THE DELIVERY OF NANOPARTICLES

Edited by **Abbass A. Hashim**

The Delivery of Nanoparticles
Edited by Abbass A. Hashim

Published by InTech

Janeza Trdine 9, 51000 Rijeka, Croatia

Publishing Process Manager Sandra Bakic

Technical Editor Teodora Smiljanic

Cover Designer InTech Design Team

Additional hard copies can be obtained from orders@intechopen.com

The Delivery of Nanoparticles, Edited by Abbass A. Hashim
p. cm.
ISBN 978-953-51-0615-9

Meet the editor

Dr Abbass A. Hashim is currently working as Head Director of Education Scheme for Training and Developments (ES4TD) British Company. The company is establishing itself in the market of supporting the scientific research and university education programs development. The company has been set up a few years ago. It is well known in the UK and Middle East and has a lot of university and company partners. Within his duties and commitments in this company, Dr. Hashim is in charge of running the research programs of Environmental Pollution Research Group (EPRG), Nanotechnology Research Group (NTRG) and Materials for Engineering Applications Research Group (MEARG). In these fields of research Dr. Hashim published many papers and research reports. He is the editor of more than six books published by InTech Open Access Publisher and a member of many journal editorial boards.

Contents

Preface

Nanoparticle is a general challenge for today's technology and the near future observations of science. Nanoparticles cover mostly all types of sciences and manufacturing technologies. The properties of this particle are flying over today scientific barriers and have passed the limitations of conventional sciences. This is the reason why nanoparticles have been evaluated for the use in many fields. A long list of science fields is inserted in the list of near future confrontation and to study the possibility of converting the conventional roles of science to explore the art beauty of the use of nanoparticles.

Researchers in this field of technology maintain to study every small system and get into the developing of even smaller devices. Professor Richard Feynman gave a talk on December 29th 1959 at the annual meeting of the American Physical Society at the California Institute of Technology (Caltech) which was first published in Caltech Engineering and Science, Volume 23:5, February 1960, pp 22-36, about the possibility of the existing of nanostructure in material. Prof. Feynman actually brought the attention of scientists when he predicted his famous speech: "There's plenty of room at the bottom". At that time tools, instruments and equipments weren't on the highly competent altitude of analyses.

Since then, the American technology institutes put forward their own potential looking for that plenty space in the bottom of the material structure. The idea was zooming inside the material to find out the novels of discoveries. In the last few years, the reality of nanotechnology has changed the dimensions of the world of science and dramatically explored the 3D piece of art of the new tiny world. This discovery confirmed that the properties of materials changed as their sizes approaches the nanoscale and as the percentage of atoms at the surface of a material becomes significantly nano-able.

A size-dependent property of this tiny space has been observed such as quantum confinement in semiconductor particles, surface plasmon resonance in some metal particles, super magnetism in magnetic materials and materials used in biomedicine. Nanoparticles often have unpredicted observable properties because of their own small size to confine their electrons and produce quantum effects. The properties of materials are absolutely diverted as their size approaches the nanoscale and where the percentage of atoms at the surface of a material becomes significantly nano. The deliverability of such nanoscale particles is another issue which has been spot out of this technology technique and put itself as the most appropriate requirement.

InTech publisher and the contributing authors of this book in nanoparticles are all overconfident to invite all scientists to read this new book. The book's potential was held until it was approached by the art of exploring the most advanced research in the field of nano-scale particles, preparation techniques and the way of reaching their destination.

25 reputable chapters were framed in this book and there were alienated into four altered sections; Toxic Nanoparticles, Drug Nanoparticles, Biological Activities and Nano-Technology.

We would like to thank all the participators and we appreciate their own potential and hard work.

Dr. Abbass A. Hashim
Head Director of ES4TD
Education Scheme for Training and Developments
UK

Section 1

Toxic Nanoparticles

1

Silver Nanoparticles

Hassan Korbekandi[1] and Siavash Iravani[2]
[1]*Genetics and Molecular Biology Department, School of Medicine, Isfahan University of Medical Sciences*
[2]*School of Pharmacy and Pharmaceutical Science, Isfahan University of Medical Sciences*
Iran

1. Introduction

Nanotechnology is an important field of modern research dealing with design, synthesis, and manipulation of particles structure ranging from approximately 1-100 nm. Tremendous growth in this emerging technology has opened novel fundamental and applied frontiers, including the synthesis of nanoscale materials and exploration or utilization of their exotic physicochemical and optoelectronic properties. Nanotechnology is rapidly gaining importance in a number of areas such as health care, cosmetics, food and feed, environmental health, mechanics, optics, biomedical sciences, chemical industries, electronics, space industries, drug-gene delivery, energy science, optoelectronics, catalysis, reorography, single electron transistors, light emitters, nonlinear optical devices, and photo-electrochemical applications (Colvin et al. 1994; Wang and Herron 1991; Schmid 1992; Hoffman et al. 1992; Hamilton and Baetzold 1979; Mansur et al. 1995).

Silver nanoparticles are of interest because of the unique properties (*e.g.*, size and shape depending optical, electrical, and magnetic properties) which can be incorporated into antimicrobial applications, biosensor materials, composite fibers, cryogenic superconducting materials, cosmetic products, and electronic components. Several physical and chemical methods have been used for synthesizing and stabilizing silver nanoparticles (Senapati 2005; Klaus-Joerger et al. 2001). The most popular chemical approaches, including chemical reduction using a variety of organic and inorganic reducing agents, electrochemical techniques, physicochemical reduction, and radiolysis are widely used for the synthesis of silver nanoparticles. Recently, nanoparticle synthesis is among the most interesting scientific areas of inquiry, and there is growing attention to produce nanoparticles using environmentally friendly methods (green chemistry). Green synthesis approaches include mixed-valence polyoxometalates, polysaccharides, Tollens, biological, and irradiation method which have advantages over conventional methods involving chemical agents associated with environmental toxicity. This chapter presents an overview of silver nanoparticle preparation by physical, chemical, and green synthesis approaches. The aim of this chapter is, therefore, to reflect on the current state and future prospects, especially the potentials and limitations of the above mentioned techniques for industries. Moreover, we discuss the applications of silver nanoparticles and their incorporation into other materials, the mechanistic aspects of the antimicrobial effects of silver nanoparticles.

2. Synthesis of silver nanoparticles

2.1 Physical approaches

Most important physical approaches include evaporation-condensation and laser ablation. Various metal nanoparticles such as silver, gold, lead sulfide, cadmium sulfide, and fullerene have previously been synthesized using the evaporation-condensation method. The absence of solvent contamination in the prepared thin films and the uniformity of nanoparticles distribution are the advantages of physical approaches in comparison with chemical processes. Physical synthesis of silver nanoparticles using a tube furnace at atmospheric pressure has some disadvantages, for example, tube furnace occupies a large space, consumes a great amount of energy while raising the environmental temperature around the source material, and requires a great deal of time to achieve thermal stability. Moreover, a typical tube furnace requires power consumption of more than several kilowatts and a preheating time of several tens of minutes to reach a stable operating temperature (Kruis et al. 2000; Magnusson et al. 1999). It was demonstrated that silver nanoparticles could be synthesized via a small ceramic heater with a local heating source (Jung et al. 2006). The evaporated vapor can cool at a suitable rapid rate, because the temperature gradient in the vicinity of the heater surface is very steep in comparison with that of a tube furnace. This makes possible the formation of small nanoparticles in high concentration. This physical method can be useful as a nanoparticle generator for long-term experiments for inhalation toxicity studies, and as a calibration device for nanoparticle measurement equipment (Jung et al. 2006).

Silver nanoparticles could be synthesized by laser ablation of metallic bulk materials in solution (Mafune et al. 2000; Mafune et al. 2001; Kabashin and Meunier 2003; Sylvestre et al. 2004; Dolgaev et al. 2002). The ablation efficiency and the characteristics of produced nano-silver particles depend upon many factors such as the wavelength of the laser impinging the metallic target, the duration of the laser pulses (in the femto-, pico- and nanosecond regime), the laser fluence, the ablation time duration and the effective liquid medium, with or without the presence of surfactants (Kim et al. 2005; Link et al. 2000; Tarasenko et al. 2006; Kawasaki and Nishimura 2006). One important advantage of laser ablation technique compared to other methods for production of metal colloids is the absence of chemical reagents in solutions. Therefore, pure and uncontaminated metal colloids for further applications can be prepared by this technique (Tsuji et al. 2002). Silver nanospheroids (20-50 nm) were prepared by laser ablation in water with femtosecond laser pulses at 800 nm (Tsuji et al. 2003). The formation efficiency and the size of colloidal particles were compared with those of colloidal particles prepared by nanosecond laser pulses. The results revealed the formation efficiency for femtosecond pulses was significantly lower than that for nanosecond pulses. The size of colloids prepared by femtosecond pulses were less dispersed than that of colloids prepared by nanosecond pulses. Furthermore, it was found that the ablation efficiency for femtosecond ablation in water was lower than that in air, while, in the case of nanosecond pulses, the ablation efficiency was similar in both water and air.

2.2 Chemical approaches

The most common approach for synthesis of silver nanoparticles is chemical reduction by organic and inorganic reducing agents. In general, different reducing agents such as sodium citrate, ascorbate, sodium borohydride ($NaBH_4$), elemental hydrogen, polyol process, Tollens reagent, N, N-dimethylformamide (DMF), and poly (ethylene glycol)-block copolymers are used for reduction of silver ions (Ag^+) in aqueous or non-aqueous solutions. The aforementioned reducing agents reduce silver ions (Ag^+) and lead to the formation of metallic silver (Ag^0), which is followed by agglomeration into oligomeric clusters. These clusters eventually lead to formation of metallic colloidal silver particles (Wiley et al. 2005; Evanoff and Chumanov 2004; Merga et al. 2007). It is important to use protective agents to stabilize dispersive nanoparticles during the course of metal nanoparticle preparation, and protect the nanoparticles that can be absorbed on or bind onto nanoparticle surfaces, avoiding their agglomeration (Oliveira et al. 2005). The presence of surfactants comprising functionalities (*e.g.*, thiols, amines, acids, and alcohols) for interactions with particle surfaces can stabilize particle growth, and protect particles from sedimentation, agglomeration, or losing their surface properties. Polymeric compounds such as poly(vinyl alcohol), poly(vinylpyrrolidone), poly(ethylene glycol), poly(methacrylic acid), and polymethylmethacrylate have been reported to be effective protective agents to stabilize nanoparticles. In one study, Oliveira et al. (Oliveira et al. 2005) prepared dodecanethiol-capped silver nanoparticles, based on Brust procedure (Brust and Kiely 2002), based on a phase transfer of an Au^{3+} complex from aqueous to organic phase in a two-phase liquid-liquid system, followed by a reduction with sodium borohydride in the presence of dodecanethiol as a stabilizing agent, binding onto the nanoparticles surfaces, thereby avoiding their aggregation and making them soluble in certain solvents. They reported that small changes in synthetic factors lead to dramatic modifications in nanoparticle structure, average size, size distribution width, stability and self-assembly patterns. Zhang et al. (2008) used a hyperbranched poly(methylene bisacrylamide aminoethyl piperazine) with terminal dimethylamine groups (HPAMAM-$N(CH_3)_2$) to produce colloids of silver. The amide moieties, piperazine rings, tertiary amine groups and the hyper-branched structure in HPAMAM-$N(CH_3)_2$ are important to its effective stabilizing and reducing abilities.

Uniform and size controllable silver nanoparticles can be synthesized using micro-emulsion techniques. The nanoparticles preparation in two-phase aqueous organic systems is based on the initial spatial separation of reactants (metal precursor and reducing agent) in two immiscible phases. The interface between the two liquids and the intensity of inter-phase transport between two phases, which is mediated by a quaternary alkyl-ammonium salt, affect the rate of interactions between metal precursors and reducing agents. Metal clusters formed at the interface are stabilized, due to their surface being coated with stabilizer molecules occurring in the non-polar aqueous medium, and transferred to the organic medium by the inter-phase transporter (Krutyakov et al. 2008). One of the major disadvantages of this method is the use of highly deleterious organic solvents. Thus large amounts of surfactant and organic solvent must be separated and removed from the final product. For instance, Zhang et al. (2007) used dodecane as an oily phase (a low deleterious and even nontoxic solvent), but there was no need to separate the prepared silver solution from the reaction mixture. On other hand, colloidal nanoparticles prepared in nonaqueous media for conductive inks are well-dispersed in a low vapor pressure organic solvent, to readily wet the surface of the polymeric substrate without any aggregation. These advantages can also be found in the applications of metal nanoparticles as catalysts to catalyze most organic reactions, which have been conducted in non-polar solvents. It is very important to transfer nanometal particles to different physicochemical environments in practical applications (Cozzoli et al. 2004).

A simple and effective method, UV-initiated photoreduction, has been reported for synthesis of silver nanoparticles in the presence of citrate, polyvinylpyrrolidone, poly(acrylic acid), and collagen. For instance, Huang and Yang produced silver nanoparticles via the photoreduction of silver nitrate in layered inorganic laponite clay suspensions which served as a stabilizing agent for the prevention of nanoparticles aggregation. The properties of the produced nanoparticles were studied as a function of UV irradiation time. Bimodal size distribution and relatively large silver nanoparticles were obtained when irradiated under UV for 3 h. Further irradiation disintegrated the silver nanoparticles into smaller sizes with a single distribution mode until a relatively stable size and size distribution was obtained (Huang and Yang 2008). Silver nanoparticles (nanosphere, nanowire, and dendrite) have been prepared by an ultraviolet irradiation photoreduction technique at room temperature using poly(vinylalcohol) (as protecting and stabilizing agent). Concentration of both poly(vinylalcohol) and silver nitrate played significant role in the growth of the nanorods and dendrites (Zhou et al. 1999). Sonoelectrochemistry technique utilizes ultrasonic power primarily to manipulate the material mechanically. The pulsed sonoelectrochemical synthetic method involves alternating sonic and electric pulses, and electrolyte composition plays a crucial role in shape formation (Socol et al. 2002). It was reported that silver nanospheres could be prepared by sonoelectrochemical reduction using a complexing agent, nitrilotriacetate to avoid aggregation (Socol et al. 2002).

Nano-sized silver particles with an average size of 8 nm were prepared by photoinduced reduction using poly(styrene sulfonate)/poly(allylamine hydrochloride) polyelectrolyte capsules as microreactors (Shchukin et al. 2003). Moreover, it was demonstrated that the photoinduced method could be used for converting silver nanospheres into triangular silver nanocrystals (nanoprisms) with desired edge lengths in the range of 30-120 nm (Jin et al. 2003). The particle growth process was controlled using dual-beam illumination of nanoparticles. Citrate and poly(styrene sulfonate) were used as stabilizing agents. In another study, silver nanoparticles were prepared through a very fast reduction of Ag^+ by α-aminoalkyl radicals generated from hydrogen abstraction toward an aliphatic amine by the excited triplet state of 2-substituted thioxanthone series ($TX{-}O{-}CH_2{-}COO^-$ and $TX{-}S{-}CH_2{-}COO^-$). The quantum yield of this prior reaction was tuned by a substituent effect on the thioxanthones, and led to a kinetic control of the conversion of Ag^+ to $Ag^{(0)}$ (Malval et al. 2010).

Electrochemical synthetic method can be used to synthesize silver nanoparticles. It is possible to control particle size by adjusting the electrolysis parameters and to improve homogeneity of silver nanoparticles by changing the composition of the electrolytic solutions. Polyphenylpyrrole-coated silver nanospheroids (3-20 nm) were synthesized by electrochemical reduction at the liquid/liquid interface. This nano-compound was prepared by transferring the silver metal ion from the aqueous phase to the organic phase, where it reacted with pyrrole monomer (Johans et al. 2002). In another study, monodisperse silver nanospheroids (1-18 nm) were synthesized by electrochemical reduction inside or outside zeolite crystals according to the silver exchange degree of the compact zeolite film modified electrodes (Zhang et al. 2002). Furthermore, spherical silver nanoparticles (10-20 nm) with narrow size distributions were conveniently synthesized in an aqueous solution by an electrochemical method (Ma et al. 2004). Poly N-vinylpyrrolidone was chosen as the stabilizer for the silver clusters in this study. Poly N-vinylpyrrolidone protects nanoparticles from agglomeration, significantly reduces silver deposition rate, and promotes silver nucleation and silver particle formation rate. Application of rotating platinum cathode effectively solves the technological difficulty of rapidly transferring metallic nanoparticles from the cathode vicinity to bulk solution, avoiding the occurrence of flocculates in vicinity of the cathode, and ensures monodispersity of particles. The addition of sodium dodecyl benzene sulfonate to the electrolyte improved the particle size and particle size distribution of the silver nanoparticles (Ma et al. 2004).

Silver nanoparticles can be synthesized by using a variety of irradiation methods. Laser irradiation of an aqueous solution of silver salt and surfactant can produce silver nanoparticles with a well defined shape and size distribution (Abid et al. 2002). Furthermore, the laser was used in a photo-sensitization synthetic method of making silver nanoparticles using benzophenone. Low laser powers at short irradiation times produced silver nanoparticles of approximately 20 nm, while an increased irradiation power produced nanoparticles of approximately 5 nm. Laser and mercury lamp can be used as light sources for the production of silver nanoparticles (Eutis et al. 2005). In visible light irradiation studies, the photo-sensitized growth of silver nanoparticles using thiophene (sensitizing dye) and silver nanoparticle formation by illumination of $Ag(NH_3)^+$ in ethanol has been accomplished (Sudeep and Kamat 2005; Zhang et al. 2003).

Microwave assisted synthesis is a promising method for the synthesis of silver nanoparticles. It was reported that silver nanoparticles could be synthesized by a microwave-assisted synthesis method employing carboxymethyl cellulose sodium as a reducing and stabilizing agent. The size of the resulting particles depended on the concentration of sodium carboxymethyl cellulose and silver nitrate. The produced nanoparticles were uniform and stable, and were stable at room temperature for 2 months without any visible change (Chen et al. 2008). The production of silver nanoparticles in the presence of Pt seeds, polyvinyl pyrrolidine and ethylene glycol was also reported (Navaladian et al. 2008). Additionally, starch has been employed as a template and reducing agent for the synthesis of silver nanoparticles with an average size of 12 nm, using a microwave-assisted synthetic method. Starch functions as a template, preventing the aggregation of the produced silver nanoparticles (Sreeram et al. 2008). Microwaves in combination with polyol process were applied in the synthesis of silver nanospheroids using ethylene glycol and poly N-vinylpyrrolidone as reducing and stabilizing agents, respectively (Komarneni et al. 2002). In a typical polyol process inorganic salt is reduced by the polyol (*e.g.*, ethylene glycol which serves as both a solvent and a reducing agent) at a high temperature. Yin et al. (Yin et al. 2004) reported that large-scale and size-controlled silver nanoparticles could be rapidly synthesized under microwave irradiation from an aqueous solution of silver nitrate and trisodium citrate in the presence of formaldehyde as a reducing agent. Size and size distribution of the produced silver nanoparticles are strongly dependent on the states of silver cations in the initial reaction solution. Silver nanoparticles with different shapes can be synthesized by microwave irradiation of a silver nitrate-ethylene-glycol-$H_2[PtCl_6]$-poly(vinylpyrrolidone) solution within 3 min (Tsuji et al. 2008). Moreover, the use of microwave irradiation to produce monodispersed silver nanoparticles using basic amino acids (as reducing agents) and soluble starch (as a protecting agent) has been reported (Hu et al. 2008). Radiolysis of silver ions in ethylene glycol, in order to synthesize silver nanoparticles, was also reported (Soroushian et al. 2005). Moreover, silver nanoparticles supported on silica aero-gel were produced using gamma radiolysis. The produced silver clusters were stable in the 2-9 pH range and started agglomeration at pH > 9 (Ramnami et al. 2007). Oligochitosan as a stabilizer can be used in a preparation of silver nanoparticles by gamma radiation. It was reported that stable silver nanoparticles (5-15 nm) were synthesized in a 1.8-9.0 pH range using this method (Long et al. 2007). Silver nanoparticles (4-5 nm) were also synthesized by γ-ray irradiation of acetic water solutions containing silver nitrate and chitosan (Cheng et al. 2007). In another study, silver nanospheroids (1-4 nm) were produced by γ-ray irradiation of a silver solution in optically transparent inorganic mesoporous silica. Reduction of silver ions within the matrix is brought about by hydrated electrons and hydroalkyl radicals generated during the radiolysis of a 2-propanol solution. The nanoparticles produced within the silica matrix were stable in the presence of oxygen for at least several months (Hornebecq et al. 2003). Moreover, silver nanoparticles (60-200 nm) have been produced by irradiating a solution, prepared by mixing silver nitrate and poly-vinyl-alcohol, with 6 MeV electrons (Bogle et al. 2006). A pulse radiolysis technique has been applied to study the reactions of inorganic and organic species in silver nanoparticle synthesis, to understand the factors controlling the shape and size of the nanoparticles synthesized by a common reduction method using citrate ions (as reducing and stabilizing agents) (Pillai and Kamat 2004), and to demonstrate the role of phenol derivatives in the formation of silver nanoparticles by the reduction of silver ions with dihydroxy benzene (Jacob et al. 2008). Dihydroxy benzene could be also used to reduce silver ions to synthesize stable silver nanoparticles (with an average size of 30 nm) in air-saturated aqueous solutions (Jacob et al. 2008).

In polysaccharide method, silver nanoparticles were prepared using water as an environmentally-friendly solvent and polysaccharides as capping/reducing agents. For instance, the synthesis of starch-silver nanoparticles was carried out with starch (as a capping agent) and β-D-glucose (as a reducing agent) in a gently heated system (Raveendran et al. 2003). The binding interactions between starch and produced silver nanoparticles were weak and could be reversible at higher temperatures, allowing for the separation of the synthesized nanoparticles. In dual polysaccharide function, silver nanoparticles were synthesized by the reduction of silver ions inside nanoscopic starch templates (Raveendran et al. 2003, 2005). The extensive network of hydrogen bands in templates provided surface passivation or protection against nanoparticle aggregation. Green synthesis of silver nanoparticles using negatively charged heparin (reducing/stabilizing agent and nucleation controller) was also reported by heating a solution of silver nitrate and heparin to 70 °C for approximately 8 h (Huang and Yang 2004). Transmission electron microscopy (TEM) micrographs demonstrated an increase in particle size of silver nanoparticles with increased concentrations of silver nitrate (as the substrate) and heparin. Moreover, changes in the heparin concentration influenced the morphology and size of silver nanoparticles. The synthesized silver nanoparticles were highly stable, and showed no signs of aggregation after two months (Huang and Yang 2004). In another study, stable silver nanoparticles (10-34 nm) were synthesized by autoclaving a solution of silver nitrate (as the substrate) and starch (as a capping/reducing agent) at 15 psi and 121 °C for 5 min (Vigneshwaran et al. 2006b). These nanoparticles were stable in solution for three months at approximately 25 °C. Smaller silver nanoparticles (≤10 nm) were synthesized by mixing two solutions of silver nitrate containing starch (as a capping agent), and NaOH solutions containing glucose (as a reducing agent) in a spinning disk reactor with a reaction time of less than 10 min (Tai et al. 2008).

Recently, a simple one-step process, Tollens method, has been used for the synthesis of silver nanoparticles with a controlled size. This green synthesis technique involves the reduction of $Ag(NH_3)_2^+$ (as a Tollens reagent) by an aldehyde (Yin et al. 2002). In the modified Tollens procedure, silver ions are reduced by saccharides in the presence of ammonia, yielding silver nanoparticle films (50-200 nm), silver hydrosols (20-50 nm) and silver nanoparticles of different shapes. In this method, the concentration of ammonia and the nature of the reducing agent play an important role in controlling size and morphology of the silver nanoparticles. It was revealed that the smallest particles were formed at the lowest ammonia concentration. Glucose and the lowest ammonia concentration (5 mM) resulted in the smallest average particle size of 57 nm with an intense maximum surface plasmon absorbance at 420 nm. Moreover, an increase in NH_3 from 0.005 M to 0.2 M resulted in a simultaneous increase in particle size and polydispersity (Kvítek et al. 2005). Silver nanoparticles with controllable sizes were synthesized by reduction of $[Ag(NH_3)_2]^+$ with glucose, galactose, maltose, and lactose (Panacek et al. 2006). The nanoparticle synthesis was carried out at various ammonia concentrations (0.005-0.20 M) and pH conditions (11.5-13.0), resulting in average particle sizes of 25-450 nm. The particle size was increased by increasing (NH_3), and the difference in the structure of the reducing agent (monosaccharides and disaccharides) and pH (particles obtained at pH 11.5 were smaller than those at pH 12.5) influenced the particle size. Polydispersity also decreased in response to decreases the pH. Produced silver nanoparticles were stabilized and protected by sodium dodecyl sulfate (SDS), polyoxyethylenesorbitane monooleate (Tween 80), and polyvinylpyrrolidone (PVP 360) (Kvitek et al. 2008; Soukupova et al. 2008).

Silver, gold, palladium, and platinum nanoparticles can be produced at room temperature, as a result of simply mixing the corresponding metal ions with reduced polyoxometalates which served as reducing and stabilizing agents. Polyoxometalates are soluble in water and have the capability of undergoing stepwise, multielectron redox reactions without disturbing their structure. It was demonstrated that silver nanoparticles were produced by illuminating a deaerated solution of polyoxometalate/S/Ag^+ (polyoxometalate: $[PW_{12}O_{40}]^{3}$, $[SiW_{12}O_{40}]^{4-}$; S:propan-2-ol or 2,4-dichlorophenol) (Troupis et al. 2002). Furthermore, green chemistry-type one-step synthesis and stabilization of silver nanostructures with Mo^{V}-Mo^{VI} mixed-valence polyoxometalates in water at room temperature has been reported (Zhang et al. 2007).

2.3 Biological approaches

In recent years, the development of efficient green chemistry methods employing natural reducing, capping, and stabilizing agents to prepare silver nanoparticles with desired morphology and size have become a major focus of researchers. Biological methods can be used to synthesize silver nanoparticles without the use of any harsh, toxic and expensive chemical substances (Ahmad et al. 2003a; Shankar et al. 2004; Ankamwar et al. 2005; Huang et al. 2007). The bioreduction of metal ions by combinations of biomolecules found in the extracts of certain organisms (*e.g.*, enzymes/proteins, amino acids, polysaccharides, and vitamins) is environmentally benign, yet chemically complex. Many studies have reported successful synthesis of silver nanoparticle using organisms (microorganisms and biological systems) (Korbekandi et al. 2009; Sastry et al. 2003; Iravani 2011). For instance, we demonstrated the bioreductive synthesis of silver nanoparticles using *F. oxysporum* (Figure 1). In this section, most of the organisms used in green synthesis of silver nanoparticles are shown.

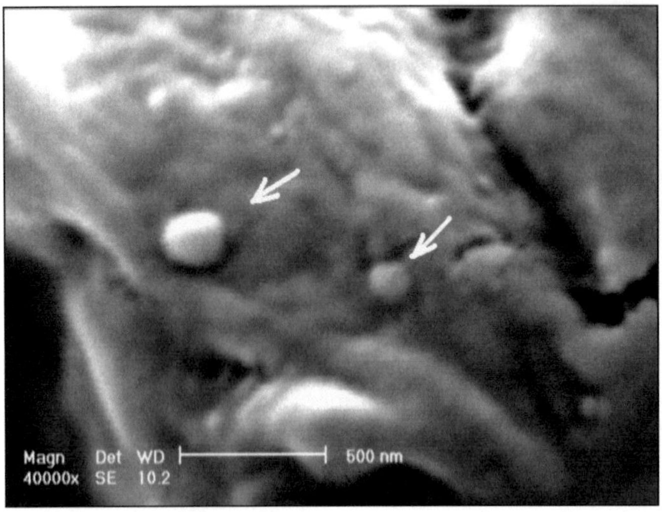

Fig. 1. SEM micrograph recorded from silver nanoparticles produced by reaction of $AgNO_3$ solution (1 mM) with *F. oxysporum* biomass.

2.3.1 Synthesis of silver nanoparticles by bacteria

It was reported that highly stable silver nanoparticles (40 nm) could be synthesized by bioreduction of aqueous silver ions with a culture supernatant of nonpathogenic bacterium, *Bacillus licheniformis* (Kalishwaralal et al. 2008b). Moreover, well-dispersed silver nanocrystals (50 nm) were synthesized using the bacterium *Bacillus licheniformis* (Kalishwaralal et al. 2008a). Saifuddin *et al.* (Saifuddin et al. 2009) have described a novel combinational synthesis approach for the formation of silver nanoparticles by using a combination of culture supernatant of *B. subtilis* and microwave irradiation in water. They reported the extracellular biosynthesis of monodispersed Ag nanoparticles (5-50 nm) using supernatants of *B. subtilis*, but in order to increase the rate of reaction and reduce the aggregation of the produced nanoparticles, they used microwave radiation which might provide uniform heating around the nanoparticles and could assist the digestive ripening of particles with no aggregation.

Silver nanocrystals of different compositions were successfully synthesized by *Pseudomonas stutzeri* AG259 (Klaus et al. 1999). The silver-resistant bacterial strain, *Pseudomonas stutzeri* AG259, isolated from a silver mine, accumulated silver nanoparticles intracellularly, along with some silver sulfide, ranging in size from 35 to 46 nm (Slawson et al. 1992). Larger particles were formed when *P. stutzeri* AG259 challenged with high concentrations of silver ions during culturing, resulted intracellular formation of silver nanoparticles, ranging in size from a few nm to 200 nm (Klaus-Joerger et al. 2001; Klaus et al. 1999). *P. stutzeri* AG259 detoxificated silver through its precipitation in the periplasmic space and its reduction to elemental silver with a variety of crystal typologies, such as hexagons and equilateral triangles, as well as three different types of particles: elemental crystalline silver, monoclinic silver sulfide acanthite (Ag_2S), and a further undetermined structure (Klaus et al. 1999). The periplasmic space limited the thickness of the crystals, but not their width, which could be rather large (100-200 nm) (Klaus-Joerger et al. 2001). In another study, rapid biosynthesis of metallic nanoparticles of silver using the reduction of aqueous Ag^+ ions by culture supernatants of *Klebsiella pneumonia, E. coli,* and *Enterobacter cloacae* (Enterobacteriacae) was reported (Shahverdi et al. 2007). The synthetic process was quite fast and silver nanoparticles were formed within 5 min of silver ions coming in contact with the cell filtrate. It seems that nitroreductase enzymes might be responsible for bioreduction of silver ions. It was also reported that visible-light emission could significantly increase synthesis of silver nanoparticles (1-6 nm) by culture supernatants of *K. pneumoniae* (Mokhtari et al. 2009). Monodispersed and stable silver nanoparticles were also successfully synthesized with bioreduction of $[Ag\ (NH_3)_2]^+$ using *Aeromonas* sp. SH10 and *Corynebacterium* sp. SH09 (Mouxing et al. 2006). It was speculated that $[Ag\ (NH_3)_2]^+$ first reacted with OH^- to form Ag_2O, which was then metabolized independently and reduced to silver nanoparticles by the biomass.

Lactobacillus strains, when exposed to silver ions, resulted in biosynthesis of nanoparticles within the bacterial cells (Nair and Pradeep 2002). It has been reported that exposure of lactic acid bacteria present in the whey of buttermilk to mixtures of silver ions could be used to grow nanoparticles of silver. The nucleation of silver nanoparticles occurred on the cell surface through sugars and enzymes in the cell wall, and then the metal nuclei were transported into the cell where they aggregated and grew to larger-sized particles.

2.3.2 Synthesis of silver nanoparticles by fungi

Silver nanoparticles (5-50 nm) could be synthesized extracellularly using *Fusarium oxysporum,* with no evidence of flocculation of the particles even a month after the reaction (Ahmad et al. 2003a). The long-term stability of the nanoparticle solution might be due to the stabilization of the silver particles by proteins. The morphology of nanoparticles was highly variable, with generally spherical and occasionally triangular shapes observed in the micrographs. Silver nanoparticles have been reported to interact strongly with proteins including cytochrome *c* (Cc). This protein could be self-assembled on citrate-reduced silver colloid surface (Macdonald and Smith 1996). Interestingly, adsorption of (Cc)-coated colloidal Au nanoparticles onto aggregated colloidal Ag resulted Ag: Cc: Au nanoparticle conjugate (Keating et al. 1998). In UV-vis spectra from the reaction mixture after 72 h, the presence of an absorption band at ca. 270 nm might be due to electronic excitations in tryptophan and tyrosine residues in the proteins. In *F. oxysporum,* the bioreduction of silver ions was attributed to an enzymatic process involving NADH-dependent reductase (Ahmad et al. 2003b). The exposure of silver ions to *F. oxysporum,* resulted in release of nitrate reductase and subsequent formation of highly stable silver nanoparticles in solution (Kumar et al. 2007). The secreted enzyme was found to be dependent on NADH cofactor. They mentioned high stability of nanoparticles in solution was due to capping of particles by release of capping proteins by *F. oxysporum*. Stability of the capping protein was found to be pH dependent. At higher pH values (>12), the nanoparticles in solution remained stable, while they aggregated at lower pH values (<2) as the protein was denatured. Kumar *et al.* (Kumar et al. 2007) have demonstrated enzymatic synthesis of silver nanoparticles with different chemical compositions, sizes and morphologies, using α-NADPH-dependent nitrate reductase purified from *F. oxysporum* and phytochelatin, *in vitro*. Silver ions were reduced in the presence of nitrate reductase, leading to formation of a stable silver hydrosol 10-25 nm in diameter and stabilized by the capping peptide. Use of a specific enzyme in *in vitro* synthesis of nanoparticles showed interesting advantages. This would eliminate the downstream processing required for the use of these nanoparticles in homogeneous catalysis and other applications such as non-linear optics. The biggest advantage of this protocol based on purified enzyme was the development of a new approach for green synthesis of nanomaterials over a range of chemical compositions and shapes without possible aggregation. Ingle *et al.* (Ingle et al. 2008) demonstrated the potential ability of *Fusarium acuminatum* Ell. and Ev. (USM-3793) cell extracts in biosynthesis of silver nanoparticles. The nanoparticles produced within 15-20 minutes and were spherical with a broad size distribution in the range of 5-40 nm with the average diameter of 13 nm. A nitrate-dependent reductase enzyme might act as the reducing agent. The white rot fungus, *Phanerochaete chrysosporium*, also reduced silver ions to form nano-silver particles (Vigneshwaran et al. 2006a). The most dominant morphology was pyramidal shape, in different sizes, but hexagonal structures were also observed. Possible involvement of proteins in synthesizing silver nanoparticles was observed in *Plectonema boryanum* UTEX 485 (a filamentous cyanobacterium) (Lengke et al. 2007).

Stable silver nanoparticles could be achieved by using *Aspergillus flavus* (Vigneshwaran et al. 2007). These nanoparticles were found to be stable in water for more than 3 months with no significant aggregation because of surface binding of stabilizing materials secreted by the fungus (Vigneshwaran et al. 2007). Extracellular biosynthesis of silver nanoparticles using *Aspergillus fumigatus* (a ubiquitous saprophytic mold) has also been investigated (Bhainsa and D'Souza 2006). The resulted TEM micrograph showed well-dispersed silver nanoparticles (5-25 nm) with variable shapes. Most of them were spherical in nature with some others having occasionally triangular shapes (Bhainsa and D'Souza 2006). Compared to intracellular biosynthesis of nanoparticles; extracellular synthesis could be developed as a simple and cheap method because of uncomplicated downstream processing and handling of biomasses.

The extracellular filtrate of *Cladosporium cladosporioides* biomass was used to synthesize silver nanoparticles (Balaji et al. 2009). It was suggested that proteins, organic acids and polysaccharides released by *C. cladosporioides* were responsible for formation of spherical crystalline silver nanoparticles. Kathiresan *et al.* (Kathiresan et al. 2009) have shown that when the culture filtrate of *Penicillium fellutanum* was incubated with silver ions and maintained under dark conditions, spherical silver nanoparticles could be produced. They also changed crucial factors such as pH, incubation time, temperature, silver nitrate concentration and sodium chloride to achieve the maximum nanoparticle production. The highest optical density at 430 nm was found at 24 h after the start of incubation time, 1 mM concentration of silver nitrate, pH 6.0, 5°C temperature and 0.3% sodium chloride. Fungi of *Penicillium* genus were used for green synthesis of silver nanoparticles (Sadowski et al.

2008). *Penicillium* sp. J3 isolated from soil was able to produce silver nanoparticles (Maliszewska et al. 2009). The bioreduction of silver ions occurred on the surface of the cells and proteins might have critical role in formation and stabilization of the synthesized nanoparticles.

Sanghi *et al.* (2009) have investigated the ability of *Coriolus versicolor* in formation of monodisperse spherical silver nanoparticles. Under alkaline conditions (pH 10) the time taken for production of silver nanoparticles was reduced from 72 h to 1 h. It was indicated that alkaline conditions might be involved in bioreduction of silver ions, water hydrolysis and interaction with protein functionalities. Findings of this study have shown that glucose was necessary for the reduction of silver nanoparticles, and S-H of the protein played an important role in the bioreduction.

2.3.3 Synthesis of silver nanoparticles by plants

Camellia sinensis (green tea) extract has been used as a reducing and stabilizing agent for the biosynthesis of silver nanoparticles in an aqueous solution in ambient conditions (Vilchis-Nestor et al. 2008). It was observed that when the amount of *C. sinensis* extract was increased, the resulted nanoparticles were slightly larger and more spherical. Phenolic acid-type biomolecules (*e.g.,* caffeine and theophylline) present in the *C. sinensis* extract seemed to be responsible for the formation and stabilization of silver nanoparticles. Black tea leaf extracts were also used in the production of silver nanoparticles (Begum et al. 2009). The nanoparticles were stable and had different shapes, such as spheres, trapezoids, prisms, and rods. Polyphenols and flavonoids seemed to be responsible for the biosynthesis of these nanoparticles.

Plant extracts from alfalfa (*Medicago sativa*), lemongrass (*Cymbopogon flexuosus*), and geranium (*Pelargonium graveolens*) have served as green reactants in silver nanoparticle synthesis. Harris *et al.* (2008) have investigated the limits (substrate metal concentration and time exposure) of uptake of metallic silver by two common metallophytes, *Brassica juncea* and *M. sativa.* They demonstrated that *B. juncea* and *M. Sativa* could be used in the phytosynthesis of silver nanoparticles. *B. juncea,* when exposed to an aqueous substrate containing 1,000 ppm silver nitrate for 72 h, accumulated up to 12.4 wt. % silver. *M. sativa* accumulated up to 13.6 wt. % silver when exposed to an aqueous substrate containing 10,000 ppm silver nitrate for 24 h. In the case of *M. sativa,* an increase in metal uptake was observed by increasing the exposure time and substrate concentration. In both cases, TEM analysis showed the presence of roughly spherical silver nanoparticles, with a mean size of 50 nm. Geranium leaf broth, when exposed to aqueous silver nitrate solution, resulted in enzymatic synthesis of stable crystalline silver nanoparticles, extracellularly (Shankar et al. 2003). Bioreduction of the metal ions was fairly rapid, occurred readily in solution, and resulted in a high density of stable silver nanoparticles (16-40 nm). The nanoparticles appeared to be assembled into open, quasilinear superstructures and were predominantly spherical in shape. It was believed that proteins, terpenoids and other bio-organic compounds in the geranium leaf broth participated in bioreduction of silver ions and in the stabilization of the nanoparticles via surface capping.

A high density of extremely stable silver nanoparticles (16-40 nm) was rapidly synthesized by challenging silver ions with *Datura metel* (*Solanaceae*) leaf extract (Kesharwani et al. 2009).

The leaf extracts of this plant contains biomolecules, including alkaloids, proteins/enzymes, amino acids, alcoholic compounds, and polysaccharides which could be used as reductant to react with silver ions, and therefore used as scaffolds to direct the formation of silver nanoparticles in the solution. Song and Kim (2008) elucidated the fact that *Pinus desiflora, Diospyros kaki, Ginko biloba, Magnolia kobus* and *Platanus orientalis* leaf broths synthesized stable silver nanoparticles with average particle size ranging from 15 to 500 nm, extracellularly. In the case of *M. kobus* and *D. kaki* leaf broths, the synthesis rate and final conversion to silver nanoparticles was faster, when the reaction temperature was increased. But the average particle sizes produced by *D. kaki* leaf broth decreased from 50 nm to 16 nm, when temperature was increased from 25 °C to 95 °C. The researchers also illustrated that only 11 min was required for more than 90% conversion at the reaction temperature of 95 °C using *M. kobus* leaf broth (Song and Kim 2008). It was further demonstrated that leaf extracts from the aquatic medicinal plant, *Nelumbo nucifera* (Nymphaeaceae), was able to reduce silver ions and produce silver nanoparticles (with an average size of 45 nm) in different shapes (Santhoshkumar et al. 2010). The biosynthesized nanoparticles showed larvicidal activities against malaria (*Anopheles subpictus*) and filariasis (*Culex quinquefasciatus*) vectors. Silver nanoparticles were biosynthesized using *Sorbus aucuparia* leaf extract within 15 min. The nanoparticles were found to be stable for more than 3 months. The sorbate ion in the leaf extract of *S. aucuparia* encapsulated the nanoparticles and this action seemed to be responsible for their stability (Dubey et al. 2010).

Studying synthesis of silver nanoparticles with isolated/purified bio-organics may give better insight into the system mechanism. Glutathione (γ-Glu-Cys-Gly) as a reducing/capping agent can synthesize water-soluble and size tunable silver nanoparticles which easily bind to a model protein (bovine serum albumin) (Wu et al. 2008). Tryptophan residues of synthetic oligo-peptides at the C-terminus were identified as reducing agents producing silver nanoparticles (Si and Mandal 2007). Moreover, silver nanoparticles were synthesized by vitamin E in Langmuir-Blodgett technique, by bio-surfactants (sophorolipids) and by L-valine-based oligopeptides with chemical structures, Z-(L-Val)$_3$-OMe and Z-(L-Val)$_2$-L-Cys (S-Bzl)-OMe (Kasture et al. 2007; Mantion et al. 2008). The sulfur content in the Z-(L-Val)$_2$-L-Cys(S-Bzl)-OMe controls the shape and size of silver nanoparticles, which suggests interaction between silver ions and thioether moiety of the peptide (Mantion et al. 2008).

Pure natural constituents could be used to bioreduce and stabilize silver nanoparticles. Kasthuri *et al.* (Kasthuri et al. 2009) have shown the ability of apiin extracted from henna leaves to produce anisotropic gold and quasi-spherical silver nanoparticles. Secondary hydroxyl and carbonyl groups of apiin were responsible for the bioreduction of metal salts. In order to control size and shape of nanoparticles, they used different amounts of apiin (as the reducing agent). The nanoparticles were stable in water for 3 months, which could be attributed to surface binding of apiin to nanoparticles. Furthermore, geraniol as a volatile compound obtained from *P. graveolens* was used for biosynthesis of silver nanoparticles (1-10 nm). They also evaluated *in vitro* cytotoxicity of silver nanoparticles against Fibrosarcoma Wehi 164 at different concentrations (1-5 g ml^{-1}) (Safaepour et al. 2009). The presence of 5 g/ml of silver nanoparticles significantly inhibited the cell line's growth (up to 60 %).

The various synthetic and natural polymers such as poly(ethylene glycol), poly-(N-vinyl-2-pyrrolidone), starch, heparin, poly-cationic chitosan (1-4-linked 2-amino-2-deoxy-β-D-glucose), sodium alginate, and gum acacia have been reported as reducing and stabilizing agents for biosynthesis of silver nanoparticles. It was reported that monodisperse spherical silver nanoparticles (3 nm) could be synthesized using gum kondagogu (non-toxic polysaccharide derived as an exudate from the bark of *Cochlospermum gossypium*) (Kora et al. 2010). It was suggested that carboxylate and hydroxyl groups were involved in complexation and bioreduction of silver ions into nanoparticles. This method was compatible with green chemistry principles as the gum serves a matrix for both bioreduction and stabilization of the synthesized nanoparticles. Due to availability of low cost plant derived biopolymer, this method could be implemented for large-scale synthesis of highly stable monodispersed nanoparticles.

Spherical silver nanoparticles (40-50 nm) were produced using leaf extract of *Euphorbia hirta* (Elumalai et al. 2010). These nanoparticles had potential and effective antibacterial property against *Bacillus cereus* and *S. aureus*. *Acalypha indica* (Euphorbiaceae) leaf extracts have produced silver nanoparticles (20-30 nm) within 30 min (Krishnaraj et al. 2010). These nanoparticles had excellent antimicrobial activity against water borne pathogens, *E. coli* and *V. cholera* (Minimum Inhibitory Concentration (MIC) = 10 μg ml^{-1}). Moreover, silver nanoparticles (57 nm) were produced when 10 ml of *Moringa oleifera* leaf extract was mixed to 90 ml of 1 mM aqueous of $AgNO_3$ and was heated at 60-80 °C for 20 min. These nanoparticles had considerable antimicrobial activity against pathogenic microorganisms, including *Staphylococcus aureus, Candida tropicalis, Klebsiella pneumoniae,* and *Candida krusei* (Prasad and Elumalai 2011). It has been reported that cotton fibers loaded with biosynthesized silver nanoparticles (~20 nm) using natural extracts of *Eucalyptus citriodora* and *Ficus bengalensis* had excellent antibacterial activity against *E. coli*. These fibers had potential for utilization in burn/wound dressings as well as in the fabrication of antibacterial textiles and finishings (Ravindra et al. 2010). *Garcinia mangostana* leaf extract could be used as reducing agent in order to biosynthesize silver nanoparticles (35 nm). These nanoparticles had high effective antimicrobial activity against *E. coli* and *S. aureus* (Veerasamy et al. 2011). It was reported that *Ocimum sanctum* leaf extract could bioreduce silver ions into crystalline silver nanoparticles (4-30 nm) within 8 min of reaction time. These nanoparticles were stable due to the presence of proteins which may act as capping agents. *O. sanctum* leaves contain ascorbic acid which may play an important role in reduction of silver ions into metallic silver nanoparticles. These nanoparticles have shown strong antimicrobial activity against *E. coli* and *S. aureus* (Singhal et al. 2011). Green synthesis of silver nanoparticles using *Cacumen platycladi* extract was also investigated. Reducing sugars and flavonoids in the extract seemed to be mainly responsible for reduction of silver ions, and their reductive capability promoted at 90 °C, leading to formation of silver nanoparticles (18.4 ± 4.6 nm) with narrow size distribution. The produced nanoparticles had significant antibacterial activity against both gram negative and gram positive bacteria (*E. coli* and *S. aureus*) (Huang et al. 2011). In case of *Cinnamon zeylanicum,* the bark extract could be used in biosynthesis of cubic and hexagonal silver nanocrystals (31-40 nm) (Sathishkumar et al. 2009). The particle size distribution varied with variation in the amount of *C. zeylanicum* bark extract. The number of particles increased with increasing dosage due to the variation in the amount of reductive biomolecules. Small nanoparticles were formed at high pH. The shape of silver nanoparticles at high pH was more spherical in nature rather than ellipsoidal. Bactericidal effect of produced nano-crystalline silver particles was tested against *E. coli* strain. As a result, the various tested concentrations of 2, 5, 10, 25, and 50 mg

L^{-1} produced inhibition of 10.9, 32.4, 55.8, 82, and 98.8 %, respectively. The minimum inhibitory concentration was found to be 50 mg L^{-1}.

3. Applications of silver nanoparticles and their incorporation into other materials

Nanoparticles are of great interest due to their extremely small size and large surface to volume ratio, which lead to both chemical and physical differences in their properties compared to bulk of the same chemical composition, such as mechanical, biological and sterical properties, catalytic activity, thermal and electrical conductivity, optical absorption and melting point (Daniel and Astruc 2004; Bogunia-Kubik and Sugisaka 2002; Zharov et al. 2005). Therefore, designing and production of materials with novel applications can be resulted by controlling shape and size at nanometer scale. Nanoparticles exhibit size and shape-dependent properties which are of interest for applications ranging from biosensing and catalysts to optics, antimicrobial activity, computer transistors, electrometers, chemical sensors, and wireless electronic logic and memory schemes. These particles also have many applications in different fields such as medical imaging, nano-composites, filters, drug delivery, and hyperthermia of tumors (Tan et al. 2006; Lee et al. 2008; Pissuwan et al. 2006). Silver nanoparticles have drawn the attention of researchers because of their extensive applications in areas such as integrated circuits (Kotthaus et al. 1997), sensors (Cao 2004), biolabelling (Cao 2004), filters (Cao 2004), antimicrobial deodorant fibres (Zhang and Wang 2003), cell electrodes (Klaus-Joerger et al. 2001), low-cost paper batteries (silver nano-wires) (Hu et al. 2009) and antimicrobials (Cho et al. 2005; Dura´n et al. 2007). Silver nanoparticles have been used extensively as antimicrobial agents in health industry, food storage, textile coatings and a number of environmental applications. Antimicrobial properties of silver nanoparticles caused the use of these nanometals in different fields of medicine, various industries, animal husbandry, packaging, accessories, cosmetics, health and military. Silver nanoparticles show potential antimicrobial effects against infectious organisms, including *Escherichia coli, Bacillus subtilis, Vibria cholera, Pseudomonas aeruginosa, Syphillis typhus*, and *S. aureus* (Cho et al. 2005; Dura´n et al. 2007). For instance, it was shown that silver nanoparticles mainly in the range of 1-10 nm attached to the surface of *E. coli* cell membrane, and disturbed its proper function such as respiration and permeability (Morones et al. 2005). It was observed that silver nanoparticles had a higher antibacterial effect on *B. subtilis* than on *E. coli*, suggesting a selective antimicrobial effect, possibly related to the structure of the bacterial membrane (Yoon et al. 2007). As antimicrobial agents, silver nanoparticles were applied in a wide range of applications from disinfecting medical devices and home appliances to water treatment (Jain and Pradeep 2005; Li et al. 2008). The synergetic activities of silver nanoparticles and antibiotics such as erythromycin, amoxicillin, penicillin G, clindamycin, and vancomycine against *E. coli* and *S. aureus* was reported (Shahverdi et al. 2007). Silver nanoparticles were also used in medical devices (*e.g.*, polymethylmetacrylate bone cement, surgical masks and implantable devices) to increase their antimicrobial activities (Figure 2). For instance, plastic catheters which were coated with silver nanoparticles using silver nitrate ($AgNO_3$), surfactant and *N,N,N ',N '*-tetramethylethylenediamine (as a reducing agent), had significant antimicrobial effects against *E. coli*, Enterococcus, *S. aureus*, *C. albicans*, Staphylococci, and *P. aeroginosa* (Roe et al. 2008). These can be useful for reducing the risk of infectious complications in patients with indwelling catheters. Silver nanoparticles/clay was reported to have significant antimicrobial effects against dermal pathogens, including *S. aureus, P. aeruginosa,* and *Streptococcus pyrogens,* as well as the methicillin- and oxacillin resistant *S. aureus* (MRSA and ORSA) (Su et al. 2009).

Antifungal effects of silver nanoparticles against fungal pathogens are mostly unknown. However, it was reported that these nanoparticles had significant antifungal activities against *T. mentagrophytes* and Candida species (such as *C. albicans, C. tropicolis, C. glabrata, C. parapsilosis,* and *C. krusei*). Silver nanoparticles disrupt fungal envelope structure and lead to significant damage to fungal cells (Kim et al. 2008; Kim et al. 2009).

Sun et al demonstrated the cytoprotective activity of silver nanoparticles (5-20 nm) toward HIV-1 infected cells (Hut/CCR5). These nanoparticles inhibited the virus replication in Hut/CCR5 cells causing HIV-associated apoptosis (Sun et al. 2005). In addition, size-dependent antiviral activities of silver nanoparticles (exclusively within the range of 1-10 nm) against HIV-1 virus have been reported. Silver nanoparticles preferentially bind to the gp120 subunit of HIV-1 viral envelope glycoprotein, and this interaction caused the virus could not bind to the host cell (Elechiguerra et al. 2005). It seems that more investigations are needed to find the mechanisms of antiviral effects of silver nanoparticles.

Pharmaceutics & Medicine	Treatment of dermatitis; Inhibition of HIV-1 replication Treatment of ulcerative colitis & acne Antimicrobial effects against infectious organisms Remote laser light-induced opening of microcapsules Silver/dendrimer nanocomposite for cell labelling Molecular imaging of cancer cells Enhanced Raman scattering (SERS) spectroscopy Detection of viral structures (SERS & Silver nanorods) Coating of hospital textile (surgical gowns, face mask) Coating of catheter for cerebrospinal fluid drainage Coating of surgical mesh for pelvic reconstruction Coating of breathing mask Patent Coating of endotracheal tube for mechanical ventilatory support Coating of driveline for ventricular assist devices Coating of central venous catheter for monitoring Additive in bone cement Implantable material using clay-layers with starch-stabilized Ag NPs Coating of intramedullary nail for long bone fractures Coating of implant for joint replacement Orthopedic stockings Superabsorbent hydrogel for incontinence material Hydrogel for wound dressing
Dentistry	Additive in polymerizable dental materials Patent Silver-loaded SiO_2 nanocomposite resin filler (Dental resin composite) Polyethylene tubes filled with fibrin sponge embedded with Ag NPs dispersion

Fig. 2. Applications of silver nanoparticles in pharmaceutics, medicine, and dentistry

In general, therapeutic effects of silver particles (in suspension form) depend on important aspects, including particle size (surface area and energy), particle shape (catalytic activity), particle concentration (therapeutic index) and particle charge (oligodynamic quality) (Pal et al. 2007). Mechanisms of antimicrobial effects of silver nanoparticles are still not fully understood, but several studies have revealed that silver nanoparticles may attach to the negatively charged bacterial cell wall and rupture it, which leads to denaturation of protein and finally cell death (Figure 3). Fluorescent bacteria were used to study antibacterial effects of silver nanoparticles. Green fluorescent proteins were adapted to these investigations (Gogoi et al. 2006). It was found that silver nanoparticles attached to sulfur-containing proteins of bacteria, and caused death. Moreover, fluorescent measurements of cell-free supernatants showed the effect of silver nanoparticles on recombination of bacteria. The attachment of silver ions or nanoparticles to the cell wall caused accumulation of envelope protein precursors resulting in immediate dissipation of the proton motive force (Lin et al. 1998). Catalytic mechanism of silver nanoparticle composites and their damage to the cell by interaction with phosphorous- and sulfur-containing compounds such as DNA have been also investigated (Sharma et al. 2009). Furthermore, silver nanoparticles exhibited destabilization of the outer membrane and rupture of the plasma membrane, thereby causing depletion of intracellular ATP (Lok 2006). Another mechanism involves the association of silver with oxygen and its reaction with sulfhydryl groups on the cell wall to form R-S-S-R bonds, thereby blocking respiration and causing cell death (Kumar et al. 2004).

Mechanisms of Antibacterial Effects of Ag Nanoparticles
Cell death due to uncoupling of oxidative phosphorylation
Cell death due to induction of free radical formation
Interference with respiratory chain at cytochrome C level Interference with components of microbial electron transport system
Interactions with protein thiol groups & membrane bound enzymes
Interaction with phosphorous- and sulfur-containing compounds such as DNA

Fig. 3. Mechanisms of antibacterial effects of silver nanoparticles

Products made with silver nanoparticles have been approved by a range of accredited bodies such as US Food and Drug Administration (FDA), US Environmental Protection Agency (EPA), and Research Institute for Chemical Industry and FITI Testing (Zhong et al. 2007; Wei et al. 2007; Jia et al. 2008). Silver nanoparticles can be exploited in medicine and pharmacy for dental materials, burn treatments, coating stainless steel materials, textile fabrics, and sunscreen lotions (Duran et al. 2007). Silver nanoparticles are widely used in surface enhanced Raman scattering (SERS) spectroscopy (surface-sensitive technique which is known for its high sensitivity due to its surface plasmon effects lacks behind at uniform substrate formation) because of high local field enhancement factor. For example, it was demonstrated that SERS and silver nanorods quickly revealed the viral structures (Shanmukh et al. 2006). It can also differentiate between respiratory viruses, virus strains, and viruses containing gene deletions without manipulating the virus. Silver nanoparticles can also be used for modification of polymer based diffraction gratings which further can be used in grating light reflection spectroscopy for sensing (Šileikaitė et al. 2007; Abu Hatab et al. 2008). Furthermore, the electrochemical properties of silver nanoparticles incorporated them in nano-scale sensors which can offer faster response times and lower detection limits (Manno et al. 2008; Hahm and Lieber 2004). Manno et al. (Hahm and Lieber 2004) electrodeposited silver nanoparticles onto alumina plates gold micro-patterned electrode which showed a high sensitivity to hydrogen peroxide (H_2O_2). Silver nanoparticles were found to catalyze the chemiluminescence from luminal-hydrogen peroxide system with catalytic activity better than Au and Pt colloid (Guo et al. 2008). The catalytic properties of silver nanoparticles varied from their size and morphology. It was reported that when the silver colloid was injected, chemiluminoscence emission from the luminal-H_2O_2 system was greatly enhanced (Guo et al. 2008). Silver nanoparticles supported halloysite nanotubes (Ag/HNTs), with silver content of approximately 11%, was applied to catalyze reduction of 4-nitrophenol with $NaBH_4$ in alkaline aqueous solutions (Liu and Zhao 2009). Nano-clusters composed of 2-8 silver atoms could be the basis for a new type of optical data storage. Moreover, fluorescent emissions from the clusters could also potentially be used in biological labels and electroluminescent displays (Berciaud et al. 2005; Kossyrev et al. 2005). The optical properties of a metallic nanoparticle depend mainly on its surface plasmon resonance, where the plasmon refers to the collective oscillation of the free electrons within the metallic nanoparticle. The plasmon resonant peaks and line widths are sensitive to the size and shape of the nanoparticle, the metallic species and the surrounding medium. The optical properties of silver nanoparticles have been taken advantage of optical chemosensors and biosensors. For instance, triangular silver nanoparticles have been used in measurement of biological binding signal between antibodies and antigens (Zhu et al. 2009).

Metal nanoparticles embedded into polymer matrix can be used as sensors, materials with solvent switchable electronic properties, optical limiters and filters, and optical data storage. They can be applied for catalytic applications and antimicrobial coatings. Production of metal/polymer nanocomposites in poly(methyl methacrylate), N-isopropylacrylamide, diallyldimethylammonium chloride, poly(vinyl alcohol), poly(vinyl acetate), poly(vinyl carbazole), and polyimide films has been reported (Boyd et al. 2006; Sakamoto et al. 2007; Sakamoto et al. 2009). Nanocomposites of polymer/silver nanoparticles have attracted great attention because of their potential applications in the fields of catalysis, bioengineering, photonics, and electronics (Kiesow et al. 2003). Polymers are considered good host materials for metallic nanoparticles as well as other stabilizing agents, including citrates, organic solvents (*e.g.*, tetrahydrofuran and tetrahydrofuran/methanol), long chain alcohols, surfactants, and organometallics (Sarkar et al. 2005; Chen et al. 2001). Various chemical and physical methods were applied to prepare metal-polymer composites (*e.g.*, fabrication approach) (Kiesow et al. 2003; Rifai et al. 2006). Fabrication approach is often referred to as the evaporation method since the polymer solvent is evaporated from the reaction mixture after nanoparticle dispersion. However, this often leads to inhomogeneous distribution of the particles in the polymer. A successful production of nanoparticles is determined by the ability to produce particles with uniform distributions and long stability, given their tendency to rapidly agglomerate in aqueous solution (Rifai et al. 2006). Another approach is a system in which simultaneous polymerization and metal reduction occur. For instance, polychrome silver nanoparticles were obtained by microwave irradiation from a solution of silver nitrate in the presence of poly(*N*-vinyl-2-pyrrolidone) without any other reducing agent. The *in-situ* reduction of silver ions in poly(N-vinyl-2-pyrrolidone) by microwave irradiation produced particles with narrow size distribution (He et al. 2002; Mohan et al. 2007). Moreover, silver nanoparticles (approximately 5 nm) were prepared by mixing equal amounts of 0.5 wt % aqueous solutions of acacia (natural polymer) and silver nitrate. In this process, acacia polymeric chains promote the reduction and act as good stabilizers over 5 months (He et al. 2002; Mohan et al. 2007).

Conventional heating approach (simultaneous polymerization-reduction method) to polymerize acrylonitrile reduces silver ions resulting in homogeneous dispersal and narrow size distributions of the silver nanoparticles in silver-polyacrylonitrile composite powders (Zhang and Han 2003). Moreover, size-controlled synthesis of silver nano-complex was recently achieved in reduction of silver nitrate by a UV-irradiated argine-tungstonsilicate acid solution (Yang et al. 2007). Other various metal-polymer nano-composites have been prepared by these reduction methods, such as poly(vinyl alcohol)-Ag, Ag-polyacrylamide, Ag-acrylonitrile, Ag_2Se-polyvinyl alcohol, Ag-polyimide, Au-polyaniline, and Cu-poly(acrylic acid) (Ma et al. 2002; Akamatsu et al. 2003). Poly(vinyl alcohol) (a water soluble and biocompatible polymer with extremely low cytotoxicity) has a lot of biomedical applications. This biologically friendly polymer can be used as a stabilizer due to its optical clarity and classified into grades of partially (85-89%) and fully (97-99.5%) hydrolyzed polymers (Badr and Mahmoud 2006). This polymer was widely used in textile, paper industries, food packaging, pharmacy, and cosmetics. Introduction of nano-sized silver into poly(vinyl alcohol) provides antibacterial activity, which is highly desired in textiles used in medicine, wound dressing applications, clothing and household products (Hong et al. 2006). For instance, poly(vinyl alcohol) nanofibers containing silver nanoparticles were prepared by electro-spinning poly(vinyl alcohol)/silver nitrate aqueous solutions, followed by short heat treatment. The produced nanofibers containing silver nanoparticles showed strong antimicrobial activities (Hong et al. 2006). Various approaches such as solvent evaporation, electron and ultraviolet radiation, *in situ* chemical reduction, thermal annealing, and sono-chemical method have been applied to synthesize poly(vinyl alcohol)-silver nanoparticles (Sharma et al. 2009). In solvent evaporation methods, synthesis of poly(vinyl alcohol)-silver nanoparticles was achieved by first reducing silver salt with $NaBH_4$, followed by mechanical dispersion of the silver colloids into the dissolved polymer, and then the solvent was evaporated resulting in final structure. Electron and ultraviolet radiation approaches involve irradiation of a Ag^+ doped polymer film which gives poly(vinyl alcohol)-silver nano-composites (Sharma et al. 2009). In one study, silver nanoparticles have been prepared in poly(vinyl alcohol) matrix by thermal treatments. The annealing time and temperature vary morphologies of poly(vinyl alcohol)-silver nanoparticles (Clemenson et al. 2007). Moreover, hydrogels of poly(vinyl alcohol)- poly(N-vinyl pyrolidone containing silver nanoparticles were produced by repeated freezing–thawing treatments. These nanosilver-containing hydrogels had excellent antibacterial effects against *E. coli* and *S. aureus* (Yu et al. 2007).

Apatite-coated Ag/AgBr/titanium dioxide (TiO_2) was prepared by deposition of silver to generate electron–hole pairs by extending the excitation wavelength to the visible-light region, AgBr, and hydroxy apatite as photosensitive material and adsorption bioceramic, respectively (Elahifard et al. 2007). In another study, nanoparticle photocatalysts based on silver, carbon, and sulfur-doped titanium dioxide were synthesized by a modified sol-gel method (Hamal and Klabunde 2007). Silver/TiO_2 nano-composites and poly(vinyl alcohol)-capped colloidal silver-TiO_2 nano-composites were prepared. The synthesized nanocomposites were effective toward destroying Gram-negative bacteria, e.g. *Escherichia coli*. (Guin et al. 2007). These silver/TiO_2 surfaces had advantageous properties, including visible light photocatalysis, biological compatibility, and antimicrobial activities (Elahifard et al. 2007; Seery et al. 2007; Hamal and Klabunde 2007). For example, Ag/TiO_2 composite films were prepared by incorporating silver in the pores of titanium dioxide films with an impregnation method via photoreduction (a versatile and convenient process with advantage of space-selective fabrication). It was reported that these composite films are promising in application of photocatalysis, antimicrobial and self-clean technologies (Liu et al. 2008). In order to synthesize silver nanoparticles on titanium dioxide surfaces, aqueous reduction, photochemical (*in-situ* photochemical fabrication approach can be used for preparation of metal nanoparticles/polymer nano-composites), liquid phase deposition, and sol-gel methods can be applied (Zheng et al. 2008; Elahifard et al. 2007; Seery et al. 2007; Liu et al. 2008; Hamal and Klabunde 2007; Cozzoli et al. 2004). Silver nanoparticles with a narrow size distribution were prepared by simple aqueous reduction from silver ions in different molar ratios of titanium dioxide suspensions and $NaBH_4$ (reducing agent) (Nino-Martinez et al. 2008). One of the photochemical reduction methods involves loading silver nanoparticles (3-5 nm) onto the surface of titanium dioxide nano-tubes first using the liquid deposition approach followed by UV or laser irradiation (Li et al. 2008).

4. Toxicity of nano-silver particles

Nano-sized particles can pass through biological membranes and penetrate even very small capillaries throughout the body (*e.g.*, pass through blood-brain and blood testes barriers). Size, morphology, and surface area are recognized as important determinants for toxicity of nanoparticles (Ji et al. 2007). Information on the toxicological implication of silver nanoparticles is limited (Chen and Schluesener 2008). Toxicity of silver nanoparticles is mostly determined *in vitro* with particles ranging from approximately 1-100 nm. The *in vitro* and *in vivo* toxicity studies revealed that silver nanoparticles have the potential ability to cause chromosomal aberrations and DNA damage, to enter cells and cause cellular damage, and are capable of inducing proliferation arrest in cell lines of zebra fish (Ji et al. 2007; Asharani PV et al. 2007; Hussain SM et al. 2005). Limited health effects of the use of nano-silver particles in humans have been reported, such as argyria (a condition in which the skin becomes blue or bluish-grey colored because of improper exposure to chemical forms of element) which was appeared to occur only after intake of large amounts of colloidal silver particles (the suspension with nano-silver of different sizes). Several cross-sectional studies reported that argyria is the most frequent adverse outcome from exposure to silver nanoparticles. For instance, prolonged ingestion of colloidal silver can change the color of skin and cause blue-grey appearance on face (the symptoms of argyria) (Chang et al. 2006). Silver nanoparticles can bind to different tissues (bind to proteins and enzymes in mammalian cells) and cause toxic effects, such as adhesive interactions with cellular membrane and production of highly reactive and toxic radicals like reactive oxygen species (ROS), which can cause inflammation and show intensive toxic effects on mitochondrial function. Potential target organs for nano-silver toxicity may involve liver and immune system. Accumulation and histopathological effects were seen in livers from rats systemically exposed to silver nanoparticles (10-15 nm). Furthermore, increased liver enzymes were reported in Sprague-Dawley rats after Twenty-eight-day oral exposure to nano-silver particles (60 nm) (Kim et al. 2008). It was reported that silver-coated dressing acticoat caused raised liver enzymes and argyria-like symptoms in burn patients (Trop et al. 2006). The *in vitro* and *in vivo* studies revealed toxicity effects of silver nanoparticles on immune system, especially cytokine excretion. Application of 1% nano-silver cream (silver nanoparticles with size range of approximately 50 nm), inhibited DNB-induced allergic contact dermatitis, and accumulation in the spleen has also been reported (Bhol and Schechter 2007; Takenaka et al. 2001). It has been suggested that silver nanoparticles are especially effective at inhibiting inflammations and may be used to treat immunologic and inflammatory diseases (Shin et al. 2007). Oral administration of silver nanoparticles (60 nm) to rats induced some local inflammatory effects (Kim et al. 2008). Moreover, in one *in vitro* study, toxicity effects of nano-silver particles on erythrocytes was reported (Garner et al. 1994), while an increase in red blood cells was seen after oral administration of silver nanoparticles (60 nm) (Kim et al. 2008), but not after inhalation of silver nanoparticles (15 nm) (Ji et al. 2007). An *in vitro* study suggests that nano-silver particles may have cytotoxic effects on mammalian (mouse) spermatogonial germ line stem cells (Braydich-Stolle et al. 2005). Silver nanoparticles (10 μg ml^{-1} and above concentration) showed dramatic changes such as necrosis and apoptosis of the cells and at 5-10 μg ml^{-1} drastically reduced mitochondrial function and cell viability. It was reported that there is no evidence available to demonstrate that silver is a cause of neurotoxic damage even though silver deposits have been identified in the region of cutaneous nerves (Lansdown 2007). The respiratory system seemed relatively unaffected by exposure to nano-silver *in vivo* in a 28 days study (Ji et al. 2007). However, cytotoxic effect of nano-silver particles on alveolar macrophages and alveolar epithelial cells was demonstrated, *in vitro* (Soto et al. 2005; Soto et al. 2007; Park et al. 2007). In a 90 days inhalation study, decrease in lung functions, including tidal volume, minute volume and peak inspiration flow as well as inflammatory lesions in lung morphology and effects on inflammatory markers were reported (Sung et al. 2008). Kim et al. (2008) performed a bone marrow micronucleus test as a part of a 28-day oral administration study in order to investigate genotoxic effects or carcinogenicity of exposure to silver nanoparticles. No significant genotoxic potential of oral exposure to silver nanoparticles (60 nm) was found in this study. An increase in (mostly local) malignant tumors was found following chronic subcutaneous administration of high doses colloidal silver (Schmaehl and Steinhoff 1960).

5. Conclusion

Silver nanoparticles have attracted the attention of researchers because of their unique properties, and proven applicability in diverse areas such as medicine, catalysis, textile engineering, biotechnology, nanobiotechnology, bioengineering sciences, electronics, optics, and water treatment. Moreover, silver nanoparticles have significant inhibitory effects against microbial pathogens, and are widely used as antimicrobial agents in a diverse range of consumer products, including air sanitizer sprays, socks, pillows, slippers, respirators, wet wipes, cosmetics, detergents, soaps, shampoos, toothpastes, air and water filters, coatings of refrigerators, vacuum cleaners, bone cement, wound dressings, surgical dressings, washing machines, food storage packaging, and cell phones. The flexibility of silver nanoparticle synthetic methods and facile incorporation of silver nanoparticles into different media have interested researchers to further investigate the mechanistic aspects of antimicrobial, antiviral and anti-inflammatory effects of these nanoparticles.

In brief, there are limited well-controlled investigations on potential toxicities of nano-silver particles, and it seems that additional long-term studies (preferably using multiple particle sizes) are needed to better characterize and understand the risk of using these particles.

Various chemical, physical and biological synthetic methods have been developed to obtain silver nanoparticles of various shapes and sizes, including laser ablation, gamma irradiation, electron irradiation, chemical reduction, photochemical methods, microwave processing, and thermal decomposition of silver oxalate in water and in ethylene glycol, and biological synthetic methods. Most of these methods are still in the development stages and the problems experienced involve the stability and aggregation of nanoparticles, control of crystal growth, morphology, size and size distribution, and occasional difficulty in the management of the synthesis, as in the case of the radiolysis technique. Moreover, the separation of produced nanoparticles for further applications is still an important issue. By using different reducing agents and stabilizers, the particle size and morphology of silver nanoparticles have been controlled. Selection of solvent medium and selection of eco-friendly nontoxic reducing and stabilizing agents are the most important issues which must be considered in green synthesis of silver nanoparticles. In biological synthetic methods, it was shown that the silver nanoparticles produced by plants are more stable in comparison with those produced by other organisms. Plants (especially plant extracts) are able to reduce silver ions faster than fungi or bacteria. Furthermore, in order to use easy and safe green methods in scale-up and industrial production of well-dispersed silver nanoparticles, plant extracts are certainly better than plant biomass or living plants. However, better experimental procedures are needed for synthesis of well-characterized nanoparticles.

6. Acknowledgment

The authors appreciate the support of Isfahan University of Medical Sciences.

7. References

Abid, J.P.; Wark, A.W.; Brevet, P.F. & Girault H.H. (2002). Preparation of silver nanoparticles in solution from a silver salt by laser irradiation. *Chem Commun*, 792-793

Abu Hatab, N.A.; Oran, J.M. & Sepaniak, M.J. (2008). Surface-enhanced raman spectroscopy substrates created via electron beam lithography and nanotransfer printing. *American Chemical Society ACS NANO*, Vol.2, pp.377–385

Ahmad, A.; Mukherjee, P.; Senapati, S.; Mandal, D.; Khan, M.I.; Kumar, R. & Sastry, M. (2003a). Extracellular biosynthesis of silver nanoparticles using the fungus *Fusarium oxysporum*, *Colloids and Surfaces B: Biointerfaces*, Vol.28, pp.313-318

Ahmad, A.; Senapati, S.; Khan, M.I.; Kumar, R.; Ramani, R.; Srinivas, V. & Sastry, M.(2003b). Intracellular synthesis of gold nanoparticles by a novel alkalotolerant actinomycete, *Rhodococcus* species. *Nanotechnology*, Vol.14, pp.824-828

Akamatsu, K.; Ikeda, S. & Nawafune, H. (2003). Site-selective direct silver metallization on surface-modified polyimide layers. *Langmuir*, Vol.19, pp.10366–10371

Ankamwar, B.; Damle, C.; Ahmad, A. & Sastry, M. (2005). Biosynthesis of gold and silver nanoparticles using *Emblica officinalis* fruit extract, their phase transfer and transmetallation in an organic solution. *J Nanosci Nanotechnol*, Vol.5, pp.1665-1671

Asharani, P.V.; Nair, G.; Zhiyuan, H.; Manoor, P. & Valiyaveettil, S. (2007). Potential health impacts of silver nanoparticles. *Abstracts of Papers, 234th ACS National Meeting, Boston, MA, USA, August*:19-23, 2007. pp:TOXI-2099.

Badr, Y. & Mahmoud, M.A. (2006). Enhancement of the optical properties of poly vinyl alcohol by doping with silver nanoparticles. *J Appl Polym Sci*, Vol.99, pp.3608-3614

Balaji, D.S.; Basavaraja, S.; Deshpande, R.; Bedre Mahesh, D.; Prabhakar, B.K. & Venkataraman, A. (2009). Extracellular biosynthesis of functionalized silver nanoparticles by strains of *Cladosporium cladosporioides* fungus. *Colloids and Surfaces B: Biointerfaces*, Vol.68, pp.88-92

Begum, N.A.; Mondal, S.; Basu, S.; Laskar, R.A. & Mandal, D. (2009). Biogenic synthesis of Au and Ag nanoparticles using aqueous solutions of Black Tea leaf extracts. *Colloids and Surfaces B: Biointerfaces,* Vol.71, pp.113–118

Berciaud, S.; Cognet, L.; Tamarat, P. & Lounis, B. (2005). Observation of intrinsic size effects in the optical response of individual gold nanoparticles. *Nano Lett*, Vol.5,pp.515–518

Bhainsa, K.C. & D'Souza, S. (2006). Extracellular biosynthesis of silver nanoparticles using the fungus *Aspergillus fumigatus. Colloids and Surfaces B: Biointerfaces*, Vol.47, pp.160-164

Bhol, K.C. & Schechter, P.J. (2007). Effects of nanocrystalline silver (NPI 32101) in a rat model of ulcerative colitis. *Digestive Dis Sci*, Vol.52, pp.2732-2742

Bogle, K.A.; Dhole, S.D. & Bhoraskar, V.N. (2006). Silver nanoparticles: synthesis and size control by electron irradiation. *Nanotechnology,* Vol.17, pp.3204

Bogunia-Kubik, K. & Sugisaka, M. (2002). From molecular biology to nanotechnology and nanomedicine. *Biosystems*, Vol.65, pp.123-138

Boyd, D.A.; Greengard, L.; Brongersma, M.; El-Naggar, M.Y. & Goodwin, D. (2006). Plasmon-assisted chemical vapor deposition. *Nano Lett,* Vol.6, pp.2592-2597

Braydich-Stolle, L.; Hussain, S.; Schlager, J.J. & Hofmann, M.C. (2005). *In vitro* cytotoxicity of nanoparticles in mammalian germline stem cells. *Toxicol Sci,* Vol.88, pp.412-419

Brust, M. & Kiely, C. (2002). Some recent advances in nanostructure preparation from gold and silver particles: a short topical review. *Colloids Surf A: Phys Eng Aspects,* Vol.202, pp.175-186

Cao, G. (2004). *Nanostructures and nanomaterials: synthesis, properties and applications*. Imperial College Press, ISBN: 1-86094-4159, London

Chang, A.L.S.; Khosravi, V. & Egbert, B. (2006). A case of argyria after colloidal silver ingestion. *J Cutan Pathol,* Vol.33, pp.809-811

Chen, J.; Wang, K.; Xin, J. & Jin, Y. (2008). Microwave-assisted green synthesis of silver nanoparticles by carboxymethyl cellulose sodium and silver nitrate. *Mater Chem Phys*, Vol.108, pp.421-424

Chen, W.; Cai, W.; Zhang, L.; Wang, G. & Zhang, L. (2001). Sonochemical Processes and Formation of Gold Nanoparticles within Pores of Mesoporous Silica. *J Colloid Interface Sci*, Vol.238, pp.291-295

Chen, X. & Schluesener, H.J. (2008). Nano-silver: A nanoproduct in medical application. *Toxicol Lett*, Vol.176, pp.1-12

Cheng, P.; Song, L.; Liu, Y. & Fang, Y.E. (2007). Synthesis of silver nanoparticles by γ-ray irradiation in acetic water solution containing chitosan. *Radiat Phys Chem*, Vol.76, pp.1165-1168

Cho, K-H.; Park, J-E.; Osaka, T. & Park, S.G. (2005). The study of antimicrobial activity and preservative effects of nanosilver ingredient. *Electrochimica Acta*, Vol.51, pp.956-960

Clemenson, S.; David, L. & Espuche, E. (2007). Structure and morphology of nanocomposite films prepared from polyvinyl alcohol and silver nitrate: Influence of thermal treatment. *J Polym Sci Part A Polym Chem*, Vol.45, pp.2657-2672

Colvin, V.L.; Schlamp, M.C. & Alivisatos, A. (1994). Light emitting diodes made from cadmium selenide nanocrystals and a semiconducting polymer. *Nature*, Vol.370, pp.354-357

Cozzoli, P.; Comparelli, R.; Fanizza, E.; Curri, M.; Agostiano, A. & Laub, D. (2004). Photocatalytic synthesis of silver nanoparticles stabilized by TiO_2 nanorods: a semiconductor/metal nanocomposite in homogeneous nonpolar solution. *J Am Chem Soc*, Vol.126, pp.3868–3879

Daniel, M.C. & Astruc, D. (2004). Gold nanoparticles: assembly, supramolecular chemistry, quantum-size-related properties, and applications toward biology, catalysis, and nanotechnology. *Chem Rev*, Vol.104, pp.293-346

Dolgaev, S.I.; Simakin, A.V.; Voronov, V.V.; Shafeev, G.A. & Bozon-Verduraz, F. (2002). Nanoparticles produced by laser ablation of solids in liquid environment. *Appl Surf Sci*, Vol.186, pp.546-551

Dubey, S.P.; Lahtinen, M.; Särkkä, H. & Sillanpää, M. (2010). Bioprospective of *Sorbus aucuparia* leaf extract in development of silver and gold nanocolloids. *Colloids and Surfaces B: Biointerfaces*, Vol.80, pp.26-33

Dura´n, N.; Marcato, P.D.; De, S.; Gabriel, I.H.; Alves, O.L. & Esposito, E. (2007). Antibacterial effect of silver nanoparticles produced by fungal process on textile fabrics and their effluent treatment. *J Biomed Nanotechnol*, Vol.3, pp.203-208

Duran, N.; Marcato, D.P.; De Souza, H.I.; Alves, L.O. & Espsito, E. (2007). Antibacterial effect of silver nanoparticles produced by fungal process on textile fabrics and their effluent treatment. *J Biomedical Nanotechnology*, Vol.3, pp.203-208

Elahifard, M.R.; Rahimnejad, S.; Haghighi, S. & Gholami, M.R. (2007). Apatite-coated Ag/AgBr/TiO_2 visible-light photocatalyst for destruction of bacteria. *J Am Chem Soc*, Vol.129, pp.9552–9553

Elechiguerra, J.L.; Burt, J.L.; Morones, J.R.; Camacho-Bragado, A.; Gao, X.; Lara, H.H. & Yacaman, M.J. (2005). Interaction of silver nanoparticles with HIV-1. *J Nanobiotechnol*, Vol.3:6

Elumalai, E.K.; Prasad, T.N.V.K.V.; Hemachandran, J.; Viviyan Therasa, S.; Thirumalai, T. & David, E. (2010). Extracellular synthesis of silver nanoparticles using leaves of *Euphorbia hirta* and their antibacterial activities. *J Pharm Sci & Res*, Vol.2, pp.549-554

Eutis, S.; Krylova, G.; Eremenko, A.; Smirnova, N.; Schill, A.W. & El-Sayed, M. (2005). Growth and fragmentation of silver nanoparticles in their synthesis with a fs laser and CW light by photo-sensitization with benzophenone. *Photochem Photobiol Sci*, Vol.4, pp.154-159

Evanoff, Jr. & Chumanov, G. (2004). Size-controlled synthesis of nanoparticles. 2. measurement of extinction, scattering, and absorption cross sections. *J Phys Chem B*, Vol.108, pp.13957-13962

Garner, M.; Reglinski, J.; Smith, W.E. & Stewart, M.J. (1994). The interaction of colloidal metals with erythrocytes. *J Inorg Biochem*, Vol.56, pp.283-290

Gogoi, K.S.; Gopina, P.; Paul, A.; Ramesh, A.; Ghosh, S.S. & Chattopadhyay, A. (2006). Green fluorescent protein expressing *Escherichia coli* as a model system for investigating the antimicrobial activities of silver nanoparticles. *Langmuir*, Vol.22, pp.9322-9328

Guin, D.; Manorama, S.V.; Latha, J.N.L. & Singh, S. (2007). Photoreduction of silver on bare and colloidal TiO_2 nanoparticles/nanotubes: synthesis, characterization, and tested for antibacterial outcome. *J Phys Chem C*, Vol.111, pp.13393–13397

Guo, J-Z.; Cui, H.; Zhou, W. & Wang, W. (2008). Ag nanoparticle-catalyzed chemiluminescent reaction between luminal and hydrogen peroxide. *J Photochem Photobiol A: Chem*, Vol.193, pp.89-96

Guo, J.; Cui, H.; Zhou, W. & Wang, W. (2008). Ag nanoparticle-catalyzed chemiluminescent reaction between luminol and hydrogen peroxide. *J Photochem Photobiol A: Chem*, Vol.193, pp.89-96

Hahm, J. & Lieber, C. (2004). Direct ultrasensitive electrical detection of DNA and DNA sequence variations using nanowire nanosensors. *Nano Lett*, Vol.4, pp.51–54

Hamal, D.B. & Klabunde, K.J. (2007). Synthesis, characterization, and visible light activity of new nanoparticle photocatalysts based on silver, carbon, and sulfur-doped TiO_2. *J Colloid Interface Sci*, Vol.311, pp.514-522

Hamilton, J.F. & Baetzold, R. (1979). Catalysis by small metal clusters. *Science*, Vol.205, pp.1213-1220

Harris, A.T. & Bali, R. (2008). On the formation and extent of uptake of silver nanoparticles by live plants. *J Nanopart Res*, Vol.10, pp.691-695

He, R.; Qian, X.; Yin, J. & Zhu, Z. (2002). Preparation of polychrome silver nanoparticles in different solvents. *J Mater Chem*, Vol.12, pp.3783-3786

Hoffman, A.J.; Mills, G.; Yee, H. & Hoffmann, M. (1992). Q-sized cadmium sulfide: synthesis, characterization, and efficiency of photoinitiation of polymerization of several vinylic monomers. *J Phys Chem*, Vol.96, pp.5546-5552

Hong, K.H.; Park, J.L.; Sul, I.H.; Youk, J.H. & Kang, T.J. (2006). Preparation of antimicrobial poly(vinyl alcohol) nanofibers containing silver nanoparticles. *J Polym Sci Part B Polym Phys*, Vol.44, pp.2468-2474

Hornebecq, V.; Antonietti, M.; Cardinal, T. & Treguer-Delapierre, M. (2003). Stable silver nanoparticles immobilized in mesoporous silica. *Chemistry of Materials*, Vol.15, pp.1993–1999

Hu, B.; Wang, S-B.; Wang, K.; Zhang, M. & Yu, S.H. (2008). Microwave-assisted rapid facile "green" synthesis of uniform silver nanoparticles: Self-assembly into multilayered films and their optical properties. *J Phys Chem C*, Vol.112, pp.11169–11174

Hu, L.; Choi, J.W.; Yang, Y.; Jeong, S.; La Mantia, F.; Cui, L-F. & Cui, Y. (2009). Highly conductive paper for energy-storage devices. Proceedings of the National Academy of Sciences of the United States of America 106:21490-21494, S21490/21491-S21490/21413

Huang, H. & Yang, X. (2004). Synthesis of polysaccharide-stabilized gold and silver nanoparticles: A green method. *Carbohydr Res*, Vol.339, pp.2627-2631

Huang, H. & Yang, Y. (2008) Preparation of silver nanoparticles in inorganic clay suspensions. *Compos Sci Technol*, Vol.68, pp.2948-2953

Huang, J.; Li, Q.; Sun. D.; Lu, Y.; Su, Y.; Yang, X.; Wang, H.; Wang, Y.; Shao, W.; He, N.; Hong, J. & Chen, C. (2007). Biosynthesis of silver and gold nanoparticles by novel sundried *Cinnamomum camphora* leaf. *Nanotechnology*, Vol.18, pp.105,104

Huang, J.; Zhan, G.; Zheng, B.; Sun, D.; Lu, F.; Lin, Y.; Chen, H.; Zheng, Z.; Zheng, Y. & Li, Q. (2011). Biogenic silver nanoparticles by *Cacumen Platycladi* extract: synthesis, formation mechanism and antibacterial activity. *Ind Eng Chem Res*, Vol.50, pp.9095-9106

Hussain, S.M.; Hess, K.L.; Gearhart, J.M.; Geiss, K.T. & Schlager, J.J. (2005). *In vitro* toxicity of nanoparticles in BRL 3A rat liver cells. *Toxicol in vitro*, Vol.19, pp.975-983

Ingle, A.; Gade, A.; Pierrat, S.; Sönnichsen, C. & Mahendra, R. (2008). Mycosynthesis of silver nanoparticles using the fungus *Fusarium acuminatum* and its activity against some human pathogenic bacteria. *Current Nanoscience*, Vol.4, pp.141-144

Iravani, S. (2011). Green synthesis of metal nanoparticles using plants. *Green Chem*, Vol.13, pp. 2638-2650.

Jacob, J.A.; Mahal, H.S.; Biswas, N.; Mukerjee, T. & Kappor, S. (2008). Role of phenol derivatives in the formation of silver nanoparticles. *Langmuir*, Vol.24, pp.528–533

Jain, P. & Pradeep, T. (2005). Potential of silver nanoparticle-coated polyurethane foam as an antibacterial water filter. *Biotechnol Bioeng*, Vol.90, pp.59-63

Ji, J.H.; Jung, J.H.; Kim, S.S.; Yoon, J.U.; Park, J.D.; Choi, B.S.; Chung, Y.H.; Kwon, I.H.; Jeong, J.; Han, B.S.; Shin, J.H.; Sung, J.H.; Song, K.S. & Yu, I.J. (2007). Twenty-eight-day inhalation toxicity study of silver nanoparticles in Sprague-Dawley rats. *Inhal Toxicol*, Vol.19, pp.857-871

Jia, X.; Ma, X.; Wei, D.; Dong, J. & Qian, W. (2008). Direct formation of silver nanoparticles in cuttlebone-derived organic matrix for catalytic applications. *Colloids Surf A:Physicochem Eng Aspects*, Vol.330, pp.234-240

Jin, R.; Cao, Y.C.; Hao, E.; Metraux, G.S.; Schatz, G.C. & Mirkin, C. (2003). Controlling anisotropic nanoparticle growth through plasmon excitation. *Nature*, Vol.425, pp.487–490

Johans, C.; Clohessy, J.; Fantini, S.; Kontturi, K. & Cunnane, V.J. (2002). Electrosynthesis of polyphenylpyrrole coated silver particles at a liquid-liquid interface. *Electrochemistry Communications*, Vol.4, pp.227–230

Jung, J.; Oh, H.; Noh, H.; Ji, J. & Kim, S. (2006). Metal nanoparticle generation using a small ceramic heater with a local heating area. *J Aerosol Sci*, Vol.37, pp.1662-1670

Kabashin, A.V. & Meunier, M. (2003) Synthesis of colloidal nanoparticles during femtosecond laser ablation of gold in water. *J Appl Phys*, Vol.94, pp.7941-7943

Kalishwaralal, K.; Deepak, V.; Ramkumarpandian, S.; Bilal, M. & Sangiliyandi G. (2008a). Biosynthesis of silver nanocrystals by *Bacillus licheniformis*. *Colloids and Surfaces B: Biointerfaces*, Vol.65, pp.150-153

Kalishwaralal, K.; Deepak, V.; Ramkumarpandian, S.; Nellaiah, H. & Sangiliyandi, G. (2008b). Extracellular biosynthesis of silver nanoparticles by the culture supernatant of *Bacillus licheniformis*. *Mater Lett*, Vol.62, pp.4411-4413

Kasthuri, J.; Veerapandian, S. & Rajendiran, N. (2009). Biological synthesis of silver and gold nanoparticles using apiin as reducing agent. *Colloids and Surfaces B: Biointerfaces*, Vol.68, pp.55-60

Kasture, M.; Singh, S.; Patel, P.; Joy, P.A.; Prabhune, A.A.; Ramana, C.V. & Prasad, B.L.V. (2007). Multiutility sophorolipids as nanoparticle capping agents: synthesis of stable and water dispersible Co nanoparticles. *Langmuir*, Vol.23, pp.11409-11412

Kathiresan, K.; Manivannan, S.; Nabeel, M.A. & Dhivya, B. (2009). Studies on silver nanoparticles synthesized by a marine fungus, *Penicillium fellutanum* isolated from coastal mangrove sediment. *Colloids and Surfaces B: Biointerfaces*, Vol.71, pp.133-137

Kawasaki, M. & Nishimura, N. (2006). 1064-nm laser fragmentation of thin Au and Ag flakes in acetone for highly productive pathway to stable metal nanoparticles. *Appl Surf Sci*, Vol.253, pp.2208-2216

Keating, C.D.; Kovaleski, K.K. & Natan, M. (1998). Heightened electromagnetic fields between metal nanoparticles: surface enhanced Raman scattering from metal-Cytochrome c-metal sandwiches. *J Phys Chem B*, Vol.102, pp.9414

Kesharwani, J.; Yoon, K.Y.; Hwang, J. & Rai, M. (2009). Phytofabrication of silver nanoparticles by leaf extract of *Datura metel*: hypothetical mechanism involvedin synthesis. *Journal of Bionanoscience*, Vol.3, pp.1-6

Kiesow, A.; Morris, J.E.; Radehaus, C. & Heilmann, A. (2003). Switching behavior of plasma polymer films containing silver nanoparticles. *J Appl Lett*, Vol.94, pp.6988-6990

Kim, K.J.; Sung, W.S.; Moon, S.K.; Choi, J.S.; Kim, J.G. & Lee, D.G. (2008). Antifungal effect of silver nanoparticles on dermatophytes. *J Microbiol Biotechnol*, Vol.18, pp.1482–1484

Kim, K.J.; Sung, W.S.; Suh, B.K.; Moon, S.K.; Choi, J.S.; Kim, J.G. & Lee, D.G. (2009). Antifungal activity and mode of action of silver nano-particles on *Candida albicans*. *Biometals*, Vol.22, pp.235-242

Kim, S.; Yoo, B.; Chun, K.; Kang, W.; Choo, J.; Gong, M. & Joo, S. (2005). Catalytic effect of laser ablated Ni nanoparticles in the oxidative addition reaction for a coupling reagent of benzylchloride and bromoacetonitrile. *J Mol Catal A: Chem*, Vol.226, pp.231-234

Kim, Y.S.; Kim, J.S.; Cho, H.S.; Rha, D.S.; Kim, J.M.; Park, J.D.; Choi, B.S.; Lim, R.; Chang, H.K.; Chung, Y.H.; Kwon, I.H.; Jeong, J.; Han, B.S. & Yu, I.J. (2008). Twenty-eight-day oral toxicity, genotoxicity, and gender-related issue distribution of silver nanoparticles in Sprague-Dawley rats. *Inhal Toxicol*, Vol.20, pp.575-583

Klaus-Joerger, T.; Joerger, R.; Olsson, E. & Granqvist, C.G. (2001). Bacteria as workers in the living factory: metal-accumulating bacteria and their potential for materials science. *Trends in Biotechnology*, Vol.19, pp.15–20

Klaus, T.; Joerger, R.; Olsson, E. & Granqvist, C.Gr. (1999). Silver-based crystalline nanoparticles, microbially fabricated. *Proc Natl Acad Sci USA*, Vol.96, pp.13611-13614

Komarneni, S.; Li, D.; Newalkar, B.; Katsuki, H. & Bhalla, A.S. (2002). Microwave-Polyol Process for Pt and Ag Nanoparticles. *Langmuir*, Vol.18, pp.5959–5962

Kora, A.J.; Sashidhar, R.B. & Arunachalam, J. (2010). Gum kondagogu (*Cochlospermum gossypium*): A template for the green synthesis and stabilization of silver nanoparticles with antibacterial application. *Carbohydrate Polymers*, Vol.82, pp.670-679

Korbekandi, H.; Iravani, S. & Abbasi, S. (2009). Production of nanoparticles using organisms. *Critical Reviews in Biotechnology*, Vol.29, pp.279-306

Kossyrev, P.; Yin, A.; Cloutier, S.; Cardimona, D.; Huang, D.; Alsing, P. & Xu, J. (2005). Electric field tuning of plasmonic response of nanodot array in liquid crystal matrix. *Nano Lett,* Vol.5, pp.1978–1981

Kotthaus, S.; Gunther, B.H.; Hang, R. & Schafer, H. (1997). Study of isotropically conductive bondings filled with aggregates of nano-sited Ag-particles. *IEEE Trans Compon Packaging Technol,* Vol.20, pp.15-20

Krishnaraj, C.; Jagan, E.G.; Rajasekar, S.; Selvakumar, P.; Kalaichelvan, P.T. & Mohan, N. (2010). Synthesis of silver nanoparticles using *Acalypha indica* leaf extracts and its antibacterial activity against water borne pathogens. *Colloids and Surfaces B: Biointerfaces,* Vol.76, pp.50–56

Kruis, F.; Fissan, H. & Rellinghaus, B. (2000). Sintering and evaporation characteristics of gas-phase synthesis of size-selected PbS nanoparticles. *Mater Sci Eng B,* Vol.69, pp.329-324

Krutyakov, Y.; Olenin, A.; Kudrinskii, A.; Dzhurik, P. & Lisichkin, G. (2008). Aggregative stability and polydispersity of silver nanoparticles prepared using two-phase aqueous organic systems. *Nanotechnol Russia,* Vol.3, pp.303-310

Kumar, S.A.; Majid Kazemian, A.; Gosavi, S.W.; Sulabha, K.K.; Renu, P.;Ahmad A. & Khan M.I. (2007). Nitrate reductase-mediated synthesis of silver nanoparticles from $AgNO_3$. *Biotechnology Letters,* Vol.29, pp.439-445

Kumar, V.S.; Nagaraja, B.M.; Shashikala, V.; Padmasri, A.H.; Madhavendra, S.S. & Raju, B.D. (2004). Highly efficient Ag/C catalyst prepared by electro-chemical deposition method in controlling microorganisms in water. *J Mol Catal A,* Vol.223, pp.313–319

Kvitek, L.; Panacek, A.; Soukupova, J.; Kolar, M.; Vecerova, R.; Prucek, R.; Holecová, M. & Zbořil, R. (2008). Effect of surfactants and polymers on stability and antibacterial activity of silver nanoparticles (NPs). *J Phys Chem C,* Vol.112, pp.5825-5834

Kvítek, L.; Prucek, R.; Panáček, A.; Novotný, R.; Hrbác, J. & Zbořil, R. (2005). The influence of complexing agent concentration on particle size in the process of SERS active silver colloid synthesis. *J Mater Chem,* Vol.15, pp.1099-1105

Lansdown, A.B. (2007). Critical observations on the neurotoxicity of silver. *Crit Rev Toxicol,* Vol.37, pp.237-250

Lee, H.Y.; Li, Z.; Chen, K.; Hsu, A.R.; Xu, C.; Xie, J.; Sun, S.; Chen, X. (2008). PET/MRI dual-modality tumor imaging using arginine-glycine-aspartic (RGD)–conjugated radiolabeled iron oxide nanoparticles, *J Nucl Med,* Vol.49, pp.1371-1379

Lengke, M.F.; Fleet, M.E. & Southam, G. (2007). Biosynthesis of silver nanoparticles by filamentous cyanobacteria from a silver(I) nitrate complex. *Langmuir,* Vol.23, pp.2694–2699

Li, H.; Duan, X.; Liu, G. & Liu, X. (2008). Photochemical synthesis and characterization of Ag/TiO_2 nanotube composites. *J Mater Sci,* Vol.43, pp.1669-1676

Li, Q.; Mahendra, S.; Lyon, D.; Brunet, L.; Liga, M.; Li, D. & Alvarez, P. (2008). Antimicrobial nanomaterials for water disinfection and microbial control: Potential applications and implications. *Water Res,* Vol.42, pp.4591-4602

Lin, Y.E.; Vidic, R.D.; Stout, J.E.; McCartney, C.A. & Yu V.L. (1998). Inactivation of *Mycobacterium avium* by copper and silver ions. *Water Res,* Vol.32, pp.1997-2000

Link, S.; Burda, C.; Nikoobakht, B. & El-Sayed, M. (2000). Laser-induced shape changes of colloidal gold nanorods using femtosecond and nanosecond laser pulses. *J Phys Chem B,* Vol.104, pp.6152–6163

Liu, P. & Zhao, M. (2009). Silver nanoparticle supported on halloysite nanotubes catalyzed reduction of 4-nitrophenol (4-NP). *Appl Surf Sci*, Vol.255, pp.3989-3993

Liu, Y.; Wang, X.; Yang, F. & Yang, X. (2008). Excellent antimicrobial properties of mesoporous anatase TiO_2 and Ag/TiO_2 composite films. *Microporous Mesoporous Mater*, Vol.114, pp.431-439

Lok, C. (2006). Proteomic analysis of the mode of antibacterial action of silver nanoparticles. *J Proteome Res*, Vol.5, pp.916–924

Long, D.; Wu, G. & Chen, S. (2007). Preparation of oligochitosan stabilized silver nanoparticles by gamma irradiation. *Radiat Phys Chem*, Vol.76, pp.1126-1131

Ma, H.; Yin, B.; Wang, S.; Jiao, Y.; Pan, W.; Huang, S.; Chen, S. & Meng, F. (2004). Synthesis of silver and gold nanoparticles by a novel electrochemical method. *Chem Phys Chem*, Vol.24, pp.68–75

Ma, X-D.; Qian, X-F.; Yin, J. & Zhu, Z-K. (2002). Preparation and characterization of polyvinyl alcohol–selenide nanocomposites at room temperature. *J Mater Chem*, Vol.12, pp.663-666

Macdonald, I.D.G. & Smith, W. (1996). Orientation of Cytochrome c adsorbed on a citrate-reduced silver colloid surface. *Langmuir*, Vol.12, pp.706

Mafune, F.; Kohno, J.; Takeda, Y.; Kondow, T. & Sawabe, H. (2001). Formation of gold nanoparticles by laser ablation in aqueous solution of surfactant. *J Phys Chem B*, Vol.105, pp.5114-5120

Mafune, F.; Kohno, J.; Takeda, Y.; Kondow, T. & Sawabe, H. (2000). Structure and stability of silver nanoparticles in aqueous solution produced by laser ablation. *J Phys Chem B*, Vol.104, pp.8333-8337

Magnusson, M.; Deppert, K.; Malm, J.; Bovin, J. & Samuelson, L. (1999). Gold nanoparticles: production, reshaping, and thermal charging. *J Nanoparticle Res*, Vol.1, pp.243-251

Maliszewska, I.; Szewczyk, K. & Waszak, K. (2009). Biological synthesis of silver nanoparticles. *Journal of Physics: Conference Series*, Vol.146, pp.1-6

Malval, J-P.; Jin, M.; Balan, L.; Schneider, R.; Versace, D-L.; Chaumeil, H.; Defoin, A. & Soppera, O. (2010). Photoinduced size-controlled generation of silver nanoparticles coated with carboxylate-derivatized thioxanthones. *J Phys Chem C*, Vol.114, pp.10396-10402

Manno, D.; Filippo, E.; Giulio, M. & Serra, A. (2008). Synthesis and characterization of starch-stabilized Ag nanostructures for sensors applications. *J Non-Cryst Solids*, Vol.354, pp.5515-5520

Mansur, H.S.; Grieser, F.; Marychurch, M.S.; Biggs, S.; Urquhart, R.S. & Furlong, D. (1995). Photoelectrochemical properties of 'q-state' cds particles in arachidic acid langmuir-blodgett films. *J Chem Soc Faraday Trans*, Vol.91, pp.665-672

Mantion, A.; Guex, A.G.; Foelske, A.; Mirolo, L.; Fromn, K.M. & Painsi, M. (2008). Silver nanoparticle engineering via oligovaline organogels. *Soft Matter*, Vol.4, pp.606-617

Merga, G.; Wilson, R.; Lynn, G.; Milosavljevic, B. & Meisel, D. (2007). Redox catalysis on "naked" silver nanoparticles. *J Phys Chem C*, Vol.111, pp.12220–12226

Mohan, Y.M.; Raju, K.M.; Sambasivudu, K.; Singh, S. & Sreedhar, B. (2007). Preparation of acacia-stabilized silver nanoparticles: A green approach. *J Appl Polym Sci*, Vol.106, pp.3375-3381

Mokhtari, N.; Daneshpajouh, S.; Seyedbagheri, S.; Atashdehghan, R.; Abdi, K.; Sarkar, S.; Minaian, S.; Shahverdi, H.R. & Shahverdi, A.R. (2009). Biological synthesis of very small silver nanoparticles by culture supernatant of *Klebsiella pneumonia*: The effects of visible-light irradiation and the liquid mixing process. *Materials Research Bulletin,* Vol.44, pp.1415-1421

Morones, J.R.; Elechiguerra, L.J.; Camacho, A.; Holt, K.; Kouri, B.J.; Ramirez, T.J. & Yocaman, J.M. (2005). The bactericidal effect of silver nanoparticles. *Nanotechnology,* Vol.16, pp.2346-2353

FU, M.; Li Q.; Sun, D.; Lu, Y.; He, N.; Xu, D.; Wang, H. & Huang, J. (2006). Rapid preparation process of silver nanoparticles by bioreduction and their characterizations. *Chinese J Chem Eng,* Vol.14, pp.114-117

Nair, B. & Pradeep, T. (2002). Coalescence of nanoclusters and formation of submicron crystallites assisted by *Lactobacillus* strains. *Crystal Growth & Design,* Vol.2, pp.293-298

Navaladian, S.; Viswanathan, B.; Varadarajan, T.K. & Viswanath, R.P. (2008). Microwave-assisted rapid synthesis of anisotropic Ag nanoparticles by solid state transformation. *Nanotechnology,* Vol.19, pp.045603

Nino-Martinez, N.; Martinez-Castanon, G.A.; Aragon-Pina, A.; Martinez-Gutierrez, F.; Martinez-Mendoza, J.R. & Ruiz, F. (2008). Characterization of silver nanoparticles synthesized on titanium dioxide fine particles. *Nanotechnology,* Vol.19, pp.065711/065711-065711/065718

Oliveira, M.; Ugarte, D.; Zanchet, D. & Zarbin, A. (2005). Influence of synthetic parameters on the size, structure, and stability of dodecanethiol-stabilized silver nanoparticles. *J Colloid Interface Sci,* Vol.292, pp.429-435

Pal, S.; Tak, Y.K. & Song, J.M. (2007). Does the antibacterial activity of silver nanoparticles depend on the shape of the nanoparticle? a study of the gram-negative bacterium *Escherichia coli. Applied and Environmental Microbiology,* Vol.73, pp.1712–1720

Panacek, A.; Kvitek, L.; Prucek, R.; Kolar, M.; Vecerova, R.; Pizurova, N.; Sharma, V.K.; Nevěná, T. & Zbořil, R. (2006). Silver colloid nanoparticles: synthesis, characterization, and their antibacterial activity. *J Phys Chem B,* Vol.110, pp.16248–16253

Park, S.; Lee, Y.K.; Jung, M.; Kim, K.H.; Eun-Kyung Ahn, N.C.; Lim, Y. & Lee, K.H. (2007). Cellular toxicity of various inhalable metal nanoparticles on human alveolar epithelial cells. *Inhalat Toxicol,* Vol.19, pp.59-65

Pillai, Z.S. & Kamat, P.V. (2004). What factors control the size and shape of silver nanoparticles in the citrate ion reduction method?. *J Phys Chem B,* Vol.108, pp.945-951

Pissuwan, D.; Valenzuela, S.M. & Cortie, M.B. (2006). Therapeutic possibilities of plasmonically heated gold nanoparticles, *Trends Biotechnol,* Vol.24, pp.62-67

Prasad, T.N.V.K.V. & Elumalai, E. (2011). Biofabrication of Ag nanoparticles using *Moringa oleifera* leaf extract and their antimicrobial activity. *Asian Pacific Journal of Tropical Biomedicine,* Vol.1, pp.439-442

Ramnami, S.P.; Biswal, J. & Sabharwal, S. (2007). Synthesis of silver nanoparticles supported on silica aerogel using gamma radiolysis. *Radiat Phys Chem,* Vol.76, pp.1290-1294

Raveendran. P.; Fu, J. & Wallen, S.L. (2003). Completely "green" synthesis and stabilization of metal nanoparticles. *J Am Chem Soc,* Vol.125, pp.13940–13941

Raveendran, P.; Fu, J. & Wallen, S.L. (2005). A simple and "green" method for the synthesis of Au, Ag, and Au–Ag alloy nanoparticles, *Green Chem,* Vol.8, pp.34-38

Ravindra, S.; Murali Mohan, Y.; Narayana Reddy, N. & Raju, K.M. (2010). Fabrication of antibacterial cotton fibres loaded with silver nanoparticles via "Green Approach". *Colloids and Surfaces A: Physicochem Eng Aspects*, Vol.367, pp.31-40

Rifai, S.; Breen, C.A.; Solis, D.J. & Swager, T.M. (2006). Facile *in situ* silver nanoparticle formation in insulating porous polymer matrices. *Chem Mater*,Vol.18, pp.21–25

Roe, D.; Karandikar, B.; Bonn-Savage, N.; Gibbins, B. & Roullet, J.B. (2008). Antimicrobial surface functionalization of plastic catheters by silver nanoparticles. *J Antimicrob Chemother*, Vol.61, pp.869-876

Sadowski, Z.; Maliszewska, I.H.; Grochowalska, B.; Polowczyk, I. & Kozlecki, T. (2008). Synthesis of silver nanoparticles using microorganisms. *Materials Science-Poland*, Vol.26, pp.419-424

Safaepour, M.; Shahverdi, A.R.; Shahverdi, H.R.; Khorramizadeh, M.R. & Gohari, A.R. (2009). Green synthesis of small silver nanoparticles using geraniol and its cytotoxicity against Fibrosarcoma-Wehi 164. *Avicenna J Med Biotech*, Vol.1, pp.111-115

Saifuddin, N.; Wong, C.W. & Nur Yasumira, A.A. (2009). Rapid biosynthesis of silver nanoparticles using culture supernatant of bacteria with microwave irradiation. *E-Journal of Chemistry*, Vol.6, pp.61-70

Sakamoto, M.; Fujistuka, M. & Majima, T. (2009). Light as a construction tool of metal nanoparticles: synthesis and mechanism. *J Photochem and Photobiol C: Photochem Reviews*, Vol.10, pp.33-56

Sakamoto, M.; Tachikawa, T.; Fujitsuka, M. & Majima, T. (2007). Photochemical formation of Au/Cu bimetallic nanoparticles with different shapes and sizes in a PVA film. *Adv Funct Mater*, Vol.17, pp.857-862

Sanghi, R. & Verma P. (2009). Biomimetic synthesis and characterisation of protein capped silver nanoparticles. *Bioresource Technology*, Vol.100, pp.501–504

Santhoshkumar, T.; Rahuman, A.A.; Rajakumar, G.; Marimuthu, S.; Bagavan, A.; Jayaseelan, C.; Zahir, A.A.; Elango, G. & Kamaraj, C. (2010). Synthesis of silver nanoparticles using *Nelumbo nucifera* leaf extract and its larvicidal activity against malaria and filariasis vectors. *Parasitol Res*, Vol.108, pp.693-702

Sarkar, A.; Kapoor, S. & Mukherjee, T. (2005). Preparation, characterization, and surface modification of silver nanoparticles in formamide. *J Phys Chem*, Vol.109, pp.7698–7704

Sastry, M.; Ahmad, A.; Khan, M.I. & Kumar, R. (2003). Biosynthesis of metal nanoparticles using fungi and actinomycete. *Current Science*, Vol.85, pp.162-170

Sathishkumar, M.; Sneha, K.; Won, S.W.; Cho, C-W.; Kim, S. & Yun, Y.S. (2009). *Cinnamon zeylanicum* bark extract and powder mediated green synthesis of nano-crystalline silver particles and its bactericidal activity. *Colloids and Surfaces B: Biointerfaces*, Vol.73, pp.332–338

Schmaehl, D. & Steinhoff, D. (1960). Studies on cancer induction with colloidal silver and gold solutions in rats. *Z Krebsforsch*, Vol.63, pp.586-591

Schmid, G. (1992). Large clusters and colloids. Metals in the embryonic state. *Chem Rev*, Vol.92, pp.1709-1727

Seery, M.K.; George, R.; Floris, P. & Pillai, S.C. (2007). Silver doped titanium dioxide nanomaterials for enhanced visible light photocatalysis. *J Photochem Photobiol A Chem*, Vol.189, pp.258-263

Senapati, S. (2005). *Biosynthesis and immobilization of nanoparticles and their applications.* University of pune, India

Shahverdi, A.R.; Minaeian, S.; Shahverdi, H.R.; Jamalifar, H. & Nohi, A. (2007). Rapid synthesis of silver nanoparticles using culture supernatants of Enterobacteria: A novel biological approach. *Process Biochemistry*, Vol.42, pp.919-923

Shahverdi, R.A.; Fakhimi, A.; Shahverdi, H.R. & Minaian, S. (2007). Synthesis and effect of silver nanoparticles on the antibacterial activity of different antibiotics against *Staphyloccocus aureus* and *Escherichia coli. Nanomed: Nanotechnol Biol Med,* Vol.3, pp.168-171

Shankar, S.S.; Absar, A. & Murali, S. (2003). Geranium leaf assisted biosynthesis of silver nanoparticles. *Biotechnol Prog,* Vol.19, pp.1627-1631

Shankar, S.S.; Rai, A.; Ankamwar, B.; Singh, A.; Ahmad, A. & Sastry, M. (2004). Biological synthesis of triangular gold nanoprisms. *Nature Materials,* Vol.3, pp.482-488

Shanmukh, S.; Jones, L.; Driskell, J.; Zhao, Y.; Dluhy, R. & Tripp, R.A. (2006). Rapid and sensitive detection of respiratory virus molecular signatures using a silver nanorod array SERS substrate. *Nano Letters*, Vol.6, pp.2630-2636

Sharma, V.K.; Yngard, R.A. & Lin, Y. (2009). Silver nanoparticles: green synthesis and their antimicrobial activities. *Advances in Colloid and Interface Science,* Vol.145, pp.83-96

Shchukin, D.G.; Radtchenko, I.L. & Sukhorukov, G. (2003). Photoinduced reduction of silver inside microscale polyelectrolyte capsules. *Chem Phys Chem,* Vol.4, pp.1101–1103

Shin, S.H.; Ye, M.K.; Kim, H.S. & Kang, H.S. (2007). The effects of nanosilver on the proliferation and cytokine expression by peripheral blood mononuclear cells. *Int Immuno pharmacol,* Vol.7, pp.1813-1818

Si, S. & Mandal, T.K. (2007). Tryptophan-based peptides to synthesize gold and silver nanoparticles: a mechanistic and kinetic study. *Chem A Eur J*, Vol.13, pp.3160-3168

Šileikaitė, A.; Puišo, J.; Prosyčevas, I.; Guobienė, A.; Tamulevičius, S.; Tamulevičius, T. & Janušas, G. (2007). Polymer diffraction gratings modified with silver nanoparticles. *Materials Science (Medžiagotyra),* Vol.13, pp.273–277

Singhal, G.; Bhavesh, R.; Kasariya, K.; Sharma, A.R. & Singh, R.P. (2011). Biosynthesis of silver nanoparticles using *Ocimum sanctum* (Tulsi) leaf extract and screening its antimicrobial activity. *J Nanopart Res*, Vol.13, pp.2981-2988

Slawson, R.M.; Van, D.M.; Lee, H. & Trevor, J. (1992). Germanium and silver resistance, accumulation and toxicity in microorganisms. *Plasmid*, Vol.27, pp.73-79

Socol, Y.; Abramson, O.; Gedanken, A.; Meshorer, Y.; Berenstein, L. & Zaban, A. (2002). Suspensive electrode formation in pulsed sonoelectrochemical synthesis of silver nanoparticles. *Langmuir*, Vol.18, pp.4736–4740

Song, J.Y. & Kim, B. (2008). Rapid biological synthesis of silver nanoparticles using plant leaf extracts. *Bioprocess Biosyst Eng*, Vol.32, pp.79-84

Soroushian, B.; Lampre, I.; Belloni, J. & Mostafavi, M. (2005). Radiolysis of silver ion solutions in ethylene glycol: solvated electron and radical scavenging yields. *Radiat Phys Chem*, Vol.72, pp.111-118

Soto, K.; Garza, K.M. & Murr, L.E. (2007). Cytotoxic effects of aggregated nanomaterials. *Acta Biomater,* Vol.3, pp.351-358

Soto, K.F.; Carrasco, A.; Powell, T.G.; Garza, K.M. & Murr, L.E. (2005). Comparative *in vitro* cytotoxicity assessment of some manufactured nanoparticulate materials characterized by transmission electron microscopy. *J Nanopart Res,* Vol.7, pp.145-169

Soukupova, J.; Kvitek, L.; Panacek, A.; Nevecna, T. & Zboril, R. (2008). Comprehensive study on surfactant role on silver nanoparticles (NPs) prepared via modified Tollens process. *Mater Chem Phys,* Vol.111, pp.77-81

Sreeram, K.J.; M N. & Nair, B.U. (2008). Microwave assisted template synthesis of silver nanoparticles. *Bull Mater Sci,* Vol.31, pp.937–942

Su, H.L.; Chou, C.C.; Hung, D.J.; Lin, S.H.; Pao, I.C.; Lin, J.H.; Huang, F.L.; Dong, R.X. & Lin, J.J. (2009). The disruption of bacterial membrane integrity through ROS generation induced by nanohybrids of silver and clay. *Biomaterials,* Vol.30, pp.5979–5987

Sudeep, P.K. & Kamat, P.V. (2005). Photosensitized growth of silver nanoparticles under visible light irradiation: a mechanistic investigation. *Chem Mater,* Vol.17, pp.5404–5410

Sun, R.W.; Chen, R.; Chung, N.P.; Ho, C.M.; Lin, C.L. & Che, C.M. (2005). Silver nanoparticles fabricated in Hepes buffer exhibit cytoprotective activities toward HIV-1 infected cells. *Chem Commun,* Vol.40, pp.5059-5061

Sung, J.H.; Ji, J.H.; Yoon, J.U.; Kim, D.S.; Song, M.Y.; Jeong, J.; Han, B.S.; Han, J.H.; Chung, Y.H.; Kim, J.; Kim, T.S.; Chang, H.K.; Lee, E.J.; Lee, J,H. & Yu, I.J. (2008). Lung function changes in Sprague-Dawley rats after prolonged inhalation exposure to silver nanoparticles. *Inhal Toxicol,* Vol.20, pp.567-574

Sylvestre, J.P.; Kabashin, A.V.; Sacher, E.; Meunier, M. & Luong, J.H.T. (2004). Stabilization and size control of gold nanoparticles during laser ablation in aqueous cyclodextrins. *J Am Chem Soc,* Vol.126, pp.7176-7177

Tai, C.; Wang, Y-H. & Liu, H.S. (2008). A green process for preparing silver nanoparticles using spinning disk reactor. *AIChE J,* Vol.54, pp.445-452

Takenaka, S.; Karg, E.; Roth, C.; Schulz, H.; Ziesenis, A.; Heinzmann, U.; Schramel, P. & Heyder, J. (2001). Pulmonary and systemic distribution of inhaled ultrafine silver particles in rats. *Environ Health Perspect,* Vol.109, pp.547-551

Tan, M.; Wang, G.; Ye, Z. & Yuan, J. (2006). Synthesis and characterization of titania-based monodisperse fluorescent europium nanoparticles for biolabeling, *Journal of Luminescence,* Vol.117, pp.20-28

Tarasenko, N.; Butsen, A.; Nevar, E. & Savastenko, N. (2006). Synthesis of nanosized particles during laser ablation of gold in water. *Appl Surf Sci,* Vol.252, pp.4439-4444

Trop, M.; Novak, M.; Rodl, S.; Hellbom, B.; Kroell, W. & Goessler, W. (2006). Silver coated dressing acticoat caused raised liver enzymes and argyria-like symptoms in burn patient. *J Trauma Injury Infect Crit Care,* Vol.60, pp.648-652

Troupis, A.; Hiskia, A. & Papaconstantinou, E. (2002). Synthesis of metal nanoparticles by using polyoxometalates as photocatalysts and stabilizers. *Angew Chem Int Ed,* Vol.41, pp.1911-1914

Tsuji, M.; Matsumoto, K.; Jiang, P.; Matsuo, R.; Hikino, S.; Tang, X-L. & Nor Kamarudin, K.S. (2008). The Role of Adsorption Species in the Formation of Ag Nanostructures by a Microwave-Polyol Route. *Bull Chem Soc Jpn,* Vol.81, pp.393-400

Tsuji, T.; Iryo, K.; Watanabe, N. & Tsuji, M. (2002). Preparation of silver nanoparticles by laser ablation in solution: influence of laser wavelength on particle size. *Appl Surf Sci,* Vol.202, pp.80-85

Tsuji, T.; Kakita, T. & Tsuji, M. (2003). Preparation of nano-Size particle of silver with femtosecond laser ablation in water. *Applied Surface Science*, Vol.206, pp.314–320

Veerasamy, R.; Xin, T.Z.; Gunasagaran, S.; Xiang, T.F.W.; Yang, E.F.C.; Jeyakumar, N. & Dhanaraj. S.A. (2011). Biosynthesis of silver nanoparticles using mangosteen leaf extract and evaluation of their antimicrobial activities. *Journal of Saudi Chemical Society*, Vol.15, pp.113-120

Vigneshwaran, N.; Ashtaputre, N.M.; Varadarajan, P.V.; Nachane, R.P.; Paralikar, K.M. & Balasubramanya, R. (2007). Biological synthesis of silver nanoparticles using the fungus *Aspergillus flavus*. *Materials Letters*, Vol.61, pp.1413-1418

Vigneshwaran, N.; Kathe, A.A.; Varadarajan, P.V.; Nachane, R.P. & Balasubramanya, R. (2006a). Biomimetics of silver nanoparticles by white rot fungus, *Phaenerochaete chrysosporium*. *Colloids and Surfaces B: Biointerfaces*, Vol.53, pp.55-59

Vigneshwaran, N.; Nachane, R.P.; Balasubramanya, R.H. & Varadarajan, P.V. (2006b). A novel one-pot 'green' synthesis of stable silver nanoparticles using soluble starch. *Carbohydr Res*, Vol.341, pp.2012-2018

Vilchis-Nestor, A.R.; Sánchez-Mendieta, V.; Camacho-López, M.A.; Gómez-Espinosa, R.M.; Camacho-López, M.A. & Arenas-Alatorre, J. (2008). Solventless synthesis and optical properties of Au and Ag nanoparticles using *Camellia sinensis* extract. *Materials Letters*, Vol.62, pp.3103–3105

Wang, Y. & Herron, N. (1991). Nanometer-sized semiconductor clusters: materials synthesis, quantum size effects, and photophysical properties. *J Phys Chem*, Vol.95, pp.525-532

Wei, H.; Li, J.; Wang, Y. & Wang, E. (2007). Silver nanoparticles coated with adenine: preparation, self-assembly and application in surface-enhanced Raman scattering. *Nanotechnology*, Vol.18, pp.175610

Wiley, B.; Sun, Y.; Mayers, B. & Xi, Y. (2005). Shape-controlled synthesis of metal nanostructures: the case of silver. *Chem Eur J*, Vol.11, pp.454-463

Wu, Q.; Cao, H.; Luan, Q.; Zhang, J.; Wang, Z.; Warner, J-H. & Watt, A.A.R. (2008). Biomolecule-assisted synthesis of water-soluble silver nanoparticles and their biomedical applications. *Inorg Chem*, Vol.47, pp.5882–5888

Yang, L.; Shen, Y.; Xie, A. & Zhang, B. (2007). Facile size-controlled synthesis of silver nanoparticles in UV-irradiated tungstosilicate acid solution. *J Phys Chem C*, Vol.111, pp.5300–5308

Yin, H.; Yamamoto, T.; Wada, Y. & Yanagida, S. (2004). Large-scale and size-controlled synthesis of silver nanoparticles under microwave irradiation. *Materials Chemistry and Physics*, Vol.83, pp.66–70

Yin, Y.; Li, Z-Y.; Zhong, Z.; Gates, B. & Venkateswaran, S. (2002). Synthesis and characterization of stable aqueous dispersions of silver nanoparticles through the Tollens process. *J Mater Chem*, Vol.12, pp.522-527

Yoon, K.Y.; Byeon, J.H.; Park, J.H. & Hwang, J. (2007). Susceptibility constrants of *Escherichia coli* and *Bacillus subtilis* to silver and copper nanoparticles. *Science of the Total Environment*, Vol.373, pp.572-575

Yu, H.; Xu, X.; Chen, X.; Lu, T.; Zhang, P. & Jiang, X. (2007). Preparation and antibacterial effects of PVA-PVP hydrogels containing silver nanoparticles. *J Appl Polym Sci*, Vol.103, pp.125-133

Zhang, G.; Keita, B.; Dolbecq, A.; Mialane, P.; Secheresse, F.; F M. & Nadjo, L. (2007). Green chemistry-type one-step synthesis of silver nanostructures based on Mo^{V}–Mo^{VI} mixed-valence polyoxometalates. *Chem Mater,* Vol.19, pp.5821-5823

Zhang, L.; Yu, J.C.; Yip, H.Y.; Li, Q.; Kwong, K.W.; A-W X. & Wong, P.K. (2003). Ambient light reduction strategy to synthesize silver nanoparticles and silver-coated TiO_2 with enhanced photocatalytic and bactericidal activities. *Langmuir,* Vol.19, pp.10372–10380

Zhang, W.; Qiao, X. & Chen, J. (2007). Synthesis of nanosilver colloidal particles in water/oil microemulsion. *Colloids Surf A: Physicochem Eng Aspects,* Vol.299, pp.22-28

Zhang, W. & Wang, G. (2003). Research and development for antibacterial materials of silver nanoparticle. *New Chem Mater,* Vol.31, pp.42-44

Zhang, Y.; Chen, F.; Zhuang, J.; Tang, Y.; Wang, D.; Wang, Y.; Dong, A. & Ren, N. (2002). Synthesis of silver nanoparticles via electrochemical reduction on compact zeolite film modified electrodes. *Chemical Communications,* Vol.24, pp.2814–2815

Zhang, Y.; Peng, H.; Huang, W.; Zhou, Y. & Yan, D. (2008). Facile preparation and characterization of highly antimicrobial colloid Ag or Au nanoparticles. *J Colloid Interface Sci,* Vol.325, pp.371-376

Zhang, Z. & Han, M. (2003). One-step preparation of size-selected and well-dispersed silver nanocrystals in polyacrylonitrile by simultaneous reduction and polymerization. *J Mater Commun,* Vol.13, pp.641-643

Zharov, V.P.; Kim, J-W.; Curiel, D.T. & Everts, M. (2005). Self-assembling nanoclusters in living systems: application for integrated photothermal nanodiagnostics and nanotherapy. *Nanomedicine: Nanotechnology, Biology and Medicine,* Vol.1, pp.326-345

Zheng, J.; Hua, Y.; Xinjun, L. & Shanqing, Z. (2008). Enhanced photocatalytic activity of TiO_2 nano-structured thin film with a silver hierarchical configuration. *Appl Surf Sci,* Vol.254, pp.1630-1635

Zhong, L.; Hu, J.; Cui, Z.; Wan, L. & Song, W. (2007). *In-Situ* loading of noble metal nanoparticles on hydroxyl-group-rich titania precursor and their catalytic applications. *Chem Mater,* Vol.19, pp.4557–4562

Zhou, Y.; Yu, S.H.; Wang, C.Y.; Li, X.G.; Zhu, Y.R. & Chen, Z.Y. (1999). A novel ultraviolet irradiation photoreduction technique for the preparation of single-crystal Ag nanorods and Ag dendrites. *Advanced Materials,* Vol.11, pp.850–852

Zhu, S.; Du, C.L. & Fu, Y. (2009). Fabrication and characterization of rhombic silver nanoparticles for biosensing. *Optical Materials,* Vol.31, pp.769-774

2

In vivo Toxicity Studies of Pristine Carbon Nanotubes: A Review

Jelena Kolosnjaj-Tabi, Henri Szwarc and Fathi Moussa
UMR CNRS 8612 and LETIAM, EA 4041, University of Paris Sud
France

1. Introduction

Discovered in 1991(Iijima, 1991), carbon nanotubes (CNTs) have attracted considerable attention in many fields of science and technology because of their unique structural, mechanical, and electronic properties. Their potential seemed paramount in the fields of materials science, including conductive and high-strength composites, energy storage and energy conversion devices, sensors, field emission displays and radiation sources, hydrogen storage, nanometer-sized semiconductor devices, probes and interconnects (Dresselhaus et al., 2004). Thus their studies progressed rapidly during the last two decades.

The chemistry of CNTs grew out from the efforts to open and fill the tubes (Ruoff et al., 1993; Sloan et al., 1998 as cited in Harris, 2009) and, indeed, to functionalize their sidewalls (J. Chen et al., 1998; Y. Chen et al., 1998). The first approach was driven partly for the study of matter in confined spaces and partly in order to use the tubes as templates for nanowires (Harris, 2009).

On the other hand, CNT functionalization was needed to disperse ("to solubilize") the tubes in aqueous media (Hirsh, 2002). Then, functionalization with biomolecules has become in vogue and many research groups have begun to investigate biological uses of these new types of nanostructures. According to some authors, CNTs could be possibly used as biosensors (Balavoine et al., 1999; Bekyarova et al., 2005; Lin et al., 2004; Richard et al., 2003; J. Wang et al., 2003; S. Wang et al., 2003), substrates for neuronal growth (Hu et al., 2004; Mattson et al., 2000; Lovat et al., 2005), supports for adhesion of liposaccharides to mimic cell membranes (X. Chen et al., 2004), delivery systems (Bianco et al., 2005), medical imaging agents (Ashcroft et al., 2007; Hartman et al., 2008; Sitharaman et al., 2005) and radiotherapeutics (Hartman et al., 2007a). Such potential uses remind those proposed for fullerenes and derivatives in the nineties (Jensen et al., 1996).

The growing use and mass production of CNTs raised concerns about their safety and environmental impact soon after the announcement of the national nanotechnology initiative by the US president in 2000 (http://www.nano.gov/). First toxicity studies addressed their safety at workplace (Oberdorster et al., 2005). Since then, the investigations of their toxicity first *in vitro* and then *in vivo* were reported in countless publications but their results remain contradictory (Kolosnjaj et al., 2007). As mentioned by most authors, this is due to several factors including surface defects, sizes, and degree of aggregation of the tested material, and exposure protocols (Oberdorster et al., 2005, Kolosnjaj et al., 2007).

In this chapter we will try to present an uptake of the current knowledge on CNT toxicity as a function of the route of administration. Because of the great number of papers devoted to this subject, we are quite aware that we will miss quoting a number of works and we apologize to forgotten colleagues.

For a better understanding of biological behaviour of CNTs we will first describe their main general characteristics.

2. General characteristics

Carbon nanotubes are mainly composed of sp^2 bonds, similar to graphite, and are categorized as single-walled nanotubes (SWNTs) and multi-walled nanotubes (MWNTs).

2.1 Structure

2.1.1 Single-walled carbon nanotubes (SWNTs)

Single-walled carbon nanotubes, composed by a rolled monolayered graphene sheet (that might be end-capped by half a C60 molecule), exist in a variety of structures corresponding to the many ways a sheet of graphite can be wrapped into a seamless tube. Each structure type has a specific diameter and chirality, or wrapping angle (α). The "armchair" structures (Fig.1 a), with α = 30°, have metallic character. The "zigzag" tubes (Fig.1 b), for which α = 0°, can be either semi-metallic or semiconducting, depending on the specific diameter. Nanotubes with chiral angles intermediate between 0 and 30° (Fig.1 c) include both semimetals and semiconductors. The terms "armchair" and "zigzag" refer to the pattern of carbon–carbon bonds along a tube's circumference (Dresselhaus et al., 2004).

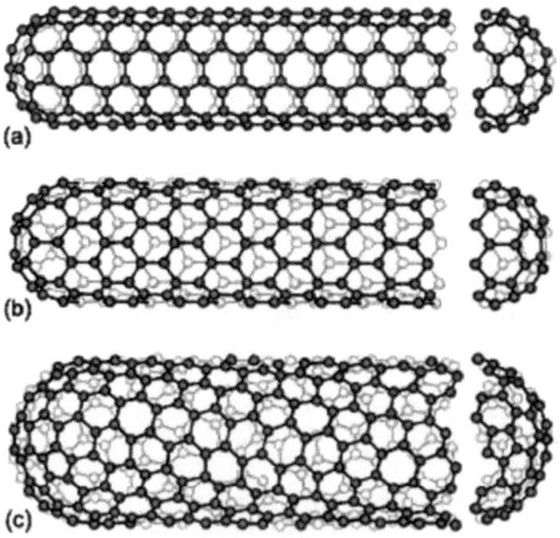

Fig. 1. Schematic representations of SWNTs in a variety of structures: (a) "armchair" (b) "zigzag" stricture (c) "intermediate" structure.

The diameter of the tubes generally varies from 0.4 to 20 nm, while the length usually reaches several micrometers. The tubes are often entangled and the ropes (Fig. 2) of SWNTs are held together by van der Waals forces (Popov et al., 2004).

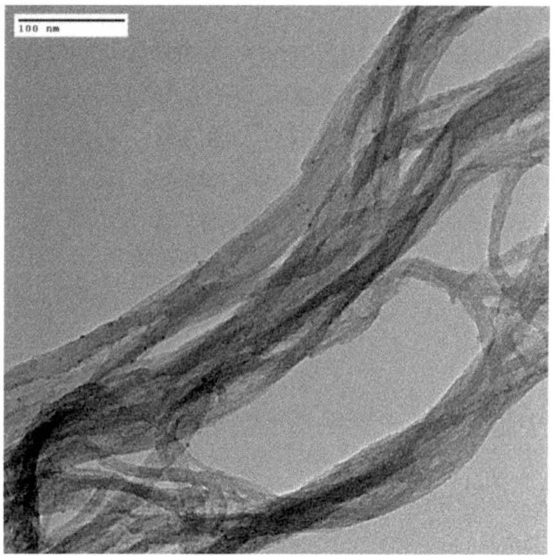

Fig. 2. Transmission electron micrograph of ropes of SWNTs

2.1.2 Multi-walled carbon nanotubes (MWNTs)

Multi-walled carbon nanotubes possess several graphitic concentric layers (Fig.3), made either by a single rolled graphene layer (resembling a scroll of parchment) or, more commonly, encased within one another (as Russian nesting dolls). The distance between each layer of graphene in a MWNT is about 0.34 nm (Iijima, 1991).

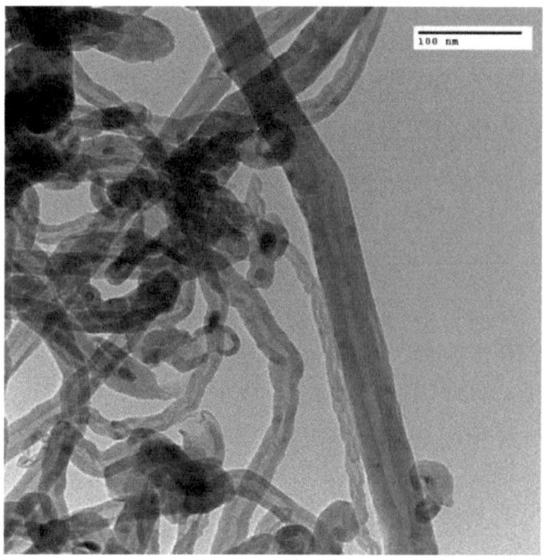

Fig. 3. Transmission electron micrograph of a bundle made of individual entangled MWNTs.

2.2 Impurities

Carbon nanotube powder often contains up to 30% metal (mainly iron and nickel) catalyst particles, as we can see on figure 2 (note the little round electron-dense particles present on nanotube ropes), and some amorphous carbon. Carbon nanotubes should be purified prior to their administration, in order to avoid any metal catalyst-related toxicity (Valko et al., 2005).

Several techniques of purification have been reviewed elsewhere (Sinha & Yeow, 2005) and the most commonly used technique appears to be the oxidation using strong acid treatments, which allows solubilisation and removal of a large part of the metallic impurities. Nevertheless, this methodology has an impact on the tubes. Strong acid conditions cut the tubes in shorter pieces and generate carboxylic functions at the tips and around the sidewalls where the curvatures of the tubes present a higher strain (Ziegler et al., 2005).

2.3 Dispersibility

Pristine (chemically unmodified) CNTs are insoluble and hardly dispersible in water. In order to disperse CNTs several groups proposed covalent and non-covalent functionalization, which will be described in this chapter in terms of *in vivo* interactions.

3. Interfacing pristine carbon nanotubes with living organisms

The interaction of CNTs with cells has been described by many research groups and has been reviewed by several authors, all converging to the conclusions that the toxicity of CNTs *in vivo* depends on the type of CNTs, including the method of production, impurities and purification process (which might affect the sidewalls), length, aggregation state, surface coating, and chemical modification. Moreover, special care is needed for the choice of reagents used to evaluate the viability of the cells (Kroll et al., 2009), as these nanoparticles may interact with assays and even dispersion agents, potentially resulting in a secondary rather than primary toxicity (Casey et al., 2007).

These physical and chemical characteristics are, indeed, important in *in vivo* studies. However, we should also consider other phenomena that may occur in complex living systems, such as interaction of CNTs with several different cells at a time, biotransformation and innate foreign body reactions, etc.

Organism-CNT interactions were first described in *in vivo* toxicity studies, which were performed in order to assess the exposure risks to CNTs at workplace. Airborne CNTs might represent a danger to people handling these materials on daily basis either by crossing their skin barrier or entering and residing in their lungs. However, in recent years, potential applications of CNTs in the biomedical field intensified the research on the *in vivo* behaviour of these materials and increased the number of studies devoted to the evaluation of their potential toxicity after non-pulmonary routes of administration.

3.1 Pilot studies on carbon nanotubes in regards to workplace safety

In 2001 the potential of CNTs to induce skin irritation was evaluated by performing two routine dermatological tests (Huczko & Lange, 2001). Initially, 40 volunteers with allergy susceptibilities were exposed for 96 h to a patch test consisting in a filter paper saturated with a water suspension of unrefined CNTs synthesized by the arc discharge process. Secondly, a modified Draize rabbit eye test using a water suspension of unrefined CNTs was conducted with four albino rabbits monitored for 72 h after exposure. Both tests showed no irritation in comparison to a CNT-free soot control and it was concluded that "no special precautions have to be taken while handling these nanostructures" (Huczko & Lange, 2001).

In a second two-part study, other investigations have been made to seek for exposure routes and toxicity of SWNTs (Maynard et al., 2004). The study was undertaken to evaluate the physical nature of the aerosol formed from SWNTs during mechanical agitation. This was complemented by a field study in which airborne and dermal exposure to SWNTs was evaluated while handling unrefined material. Although laboratory studies indicated that unrefined SWNT material could release fine particles into air under sufficient agitation, concentrations generated while handling material in the field were very low. Estimates of the airborne concentrations of nanotube materials generated during handling suggest that concentrations were lower than 53 $\mu g/m^3$ in all instances. In another way, glove deposits of SWNTs during handling were estimated at between 0.2 mg and 6 mg per hand (Maynard et al., 2004).

3.2 Respiratory exposure: Pulmonary toxicity

Carbon nanotubes are very light and could become airborne and potentially reach the lungs; therefore the earliest *in vivo* studies tried to assess their pulmonary toxicity (Lam et al., 2004; Warheit et al., 2004).

For this purpose three kinds of SWNTs were studied, namely raw and purified HiPco and CarboLex CNTs. The first material is rich in iron impurities and the last one contains nickel and yttrium impurities. The particles were dispersed by brief shearing (2 min in a small glass homogenizing tube) and subsequent sonication (0.5 min) in heat-inactivated mouse serum. Mice were then intra-tracheally instilled with 0, 0.1, or 0.5 mg of CNT or carbon black or quartz particles used as negative and positive control, respectively. Seven and 90 days after this single treatment the animals were sacrificed for histopathological examination of the lungs. All CNT treatments induced dose-dependent epithelioid granulomas and, in some cases, interstitial inflammation in the animals euthanized after 7 days (Lam et al., 2004). These lesions persisted and were more pronounced in the group euthanized after 90 days. The lungs of some animals also revealed peri-bronchial inflammation and necrosis that had extended into the alveolar septa. The lungs of mice treated with carbon black were normal, whereas those treated with high-dose quartz revealed mild to moderate inflammation. These results show that, under these conditions and on an equal- weight basis, if carbon nanotubes reach the lungs, they are much more toxic than carbon black and can be more toxic than quartz, which is considered a serious occupational health hazard in chronic inhalation exposures (Lam et al., 2004).

In a similar way, a parallel pulmonary toxicity assessment of pristine SWNTs was described (Warheit et al., 2004). The aim of the study was to evaluate the acute lung toxicity of intratracheally instilled SWNTs in rats. The applied CNTs were produced by laser ablation and contained about 30 to 40% amorphous carbon (by weight) and 5% each of nickel and cobalt. The lungs of rats were instilled either with 1 or 5 mg/kg of the following control or particle types: SWNTs, quartz particles (positive control), carbonyl iron particles (negative control), and the vehicle - phosphate buffered saline (PBS) and 1% Tween 80, or graphite particles (Warheit et al., 2004). Following exposure, the lungs of treated rats were assessed using bronchoalveolar fluid biomarkers and cell proliferation methods, as well as by histopathological examination of lung tissue at 24 h, 1 week, 1 month, and 3 months post-instillation. Exposures to high-dose (5 mg/kg) of SWNT produced mortality in approximately 15% of the instilled rats within 24 h post-instillation. This mortality resulted from mechanical blockage of the upper airways by the instilled particulate SWNT. In the surviving animals, SWNT produced temporary inflammatory and cell injury effects. Results from the lung histopathology indicated that pulmonary exposures to SWNT in rats produced a non-dose-dependent series of multifocal granulomas, which were evidence of a foreign tissue body reaction. However, they were non-uniform in distribution and not progressive beyond one month of post-exposure. The observation of SWNT-induced multifocal granulomas was inconsistent with the following: lack of lung toxicity by assessing lavage parameters, lack of lung toxicity by measuring cell proliferation parameters, apparent lack of a dose response relationship, non-uniform distribution of lesions, the paradigm of dust-related lung toxicity effects, and possible regression of effects over time. The observation of granulomas, in the absence of adverse effects measured by pulmonary endpoints was surprising, and did not follow the normal inflammogenic/fibrotic pattern produced by fibrogenic dusts, such as quartz, asbestos, and silicon carbide whiskers (Warheit et al., 2004).

While the first authors (Lam et al., 2004) concluded that SWNT were more toxic than quartz nanoparticles and crystalline silica particles, the second ones (Warheit et al., 2004) observed only a transient pulmonary inflammation and granuloma formation after SWNT exposure, contrarily to sustained lung inflammation, cytotoxicity, enhanced lung cell proliferation, foamy macrophage accumulation and lung fibrosis after exposure to quartz particles. The differences between these findings may be related in part to species differences (mouse vs. rat), but are more likely due to the differences in the experimental designs of the two studies (Warheit, 2006).

Respiratory toxicity of MWNTs has been also evaluated after intra-tracheal administration of MWNTs or ground MWNTs suspended and sonicated in sterile 0.9 % saline containing 1 % of Tween 80, at doses of 0.5, 2.0 or 5.0 mg, corresponding to approximately 2.2 mg/kg, 8.9 mg/kg and 22.2 mg/kg body-weight (bw) to Sprague-Dawley rats (Muller et al. 2005). The applied CNTs were still present in the lungs after 60 days (80% and 40% of the lowest dose) and both induced inflammatory and fibrotic reactions (Muller et al. 2005). At 2 months, pulmonary lesions induced by MWNTs were characterized by the formation of collagen-rich granulomas protruding in the bronchial lumen, in association with alveolitis in the surrounding tissues. These lesions were caused by the accumulation of large MWNT agglomerates in the airways. Ground CNTs were better dispersed in the lung parenchyma and also induced inflammatory and fibrotic responses. Both MWNTs and ground MWNTs stimulated the production of TNF-α in the lung of treated animals (Muller et al. 2005).

The physiological relevance of intra-tracheal instillation of CNTs is debatable since physiologically inspired particles would probably encounter several barriers in the upper respiratory tract before reaching the trachea and the lungs. Nevertheless, purified SWNTs elicited inflammation, fibrosis and granulomas formation in C57BL/6 mice even when administered by pharyngeal aspiration (Shvedova et al., 2005). The nanotubes used in this study were produced by HiPco and where further purified by acidic treatment. The analysis also proved that CNTs accounted for more than 99% of carbon. The animals were treated with either SWNT (0, 10, 20, 40 μg/mouse) or two reference materials (ultrafine carbon black or SiO2 at 40 μg/mouse). The animals were sacrificed at 1, 3, 7, 28, and 60 days following exposures. A rapid progressive fibrosis found in mice exhibited two distinct morphologies: 1- SWNT-induced granulomas mainly associated with hypertrophied epithelial cells surrounding dense micrometer-scale SWNT aggregates and 2- diffuse interstitial fibrosis and alveolar wall thickening likely associated with dispersed SWNTs. These differences in fibrosis morphology were attributed to the distinct particle morphologies of compact aggregates and dispersed SWNT structures. Importantly, deposition of collagen and elastin was also observed in both granulomatous regions as well as in the areas distant from granulomas. Increased numbers of alveolar type II (AT- II) cells, the progenitor cells that replicate following alveolar type I (AT-I) cell death, were also observed as a response to SWNT administration. Moreover, functional respiratory deficiencies and decreased bacterial clearance (Listeria monocytogenes) were found in mice treated with SWNT (Shvedova et al., 2005).

In a mechanistically oriented study, the physicochemical determinants of the MWNTs' toxicity mechanism were investigated (Muller et al., 2008). In this experiment the toxicity of MWNTs was evaluated after the tubes were heated at 600°C (which allowed loss of oxygenated carbon functionalities and reduction of oxidized metals) or at 2400°C (which annealed structural defects and eliminated metals) or after the MWNTs heated at 2400°C were grinded (introduction of structural defects in a metal-deprived framework). The MWNTs were suspended in 1% Tween 80 and physiological saline and administered intra-tracheally (2 mg/rat). The results show that pulmonary toxicity (and genotoxicity of MWNTs, determined *in vitro*) were reduced upon heating but restored upon grinding, indicating that the intrinsic toxicity of the tubes was mainly mediated by the presence of defective sites in their carbon framework (Muller et al., 2008).

Finally, in order to check the hypothesis linking lung toxicity to CNT aggregates (Mutlu, Budinger et al., 2010), the authors instilled intratracheally unpurified aggregated or highly dispersed SWNTs in 1% Pluronic F 108NF to mice. As-produced HiPco SWNTs were either dispersed in PBS or highly dispersed in Pluronic solution in a dose of 40 μg, which was chosen to match or exceed those previously reported to cause pulmonary fibrosis in mice (Mutlu, Budinger et al., 2010; Shvedova et al., 2005). According to the authors, lung inflammation induced by SWNTs is minimal compared to that induced by urban particulate matter or asbestos fibers (used as positive control). Aggregated SWNTs in PBS caused areas of chronic inflammation, while highly dispersed SWNTs do not cause any inflammation or fibrosis. Moreover, nanoscale dispersed SWNTs are taken up by alveolar macrophages and cleared from the lung over time (Mutlu, Budinger et al., 2010). Besides, by administering unpurified CNTs, the authors (Mutlu, Budinger et al., 2010) avoided a potential effect due to surface defects of tube sidewalls (Ziegler et al., 2005), which might contribute to an increase in collagen deposition (Mercer et al., 2008).

In a study where rats were instilled with 0.04, 0.2, or 1 mg/kg of individually dispersed MWNTs in Tween 80 (Kobayashi et al., 2010), it has been observed that pulmonary inflammatory responses occur only in the lungs of the group treated with the highest dose. Moreover, the authors did not find any evidence of chronic inflammation, such as angiogenesis or fibrosis, induced by MWNT instillation (Kobayashi et al., 2010). Light microscopic examination indicated that MWNT aggregates deposited in the lungs were phagocytized by alveolar macrophages and were accumulated in the lungs until 6-month post-exposure. These aggregates were located in the alveolar or interstitial macrophages, but individual MWNTs were not present in the cells of the interstitial tissue (Kobayashi et al., 2010). However, in the light micrograph panels, provided by the authors (Kobayashi et al., 2010) MWNTs seem to extend outside the macrophages in several directions, which is commonly referred to as incomplete or frustrated phagocytosis that is known to be a pro-inflammatory condition (Balkwill & Mantovani, 2001). According to the authors, the MWNTs used in this study were less than 20 µm long, but after measuring the tubes inside the alveolar macrophages, they concluded that median length was approximately 1.5 µm, although some tubes were measuring up to 6 µm (Kobayashi et al., 2010).

In contrast, a recent study reported that highly dispersed MWNTs could, depending on the way of administration (i.e. intratracheal instillation or inhalation) and dose, produce pulmonary lesions (Morimoto et al., 2011). The MWNTs that were used in this study were ground in a fructose mold - the fructose was rinsed afterwards with water and hydrogen peroxide. According to the authors this process slightly oxidized the tubes, which were subsequently dispersed in a 0.05% Triton X-100 solution (Morimoto et al., 2011). Triton X-100 is often used in cell biology to digest the cell membrane and cytoplasm to access the cell nucleus (http://fr.wikipedia.org/wiki/Triton_X-100). In the experiment of intratracheal instillation, two single doses (0.2 mg or 1 mg, 1.1 µm of mean particle length) were administered to rats and the study was conducted up to 6 months. While only a transient infiltration of inflammatory cells was observed for 0.2 mg treated animals, the high dose caused small granulomatous lesions and transient collagen depositions. In parallel, the authors conducted an inhalation study of dispersed MWNTs in a daily average mass of 0.37 ± 0.18 mg/m^3 (Morimoto et al., 2011). The rats were exposed to aerosol particles for 6h per day, 5 days a week for 4 weeks. At the end of the experiment, the dispersed MWNTs with the average length of 1.1 µm caused only a minimal transient inflammation, which did not cause neutrophil infiltration into alveolar space. Moreover granulomatous lesions or collagen depositions were not observed (Morimoto et al., 2011).

In conclusion, the studies performed thus far indicate that due to van der Waals interactions individual SWNTs are prone to form large aggregates, in air or in aqueous solutions, which can be more than one hundred micrometers in diameter. While accidental industrial exposure is the most probable risk and might have a serious impact at workplace, toxicity was not observed after the intratracheal instillation of nanoscale dispersed SWNTs at a dose of 1.6 mg/kg (Mutlu, Budinger et al., 2010), or nanoscale dispersed MWNTs at a dose of 0.66 mg/kg (Morimoto et al., 2011). It is worthy to note that these doses would be the equivalent to a single instilled dose of approximately 112 g or 46 g, respectively, in an average weighting human.

3.2.1 Respiratory exposure and health risks

The analogy between CNTs and asbestos fibres was pointed out in the late nineties (Service, 1998). The term asbestos refers to a variety of fibrous silicates, which were exploited commercially in past for their desirable physical properties, such as sound absorption, average tensile strength, and resistance to fire, heat, electrical and chemical damage (http://en.wikipedia.org/wiki/Asbestos). Asbestos fibres have a high aspect ratio and they are characterized by high chemical stability in physiological environment (Kane & Hurt, 2008). The pathologies related to asbestos exposure, especially lung fibrosis (asbestosis) and lung cancer (mesothelioma) that most often originates in the pleura, the outer lining of the lungs, have caused a major worldwide occupational health disaster (Donaldson & Poland, 2009) and founded reasonable fear of these airborne fibres.

Each lung is invested by a membrane, the pleura, which is arranged in the form of a closed invaginated sac. The membrane lining on the lung with its 'visceral' mesothelial layer is the visceral pleura and the membrane attached to the chest wall, covered by a continuous 'parietal' mesothelial cell layer is the parietal pleura. The pleural mesothelium is the primary mesothelial target for inhaled fibres (Donaldson et al, 2010). The space between the visceral and the parietal pleura contains the pleural fluid and a population of pleural macrophages. Pleural liquid is derived mainly from capillaries in the parietal pleura and is principally removed by lymphatic stomata in the parietal pleura (Lai-Fook, 2004), which drains the pleural fluid to the lymph nodes. This turnover is important for clearance of particles and fibres that reach the pleural space (Donaldson et al, 2010). While the exact mechanism of fibre deposition in the pleural mesothelium remains unclear, research indicates that retention of biopersistent fibres at the parietal pleura initiates mesothelial injury and inflammation that, overtime, lead to mesothelioma. When the fibre diameter is small, the fibre will align with the flow and deeply penetrate the lungs. The fibres are more or less cleared by macrophages, depending on their length. If the fibers are too long they cannot be entirely phagocytized. This unachieved - 'frustrated' phagocytosis is pro-inflammatory condition, characterized by the release of inflammatory mediators into the environment. These mediators may recruit other cells (for example collagen synthesizing fibroblasts) or cause DNA damage and mutations to proliferating cells, which may in term cause tumour development (Balkwill & Mantovani, 2001). Carbon nanotubes are fiber-shaped, however, for what concerns SWNTs, they are flexible and bendable, and often entangled. These particle-sized tangles would not obey the fibre toxicity paradigm because of their non-fibrous geometry (Donaldson et al, 2010; Kane & Hurt, 2008). Multi-walled carbon nanotubes, on the other hand, are much stiffer and generally less entangled; therefore, if long enough, they might present a risk (Donaldson et al, 2010).

A study performed with nickel containing milled pluronic-suspended MWNTs (Ryman-Rasmussen et al., 2009) with a length ranging between 100 nm and 10 µm showed that the nanotubes are observed inside the sub-pleural tissue macrophages after a single 6 h inhalation exposure of 30 mg/m^3. Fibrotic lesions, which increase 2 and 6 weeks after exposure, remain focal and regional. This effect did not occur after exposure to a dose of 1 mg/m^3, which according to the authors corresponds to 0.2 mg/kg. While the authors did not find MWNTs-loaded macrophages inside the pleura, they did notice an increased number of pleural mononuclear cells.

A study published about the same time reported that MWNTs could reach the pleura after pharyngeal aspiration (Porter et al., 2010). The inflammation extended from lungs to pleura in half of the MWNT-exposed mice. At 56 days post exposure, MWNTs penetrated the pleura in two out of four mice treated with the highest MWNT dose. The inflammation induced by the nanotubes was transient at low doses but persistent through day 56 at a dose of 40 µg.

In another study performed by the same group (Mercer et al., 2010), the authors reported an initial high density of penetrations into the sub-pleural tissue and the intra-pleural space one day following aspiration of MWNTs (80 µg per mice). The kinetics of penetration decreased due to the clearance by alveolar macrophages by day 7 and reached steady state levels in the sub-pleural tissue and intra-pleural space from days 28 to 56. The majority of the tubes (62% of the dose) resided in alveolar macrophages, while 0.6% of tubes, reached the visceral pleura region (sub-pleura and intra-pleural space).

As it has been already emphasized (Donaldson et al, 2010), the question that we should ask with regard to any fibre in relation to mesothelioma is not "Do fibres reach the pleura?" but "Are fibres retained in the parietal pleura", which is the site of origin of pleural mesothelioma.

When long and short CNTs (as well as long and short asbestos fibres) were injected directly into the pleural space, the authors found (Murphy et al., 2011) evidence of length-related inflammation, with no significant inflammation when short tubes (fibres) were administered. While no short samples were visible at day 1 or day 7, the mesothelium, that was thicker on day 1, returned to its normal thickness by day 7. While short tubes and fibres cleared from the pleura through stomatal openings, long tubes and fibres remained inside the pleura near stomata, where they persisted and caused inflammation and progressive fibrosis.

3.3 Effects of carbon nanotubes after intra-peritoneal administration

While pulmonary toxicity studies clearly indicate that inhalation of CNTs aggregates represents a possible occupational health hazard, the toxicity of CNTs after *in vivo* administration through bio-medically relevant routes is still a matter of debate. Among different routes of administration, the intra-peritoneal way has several advantages, firstly it offers the possibility to administer larger doses of suspended nanoparticles and secondly, the peritoneal cavity has a recognized particle-clearance mechanism. Particles leaving the peritoneal cavity pass *via* the retrosternal route through stomata (pore like structures) to the parathymic (mediastinal) nodes to the upper terminal thoracic duct or right lymphatic duct (Abu-Hijleh et al., 1995). Moreover, the peritoneal cavity and its mesothelium-covered viscera were recognized as a convenient substitute for pleural cavity mesothelium in fibre toxicity studies (Donaldson et al., 2010).

The first study of *in vivo* toxicity of CNTs after intra-peritoneal administration was conducted in our laboratory in collaboration with the Department of Chemistry of Rice University (Hartmann et al., 2007b). We compared the acute toxicity of full-length and ultra-short CNTs suspended in a Tween 80 aqueous solution, under the same conditions we used since 1996 to study the acute and sub-acute toxicity of [60]fullerene in mice (Moussa, F. et al, 1996). Our preliminary results showed that irrespective of the length of the administered CNT material, CNT aggregates induced a granulomatous response inside the organs like that which occurs in lungs after inhalation or intra-tracheal instillation (Hartmann et al. 2007).

One year later, in a comparative study of MWNTs and asbestos fibres, it was reported that MWNTs exhibit a length-dependent pathogenic behaviour (Poland et al., 2008) including granuloma formation and inflammation. In order to assess the role of fibre length, samples of long and short asbestos fibres and MWNTs with length ranges less than 5 µm, less than 20 µm- referred to as short, tangled MWNTs; and long tubes of the mean of 13 µm (24% of them were larger than 15 µm) and maximum 56 µm have been suspended in bovine serum albumin and saline and administered intra-peritoneally to mice in a dose of 50 µg per mouse. The MWNTs samples differed in the source, preparation and purification method (the short ones being purified by acid treatment). At day 7 after injection, the authors reported that only the samples containing long fibres (asbestos or MWNTs) caused significant polymorphonuclear leukocyte infiltration, protein exudation and granulomas. However, the mesothelial lining on the pleural side of the diaphragm was normal in every case (Poland et al., 2008) and the inflammation decreased by day 7. Short fibres of any kind did not cause significant inflammation, neither at day 1 nor at day 7 after administration, except for one mouse out of three in the group treated with short tangled MWNTs of the length < 20 µm. The overall conclusion of the study was that short MWNTs do not mimic the behaviour of long asbestos, but that their data cannot preclude the possibility that short MWNTs may be by some other mechanism that was not assessed in this study. Long MWNTs produced inflammation, foreign body giant cells and granulomas that were qualitatively and quantitatively similar to the foreign body inflammatory response caused by long asbestos. However, for the specimen treated with shorter MWNTs that did exhibit granuloma, the authors concluded that it was maybe due to the fact that the sample they injected was contaminated with long fibres, caused by some other unidentified component specific for the precise MWNTs sample, or the granulomas could have arisen spontaneously by chance (Poland et al., 2008).

In our laboratory (Kolosnjaj-Tabi et al., 2010) we administered Tween-suspended ultra-short (20-80 nm long) and full length SWNTs in a dose up to 1000 mg/kg. Our results indicated that regardless of the administered dose (50-1000 mg/kg b.w.), length, or surface state of the administered material, large aggregates of CNTs (>10 µm) irremediably induce granuloma formation (Fig. 4).

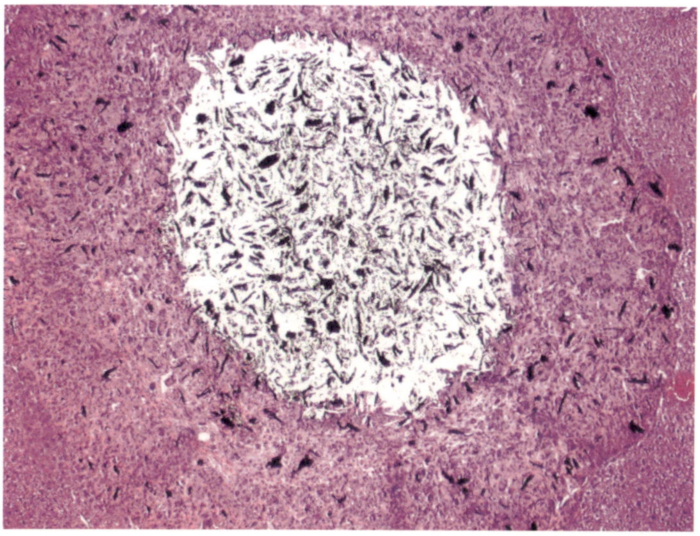

Fig. 4. Light micrograph after hematoxylin-eosin staining of a liver section from mice i.p. injected with a single dose of ultra-short SWNTs at 90 days post-administration showing a US-tube-laden granuloma. (Magnification = 10).

The administered doses were high, yet necessary to ascertain a sufficient circulating dose of administered material. The bolus dose was responsible, in set terms, for granulomas that were formed after aggregation of intra-peritoneally administered tubes. Smaller agglomerates (< 300 nm), on the other hand, did not induce granuloma formation nor did they cause any major life-threatening condition under our experimental conditions. A large portion of well-dispersed CNTs was eliminated through the kidney and the bile ducts. However, the aggregated part of the administered dose was not cleared from the body and persisted inside cells 5 months after administration (Fig. 5).

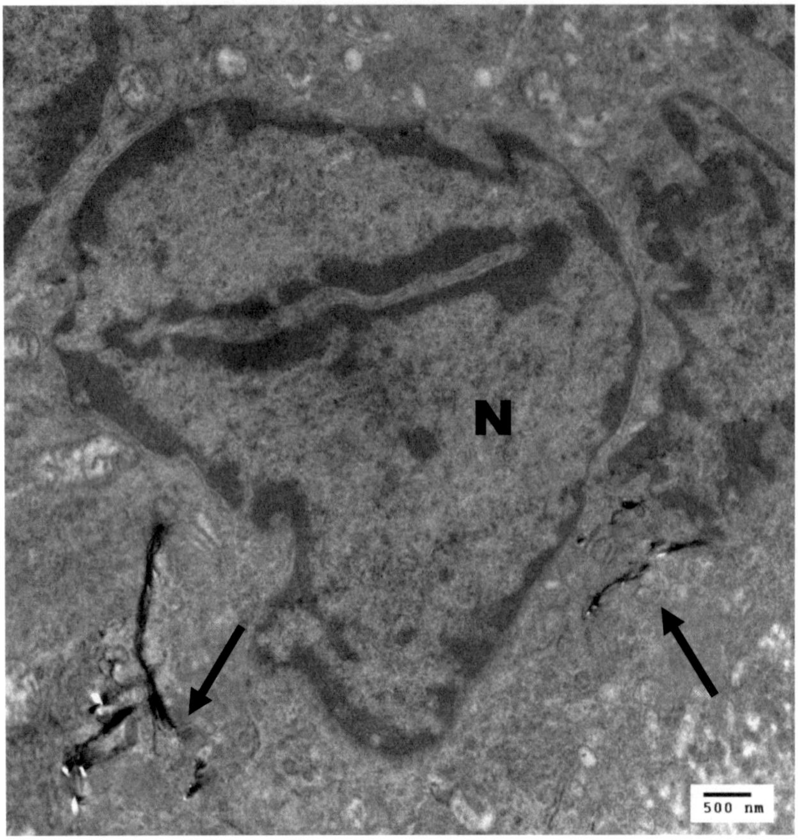

Fig. 5. Transmission electron micrograph of bundles of SWNTs in a Kuppfer cell, found in the liver of a mouse 5 months after treatment

The persistence of the nanotube aggregates inside the cells was probably due to the slow disaggregation and slow elimination of larger aggregates.

3.3.1 Carbon nanotubes and mesothelioma

The ability of MWNNTs to induce mesothelioma, a deadly cancer, in experimental models in rodents is still a matter of debate. This cancer is a highly specific response to bio-persistent fibres and may occur in the pleura (outer lining of the lungs and internal chest wall), peritoneum (the lining of the abdominal cavity), pericardium (a sac that surrounds the heart), or tunica vaginalis (a sac that surrounds the testis) (Moore et al., 2008).

In 2008, a Japanese team observed for the first time that MWNTs could induce mesothelioma after intra-peritoneal administration to p53 heterozygous mice (Takagi et al., 2008a). This type of mice are, however, more sensitive and have a shorter tumour onset rate than standard wild-type rodents. Indeed, different groups contested this study (Donaldson et al., 2008, Ichihara et al., 2008), but Takagi et al. explained and justified their experimental choices letting other studies to confirm mesothelium threats (Takagi et al., 2008b, c). The authors administered intra-peritoneally 3 mg of CNTs per mouse suspended in 1% Tween 80 and 0.5% methyl cellulose- aqueous solution. While the samples of crocidolite asbestos were evenly dispersed, MWNTs contained aggregates and fibres. The mice were monitored until one of the groups (MWNTs treated group) reached 100% mortality, which happened on day 180. The mice treated with MWNTs exhibited moderate to severe fibrous peritoneal adhesions, slight ascites, fibrous peritoneal thickening and a high incidence of macroscopic peritoneal tumours. Similar, but less severe findings were noted in the asbestos group. As the authors emphasized, it is important to limit the significance of this study to the monitoring of biological activity of a compartment of MWNTs longer that 5 µm. There is no information that this study method would be sensitive to pure nanometer-sized particles within the same timeframe (Takagi et al. 2008a).

An analogous study was made, in time, on wild-type rats (Muller et al., 2009) in a two years exposure period, where up to 20 mg of MWNTs with and without sidewall defects (induced by grounding of the raw material) suspended in phosphate buffer were intraperitoneally injected to a large number of rats. After 24 months, crocidolite induced mesothelioma in 34.6% animals whereas mesothelioma occurred only in 3.8% in the vehicle-treated rats. MWNTs with or without structural defects did not induce significant mesothelioma in this study, as mesothelioma occurrence was detected in up to 6% of MWNT-treated rats, which is in line with the spontaneous incidences of mesothelioma in rats (up to 6%). The incidence of tumours other than mesothelioma was not significantly increased across the groups (Bignon et al., 1995).

In contrast to what was observed after intra-peritoneal administration to wild type rats (Muller et al., 2009), MWNTs injected in a single intra-scrotal dose to rats, induced mesothelioma 37 to 40 weeks after treatment (Sakamoto et al., 2009). In the latter experiment, MWNTs were suspended in a 2% carboxymethyl cellulose aqueous solution and administered to rats at a dose of 0.24 mg/animal. According to the authors this dose corresponds to the maximal value recommended by the guideline for man-made mineral fibers (Bernstein & Riego Sintes, 1999, as cited in Sakamoto et al., 2009).

In conclusion, these findings also show that under some experimental conditions, MWNTs may induce mesothelioma formation. Thus, further investigations are urgently needed.

3.4 Effects of carbon nanotubes after sub-cutaneous administration

Subcutaneous implantations of clusters of MWNTs of different lengths (220 nm and 825 nm) in rats showed that the degree of inflammatory response around the shorter MWNTs was slighter than that observed around the longer ones, thus indicating that macrophages could envelop the shorter nanotubes more readily than the longer ones (Sato et al., 2005). However, no severe inflammatory response such as necrosis, degeneration or neutrophils infiltration was observed around both types of MWNTs.

These biological responses have also been described after subcutaneous implantations in mice of 2 mg/animal of SWNTs, two different types of MWNTs (20 and 80 nm of average diameter) and cup-stacked carbon nanotubes (CSNTs made with stacked truncated cones), for up to 3 months (Koyama et al., 2006). The nanotubes used in this study were purified by thermal treatment process during which the metal particles evaporate. After 1, 2, 3 weeks, 1 month, 2 months and 3 months post-implantation, the animals were sacrificed, blood was collected for CD4+ and CD8+ T-cells counting by flow-cytometry and tissue of skin (including muscle layers) were collected for histo-pathological examination. All mice survived, and no large changes in their weights were observed during the experimental period. After one week of implantation, only SWNTs activated the major histocompatibility complex (MHC) class I pathway of antigen-antibody response system (higher CD4+/CD8+ value), leading to the appearance of an oedematous aspect. After two weeks, significantly high values of CD4+ without changes in CD8+ signified the activation of MHC class II for all samples. The authors noted that antigenic mismatch becomes less evident with time, notably one month post-implantation, indicating an establishment of granuloma formation. Furthermore, the toxicological response of CNTs was absolutely lower than that of asbestos.

3.5 Effects of carbon nanotubes after intra-venous administration

3.5.1 Effects of SWNTs

The first report on the effects of SWNTs after intravenous administration to mice (3 mg/kg) monitored over four months indicated normal blood chemistries and normal histological examinations (Schipper et al., 2008). The animals did not show significant inflammatory lesions and SWNTs accumulated in liver and spleen as evidenced by Raman spectroscopy. The CNTs used in this study were highly dispersed with polyethylene glycol (PEG). CNT aggregates were eliminated with ultracentrifugation before administration to the animals.

Another study used highly dispersed pristine HiPco SWNTs with different PEGylated phospholipids (Liu et al., 2008). Big bundles and impurities were removed by centrifugation, and individually suspended tubes or small bundles were administered intravenously at a dose of approximately 20 µg or 100 µg per mouse. Blood, tissue and organ distribution and elimination in urines and faeces were evaluated by Raman spectroscopy, by assessment of the tangential graphite-like phonon mode (G band). Administered SWNTs accumulated mainly in liver and spleen, however the quantity decreased over a 3-month period. The authors concluded that SWNTs were mainly eliminated by the biliary pathway; only a small portion of short tubes (< 50 nm in length) was eliminated in the urines. Finally, the authors did not report any obvious sign of toxicity.

Other authors also reported low toxicity of SWNTs in mice over a 3-month period (Yang et al., 2008). Purified SWNTs were suspended in 1% Tween 80 aqueous solution and sonicated for 30 minutes prior administration to animals at various doses from 40 µg to 1 mg per mouse. Some of the serum biochemical parameters (ALT, AST and LDH) were higher in SWNT-treated animals compared to the control group 90 days post-exposure, indicating that induced hepatic injury and tissue breakdown were dose-dependent. The long-term accumulation of aggregated SWNTs was evidenced in histological sections of lungs, livers and spleens and was confirmed by Raman spectroscopy and transmission electron microscopy in organ lysates. However, no fibrosis was detected in the organs.

Embriotoxicity was recently reported as an effect caused by intravenous administration of SWNTs in mice (30 µg/mouse) (Pietroiusti et al., 2011). The authors used pristine, oxidized or ultra-oxidized by acid treatment SWNTs. Cobalt was the only impurity that was released in the medium in which the tubes were dispersed (DMEM cell culture medium with fetal bovine serum). Before the end of gestation, the animals were sacrificed, and uteri, placentas, and foetuses were examined. A high percentage of early miscarriages and foetal malformations were observed in females exposed to oxidized SWNTs, while lower percentages were found in animals exposed to the pristine material. The lowest effective dose was 100 ng/mouse. Extensive vascular lesions and increased production of reactive oxygen species (ROS) were detected in placentas of malformed but not of normally developed foetuses. Increased ROS levels were likewise detected in malformed foetuses. No increased ROS production or evident morphological alterations were observed in maternal tissues (Pietroiusti et al., 2011).

3.5.2 Effects of MWNTs

MWNTs dispersed in mouse-serum (10 min of sonication) and injected to mice at a dose of 200 or 400 µg per mouse showed no severe acute response, 24 hours after administration (Lacerda et al., 2008). However, mice treated with aggregates of pristine MWNTs exhibited subdued behavior, hunched posture, and signs of respiratory distress. While serum biochemistry data did not show significant increase, optical microscopy revealed aggregate accumulation, mostly in livers and lung vessels, which were probably responsible for respiratory distress. Nevertheless, no tissue degeneration, inflammation, necrosis or fibrosis occurred 24 hours after injection.

The effect of Tween-suspended pristine MWNTs and PBS-suspended acid oxidized MWNTs up to 2 months were investigated after administration to rats at a dose level of 10 or 60 mg/kg (Ji et al., 2009). The authors reported an impact on body weight gain of the highest dose. Severe inflammatory cell infiltration in the portal region, cellular necrosis and focal necrosis were seen at a dose of 60 mg/kg in the MWNT-treated group 15 and 60 days following the treatment. Moreover, severe mitochondrial swelling, bile canaliculi expansion, mitochondrial destruction, loss and lysis of mitochondrial crest were also observed. In the acid treated (oxidized) MWNTs group only slight inflammatory cell infiltration was observed after 2 months. A slight increase of AST activity used as biochemical marker of liver injury was also reported in treated animals, but it did not increase more than twofold. Interestingly, 329 genes were up-regulated and 31 genes that were down-regulated more than twofold in MWNT-treated mice and 1139 genes were up-regulated and 505 genes were down-regulated over twofold in the mice treated with oxidized MWNTs.

In order to avoid mechanical blockage by the administered material, it is of capital importance to administer only individually suspended, short CNTs. Thus, further studies with individually suspended CNTs have to be made in order to confirm the direct effect of these materials after administration by the intravenous route.

3.6 Effects of carbon nanotubes after oral administration

Oral administration of 1000 mg/kg of body weight of SWNTs to mice (Kolosnjaj-Tabi et al.) resulted in neither animal death nor behavioral abnormalities. Compound-colored stool was found 24 h after gavage in all treated groups. Two weeks after treatment, regardless of the length or of the iron content, the nanotube materials did not induce any abnormalities after pathological examination, indicating that under these conditions, the lowest lethal dose (LDLo) is greater than 1000 mg/kg b.w. in Swiss mice.

The potential effects of MWNTs after oral administration were also investigated on pregnant dams and embryo-fetal development in rats (Lim et al., 2011). MWNTs were administered to pregnant rats by gavage at 0, 40, 200, and 1,000 mg/kg/day. All dams were subjected to Cesarean section on day 20 of gestation, and the foetuses were examined for morphological abnormalities. A decrease in thymus weight was observed in the high dose group in a dose-dependent manner. However, maternal body weight, food consumption, and oxidant-antioxidant balance in the liver were not affected by treatment with MWNTs. No treatment-related differences in gestation index, foetal deaths, foetal and placental weights, or sex ratio were observed between the groups. Morphological examinations of the foetuses demonstrated no significant difference in incidences of abnormalities between the groups.

Intriguingly, it has also been reported that CNTs may be involved in oxidative stress with oxidative damage of DNA in the colon mucosa, liver, and lung of rats after oral administration of SWNTs in a dose of 0.064 or 0.64 mg/kg b.w. suspended in saline solution or corn oil (Folkmann et al., 2009). Suspensions of particles in saline solution or corn oil yielded a similar extent of genotoxicity. However, corn oil per se generated more genotoxicity than the particles (Folkmann et al., 2009).

4. Conclusion

Considered together, these diverging results highlight the difficulties in evaluating the toxicity of CNT materials. While the toxicity is certainly governed by the state of aggregation, length and stiffness of CNTs, other parameters might be involved. Most of them probably depend of the method of production of the sample, the method of purification and the method of preparation of the tested formulations. As the reactivity and the general behaviour of CNTs in biological media are not completely understood, assessing the safety of these nanoparticles should also include a careful selection of appropriate experimental methods. Thus, more studies are needed in order to determine the safety of CNTs. For the time being, precaution is necessary notably in case of CNT-exposure at workplace.

5. References

Abu-Hijleh, M.F., Habbal, O.A. & Moqattash, S.T. (1995). The role of the diaphragm in lymphatic absorption from the peritoneal cavity. *Journal of Anatomy, 18*, 453-467.

Ashcroft, J. M., Hartman, K. B., Kissell, K. R., Mackeyev, Y., Pheasant, S., Young, S., . . . Wilson, L. J. (2007). Single Molecule I2@ US Tube Nanocapsules: A New X ray Contrast Agent Design. *Advanced Materials, 19*(4), 573-576.

Balavoine, F., Schultz, P., Richard, C., Mallouh, V., Ebbesen, T. W., & Mioskowski, C. (1999). Helical crystallization of proteins on carbon nanotubes: a first step towards the development of new biosensors. *Angewandte Chemie International Edition, 38*(13/14), 1912-1915.

Balkwill, F., & Mantovani, A. (2001). Inflammation and cancer: back to Virchow? *The Lancet, 357*(9255), 539-545.

Bekyarova, E., Ni, Y., Malarkey, E. B., Montana, V., McWilliams, J. L., Haddon, R. C., & Parpura, V. (2005). Applications of carbon nanotubes in biotechnology and biomedicine. *Journal of biomedical nanotechnology, 1*(1), 3.

Bernstein, D., & Riego-Sintes, J. (1999). Methods for the determination of the hazardous properties for human health of man made mineral fibers (MMMF). Vol. EUR 18748 EN, April. 93.

Bianco, A., Kostarelos, K., & Prato, M. (2005). Applications of carbon nanotubes in drug delivery. *Current Opinion in Chemical Biology, 9*(6), 674-679.

Bignon, J., Brochard, P., Brown, R., Davis, J., Vu, V., Gibbs, G., . . . Sebastien, P. (1995). Assessment of the toxicity of man-made fibres. A final report of a workshop held in Paris, France 3-4 February 1994. *Annals of Occupational Hygiene, 39*(1), 89-106.

Casey, A., Herzog, E., Davoren, M., Lyng, F., Byrne, H., & Chambers, G. (2007). Spectroscopic analysis confirms the interactions between single walled carbon nanotubes and various dyes commonly used to assess cytotoxicity. *Carbon, 45*(7), 1425-1432.

Chen, J., Hamon, M. A., Hu, H., Chen, Y., Rao, A. M., Eklund, P. C., & Haddon, R. C. (1998). Solution properties of single-walled carbon nanotubes. *Science, 282*(5386), 95.

Chen, X., Lee, G. S., Zettl, A., & Bertozzi, C. R. (2004). Biomimetic engineering of carbon nanotubes by using cell surface mucin mimics. *Angewandte Chemie International Edition, 43*(45), 6111-6116.

Chen, Y., Haddon, R., Fang, S., Rao, A., Eklund, P., Lee, W., . . . Chavan, A. (1998). Chemical attachment of organic functional groups to single-walled carbon nanotube material. *J. Mater. Res, 13*(9), 2423-2431.

Donaldson, K., Murphy, F. A., Duffin, R., & Poland, C. A. (2010). Asbestos, carbon nanotubes and the pleural mesothelium: a review of the hypothesis regarding the role of long fibre retention in the parietal pleura, inflammation and mesothelioma. *Part Fibre Toxicol, 7*(5).

Donaldson, K., & Poland, C. A. (2009). Nanotoxicology: new insights into nanotubes. *Nature Nanotechnology, 4*(11), 708-710.

Donaldson, K., Stone, V., Seaton, A., Tran, L., Aitken, R., & Poland, C. (2008). Re: Induction of mesothelioma in p53+/-mouse by intraperitoneal application of multi-wall carbon nanotube. *The Journal of toxicological sciences, 33*(3), 385.

Dresselhaus, M., Dresselhaus, G., Charlier, J., & Hernandez, E. (2004). Electronic, thermal and mechanical properties of carbon nanotubes. *Philosophical Transactions of the Royal Society of London. Series A: Mathematical, Physical and Engineering Sciences, 362*(1823), 2065.

Folkmann, J. K., Risom, L., Jacobsen, N. R., Wallin, H., Loft, S., & M¯ller, P. (2009). Oxidatively damaged DNA in rats exposed by oral gavage to C60 fullerenes and single-walled carbon nanotubes. *Environmental health perspectives, 117*(5), 703.

Harris, PJF. (2009). *Carbon nanotube science Synthesis, properties and applications* (1st Edition), Cambridge University Press, ISBN 9780521828956, Cambridge

Hartman, K. B., Hamlin, D. K., Wilbur, D. S., & Wilson, L. J. (2007). 211AtCl@ US Tube Nanocapsules: A New Concept in Radiotherapeutic Agent Design. *Small, 3*(9), 1496-1499.

Hartman, K., Kolosnjaj, J., Gharbi, N., Boudjemaa, S. Wilson, L. J. and Moussa, F. (2007). Comparative In vivo Toxicity Assessment of Singlewalled Carbon Nanotubes in Mice. The 211th Meeting of The Electrochemical Society, Chicago, USA, May 6-10

Hartman, K. B., Laus, S., Bolskar, R. D., Muthupillai, R., Helm, L., Toth, E., . . . Wilson, L. J. (2008). Gadonanotubes as ultrasensitive pH-smart probes for magnetic resonance imaging. *Nano letters, 8*(2), 415-419.

Hirsch, A. (2002). Functionalization of single walled carbon nanotubes. *Angewandte Chemie International Edition, 41*(11), 1853-1859.

Hu, H., Ni, Y., Montana, V., Haddon, R. C., & Parpura, V. (2004). Chemically functionalized carbon nanotubes as substrates for neuronal growth. *Nano letters, 4*(3), 507-511.

Huczko, A., & Lange, H. (2001). Carbon nanotubes: experimental evidence for a null risk of skin irritation and allergy. *Fullerene Science and Technology, 9*(2), 247-250.

Ichihara, G., Castranova, V., Tanioka, A., & Miyazawa, K. (2008). Letter to the editor. *The Journal of toxicological sciences, 33*(3), 381.

Iijima, S. (1991). Helical microtubules of graphitic carbon. *Nature, 354*(6348), 56-58.

Jensen, A. W., Wilson, S. R., & Schuster, D. I. (1996). Biological applications of fullerenes. *Bioorganic & medicinal chemistry, 4*(6), 767-779.

Ji, Z., Zhang, D., Li, L., Shen, X., Deng, X., Dong, L., . . . Liu, Y. (2009). The hepatotoxicity of multi-walled carbon nanotubes in mice. *Nanotechnology, 20*, 445101.

Kane, A. B., & Hurt, R. H. (2008). Nanotoxicology: The asbestos analogy revisited. *Nature Nanotechnology, 3*(7), 378-379.

Kobayashi, N., Naya, M., Ema, M., Endoh, S., Maru, J., Mizuno, K., & Nakanishi, J. (2010). Biological response and morphological assessment of individually dispersed multi-wall carbon nanotubes in the lung after intratracheal instillation in rats. *Toxicology*, *276*(3), 143-153.

Kolosnjaj, J., Szwarc, H., & Moussa, F. (2007). Toxicity studies of carbon nanotubes. *Bio-Applications of Nanoparticles*, 181-204.

Kolosnjaj-Tabi, J., Hartman, K. B., Boudjemaa, S., Ananta, J. S., Morgant, G., Szwarc, H., . . . Moussa, F. (2010). In vivo behavior of large doses of ultrashort and full-length single-walled carbon nanotubes after oral and intraperitoneal administration to Swiss mice. *Acs Nano, 4*(3), 1481-1492.

Koyama, S., Endo, M., Kim, Y. A., Hayashi, T., Yanagisawa, T., Osaka, K., . . . Kuroiwa, N. (2006). Role of systemic T-cells and histopathological aspects after subcutaneous implantation of various carbon nanotubes in mice. *Carbon, 44*(6), 1079-1092.

Kroll, A., Pillukat, M. H., Hahn, D., & Schnekenburger, J. (2009). Current in vitro methods in nanoparticle risk assessment: Limitations and challenges. *European Journal of Pharmaceutics and Biopharmaceutics, 72*(2), 370-377.

Lacerda, L., Ali-Boucetta, H., Herrero, M. A., Pastorin, G., Bianco, A., Prato, M., & Kostarelos, K. (2008). Tissue histology and physiology following intravenous administration of different types of functionalized multiwalled carbon nanotubes. *Nanomedicine, 3*(2), 149-161.

Lai-Fook, S. J. (2004). Pleural mechanics and fluid exchange. *Physiological reviews, 84*(2), 385.

Lam, C. W., James, J. T., McCluskey, R., & Hunter, R. L. (2004). Pulmonary toxicity of single-wall carbon nanotubes in mice 7 and 90 days after intratracheal instillation. *Toxicological Sciences, 77*(1), 126.

Lim, J. H., Kim, S. H., Shin, I. S., Park, N. H., Moon, C., Kang, S. S., . . . Kim, J. C. (2011) Maternal exposure to multi wall carbon nanotubes does not induce embryo-fetal developmental toxicity in rats. *Birth Defects Research Part B: Developmental and Reproductive Toxicology. 92*(1),69-76.

Lin, Y., Taylor, S., Li, H., Fernando, K. A. S., Qu, L., Wang, W., . . . Sun, Y. P. (2004). Advances toward bioapplications of carbon nanotubes. *Journal of Materials Chemistry, 14*(4), 527-541.

Liu, Z., Davis, C., Cai, W., He, L., Chen, X., & Dai, H. (2008). Circulation and long-term fate of functionalized, biocompatible single-walled carbon nanotubes in mice probed by Raman spectroscopy. *Proceedings of the National Academy of Sciences, 105*(5), 1410.

Lovat, V., Pantarotto, D., Lagostena, L., Cacciari, B., Grandolfo, M., Righi, M., . . . Ballerini, L. (2005). Carbon nanotube substrates boost neuronal electrical signaling. *Nano letters, 5*(6), 1107-1110.

Mattson, M. P., Haddon, R. C., & Rao, A. M. (2000). Molecular functionalization of carbon nanotubes and use as substrates for neuronal growth. *Journal of Molecular Neuroscience, 14*(3), 175-182.

Maynard, A., Baron, P., Foley, M., Shvedova, A., Kisin, E., & Castranova, V. (2004). Exposure to carbon nanotube material: aerosol release during the handling of unrefined single-walled carbon nanotube material. *Journal of Toxicology and Environmental Health Part A, 67*(1), 87-107.

Mercer, R., Scabilloni, J., Wang, L., Kisin, E., Murray, A., Schwegler-Berry, D., . . . Castranova, V. (2008). Alteration of deposition pattern and pulmonary response as a result of improved dispersion of aspirated single-walled carbon nanotubes in a mouse model. *American Journal of Physiology-Lung Cellular and Molecular Physiology, 294*(1), L87.

Mercer, R. R., Hubbs, A. F., Scabilloni, J. F., Wang, L., Battelli, L. A., Schwegler-Berry, D., . . . Porter, D. W. (2010). Distribution and persistence of pleural penetrations by multi-walled carbon nanotubes. *Particle and Fibre Toxicology, 7*(1), 28.

Moore, A. J., Parker, R. J., & Wiggins, J. (2008). Malignant mesothelioma. *Orphanet J Rare Dis, 3*(1), 34.

Morimoto, Y., Hirohashi, M., Ogami, A., Oyabu, T., Myojo, T., Todoroki, M., . . . Lee, B. W. (2011). Pulmonary toxicity of well-dispersed multi-wall carbon nanotubes following inhalation and intratracheal instillation. *Nanotoxicology*(0), 1-15.

Moussa, F., Trivin, F., Céolin, R., Hadchouel, M., Sizaret, P.-Y., Greugny, V., Fabre, C., Rassat, A. and Szwarc, H. (1996). Early effects of C60 administration in Swiss Mice: a preliminary account for *in vivo* C60 toxicity. *Fullerenes Science & Technology*, 4, 21–29

Muller, J., Delos, M., Panin, N., Rabolli, V., Huaux, F., & Lison, D. (2009). Absence of carcinogenic response to multiwall carbon nanotubes in a 2-year bioassay in the peritoneal cavity of the rat. *Toxicological Sciences, 110*(2), 442.

Muller, J., Huaux, F., Fonseca, A., Nagy, J. B., Moreau, N., Delos, M., . . . Fenoglio, I. (2008). Structural defects play a major role in the acute lung toxicity of multiwall carbon nanotubes: Toxicological aspects. *Chemical research in toxicology, 21*(9), 1698-1705.

Muller, J., Huaux, F., Moreau, N., Misson, P., Heilier, J. F., Delos, M., . . . Lison, D. (2005). Respiratory toxicity of multi-wall carbon nanotubes. *Toxicology and Applied Pharmacology, 207*(3), 221-231.

Murphy, F. A., Poland, C. A., Duffin, R., Al-Jamal, K. T., Ali-Boucetta, H., Nunes, A., . . . Li, S. (2011). Length-Dependent Retention of Carbon Nanotubes in the Pleural Space of Mice Initiates Sustained Inflammation and Progressive Fibrosis on the Parietal Pleura. *The American Journal of Pathology, 178*(6), 2587-2600.

Mutlu, G. M., Budinger, G. R. S., Green, A. A., Urich, D., Soberanes, S., Chiarella, S. E., . . . Hersam, M. C. (2010). Biocompatible nanoscale dispersion of single-walled carbon nanotubes minimizes in vivo pulmonary toxicity. *Nano letters, 10*(5), 1664-1670.

Oberdorster, G., Maynard, A., Donaldson, K., Castranova, V., Fitzpatrick, J., Ausman, K., . . . Lai, D. (2005). ILSI Research Foundation/Risk Science Institute Nanomaterial Toxicity Screening Working Group. Principles for characterizing the potential human health effects from exposure to nanomaterials: Elements of a screening strategy. *Part Fibre Toxicol, 2*(8).

Pietroiusti, A., Massimiani, M., Fenoglio, I., Colonna, M., Valentini, F., Palleschi, G., . . . Bergamaschi, A. (2011) Low Doses of Pristine and Oxidized Single Wall Carbon Nanotubes Affect Mammalian Embryonic Development. *Acs Nano*. 5(6), 4624-4633.

Poland, C. A., Duffin, R., Kinloch, I., Maynard, A., Wallace, W. A. H., Seaton, A., . . . Donaldson, K. (2008). Carbon nanotubes introduced into the abdominal cavity of mice show asbestos-like pathogenicity in a pilot study. *Nature Nanotechnology, 3*(7), 423-428.

Popov, V. N. (2004). Carbon nanotubes: properties and application. *Materials Science and Engineering: R: Reports, 43*(3), 61-102.

Porter, D. W., Hubbs, A. F., Mercer, R. R., Wu, N., Wolfarth, M. G., Sriram, K., . . . Friend, S. (2010). Mouse pulmonary dose-and time course-responses induced by exposure to multi-walled carbon nanotubes. *Toxicology, 269*(2-3), 136-147.

Richard, C., Balavoine, F., Schultz, P., Ebbesen, T. W., & Mioskowski, C. (2003). Supramolecular self-assembly of lipid derivatives on carbon nanotubes. *Science, 300*(5620), 775.

Ruoff, R. S., Lorents, D. C., Chan, B., Malhotra, R., & Subramoney, S. (1993). Single crystal metals encapsulated in carbon nanoparticles. *Science, 259*(5093), 346.

Ryman-Rasmussen, J. P., Cesta, M. F., Brody, A. R., Shipley-Phillips, J. K., Everitt, J. I., Tewksbury, E. W., . . . Andersen, M. E. (2009). Inhaled carbon nanotubes reach the subpleural tissue in mice. *Nature Nanotechnology, 4*(11), 747-751.

Sakamoto, Y., Nakae, D., Fukumori, N., Tayama, K., Maekawa, A., Imai, K., . . . Ogata, A. (2009). Induction of mesothelioma by a single intrascrotal administration of multi-wall carbon nanotube in intact male Fischer 344 rats. *The Journal of toxicological sciences, 34*(1), 65-76.

Sato, Y., Yokoyama, A., Shibata, K., Akimoto, Y., Ogino, S., Nodasaka, Y., . . . Uo, M. (2005). Influence of length on cytotoxicity of multi-walled carbon nanotubes against human acute monocytic leukemia cell line THP-1 in vitro and subcutaneous tissue of rats in vivo. *Mol. BioSyst., 1*(2), 176-182.

Schipper, M. L., Nakayama-Ratchford, N., Davis, C. R., Kam, N. W. S., Chu, P., Liu, Z., . . . Gambhir, S. S. (2008). A pilot toxicology study of single-walled carbon nanotubes in a small sample of mice. *Nature Nanotechnology, 3*(4), 216-221.

Service, R. (1998). Chemistry: Nanotubes: The Next Asbestos? *Science, 281*(5379), 941-941.

Shvedova, A. A., Kisin, E. R., Mercer, R., Murray, A. R., Johnson, V. J., Potapovich, A. I., . . . Schwegler-Berry, D. (2005). Unusual inflammatory and fibrogenic pulmonary responses to single-walled carbon nanotubes in mice. *American Journal of Physiology-Lung Cellular and Molecular Physiology, 289*(5), L698.

Sinha, N., & Yeow, J. T. W. (2005). Carbon nanotubes for biomedical applications. *NanoBioscience, IEEE Transactions on, 4*(2), 180-195.

Sitharaman, B., Kissell, K. R., Hartman, K. B., Tran, L. A., Baikalov, A., Rusakova, I., . . . Chiu, W. (2005). Superparamagnetic gadonanotubes are high-performance MRI contrast agents. *Chem. Commun.*(31), 3915-3917.

Sloan, J., Hammer, J., Zwiefka-Sibley, M., & Green, M. L. H. (1998). The opening and filling of single walled carbon nanotubes (SWTs). *Chem. Commun.*(3), 347-348.

Takagi, A., Hirose, A., Nishimura, T., Fukumori, N., Ogata, A., Ohashi, N., Kitajima S. & Kanno, J. (2008). Induction of mesothelioma in p53+/-mouse by intraperitoneal application of multi-wall carbon nanotube. *The Journal of toxicological sciences, 33*(1), 105-116.

Takagi, A., Hirose, A., Nishimura, T., Fukumori, N., Ogata, A., Ohashi, N., Kitajima S. & Kanno, J. (2008b) Letter to the editor. *The Journal of toxicological sciences, 33*(3), 382-384.

Takagi, A., Hirose, A., Nishimura, T., Fukumori, N., Ogata, A., Ohashi, N., Kitajima S. & Kanno, J. (2008c) Letter to the editor. *The Journal of toxicological sciences, 33*(3), 386-388.

Valko, M., Morris, H., & Cronin, M. (2005). Metals, toxicity and oxidative stress. *Current Medicinal Chemistry, 12*(10), 1161-1208.

Wang, J., Musameh, M., & Lin, Y. (2003). Solubilization of carbon nanotubes by Nafion toward the preparation of amperometric biosensors. *Journal of the American Chemical Society, 125*(9), 2408-2409.

Wang, S., Humphreys, E. S., Chung, S. Y., Delduco, D. F., Lustig, S. R., Wang, H., . . . Chiang, Y. M. (2003). Peptides with selective affinity for carbon nanotubes. *Nature Materials, 2*(3), 196-200.

Warheit, D. (2006). What is currently known about the health risks related to carbon nanotube exposures? *Carbon, 44*(6), 1064-1069.

Warheit, D. B., Laurence, B., Reed, K. L., Roach, D., Reynolds, G., & Webb, T. (2004). Comparative pulmonary toxicity assessment of single-wall carbon nanotubes in rats. *Toxicological Sciences, 77*(1), 117.

Yang, S. T., Wang, X., Jia, G., Gu, Y., Wang, T., Nie, H., Liu, Y. (2008). Long-term accumulation and low toxicity of single-walled carbon nanotubes in intravenously exposed mice. *Toxicology letters, 181*(3), 182-189.

Ziegler, K. J., Gu, Z., Peng, H., Flor, E. L., Hauge, R. H., & Smalley, R. E. (2005). Controlled oxidative cutting of single-walled carbon nanotubes. *Journal of the American Chemical Society, 127*(5), 1541-1547.

3

Cytotoxicity of Tamoxifen-Loaded Solid Lipid Nanoparticles

Roghayeh Abbasalipourkabir, Aref Salehzadeh and Rasedee Abdullah
[1]*Hamedan University of Medical Science*
[2]*Universiti Putra Malaysia*
[1]*Iran*
[2]*Malaysia*

1. Introduction

Breast cancer is one the most important health concerns of the modern society (Ferlay *et al.*, 2007). Worldwide, it is estimated that over one million new cases of breast cancer are diagnosed every year, and more than 400 thousands will die from the breast cancer (Coughlin & Ekwueme, 2009) The life-time risk in women contracting breast cancers is estimated to be 1 in 8, which is the highest among all forms of cancers. (DevCan, 2004). Although the mortality rates from breast cancers have decreased in most developed countries because more frequent mammographic screening and extensive use of tamoxifen, it still remains the second highest in women (Clark, 2008). Breast cancer incidence rates were reported to have doubled or tripled in developing countries in the past 40 years (Anderson *et al.*, 2008; Porter, 2008). The main options for breast cancer treatment include surgery, radiation therapy and chemotherapy (Mirshahidi & Abraham, 2004). Surgical procedures usually lead to significant morbidity such as lymph edema, muscle wasting, neuropathy and chronic pain (Paci *et al.*, 1996). Radiation therapy is useful for cancer which is more localized, but it also carry a number of acute and chronic side- effects such as nausea, diarrhea, pain and fatigue (Ewesuedo & Ratain, 2003). Endocrine therapy may be used as a supplementary treatment. This method of therapy is applied to specific group of patients, e.g. women after menopause with hormone-responsive disease (Gradishar, 2005). In hormone-sensitive cancer patients receive chemotherapy with cytotoxic drugs. The cytotoxic drugs treat cancers by causing cell death or growth arrest. Effective cancer chemotherapy is able either to shrink a tumor or to help destroy cancer cells (Ewesuedo & Ratain, 2003). A number of obstacles such as drug toxicity, possible undesirable drug interactions and various forms of drug resistance have to be overcome to achieve effective chemotherapy (Cardosa *et al.*, 2009). Drug resistance is a general problem in the chemotherapy of several cancers including breast cancers (Wong *et al.*, 2006). Failures in treatment of cancers are common. Development of new drugs is also slow to progress. Among the reasons contributing to this are weak absorption, high rate of metabolism and elimination of drugs per oral administration resulting in less or variable concentrations in blood, poor drug solubility, unpredictable bioavailability of oral drugs due to food, and tissue toxicity (Sipos *et al.*, 1997). Thus alternative methods of drug administration like appropriate drug carrier system is needed to overcome this problem. Depending on the route of administration, the size of drug carriers may range from a few nanometers (colloidal carriers), to micrometers (microparticles) and to several millimeters (implants). Among these carriers, nanoparticles had shown great promise for parenteral application of chemotherapeutic drugs (Mehnert *et al.,* 2001). Targeting of unhealthy tissues and organs of the body is one of the important challenges of the drug delivery systems. Nanoparticles seem to show promise as a drug targeting systems supplying drug to target tissues at the right time (Kayser *et al.,* 2005). The main objective of new drug delivery systems is to improve the anti-tumor efficacy of drug and reduce their toxic effects on normal tissues. Nanoparticle is expected to be able to diminish toxicity of chemotherapy drug. Nanoparticles based on lipids that are solid at room temperature, namely solid lipid nanoparticle (SLN) using physiological well-tolerable lipids have potentially wide application (Siekmann & Westesen, 1992; Müller *et al.,* 1995; Müller *et al.,* 2000). The SLN is a drug delivery system that loads lipophilic or chemically unstable drugs (Fig. 1). Among the advantages of SLN are high potential for management of drug release and drug targeting, high stability for drug loading and high capacity for drug payload. This delivery system makes possible the encapsulation of lipophilic and hydrophilic drugs without the toxic effect of the carriers. This system also avoids the use

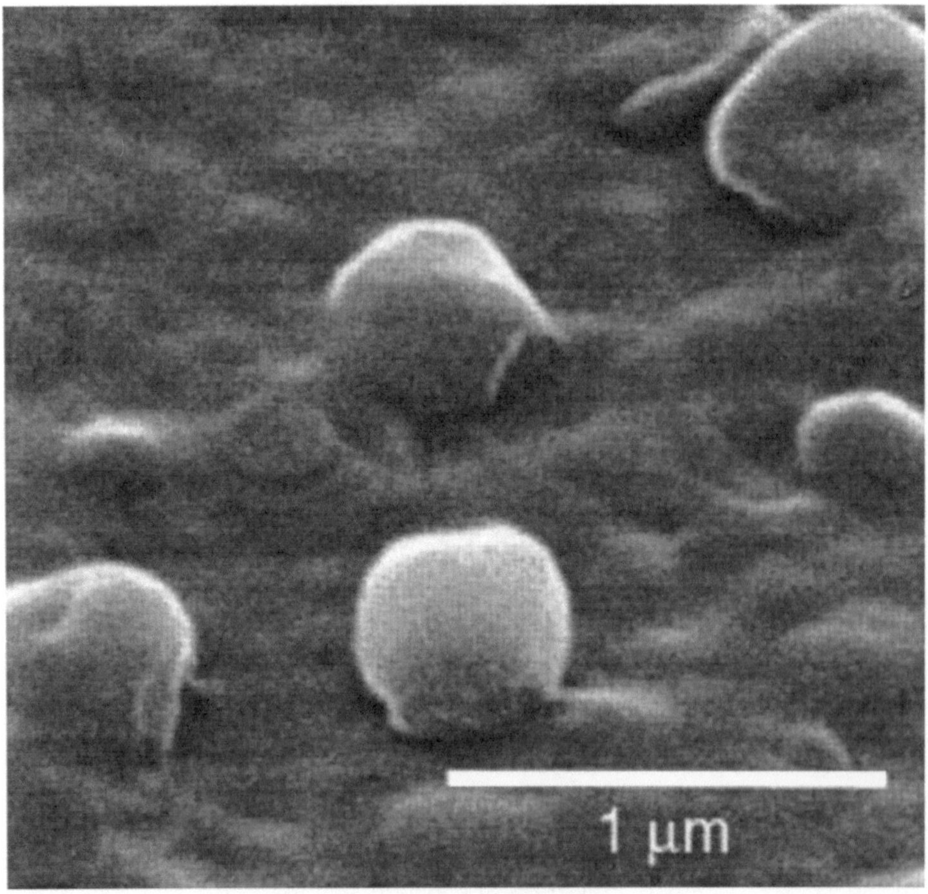

Fig. 1. Electron microscopy picture of solid lipid nanoparticles made from Compritol® stabilized with poloxamer 188, diameter 400nm (Adapted from Müller and Mäder 2000)

organic solvents, and has potential for large scale production. However, several disadvantages are associated with SLNs to include particle growth, particle aggregation, unpredictable gelation tendency, polymorphic transitions, burst drug release and inherently low incorporation capacities due to the crystalline structure of the solid lipid (Mehnert & Mäder, 2001).

The most common production technique of SLNs are high-pressure homogenization (HPH), high-shear homogenization combined with ultrasound, solvent emulsification/evaporation and microemulsion techniques. HPH is the predominant production method because it is easy to handle and scale-up. In this method, drug incorporation is achieved by dissolving or dispersing the drug in the melted lipid (He *et al.*, 2007). The drug can be encapsulated in the matrix or attached to the particle surface. In spite of improved researches in production of high quality SLN, it is still not routinely used clinically.

2. Tamoxifen-encapsulated solid lipid nanoparticles

The chemical name of tamoxifen is trans-2-[4-(1,2-diphenyl-1-butenyl)phenoxyl] N,Ndimethylethylamine (Fig. 2). Tamoxifen, an antiestrogen molecule and strong hydrophobic drug (water solubility, 0.04 µg/mL at 37°C), is widely administered in breast cancer and high risk patients (McGregor & Jorda, 1998). Although tamoxifen was primarily used as a drug against hormone-dependent breast cancers (Wyld *et al.*, 1998), it has also been used in the treatment of hormone-insensitive estrogen receptor-negative breast cancers (Jordan, 1994). Tamoxifen inhibits cell proliferation and induces apoptosis in breast cancer cells (MCF-7, MDA-MB231 and BT-20) (Mandlekar & Kong, 2001; Mandlekar *et al.*, 2000). In spite of being high effective, tamoxifen has harmful dose-dependent long-term side-effects such as development of endometrial cancer (Brigger *et al.*, 2001), hyperplasia, polyps, carcinoma, sarcoma (Peters-Engl *et al.*, 1999; Cohen, 2004) vaginal hemorrhage, blazes and liquid retention in postmenopausal breast cancer patients (Mourits *et al.*, 2001; Delima *et al.*, 2003). Formulations with the encapsulation of low-dose tamoxifen in colloidal delivery systems have been effective. Tamoxifen has been formulated in nanospheres such as poly-ε-caprolactone nanoparticles (Chawla & Amiji, 2003) and long circulating Poly(MePEGcyanoacrylate-co-hexadecylcyanoacrylate) nanoparticles in the form of free base (Brigger, *et al.*, 2001). Tamoxifen, as a nonsteroidal antiestrogen drug was recently encapsulated in SLNs and was shown to be effective on induced mammary tumor gland in Sprague-Dawley rats (Abbasalipourkabir *et al.*, 2010,1) in parenteral administration. The SLN

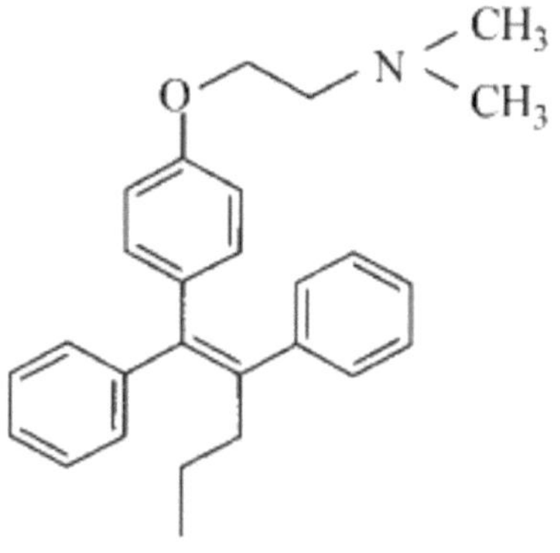

Fig. 2. Chemical Structure of Tamoxifen (Adapted from Christov *et al.*, 2007)

systems offer a sustained release of the drug in its intact form (Fontana *et al.*, 2005 Using human breast cancer cell line, MCF7, some *in vitro* studies have shown that drug release from the tamoxifen-incorporated SLN has the same antitumoral activity as the free drug (Abbasalipourkabir *et al.*, 2011). Therefore the tamoxifen-loaded SLN as a carrier system has excellent potential in prolonged drug release in breast cancer therapy (Fundaro *et al.*, 2000).

2.1 Preparation of tamoxifen-loaded SLN

Drug-loaded SLNs can be prepared using the high-pressure homogenization technique (Abbasalipourkabir *et al.*, 2011). A mixture of Hydrogenated palm oil (Softisan 154 or S154) and Hydrogenated soybean lecithin (Lipoid S100-3, containing 90% phosphatidylcholine, including 12–16% palmitic acid, 83–88% stearic acid, oleic acid and isomers, and linoleic acid] at a ratio of 70:30 is grounded in a ceramic crucible. The mixture is heated to 65–70°C while being stirred with a PTFE-coated magnet until a clear-yellowish lipid matrix (LM) solution is obtained. A solution containing 1 mL oleyl alcohol, 0.005 g thimerosal, 4.75 g Sorbitol, and 89.25 mL bidistilled water (all w/w) at the same temperature is added to 5 g of LM. A pre-emulsion of SLN is obtained using the homogenizer (Ultra Turrax, Ika) at 13,000 rpm for 10 min and high-pressure homogenizer (EmulsiFlex-C50 CSA10, Avestin) at 1000 bar, 20 cycles, and 60 °C. The lipophilic drug tamoxifen (1 mg) is dissolved in oleyl alcohol and mixed with 5 mg of SLN pre-emulsion using the Ultra Turrax homogenizer at 13,000 rpm for 10 min. This mixture is then incubated overnight at 50–60 °C, stirred periodically with a PTFE-coated magnet at 500 rpm, and finally will expose to air to solidify.

2.2 *In Vitro* antitumoral activity of Tamoxifen-loaded solid lipid nanoparticle

Cell death basically can occur in two ways. The first is through the necrosis pathway, where traumatic injuries cause cells damage in particular cell enlarges, bursts and liberate its intracellular components into the surrounding environment. The second pathway is programmed cell death or apoptosis, which is a molecular signaling cascade, inducing a disturbance in the organization and package of the cell causing death (Fadok, 1999; Messmer & Pfeilschifter, 2000). Other mode of cell death has also been suggested, for example mitotic cell death, which plays an important role in cell death caused by ionizing radiation (Steel, 2001). Breast cancer is the most common malignancy (18% of all malignancies) in women worldwide and its occurrence is slowly increasing (Salami & Karami-Tehrani, 2003). Like many cancers, breast cancer appears to be a result of high genetic damage that caused uncontrolled cellular proliferation and unusual apoptosis. These phenomena activate proto-oncogenes and inactivate tumor suppressor genes. These events can be activated by exposure of living cells to environmental, physical, chemical and/or biological carcinogens (Russo & Russo, 2002). The antiestrogen molecule, Tamoxifen (TAM) or trans-2-[4-(1,2-diphenyl-1-butenyl) phenoxyl]-N, N-dimethylethylamine, has been widely applied in treatment of breast cancer and high risk patients. Tamoxifen can reduce the occurrence of contralateral breast cancers by at least 40% (Fontana *et al.*, 2005). Tamoxifen exhibits anti-estrogenic activity by binding to the intracellular estrogen receptor. The tamoxifen-estrogen receptor complex binds with DNA and can subsequent inhibit mRNA transcription and lead to cellular apoptosis (Chawla & Amiji, 2003). Recently nanoparticulate delivery systems in the form of nanospheres like poly- caprolactone nanoparticles and long-circulating PEG-coated poly (MePEGcyanoacrylate-co-hexadecylcyanoacrylate) nanoparticles in the form of free base have been used for tamoxifen encapsulation. The basis of this formulation is to obtain the necessary dose of drug at tumor location for a known period of time and reducing adverse effects on normal organs in the body (Chawla & Amiji, 2003). In recent years Delivering Tamoxifen within Solid Lipid Nanoparticles have been recommended. Animal models play an important role in cancer chemotherapy (Abbasalipourkabir *et al.,* 2010,2). However, today there is increasing acceptation for *in vitro* tests as the method for determining cytotoxicity and viability of chemotherapeutic drugs. The reason for change lies partly in the limitations of animal models, to include financial considerations, time, and differences between animal and human metabolism. Finally, there is the moral pressure to reduce animal experimentation. *In vitro* tests are more likely to be reproducible. In general, the procedure involves the exposure of cells to a range of concentrations of the chemicals under test for a defined time and then to test for cell viability. Such tests are most easily performed in microtitre plates, which allow rapid quantitation of the results using a micrometer plate reader (Adams, 1990). The responses of breast cancer cell lines are determined by cytotoxicity assay, cellular and nuclear morphology, apoptosis and cell cycle distribution.

2.3 Cytotoxicity effect of TAM-loaded SLN on human breast cancer cells

The TAM-loaded SLN has an equally efficient cytotoxic activity as free tamoxifen. Therefore TAM-loaded SLN preserves the antitumoral activity of the free drug. When TAM is incorporated into the SLN carrier system, its antitumoral activity is still maintained and formulating TAM by incorporating into SLN will potentially enhance the solubility of the drug through inclusion into the lipid phase and facilitating the entrapment of greater amounts of the drug in the SLN, suggesting that SLN is a good carrier for the drug (Fig. 3. & Fig. 4).

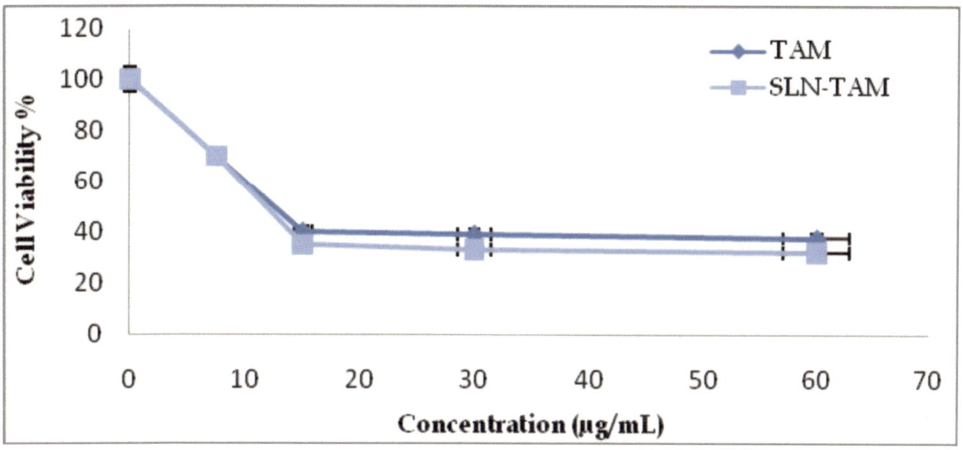

Fig. 3. MCF-7 cells viability after 72 hours incubation with TAM and TAM-Loaded SLN formulation. The percentage of cell viability is expressed as a ratio of treated cells to the untreated control cells. Each point represents the mean ± standard deviation of 5 wells. (Adapted from Abbasalipourkabir, 2010).

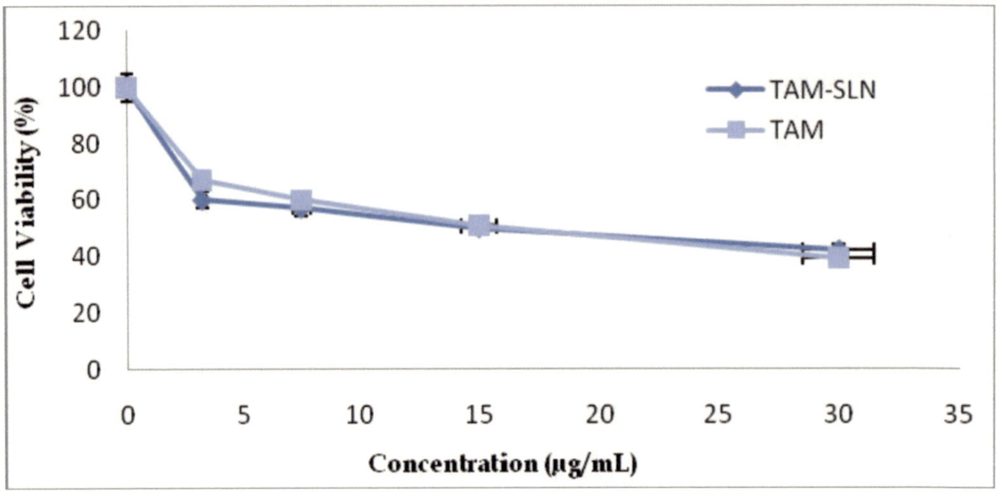

Fig. 4. MDA-MB231 cells viability after 72 hours incubation with TAM and TAM-Loaded SLN formulation. The percentage of cell viability is expressed as a ratio of treated cells to the untreated control cells. Each point represents the mean ± standard deviation of 5 wells. (Adapted from Abbasalipourkabir, 2010).

The IC_{50} of TAM and TAM-loaded SLN for MDA-MB231 cells (ER-negative or ER-independent) is higher than for MCF-7 cells (ER-positive or ER-dependent), (Tables 1-2).

Treatment	IC_{50} (µg/mL)		
	24h	48h	72h
TAM	13.45±0.46	13.00±0.98	12.50±0.91
TAM-SLN	13.18^{a}±0.66	12.50^{a}±1.50	11.78^{b}±0.18

All value represent the means ± std. dev., (n=5) ^{A,b}means in each row with different superscripts are significantly different TAM = Tamoxifen; TAM-SLN = Tamoxifen-loaded solid lipid nanoparticle (Adapted from Abbasalipourkabir, 2010).

Table 1. The IC_{50} of TAM and TAM-loaded SLN on MCF-7 cells after 24, 48 and 72h.

Treatment	IC_{50} (µg/mL)		
	24h	48h	72h
TAM	17.21±1.44	16.87±1.97	15.97±0.86
TAM-SLN	16.93±0.82	16.00±0.10	15.80±0.69

All value represent the means ± std. dev., (n=5) TAM = Tamoxifen; TAM-SLN = Tamoxifen-loaded solid lipid nanoparticle (Adapted from Abbasalipourkabir, 2010).

Table 2. The IC_{50} of TAM and TAM-loaded SLN on MDA-MB231 cells after 24, 48 and 72h.

The mechanisms of ER-independent, TAM-induced apoptosis may be through the inhibition of protein kinase C. The IC_{50} value of tamoxifen for protein kinase C inhibition is 4 to 10 times the concentration for ER inhibition in ER-positive cells. Therefore, the dose of tamoxifen for treatment of patients with ER-positive breast cancer would have to be increased over the usual 20 mg per day used. High dose of tamoxifen might decrease the therapeutic index by increasing toxicity (Gelmann, 1996). It seems that improved cytotoxicity of incorporated drug is not dependent of the composition on the SLN. In fact it was reported that the IC_{50} value of drug-loaded SLN composed of different materials were lower than that of free drug solution (Yuan *et al*, 2008). There are at least two mechanisms that have been associated with the cytotoxicity of drug-loaded SLN. Using Doxorubicin (DOX)-loaded SLN, it was suggested that the first mechanism involves the release of DOX from DOX-SLN outside the cells, and the cytotoxicity of DOX is increased by the nanoparticles. The second mechanism suggested was, release of the drug inside the cell and thus produces greater cytotoxicity (Wong *et al.*, 2006).

2.4 Morphological changes of TAM-loaded SLN on human breast cancer cells

Apoptotic cell death can be recognized under phase contrast and fluorescence inverted microscope after staining. This is the most practical method to identify cell morphological changes attributed to apoptotic cell death.

2.4.1 Phase contrast microscopy

TAM-loaded SLN treatments at concentrations equal to IC_{50}, causes detachment of MCF-7 and MDA-MB231 cells and loss of colony formation ability. These cells appear rounded-up and lose contact with neighboring cells (Fig. 5). The normal untreated MCF-7 and MDA-MB-231 cells, however, appear healthy and exhibiting epithelial-like features and forming a monolayer on the surface of the culture flask. In the presence of TAM and TAM-loaded SLN, the viability of the both cells diminishes and the cancer cells loss their normal morphological characteristics, detaches, aggregates, and later develops apoptotic bodies. The detachment of cells in the presence of free TAM and TAM-loaded SLN suggests that tamoxifen is cytotoxic, even when incorporated in the SLN.

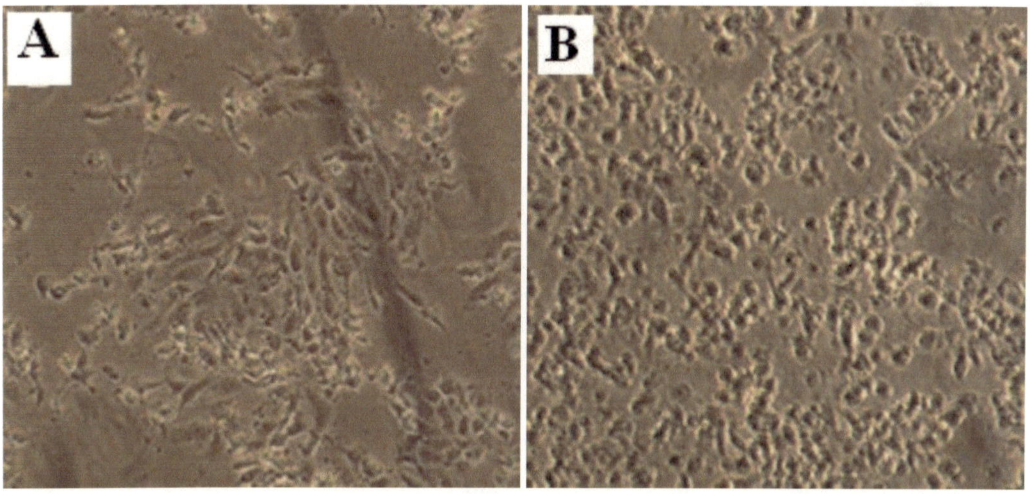

Fig. 5. Phase contrast micrographs of MCF-7 cell (A) and MDA-MB231 cell (B) treated with TAM loaded SLN (12 µg/mL) (magnification x10). (Adapted from Abbasalipourkabir, 2010).

2.4.2 Nuclear morphology

Cell death is either by physiological or pathological means. Physiological cell death is distinguished by apoptotic morphology, including chromatin condensation, membrane blebbing, internucleosomal degradation of DNA, and apoptotic body formation. Pathological cell death or necrosis is associated with cellular swelling and collapse, without severe damage to nuclei or breakdown of the DNA. In apoptosis, several cellular and molecular biological features, including cell shrinkage and DNA fragmentation are exhibited (Yu *et al.*, 2010). To characterize the cell death, the nuclear morphology of dying cells can be examined under Hoechst dye 33258 staining. The Hoechst dye 33258 is a bis-benzimide derivatives and a fluorescent DNA-binding agent. This dye is useful for cell cycle analysis because it can be used in low concentrations, and thus minimizing the problem of toxicity. According to Latt & Stetten, (1976) the Hoechst dye binds to AT-rich regions of the DNA and when excited with an ultraviolet light produces bright fluorescence at 465 nm. TAM-loaded SLN induce death of MCF-7 and MDA-MB231 cells by apoptosis. This is evident by the typical apoptotic changes showing clear condensation of cell nuclei, nuclear fragmentation and apoptotic bodies (Fig. 6).

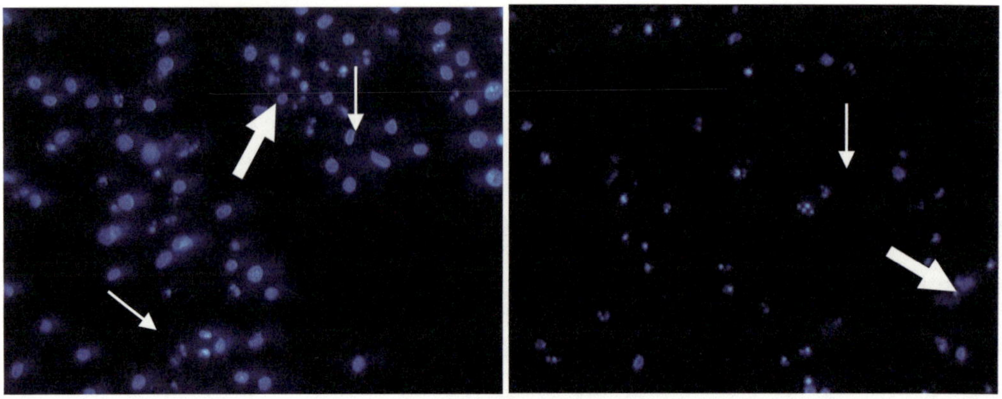

Fig. 6. Fluorescence microscopy of MCF-7 cells (A) and MDA-MB-231 cells (B) treated with TAM-loaded SLN. Cell shrinkage (thin arrow) and apoptotic cells (thick arrow) are evident (magnification x40). (Adapted from Abbasalipourkabir, 2010).

Tamoxifen-loaded SLN like free TAM display antitumoral activity against human breast cancer cells. The biological availability of drug is not affected when incorporated into SLN. Therefore SLN could be applied as a drug delivery system for cancer treatments. In conclusion, the TAM-loaded SLN, because of its small size, could not be easily phagocytosed by macrophages and therefore the nanoparticles could be potentially used in long-term circulating carrier system for breast cancer therapy.

3. Conclusion

The main challenge in cancer chemotherapy is toxic side-effects induced by chemotherapeutic drugs. Single dose or short-time application (1-2 weeks) will probably causes serious health problems, but the use of biodegradable nano-sized particles for long-term or life-time therapy may produce other serious side-effects. Increasing the encapsulation efficiency of poorly water-soluble molecules will lead to the development of improved SLN formulations. In the near future, it is expected more studies will focus on improving SLN and drug-loaded SLN formulations to increase the efficacy and reduce the side-effects of chemotherapeutic drugs for anticancer treatment. These studies should include preparation of formulations with different particle size and distributions, different matrix lipids and additional ingredients. Thus, if nanoparticulate drug delivery systems to be used effectively and routinely, the matter of toxicity of the components of nanoparticles must be addressed. Indeed, SLN requires further development before it can be used as a new drug delivery system for chemotherapy drugs in treatment of human cancers.

4. References

Abbasalipourkabir, R. Dehghan, A. Salehzadeh, A. Shams Abadi, F. and Rasedee Abdullah (2010,2) "Induction of mammary gland tumor in female Sprague-Dawley rats with LA-7 cells." *African Journal of Biotechnology*, 9(28):4491-4498

Abbasalipourkabir, R. 2010. Development, characterization, cytotoxicity and antitumor effect of tamoxifen-loaded solid lipid nanoparticles. Ph.D. thesis. Universiti Putra Malaysia. Malaysia

Abbasalipourkabir, R. Salehzadeh, A. and Rasedee Abdullah (2010,1). "Antitumor activity of tamoxifen loaded solid lipid nanoparticles on induced mammary tumour gland in Sprague-Dawley rats." *African Journal of Biotechnology*, 9(43):7337-7345

Abbasalipourkabir, R. Salehzade, A. and Rasedee A. 2011. *Pharmaceutical technology* 35(4): 74-79

Adams, R.L.P. 1990. Cell structure for biochemists. 2nd Edition. Elsevier, Amsterdam, pp. 1-22

Anderson B.O. Yip C.H. Smith R.A. Shyyan R. Sener S.F. Eniu A. Carlson R.W. Azavedo, E. and Harford, J. 2008. Guideline implementation for breast health care in low-income and middle-income countries: overview of the breast health global initiative global summit 2007. *Cancer* 113:2221–43.

Brigger, I. Chaminade, P. Marsaud, V. Appel, M. Besnard, M. Gurny, R. Renoir, M. and Couvreur, P. 2001. Tamoxifen encapsulated within polyethylene glycol-coated nanospheres. A new antiestrogen formulation. *International Journal of Pharmaceutics* 214:37-42.

Cardosa, F. Bedard, P.L. Winer, E.P. Pagani, O. Senkus-Konefka, E. Fallowfield, L.J. Kyriakides,S. Costa, A. Cufer, T. and Albain, K.S. 2009. International Guidelines for Management of Metastatic Breast Cancer: Combination vs Sequential Single-Agent Chemotherapy. *Journal of National Cancer Institute* 101: 1174-1181

Clark M.J. 2008. WITHDRAWN: Tamoxifen for early breast cancer. *Cochrane Database of Systematic Reviews* Oct 8(4):CD000486.

Cohen, I. 2004. Endometrial pathologies associated with postmenopausal tamoxifen treatment. *Gynecology & Oncology*. 94:256-266.

Coughlin, S.S. and Ekwueme, D.U. 2009. Breast cancer as a global health concern. *Cancer Epidemiology* 33:315-318.

Chawla, J.S. and Amiji, M.M. 2003. Cellular uptake and concentrations of tamoxifen upon administration in poly(-caprolactone) nanoparticles. *The American Association of Pharmaceutical Scientists Journal* 5:28-34.

Christov, K. Grubbs, C.J. Shilkaitis, A. Juliana, M.M. and Lubet. R.A. 2007. Short-Term Modulation of Cell Proliferation and Apoptosis and Preventive/Therapeutic Efficacy of Various Agents in a Mammary Cancer Model. *Clinical Cancer Research* 13:5488-5496.

Delima, G.R. Facina, G. Shida, J.Y. Chein, M.B.C. Tanaka, P. Dardes, R.S. Jordan, V.C. and Gebrim, L.H. 2003. Effects of low dose tamoxifen on normal breast tissue from premenopausal women. *European Journal of Cancer* 39:891-898.

DevCan, 2004. Probability of Developing or Dying of Cancer Software, Version 5.2. Statistical Research and Applications Branch, National Cancer Institute. http://srab.cancer.gov/devcan accessed on 11 January 2010.

Ewesuedo, RB. and Ratain, MJ. 2003. Principles of cancer therapeutics. In: Vokes EE, Golomb HM (eds) Oncologic therapies. Springer, Secaucus, NJ, pp 19–66.

Fadok, VA. 1999. Clearance: the last and often forgotten stage of apoptosis. *Journal of Mammary Gland Biology and Neoplasia* 4:203–211.

Ferlay, J. Autier, P. Boniol, M. Heanue, M. Colombet, M. and Boyle, P. 2007. Estimates of the cancer incidence and mortality in Europe in 2006. *Annals of Oncology* 18:581-592.

Fontana, G. Maniscalco, L. Schillaci, D. and Cavallaro, G. 2005. Solid Lipid Nanoparticles Containing Tamoxifen. Characterization and *in vitro* Antitumoral Activity. *Drug Delivery,* 12:385–392.

Fundaro, A. Cavalli, R. Bargoni, A. Vighetto, D. Zara, G.P. and Gasco, M.R. 2000. Non-stealth and stealth solid lipid nanospheres carrying doxorubicin; pharmacokinetics and tissue distribution after I.V. administration to rats. *Pharmacology* 42:337-343.

Gelman, E.P. 1996. Tamoxifen induction of apoptosis in estrogen receptor-negative cancers: New tricks for an old dog? *Journal of the National Cancer Institute* 88:224-226.

Gradishar, W.J. 2005. Safety considerations of adjuvant therapy in early breast cancer in postmenopausal women. *Oncology* 69:1-9.

He, J. Hou, S. Lu, W. Zhu, L. and Feng, J. 2007. Preparation, pharmacokinetics and body distribution of Silymarin-loaded Solid Lipid Nanoparticles after oral administration. *Journal of Biomedical Nanotechnology* 3:195-202

Jordan, V.C. 1994. Molecular mechanisms of antiestrogen action in breast cancer. *Breast Cancer Research and Treatment* 31:41-52.

Kayser, O. lemke, A. and Hernandez-Trejo, N. 2005. The impact of nanobiotechnology on the development of new drug delivery systems. *Current Pharmaceutical biotechnology* 6:3-5.

Latt, S.A. and Stetten, G. 1976. Spectral studies on 33258 Hoechst and related bisbenzimidazole dyes useful for fluorescent detection of deoxyribonucleic acid synthesis. *Journal of Histochemistry and Cytochemistry* 24:24-33.

Mandlekar, S. and Kong, A.N. 2001. Mechanisms of tamoxifen-induced apoptosis. *Apoptosis* 6:469-477.

Mandlekar, S. Yu, R. Tan, TH. and Kong, AN. 2000. Activation of caspase-3 and c-Jun NH2-terminal Kinase-1 signaling pathways in tamoxifen-induced apoptosis of human breast cancer cells. *Cancer Research* 60:5995-6000.

McGregor, J. and Jorda, V. 1998. Basic guide to the mechanisms of antiestrogen action. *Pharmacological Reviews* 50:151-196.

Mehnert, W. and Mäder, K. 2001. Solid lipid nanoparticles Production, characterization and applications. *Advanced Drug Delivery Reviews* 47:165–196.

Messmer, U.K. and Pfeilschifter J. 2000. New insights into the mechanism for clearance of apoptotic cells. *Bio Essays* 22:878–881.

Mirshahidi, H.R. and Abraham, J. 2004. Managing early breast cancer: prognostic features guide choice of therapy. *Postgraduate Medicine* 116:23-34.

Mourits, M.J.E. De Vries, E.G.E. Willemse, P.H.B. Ten Hoor, K.A. Hollema, H. and Van Der Zee, A.G.J. 2001. Tamoxifen treatment and gynecologic side effects: a review. *Obstetrics and Gynecology* 97:855-866.

Müller, R.H. Mäder, K. and Gohla, S. 2000. Solid lipid nanoparticles (SLN) for controlled drug delivery - a review of the state of the art. *European Journal of Biopharmaceutics* 50:161-177

Müller, R.H. Mehnert, W. Lucks, J.S. Schwarz, C. Mühlen, A. Z. Weyhers, H. Freitas, C. and Rühl, D. 1995. Solid lipid nanoparticles (SLN) - an alternative colloidal carrier system for controlled drug delivery. *European Journal of Biopharmaceutics* 41:62-69.

Paci, E. Cariddi, A. Barchielli, A. Bianchi, S. Cardona, G. Distante, V. Giorgi, D. Pacini, P. Zappa, M. and Del Turco, MR. 1996. Long-term sequelae of breast cancer surgery. *Tumori* 82:321-4.

Peters-Engle, C. Frank, W. Danmayr, E. Friedl, H.P. Leodolter, S. and Medl. M. 1999. Association between endometrial cancer and tamoxifen treatment of breast cancer. *Breast Cancer Treatment* 54:255-260.

Porter P. 2008. "Westernizing" women's risks? Breast cancer in lower-income countries. *The New England Journal of Medicine* 358:213–6.

Russo, J. and Russo, IH. 2002. Mechanisms involved in carcinogenesis of the breast. In *Breast cancer*, ed. J.R. Pasqualini, Marcel Dekker. New York. pp. 1-2.

Salami, S. and Karami-Tehrani, F. 2003. biochemical studies of apoptosis induced by tamoxifen in estrogen receptor positive and negative breast cancer cell lines. *Clinical Biochemistry* 36:247-253.

Siekmann, B. and Westesen. K. 1992. Submicron-sized parenteral carrier systems based on solid lipids. *Pharmaceutical and Pharmacological Letters* 1:123-126.

Sipos, E.P. Tyler, B. Piantadosi, S. Burger, P.C. and Brem, H. 1997. Optimizing interstitial delivery of BCNU from controlled release polymers for the treatment of brain tumors, *Cancer Chemotherapy and Pharmacology* 39:383-389.

Wong, H.L. Rauth, A.M. Bendayan, R. Manias, J.L. Ramaswamy, M. Liu, Z. Erhan, S.Z. and Wu, X.Y. 2006. A New Polymer-Lipid Hybrid Nanoparticle system Increases Cytotoxicity of Doxorubicin Against Multidrug-Resistaant Human Breast Cancer Cells. *Pharmaceutical Research* 23:1574-1584.

Wyld, D.K. Chester, J.D. and Perren, T.J. 1998. Endocrine aspect of the clinical management of the breast cancer-current issue. *Endocrine-Related Cancer* 58:97-110.

Yu, T, Lee, J. Lee, Y.G. Byeon, S.E. Kim, M.H. Sohn, E,H. Lee, Y.J. Lee, S.G. and Youl, j. 2010. *In vitro* and *in vivo* anti-inflammatory effects of ethanol extract from Acer tegmentosum. *Journal of Ethnopharmacology* 128(1):139-47).

Yuan, H. Miao, J. Du, YZ. You, J. Hu, FQ. and Su Z. 2008. Cellular uptake of solid lipid nanoparticles and cytotoxicity of encapsulated paclitaxel in A549 cancer cells. *International Journal of Pharmaceutics* 348:137-145.

4

In-Situ Versus Post-Synthetic Stabilization of Metal Oxide Nanoparticles

Georg Garnweitner
Technische Universität Braunschweig
Germany

1. Introduction

Within the last two decades, the synthesis and application of nanoparticles has evolved as one of the most active fields of research and development, promising solutions to some of humanity's most pressing needs in cases as diverse as cancer therapy, water purification, or energy storage (Duget et al., 2006; Bazito & Torresi, 2006; Centi & Perathoner, 2009; Feldmann & Goesmann, 2010; Goyal et al., 2011; Kim & van der Bruggen, 2010; Li et al., 2008; Moghimi et al., 2005; Prandeep & Anshup, 2009; Sanvicens & Pilar, 2008; Sides et al., 2002; Theron et al., 2007). Today, nanoparticles are predominantly being applied as particulate materials with enhanced properties as compared to the standard bulk material. One however envisions nanotechnology already in the near future to strongly increase in complexity, leading to the development of hierarchical nanosystems being capable of computing and robotic or even self-replicating tasks (Mallouk & Sen, 2009; Rasmussen et al., 2003; Requicha, 2003; Sánchez & Pumera, 2009). Most future applications are expected to be eventually realised via the so-called bottom-up approach, emanating from nanoparticles as tiny building blocks that self-assemble in a controlled and organised manner to a higher level of hierarchy (Shenhar & Rotello, 2003; Dong et al., 2007). The primary prerequisite for the realisation of this strategy is the availability of high-quality nanoparticles that can act as such building blocks, possessing a predefined and uniform size and shape. Consequently, research on the controlled preparation of nanoparticles has greatly intensified within the last two decades, mainly focusing on carbon-based nanostructures (Mostofizadeh et al., 2011), metals (Cushing et al., 2004; Guo & Wang, 2011), metal oxides (Cushing et al., 2004; Chen & Mao, 2007; Pinna & Niederberger, 2008), semiconductor nanostructures (Trindade et al., 2001) and organic or hybrid nanostructures (Biswas & Ray, 2001; Ballauff, 2003). Remarkable progress has been made in the synthesis of nanoparticles with defined and even complex shape for certain materials (Jun et al., 2006), although a general unterstanding of the formation processes of nanostructured materials appears to remain a long way off.

An aspect that however is often not treated accordingly is the second step in the fabrication chain: the control of the surface interactions of the nanoparticles to determine their secondary structure. Even though by now a large number of examples have been reported on the preparation of highly ordered superstructures for diverse nanoparticle systems (Gao & Tang, 2011), the truly rational and controlled self-assembly, which would be vital for the large-scale feasibility of many of the envisioned applications, appears still to be long out of reach. Even the first level, the control of the interaction forces between the individual nanosized building blocks to avoid uncontrolled agglomeration, is neither realised nor understood in most commercial nanotechnology products today, and one is only beginning to develop general concepts aimed to describe the interactions between ultra-small particles in the low nanometer regime, even though the models of particle interaction for larger particles have been established in colloid science almost a century ago. One concept that is particularly promising for the control of nanoparticle interaction is the use of small organic ligands - in contrast to the classical models of purely electrostatic stabilisation on the one hand and polymer-based steric stabilisation on the other hand - which in principle allows long-term stabilisation with minimum stabiliser content. This Chapter is aimed to introduce the advantages and challenges of the small molecule-based stabilisation, and will present a number of select examples for the successful application of this strategy. A special focus is set on comparing the different options of introducing the stabiliser into the system: on the one hand, the so-called in-situ stabilisation strategy, where ligands are already introduced into the system prior to or during particle formation, and on the other hand the post-synthetic stabilisation, where the ligands are added in a separate step following the synthesis, in order to give the reader significant insight into this field to allow individual judgement on the benefits and disadvantages of each strategy. The Chapter will be concluded with an outlook to current and possible future developments in the field of the small-molecule stabilisation of nanoparticles.

2. The stabilisation of nanoparticles

The systematic study of the interactions between small particles began in the first decades of the 20th century within the rapidly growing field of colloids science, following the discovery that colloidal systems consist of small units below 500 nm in size dispersed in a liquid medium. In the pioneering works of Ostwald, the inherent thermodynamic instability of such colloidal systems was recognised (Ostwald, 1915), which causes a general tendency of such systems to agglomerate and aggregate, in order to minimise the particle-fluid interface area and thus the interface free energy. A number of theories were proposed in the following decades to explain the stability of colloidal systems (Everett, 1988), providing different possibilities to change and enhance the stability, which shall be briefly summarised in the following.

2.1 Colloidal theories of particle stabilisation and their application to nanoparticles

A few decades after Graham and Ostwald, first theories to explain the stability of aqueous colloidal systems in the presence of charged species were formulated, leading to the well-known DLVO theory (Derjaguin & Landau, 1941; Verwey & Overbeek, 1948). This theory is based on previous works by Hamaker on the attractive interactions of colloidal particles, which he explained as the combined van-der-Waals attractions of the individual atoms (Hamaker, 1937). These attractive forces are balanced by the electrostatic repulsion of likely charged particles, with charges not only arising from the surface potential of the particles, but also the electrochemical double layer surrounding them. Also the Born repulsion, effective only at small distances when the electron shells of atoms start to penetrate each other, is taken into account. The DLVO theory shows that the stabilisation of particles can be easily achieved by electrostatic repulsive forces, which however are highly labile and depend on conditions such as the pH and ionic strength in the dispersion medium. Moreover, this mechanism cannot be applied to hydrophobic systems that often are preferred for the handling and manipulation of nanostructured materials, e.g. due to their lower surface tension and lower adsorption tendency to the nanoparticle surface.

Another possibility is the stabilisation by coverage of the particle surface with organic polymers. This steric stabilisation concept was investigated theoretically already briefly after the DLVO theory (Mackor, 1951). The polymer chains adsorb to the particle surface to create an organic shell around the particle. As two particles approach each other, their organic shells touch and eventually interpenetrate. At that point, the solvation shell of the polymer is disrupted, as the polymer density increases upon penetration, which creates a counteracting osmotic pressure of the solvent that forces the particles apart (Napper, 1983). Good solvent compatibility of the polymer is therefore required for an effective steric stabilisation. The electrosteric stabilisation concept involves the use of polyelectrolytes, i.e. polymers with charged groups, to attach to the particle surface, thereby combining the electrostatic and steric stabilisation concepts (Hunter, 2001).

These classic colloidal theories have been proven successful for the description of colloidal systems in numerous cases and hence, also their application to nanoparticle dispersions seems obvious. The DLVO theory has thus been applied in a number of cases to describe the behaviour of nanoparticulate systems (Schwarz & Safran, 2000; Tadmor & Klein, 2001; Reindl & Peukert, 2008; Marczak et al., 2009; Segets et al., 2011). However, some special aspects need to be taken into account when attempting to describe dispersions of nanoparticles below 20 nm in size by means of DLVO theory: On the one hand, the distance dependence of the attractive and repulsive forces is strongly dependent on particle size already in the classical DLVO concept (Verwey & Overbeek, 1948). The electrostatic interaction is simplified in colloidal models as the interaction between two planar electrochemical double layers, because the interparticle distance becomes much smaller than the particle size as the particles approach each other. For small nanoparticles, this model is no longer valid, and thus spherical double layers must be included in the calculation (Mulvaney, 1998). The interaction of small nanoparticles was studied by Wiese and Healy in 1969. They reported two significant effects when reducing the particle size to the low nanometre regime: on the one hand, the repulsive forces become very small, but on the other hand also the energy minimum becomes very shallow (Wiese & Healy, 1969). The application of the DLVO concept to such small particles generally leads to interaction potentials of only a few $k{\cdot}T$, which results in only small changes between a stable and an unstable system, making the prediction of the stability of a system a challenging task (Marczak et al., 2009; Segets et al., 2011). Moreover, the minimum is also shifted in position towards smaller interparticle distances and thus, the description of the particles as solid, spherical systems with defined interface becomes inappropriate. A molecular layer of organics adsorbed to the particle surface that may be neglected for large particles can become important for the stability of nanoparticle systems, as it may be sufficient to prevent the particles from reaching the critical distance. Additionally, inhomogeneities within the particles, such as an oxide shell of metal nanoparticles, can suddenly play an important role.

This is nicely exemplified in the work of Reindl and Peukert, who investigated the stabilisation of silicon nanoparticles in organic solvents, taking into account an oxidised silica shell around the particles as well as the adsorption of solvent molecules, by applying a core-shell-adsorbate model based on the DLVO theory (Reindl & Peukert, 2008). Figure 1 shows the used models as well as results of the calculation of the total energy barrier. Whilst at higher concentrations, the system would be unstable for pure Si nanoparticles, the oxide shell provides high stability against agglomeration. Moreover, it is clearly visible that the adsorbate layer formed by solvent adsorption also has a noticeable effect on the energy barrier despite the relatively large particle size of 100 nm, further increasing the stability of the system. Such core-shell models have been applied also to other systems in the meantime (Marczak et al., 2009; Segets et al., 2011).

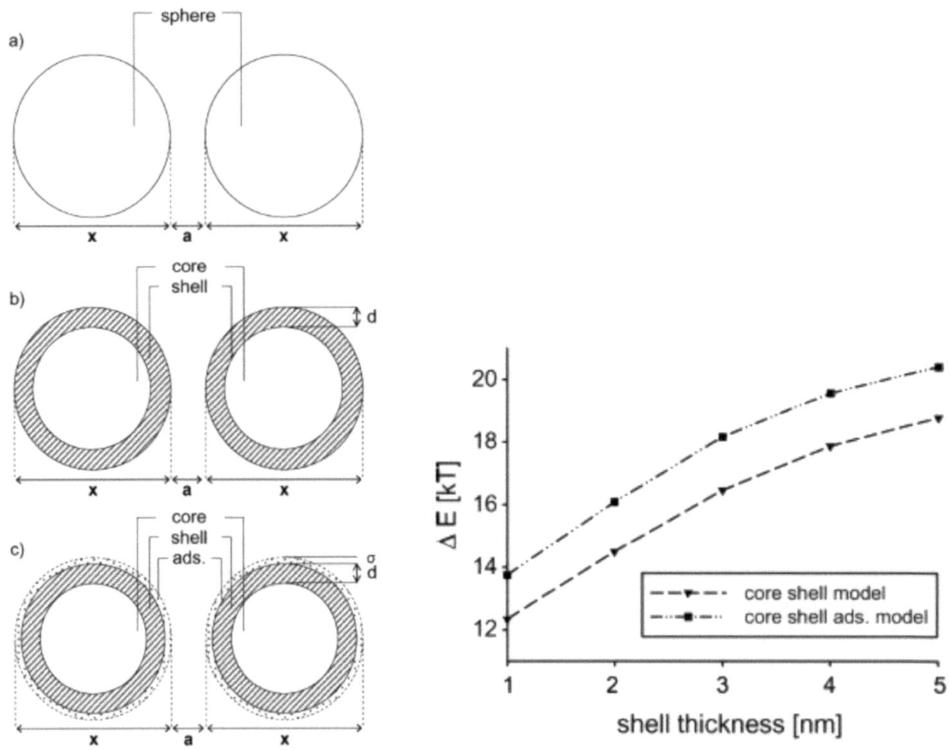

Fig. 1. Left: schematic of the model systems used for DLVO calculations on Si nanoparticles: (a) solid sphere model, (b) core-shell model, (c) core-shell-adsorbate model; right: calculated energy barrier ΔE of Si nanoparticles for the model systems in dependence of shell thickness; reproduced from (Reindl & Peukert, 2008) with kind permission of Elsevier.

On the other hand, an extension of the DLVO theory has been proposed, leading to the so-called XDLVO concept, in the works by van Oss et al. (van Oss et al., 1987, 2003). Here, Lewis acid-base interactions are additionally taken into account, and this model was shown to allow a more accurate prediction of the stability of dispersions of hectorite and chrysotile particles in various media (Wu et al., 1999). Also this concept has been applied to nanoparticle systems, being able to explain differences in the stability of iron oxide nanoparticle dispersions when adding fatty acids as stabilisers (Gyergyek et al., 2010).

The precise modelling of the steric stabilisation of nanoparticles, on the other hand, remains complex, which prevented widespread attempts to calculate particle stability by adsorption of polymers. To allow a practical application, several simplified models were presented, such as the theory of Vincent and Edwards (Vincent et al., 1986). Here, a uniform polymer segment density is assumed within the polymer shell, thus requiring a smaller number of parameters. This model has also been applied to the steric stabilisation of nanoparticles. For example, K. Lu investigated the interaction energies of Al_2O_3 nanoparticles of 20-45 nm in size stabilised with polymeric shells of different thickness (Lu, 2008). Figure 2 shows the results of these calculations, clearly revealing that with increasing layer thickness, the stability of the system clearly increases, and that a critical layer thickness is required to prevent agglomeration (in this case, amounting to 2.5 nm), according to the model.

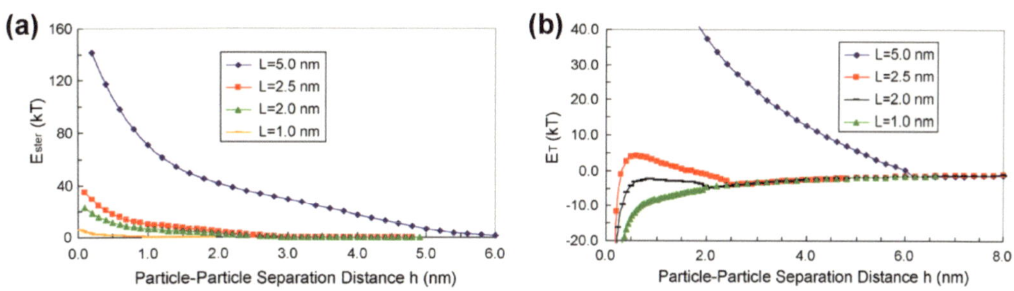

Fig. 2. Calculation of the distance-dependent steric interaction energy E_{ster} (a), and the total interaction energy E_T (b) for the stabilisation of Al_2O_3 nanoparticles protected by polymer shells of different thickness (Reproduced from (Lu, 2008) with kind permission of Elsevier).

2.2 The small-molecule stabilisation concept

Whilst the classical colloidal models of electrostatic and steric stabilisation have thus been known for decades, the treatment of nanoparticles to obtain stable dispersions in practice has always followed an empirical approach, with the well-known gold nanoparticle sols serving as typical examples that small organic molecules are sufficient to act as long-term effective stabilisers to prevent agglomeration (Turkevich et al., 1953). In many cases, the precise mechanism of stabilisation is unclear, and it is generally still being disputed whether organic ligands bound to the surface in most cases bear charges that result in electrostatic stabilisation, or whether steric repulsion effects (Peyre et al., 1997; Shah et al., 2001; Arita et al., 2009) or in fact some combined effects (Marczak et al.; 2009, Segets et al., 2011) determine the stability of the system.

Nonetheless, it has become clear that this concept of adsorption of small molecules, resulting in very thin organic layers (termed adlayers) around nanoparticles, has significant advantages over a conventional steric stabilisation with polymers: The packing of small particles is strongly influenced by the adlayer thickness Δ, basically equivalent to the chain length of the surface-bound ligands, as shown in Figure 3. For long-chain ligands, such as a polymer coating, the large adlayer thickness results in a much lower packing density of the particles as compared to short-chain ligands.

This can also be expressed as the effective volume fraction Φ_{eff} which includes the adlayer in the particle volume and is related to the true volume fraction Φ of the particles by

$$\Phi_{eff} = \Phi\left(1 + \frac{\Delta}{r}\right)^3 \qquad (1)$$

r referring to the particle radius (Bell et al., 2005). Hence, the particles can be packed less densely, which limits the maximum achievable solids volume fraction in concentrated dispersions, leads to higher viscosity for a given volume fraction, and results in a "dilution" of core particle properties in applications such as nanocomposites.

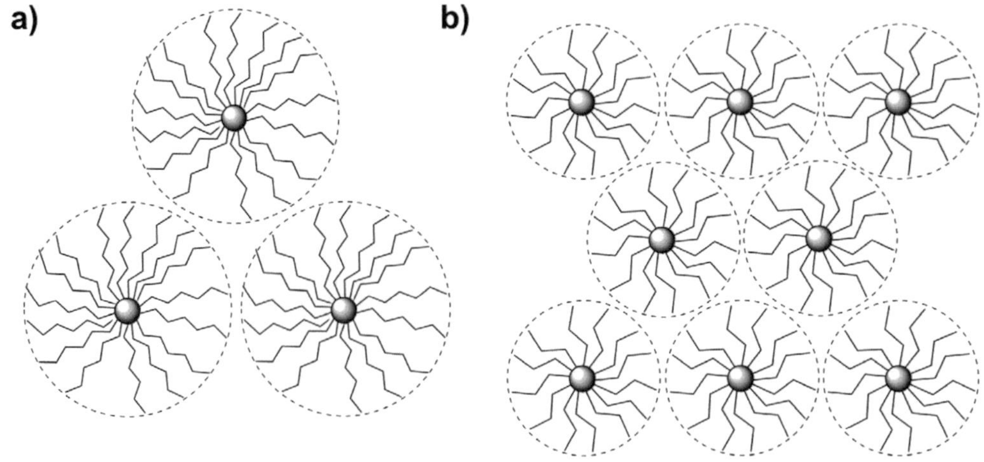

Fig. 3. Schematic (not to scale) of the packing of spherical particles with a stabilising adlayer comprised of long-chain ligands (a) and short-chain ligands (b).

3. In-situ stabilisation of nanoparticles

The stabilisation of nanoparticles can be realised experimentally most easily via the in-situ stabilisation concept. This involves the addition of stabilisers or ligands to the reaction system already during or even prior to nanoparticle formation. Hence, it is straightforward, and in principle can lead to optimum stabilisation, as it prevents particle agglomeration *a priori*, but on the other hand bears the imminent risk of an alteration of the nanoparticle formation process - an aspect that in some cases may even be desired, e.g. to obtain smaller and more uniform particles, but in many other cases is unwanted, as it makes the formation mechanism even more complex, or results in a degradation of intrinsic properties of the nanoparticles. Conceptually, the in-situ stabilisation of nanoparticles can be performed with optimised stability in the reaction medium when the processing or application of the nanoparticles in that medium is desired or feasible. In other cases, an intermediate precipitation step is performed to facilitate a change in dispersion medium, for example when a more volatile or more hydrophilic/hydrophobic medium is required than was used for the synthesis, and thus, the stabiliser is targeted to result in optimised stability in the final medium rather than the reaction solvent.

The in-situ stabilisation strategy has found widespread application, as the addition of surfactants or surface-modifying agents to a reaction mixture is commonly employed in the synthesis of metal nanoparticles (Turkevich et al., 1953; Murphy et al., 2005) as well as in the so-called hot injection method, where the precursors are injected into a hot solvent/surfactant mixture, resulting in highly uniform and well-dispersible nanocrystals (Murray et al., 1993). This method has also been extended to transition metal oxides (Rockenberger et al., 1999; Sun et al., 2004; Epifani et al., 2008; Gilstrap Jr. et al., 2008). Despite the widespread use of in-situ stabilisation, the precise effect of the ligands on the properties of the resulting products has only been investigated in a small number of studies. Only a few studies were directed towards a systematic investigation of the effect of ligands on the particle properties, whilst in most reports the influence of the chemistry of the ligand has not been studied specifically. In the following section, some recent investigations of ligand effects on the particle properties are presented for different metal oxide nanoparticle systems.

3.1 Effects on particle crystallinity

Most ligands possess Lewis-basic groups that act as electron donors to metal centres on the nanoparticle surface, thereby enabling a firm binding of the ligands. Naturally, these groups also coordinate to molecular complexes of the same metal and hence, also to the precursor species. If the ligands are coordinated more strongly than the organics in the precursor complex, the reaction rates during the particle formation are decreased drastically, a fact that is utilised in aqueous sol-gel chemistry to deliberately decrease the rates of condensation in many cases. For strongly coordinating ligands, this may even lead to an incomplete formation, resulting in substantial organics content or a decrease in crystallinity of the resultant nanoparticles, a loss of yield, or may even impede the formation of nanoparticles altogether.

The effect of ligands on the crystallinity of in-situ stabilised titania nanoparticles has been investigated in a number of studies, because the crystallinity and phase of the particles is crucial to their performance in photocatalytic applications (Chen & Mao, 2007). Thereby, the high suitability of catechol ligands for the in-situ surface modification of TiO_2 nanoparticles, due to a very strong, covalent binding, has been reported (Niederberger et al., 2004). If a certain, critical concentration of these ligands however is exceeded and the molar Ti:ligand ratio becomes smaller than 10, a substantial decrease in crystallinity is observed (Niederberger et al., 2004). Therefore, the dopamine ligands cannot be used in high amounts, which results in the additional presence of the reaction medium benzyl alcohol on the particle surface after the synthesis (Niederberger et al., 2004). In this example, the in-situ stabilisation process did not lead to stability in benzyl alcohol but in various common organic solvents after centrifugation, washing and redispersion steps. Furthermore, different substituted catechols were investigated. Depending on the precise chemistry of the catechol side group, the particles could be made more hydrophilic or hydrophobic, even allowing stability in water when using dopamine hydrochloride as ligand (Niederberger et al., 2004).

In a more recent study, we have taken a closer look at the precise influence of the chemistry of the ligands on the crystallinity (Garnweitner et al., 2010). Different ligands with amine, alcohol and/or carboxylic acid functional groups were added prior to the synthesis of TiO_2 nanoparticles, again in benzyl alcohol as reaction medium. Interestingly, strong differences were found with respect to stability of the resulting nanoparticles, as well as their crystallinity; the differences in these properties however could not be directly correlated. Figure 4 (left) shows a series of X-ray diffraction (XRD) patterns measured from products obtained in presence of different ligands. Under the chosen experimental conditions, in the ligand-free system small anatase nanoparticles with good crystallinity and a size of about 3 nm are obtained. Whilst for glycine and malonic acid, no negative effects on crystallinity could be observed, for malic acid and glycerol a strongly decreased crystallinity was found. This could be correlated to the binding of the ligands to the particle surface by means of thermogravimetric analysis (TGA), showing that malic acid and especially glycerol bind to the nanoparticles in substantially higher quantities than the other ligands, equivalent to a higher mass loss (Figure 4, right).

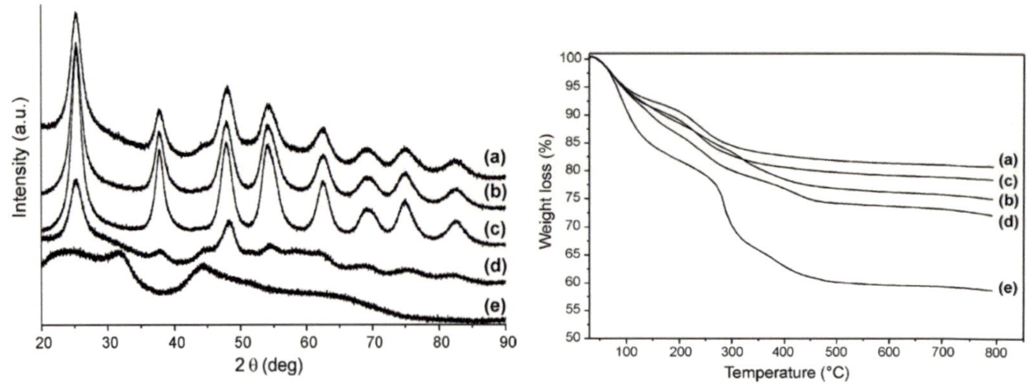

Fig. 4. X-ray diffractograms (left) and thermograms (right) of samples obtained without any present ligands (a), and in the presence of glycine (b), malonic acid (c), malic acid (d) and glycerol (e). Images reproduced from (Garnweitner et al., 2010) with kind permission of Elsevier.

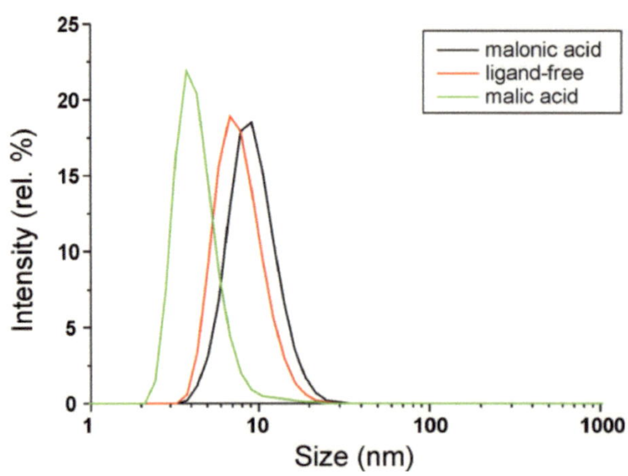

Fig. 5. Particle size distribution of TiO_2 nanoparticles modified in-situ by addition of various ligands, as determined by dynamic light scattering. Reproduced from (Garnweitner et al., 2010) with kind permission of Elsevier.

Surprisingly, the stability of dispersions of the nanoparticles could not be directly related to these results. The stability was investigated after precipitation of the particles in diethyl ether and redispersion by simple addition of de-ionised water to the precipitate. Best stability – in terms of highest solid content for minimal water addition – was observed for the glycine and malic acid systems (Garnweitner et al., 2010). However, good stability was even achieved for the ligand-free system, as in this case, the reaction of $TiCl_4$ with benzyl alcohol causes the formation of HCl as byproduct, resulting in a highly positive zeta potential. The secondary particle size distribution, as determined by dynamic light scattering (DLS), did not reveal substantial differences between the samples (Figure 5), which - in combination with the TGA results - proves that the reaction medium coordinates to the particle surface in addition to any ligand, and that for most systems, the ligand merely replaces the reaction medium rather than forming an additional shell.

3.2 Effects on particle size and morphology

Although the effect of in-situ stabilisers on the morphology of metal oxide nanoparticles has been studied in a number of publications, as these stabilisers were often purposely used to modify particle size, consistent knowledge on the precise influence of the stabilisers has not been reached, probably due to the complex interaction and the necessity of a thorough analysis of the underlying chemistry to be able to understand the effects. Gaynor et al. investigated the effect of soluble hydroxypropylcellulose on the particle size of TiO_2 nanoparticles, reporting a rather complex effect (Gaynor et al., 1997). The in-situ stabilisation of magnetite nanoparticles with L-arginine was investigated by Theerdhala et al., who showed that the particle size decreased steadily with decreasing precursor:stabiliser ratio, with however rather broad size distributions and additional changes in aspect ratio of the particles (Theerdhala et al., 2010).

In other cases, the addition of ligands even in small amounts was shown to result in dramatic changes in morphology. Again, TiO_2 nanoparticles are among the best studied systems: The influence of amino acids used as in-situ ligands in a "gel-sol" synthesis of anatase nanoparticles was investigated by Kanie and Sugimoto, showing that for certain amino acids, the initial particle-type morphology changes to rod-like due to strong adsorption of the amino acids to specific crystal faces (Kanie & Sugimoto, 2004). The synthesis of TiO_2 nanorods by hydrolysis of titanium tetraisopropoxide in presence of oleic acid was investigated by Cozzoli et al., who even proposed a model to explain different morphologies obtained for a "fast" and "slow"-type hydrolysis, again involving selective adsorption of the ligand (Cozzoli et al., 2003).

As another example, Polleux et al. investigated the synthesis of tungsten oxide nanostructures in a nonaqueous system, using benzyl alcohol as reaction medium (Polleux et al., 2006). In the pristine system, comprising only the precursor and solvent, nanoplatelets of about 5-10 nm in thickness and 30-100 nm lateral size were obtained (Figure 6, a), but after addition even of small amounts of 4-*tert*-butylcatechol, anisotropic rod-like structures with diameters of 35-40 nm and lengths of 50-800 nm were observed (Figure 6, b-d), which in fact are comprised of a highly ordered lamellar organic-inorganic hybrid nanostructure (Polleux et al., 2006). The ligand was hence inferred to act in manifold ways, not only controlling the crystal growth through selective binding to high surface energy edges but also the interaction of the individual platelets to enable their assembly (Polleux et al., 2006). Although in this example, the dispersibility of the obtained nanostructures in various organic solvents was not specifically investigated, the tremendous effect of organic ligands on the morphology of inorganic nanostructures is nicely illustrated. A broader review of these effects can be found in our earlier publication (Garnweitner & Niederberger, 2008).

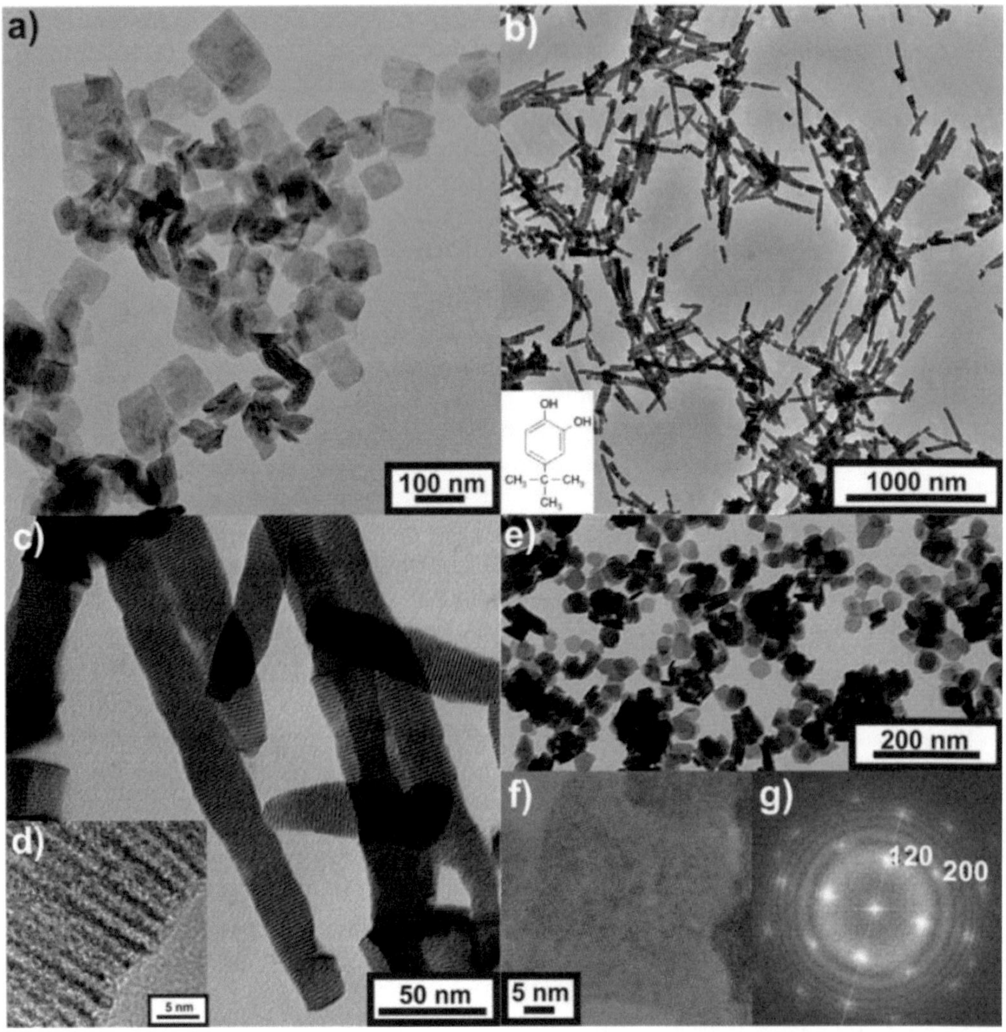

Fig. 6. TEM images of tungstite nanoparticles obtained without 4-*tert*-butylcatechol (a), and of the hybrid nanostructures obtained in the presence of 4-*tert*-butylcatechol, as-prepared (b-d) and after redispersion in water (e-g). The chemical structure of the ligand is shown in (b, Inset) (Polleux et al., 2006). Reproduced by permission of the Royal Society of Chemistry.

4. Post-synthetic stabilisation

The biggest advantage of the post-synthesis stabilisation approach is that the original synthesis is preserved, with particle properties such as size and crystallinity being unaffected by the successive stabilisation treatment that is performed at much more lenient conditions than the synthesis. Under these conditions, a possible unwanted reaction or decomposition of the stabiliser can also be excluded, thus allowing higher flexibility in terms of ligand structure and chemistry. On the other hand, after the synthesis already rather strong agglomerates may be present, and organic species from the reaction may be strongly adsorbed to the particle surface, thus limiting or preventing the stabilisation.

The post-synthetic modification of metal oxide nanoparticles has been investigated in a large number of cases, and with very different goals, not only for an enhancement of colloidal stability of the nanoparticles, but also to adjust many other properties, for example the optical and photocatalytic performance (Rehm et al., 1996; Rajh et al., 1999, 2002; Stowdam & van Veggel, 2004). Different classes of surface modifiers that have shown to be especially suitable for metal oxides include alkoxysilanes and chlorosilanes (Sanchez et al., 2001), carboxylic acids (Bourlinos et al., 2002; Arita et al., 2010), or phosphonic/phosphoric acid derivatives (Rill et al., 2007; Lomoschitz et al., 2011). Interestingly, whereas in some systems, the binding of ligands can be perceived as an exchange process of previously surface-coordinated organics (Shenhar & Rotello, 2003), in other systems a grafting of ligands to a formerly "bare" surface was observed (Arita et al., 2010). Many other groups use the term "surface modification" without further specification of the chemical surface processes. Apparently, differences between the individual systems do exist, and in contrast to the well-explored metal and semiconductor nanoparticle systems, further investigations will be required to fully understand and explain the processes during post-synthetic stabilisation of metal oxide nanostructures. Within the scope of this Chapter, only some select examples can be shown to illustrate the potentials of post-synthetic stabilisation strategies.

4.1 Exchange chemistry of metal oxide nanoparticles

The exchange chemistry of γ-Fe_2O_3 nanoparticles has been investigated in detail by Boal et al. (Boal et al., 2002). Here, alkylamine-protected maghemite nanoparticles were subjected to different tailor-made long-chain ligands (Figure 7). The amines were exchanged at least partially by all alcohol ligands used; however substantial differences in the stability of the modified particles were observed: whilst dodecanol (**1**)-modified nanoparticles agglomerated within a few days both when being in dispersion and in a dried state, modification of the nanoparticles with pentacosane-13-ol (**2**) and 13-dodecylpentacosane-13-ol (**3**), bearing two and three C_{12}-chains, respectively, resulted in high stability of the nanoparticles in the solid state after initial precipitation. Long-term stability in dispersion, however, was only obtained when using ligands with bidentate diol anchoring groups. Again, ligands with only a single C_{12} chain (**4**) showed low stability after drying, whilst double-chain bidentate ligands (**5**) proved most effective, leading to long-term stability both in dispersion and in the dried state (Boal et al., 2002). The authors concluded that according to their findings, a certain density of the monolayer as well as "kinetic stabilisation", in the form of multivalent ligand binding, are required in order to achieve stability, with the effect of kinetic stabilisation being especially pronounced in liquid dispersions.

In other studies, the surface modification was targeted specifically in order to adjust the polarity of the nanoparticles, allowing their redispersion in a desired organic solvent or water. For example, TiO_2 aqueous sols were modified successfully with hydrophobic DTMS and hydrophilic APTMS to make the nanoparticles redispersible in solvents of different polarity, depending on the ratio of the used silanes (Iijima et al., 2009).

4.2 Instant stabilisation through surface modification

In some cases, the addition of ligands to an agglomerated particle system can prove sufficient to result in instant stabilisation. As an example, our earlier studies on the

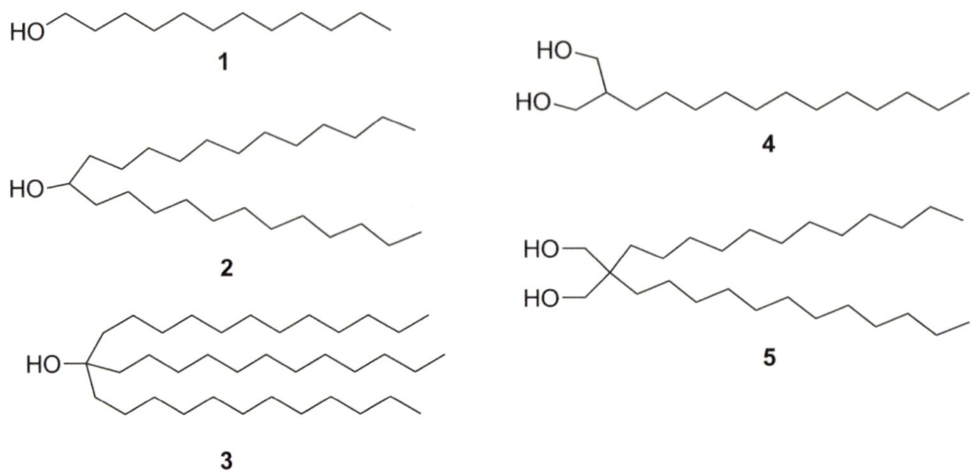

Fig. 7. Chemical structures of different ligands utilised for the modification of γ-Fe_2O_3 nanoparticles (after Boal et al., 2002).

stabilisation of ZrO_2 nanoparticles are presented (Garnweitner et al., 2007; Tsedev & Garnweitner, 2008). The nanoparticles were prepared in benzyl alcohol, which binds to the particle surface during the synthesis but is not capable of a full stabilisation. Hence, large agglomerates are obtained. After retrieval of the nanoparticles from the solvent by centrifugation and washing, solutions of fatty acid stabilisers were added, which resulted in instant stabilisation, visible as an immediate optical change of the mixture from turbid to transparent (Figure 8, left). TEM investigations of the nanoparticles after addition of the stabiliser show that the nanoparticles are fully stabilised, with no agglomeration occurring even during the drying process on the sample grid (Figure 8, right).

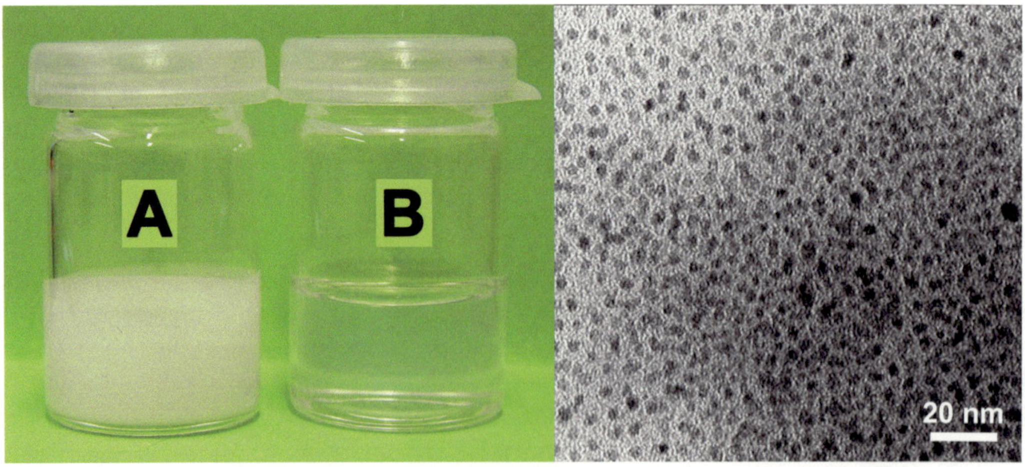

Fig. 8. Post-synthetic stabilisation of ZrO_2 nanoparticles by addition of fatty acids.
Left: digital images of two samples before (A) and after (B) the stabilisation treatment; right: TEM image of oleic acid-stabilised nanoparticles.

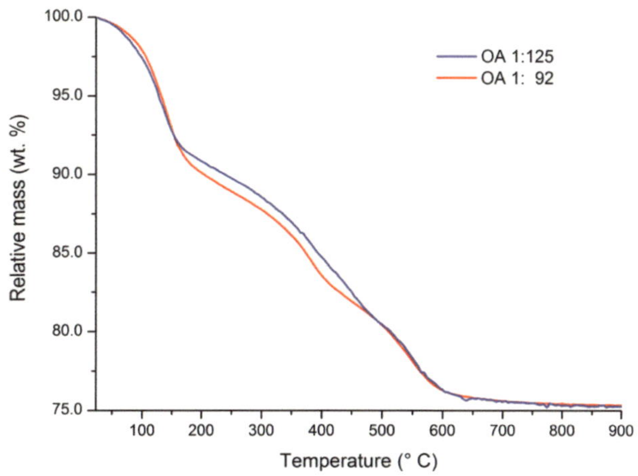

Fig. 9. Thermogravimetric analysis (TGA) of fatty acid-stabilised zirconia nanoparticles, with the stabiliser oleic acid (OA) added in different molar ratios.

In order to elucidate the binding of the fatty acid stabilisers in further detail, we performed thermogravimetric analysis of the nanoparticles before and after the stabilisation treatment. Even the as-prepared samples show a substantial amount of adsorbed organics, which can be attributed to benzyl alcohol coordinating to the particle surface during the synthesis (Niederberger et al., 2004). Figure 9 shows a comparison of two samples obtained when adding different concentrations of oleic acid (OA) as stabiliser, in a molar OA: ZrO_2 ratio of 1:125 (shown in blue), and 1:92 (shown in red). Both samples were fully stable against agglomeration. The thermograms show the same total mass loss of 24.7 wt. % at temperatures exceeding 800 °C for both samples, with however some differences in the decomposition behaviour at lower temperatures. Therefore, the total amount of organic and volatile species bound to the surface is equal, and does not rise even when increasing the amount of added stabiliser. By comparing the decomposition profile with samples measured before stabilisation and performing NMR spectroscopy (not shown), the observed differences could be attributed to a different ratio of adsorbed stabiliser to benzyl alcohol. Hence, when added above a certain concentration, the stabiliser does not increase the overall organics content of the nanoparticles, but replaces previously bound benzyl alcohol groups (Garnweitner et al., 2007).

In other investigations, however, the ligands were utilised rather to bind to a previously "bare" nanoparticle surface. This strategy is especially employed in cases of metal oxide nanoparticles synthesised by aqueous precipitation, where the resulting dispersions are charged-stabilised and highly sensitive to pH changes (Sehgal et al., 2005). Here, the binding of organic ligands can be utilised to considerably extend the range of pH stability, or achieve stability under completely different conditions. For example, the binding of poly(acrylic acid) (PAA) to CeO_2 nanoparticles was utilised to render the nanoparticles stable against agglomeration under alkaline conditions, which was attributed to an electrosteric mechanism (Sehgal et al., 2005). Figure 10 (a) shows a photograph of dispersions at different pH, proving the high stability in the alkaline regime. Additionally, DLS measurements performed by the group nicely illustrate the stabilisation process (Figure 10 (b)), with a measurable steady decrease in particle size being the result of the stabilisation process upon pH increase (Sehgal et al., 2005). In a subsequent report, Qi et al. investigated the binding of phosphonated polyethylene glycol (PPEG) oligomers on ceria nanoparticles in a comprehensive binding study (Qi et al., 2008).

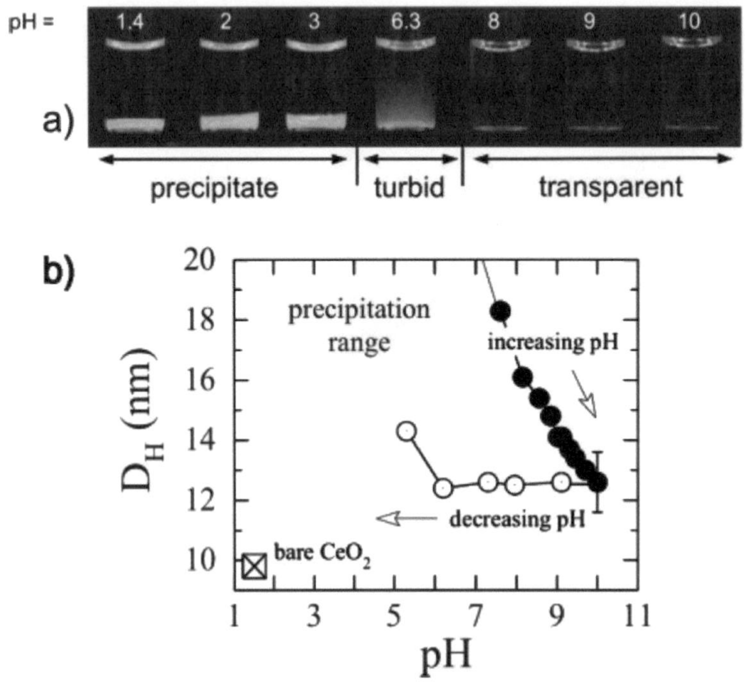

Fig. 10. Photographs of CeO_2-PAA solutions prepared at different pH, adjusted by NH_4OH (a); hydrodynamic particle diameters measured by DLS at different pH (b). Reprinted with permission from (Sehgal et al., 2005). Copyright 2005 American Chemical Society.

5. Conclusion

In this Chapter, a short review of the possibilities for the stabilisation of metal oxide nanoparticles using molecular organic stabilisers has been provided, showing select examples of the potentials of this strategy. In particular, the in-situ stabilisation technique has been compared to the post-synthetic stabilisation strategy. On the one hand, it was shown that the in-situ stabilisation strategy bears several drawbacks, as the particle properties, especially their crystallinity and morphology, may be adversely affected by the added ligands, that these problems however can be overcome by choice of the "right" ligand. The ideal in-situ stabiliser, therefore, needs to bind to the particle surface but not form too strong bonds, as this could interfere with the synthesis mechanism. For some metal oxides, such as TiO_2, first insights into the precise effect of different functional groups on the synthesis and stabilisation have been gained, and it can be expected that in the future, a toolbox will be available for different metal oxide systems, allowing the precise choice of an in-situ ligand for optimum effect. Then, the benefits of in-situ stabilisation can fully be exploited, such as the experimental simplicity and best stability.

Only in rare cases, a direct comparison of in-situ versus post-synthetic stabilisation has been performed, but these results clearly point to a higher efficiency of in-situ stabilisation approaches. For example, in the two-step stabilisation of magnetite nanoparticles, Theerdhala et al. achieved long-time stable dispersions only for the in-situ approach (Theerdhala et al., 2010). On the other hand, there are many cases where the initial reaction medium is not the perfect choice for further handling and storage of the nanoparticles, e.g. due to its toxicity or a too high or too low boiling point. In these cases, the post-synthetic stabilisation of nanoparticles will always be an attractive strategy, as the lack of stability of the nanoparticles allows for their simple and fast separation and recovery, and the presented examples clearly show that at least in some cases, a very simple and fast post-synthetic step is sufficient to achieve high stability of the nanoparticles, without any agglomeration even upon drying. Also in this case, it can be expected that within the next decade research on the chemical interaction of stabilisers and different metal oxide nanoparticle systems will lead to a much broader understanding than today, with general strategies available to enable the *a priori* choice of ligands at least for the most common metal oxides and reaction systems to achieve best stability with minimum addition of organics. Additionally, the increasing complexity of self-assembly tasks will also demand for structurally more complex ligands, which most likely will have to be introduced in post-synthetic steps due to their sensitivity to temperature and harsh chemical conditions. Hence, it appears clear that both the in-situ and post-synthetic approaches to the surface modification and stabilisation of nanoparticles with be applied in the future, will a more detailed knowledge being a fundamental prerequisite for the precise and rational application of experimental strategies to achieve optimum efficiency.

6. Acknowledgment

The author gratefully acknowledges funding by the Deutsche Forschungsgemeinschaft (DFG), grant GA 1492/4-2, within the framework of the priority programme SPP 1273 "Kolloidverfahrenstechnik". Furthermore, Tarik A. Cheema and Rona Pitschke, Max Planck Institute of Colloids and Interfaces, Potsdam/Germany, are acknowledged for Fig. 8, and Mario Raab is acknowledged for help in formatting the manuscript.

7. References

Arita, T.; Yoo, J.; Ueda, Y. & Adschiri, T. (2009). Size and size distribution balance the dispersion of colloidal CeO_2 nanoparticles in organic solvents. *Nanoscale*, Vol.2, No.5, (February 2010), pp. 689-693, ISSN 2040 3364

Ballauff, M. (2003). Nanoscopic Polymer Particles with a Well-Defined Surface: Synthesis, Characterization, and Properties. *Macromolecular Chemistry and Physics*, Vol.204, No. 2, (February 2003), pp. 220-234, ISSN 1022-1352

Bazito, F.F.C. & Torresi, R.M. (2006). Cathodes for Lithium Ion Batteries: The Benefits of Using Nanostructured Materials. *Journal of the Brazilian Chemical Society*, Vol.17, No.4 , (May 2006), pp. 627-642, ISSN 0103-5053

Bell, N.S.; Schendel, M.E. & Piech, M. (2005). Rheological Properties of Nanopowder Alumina coated with adsorbed Fatty Acids. *Journal of Colloid and Interface Science*, Vol.287, No.1, (July 2005), pp. 94-106, ISSN 0021-9797

Biswas, M. & Ray, S.S. (2001). Recent Progress in Synthesis and Evaluation of Polymer-Montmorillonite Nanocomposites. *Advances in Polymer Science*, Vol.155, (2001), pp. 167-221, ISSN 0065-3195

Boal, A.K.; Das, K.; Gray, M. & Rotello, V.M. (2002). Monolayer Exchange Chemistry of γ-Fe_2O_3 Nanoparticles. *Chemistry of Materials*, Vol.14, No.6, (May 2002), pp. 2628-2636, ISSN 0897-4756

Bourlinos, A.B.; Bakandritsos, A.; Georgakilas, V. & Petridis, D. (2002). Surface Modification of Ultrafine magnetic Iron Oxide Particles. *Chemistry of Materials*, Vol.14, No.8, (August 2002), pp. 3226-3228, ISSN 0897-4756

Centi, G. & Perathoner, S. (2009). The Role of Nanostructure in Improving the Performance of Electrodes for Energy Storage and Conversion. *European Journal of Inorganic Chemistry*, Vol.2009, No.26, (September 2009), pp. 3851-3878, ISSN 1099-0682

Chen, X. & Mao, S.S., (2007). Titanium Dioxide Nanomaterials: Synthesis, Properties, Modifications and Applications. *Chemical Reviews*, Vol.107, No.7, (July 2007), pp. 2891-2959, ISSN 0009-2665

Cozzoli, P.D.; Kornowski, A. & Weller, H. (2003). Low-Temperature Synthesis of Soluble and Processable Organic-Capped Anatase TiO_2 Nanorods. *Journal of the American Chemical Society*, Vol.125, No.47, (November 2003), pp. 14539-14548, ISSN 0002-7863

Cushing, B.L.; Kolesnichenko, V.L. & O'Connor, C.J. (2004). Recent Advances in the Liquid-Phase Syntheses of Inorganic Nanoparticles. *Chemical Reviews*, Vol.104, No.9, (September 2004), pp. 3893-3946, ISSN 0009-2665

Derjaguin, B. V.& Landau, L. D. (1941). Theory of the Stability of Strongly charged Lyophobic Sols and the Adhesion of strongly charged Particles in Solutions of Electrolytes. *Progress in Surface Science*, Vol.43, No.1-4, (May 1993), pp. 30-59, ISSN 0079-6816

Dong, L.; Subramanian, A.; Nelson, B.J. (2007). Carbon Nanotubes for Nanorobotics. *Nano Today*, Vol.2, No.6, (December 2007), pp. 12-21, ISSN 1748-0132

Duget, E.; Vasseur, S.; Mornet, S. & Devoisselle, J.M. (2006), Magnetic nanoparticles and their applications in medicine. *Nanomedicine*, Vol.1, No.2, (August 2006), pp.157-168, ISSN 1748-6963

Epifani, M.; Arbiol, J.; Andreu, T. & Morante, J.R. (2008). Synthesis of Soluble and Size-Controlled SnO_2 and CeO_2 Nanocrystals: Application of a General Concept for the Low-Temperature, Hydrolytic Synthesis of Organically Capped Oxide Nanoparticles. *European Journal of Inorganic Chemistry*, No.6, (February 2008), pp. 859-862, ISSN 1434-1948

Everett, D. H. (1988). *Basic Principles of Colloid Science*. Royal Society of Chemistry, ISBN 0-85186-443-0, London, England

Feldmann C. & Goesmann H. (2010). Nanoparticulate Functional Materials. *Angewandte Chemie International Edition*, Vol.49, No. 8, (February 2010), pp. 1362-1395, ISSN 1521-3773

Gao, Y. & Tang, Z. (2011). Design and Application of Inorganic Nanoparticle Superstructures: Current Status and Future Challenges. *Small*, Vol.7, No.15, (August 2011), pp. 2133-2146, ISSN 1613-6810

Garnweitner, G.; Goldenberg, L.M.; Sakhno, O.V.; Antonietti, M.; Niederberger, M. & Stumpe, J. (2007). Large-Scale Synthesis of Organophilic Zirconia Nanoparticles and their Application in Organic-Inorganic Nanocomposites for Efficient Volume Holography. *Small*, Vol.3, No.9, (September 2007), pp. 1626-1632, ISSN 1613-6810

Garnweitner, G. & Niederberger, M. (2008). Organic Chemistry in Inorganic Nanomaterials Synthesis. *Journal of Materials Chemistry*, Vol.18, No.11, (March 2008), pp. 1171-1182, ISSN 0959-9428

Garnweitner, G.; Ghareeb, H.O. & Grote C. (2010), Small-molecule in situ Stabilization of TiO_2 Nanoparticles for the facile Preparation of stable colloidal Dispersions. *Colloids and Surfaces A: Physicochemical and Engineering Aspects*, Vol.372, No.1-3, (December 2010), pp. 41-47, ISSN 0927-7757

Gaynor, A.G.; Gonzalez, R.J.; Davis, R.M. & Zallen, R. (1997). Characterization of Nanophase Titania Particles Synthesized Using In Situ Steric Stabilization. *Journal of Materials Research*, Vol.12, No. 7, (July 1997), pp. 1755-1765, ISSN 0884-2914

Gilstrap Jr., R.A.; Capozzi, C.J.; Carson, C.G.; Gerhardt, R.A. & Summers, C.J. (2008). Synthesis of a nonagglomerated Indium Tin Oxide Nanoparticle Dispersion. *Advanced Materials*, Vol.20, No.21, (November 2008), pp. 4163-4166, ISSN 0935-9648

Goyal, A.K.; Johal, E.S. & Rath, G. (2011) Nanotechnology for Water Treatment. *Current Nanoscience*, Vol.7, No. 4, (August 2011), pp. 640-654, ISSN 1573-4137

Guo, S. & Wang, E. (2011). Noble Metal Nanomaterials: Controllable Synthesis and Application in Fuel Cells and Analytical Sensors. *Nano Today*, Vol.6, No.3, (June 2011), pp. 240-264, ISSN 1748-0132

Gyergyek, S.; Makovec, D. & Drofenik, M. (2010). Colloidal Stability of Oleic- and Ricinoleic-Acid-Coated Magnetic Nanoparticles in Organic Solvents *Journal of Colloid and Interface Science*, Vol.354, No.2, (February 2011), pp. 498-505, ISBN 0021-9797

Hamaker, H. C. (1937) The London-van der Waals Attraction between Spherical Particles. *Physica*, Vol.4, No.10, (October 1937), pp. 1058-1072, ISSN 0031-8914

Heller, W. & Pugh, T.L. (1954). ``Steric Protection'' of Hydrophobic Colloidal Particles by Adsorption of Flexible Macromolecules. *Journal of Chemical Physics*, Vol.22, No.10, (1954) pp. 1778, ISSN 1089-7690

Hunter, R.J. (2001). *Foundations of Colloid Science*, Oxford University Press, ISBN 978-0-19-850502-0, Oxford, England

Iijima, M.; Kobayakawa, M. & Kamiya, H. (2009). Tuning the Stability of TiO_2 Nanoparticles in various Solvents by mixed Silane Alkoxides. *Journal of Colloid and Interface Science*, Vol.337, No.1, (September 2009), pp. 61-65, ISSN 0021-9797

Jun, Y.-W.; Choi, J.-S. ; Cheon, J. (2006). Shape Control of Semiconductor and Metal Oxide Nanocrystals through Nonhydrolytic Colloidal Routes. *Angewandte Chemie International Edition*, Vol.45, No.21, (May 2006), pp. 3414-3439, ISSN 1521-3773

Kanie, K. & Sugimoto, T. (2004). Shape Control of Anatase TiO_2 Nanoparticles by Amino Acids in a Gel-Sol System. *Chemical Communications*, Vol.2004, No.14, (July 2004), pp. 1584-1585, ISSN 1359-7345

Kim, J. & van der Bruggen, B. (2010), The Use of Nanoparticles in Polymeric and Ceramic Membrane Structures: Review of Manufacturing Procedures and Performance Improvement for Water Treatment, *Environmental Pollution*, Vol. 158, No. 7, (July 2010), pp. 2335-2349, ISSN 0269-7491

Li, Q.; Mahendra, S.; Lyon, D.Y.; Burnet, L.; Liga, M.V.; Li, D. & Alvarez, J.J. (2008). Antimicrobial nanomaterials for water disinfection and microbial control: Potential applications and implications. *Water Research*, Vol.42, No.18, (November 2008), pp 4591-4602, ISSN 0043-1354

Lomoschitz, C.J.; Feichtenschlager, B.; Moszner, N.; Puchberger, M.; Müller, K.; Abele, M. & Kickelbick, G. (2011). Directing Alkyl Chain Ordering of Functional Phosphorus Coupling Agents on ZrO_2. *Langmuir*, Vol.27, No.7, (April 2011), pp. 3534-3540, ISSN 0743-7463

Lu, K. (2008). Theoretical Analysis of Colloidal Interaction Energy in Nanoparticle Suspensions. *Ceramics International*, Vol.34, No.6, (August 2008) , pp.1353-1360, ISSN 0272-8842

Mackor, E. L. (1951) A Theoretical Approach of the Colloid-Chemical Stability of Dispersions in Hydrocarbons. *Journal of Colloid Sciences*, Vol.6, No.5, (October 1951), pp. 492-495, ISSN 0021-9797

Mallouk, T.E & Sen, A. (2009). Powering nanorobots: Catalytic Engines Enable Tiny Swimmers to Harness Fuel From Their Environment and Overcome the Weird Physics of the Microscopic World. *Scientific American*, Vol. 300, No.5 (May 2009), pp.72-77, ISSN 0036-8733

Marczak, R.; Segets, D.; Voigt, M. & Peukert W. (2009). Optimum between Purification and Colloidal Stability of ZnO Nanoparticles. *Advanced Powder Technology*, Vol.21, No.1, (January 2010), pp. 41-49, ISSN 0921-8831

Moghimi, S.M.; Hunter, A.C. & Murray J.C. (2005), Nanomedicine: Current Status and Future Prospects. *The FASEB Journal*, Vol. 19, No.3, (March 2005), pp. 311-330, ISSN 1530-6860

Mostofizadeh, A.; Li, Y.; Song, B. & Huang, Y. (2011). Synthesis, Properties, and Applications of Low-Dimensional Carbon-Related Nanomaterials. *Journal of Nanomaterials*, Vol.2011, Art. No. 685081, (2011), 21 p. , ISSN 1687-4110

Mulvaney, P. (1998) Zeta Potential and Colloid Reaction Kinetics. In: *Nanoparticles and Nanostructured Films*, J.H. Fendler, (Ed.), Wiley-VCH, ISBN 3-527-29443-0, Weinheim, Germany

Murphy, C.J.; Sau, T.K.; Gole, A.M.; Orendorff, C.J.; Gao, J.; Gou, L.; Hunyadi, S.E.; Li, T. (2005). Anisotropic Metal Nanoparticles: Synthesis, Assembly, and Optical Applications. *Journal of Physical Chemistry B*, Vol.109, No.29, (July 2005), pp. 13857-13870, ISSN 1520-6106

Murray, C.B.; Norris, D.J.; Bawendi, M.G. (1993). Synthesis and Characterization of Nearly Monodisperse CdE (E = Sulfur, Selenium, Tellurium) Semiconductor

Nanocrystallites. *Journal of the American Chemical Society*, Vol.115, No. 19, (September 1993), pp. 8706-8715, ISSN 1520-1526

Napper, D. (1983) *Polymeric Stabilization of Colloidal Dispersions*, Academic Press, ISBN 0-12513-980-2 London, England

Niederberger, M.; Garnweitner, G.; Krumreich, F.; Nesper, R.; Cölfen, H. & Antionetti, M. (2004). Tailoring the Surface and Solubility Properties of Nanocrystalline Titania by a Nonaqueous In Situ Functionalization Process. *Chemistry of Materials*, Vol.16 , No.7, (February 2004), pp. 1202-1208, ISSN 0897-4756

Ostwald, W. (1915). *Die Welt der vernachlässigten Dimensionen: Eine Einführung in die Kolloidchemie mit besonderer Berücksichtigung ihrer Anwendungen*. Steinkopff Verlag, Dresden, Germany

Peyre, D.; Spalla, O.; Belloni, L. & Nabavi, M. (1997). Stability of a Nanometric Zirconia Colloidal Dispersion under Compression: Effect of Surface Complexation by Acetylacetone. *Journal of Colloid and Interface Science*, Vol.187, No.1, (March 1997), pp. 184-200, ISSN 0021-9797.

Pinna, N. & Niederberger, M. (2008). Surfactant-free Nonaqueous Synthesis of Metal Oxide Nanostructures. *Angewandte Chemie International Edition*, Vol.47, No.29, (July 2008), pp. 5292-5304, ISSN 1433-7851

Polleux, J.; Antonietti, M. & Niederberger, M. (2006). Ligand and Solvent Effects in the Nonaqueous Synthesis of Highly Ordered Anisotropic Tungsten Oxide Nanostructures. *Journal of Materials Chemistry*, Vol.16, No.40, (July 2006), pp. 3969-3975, ISSN 0959-9428

Prandeep, T. & Anshup (2009), Noble Metal Nanoparticles for Water Purification: A Critical Review. *Thin Solid Films*, Vol.517, No.24, (October 2009), pp. 6441-6478, ISSN 1879-2731

Qi, L.; Sehgal, A.; Castaing, J.-C.; Chapel, J.-P.; Fresnais, J.; Berret, J.-F. & Cousin, F. (2008). Redispersible Hybrid Nanopowders: Cerium Oxide Nanoparticle Complexes with Phosphonated-PEG Oligomers. *ACS Nano*, Vol.2, No.5, (May 2008), pp. 879-888, ISSN 1936-0851

Rajh, T.; Nedeljkovic, J.M.; Chen, L.X.; Poluektov, O. & Thurnauer M.C. (1999). Improving Optical and Charge Separation Properties of Nanocrystalline TiO_2 by Surface Modification with Vitamin C. *Journal of Physical Chemistry B*, Vol.103, No.18, (April 1999), pp. 3515-3519, ISSN 1089-5647

Rajh, T.; Chen, L.X.; Lukas, K.; Liu, T.; Thurnauer, M.C. & Tiede, D.M. (2002). Surface Restructuring of Nanoparticles: An efficient Route for Ligand-Metal Oxide Crosstalk. *Journal of Physical Chemistry B*, Vol.106, No.41, (October 2002), pp. 10543-10552, ISSN 1089-5647

Rasmussen, S.; Chen, L.; Nilsson, M. & Abe, S. (2003). Bridging Nonliving and Living Matter. *Artificial Life*, Vol.9, No.3, (June 2003), pp. 269-316, ISSN 1064-5462

Rehm, J.M.; McLendon, G.L.; Nagasawa, Y.; Yoshihara, K.; Moser, J. & Grätzel, M. (1996). Femtosecond Electron-Transfer Dynamics at a Sensitizing Dye−Semiconductor (TiO_2) Interface. *Journal of Physical Chemistry*, Vol.100, No.23, (June 1996) pp. 9577-9578, ISSN 0022-3654

Reindl, A. & Peukert, W. (2008). Intrinsically stable Dispersions of Silicon Nanoparticles. *Journal of Colloid and Interface Science*, Vol.325, No.1, (September 2008), pp. 173-178, ISSN 0021-9797

Renger, C.; Kuschel, P.; Kristoffersson, A.; Clauss, B.; Oppermann, W. & Sigmund, W. (2006). Rheology Studies on Highly Filled Nano-Zirconia Suspensions. *Journal of the European Ceramic Society*, Vol.27, No.6, (November 2006), pp. 2361-2367, ISSN 0955-2219

Requicha, A.G.G (2003). Nanorobots, NEMS, and Nanoassembly. *Proceedings of the IEEE*, Vol.91, No.11, (November 2003), pp. 1922-1933, ISSN 0018-9219

Rill, C.; Ivanovici, S. & Kickelbick, G. (2007). Hybrid Nanoparticles Prepared by In-Situ and Post-Synthetic Surface Modification of Lanthanide-based Nanoparticles with Phosphonic Acid Derivatives, *Proceedings of Materials Research Society Symposium 1007*, pp. 207-212, ISBN: 978-155899967-1, San Francisco, USA, California, USA, April 9-13, 2007

Rockenberger, J.; Scher, E.C. & Alivisatos, A.P. (1999), A New Nonhydrolytic Single-Precursor Approach to Surfactant-Capped Nanocrystals of Transition Metal Oxides, *Journal of the American Chemical Society*, Vol.121, No.49, (December 1999), pp. 11595-11596, ISSN 0002-7863

Sanchez, C.; Soler-Illia, G.J.D.A.A.; Ribot, F.; Lalot, T.; Mayer, C.R. & Cabuil, V. (2001). Designed Hybrid Organic-Inorganic Nanocomposites from Functional Nanobuilding Blocks. *Chemistry of Materials*. Vol.13, No.10, (October 2001), pp. 3061-3083, ISSN 0897-4756

Sánchez, S. & Pumera, M. (2009). Nanorobots: The Ultimate Wireless Self-Propelled Sensing and Actuating Devices, *Chemistry-An Asian Journal*, Vol.4, No.9, (September 2009), pp. 1402-1410, ISSN 1861-471X

Sanvicens, N.& Pilar, M. (2008) Multifunctional nanoparticles - properties and prospects for their use in human medicine. *Trends in Biotechnology*, Vol.26, No.8, (August 2008), pp. 425-433, ISSN 0167-7799

Schwarz, U.S. & Safran,S.A. (2000) Phase Behavior and Material Properties of Hollow Nanoparticles, *Physical Review E*, Vol.62, No.5, (November 2000) pp. 6957–6967, ISSN 1539-3755

Segets, D.; Marczak, R.; Schäfer, S.; Paula, C.; Gnichwitz, J.F.; Hirsch, A. & Peukert, W. (2011). Experimental and Theoretical Studies of the Colloidal Stability of Nanoparticles−A General Interpretation Based on Stability Maps. *ACS Nano*, Vol.5, No.6, (May 2011), pp. 4658–4669, ISSN 1936-0851

Sehgal, A.; Lalatonne, Y.; Berret, J.-F.; Morvan, M. (2005). Precipitation-Redispersion of Cerium Oxide Nanoparticles with Poly(acrylic acid): Toward Stable Dispersions. *Langmuir*, Vol.21, No.20, (September 2005), pp. 9359-9364, ISSN: 0743-7463.

Shah, P.S.; Holmes, J.D.; Johnston, K.P. & Korgel, B.A. (2001). Size-Selective Dispersion of Dodecanethiol-coated Nanocrystals in Liquid and Supereritical Ethane by Density Tuning. *Journal of Physical Chemistry B*, Vol.106, No. 10, (March 2002), pp. 2545-2551, ISSN 1089-5647

Shenhar, R. & Rotello, V.M (2003). Nanoparticles: Scaffolds and Building Blocks. *Accounts of Chemical Research*, Vol.36, No.7, (July 2003), pp. 549-561, ISSN 0001-4842

Sides, C.R.; Li, N.; Patrissi, C.J.; Scrosati, B. & Martin, C. R. (2002). Nanoscale Materials for Lithium-Ion Batteries. *MRS Bulletin*, Vol.27, No.8, (August 2002), pp. 604-607, ISSN 0883-7694

Stowdam, J.W. & van Veggel, F.C.J.M. (2004). Improvement in the Luminescence Properties and Processability of LaF_3/Ln and $LaPO_4$/Ln Nanoparticles by Surface Modification. *Langmuir*, Vol.20, No.26, (November 2004), pp. 11763-11771, ISSN 0743-7463.

Sun, S.; Zeng, H.; Robinson, D.B.; Raoux, S.; Rice, P.M.; Wang, S.X.; Li, G. (2004). Monodisperse MFe_2O_4 (M = Fe, Co, Mn) Nanoparticles. *Journal of the American Chemical Society*, Vol.126 No.1, pp. 273-279, ISSN 0002-7863

Tadmor, R. & Klein, J.H. (2001). Additional Attraction Between Surfactant-Coated Surfaces, *Journal of Colloid and Interface Science*, Vol.247, No.2, (March 2002), pp. 321–326, ISSN 0021-9797

Theerdhala, S.; Bahadur, D.; Vitta, S.; Perkas, N.; Zhong, Z.; Gedanken, A. (2010). Sonochemical Stabilization of Ultrafine Colloidal Biocompatible Magnetite Nanoparticles using Amino Acid, L-Arginine, for Possible Bio Applications. *Ultrasonics Sonochemistry*, Vol.17, No.4, (April 2010), pp. 730-737, ISSN 1350-4177

Theron, J.; Walker, J.A. & Cloete (2008), T.E. Nanotechnology and Water Treatment: Applications and Emerging Opportunities, *Critical Reviews in Microbiology*, Vol. 34, No.1, (February 2008), pp. 43-69, ISSN 1549-7828

Traina, C.A. & Schwartz, J. (2007). Surface Modification of Y_2O_3 Nanoparticles. *Langmuir*, Vol.23, No.18, (August 2007), pp. 9158-9161, ISSN 0743-7463

Trindade, T.; O'Brien, P. & Pickett, N.L. (2001). Nanocrystalline Semiconductors: Synthesis, Properties, and Perspectives. *Chemistry of Materials*, Vol.13, No.11, (October 2001), pp. 3843-3858, ISSN 0897-4756

Tsedev, N. & Garnweitner, G. (2008). Surface Modification of ZrO_2 Nanoparticles as Functional Component in Optical Nanocomposite Devices, *Materials Research Society Symposium Proceedings*, pp. 175-180, ISBN 978-160511046-2, San Francisco, California, USA, March 25-27, 2008

Turkevich, J.; Stevenson, P.C. & Hillier, J. (1953). The Formation of Colloidal Gold. *Journal of Physical Chemistry*, Vol.57, No.7, (July 1953), pp. 670-673, ISSN 00223654

van Oss, C.J.; Chaudhury, M.K.& Good, R.J. (1987). Monopolar Surfaces. *Advances in Colloid and Interface Science*, Vol.28, No.C, (1987)pp. 35-64, ISSN 1873-3727

van Oss, C.J. (2003). Long-Range and Short-Range Mechanisms of Hydrophobic Attraction and Hydrophilic Repulsion in Specific and Aspecific Iinteractions. *Journal of Molecular Recognition*, Vol.16, No.4, (July/August 2003), pp. 177-190, ISSN 1099-1352

Verwey, E. J. W. & Overbeek, J. T. G. (1948). *Theory of the Stability of Lyophobic Colloids*, Elsevier, ISBN 978-0-48640-929-0, New York, USA

Vincent, B.; Edwards, J.; Emmett, S. & Jones, A. (1986). Depletion Flocculation in Dispersions of sterically-stabilized Particles ("Soft Spheres"). *Colloids and Surfaces*,Vol.18, No.2-4, (June 1986), pp. 261-281, ISSN 0166-6622

Wiese, G.R. & Healy, T.W. (1969). Effect of Particle Size on Colloid Stability. *Transactions of the Faraday Society,* Vol.66, No.0, (January 1970), pp. 490-499, ISSN 0014-7672

Wu, W.; Giese, R.F. & van Oss, C.J. (1999). *Colloids and Surfaces B: Biointerfaces,* Vol.14, No.1-4, (August 1999), pp. 47-55, ISSN 0927-7765

Section 2

Drug Nanoparticles

5

The Development of Magnetic Drug Delivery and Disposition

Michał Piotr Marszałł
Department of Medicinal Chemistry, Collegium Medicum in Bydgoszcz,
Nicolaus Copernicus University
Poland

1. Introduction

The process of drug delivery and disposition in the modern scientific aspect is very complex. Advances in many fields are converging to make the commercialisation of advanced drug delivery concepts possible. It integrates many disciplines, including biotechnology, medicine and pharmacology. Innovative devices should protect labile active ingredients, precisely control drug release kinetics and minimise the release of the drug to non-target sites. Rapid advances in these areas led to the revolutionary change in discovery of new methods of drug delivery and disposition.

In the last two decades, interest in nanometre-size particles such dendrimers, liposomes, micelles and polymeric particles has increased significantly. As a new drug delivery system, nanoparticles resolved many drug pharmacokinetics problems related with precisely control drug release, poor stability and toxicity of active ingredient. Crucially, the nano-sized particles have very important advantage in intravenous application (Ilium et al., 1982). Generally, their circulation time is limited by the smallest blood capillaries in human vascular system (Table 1) (Arruebo et al., 2007; Yoo et al., 2011). Therefore, the particles size smaller than 1000 nm are usually objects of interest as a new potential carriers for efficient drug delivery (Figure 1). Also, the other parameters including shape and surface activity influence on their distribution and elimination from vascular system.

In the last years, a new drug targeting employing magnetic particles as a drug carriers is a promising tool for the more effective and safe chemotherapy. Initially, the micro- and nano-sized magnetic particles (MMPs; MNPs) provided an original modern technology for bioseparations, especially for ligand "fishing", protein, enzyme, DNA, RNA and cell isolation or purification (Corchero & Villaverde, 2009). Because of the superparamagnetic properties, magnetic particles can be used to simply isolate any target and can be linked with diverse manual and automated applications (Marszałł et al., 2008, 2011a).

The stability of captured protein/enzyme/drug depends on the morphology of the magnetic microspheres. There are many materials available with magnetic properties. However, the iron oxide (Fe_3O_4) or meghemite (γ-Fe_2O_3) are most commonly employed for in vivo applications. The materials, such as cobalt, chromium, nickel and other iron-based metal oxides, such as $CoFe_2O_4$, $NiFe_2O_4$, $MnFe_2O_4$ should be limited for biomedical applications due to their high toxicity and long-term changes in enzyme kinetics. The ferrite nanoparticles have large surface area to volume ratio and can adsorb plasma proteins and other biomolecules as well as agglomerate in vivo (Douziech-Eyrolles et al., 2007). Additionally, the magnetic nanoparticles without coating are rapidly cleared by macrophages in the reticulo-endothelial system. Because of the high physicochemical activity of the surface of magnetic particles and risk of corrosion process they are often encapsulated or coated with protective material: carbohydrates, lipids, gold, proteins and synthetic polymers.

Classes of blood capillaries	Permeability (size)	Location
Tight-junction capillaries	< 1nm	Blood-brain barrier (BBB)
Continuous capillaries	~ 6 nm	Muscle, lung, skin
Fenestrated capillaries	~ 5-60 nm	Kidney, intestine and some endocrine and exocrine glands
Sinusoid capillaries	~ 100-1000 nm	Liver, spleen, bone narrow

Table 1. Different classes of blood capillaries in human vascular system (Arruebo et al., 2007).

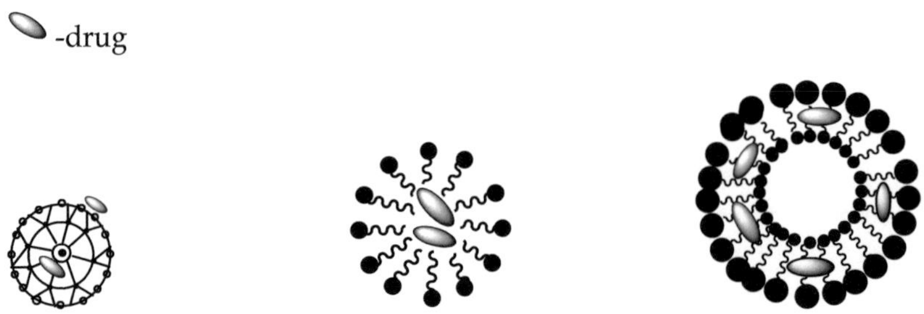

Fig. 1. The most popular nano-sized drug delivery systems for in vitro and in vivo applications.

With the development of improved syntheses and techniques for determining the particle-size distribution the new name of MNPs was defined as a superparamagnetic iron oxide naoparticles (SPIONs). The main principles of SPIONs compared to ferromagnetic particles is lack of permanent magnetic dipole. The promising perspectives for biomedical application of SPIONs is that the moment of each particles fluctuates between different directions at the rate given by the temperature and the applied magnetic field (Woodward et al., 2007). This phenomenon allows for the accumulation of local chemotherapeutic agent in vivo with the use of external magnetic field which can be controlled by magnetisation expressed as:

$$M = c\mu L(\xi) = c\mu[coth(\xi) - (1/\xi)]$$

where c is particle concentration, μ is the magnetic moment of the particle and L(ξ) is the Langevin function in which $\xi = \mu H/kT$, where H is the applied magnetic field, k *is* the Boltzman constant and T is the temperature in Kelvin. μ is the magnetic moment of the particle which is related to the particle volume (v) by $\mu = vM_s$ where M_s is the saturation magnetization.

SPIONs have interesting properties such as high field irreversibility, high saturation field and they do not show magnetic interaction after the external magnetic field is removed (Mahmoudi et al., 2011). The advantage of SPIONs compared to ferromagnetic particles is that the standard particles tend to agglomerate due to their permanent magnetic dipoles. Thus, the suspension of SPIONs produce very stable ferrofluids which in vivo application are still investigated.

The use of solid polymer matrix or core-shell type silica-based superpara- and paramagnetic particles as a carriers for binding proteins, enzymes and drugs received increasing attention in drug delivery studies. The external multifunctional layer of magnetic beads provides reactive groups that are able to covalently conjugate aldehyde, amine, carboxy, hydroxyl, ketone, sulfhydryl and organic linkers for formation of stable linkage (Figure 2).

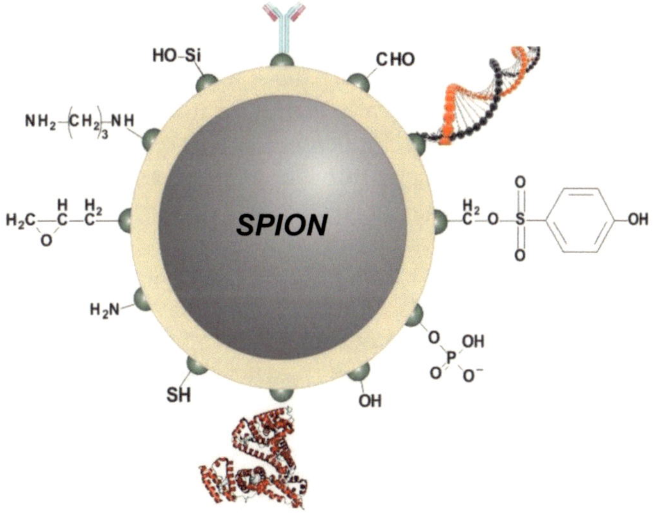

Fig. 2. The core-shell type silica-based superparamagnetic iron oxide nanoparticle (SPION) with different functional groups on the surface.

The pioneering "medical" application in the treatment of lymphatic nodes and metastases based on injecting "metallic particles" preheated in a magnetic field was first published in 1957 (Gilchrist et al., 1957). However, the potential medical application of new finding was limited by the poor performance of particles. In recent years, the new nanotechnologies allowed for the precise manufacturing of nanoparticles with specific size, surface area and biocompatibility with the view of in vivo application. Moreover, the magnetic nanoparticles offer many exciting possibilities for development not only a new potential drug carriers but also are being investigated as a magnetic contrast agents in magnetic resonance imaging (MRI) and hyperthermia agents in hyperthermia cancer treatment.

2. Magnetic drug delivery system in cancer therapy

2.1 Ionic binding of the drug to the magnetic nanoparticles

In 1978, magnetic particles were first applied as a new class of drug target (Widder et al., 1978). The superparamagnetic properties of the magnetic micro- and nanoparticles have opened promising new perspectives for in vivo application. The magnetic nanoparticles as drug carriers provide huge opportunities in cancer treatment. The use of such carriers in targeted therapy considerably reduces the side effects of conventional chemotherapy. A novel carrier system allows for intravenous drug delivery and the local accumulation of chemotherapeutic agent which is comparable to that achieved by the administration of a 100-fold higher dose of the drug. The magnetic drug targeting enables the fast and precise location of drug in the body with the use of external magnetic field. After the vascular injection, the particles can be transported and concentrated at desired location with the help of magnet (Figure 3). For drug delivery application the optimal size of nanoparticles should be in the range from 10 to 200 nm. The micrometer size below 200 allows for systematic administration in circulation system and also into targeted tissue as well as enhance the ability of nanoparticles to evade the biological particulate filters, such as the reticulo-endothelial system (Tran & Webster, 2010). To date, magnetic drug targeting has been studied mostly in pre-clinical models for cancer therapy with intravascular administration of chemotherapeutics.

The Phase I clinical trial demonstrated that the epirubicin attached to magnetic fluids (ferrofluids) could be concentrated in locally advanced tumor by magnetic field. The drug was bound to phosphate groups of starch polymers which covered the iron oxide core with particle size of 100 nm. The "magnetic epirubicin" with doses in the range of 5-100mg/m^2 was infused i.v. over 15 min into a vein located contra laterally to the tumor in patients with advanced and unsuccessfully pretreated metastatic breast cancer, chondrosarcoma and squamous cell carcinoma. The neodymium magnets were large 8x4x2cm or 3x3x1cm and were kept at a distance of 0.5 cm to tumor surface. The clinical trials provided the important information about the conditions of drug release, distribution and mechanism of action. The researchers concluded that the size of magnetic particles should be close to 1µm to increase of their accumulation in tumor site and, consequently, the concentration of the drug (Lübbe et al., 1996). The better optimization of ionic binding epirubicin on to the surface of magnetic nanoparticles also can improve the drug release in physiological parameters. Hence, a new group of novel nano- and micro magnetic particles consisting of various synthetic and natural matrices were investigated (Bergemann et al., 1999).

The idea for ionic binding of the drug to magnetic particles was continued because of the ligand/drug can be easily released from the matrix by simply changing either ionic strength or the pH. The surface of polyvinylalcohol, polyacrylates and polyacharide matrices consisted of cationic exchange groups (i.e. sulphate, phosphate) is able to ionically bind pharmaceutical active compound or biological substances. Furthermore, the modified starch nanoprticles with diethylaminoethyl (DEAE) groups can bind to negatively charged cell membrane. This tool was successful used for separation cells and gene transfer by magnetic transfection of cytokine-induced killer cells (CIK-cells) with plasmid DNA. The representative cationic and anionic nanopartcles are presented in Figure 4 (Bergemann et al., 1999).

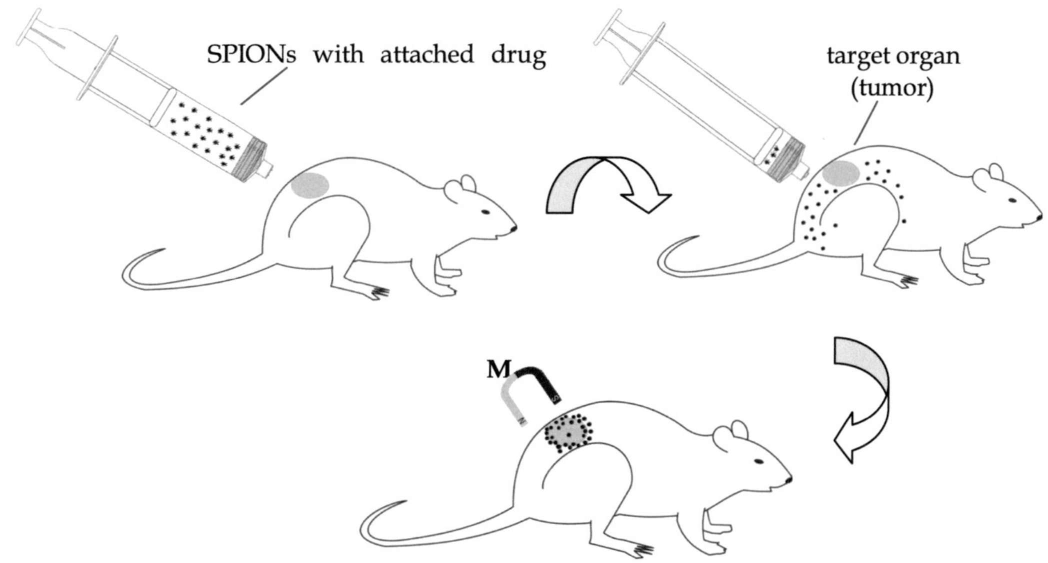

Fig. 3. The scheme of in vivo application of magnetic drug delivery system after the vascular injection; superparamagnetic iron oxide nanoparticles (SPIONs), external magnet (M).

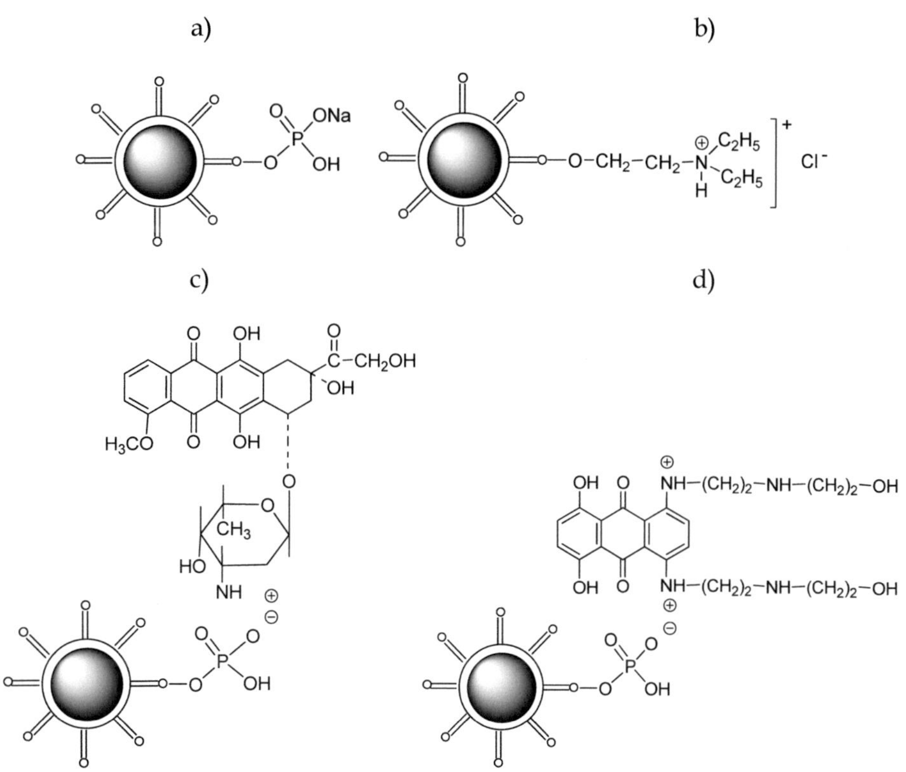

Fig. 4. The representative ion-exchange-active magnetic nanoparticles: cationic (a), anionic (b) as well as loaded with cytostatic drugs: epirubicin (c) and mitoxantrone (d).

The magnetic dug targeting approach offers a new opportunity to treat malignant tumors locoregionally. The treatment of squamous cell carcinoma in rabbits with nanoparticles covered with modified starch to which the mitoxantrone was ionically bound, caused complete and permanent remission of the cancer compared with control group (Alexiou et al., 2000). The advantage of ionically bound mitoxantrone is that the anticancer agent is able to desorb from the magnetic caries after the 30 min (half-time). Determination of time of desorption is very important because the ferrofluids have to be transferred to the tumor region by the magnetic field. Next, the drug must dissociate to act within the tumor. Generally, the total release of drug from the magnetic carriers is recommended at less than 1 h. The 100 nm particles size and strong magnetic field (1.7 Tesla) are optimal for efficient treatment of smaller animals such as mouse or rat. However, the appropriate magnetic field strength and particle size for treatment of deep body cavities and human cancer has to be optimised. The extensive biodistribution study with Iod^{123} - labeled ferrofluids demonstrated that magnetic flux density is an important factor in magnetic drug targeting (Alexiou et al., 2005).

2.2 Covalent binding of the drug to the magnetic nanoparticles

On the contrary to ionic binding, the covalent binding of the drug to magnetic particles prevents unwanted drug release in a physiological environment. Hence, the new strategy is based on the covalent coupling of antibodies, nucleic acids, proteins and active compounds on to the surface of functional magnetic particles coated with polymer or silica. The most popular surface groups such as amine, carboxy and aldehyde allow for a covalent conjugate almost any custom ligand (drug) (Figure 5). The coupling efficiency depends on the functional group density and type of linker. The magnetic particles with primary amine or aldehyde functional groups on the surface are used to covalently conjugate primary amine - containing ligands. The particles coated with carboxyl functional groups on the surface can be covalently attached to primary amine- containing ligands via a stable amide bond.

This strategy was proposed for methotrexate-modified superparamagnetic nanoparticles as a drug carrier in controlled drug delivery, targeted at cancer diagnostics and therapeutics (Kohler et al., 2005). The nanoparticles were modified with (3-aminopropyl)-trimethoxysilane (APS) to form aminopropyl layer and subsequently conjugated with methotrexate through amidation using the 1-ethyl-3-[3-dimethylaminopropyl]carbodiimide hydrochloride (EDC) as a crosslinking agent.

The in vitro studies in both human breast cancer (MCF-7) and human cervical cancer (HeLa) cells demonstrated the successful internalization of used methotrexate conjugates into lysosomes as a promising tool for anticancer activity. Methotrexate is an analogue of folic acid. Folic acid is generally recognized as an effective targeting agent which receptors are overexpressed on the cell membranes of many cancer cells. Hence, the methotrexate is used as a therapeutic agent for the treatment of several forms of cancer (carcinomas, lymphomas, breast, head, leukemia, neck cancer) (Messmann & Allegra, 2001). The proposed methotrexate-modified superparamagnetic nanoparticles enable real-time monitoring of drug carriers by MRI after intravenous drug delivery. Moreover, the methotrexate is not released from magnetic carriers due to the high stability binding of drug-particles in physiological parameters. Curiously, the covalent-amide bond is broken under conditions

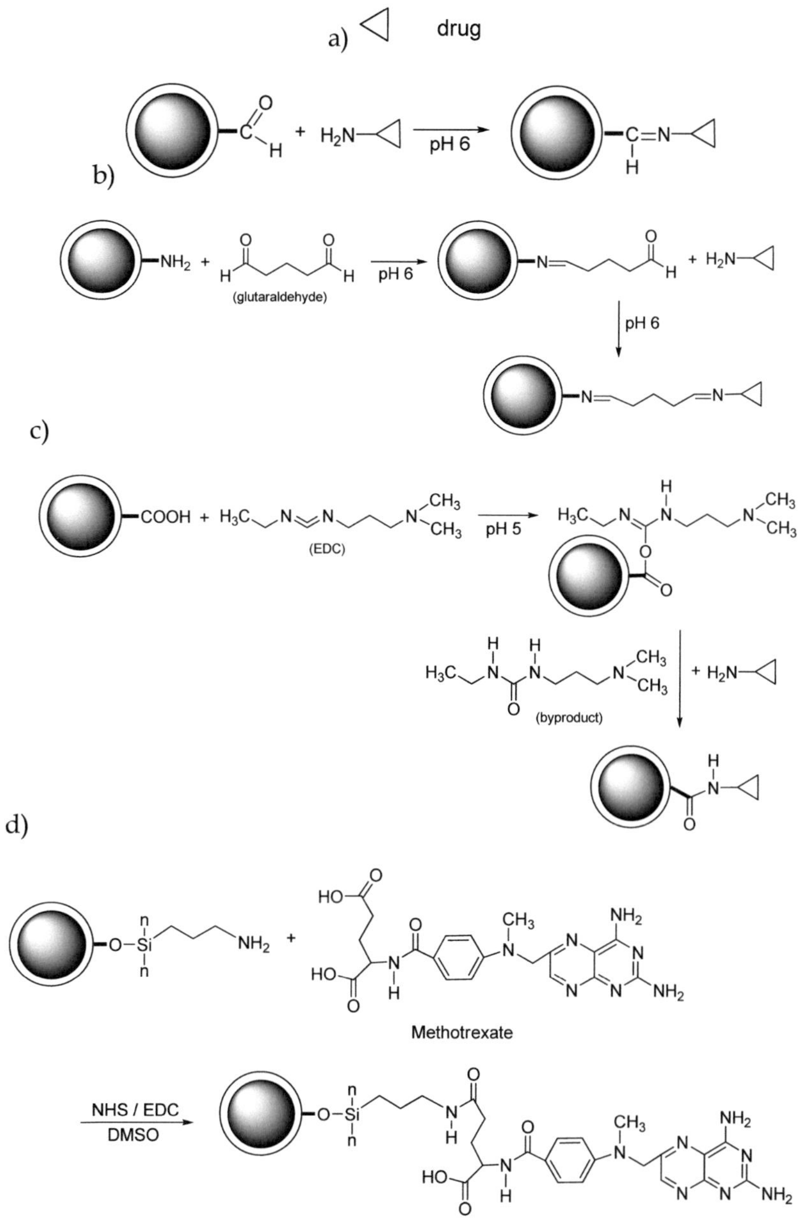

Fig. 5. The covalent immobilization of drugs on to the (a) aldehyde-, (b) amine- or (c) carboxy-terminated magnetic nanoparticles surface and metotrexate on to the surface modified with (3-aminopropyl)-trimethoxysilane (APS) (d).

present in the lysosomal compartment. The intracellular uptake of the metohrexate-magnetic nanoparticle conjugates is assumed to depend on a receptor-mediated endocytosis (Figure 6) (Kohler et al., 2005). Next, they are transported as a endosomes and fused with lyposomes containing proteases and low pH. In this condition, the peptide bond between the methotrexate and magnetic carriers is broken and anticancer drug is released inside the target cell.

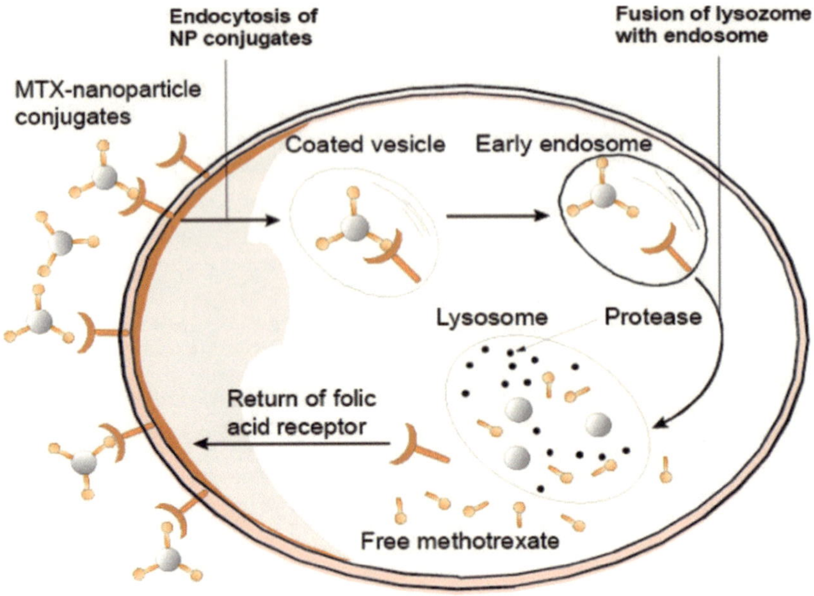

Fig. 6. The intracellular model of the uptake of methotrexate (MTX) - modified nanoparticles into breast cancer cells (Kohler et al., 2005).
(Adapted with permission from @ 2005 American Chemical Society).

2.3 Magnetic nanoparticles for tumor imaging and therapy

Magnetic resonance imaging (MRI) is widely used as a screening non-invasive method of the human body. It is also used to monitor cell migration to targets tissue in cell-based therapy. The fact that the magnetic nanoparticles with labeled cells or as a drug targeting can be visualized using MRI, they are a new alternatives and noninvasive imaging techniques for monitoring of cell or drug migration to target tissue. For in vitro application, their advanced development in cell manipulation/therapy, biomolecule separation, selection and purification was found (Gijs, 2004). The ability to produce a distortion in magnetic field monitored by MRI, allowed for the increasing application of magnetic beads in vivo. The different strategies of the use of anionic magnetic nanoparticles (AMNPs), ultrasmall paramagnetic iron oxides (USPIOs) and superparamagnetic iron oxide nanoparticles (SPIONs) have been demonstrated as a contrast agents to identify magnetically labelled cells during MRI monitoring of cellular therapies (Wilhelm & Gazeau, 2008; Modo et al., 2005). The demonstrated studies confirm the efficacy of labelling with magnetic particles for a wide variety of mammalian cells, including non- and phagocytic cells, different species, cell size, types and culture properties. The MRI contrast is a result of different signal intensities of tissue, produced in response to applied radio frequency pulses (Gijs, 2004). Labelled-specific magnetic particles provide a suitable source of contrast and convenient tool for the non-invasive study of biological processes, such a tumor imaging and therapy with the use of MRI.

The use of magnetic particles can significantly improve hyperthermia cancer treatment (Marszałł, 2011b). This therapy involves raising the temperature of the target tissue to 43-46°C. In this conditions its sensitivity to chemo- and radiotherapy increases and may additionally stimulate activities of the host immune system (Ang et al., 2007). The problem with hypethermia therapy is the heating the large area of tissue or body in general, not only the tumor region. The healthy tissues adsorb microwave, laser and ultrasound energy which can cause burns and blisters (Phillips & Johnson, 2005). Magnetic hyperthermia is one of the anti-cancer approches based on the introduction of ferro- or superparamagnetic particles into the tumor tissue. The main advantage of magnetic particle hyperthermia is that the particles can heat previously localized target issue by external magnetic field. Next, under an applied magnetic field, energy is converted to thermal energy in tumor region which destroys cancer tissues (Figure 7) (Cole et al., 2011).

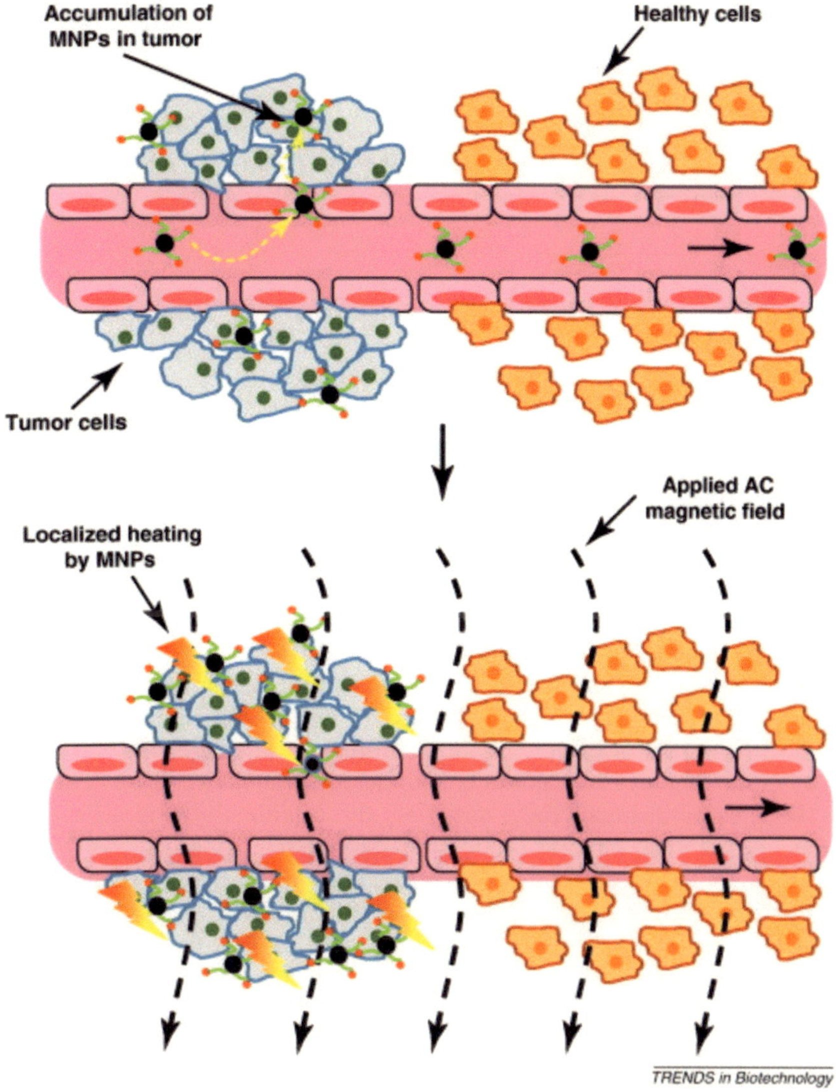

Fig. 7. Scheme of magnetic hyperthermia treatment of affected tissue; (a) accumulation of magnetic nanoparticles (MNPs) by the magnet at the tumor site; (b) exposition of tumor cells to an altering current (AC) magnetic field (Cole et al., 2011). (Adapted with permission from @ 2011 Elsevier).

Cancer-specific binding agents such as antibodies, hormones and other exo- and endogenous substances attached to magnetic nanoparticles has promising implications for magnetic fluid hyperthermia treatment (MFH) of breast, prostate and thyroid cancer or in vivo magnetic resonance imaging of acute brain inflammation (Phillips & Johnson, 2005; McAteer et al., 2007). MFH cancer treatment is based on the injecting a fluid containing magnetic nanoparticles directly into cancer region and the use of altering magnetic filed to generate the heat and destroy the tumor. Alternatively, the fluid can be injected into an artery that supplies the tumor with blood. However the location of the magnetic fluid should be precisely to minimize the effect of MFH. Attaching other cancer specific agents onto the surface of magnetic particles such as monoclonal antibodies or viruses has also promising applications. The agents can be modified to selectively bind cancerous cells. For instance, the genetic material in the viruses can be replaced with anticancer drug and than can be precisely released in cancer area at elevated temperature. The promising benefits of hyperthermia cancer treatment open a new perspectives for application of magnetic particles in cancer treatment.

3. Nanomagnetosols

Aerosol drug delivery system allows for the pulmonary drug administration delivery of therapeutic agents. The non-invasive drug delivery is mainly used for treatment of lung disorders such as asthma, chronic obstructive pulmonary disease and lung cancer. Pulmonary delivered drugs are rapidly absorbed alike to other mucosal surfaces. The appropriate size of aerosols droplets determines the passive targeting by their deposition in different lung regions. However the exact targeting to specific lung regions other than airways or the lung periphery has not been achieved to date (Dames et al., 2007). The high and effective drug concentration at disease site in standard chemotherapy in lung cancer with cytotoxic drugs is particularly difficult. The problem is that the cytotoxic potency of chemoterapeutics is not limited to cancer region.

The innovative study reports for the first time target aerosol delivery to the lung achieved with aerosol droplets comprising superparamagnetic iron oxide nanoparticles in combination with a target-directed magnetic gradient field. The high efficiency of aerosol droplets comprising SPIONs in combination with external magnetic gradient field was confirmed by computer aided simulation and demonstrated experimentally in mice. In contrast to intravenous magnetic drug targeting, the pulmonary drug administration has less limitations of the drug binding capacity of the nanoparticles. The new drug carriers, also called as a nanomagnetosols, offer high a flexibility of the system (Dames et al., 2007). The main advantage is that drug dose can be easily adjusted by the changing the drug concentration in the magnetic particles solution. Additionally, different drugs can be attached in different manner to the magnetic particles and also they can be co-delivered with other nanocarriers such as liposomes.

The successful magnetic aerosol targeting in vivo with plasmid DNA (pDNA) was achieved in intact mouse. The electromagnet with the iron circuit and the tip comprising an iron-cobalt alloy allows for accurate positioning of the SPIONs in selected site of the lung. The amount of deposited pDNA did not differ between the left and right lung in the absence of the external magnetic field. A twofold higher amount of pDNA was evaluated in the magnetized right lung than in unmagnetized left lung (Dames et al., 2007). Moreover, the authors conclude that the nanomagnetosol droplets with SPIONs are responsible for their transport, deposition the accumulation of the attached ligand/drug in affected tissue and not the single SPION as was supposed previously.

The anatomy of human lung also allows for the implementation of magnetosols in targeted therapy. The concept of non-invasive targeted delivery of magnetic aerosol droplets to the lung is presented in Figure 8. In last decade, the use of new drug delivery approach in the small animal were often demonstrated and well elaborated. The results confirm the higher deposition (~12%) of superparamagnetic nanoparticles (of ~50 nm diameter) on the magnetized airway surface and only ~4% on the unmagnetized airway area (Dames et al.,

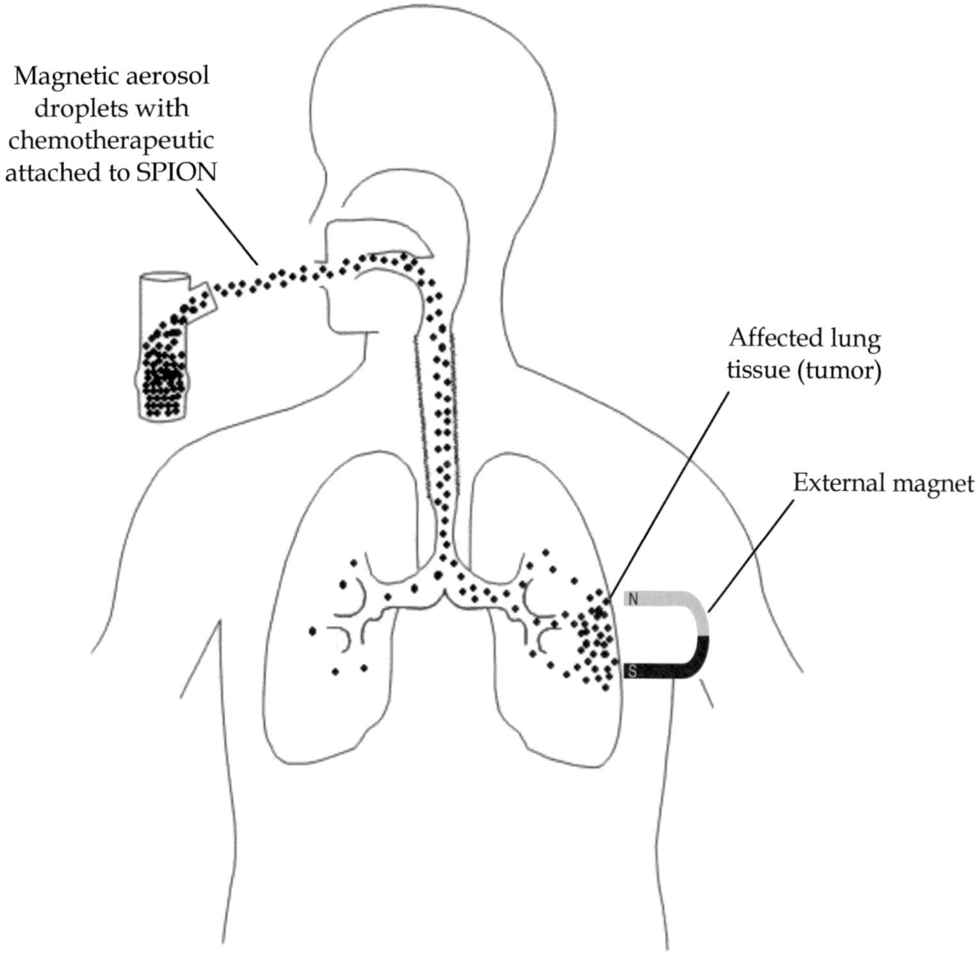

Fig. 8. The concept of non-invasive delivery of magnetic aerosol droplets to the human lungs and concentrated to the target site (tumor) with the help of magnet.

2007). Crucially, the non-deposited aerosol droplets can be transferred to further lung zones or exhaled. Regardless of the advantages for aerosol droplets with magnetic particles, the objective risk assessment studies have to be performed. Many of studies confirmed that the metal oxide particles are very toxic, especially in the pulmonary system (Machado et al., 2010). Nowadays, the in vitro studies are focused on the characterization and cytotoxic assessment of aerosol particulates, especially the nanoparticles based on heavy alloys W-Fe-Ni and W-Fe-Co. The exposure to Co, Ni, W and Fe can cause pulmonary fibrosis asthma, oedema and pneumonia among other effects.

The cytotoxic assessment studies of aerosol particulates are usually performed with the use of model for lung tissue – human epithelial cells (A549). The use of different filter exposure cytotoxicity assays allow to assess the inflammatory and related respiaratory health effects. Generally, the inhaled ~1μm particulates are moved by cilia of the bronchial epithelial cells towards the upper respiratory tract. The nanoparticles at size of <100nm can accumulate, aggregate and even cause oxidative stress and inflammation. So far it is proved that the most important parameters that can affect cytotoxicity are chemical composition, size and time of deposition of magnetic nanoparticles. Because of a lack of detailed limitations regarding the in vivo use of nanomagnetosols, it is necessary to continue the risk assessment studies and establish the well-defined regulations.

4. Magnetic drug delivery system in antimicrobial and antiviral therapy

The magnetic nanoparticles are also promising carries for antibacterial agents such copper and silver and can provide an alternative treatment for bacterial infections. As known, silver is distinguished by extraordinary inhibitory and bactericidal properties for a broad spectrum of bacterial strains. Antibacterial coatings based on hydrogen bonded multilayer containing in situ synthesized silver magnetic nanoparticles can be delivered to a specific region to localize a high concentration of antibacterial silver while maintaining a low concentration in general (Lee et al., 2005). The silver multilayer assembled on spherical support such as magnetic micro- or nanospheres were localized and focused by external magnetic field and showed excellent antibacterial properties against the Gram-positive strain *Staphylococcus epidermidis* and the Gram-negative strain *Escherichia coli*. It is expected that the modification of silver particles multilayer by the increase of surface-volume ratios can improve the antibacterial treatment (Tran & Webster, 2010).

The other studies demonstrates that the magnetic formulation of 3′-azido-3′-deoxythymidine-5′-triphosphate (AZTTP) is a new potential drug carries for targeted delivery across the blood brain barrier (Saiyed et al., 2009, 2010). Nucleoside and nucleotide analog reverse transcriptase inhibitors (NRTIs) are important part of the antiretroviral therapy (ART). Their inefficient cellular phosphorylation cause a limitation for treatment of human immunodeficiency virus (HIV). Main assumption of the in vitro study is that active NRTIs can directly bind to magnetic nanoparticles by ionic interaction and can inhibit HIV-1 replication. The development of magnetic AZTTP liposomal nanoformulation (150 nm) allows to cross blood brain barrier model by direct transport or via monocyte-mediated transport by external magnetic field. Hence, the magnetic carriers give a new opportunity to for treatment of neurological disorders.

5. Conclusion

The micro- and nanoparticle technology is highly novel and offers many possibilities for future development of new drug delivery systems. The innovative methods for drug targeting and delivery based on the micro- and nano-sized magnetic support provide a numerous advantages compared to conventional drug delivery systems. Presently, only molecular targeting ligands coupled to magnetic particles are successfully used and commercial available as contrast agents in MRI. Although the magnetic particles possess unique and useful properties for biomedical and pharmaceutical applications, they also carry potential health risks. The small size and high surface area to volume ratio of magnetic nanoparticles may have important implications due to higher biological activity per given mass compared to larger particulate forms (Helland et al., 2008). That may cause their remarkable activity and toxicity effect. Therefore, the material used for surface coating of the magnetic particles for in vivo application must not only be nontoxic but also biocompatible. So far there are no criteria and tests for evaluation of toxicity level, parameters of release as well as uptake magnetically targeted drugs. Regarding the in vivo application, it will have to be proved, that the new drug carriers possess not only useful properties but also are quit safe for environment and human health. Hence, the Food and Drug Administration issued a report to consider developing guidance for regulation of nanotechnology products and their adaptation from science to biomedical application.

6. Acknowledgment

Thanks are due to Mr. Tomasz Sióḋmiak, Mr. Wiktor Sroka and Ms. Beata Kochanek for technical assistance in preparing this manuscript.

7. References

Alexiou, Ch.; Arnold, W.; Klein, R.J. et al. (2000). Locoregional Cancer Treatment with Magnetic Drug Targeting. *Cancer Research*, Vol.60, (December 2000), pp. 6641-6648, ISSN 1538-7445.

Alexiou, Ch.; Jurgons, R.; Schmid, R.; Hilpert, A.; Bergemann, Ch.; Parak, F.; Iro, H. (2005). In vitro and in vivo investigations of targeted chemotherapy with magnetic nanoparticles. *Journal of Magnetism and Magnetic Materials*, Vol.293, (March 2005), pp. 389-393, ISSN 0304-8853

Ang, K.L; Venkatraman, S; Ramanujan, R.V. (2007). Magnetic PNIPA Hydrogels for Hyperthermia Applications in Cancer Therapy. *Materials Science and Engineering C.* Vol.27, No.3 (April 2007), pp. 347–351, ISSN 0928-4931

Arruebo, M.; Fernández-Pacheco, R.; Ibarra, M. R. & Santamaría. J. (2007). Magnetic nanoparticles for drug delivery. *Nanotoday*, Vol.2, No.3, (June 2007), pp. 22-32, ISSN 1748-0132

Bergemann, C.; Müller-Schulte, D.; Oster, J.; Brassard, L.; Lübbe, A.S. (1999). Magnetic ion-exchange nano- and microparticles for medical, biochemical and molecular biological applications. *Journal of Magnetism and Magnetic Materials*, Vol.194, (April 1999), pp.45-52, ISSN 0304-8853

Cole, A.J., Yang, V.C., David, A.E. (2011). Cancer theranostics: the rise of targeted magnetic nanoparticles. *Trends in Biotechnology*, Vol.29, No.7, (July 2011) pp. 323-332, ISSN 0167-7799

Corchero, J.L. & Villaverde, A. (2009). Biomedical application of distally controlled magnetic nanoparticles. *Trends in Biotechnology*, Vol.27, No.8, (June 2009) pp. 468-476, ISSN 0167-7799

Dames, P.; Gleich, B.; Flemmer, A. et al. (2007). Targeted Delivery of Magnetic Aerosol Droplets to the Lung. *Nature Nanotechnology*. Vol.2, No.8 (July 2007), pp. 495–499, ISSN 1748-3395

Douziech-Eyrolles, L.; Marchais, H.; Hervé, K.; Munnier, E. & Soucé, M. (2007). Nanovectors for anticancer agents based on superparamagnetic iron oxide nanoparticles. *International Journal of Nanomedicine*, Vol.2, No.4, pp.541-550, ISSN 1178-2013

Gijs, M.A.M. (2004).Magnetic Beads Handling on-chip: New Opportunities for Analytical Applications. *Macrofluid Nanofluid* Vol.1, No.1 (November 2004), pp. 22-40, ISNN 1613-4982

Gilchrist, R.D.; Medal, R.; Shorey, W.D.; Hanselman, R.C.; Parrott, J.C. & Taylor, C.B. (1957). Selective inductive heating of lymph nodes. *Annals of Surgery*. Vol.146, No.4, (October 1957), pp. 596-606, ISSN 1528-1140

Helland, A.; Scheringer, M.; Siegrist, M.; Kastenholz, H.G.; Wiek, A.; Scholz, R.W. (2008). Risk Assessment of Engineered Nanomaterials: A Survey of Industrial Approaches. *Environmental Science & Technology*, Vol.42, No.2 (January 2008), pp. 640-646, ISSN 1520-5851

Ilium, L.; Davis, S.; Wilson, C.; Thomas, N.; Frier, M. & Hardy, J. (1982). Blood clearance and organ deposition of intravenously administered colloidal particles. The effects of particle size, nature and shape. *International Journal of Pharmaceutics*, Vol.12, No 2–3, (October 1992) pp. 135–146, ISSN 0378-5173

Kohler, N.; Sun, C.; Wang, J.; Zhang, M. (2005). Methotrexate-modified superparamagnetic nanoparticles and their intracellular uptake into human cancer cells. *Langmuir*, Vol.21, No.19 (June 2005), pp. 8858–8864, ISSN 1520-5827

Lee, D.; Cohen, R.E.; Rubner, M.F. (2005). Antibacterial Properties of Ag Nanoparticle Loaded Multilayers and Formation of Magnetically Directed Antibacterial Microparticles, *Langmuir* Vol.21, No.21 (October 2005), pp. 9651-9659, ISSN 1520-5827

Lübbe, A.S.; Bergemann, Ch.; Riess, H. et al. (1996). Preclinical experiences with magnetic drug targeting: tolerance and efficacy. *Cancer Research,* Vol.56, (October 1996), pp. 4686-4693, ISNN 1538-7445

Mahmoudi, M.; Sant, S.; Wang, B.; Laurent, S. & Sen, T. (2011). Superparamagnetic iron oxide nanoparticles (SPIONs): Development, surfach modification and applications in chemotherapy. *Advanced Drug Delivery Reviews,* Vol.63, pp. 24–46, ISSN 0169-409X

Machado, B.I.; Murr, L.E.; Suro, R.M.; Gaytan, S.M.; Ramirez, D.A.; Garza, K.M.; Schuster, B.E. (2010). Characterization and Cytotoxic Assessment of Ballistic Aerosol Particulates for Tungsten Alloy Penetrators into Steel Target Plates. *International*

Journal of Environmental Research and Public Health. Vol.7, No.9 (September 2010), pp. 3313-3331, ISSN 1660-4601

Marszałł, M.P.; Moaddel, R.; Kole, S.; Gandhari, M.; Bernier, M. & Wainer, I.W. (2008). Ligand and protein fishing with heat shock protein 90 coated magnetic beads. *Analytical Chemistry,*Vol.80, No.19, (October 2008), pp. 7571-7575, ISSN 0003-2700

Marszałł, M.P.; Buciński, A.; Kruszewski, S.; Ziomkowska B. (2011a). A New Approach to Determine Camptothecin and Its Analogues Affinity to Human Serum Albumin. *Journal of Pharmaceutical Science,,* Vol.100, No.3, (March 2011) pp. 1142-1146, ISSN 1520-6017

Marszałł, M.P. (2011b). Application of Magnetic Particles in Pharmaceutical Sciences. *Pharmaceutical Research,* Vol.28, No.3 (March 2005), pp. 480–483, ISSN 1573-904X

McAteer, M.A.; Sibson, N.R.; von zur Muhlen, C.; Schneider, J.E.; Lowe, A.S.; Warrick, N.; Channon, K.M.; Anthony, D.C.; Choudhury R.P. (2007). *In vivo* magnetic resonance imaging of acute brain inflammation using microparticles of iron oxide. *Nature Medicine.*Vol.13, No.10 (October 2007), pp. 1253–1258, ISSN 1078-8956

Messmann, R.& Allegra, C. (2001) Antifolates. In: *Cancer Chemotherapy& Biotherapy,* Chabner, B., Longo, D. (3 ed.), ISSN 0036-8075, 139-184, Eds.; Lippincott Williams & Wilkins, Philadelphia

Modo, M; Hoehn, M; Bulte, J.W. (2005). Cellular MR Imaging. *Molecular Imaging.* Vol.4, No.3 (August 2005), pp. 143–164, ISSN 1535-3508Phillips, J. & Johnson, D.T. (2005). *Magnetic Fluid Hyperthermia:* A Topical Review, *The Journal of Science and Health at The University of Alabama.*Vol.3, (August 2005), pp. 14–18

Saiyed, Z.M.; Gandhi, N.H.; Nair, M.P.N. (2009). AZT 5′-triphosphate nanoformulation suppresses HIV-1 replication in peripheral blood mononuclearcells. *Journal of Neurovirology.* Vol.15, No.4 (July 2009), pp. 343-347, ISSN 1355-0284

Saiyed, Z.M.; Gandhi, N.H.; Nair, M.P.N. (2010). Magnetic nanoformulation of azidothymidine 5′-triphosphate for targeted delivery across the blood-brain barrier. *International Journal of Nanomedicine* Vol.5 (March 2010), pp. 157-166, ISSN 1178-2013

Tran, N. & Webster, T.J. (2010). Magnetic nanoparticles: biomedical applications and challenges. Journal of Materials Chemistry, Vol.20, No.40 2010, pp. 8760–8767, ISNN 0959-9428

Widder, K.J; Senyei, A.E. & Scarpelli, D.G. (1978). Magnetic microspheres: a model system of site specific drug delivery in vivo. *Proceedings of the Society for Experimental Biology and Medicine,* Vol.158, No.2 (June 1978) pp. 141–146, ISSN 0037-9727

Wilhelm, C.; Gazeau, F. (2008) Universal Cell Labelling with Anionic Magnetic Nanoparticles. *Biomaterials,* Vol.29, No.22 (August 2008), pp. 3161–3174, ISSN 0142-9612

Woodward, R.C.; Heeris, J.; Pierre, T.G.St.; Saunders, M.; Gilbert, E.P.; Rutnakornipituk, M.; Zhang. Q.; Riffle J.S. (2007). A comparison of methods for the measurement of the particle-size distribution of magnetic nanoparticles. *Journal of Applied Crystallography,* Vol.40, (January 2007) ISSN 0021-8898

Yoo, J-W.; Doshi, N.& Mitragotri, S. (2011) Adaptive micro and nanoparticles: Temporal control over carrier properties to facilitate drug delivery, *Advanced Drug Delivery Reviews*, in press, doi:10.1016/j.addr.2011.05.004, ISNN 0169-409X

6

Drug Nanoparticles – An Overview

Vijaykumar Nekkanti, Venkateswarlu Vabalaboina and Raviraj Pillai*
Dr. Reddy's Laboratories Limited, Hyderabad, India

1. Introduction

Advances in drug discovery technologies and combinatorial chemistry techniques have led to identification of a number of compounds with good therapeutic potential. However, because of their complex chemistry majority of these compounds have poor aqueous solubility resulting in reduced and variable bioavailability (Lipinski et al., 2002). The variability in systemic exposure observed often makes it difficult for dose delineation, results in fed and fast variability and in slower onset of action. These issues may lead to sub-optimal dosing and concomitantly poor therapeutic response. For compounds with poor aqueous solubility that are ionizable, preparation of salts to improve solubility/dissolution rate is a commonly used approach that had limited success. From a product development standpoint, generally a crystalline salt is preferred due to potential physical and chemical stability issues associated with the amorphous form. Identification of a crystalline salt with adequate aqueous solubility requires screening various counter-ions and solvents/crystallization conditions and at times isolation of a crystalline material is difficult. In some instances the salt formed is extremely hygroscopic posing product development and manufacturing challenges (Elaine et al., 2008).

Currently there are limited formulation approaches for compounds with poor aqueous solubility. The most commonly used approaches are micronisation and solid dispersions of the drug in water-soluble careers for filling into hard or soft gelatin capsules. Micronisation results in particles that are < 5 µm with a very small fraction that is in the sub-micron range. The decrease in particle size results in a modest increase in surface area that may not change the dissolution rate or saturation solubility to significantly impact bioavailability (Jens-Uwe et al., 2008).

Solid dispersion compositions comprise of molecular dispersion of the drug in water soluble and lipid-based surface-active carriers that can emulsify upon contact with the dissolution medium. Formation of molecular dispersions (solid solution) provides a means of reducing the particle size of the compounds to nearly molecular levels (i.e., there are no visible particles). As the carrier dissolves, the compound is exposed to the dissolution media as fine particles that are amorphous, which can dissolve rapidly and concomitantly absorbed. These formulations are filled in soft or hard gelatin capsules. There are several products using this approach in the market, e.g.,

* Corresponding Author

Sandimmune®/Neoral® (cyclosporin microemulsion), Norvir® (Ritnovir) and Fortovase® (Saquinavir). This approach is generally suitable for highly potent compounds and thus not applicable for moderately potent compounds where the dose requirement may be high (Merisko-Liversidge et al., 2003).

In recent years an area that is gaining popularity with formulation scientists for developing a viable dosage form for poorly soluble compounds that are moderately potent is to develop a formulation incorporating drug nanoparticles, usually less than 1 μm in diameter. For example, when the particle size of the drug is reduced from 8 μm to 200 nm there is 40-fold increase in the surface area to volume ratio. This increase in surface area can provide substantial increase in the dissolution rate if the formulation disperses into discrete particles (Liversidge et al., 1995). The nanoparticle formulation approach is proven to be very useful and invaluable in all stages of the drug product development and has opened opportunities for revitalizing marketed products with suboptimal delivery.

Nanoparticle formulation technologies have provided the pharmaceutical industry with options for addressing solubility and bioavailability issues associated with poorly soluble compounds. In new chemical entities (NCE) development, the technology has been of great value when it is used as a screening tool during preclinical efficacy and / or safety assessment studies in the early development phase. For marketed products requiring life-cycle extension opportunities, nanoparticle formulation strategies provide a means to develop a new drug-delivery platform with improved therapeutic outcome incorporating the existing drug, thus creating new avenues for addressing unmet medical needs.

2. History

Nanotechnology has a long development and application history. However, the most important scientific advancements have only taken place in the last two decades. Heterogeneous catalysts were among the first examples, developed in the early 19th century (Rogers et al., 2001). The earliest example of pharmaceutical application was Danazol that was milled using a bead mill to obtain a median particle size of 169 nm (Robertson, 1983). The Danazol nanosuspension showed enhanced oral bioavailability (82.3 ± 10.1%) as compared to the drug suspension using the "as-is" drug (5.1 ± 1.9%).

The first nanoparticle technology based product approved by FDA was Rapamune® (Sirolimus) - an immunosuppressant developed by Wyeth Pharmaceuticals (now Pfizer). The second product approved by FDA was Tricor® by Abbott Laboratories, an improved formulation of Fenofibrate (for hypercholesterolemia) incorporating drug nanoparticles that reduced the fed-fast variability resulting in no dosing restriction that allowed co-administration with other drugs used for treating lipid disorders. Another product containing Fenofibrate nanoparticles is Triglide®. The product was developed by Skye Pharma using their patented IDD-P® technology and marketed by Sciele Pharma Inc. (Atlanta, USA). Antiemetic drug, Emend® (Aprepitant) was approved by the FDA in March 2003 and launched in the United States by Merck in April 2003. Emend is a capsule containing 80 or 125 mg of Aprepitant formulated as drug nanoparticles using Elan's drug NanoCrystal® technology (Mary et al., 2005). Megace ES® (ES stands for enhanced solubility) is another product containing drug nanoparticles that was developed by Par Pharmaceutical Inc. (USA). It is an aqueous suspension of Megestrol Acetate (a synthetic progestin, anti anorexic) with a dose of

625 mg / 5 mL. The drug nanosuspension reduced the fed and fast variability similar to Tricor®. The product in nanosuspension demonstrated that aqueous nanosuspension can be produced with adequate physical stability with acceptable shelf life using this technology. A list of products developed using nanoparticle technology (Ranjita Shegokar et al., 2010; Rajesh Dubey, 2006) currently available in the market is summarized in Table 1.

Brand	Generic Name	Indication	Drug Delivery Company	Innovator	Status
Rapamune®	Rapamycin, Sirolimus	Immunosuppressant	Elan Nanosystems	Wyeth	Marketed
Emend®	Aprepitant	Anti-emetic	Elan Nanosystems	Merck & Co.	Marketed
Tricor®	Fenofibrate	Hypercholesterolemia	Abbott Laboratories	Abbott Laboratories	Marketed
Megace ES®	Megestrol	Anti-anorexic	Elan Nanosystems	Par Pharmaceuticals	Marketed
Triglide®	Fenofibrate	Hypercholesterolemia	IDD-P Skyepharma	Sciele Pharma Inc.	Marketed
Avinza®	Morphine Sulphate	Phychostimulant	Elan Nanosystems	King Pharmaceuticals	Marketed
Focalin	Dexmethyl-Phenidate HCl	Attention Deficit Hyperactivity Disorder (ADHD).	Elan Nanosystems	Novartis	Marketed
Ritalin	Methyl Phenidate HCl	CNS Stimulant	Elan Nanosystems	Novartis	Marketed
Zanaflex Capusules TM	Tizanidine HCl	Muscle Relaxant	Elan Nanosystems	Acorda	Marketed

Table 1. Overview of nanoparticle technology based products

3. Formulation theory

The basic principle of micronisation and nanonisation is based on increase in surface area leading to enhancement in dissolution rate according to Noyes-Whitney equation (Muller et al., 2000). Poor aqueous solubility correlates with slower dissolution and decreasing particle size increases the surface area with concomitant increase in the dissolution rate.

Dissolution kinetics is the primary driving force behind the improved pharmacokinetic properties of nanoparticle formulations of poorly water soluble compounds. Dissolution rate of a drug is a function of its particle size and intrinsic solubility. For drugs with poor aqueous solubility, surface area of the drug particles drives dissolution. As described by the Nernst-Brunner and Levich modification of Noyes-Whitney model the rate of drug dissolution is directly proportional to surface area;

$$dx/dt = (A \times D/\delta) \times (C - X/V) \quad (1)$$

Where X is the amount of drug in solution, t is time, A is the effective surface area, D is the diffusion coefficient of the drug, δ is the effective diffusion boundary layer, C is the saturation solubility of the drug, and V is the volume of dissolution medium.

Saturation solubility usually is a compound specific constant that depends on temperature. This understanding is true for regular particles that are above the micron range however, different for drug nanoparticles. This is because the dissolution pressure is a function of the curvature of the surface that means it is much stronger for a curved surface of nanoparticles. Below a particle size of approximately 2 µm, the dissolution pressure increases distinctly leading to an increase in the saturation solubility. In addition, the diffusional distance on the surface of drug nanoparticles is decreased, thus leading to an increased concentration gradient. The increase in surface area and concentration gradient lead to much more pronounced increase in dissolution velocity and saturation solubility compared to products containing micronized particles concomitantly resulting in improved bioavailability (Keck et al., 2006).

Increased solubility near the particle surface results in enhanced concentration gradient between the surface and the bulk solution. The high concentration gradient according to Fick's law must lead to an increased mass flux away from the particle surface (Dressman et al., 1998). As the particle diameter decreases, its surface area to volume ratio increases inversely, further leading to an increased dissolution rate. Under sink conditions in which the drug concentration in the surrounding medium approaches zero, rapid dissolution could theoretically occur.

4. Production of drug nanoparticles

There are several techniques used to produce drug nanoparticles. The existing technologies can be divided into two categories; 'bottom up' and 'top down'. The bottom-up technologies involves controlled precipitation/crystallization by adding a suitable non-solvent. The top down technologies include milling or homogenization. However, combination techniques that involves pretreatment step followed by size reduction are also being used to produce nanoparticles with the desired size distribution.

4.1 Bottom-up technologies (Precipitation methods)

Precipitation has been applied for many years for preparation of fine particles, particularly in the development of photographic film, and lately for preparation of sub-micron (nano) particles for pharmaceutical applications (Otsuka et al., 1986; Illingworth, 1972). Examples for precipitation techniques are hydrosols developed by Sucker (Sandoz, presently Novartis) and Nanomorph developed by Soliqs/Abbott (Musliner, 1974; Sjostrom et al., 1993; Gassmann et al., 1994; List et al., 1988; Sucker et al., 1994).

In this process, the drug is dissolved in a suitable solvent and the solution is subsequently added to a non-solvent. This results in high super saturation, rapid nucleation and the formation of many small nuclei. Upon solvent removal, the suspension is sterile filtered and lyophilized (Kipp et al. 2003). The mixing processes may vary considerably. Through careful control of this addition process it is possible to obtain a particle with a narrow size distribution. In the case of Nanomorph, amorphous drug nanocrystals are produced to further enhance dissolution velocity and solubility (Muller et al., 2001a).

Simple precipitation methods, however, have numerous limitations; it is very difficult to control nucleation and crystal growth to obtain a narrow size distribution. Often a metastable solid, usually amorphous, is formed which is converted to more stable crystalline forms (Violante et al., 1989; Bruno et al., 1992). Furthermore, non-aqueous solvents utilized in the precipitation process must be reduced to toxicologically acceptable levels in the end product and due to the fact that many poorly soluble drugs are sparingly soluble not only in aqueous but also in organic media. Considering these limitations, the "bottom up" techniques are not widely used for production of drug nanocrystals. Instead, "top down" technologies that include homogenization and milling techniques are more frequently used.

4.2 Top-down technologies

The two top down technology frequently used for producing drug nanoparticles include;

a. High pressure homogenization
b. Milling

a. High pressure homogenization methods

One of the disintegration method used for size reduction is high-pressure homogenization. The two-homogenization principles/homogenizer types used are;

1. Microfluidisation (Microfluidics, Inc.)
2. Piston-gap homogenizers (e.g. APV Gaulin, Avestin, etc.)

b. Microfluidisation for production of drug nanoparticles

Microfluidisation works on a jet stream principle where the suspension is accelerated and passes at a high velocity through specially designed interaction chambers. Frontal collision of fluid streams under high pressures (up to 1700 bar) inside the interaction chamber generates shear forces, particle collision, and cavitation forces necessary for particle size reduction. The Microfluidizer processor keeps a constant feed stream that gets processed by a fixed geometry which produces high shear and impact necessary to break down larger particles. This process yields smaller particles with narrow particle size distribution with repeatability and scalability.

The interaction chamber's exterior and interior is either made of stainless steel, poly-crystalline diamond (PCD) or aluminum oxide. The poly-crystalline diamond chambers typically have a lifetime 3 - 4 times longer than the aluminum oxide ceramic chambers. Single slotted interaction chambers are used for lab-scale manufacturing and multi-slotted chambers for commercial scale. Multi-slotted chambers are comprised of multiple single slots in parallel for processing larger volumes of the products. There are two types of interaction chambers: Y chamber is useful for liquid-liquid emulsions and finds application in preparing liposomes while Z-chamber is typically used for cell disruption and nanodispersion. A schematic representation of mechanism of particle size reduction in high pressure homogenizers is shown in Fig. 1. The selection of correct chamber depends upon the feed particle size, the application, and the amount of shear and impact required to carryout the operation. The Insoluble Drug Delivery - Particles (IDD-P™) technology developed by SkyePharma Canada Inc. use the Microfluidizer (Jens-Uwe et al., 2008).

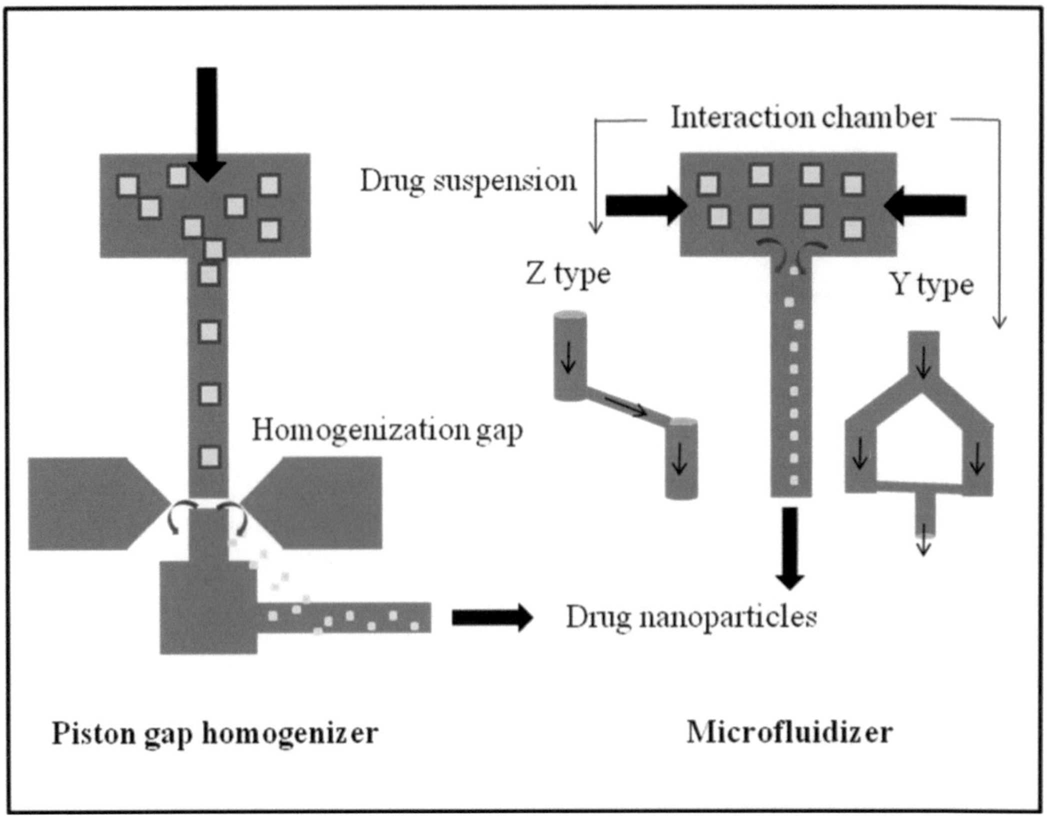

Fig. 1. Schematic representation of mechanism of particle size reduction in high pressure homogenizers

4.2.1 Process parameters affecting particle size

Studies on particle size reduction of a sparingly soluble drug (BCS class II) using the Microfluidizer (Model - Microfluidics M110-P) in our laboratory indicated that particle size reduction depends on various process parameters viz., number of homogenization cycles, homogenization pressure and, stabilizer concentration. At a constant homogenization pressure (30,000 psi) the value of mean particle size d50 decreased with increasing number of cycles from 5 to 60 (Fig. 2). Homogenization pressure has a significant effect on particle size distribution as shown in Fig. 3. At high homogenization pressure (30,000 psi) particle size reduction was significantly higher than at low homogenization pressure (10,000 psi) after 60 homogenization cycles. Surfactant concentration also plays an important role in particle size reduction through particle stabilization by forming a thin layer around the newly formed surface as evident based on the observation that at constant homogenization pressure and homogenization cycles, particle size reduced with increase in surfactant concentration from 10 mg/mL to 12 mg/mL (Fig. 4).

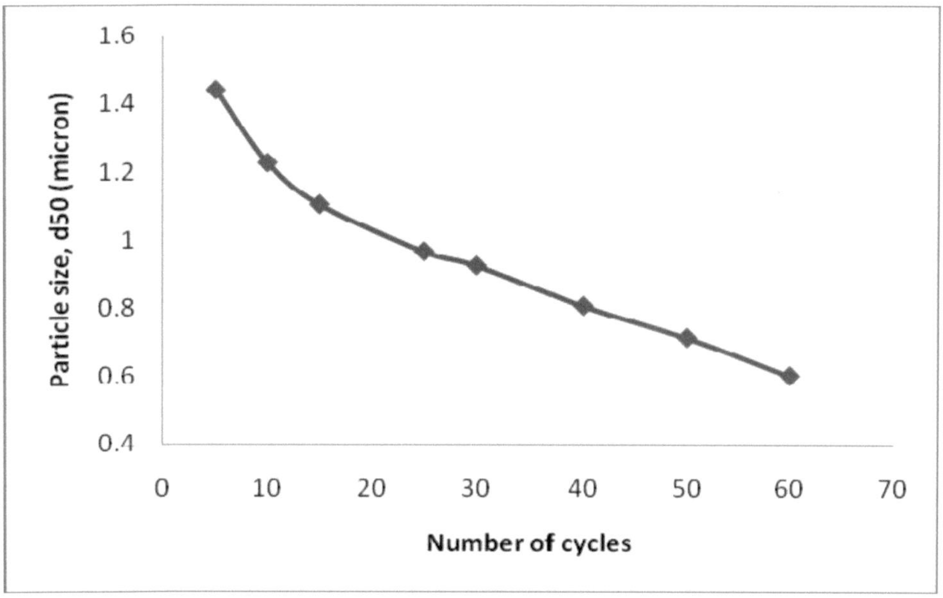

Fig. 2. Effect of number of cycles on mean particle size at constant homogenization pressure

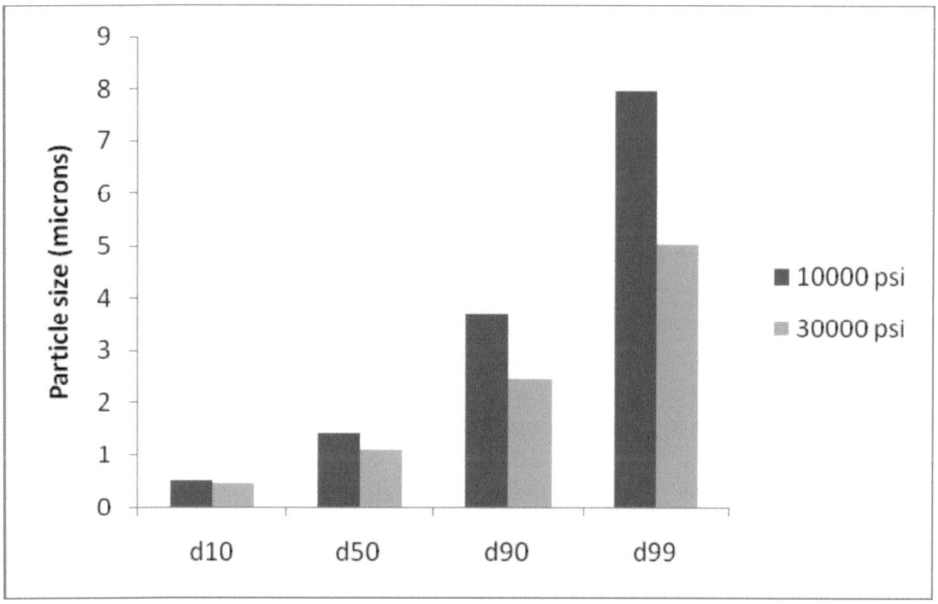

Fig. 3. Effect of homogenization pressure on particle size distribution

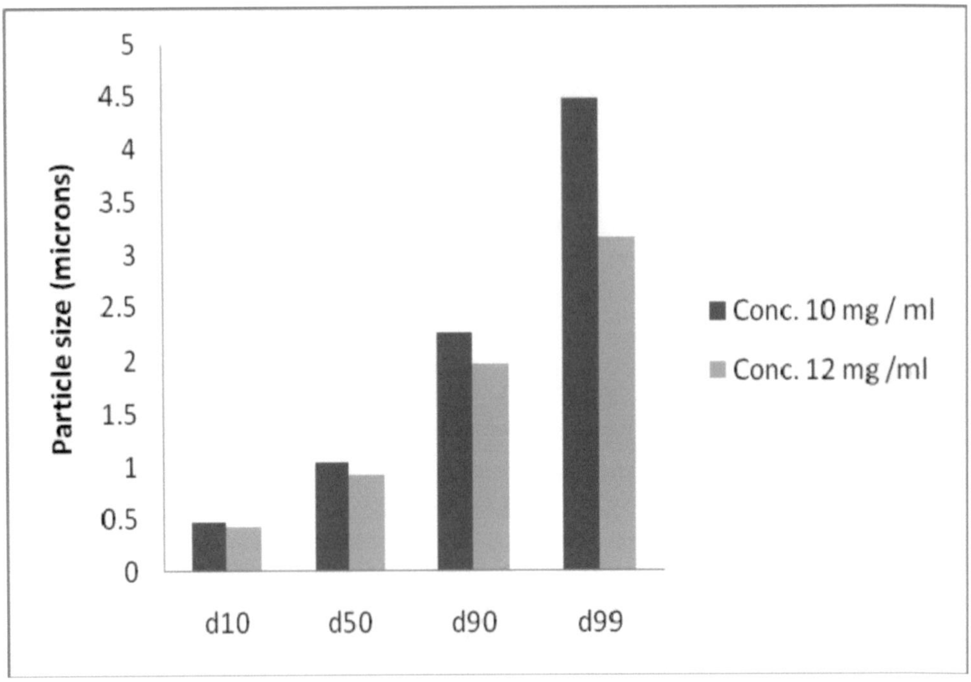

Fig. 4. Effect of surfactant concentration on particle size distribution

4.3 Piston-gap technologies

Using the microfluidisation principle, an alternative technology based on piston-gap homogenizers was developed in the middle of the 1990's for production of drug nanoparticles. Homogenization can be performed in water (DISSOCUBES®) or alternatively in non-aqueous media or water reduced media (NANOPURE®). Dissocubes® technology employs piston-gap homogenizers in which drug powder is dispersed in an aqueous surfactant solution and subsequently forced by a piston through the tiny homogenization gap (5 µm - 20 µm depending upon the viscosity of the suspension and the applied pressure) at a very high pressure (up to 4000 bar). Prior to entering the gap, the suspension is contained in a cylinder with a relatively large diameter compared to the width of the following gap. The resulting high streaming velocity of the suspension causes formation and implosion of the gas bubbles also known as cavitation which results in generation of shockwaves. The drug particle gets reduced by these high shear forces, turbulent flow and powerful shockwaves. Another approach viz., Nanopure® technology (by Pharma-Sol GmbH) is useful for particle size reduction of thermolabile drugs because it use low vapor pressure dispersion media for homogenization that helps in processing at low temperatures due to very little cavitation in the homogenization gap (Muller et al., 1999, 2001b, 2003; Muller RH & Moschwitzer JP, 2005; Jens-Uwe et al., 2008). In addition, there is also a combination process of precipitation followed by a second high-energy homogenization step (NANOEDGE®). The major limitation of this method is that nanoparticulate dispersion of low solid content (usually < 10% w/v) is produced that may be difficult for conversion to solid intermediates required for capsule filling or tableting.

4.4 Milling methods

Conventional milling and precipitation processes generally result in particles much greater than 1 μm. Milling techniques were later refined to enable milling of solid drug particles to sub-micron range. Ball mills are already known from the first half of the 20th century for the production of fine suspensions. In this method, the suspension comprising of drug and stabilizers along with milling media are charged into the grinding chamber. The reduction of particle size occurs because of the shear forces generated due to impaction of milling media. In contrast to high pressure homogenization, this is a low energy technique. Smaller or larger beads can be used as milling or attrition media. The milling media comprise of ceramics (cerium or yttrium stabilized zirconium dioxide), stainless steel or highly cross linked polystyrene resin-coated beads. Potential for erosion of the milling media during the milling process resulting in product contamination is one of the drawbacks of this technology. To overcome this issue, the milling media are often coated (Merisko-Liversidge et al., 2003). Another problem with milling process is the adherence of product to the inner surface of the mill (consisting mainly of the surface of milling media and the inner surface of milling chamber). There are two basic milling principles - either the milling medium is moved by an agitator or the complete container is moved in a complex direction leading to movement of the milling media to generate the shear forces required to fracture the drug crystals. The milling time depends on many factors such as solid content, surfactant concentration, hardness, suspension viscosity, temperature, energy input and, size of the milling media. The milling time may vary from minutes to hours or days depending on the particle size desired (Jens-Uwe et al., 2008).

In the bead milling process used for production of drug nanosuspension, the drug suspension is passed through a milling chamber containing milling media ranging from 0.2 to 3 mm. These media may be composed of glass, zirconium salts, ceramics, plastics (e.g., cross-linked polystyrene) or special polymers such as hard polystyrene derivatives. The drug concentration in the suspension may range from 5 - 40% w/v. Stabilizers such as polymers and/or surfactants are used to aid the dispersion of particles. To be effective the stabilizers must be capable of wetting the drug particles and providing steric and ionic barrier. In the absence of appropriate stabilizers, the high surface energy of the nanometer-sized particles would lead to agglomeration or aggregation of drug crystals. The concentration of polymeric stabilizers can range from 1 - 10% w/v and the concentration of surfactants is generally < 1 % w/v. If required other excipients such as buffers, salts and diluents like sugar can be added to the dispersion to enhance stability and aid further processing (Keck et al., 2006).

The milling chamber has a rotor fitted with disks that can be accelerated at the desired speed (500 - 5000 RPM). The rotation of the disk accelerates the milling media radially. The product flows axially through the milling chamber where the shear forces generated and/or forces generated during impaction of the milling media with the drug provides the energy input to fracture the drug crystals into nanometer-sized particles. The temperature inside the milling chamber is controlled by circulating coolant through the outer jacket. The process can be performed either in a batch mode or in a recirculation mode. The milled product is subsequently separated from the milling media using a separation system. A schematic of the bead milling process is shown in Fig. 5.

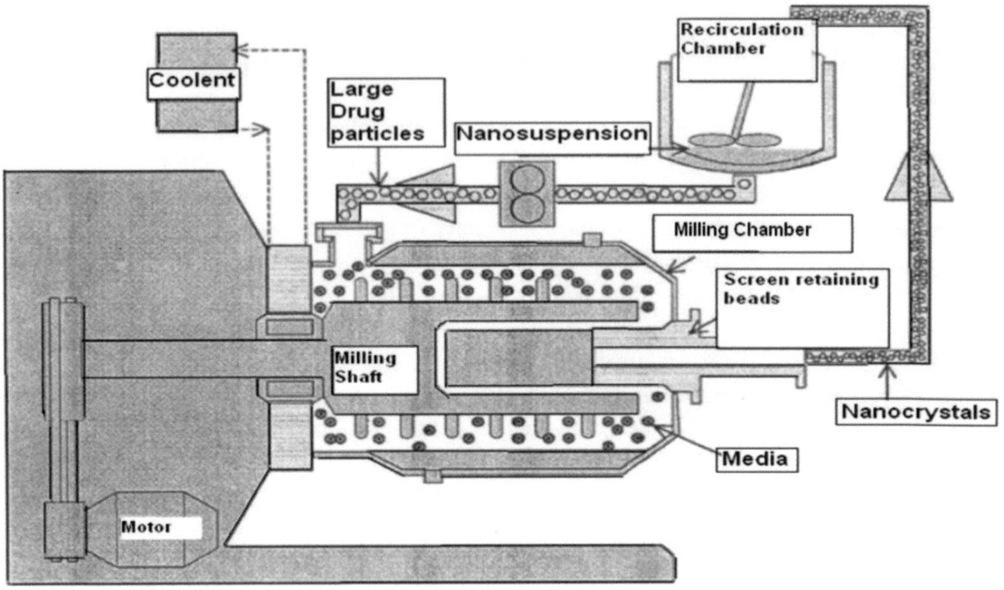

Fig. 5. Schematic of wet bead milling process used for production of drug nanoparticles

Scaling up the bead milling process is relatively easy and convenient because the process variables are scale independent. The batch size can be increased above the void volume (volume in between the hexagonal packaging of the beads) using the mill in a recirculation mode. The suspension is contained in the product container and is continuously pumped through the mill in a circular motion. This increases the batch size with concomitant increase in the milling time because the required exposure time of the drug particles per unit mass to the milling material remains unchanged.

Surfactants or stabilizers have to be added to ensure physical stability of the nanosuspensions. In the manufacturing process the drug substance is dispersed by high speed stirring or homogenizer in a surfactant/ stabilizer solution to yield a macro suspension. The choice of surfactants and stabilizers depends not only on the physical principles (electrostatic versus steric stabilization) and the route of administration. In general, steric stabilization is recommended because it is less susceptible to electrolytes in the gut or blood. Electrolytes if added can reduce the zeta potential and subsequently impair the physical stability, especially of ionic surfactants. In many cases an optimal approach is the combination of a steric stabilizer with an ionic surfactant, i.e, a combination of steric and electrostatic stabilization. There is a wide variety of bead mills available in the market, ranging from laboratory-scale to industrial-scale volumes. The ability for large-scale production is an essential prerequisite for introduction of product into market. In general, bead milling offers a convenient process for production of drug nanoparticle at high concentrations necessary for solid dosage form processing with ease of scale-up for commercial manufacturing.

5. Process optimization for the production of drug nanoparticles

Experimental design has been applied widely to formulation development, and is useful in process optimization and process validation (Fisher RA, 1926). A manufacturing process optimized using design of experiments (DOE) should result in a robust process amenable for seamless scale-up and validation (Dhananjay et al, 2010; Nekkanti et al, 2009a). The process variables in media milling can be optimized using design of experiments (DOE) to understand the effect on particle size, milling time and percentage yield (Nekkanti, et al., 2010). Though a number of statistical designs are reported, a face centered centre composite design (CCD) is often used because it provides information on direct effects, pair wise interaction effects and curvilinear variable effect (Billon et al., 2000; Vaithiyalingam & Khan, 2002; Tagne et al., 2006). For example, a design matrix prepared based on 3 variable factors at three levels (-1, 0, +1) to compute the design using statistical software program Design Expert (version No. 7.3.1) is summarized in Table 2.

S. No	Process Parameters	Level		
		Low (-1)	Center (0)	High (+1)
1	Disk Speed (RPM)	2000	2350	2750
2	Pump Speed (RPM)	40	50	60
3	Bead Volume in Milling Chamber (%)	60	70	80

Table 2. Process variables (factors) and levels

A stepwise regression can be used to generate quadratic equations for each response variable. Analysis of variance (ANOVA) and regression is used to evaluate the significant effects and model building for each response variable. Each response is then fitted to a second-order polynomial model and, the regression coefficients for each term in the model can be estimated along with R^2 and adjusted R^2 of regression model to understand how these parameters effect the critical product attributes either through non-linear, quadratic or interaction effects.

The interaction effect of pump and disc speed on milling time is shown in Fig. 6. The plot indicates that at lower disk speed the milling time (to achieve the desired particle size) increases. This may be attributed to the fact that at low disk speed the shear forces generated by accelerating beads may not be sufficient to fracture the drug crystals into smaller particles. The milling efficiency was high when the disk and pump were run at moderate speeds.

The interaction effect of pump speed and bead volume on particle size is shown in Fig. 7. The plot indicates that increase in pump speed and bead volume resulted in larger particles where as, their interaction resulted in a decrease in particle size. Both pump speed and bead volume have an effect on particle size with bead volume having a significant impact in controlling the drug particle size due to increased probability of impaction.

The interaction effect of disk speed and bead volume on yield is shown in Fig. 8. The plot indicates that the process yields obtained was significantly affected by disk speed and bead volume. At lower disk speed and higher bead volume there was a decrease in yield; this may be attributed to loss in the milling chamber due to sticking.

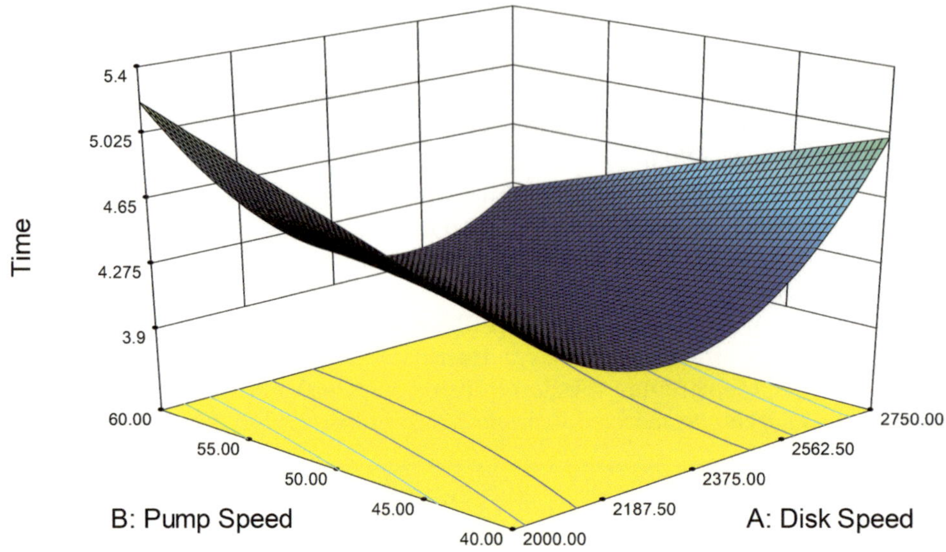

Fig. 6. Effect of disk and pump speeds on milling time

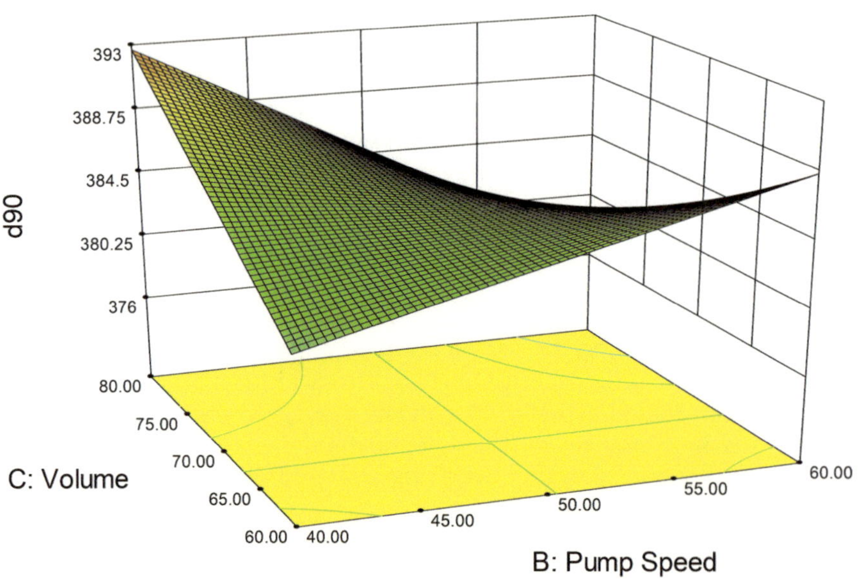

Fig. 7. Effect of pump speed and bead volume on particle size

The robustness of the model used can be validated based on confirmatory trials to ascertain difference between predicted and experimental values. The use of DOE for process optimization will result in a robust scalable manufacturing process with design space established for critical process parameters that can balance milling time, particles size and yield.

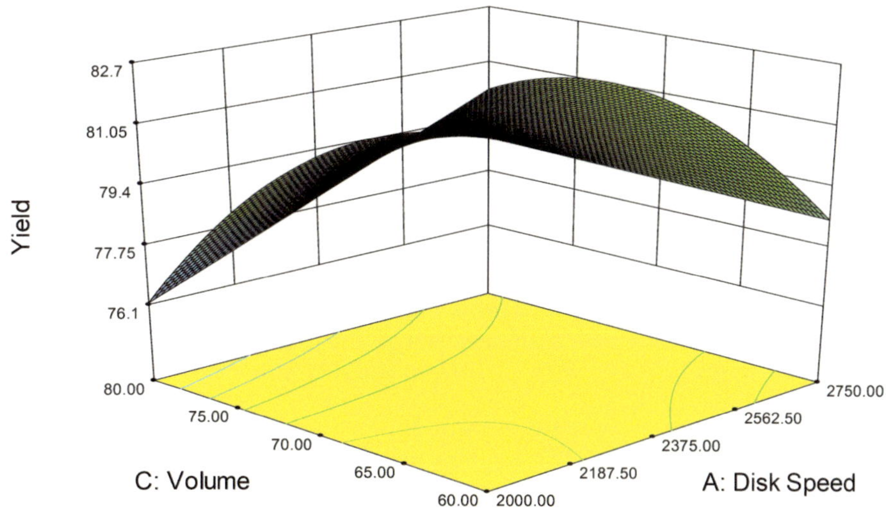

Fig. 8. Effect of disk speed and bead volume on yield

6. Conversion of nanosuspension into solid intermediate

For production of solid intermediate, the water has to be removed from the drug nanosuspension to obtain a dry powder required for tabletting or capsule filling. The objective of solid nanoparticle system is to release the drug nanoparticles in the gastrointestinal (GI) fluids as fine non-aggregated suspension and ensure physical stability upon long term storage. The solvent from nanosuspension can be removed using drying processes such as fluid bed coating / granulation, spray drying and freeze drying. Freeze drying is considered as a complex and cost-intensive process leading to a highly sensitive product. The main challenge is to preserve the re-dispersibility of the nanoparticles upon reconstitution in aqueous media and gastrointestinal fluids. The re-dispersants must be incorporated in the nanosuspensions prior or during the drying step. Commonly used re-dispersants include sugars such as lactose, sucrose and mannitol (Chu, 2000). Generally, re-dispersibility depends on the choice of re-dispersants, surfactants and polymeric stabilizers. The loading capacity of solid intermediate with drug nanoparticles can be adjusted by varying excipient concentrations.

In spray drying process the nanosuspension is atomized using a rotary or air-jet atomizer. In this process fast drying of the liquid feed happens due to the large surface area created by the atomization of the liquid feed into fine droplets and high heat transfer coefficients generated. The short drying time and consequently fast stabilisation of feed material at moderate temperatures make spray drying suitable for producing nanoparticles of drugs that are thermolabile. The spray drying process in general comprise of the following steps;

- **Atomization:** The liquid feed in the form of drug suspension is atomized into droplets by means of a nozzle or rotary atomizer. Nozzles use pressure or compressed gas to atomize the liquid feed while rotary atomizers employ an atomizer wheel rotating at high speed.

- **Drying:** Hot process gas (air or nitrogen) is brought into contact with the atomized feed using a gas disperser for drying. The balance between temperatures, feed flow rate and droplet size controls the drying process.
- **Particle formation:** As the liquid evaporates from the droplet surface the solid nanoparticle that is formed falls into the bottom of the drying chamber.
- **Recovery:** The dried nanoparticles are separated from the exhaust gas using a cyclone separator.

Depending on the spray conditions and nature of formulation, the resulting powder may be filled into hard gelatin capsules or blended with extra granular excipients and compressed in to tablets. In the case of drugs which are acid liable, the capsule or tablet can be coated with enteric polymers to protect acid labile drug from gastric fluids.

An alternative way to convert nanosuspension into solid intermediate is suspension layering onto water soluble carriers. The binders that are necessary for layering must be added before the milling process. The suspension is layered at a predetermined rate on to the water-soluble carriers using a top spray fluid bed process. Top-spraying is the most well known process for drug layering. A top-spray fluid bed processor (FBP) has three components;

- An air-handling system, which can be equipped with humidification or de-humidification and dew-point control
- A product container and expansion chamber
- An exhaust system

The nanosuspension is sprayed into the fluid bed from the top against the air flow (counter current). The granules are dried as they move upward in the fluid bed, small droplets and low viscosity of the spray medium ensures that distribution is uniform resulting in granules with a narrow size distribution (Nekkanti et al., 2008; Basa et al., 2008). The critical process variables of the top-spray layering method include the suspension spray rate, inlet air temperature, fluidization air volume, process air humidity, and the atomization air pressure (Gu et al., 2004).

7. Characterization of drug nanoparticles

There are various techniques used for characterization of drug nanoparticles. There is no single method that can be selected as the "best" for analysis. Most often the method is chosen to balance the restriction on sample size, information required, time constraints and the cost of analysis. Following methods are used commonly for characterization of drug nanoparticles.

7.1 Particle size and size distribution

The characterization of particle size of nanosuspensions is done to obtain information about its average size, size distribution and change upon storage (e.g. crystal growth and/or agglomeration). Particle size distribution of drug nanoparticles can be measured using the following techniques;

7.1.1 Spectroscopy

As nanosuspensions usually comprise of submicron particles, the appropriate method used to evaluate particle size distribution is photon correlation spectroscopy (PCS). In PCS or dynamic light scattering analyses scattered laser light from particles diffusing in a low viscosity dispersion medium (e.g. water). PCS analyze the fluctuation in velocity of the scattered light rather than the total intensity of the scattered light. The detected intensity signals (photons) are used to measure the correlation function. The diffusion coefficient D of the particles is obtained from the decay of this correlation function. Applying Stokes-Einstein equation, the mean particle size (called z-average) can be calculated. In addition, a polydispersity index (PI) is obtained as a measure for the width of the distribution. The PI value is 0 in case particles are monodisperse. Incase of narrow distribution, the PI values vary between 0.10 - 0.20, values of 0.5 and higher indicate a very broad distribution (polydispersity). From the values of z-average and PI, even small increases in size of drug nanoparticles can be evaluated. The extent of increase in particle size upon storage is a measure of instability. Therefore, PCS is considered as a sensitive instrument to detect instabilities during long-term storage (Kerker, 1969).

7.1.2 Laser Diffraction

Laser Diffractometry (LD) developed around 1980 is a very fast and used routinely in many laboratories. The instrument is also used for quantifying the amount of microparticles present, which is not possible using PCS. LD analyses the Fraunhofer diffraction patterns generated by particles in a laser beam. The first instruments were based on the Fraunhofer theory which is applicable for particle sizes 10 times larger than the wavelength of the light used for generating the diffraction pattern. For particle less than 6.3 µm (in case of using a helium neon laser, wavelength 632.8 nm) in size, the Mie theory is used to obtain the correct particle size distribution. The Mie theory requires knowledge of the actual refractive index of particles and their imaginary refractive index (absorbance of the light by the particles). Unfortunately, for most of pharmaceutical solids the refractive index is unknown. However, laser diffractometry is frequently used as a preferred characterization method for nanosuspensions because of its "simplicity" (Zhang et al., 1992; Calvo et al., 1996).

7.2 Microscopy

Microscopy based techniques can be used to study a wide range of materials with a broad distribution of particle sizes, ranging from nanometer to millimeter scale. Instruments used for microscopy based techniques include optical light microscopes, scanning electron microscopes (SEM) transmission electron microscopes (TEM) and atomic force microscopes (AFM). The choice of instrument for evaluation is determined by the size range of the particles being studied, magnification, and resolution. However, the cost of analysis is also observed to increase as the size of the particles decreases due to requirements for higher magnification, improved resolution, greater reliability and, reproducibility. The cost of size analysis also depends upon the system being studied, as it dictates the technique used for specimen preparation and image analysis. Optical microscopes tend to be more affordable and comparatively easier to operate and maintain than electron microscopes but have limited magnification and resolution (Molpeceres et al., 2000; Cavalli et al., 1997).

The surface morphology of 'as-is' drug and spray dried nanoparticles for a sparingly soluble drug, Candesartan cilexetil, examined using scanning electron microscope (Hitachi S-520 SEM, Tokyo, Japan) is shown in Fig. 9. The scanning electron micrographs of "as-is" drug and drug nanoparticles as shown in these Figures illustrate the recrystallization of water-soluble carrier around the drug creating a highly hydrophilic environment preventing particle interaction and aggregation.

Fig. 9. SEM micrographs of "as-is" drug (left); spray-dried drug nanoparticles (right).

7.3 Solid-state properties

7.3.1 Differential Scanning Calorimetry (DSC)

Differential scanning calorimetry (DSC) is used to determine the crystallinity of drug nanoparticles by measuring its glass transition temperature, melting point and their associated enthalpies. This method along with X-ray powder diffraction (XRPD) described below is used to determine the extent to which multiple phases exist in the interior and their interaction following the milling process.

7.3.2 X-ray powder diffraction (XRPD)

X-ray powder diffraction (XRD) is a rapid analytical technique primarily used for phase identification of a crystalline material and can provide information on unit cell dimensions. X-ray diffraction is based on constructive interference of monochromatic X-rays and a crystalline sample. These X-rays generated by a cathode ray tube are filtered to produce monochromatic radiation, collimated to concentrate, and directed toward the sample. The interference obtained is evaluated using Bragg's Law to determine various characteristics of the crystal or polycrystalline material (Hunter et al., 1981).

7.4 Saturation solubility

Saturation solubility evaluations to ascertain drug nanoparticles are usually carried out in buffer media at different pH conditions using a shake flask method. In this method excess amount (100 mg/mL) of drug ("as-is" and dried suspension containing microparticles or nanoparticles) is added to 25 mL of buffer medium maintained at 37°C and shaken for a period up to 24 hours. The samples are filtered using 0.10 µm pore size Millex-VV PDVF filters (Millipore Corporation, USA) prior to analysis and concentrations determined using an HPLC method. The results from saturation solubility for "as-is", micronized and spray dried nanoparticles of Candesartan cilexetil used as a model drug is summarized in Table 3 to demonstrate the impact of particle size on saturation solubility.

Solvents	Solubility (mg/mL)		
	"as-is" drug*	Micronized drug*	Spray dried drug nanoparticles
0.1 N HCl	0.011	0.016	0.134
Acetate buffer pH 4.5	0.001	0.014	0.106
Phosphate buffer pH 6.8	0.001	0.012	0.105
Water	0.000	0.001	0.073

*Solubility was tested in respective solvents containing surfactant and Stabilizer.

Table 3. Saturation solubility of "as-is", micronized and nanoparticles of Candesartan cilexetil

The saturation solubility of Candesartan cilexetil nanoparticles is significantly higher than jet-milled particles and "as-is" drug at all pH conditions. These results clearly demonstrate that reduction in particle size to sub-micron or nanometer range affects saturation solubility resulting in enhancement of dissolution rate.

The effect of particle size of Candesartan cilexetil following oral administration in male Wistar is shown in Fig. 10. As seen there is a significant enhancement in the rate and extent of drug absorption for nanosuspension. The rate and extent of drug absorption showed a 2.5-fold increase in the area under the plasma concentration - time curve (AUC_{0-t}) and a 1.7-fold increase in the maximum plasma concentration (C_{max}) and, significant reduction in the time required (1.81 hours as compared to 1.06 hours) to reach maximum plasma concentration (T_{max}) when compared to the micronized suspension (Nekkanti et al., 2009b).

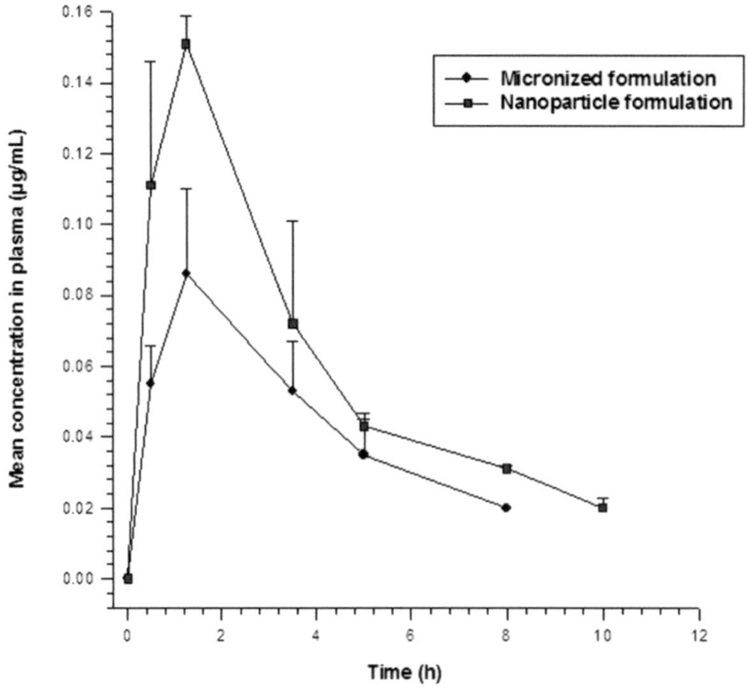

Fig. 10. Plasma concentration–time profiles following oral administration of micronized suspension and drug nanosuspension to male Wister rats

8. Conclusion

Enhancing solubility and dissolution rate of poorly soluble compounds correlates with improved pharmacokinetic (PK) profile. The approach herein can be extended to other BCS class II compounds where absorption is either solubility and/or dissolution limited. The manufacturing process used is relatively simple and scalable indicating general applicability of the approach to develop oral dosage forms of poorly soluble drugs. The enhanced bioavailability should translate into reduced dose, mitigate food effects, offer better dose delineation and result in faster onset of action that may translate into improved therapeutic outcome.

9. References

Basa, S.; Karatgi, P.; Muniyappan, T.; Raghavendra, P. & Pillai, R. (2008). Production and in vitro characterization of solid dosage form incorporating drug nanoparticles. *Drug Development and Industrial Pharmacy,* (11): 24–28.

Billon, A.; Bataille, B.; Cassanas, G. & Jacob, M. (2000). Development of spray-dried acetaminophen microparticles using experimental designs. *International Journal of Pharmaceutics,* 203, 159-168.

Bruno, JA.; Doty, BD. & Gustow, E. (1992). Method of grinding pharmaceutical substances. *US Patent* 5518187.

Calvo, P.; Vila-Jato, JL. & Alonso, MJ. (1996). Comparative in vitro evaluation of several colloidal systems, nanoparticles, nanocapsules, and nanoemulsions, as ocular drug carriers. *Journal of Pharmaceutical Sciences,* 85, 530–536.

Cavalli, R.; Caputo, O.; Carlotti, E.; Trotta, M.; Scarnecchia, C. & Gasco, MR. (1997). Sterilization and freeze-drying of drug-free and drug-loaded solid lipid nanoparticles. *International Journal of Pharmaceutics,* 148:47–54.

Chu, B. & Liu, T. (2000). Characterization of nanoparticles by scattering techniques. *Journal of Nanoparticle Research,* 2, 29–41.

Dhananjay, S.; Seshasai, M.; Gowtamrajan, K.; Giriraj, T.; Rajesh, V. & Srinivasarao, P. (2010). Optimization of formulation and process variable of nanosuspension: An industrial prospective. *International Journal of Pharmaceutics,* 402, 213–220.

Dressman, JB.; Amidon, GL.; Reppas, C. & Shah, VP. (1998). Dissolution testing as a prognostic tool for oral drug absorption: immediate release dosage forms. *Pharmaceutical Research,* 15 (1), 11-22.

Elaine, M.; Merisko-Liversidge. & Gary, L. (2008). Drug nanoparticles: Formulating poorly water-soluble compounds. *Toxicologic Pathology,* 36, 43-48.

Fisher RA. (1926). The Design of Experiments, *Oliver & Boyd,* London.

Gassmann, P.; List, M. Schweitzer, A. & Sucker, H. (1994). Hydrosols - alternatives for the parenteral application of poorly water-soluble drugs. *European Journal of Pharmaceutics and Biopharmaceutics,* 40, 64–72.

Gu, L.; Liew, CV. & Heng, PW. (2004). Wet spheronization by rotary processing: a multistage single-pot process for producing spheroids. *Drug Development and Industrial Pharmacy,* 30, 111-123.

Hunter RJ, Ed. (1981). Colloid Science: Zeta Potential in Colloid Science: Principles and Applications. London: *Academic Press.*

Illingworth, BD. (1972). Preparation of silver halide grains, *US Patent 3,655,394.*

Jens-Uwe, AH.; Junghanns. & Rainer, H. Muller. (2008). Nanocrystal technology, drug delivery and clinical applications. *International journal of Nanomedicine,* 3(3), 295-309.

Keck, CM. & Müller, RH. (2006). "Drug nanocrystals of poorly soluble drugs produced by high pressure homogenisation." *European Journal of Pharmaceutics and Biopharmaceutics,* 62(1), 3-16

Kerker, M. (1969). The scattering of light and other electromagnetic radiation. New York: *Academic Press.*

Kipp, JE.; Wong, JCT.; Doty, MJ. & Rebbeck, CL. (2003). Micro precipitation Method for preparing sub micron suspensions. *United States Patent 6,607,784,* Baxter International Inc. (Deerfield, IL), USA.

Lipinski, C. (2002). Poor aqueous solubility: an industry wide problem in drug discovery. *American Pharmaceutical Review,* 5, 82-5.

Lipinski, CA. (2004). Avoiding investment in doomed drugs, is poor solubility an industry wide problem?. *Current Drug Discovery,* 17–19.

List, M. & Sucker, H. (1988). Pharmaceutical colloidal hydrosols for injection. *GB Patent 2200048,* Sandoz Ltd.

Liversidge, GG. & Cundy, KC. (1995). Particle size reduction for improvement of oral bioavailability of hydrophobic drugs. *International Journal of Pharmaceutics,* 125, 91–97.

Mary C.; Till, Michele M.; Simkin. & Stephen Maebius. Nanotech Meets the FDA (2005): A success Story about the First Nanoparticulate drugs Approved by the FDA, *NANOTECHNOLOGY LAW & BUSINESS,* Volume 2.2

Merisko-Liversidge, E.; Liversidge, GG. & Copper, ER. (2003). Nanosizing: a formulation approach for poorly-water-soluble compounds. *European Journal of Pharmaceutical Sciences,* 18, 113–20.

Molpeceres, J.; Aberturas, MR. & Guzman, M. (2000). Biodegradable nanoparticles as a delivery system for cyclosporin: preparation and characterization. *Journal of Microencapsulation,* 17, 599–614.

Muller, RH. & Bohm, BHL. (2001a). *Dispersion techniques for laboratory and industrial scale processing, Wissenschaftliche Verlagsgesellschaft, Stuttgart.*

Müller, RH.; Becker, R. & Kruss, B. (1999). Pharmaceutical nanosuspensions for medicament administration as systems with increased saturation solubility and rate of solution. *US Patent 5858410.* USA.

Muller, RH.; Dingler, A.; Schneppe, T. & Gohla, S. (2000). Large-scale production of solid lipid nanoparticles (SLN) and nanosuspensions (DISSOCUBES). In:Wise, D.L. (Ed.), *Handbook of Pharmaceutical Controlled Release Technology. Marcel Dekker,* 359–376.

Muller, RH.; Jacobs, C. & Kayser, O. (2001b). Nanosuspensions as particulate drug formulations in therapy: Rationale for development and what we can expect for the future. *Advanced Drug Delivery Reviews,* 471, 3–19.

Müller, RH.; Jacobs, C. & Kayser, O. (2003). DissoCubes - a novel formulation for poorly soluble and poorly bioavailable drugs. In Rathbone MJ, Hadgraft J, Roberts MS (Eds). Modified-release drug delivery systems. New York: *Marcel Dekker,* 135–49.

Müller, RH. & Moschwitzer, JP. (2005). Method and apparatus for the production of ultrafine particles and coating of such particles. DE 10 2005 053 862.2 Application, Germany.

Musliner, WJ. (1974). Precipitation of metal salts. *US Patent* 3,790,387.

Nekkanti, V.; Karatgi, P.; Joshi, M. & Pillai, R. (2008) Developing nanoparticle formulations of poorly soluble drugs. *Pharmaceutical Technology Europe.* (11), 24–28.

Nekkanti, V.; Marella S.; Rudhramaraju, R. & Pillai, R. (2010). Media milling process optimization for manufacture of drug nanoparticles using design of experiments (DOE). *AAPS*, USA, Nov-10.

Nekkanti, V.; Muniyappan, T.; Marella, S.; Karatgi, P. & Pillai, R. (2009a). Spray-drying process optimization for manufacture of drug-cyclodextrin complex powder using design of experiments. *Drug Development and Industrial Pharmacy*, 5(10), 1219-29.

Nekkanti, V.; Pillai, R.; Venkateswarlu, V. & Harisudhan, T. (2009b). Development and characterization of solid oral dosage form incorporating candesartan nanoparticles. *Pharmaceutical Development Technology*, 14 (3), 290–298.

Otsuka, M. & Kaneniwa, N. (1986). Effect of seed crystals on solid state transformation of polymorphs of chloramphenicol palmitate during grinding. *Journal of Pharmaceutical Sciences*, 75, 506–511.

Rajesh Dubey. (2006). Pure Drug Nanosuspensions: Impact of nanosuspension technology on drug discovery & development. *Drug Delivery Technology*, Vol. 6, No. 5; 65-71.

Ranjita, S. & Rainer, H. Muller. (2010). Nanocrystals: Industrially feasible multifunctional formulation technology for poorly soluble actives, *International Journal of Pharmaceutics*, 399, 129-139.

Robertson, AJB. (1983). The development of ideas on heterogeneous catalysis. Progress from Davy to Langmuir. *Platinum Metals Review*, 27, 31–39.

Rogers, TL.; Johnston, KP. & Williams, III R.O. (2001). A comprehensive review: solution-based particle formation of pharmaceutical powders by supercritical or compressed fluid CO2 and cryogenic spray-freezing technologies. *Drug Development and Industrial Pharmacy*, 27, 1003–1015.

Sjostrom, B.; Kronberg, B. & Carlfors, J. (1993). A method for the preparation of submicron particles of sparingly water-soluble drugs by precipitation in oil-in-water emulsions. *Journal of Pharmaceutical Sciences*, 82, 579–583.

Sucker, H. & Gassmann, P. *(1994).* Improvements in pharmaceutical compositions, *GB-Patent 2269536A, Sandoz LTD. CH, GB.*

Tagne, PT.; Briancon, S. & Fessi, H. (2006). Spray-dried microparticles containing polymeric nanocapsules: formulation aspects, liquid phase interactions and particle characteristics. *International Journal of Pharmaceutics*, 325, 63-74.

Vaithiyalingam, S. & Khan, MA. (2002). Optimization and characterization of controlled release multi-particulate beads formulated with a customized cellulose acetate butyrate dispersion. *International Journal of Pharmaceutics*, 234, 179-193.

Violante, M.R. & Fischer, H.W. (1989). Method for making uniformly sized particles from water-insoluble organic compounds. *US Patent 4,826,689.*

Zhang, H. & Xu, G. (1992). The effect of particle refractive index on size measurement. *Powder Technology*, 70, 189-192.

7

Aptamer-Nanoparticle Bioconjugates for Drug Delivery

Veli C. Özalp and Thomas Schäfer
University of the Basque Country
Spain

1. Introduction

Drug delivery systems traditionally relied on passive diffusion mechanisms for targeting and releasing of therapeutically active molecules. The major problems associated with traditional delivery are poor specificity and dose-limited toxicity. Nanoparticles have found applicability in the development of novel drug delivery systems by easily overcoming toxicity problem. However, specificity of delivery has remained as a challenge. Developments in the methods of reaching to targeted tissue have lead to new and improved drug delivery platforms. Recently, active targeting has been incorporated by cell specific ligands such as antibodies, lectins, growth factor receptors. More recently, aptamers gained popularity in construction of novel actively targeted drug delivery systems (Ozalp *et al.*, 2011). Considerable proportions of aptamer-based delivery systems have been incorporated to a variety of nanomaterials in order to improve their specific targeting properties (Chen *et al.*, 2011; Zhou *et al.*, 2011).

A successful therapy starts with diagnosis and then continues with application of right dose of therapeutic molecules to the site of diseased tissues or cells. Specific detection of cancer cells is essential for early diagnosis. Cancer cell types are diverse even within the same cancer category. Fast and accurate diagnosis requires multiple specific probes which can profile more than one type of cancer cell at a given time on the same sample. In that respect, cell-based aptamer selection can supply multiple biorecognition probe molecules for any desired number of target cell type. Early diagnosis of cancer relies largely on the sensitivity of detection methods. Nanoparticle-bioconjugates has been designed to develop high sensitivity sensors. In this chapter, we will summarize the recent developments in selecting cell specific aptamer selection and development of aptamer-targeted nanoparticles for medical applications.

2. Selection of cell specific aptamers

Aptamers are increasingly recognized as the future affinity molecules that can readily incorporated in nanodevices for medical applications. The recent progress in cell-specific selection methodology reconfirms this premise by supplying highly specific and high affinity aptamers for any type of diseased cells. Aptamers have generally been shown to have affinities and specificity that are comparable to antibodies. They have the advantage of high stability at a variety of extreme physical conditions such as high temperature, high salt and ionic conditions with reversible denaturation properties (Deng et al., 2001; Liss et al., 2002). Cell-specific apramers have been obtained by following two types of Selex procedure; by using the traditional approach i) against purified protein targets or by new approaches ii) against whole cells. Using purified proteins as targets has the advantage of technical control over assessments of enrichment during Selex procedure. However, the second approach has a clear advantage when the marker target is unknown.

2.1 Aptamers selected for cell-specific proteins

Aptamers are selected through a combinatorial procedure to identify short nucleic acid sequences that recognize specific targets. The procedure is based on a repeated cycle of affinity and separation. First target molecules are brought together with a library of nucleic acids. The usual number of members in such libraries is 10^{14}-10^{18} random sequences. The library members will bind to target molecules if there is any recognition. The next step is the separation of bound and unbound library members through various chromatography techniques. It is commonly required to a cycle of 8 to 20 to enrich enough aptamer sequences which can be identified through sequencing since unspecific or common binder sequences will be favoured under the binding conditions. Detailed procedures on SELEX methods can be found elsewhere (Mayer, 2009). The simplest and direct way to obtain cell-specific aptamers is to isolate a marker protein that exists on the surface of target cell type and to use that protein as a target molecule in SELEX procedure. Many cell-specific aptamers have been selected using this strategy (Janas, 2011). The critical parameters for a successful selection are to identify cell-specific markers and to obtain the marker as pure starting material. Furthermore, the target that will be used in selection should be processed in such way that cell surface exposed regions of the molecule can be used during selection procedure. Several aptamers against cancer-related cells have been selected by using prufied proteins, including platelete-drived growth factor (PDGF) (Green *et al.*, 1996), vascular endothelial growth factor (VEGF) (Ruckman *et al.*, 1998), epidermal growth factor receptor 3 (HER3) (Chen *et al.*, 2003), transcription factor NFkB (Cassiday *et al.*, 2001), tenascin-C (Hicke *et al.*, 2001), prostate-specific membrane antigen (PSMA) (Lupold *et al.*, 2002), and ErbB in breast cancer (Kim *et al.*, 2011).

The aptamer against ErbB is a typical selex procedure for selecting cell specific aptamers by using purified surface proteins. Epidermal growth factor receptor family (ErbB family) of receptor tyrosine kinases plays major roles in formation and progression of human cancers. Under normal physiological, ErbB receptors play important roles in some basic cellular processes such as proliferation, differentiation, motility and apoptosis. Upon ligand binding, ErbB receptors form homo- and heterodimers leading to the activation of their tyrosine kinase domain and subsequent autocatalytic phosphorylation of specific tyrosine residues in the cytoplasmic region. Amplification and overexpression of ErbB tyrosine kinases is associated with many types of solid tumors including breast, ovarian, head and neck, gastric, bladder, colorectal, salivary, renal, prostate and lung cancers (Salomon et al., 1995). ErbB receptors consist of an extracellular region, a single transmembrane region and a cytoplasmic tyrosine kinase domain (Cho et al., 2002). Thus, ErbB has been a target in cancer treatment. 2′-fluorine-modified RNA aptamers for the extracellular region of the receptor has been recently selected for their recognition of breast cancer cells (Kim and Jeong, 2011).

2.2 Aptamers selected for whole cells (cell-Selex)

Cell specific aptamers can also be selected by using whole cells as target. Instead of selecting for a purified and defined target, aptamers can be selected by subtracting the background binding sequences after a selection cycle against the target. Whole-cell approach has been increasingly adapted more commonly for selecting new aptamers. Selection using live cell target means that the cell surface proteins will be in their native conformations by the right exposure as it would be in a real application. A panel of aptamers were selected for leukemia, myeloid leukemia, liver cancer, and lung cancer by using cell-Selex procedures. The aptamers selected by cell-selex resulted in high affinity specific aptamers, but more selection rounds compared to traditional selex were required (Fang et al., 2010). Figure 1 is a schematic representation of the cell-Selex approach. Complex targets such as tumor cells have been demonstrated to be compatible with the cell-Selex approach. The cell-selex procedure is easy to implement and faster than traditional Selex based methodologies. Most cell targets like cancer cells usually lack a clear marker protein which is highly specific for the type of application. Another important advantage of cell-selex is that the target

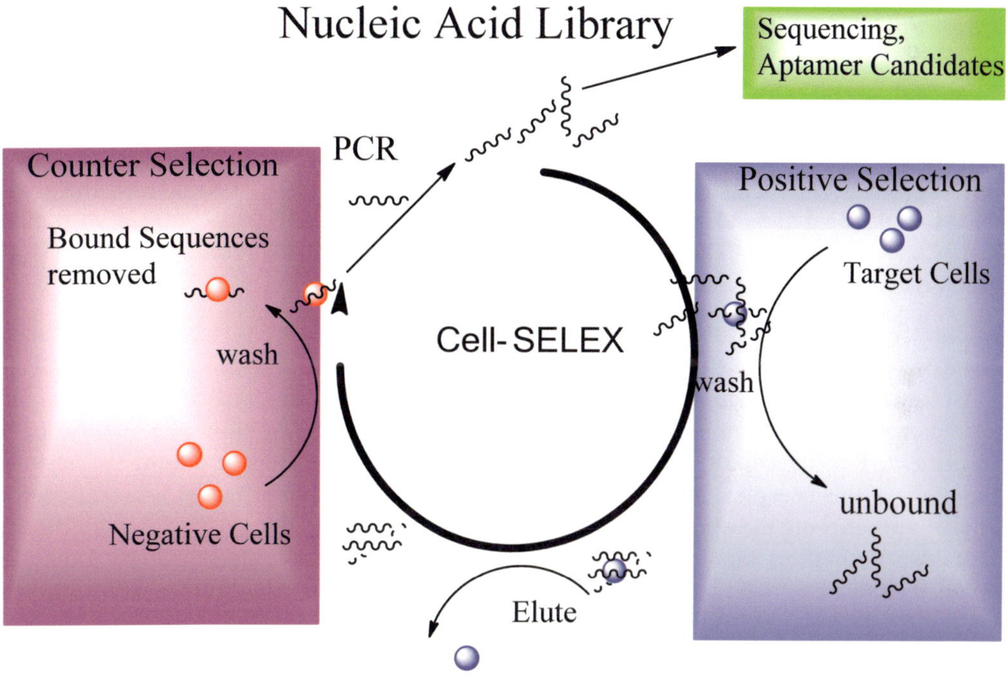

Fig. 1. Steps in a typical cell-Selex procedure. A single-stanraded DNA library was first incubated with target cells. Nano-binding sequences were washed away and the sequences bound on target cells were eluted and isolated. The recovered sequences were incubated with control cells (negative cells) to eliminate the sequences which bind to common epitopes present on both target and control cell surfaces. This step ensures the enrichment of target cell specific aptamers. The sequences specifically binding to target cells were then amplified to obtain the pool for the next round of selex. The enrichment of target cell binding sequences were evaluated using a binding assay and the sequences from last round were cloned and sequenced.

molecules are at their optimum exposed form in vivo. A traditional selex procedure using purified marker protein can select aptamers recognizing any subregion of the protein, which can be cytoplasmic site, transmembrane regions or the regions of target proteins can be hidden when they are sitting in membrane. Therefore, cell-Selex allows a comparative strategy to identify differences at molecular levels. In this section, cell-Selex will be explained by using specific examples. However, recent reviews on cell-selex can be referred for a more detailed information (Guo *et al.*, 2008; Fang and Tan, 2010).

2.2.1 Aptamers targeting tumor cells

Cancer cell types have been commonly a target for cell-Selex procedures in multiple example. An aptamer against RET receptors was selected by using cell line engineered for expressing RET tyrosine kinase receptors and background subtraction against non-expressing cell line (Cerchia et al., 2005). Another advantage is that cell-selex can be carried out without knowing identity of surface proteins in a target cell. Moreover, it is possible to select a panel of aptamers recognizing surface proteins which does not exist in counter selection target. Cultured T cell acute lymphoblastic leucemia (CCRF-CEM) cells was used as target and human Burkitt's lymphoma B-cells for counter-selex (Shangguan et al., 2006). Flow cytometry has been used to monitor the enrichment of aptamers after each cycle and several high affinity aptamers (Kd in nanomolar ranges) have been obtained with specific recognition properties for CCRF-CEM cells in a mixture of cells. The same procedure was also adapted to adherent liver tumor cells by first detaching the cells with a non-enzymatic solution (Shangguan et al., 2008). Six aptamers for Toledo and CCRF-CEM cell lines were tested for their ability to differentiate between real samples from leucemia patients (Shangguan et al., 2007). The results showed the ability of these aptamers to specifically detect molecular differences between patients.

The identity of target on the surface of cell to which an aptamer are selected using cell-selex procedures can be determined by using post-Selex procedures. TD05 is an aptamer selected against Ramos cells by using cell-Selex methodology. The aptamer was determined to bind to Ramos cell through immunoglobulin heavy mu chain which is the heavy portion of the IgM protein. B-cell receptor complex are expressed in Ramos cells and IgM is the major component of this complex. The identification of identity of aptamer target was achieved by chemically modifying the aptamers with a photoactive uracil derivative for covalent binding of aptamer with their targets. Once aptamers are bound to targets covalently, they are isolated through magnetic extraction and the target protein was determined through mass spectroscopy techniques (Mallikaratchy et al., 2007).

2.2.2 Aptamers targeting stem cells

Aptamers can be used for recognition of any given cell population. The aptamer-based capture of circulating cells has been proposed for applications in regenerative medicine and tissue engineering. Mesechymal stem cells from whole bone marrow were used to generate aptamers for isolation purposes by local injection methods (Guo *et al.*, 2006; Schaefer *et al.*, 2007). The aptamers recognizing osteoblasts were used to enrich them quickly and efficiently on titanium surfaces (Guo *et al.*, 2007). Stents coated with capturing agents for endothelial proginator cells (EPC) from circulating blood stream have been proposed to achieve fast enthelialization of the implanted stents and hence acceleration of the healing process of vascular tissue (Aoki *et al.*, 2005; Szmitko *et al.*, 2006). The capturing techniques was demonstrated with antibodies, peptides and aptamers (Wendel *et al.*, 2010). In one example, aptamers for high affinity for circulating porcine EPC was selected for EPC fishing (Hoffmann *et al.*, 2008). The application of EPC capture stents is a promising therapeutic approach to rapidly create an in vivo endothelialization of artificial surfaces. However, antibody based application has the problems due to the heterogeneity of EPC populations and the poor definition of EPC surface molecules. Therefore, the success of this approach depends on a better characterization and understanding of EPC biology and aptamers can offer the needed biorecognition molecules for the desired cell population for using in fishing approach as a successful therapy. Aptamer-based biological coatings which mimic the self healing potential of the patients' stem cell pool for vascular wall regeneration offer excitating new therapy strategies (Wendel, Avci-Adali et al., 2010). Aptamers recognizing EPC cells have also been demonstrated as one of the cell sources useful in cardiac stem cell therapy because they can contribute to neoangiogenesis and regenerate infarcted myocardium. This new therapeutic method based on aptamers was demonstrated that the transplantation of aptamer-isolated EPCs after myocardial infarction improves angiogenesis (Sobolewska *et al.*, 2010).

2.2.3 Aptamers targeting microbial cells

Another group of cell specific aptamers has been selected for microbial cells. Aptamer-based detection of pathogens is an effective diagnostic tool in food industry and medicine. Aptamers targeting viruses, protozoa, or bacteria have been selected for biosensor development (Karkkainen *et al.*, 2011). A subtype of human influenza virus has been detected by aptamers demonstrating the feasibility of using aptamers in detecting pathogenic viruses in contaminated environmental and food matrixes (Gopinath *et al.*, 2006). African trypanosomes are a specific class of protozoan organisms responsible for the parasitic disease sleeping sickness. Aptamers specific for the African trypanosomes were reported for a surface protein located within the flagellar pocket of the parasite (Hoffmann, Paul et al., 2008) and for a surface glycoprotein of the Trypnosoma brucei subsp. Brucei (Lorger *et al.*, 2003). Aptamers can target against some of the surface proteins of micro-organism, and the growth of the bacteria can be inhibited. The only example of therapeutic aptamer with antimicrobial properties has been demonstrated with virulent *Mycobacterium tuberculosis* (Chen *et al.*, 2007).

3. Drug delivering nanoparticles

Nanoparticles have many potential applications in medicine ranging from bioimaging to drug delivery. They have been synthesized from a variety of materials that could be degradable or stable, depending on the type of application. Although developing new types of materials is an exciting research area for future applications, pharmaceutical applications mostly rely on materials with history in human use to reduce the risks involved (Grama et al., 2011). Exposing malignant tumour cells that migrate to adjacent tissues or circulate in bloodstream is critical for early detection and effective therapy. Passive delivery refers to nanoparticles loaded with drugs are administered into body without any specific targeting. The improvement over conventional methods has been obtained by adjusting the release properties of the nanoparticles by selecting special materials during the synthesis of particles. On the other hand, increasing surface area to volume ratio for a delivery vehicle has huge benefits in improving solubility of biologically available drugs. Simple entrapment of bioactives inside nanoparticles has been shown to result in significantly higher bioavailability compared to their conventional administration methods (Mittal et al., 2007). Designed formulations for adjusting release rate are pursued with success in multiple examples (Cozar-Bernal et al., 2011; Tsai et al., 2011). For example, PLGA nanoparticles are biocompatible, synthesized at different grades, and safe (used for long time with humans). The release, degradation and elimination rates of PLGA nanopartiles can be tailored to desired parameters. On the other hand, nanoparticles can be functionalized for targetd delivery. Active delivery refers to incorporating a specific targeting mechanism in drug-loaded nanoparticle delivery vehicle.

Aptamers have been considered as potentially interesting targeting molecules for nanoparticles. The main reason is that conjugation of aptamers to other molecules or nanoparticles is a straight-forward procedure and the attachment does not change the affinity or specificity properties of aptamers. Early diagnosis can be achieved by systems of aptamer-nanoparticle conjugates because high affinity aptamers can be obtained for tumour cells. There have been many approaches in for detection of tumor cells that is based on mechanical forces, immunohistochemistry, magnetic cell sorting, and flow cytometry. The detection based on affinity interactions is expected to yield higher efficiency and sensitivity. Antibody based affinity sensors are often found to be high levels of cross-reactivity. There are also technical challenges in crosslinking antibodies on the surface of nanodevices (Wan et al., 2010). Thus, there is an increasing recognition for aptamers as promising utility in cancer diagnosis.

Three critical steps should be considered in designing an efficient aptamer-targeted nanoparticle drug therapy; targeting, internalization and degredation of the targeted cells by release of loaded drugs (Fig. 2).

4. Aptamer-based targeting of nanoparticles

Aptamers are biorecognition molecules with multiple properties that make them attractive targeting elements. Aptamers can bind to their targets with high affinity and specificity, comparable to antibodies. The *in vitro* selection methodology is the major advantage ensuring availability for any desired application. Moreover, nucleic acids have unique properties that are frequently exploited in construction of nanomachines . A combination of all of the above-mentioned properties makes aptamers a desirable targeting agent for novel nanoparticle-based drug delivery systems. Recent developments in cell based Selex technology have raised expectations for medical applications required targeting of whole living cells. Two major medical applications of aptamer-based targeting of nanoparticles are bioimaging and drug delivery.

4.1 Bioimaging applications

In vivo imaging is a valuable tool in clinical diagnostics and critical to develop ultrasensitive methods (Soontornworajit et al., 2011). Many techniques of imaging from whole organism (e.g. NMR) to specific molecular imaging (e.g. Fluorescence) are available and they are envisioned to be on the focus for developing cancer diagnosis, drug delivery, guided stem

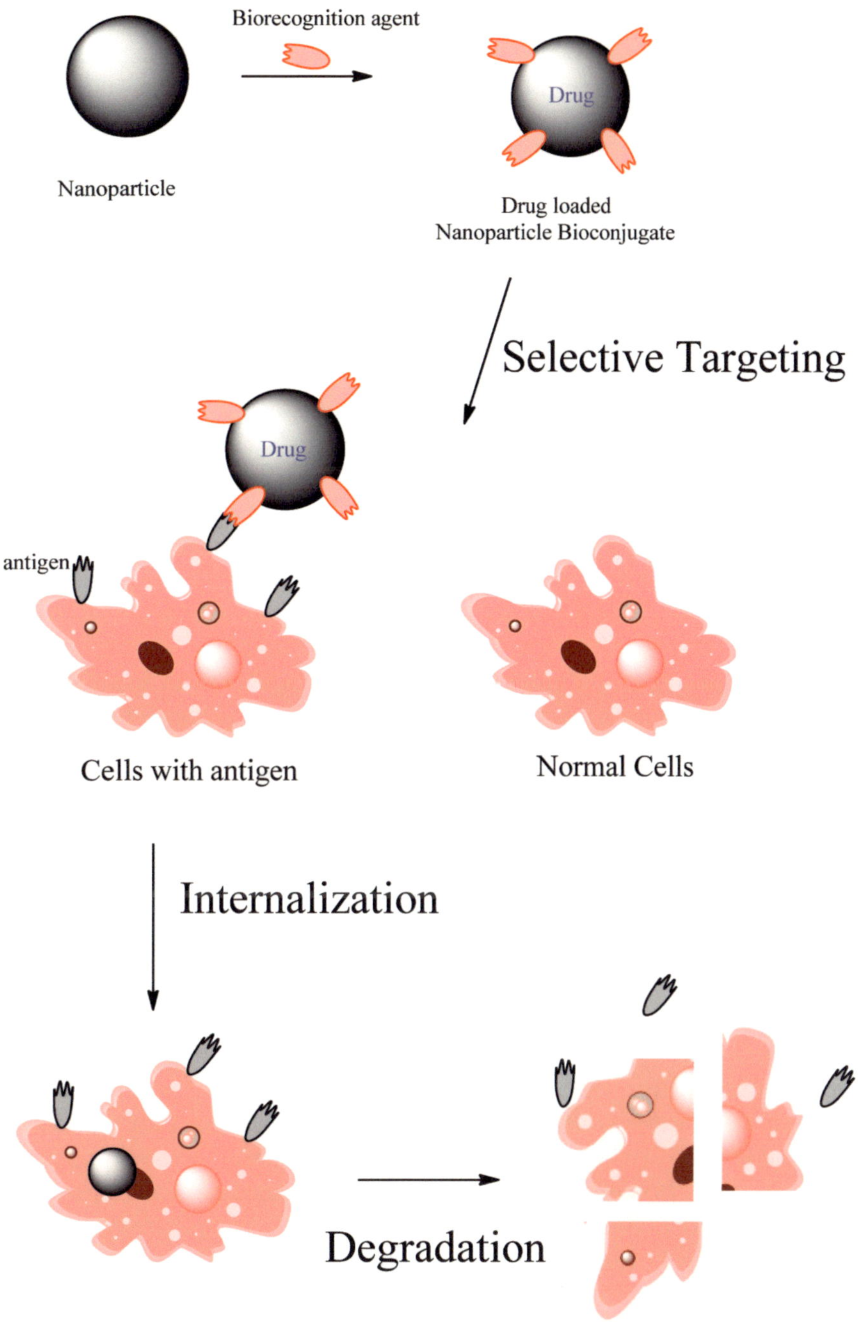

Fig. 2. A scheme of targeted Nanoparticle drug delivery. Drugs are loaded inside the nanoparticles that are conjugated to antigen-binding recognition element which bind specifically to antigen expressing cells. The nanoparticles are rapidly internalized once bound on the surface of diseased cells. The drugs encapsulated in the nanoparticles are released inside the cells and destroys the integrity of cells.

cell therapies, imaging of gene expression to monitor disease development, gene therapy, image guided surgery (Hahn et al., 2011). Nanoparticles for bioimaging should fullfill several requirements; being dispersable and stable in vivo, low non-specific binding, and high selectivity for their targets. There are four types of nanoparticles that are commonly used in conjugation with aptamers for biological imaging applications; 1) gold nanoparticles, 2) quantum dots, 3) silica nanoparticles and 4) magnetic nanoparticles.

A nonoparticle imaging and drug delivery system has been developed by conjugating A10 aptamers to quantum dots (Bagalkot et al., 2007). Fig. 3 shows a schematic representation of aptamer-quantum dot bioconjugate for drug delivery and imaging of delivered drugs as reported. The chemotherapeutic drug doxorubicin is an nucleic acid intercalating molecules. A bi-FRET system was prepared by attaching PSMA aptamers on the surface of quantum dot nanoparticles and doxorubicin was immobilized on the aptamer molecules by intercalation. Doxorubucin is a fluorescent molecule, but quenched by gold and nucleic acids. Therefore, the system targets the doxorubicin in non-fluorescent form to the surface of prostate cell. When the drug carrier system is internalized by the cells, doxorubicin is free from aptamer and quantum dots and become fluorescent, signalling the delivery of drug into cancer cells. Another example of quantum dot-aptamer conjugates in bioimaging was to identify and isolate subpopulations of tumor cells with high specificity. In another example, quantum dots were conjugated to liver hepatoma cell line aptamers through a sterptavidin-biotin binding. The quantum-dor-aptamer complexes were shown to specifically recognize MEAR cells specifically comapered to HeLa cells and BNL cells (Zhang et al., 2010).

Aptamer conjugated magnetic nanoparticles are proposed for in vivo isolation of specific cells and also for magnetic resonance imaging (MRI) techniques. Aptamer-magnetic nanoparticles were used to isolate mesenchymal stem cells and endothelian progenitor cells for isolating "ready to transplant" cells as promising methods (Schaefer, Wiskirchen et al., 2007; Sobolewska, Avci-Adali et al., 2010). Prostate cancer specific A10 aptamers conjugated to magnetic nanoparticles were used as magnetic resonance contrast agents for lower toxicity and stronger binding properties (Wang et al., 2008).

Silica nanoparticles can be readily loaded with fluorescent molecules and highly fluorescent, stable nanoparticles can be obtained. They are stable, hydrophilic, biocompatible, chemically inert. Another advantage is that organic dyes can be encapsulated in the silica with advantages of less photobleaching. Furthermore, aptamers can be easily attached to the surface of silica. Up to 100 fold increase in sensitivity has been obtained in flow cytometry applications with dye-doped silica nanopartiles (Estevez et al., 2009). In an effort to develop sensitive monitoring of multiple cancer cell lines, three different aptamers were conjugated to fluorescent doped silica nanoparticles (Chen et al., 2009). Human acute lymphoblastic leukemia cells specific aptamer, Ramos cells specific aptamer, and Toledo cells specific aptamer were used in developing multiple assays. Nanomaterials are often used for multivalent ligand binding purposes to increase the signal and sometimes enhance the binding affinity. For example, gold-silver nanorods were used as a platform for multiple aptamer immobilization. Up to 80 aptamers could be attached to rods for obtaining 26-fold higher affinity and 300-fold higher fluorescence (Huang et al., 2008).

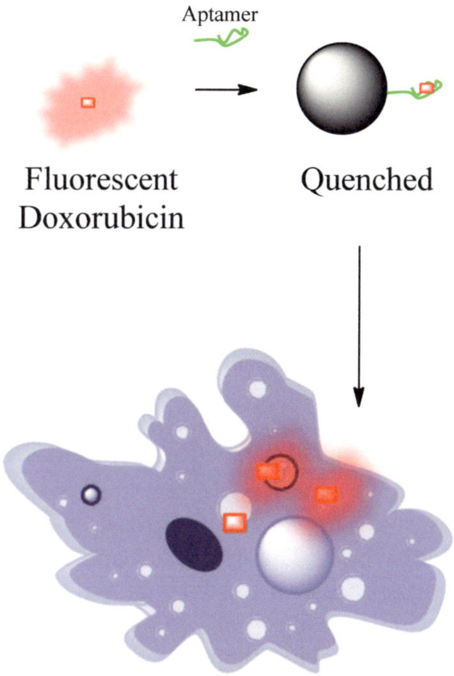

Fig. 3. Represantation of nanoparticle-aptamer based bi-FRET system for drug delivery and bioimaging. Doxorubicin is a fluophore and intercalates into nucleic acids. The fluorescence of doxorubicin is quenched by both gold and the aptamers. The quantum dot-aptamer-doxorubicin complex is targeted to tumor cells and internalized, and doxorubicin is released inside the cell, removing the quenching. The system targets the drug into diseased cells and the release of the drug can be monitored once it is delivered into desired site.

4.2 Aptamer-targeted drug delivery in nanoparticles

Aptamer-targeted delivery of drugs by using nanoparticles has similar advantages to bioimaging applications. The nanoparticles are specifically targeted to cell-types that are required for therapeutic molecules to act on. The drug delivery through nanoparticle encapsulation provides a vehicle for controlling the precise level and location of drugs in the body, reducing unwanted toxicity to healthy cells, lowering the doses needed. For example, drug encapsulated nanoparticles with a controlled release system can be developed and targeting to diseased cells can maximize the therapeutic efficacy and reduce undesired side effects (Langer, 2001).

Proof-of-concept nanoparticle-based vehicles have been developed for targeted drug delivery to prostate cancer cell surfaces with high specificity and efficiency. Once bound to tumor antigens, nanoparticles will be internalized through endocytosis. The toxic cargo of the nanoparticles will be released inside the tumour cells. Targeted delivery achieves a safe vehicle to destroy selectively the cancer cells. During the research of similar proof-of-concept research, teh nanoparticles commonly loaded with fleorescent molecules to study the pathways, the success of targeting and release rate of cargo.

4.2.1 Prostate cancer as model system

Aptamers potentially can direct nanoparticles to cancer cell antigens present on the surface of tumour cells. Therapeutic nanoparticles are specifically designed to encapsulate drugs and release the drugs in preregulated and predetermined way. Prostate cancer has been the first model disease for aptamer mediated nanoparticle targeting. Prostate specific membrane antigen protein (PMSA) is a well-known marker antigen which is over expressed on tumorigenic prostatic epithelial cells. Two RNA aptamer binding to extracellular portions of PSMA has been selected and shown to adhere selectively to PMSA positive cells (Lupold, Hicke et al., 2002). Later, an in vitro analysis device was developed in order to study the specific targeting properties of PMSA aptamer-nanoparticle conjugates (Farokhzad et al., 2005). The device was a model of prostate cancer targeting using a microfluidic cell under physiological fluid conditions similar to microvasculature. In this experiments, the microchannels were seeded with protate cancer epithelial cells expressing PMSA or non-expressing cells lines. Pegylated poly(lactic acid) (PLA) particles conjugated to A10 PMSA aptamer were tested for various experimental parameters mimicking the intraveneous administration of nanoparticles. The results confirmed that the particles were rapidly internalized into PMSA positive cells. The optimized aptamer-nanoparticle vehicle loaded with the chemotherapeutic drug docetaxel was later shown to reduce tumors upon injection into mice (Farokhzad et al., 2006; Cheng et al., 2007). This example was the first targeted delivery using nanoparticle-aptamer bioconjugates. The drug delivery was achieved to be highly specific for PMSA positive cells under *in vivo* contions. The same system of nanoparticle-aptamer bioconjugate was shown to deliver cisplatin drug to prostate cancer cells by improving simultaneously both tolerance and efficacy in mice (Kolishetti et al., 2010).

Nanomaterials for drug delivery are required to be stable and biocompatible. Current technology can supply materials that are capable of partially fulfilling these properties. Thus, combinations of surface coatings with complementary properties are proposed for developing better functioning delivery systems. In one example, gold nanorods were encapsulated within thin and uniform layer of silica shell for enhanced stability and and a layer of PEG was used for biocompatibility. The composite nanoparticles were functionalized with PSMA aptamers and tested for their ability to target the nanopartciles to prostate cancer cells (Hu et al., 2011).

4.2.2 Other examples

Vasculature targeted delivery is an attractive strategy due to the phenotypic changes on the endothelial cell surface associated with pathological conditions such as inflammation and angiogenesis. E-selectin is expressed in inflammed vaculator in advanced tumors and it can be used as marker for pathological vasculator. A high affinity thio-backbone modified aptamers against E-selectin were recently selected by using recombinant-expressed protein (Mann et al., 2010). E-selection aptamers were incorporated in liposomes, the targeting efficiency and pharmokinetic properties of delivery were investigated (Mann et al., 2011). CCRF-CEM recognizing aptamers (sgc8c) were conjugated to doxorubucin-doped silica nanoparticles and delivered leukemia cells with high specificity and efficiency (He et al., 2011).

4.2.3 Aptamer-based gating systems

The efficacy of drug therapy is determined by two parameters; efficient targeting of drug to the diseased site and the controlled release of the drug. The potential of aptamer-based targeting have been demonstrated in multiple examples as presented in above sections in order to improve specific delivery properties of drug carrying nanoparticles. The bioactive molecules in the nanoparticles were released with traditional mechanism such as internalization, endocytosis and degradation of nanomaterial once the nanoparticles attach to the surface of diseased cell surface. However, aptamers can be also exploited in controlling the release of drugs from nanoparticles. Two recent reports indicate to future potential of aptamers as targeting and gating agents in nanoparticle-based drug delivery. A snap-top type molecular gate using aptamer-gold nanoparticle complex to control the release of cargo of silica nanoparticles was shown to respond to aptamer target for stimuli-responsive release (Zhu *et al.*, 2011). In another report, switchable control of the gating in nanocarriers through aptamers was shown to control the release rate of active molecules in response to taptamer target stimuli (Ozalp *et al.*, 2011).

5. Conclusion

Nanoparticle-aptamer bioconjugates provide exciting prospects in medical nanotechnology for future disease treatments. The advancements in nanomaterial field together with cell-Selex procedures offer the controlled release polymer systems conjugated to aptamers tweaked to the any target diseased cell. Therefore, it is possible to produce a diverse range of specific and selective nanoparticle-aptamer bioconjugates. Drug delivery vehicles can improve the therapy of a myriad of important human diseases by using aptamer-nanoparticle bioconjugates. Aptamer-based targeting has already achieved progress in demonstrating the usability nanoparticle-aptamer bioconjugate systems to improve nanomaterial drug delivery vehicles in model animals. Next critical step for realization of potentials of bioconjugated nanoparticles will be clinical realization studies in near future. It is worth to indicate that with aptamers being in principle available for any kind of target (e.g., small metabolites, proteins, lipids, ions), it can be anticipated a straightforward transfer of proof-of-concept results to other applications when developing efficient drug-delivery systems. Finally, aptamer-based molecular gating of nanoparticle cargo can bring new approaches in drug delivery field by combining targeting and controlled release in the same mechanism.

6. Acknowledgment

A starting-grant from the European Research Council (No. 209842; "MATRIX") is gratefully acknowledged.

7. References

Aoki, J., P. W. Serruys, H. van Beusekom, A. T. Ong, E. P. McFadden, G. Sianos, W. J. van der Giessen, E. Regar, P. J. de Feyter, H. R. Davis, S. Rowland and M. J. Kutryk (2005). "Endothelial progenitor cell capture by stents coated with antibody against

CD34: the HEALING-FIM (Healthy Endothelial Accelerated Lining Inhibits Neointimal Growth-First In Man) Registry. J Am Coll Cardiol 45(10): 1574-9.

Bagalkot, V., L. Zhang, E. Levy-Nissenbaum, S. Jon, P. W. Kantoff, R. Langer and O. C. Farokhzad (2007). "Quantum dot - Aptamer conjugates for synchronous cancer imaging, therapy, and sensing of drug delivery based on Bi-fluorescence resonance energy transfer. Nano Letters 7(10): 3065-3070.

Cassiday, L. A. and L. J. Maher, 3rd (2001). "In vivo recognition of an RNA aptamer by its transcription factor target. Biochemistry 40(8): 2433-8.

Cerchia, L., F. Duconge, C. Pestourie, J. Boulay, Y. Aissouni, K. Gombert, B. Tavitian, V. de Franciscis and D. Libri (2005). "Neutralizing aptamers from whole-cell SELEX inhibit the RET receptor tyrosine kinase. Plos Biology 3(4): 697-704.

Cozar-Bernal, M. J., M. A. Holgado, J. L. Arias, I. Munoz-Rubio, L. Martin-Banderas, J. Alvarez-Fuentes and M. Fernandez-Arevalo (2011). "Insulin-loaded PLGA microparticles: flow focusing versus double emulsion/solvent evaporation. Journal of Microencapsulation 28(5): 430-441.

Chen, C. H. B., G. A. Chernis, V. Q. Hoang and R. Landgraf (2003). "Inhibition of heregulin signaling by an aptamer that preferentially binds to the oligomeric form of human epidermal growth factor receptor-3. Proceedings of the National Academy of Sciences of the United States of America 100(16): 9226-9231.

Chen, F., J. Zhou, F. Luo, A. B. Mohammed and X. L. Zhang (2007). "Aptamer from whole-bacterium SELEX as new therapeutic reagent against virulent Mycobacterium tuberculosis. Biochem Biophys Res Commun 357(3): 743-8.

Chen, T., M. I. Shukoor, Y. Chen, Q. A. Yuan, Z. Zhu, Z. L. Zhao, B. Gulbakan and W. H. Tan (2011). "Aptamer-conjugated nanomaterials for bioanalysis and biotechnology applications. Nanoscale 3(2): 546-556.

Chen, X., M. C. Estevez, Z. Zhu, Y.-F. Huang, Y. Chen, L. Wang and W. Tan (2009). "Using Aptamer-Conjugated Fluorescence Resonance Energy Transfer Nanoparticles for Multiplexed Cancer Cell Monitoring. Analytical Chemistry 81(16): 7009-7014.

Cheng, J., B. A. Teply, I. Sherifi, J. Sung, G. Luther, F. X. Gu, E. Levy-Nissenbaum, A. F. Radovic-Moreno, R. Langer and O. C. Farokhzad (2007). "Formulation of functionalized PLGA-PEG nanoparticles for in vivo targeted drug delivery. Biomaterials 28(5): 869-876.

Cho, H. S. and D. J. Leahy (2002). "Structure of the extracellular region of HER3 reveals an interdomain tether. Science 297(5585): 1330-1333.

Deng, Q., I. German, D. Buchanan and R. T. Kennedy (2001). "Retention and separation of adenosine and analogues by affinity chromatography with an aptamer stationary phase. Analytical Chemistry 73(22): 5415-5421.

Estevez, M. C., M. B. O'Donoghue, X. Chen and W. Tan (2009). "Highly Fluorescent Dye-Doped Silica Nanoparticles Increase Flow Cytometry Sensitivity for Cancer Cell Monitoring. Nano Research 2(6): 448-461.

Fang, X. H. and W. H. Tan (2010). "Aptamers Generated from Cell-SELEX for Molecular Medicine: A Chemical Biology Approach. Accounts of Chemical Research 43(1): 48-57.

Farokhzad, O. C., J. J. Cheng, B. A. Teply, I. Sherifi, S. Jon, P. W. Kantoff, J. P. Richie and R. Langer (2006). "Targeted nanoparticle-aptamer bioconjugates for cancer chemotherapy in vivo. Proceedings of the National Academy of Sciences of the United States of America 103(16): 6315-6320.

Farokhzad, O. C., A. Khademhosseini, S. Y. Yon, A. Hermann, J. J. Cheng, C. Chin, A. Kiselyuk, B. Teply, G. Eng and R. Langer (2005). "Microfluidic system for studying nanoparticles and microparticles the interaction of with cells. Analytical Chemistry 77(17): 5453-5459.

Gopinath, S. C., T. S. Misono, K. Kawasaki, T. Mizuno, M. Imai, T. Odagiri and P. K. Kumar (2006). "An RNA aptamer that distinguishes between closely related human influenza viruses and inhibits haemagglutinin-mediated membrane fusion. J Gen Virol 87(Pt 3): 479-87.

Grama, C. N., D. D. Ankola and M. N. V. R. Kumar (2011). "Poly(lactide-co-glycolide) nanoparticles for peroral delivery of bioactives. Current Opinion in Colloid & Interface Science 16(3): 238-245.

Green, L. S., D. Jellinek, R. Jenison, A. Ostman, C. H. Heldin and N. Janjic (1996). "Inhibitory DNA ligands to platelet-derived growth factor B-chain. Biochemistry 35(45): 14413-24.

Guo, K.-T., G. Ziemer, A. Paul and H. Wendel (2008). "CELL-SELEX: Novel Perspectives of Aptamer-Based Therapeutics. International Journal of Molecular Sciences 9(4): 668-678.

Guo, K. T., R. SchAfer, A. Paul, A. Gerber, G. Ziemer and H. P. Wendel (2006). "A new technique for the isolation and surface immobilization of mesenchymal stem cells from whole bone marrow using high-specific DNA aptamers. Stem Cells 24(10): 2220-31.

Guo, K. T., D. Scharnweber, B. Schwenzer, G. Ziemer and H. P. Wendel (2007). "The effect of electrochemical functionalization of Ti-alloy surfaces by aptamer-based capture molecules on cell adhesion. Biomaterials 28(3): 468-74.

Hahn, M. A., A. K. Singh, P. Sharma, S. C. Brown and B. M. Moudgil (2011). "Nanoparticles as contrast agents for in-vivo bioimaging: current status and future perspectives. Analytical and Bioanalytical Chemistry 399(1): 3-27.

He, X. X., L. Hai, J. Su, K. M. Wang and X. Wu (2011). "One-pot synthesis of sustained-released doxorubicin silica nanoparticles for aptamer targeted delivery to tumor cells. Nanoscale 3(7): 2936-2942.

Hicke, B. J., C. Marion, Y. F. Chang, T. Gould, C. K. Lynott, D. Parma, P. G. Schmidt and S. Warren (2001). "Tenascin-C aptamers are generated using tumor cells and purified protein. J Biol Chem 276(52): 48644-54.

Hoffmann, J., A. Paul, M. Harwardt, J. Groll, T. Reeswinkel, D. Klee, M. Moeller, H. Fischer, T. Walker, T. Greiner, G. Ziemer and H. P. Wendel (2008). "Immobilized DNA aptamers used as potent attractors for porcine endothelial precursor cells. J Biomed Mater Res A 84(3): 614-21.

Hu, X. and X. Gao (2011). "Multilayer coating of gold nanorods for combined stability and biocompatibility. Physical Chemistry Chemical Physics 13(21): 10028-10035.

Huang, Y.-F., H.-T. Chang and W. Tan (2008). "Cancer cell targeting using multiple aptamers conjugated on nanorods. Analytical Chemistry 80(3): 567-572.

Janas, T. (2011). "The Selection of Aptamers Specific for Membrane Molecular Targets. Cellular & Molecular Biology Letters 16(1): 25-39.

Karkkainen, R. M., M. R. Drasbek, I. McDowall, C. J. Smith, N. W. G. Young and G. A. Bonwick (2011). "Aptamers for safety and quality assurance in the food industry: detection of pathogens. International Journal of Food Science and Technology 46(3): 445-454.

Kim, M. Y. and S. Jeong (2011). "In Vitro Selection of RNA Aptamer and Specific Targeting of ErbB2 in Breast Cancer Cells. Oligonucleotides 21(3): 173-178.

Kolishetti, N., S. Dhar, P. M. Valencia, L. Q. Lin, R. Karnik, S. J. Lippard, R. Langer and O. C. Farokhzad (2010). "Engineering of self-assembled nanoparticle platform for precisely controlled combination drug therapy. Proceedings of the National Academy of Sciences of the United States of America 107(42): 17939-17944.

Langer, R. (2001). "Perspectives: Drug delivery - Drugs on target. Science 293(5527): 58-59.

Liss, M., B. Petersen, H. Wolf and E. Prohaska (2002). "An aptamer-based quartz crystal protein biosensor. Analytical Chemistry 74(17): 4488-4495.

Lorger, M., M. Engstler, M. Homann and H. U. Goringer (2003). "Targeting the variable surface of African trypanosomes with variant surface glycoprotein-specific, serum-stable RNA aptamers. Eukaryot Cell 2(1): 84-94.

Lupold, S. E., B. J. Hicke, Y. Lin and D. S. Coffey (2002). "Identification and characterization of nuclease-stabilized RNA molecules that bind human prostate cancer cells via the prostate-specific membrane antigen. Cancer Research 62(14): 4029-4033.

Mallikaratchy, P., Z. Tang, S. Kwame, L. Meng, D. Shangguan and W. Tan (2007). "Aptamer directly evolved from live cells recognizes membrane bound immunoglobin heavy mu chain in Burkitt's lymphoma cells. Molecular & Cellular Proteomics 6(12): 2230-2238.

Mann, A. P., R. C. Bhavane, A. Somasunderam, B. L. Montalvo-Ortiz, K. B. Ghaghada, D. Volk, R. Nieves-Alicea, K. S. Suh, M. Ferrari, A. Annapragada, D. G. Gorenstein and T. Tanaka (2011). Thioaptamer Conjugated Liposomes for Tumor Vasculature Targeting.

Mann, A. P., A. Somasunderam, R. Nieves-Alicea, X. Li, A. Hu, A. K. Sood, M. Ferrari, D. G. Gorenstein and T. Tanaka (2010). "Identification of Thioaptamer Ligand against E-Selectin: Potential Application for Inflamed Vasculature Targeting. Plos One 5(9).

Mayer, G. (2009). Nucleic Acid and Peptide Aptamers, Humana Press.

Mittal, G., D. K. Sahana, V. Bhardwaj and M. N. V. R. Kumar (2007). "Estradiol loaded PLGA nanoparticles for oral administration: Effect of polymer molecular weight and copolymer composition on release behavior in vitro and in vivo. Journal of Controlled Release 119(1): 77-85.

Ozalp, V. C., F. Eyidogan and H. A. Oktem (2011). "Aptamer-Gated Nanoparticles for Smart Drug Delivery. Pharmaceuticals 4(8): 1137-1157.

Ozalp, V. C. and T. Schafer (2011). "Aptamer-Based Switchable Nanovalves for Stimuli-Responsive Drug Delivery. Chemistry - A European Journal: DOI: 10.1002/chem.201101403.

Ruckman, J., L. S. Green, J. Beeson, S. Waugh, W. L. Gillette, D. D. Henninger, L. Claesson-Welsh and N. Janjic (1998). "2 '-fluoropyrimidine RNA-based aptamers to the 165-amino acid form of vascular endothelial growth factor (VEGF(165)) - Inhibition of receptor binding and VEGF-induced vascular permeability through interactions requiring the exon 7-encoded domain. Journal of Biological Chemistry 273(32): 20556-20567.

Salomon, D. S., R. Brandt, F. Ciardiello and N. Normanno (1995). "Epidermal Growth Factor-Related Peptides and Their Receptors in Human Malignancies. Critical Reviews in Oncology/Hematology 19(3): 183-232.

Schaefer, R., J. Wiskirchen, K. Guo, B. Neumann, R. Kehlbach, J. Pintaske, V. Voth, T. Walker, A. M. Scheule, T. O. Greiner, U. Hermanutz-Klein, C. D. Claussen, H. Northoff, G. Ziemer and H. P. Wendel (2007). "Aptamer-based isolation and subsequent imaging of mesenchymal stem cells in ischemic myocard by magnetic resonance imaging. Rofo-Fortschritte Auf Dem Gebiet Der Rontgenstrahlen Und Der Bildgebenden Verfahren 179(10): 1009-1015.

Shangguan, D., Z. C. Cao, Y. Li and W. Tan (2007). "Aptamers evolved from cultured cancer cells reveal molecular differences of cancer cells in patient samples. Clinical Chemistry 53(6): 1153-1155.

Shangguan, D., Y. Li, Z. Tang, Z. C. Cao, H. W. Chen, P. Mallikaratchy, K. Sefah, C. J. Yang and W. Tan (2006). "Aptamers evolved from live cells as effective molecular probes for cancer study. Proceedings of the National Academy of Sciences of the United States of America 103(32): 11838-11843.

Shangguan, D., L. Meng, Z. C. Cao, Z. Xiao, X. Fang, Y. Li, D. Cardona, R. P. Witek, C. Liu and W. Tan (2008). "Identification of liver cancer-specific aptamers using whole live cells. Analytical Chemistry 80(3): 721-728.

Sobolewska, B. W., M. Avci-Adali, B. Neumann, T. O. Greiner, A. Stolz, D. Bail, T. Walker, A. Scheule, G. Ziemer and H. P. Wendel (2010). "A novel method for isolation of endothelial progenitor cells for cardiac stem cell therapy. Kardiochirurgia I Torakochirurgia Polska 7(1): 61-65.

Soontornworajit, B. and Y. Wang (2011). "Nucleic acid aptamers for clinical diagnosis: cell detection and molecular imaging. Analytical and Bioanalytical Chemistry 399(4): 1591-1599.

Szmitko, P. E., M. J. Kutryk, D. J. Stewart, M. H. Strauss and S. Verma (2006). "Endothelial progenitor cell-coated stents under scrutiny. Can J Cardiol 22(13): 1117-9.

Tsai, Y.-M., W.-C. Jan, C.-F. Chien, W.-C. Lee, L.-C. Lin and T.-H. Tsai (2011). "Optimised nano-formulation on the bioavailability of hydrophobic polyphenol, curcumin, in freely-moving rats. Food Chemistry 127(3): 918-925.

Wan, Y. A., Y. T. Kim, N. Li, S. K. Cho, R. Bachoo, A. D. Ellington and S. M. Iqbal (2010). "Surface-Immobilized Aptamers for Cancer Cell Isolation and Microscopic Cytology. Cancer Research 70(22): 9371-9380.

Wang, A. Z., V. Bagalkot, C. C. Vasilliou, F. Gu, F. Alexis, L. Zhang, M. Shaikh, K. Yuet, M. J. Cima, R. Langer, P. W. Kantoff, N. H. Bander, S. Jon and O. C. Farokhzad (2008). "Superparamagnetic iron oxide nanoparticle-aptamer bioconjugates for combined prostate cancer imaging and therapy. Chemmedchem 3(9): 1311-1315.

Wendel, H. P., M. Avci-Adali and G. Ziemer (2010). "Endothelial progenitor cell capture stents - hype or hope? International Journal of Cardiology 145(1): 115-117.

Zhang, J., X. Jia, X.-J. Lv, Y.-L. Deng and H.-Y. Xie (2010). "Fluorescent quantum dot-labeled aptamer bioprobes specifically targeting mouse liver cancer cells. Talanta 81(1-2): 505-509.

Zhou, J. H. and J. J. Rossi (2011). "Cell-Specific Aptamer-Mediated Targeted Drug Delivery. Oligonucleotides 21(1): 1-10.

Zhu, C. L., C. H. Lu, X. Y. Song, H. H. Yang and X. R. Wang (2011). "Bioresponsive Controlled Release Using Mesoporous Silica Nanoparticles Capped with Aptamer-Based Molecular Gate. Journal of the American Chemical Society 133(5): 1278-1281.

8

Nanoparticles Based on Modified Polysaccharides

Hassan Namazi[1,2,*], Farzaneh Fathi[2] and Abolfazl Heydari[2]
[1]*Research Center for Pharmaceutical Nanonotechnology, Tabriz University of Medical Science, Tabriz,*
[2]*Research Laboratory of Dendrimers and Nanopolymers, University of Tabriz, Tabriz*
Iran

1. Introduction

Nanoparticles may be comprised of several kind materials being classified as non-degradable and biodegradable. Biodegradable systems have an advantage over non-degradable systems in that they are non-toxic, biotolerabl, biocompatible, biodegradable, and water-soluble. Among these systems, the role of natural polysaccharides in developing prepared nanoparticles has significantly increased (Zhang *et al.*, 2011; Yang *et al.*, 2008a; Aumelas *et al.*, 2007; Leonard *et al.*, 2003).

On the other hand, polysaccharides are the most abundant macromolecules in the biosphere. The complex carbohydrates constituted of monosaccharides joined together by glycosidic bonds are often one of the main structural elements of plants and animals exoskeleton (cellulose, carrageenan, chitosan, chitin, etc.) or have a key role in the plant energy storage (starch, paramylon, etc.) (Aminabhavi *et al.*, 1990). Polysaccharides have a large number of reactive groups, a wide range of molecular weight, varying chemical composition, which contribute to their diversity in structure and in property. The amphiphilic nature imparted upon polysaccharides after modification gives them a wide and interesting application spectrum, for instance as rheology modifiers, emulsion stabilizers, surface modifiers for liposomes and nanoparticles and as drug delivery vehicles (Sinha and Kumria, 2001; Gurruchaga *et al.*, 2009; Chen *et al.*, 2003a; Durand *et al.*, 2002; Gref *et al.*, 2003). Recently, the hydrophobically modification of polysaccharides has been received increasing attention because they can form self-assembled nanoparticles for biomedical uses. In the aqueous phase, the hydrophobic cores of polymeric nanoparticles are surrounded by hydrophilic outer shells. Thus, the inner core can serve as a nano-container for hydrophobic drugs. Starch, chitosan, dextran, cyclodextrin, cellulose and pullulan are polysaccharides that have been modified with various reactants and after the modification step the nanoparticles based on modified polysaccharides were prepared with using various methods (Onyuksel *et al.*, 2003; Aumelas *et al.*, 2007; Ragauskas *et al.*, 2007; Kwon, 2003; Namazi and Dadkhah, 2010; Namazi. and Mosadegh, 2011).

Nanoparticles are defined as particulate dispersions or solid particles with a size in the range of 10-1000nm (P., 1988; Hamidi *et al.*, 2008). Depending upon the method of preparation, nanoparticles, nanospheres or nanocapsules can be obtained. These nano-sized objects, e.g., "nanoparticles", take on novel properties and functions such small size, modified surface, improved solubility and multi-functionality. The drug is dissolved, entrapped, encapsulated or attached to a nanoparticle matrix. Nanoparticles based on modified polysaccharides have been prepared most frequency by these methods: solvent evaporation method, spontaneous emulsification or solvent diffusion method, self-assembly of hydrophobically modified and dialysis method (Kim *et al.*, 2001; Aumelas *et al.*, 2007; Sun *et al.*, 2006; Couvreur, 1998). Modified polysaccharide could be used as stabilizers to produce stable hydrophilic nanoparticles by the o/w emulsion/evaporation technique. Modified polysaccharides were shown to exhibit surface active properties and to act as efficient emulsion stabilizers. Surface modified colloidal carriers such as nanoparticles are able to modulate the biodistribution of the loaded drug when given intravenously, but also to control the absorption of drugs administered by other routes (Durand *et al.*, 2004).

This review presents the several mechanisms to prepare polysaccharides-based nanoparticles after discusses about modification of polysaccharides with various agents. Also characterization of nanoparticles such as size particles, surface coverage, colloidal stability and enzyme degradability have been described and also provided are examples of use of the polysaccharide nanoparticles and their derivatives as medical applications.

2. Polysaccharides

Polysaccharides with polymeric carbohydrate structures, formed from repeating units joined together with glycosidic bonds. Their structures are often linear, but may contain various degrees of branching. In nature, polysaccharides have various resources from algal origin, plant origin, microbial origin and animal origin .Polysaccharides have a general formula of $C_x(H_2O)_y$ where x is usually a large number between 200 and 2500. Considering that the repeating units in the polymer backbone are often six-carbon monosaccharides, the general formula can also be represented as $(C_6H_{10}O_5)_n$ where $40 \leq n \leq 3000$.(Aminabhavi *et al.*, 1990)

2.1 Starch

Starch is made up of two types of polymers: amylose and amylopectin. Amylose is a linear homopolymer of α-1,4-linked glucose. Amylose may have a low level of branching with a α-1,6-linkage (Fig 1). Amylose makes up ~35% of starch. In solution amylose forms hydrogen bound with other amylase molecules to yield rigid gels. Amylopectin is highly branched form of "amylose". The linear α-1,4-linked glucose backbone is branched at every ~20 residues by an α-1,6-linkage which is extended by α-1,4-linked linkages (Namazi and Dadkhah, 2008; Della Valle *et al.*, 1998; Namazi *et al.*, 2009; Namazi and Dadkhah, 2010)

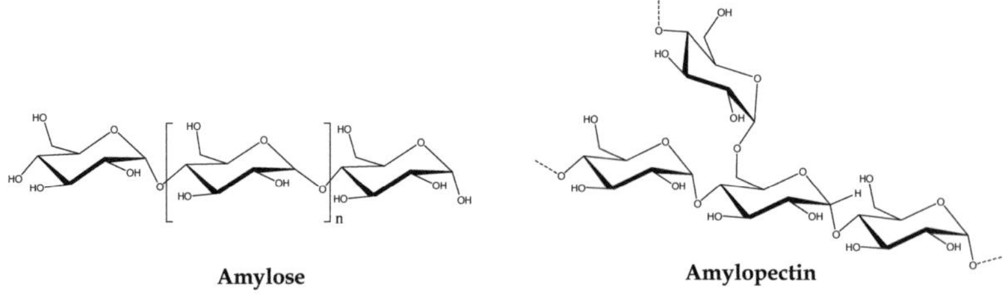

Fig. 1. Chemical structure of the starch

2.2 Chitosan and chitin

Chitosan is a linear polysaccharide composed of randomly distributed β-(1-4)-linked D-glucosamine (deacetylated unit) and N-acetyl-D-glucosamine (acetylated unit) (Fig 2). It has a number of commercial and possible biomedical uses. Chitosan is produced commercially by deacetylation of chitin, which is the structural element in the exoskeleton of crustaceans (such as crabs and shrimp) and cell walls of fungi (Thanou *et al.*, 2005; Tharanathan and Ramesh, 2003; Yuan and Zhuangdong, 2007). Chitin $(C_8H_{13}O_5N)_n$ is a long-chain polymer of a *N*-acetylglucosamine, a derivative of glucose (Fig 2), and is found in many places throughout the natural world. It is the main component of the cell walls of fungi, the exoskeletons of arthropods such as crustaceans (e.g., crabs, lobsters and shrimps) and insects, the radulas of mollusks, and the beaks of cephalopods, including squid and octopuses. In terms of structure, chitin may be compared to the polysaccharide cellulose and, in terms of function, to the protein keratin. Chitin has also proven useful for several medical and industrial purposes (Kumar, 2000; Kurita, 2001).

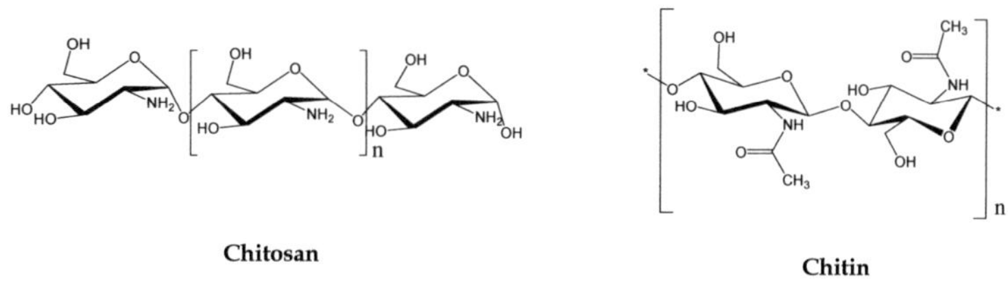

Fig. 2. Chemical structure of the chitosan and chitin

2.3 Dextran

Dextran is a polysaccharide consisting of glucose molecules coupled into long branched chains, mainly through a 1,6- and some through a 1,3-glucosidic linkages as shown in Fig 3. Dextrans are colloidal, hydrophilic and water-soluble substances, inert in biological systems. It is used medicinally as an antithrombotic (anti-platelet), to reduce blood viscosity, and as a volume expander in anemia (Bertholon *et al.*, 2006; Durand *et al.*, 2004).

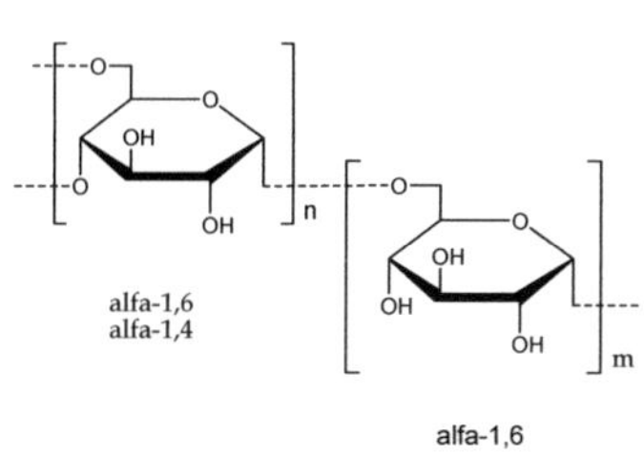

Fig. 3. Chemical structure of the dextran

2.4 Pullulan

Pullulan is a polysaccharide polymer consisting of maltotriose units, also known as α-1,4- ; α-1,6-glucan (Fig 4). Three glucose units in maltotriose are connected through an α-1,4-glycosidic bond, whereas consecutive maltotriose units are connected to each other by an α-1,6 glycosidic bond. Pullulan is produced from starch by the fungus Aureobasidium pullulans (Bataille *et al.*, 1997; Glinel *et al.*, 1999).

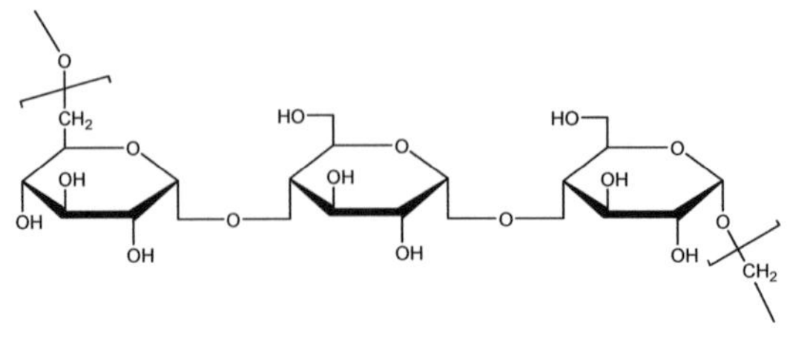

Fig. 4. Chemical structure of the pullulan

2.5 Cyclodextrins

Cyclodextrins (CDs), also using the name cycloamyloses, cyclomaltoses, or Schardinger dextrins, are natural macrocycles connected through α-(1-4)-linked glucose units in a rigid $^4C_1$ chair conformation. CDs can be produced through the enzymatic degradation of starch derived from potatoes, corn, rice or other sources. The number of glucose units per CD ring varies from 6-13, (Saenger *et al.*, 1998; Larsen, 2002; Ueda, 2002; Hennink *et al.*, 2009; Namazi and Kanani, 2009) as the enzyme produces a range of oligosaccharides. Because of steric factors, cyclodextrins constructed from less than six glucose units such as the five-membered cyclic oligomer, cyclomaltopentaose, has been obtained by chemical synthesis in small quantities.(T. Nakagawa *et al.*, 1994) A chemical synthesis for other CDs has been reported, but it is too tedious for commercial production of cyclodextrins.(Ogata and Takahashi, 1995) The most common CDs contain 6, 7, and 8 D-glucose units and are known as αCD, βCD, and γCD, respectively,(Saenger, 1980) (Figure 5), while greater cyclodextrins have been reported as well.(Larsen *et al.*, 1998; French *et al.*, 1965; Fujiwara *et al.*, 1990; Miyazawa *et al.*, 1995)

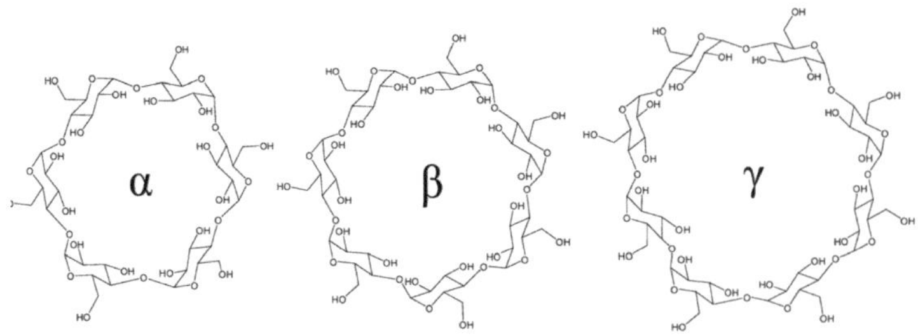

Fig. 5. Chemical structure of the cyclodextrins

2.6 Cellulose

Cellulose is an organic compound with the formula $(C_6H_{10}O_5)_n$, a polysaccharide consisting of a linear chain of several hundred to over ten thousand β(1→4) linked D-glucose units (fig. 6). Cellulose is the structural component of the primary cell wall of green plants, many forms of algae and the oomycetes. Some species of bacteria secrete it to form biofilms. Cellulose is the most common organic compound on Earth (Hinrichsen *et al.*, 2000; Riedel and Nickel, 1999; Gassan and Bledzki, 1999).

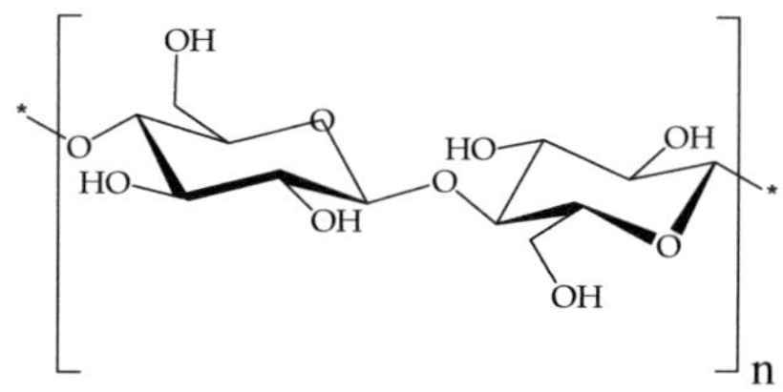

Fig. 6. Chemical structure of the cellulose

3. Modified polysaccharides (MP) for preparation of their nanoparticles

Amphiphillic polysaccharides consisting of hydrophilic and hydrophobic fragments have been modified because they can form self-assembled nanoparticles and they show unique physicochemical characteristics such as a nanoparticle structure and thermodynamic stability. Natural biopolymers have various advantages, such as availability from replenish able agricultural or marine food resources, biocompatibility, and biodegradability, therefore leading to ecological safety and the possibility of preparing a variety of chemically or enzymatically modified derivatives for specific end uses. Recently, there has been considerable interest in developing modified derivatives of polysaccharides for biodegradable nanoparticles. These nanoparticles have shown the following advantages for biomedical applications such as drug protection and ability to control the drug release. Polysaccharides have a number of positive characteristics such biotolerability, biodegradability, protein rejecting ability, receptor interaction through specific sugar moieties, and abundance of functional groups for modification or functionalization (Couvreur *et al.*, 2004). The amphiphilic character imparted upon polysaccharides after hydrophobic modification gives them a wide and interesting use spectrum, for instance as rheology modifiers, emulsion stabilizers (Chen *et al.*, 2003a; Durand *et al.*, 2002), surface modifiers for liposomes and nanoparticles (Vyas and Sihorkar, 2001) and as drug delivery vehicles (Rodrigues *et al.*, 2003; Leonard *et al.*, 2003).

3.1 Modified starch

Starch is one of the polysaccharide that it has been modified with various reactants for preparation of nanoparticles. The use of starch nanoparticles is receiving a significant amount of notice because of the plentiful availability of natural polymer, inexpensive, renewability, biocompatibility, biodegradability and nontoxicity. Chemical modification of starch has been widely studied for producing modified starch by way of chemical reaction with hydroxyl groups in the starch molecule. Starch esters are a kind of modified starches which are synthesized with various reactants such as acid anhydrides octenyl succinic anhydride (OSA), dodecenyl succinic anhydride (DDSA) fatty acids and fatty acid chlorides (Tukomane and Varavinit, 2008; Wang *et al.*, 2007a; Borredon *et al.*, 1999; Fowler *et al.*, 2002). Hydroxyethyl starch was esterified with the long chain fatty acids under mild reaction conditions using DCC and DMAP (Mader *et al.*, 2007). The synthesis of modified hydrophobic starch using fatty acids was done by means of potassium persullphate as catalyst in DMSO (Abraham and Simi, 2007). Several substituted starches were prepared by acylation of starch with fatty acid chlorides in organic solvents, such as pyridine or dimethylacetamide (Kapusniak and Siemion, 2007; Wang *et al.*, 2008). Hydrophilic

Starch	Grafting agent	References
Amylopectin (from waxy corn)	Lactic acid	(Hong-Wei Lua and Li-Ming Zhanga, 2011)
Modification: Amylopectin and aqueous lactic acid (LA) were added to a three-necked flask equipped. After the stirring at 75 ^{o}C for 30 min, the temperature of the reaction system was thermostated to be 100 ^{o}C.Then a required amount of $Sn(Oct)_2$ was added to the flask. Then the product was further purified by Soxhlet extraction to remove completely the unreacted LA monomer as well as PLA homopolymer that may be formed during the reaction.		
Amylopectin-rich waxy maize starch	Stearic acid 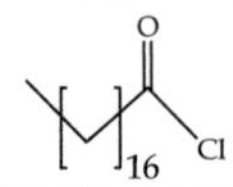	(Dufresne *et al.*, 2004; Dufresne *et al.*, 2006)
Modification: Chemical modification of the nanoparticles was performed in a round-bottomed reaction flask under a nitrogen atmosphere while constantly stirring with amagnetic stir bar. The stearate modification was performed by the reaction of dry starch nanocrystals with stearic acid chloride in methyl ethyl ketone.		
Amylomaize starch	n-Butanol	(Lim and Kim, 2009)
Modification: The amylomaize starch (0.5%, w/v) was dispersed DMSO solution with heating and stirring in a boiling water bath, and then magnetic-stirred at room temperature for 24 h. An aliquot of the starch solution was allowed to gravimetrically pass through a membrane filter into the bottom compartment filled with n-butanol. The precipitate in the butanol layer was collected by centrifugation, and then washed three timesin the n-butanol.		
Cassava starch	Monochloroacetic acid (MAC) $ClCH_2COONa$	(Wu *et al.*, 2011)
Modification: Cassava starch in anhydrous ethanol was placed in a glass reactor. An aqueous solution of sodium hydroxide was added drop wise to the starch–solvent mixture under stirring until the whole amount of sodium hydroxide were added. Then, the solution of MAC was added drop wise to the starch–solvent–sodium hydroxide mixture under ultrasonic irradiation.		
Waxy corn starch	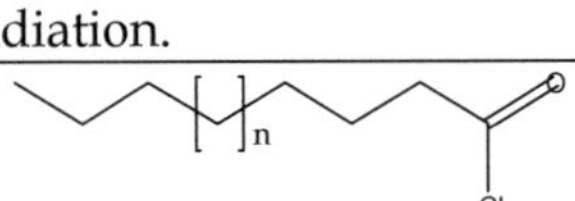	(Fowler *et al.*, 2004; Namazi and Dadkhah, 2010)
Modification: Starch esterification was carried out in two steps. In the first step, starch nanocrytstals dispersed in the reaction medium were alkali treated at room temperature with mechanical stirring under an atmosphere of N_2 for 10 min and in the second step, 0.5 mol equivalents of the required acid chloride was added drop wise and the reaction mixture was stirred for 20 min.		
Cassava starch	$CH_3(CH_2)_7CH{=}CH(CH_2)_7COOH$	(Abraham and Simi, 2007)
Modification: For the graft copolymerization, about 1g starch was dissolved in 10 ml DMSO and was taken in a round bottom flask. Oleic acid, weighing was added and potassium per sulphate was the catalyst.		

Hydroxyethyl starch (HES)	Fatty acid	(Mader *et al.*, 2007)
Modification: HES was dried for 2 h before dissolving in 20 mL of dry DMSO. To the solution were added the fatty acid, DCC, and DMAP, and they dissolved for 24 h. The formed precipitate (dicyclohexyl urea, DCU) was removed by filtration, and the filtrate was added to 200 mL of precipitating solvent mixture.		
Potato starch	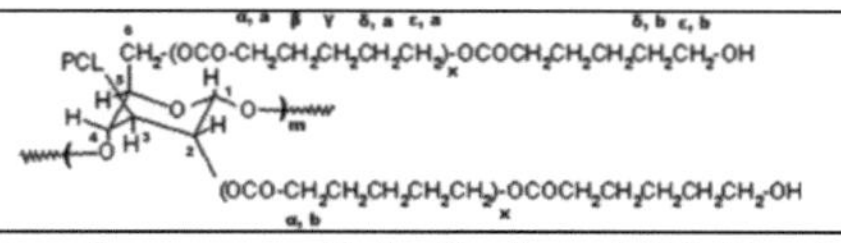	(Namazi and Dadkhah, 2008; Dufresne *et al.*, 1996)
Modification: A mixture of starch nanoparticle (1 g) and CL (2 mL) was first added to flask. A determined amount of $Sn(Oct)_2$ of total amount of reagents was then introduced via a conditioned syringe. Polymerization was stopped by fast cooling to room temperature.		

Table 1. Functional molecules for modification of starch

amylopectin was modified by grafting hydrophobic poly (lactic acid) chains (Hong-Wei Lua and Li-Ming Zhanga, 2011). Since 1950, considerable effort has gone into hydrophobically modified derivatives of hydrophilic polysaccharides(Namazi *et al.*, 2011).

Recent studies have been carried out to investigate the synthesis and the application of polysaccharide-based nanoparticles. In Table 1 functional molecule that used for modification of starch are listed which have been used for preparation of their nanoparticles.

3.2 Modified chitosan and chitin

Biopolymer chitosan with a lot of primary amino groups is a polysaccharide derived from deacetylation of chitin. Due to the excellent film-forming ability, biocompatibility, nontoxicity, high mechanical strength, cheapness of chitosan, it is used for synthesis and the application of polysaccharide-based nanoparticles (Payne *et al.*, 2005; Kwon *et al.*, 2003). Chitosan is one of the polysaccharides that modified with various groups such as 5β-cholanic acid, linoleic acid, Monomethoxy poly (ethyleneglycol) and etc. After modification process, modified chitosan are used for preparation of their nanoparticles. These groups are listed in Table 2.

Grafted chitosan has been studied by many researchers. These studies have been intensified since 1992 because chitin and chitosan show excellent biological properties such as biodegradation in the human body. Modification can marginally improve the solubility of chitosan. As a polymeric amphiphile, grafted-chitosan with monomethoxy poly (ethyleneglycol) can aggregate into core–shell nanoparticles in aqueous media because in the aqueous phase, the hydrophobic cores of chitosan nanoparticles are encircled by hydrophilic outer shells. Thus, the internal core can serve as a nano-container for hydrophobic drugs. Modified chitosan is appropriate for decreasing severe side effects such as cytotoxicity in usual tissue (Fang *et al.*, 2006; Gorochovceva *et al.*, 2005; Opanasopit *et al.*, 2006).

Chitosan	Grafting agent	References
Glycol chitosan	5β-Cholanic acid	(Kwon *et al.*, 2006; Kwon *et al.*, 2004; Kwon *et al.*, 2003)
Modification: Glycol chitosan was hydrophobically modified with cholanic acid in methanol/water. To activate the carboxylic acid groups of cholanic acid, equal amounts of 1-ethyl-3-(3-dimethylaminopropyl) - carbodiimide ydrochloride and N-hydroxysuccinimide were added.		
Chitosan of 100 mesh	Linoleic acid	(Lu *et al.*, 1994; Ichinose *et al.*, 2000)
Modification: Chitosan was dissolved in aqueous acetic acid solution and diluted of methanol. LA was added to the chitosan solution glucosamine residue of chitosan followed by a dropwise addition of 15 mL of EDC methanol solution (0.07 g/L) while stirring.		
chitosan	α-Cyclodextrin	(Sakairi *et al.*, 1998; Martel *et al.*, 2001; Aoki *et al.*, 2003)
Modification: Sakairi prepared α-CD linked chitosan using 2-O-formylmethyl-α-CD by reductive N-alkylation and confirmed the host-guest complex with p-nitrophenol.		
Chitosan	ε-Caprolactone 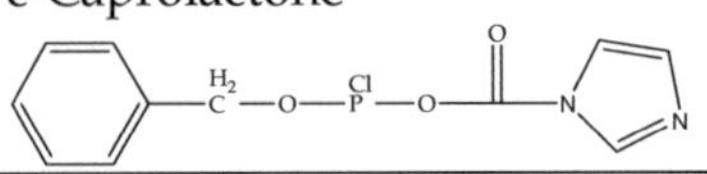	(Albertsson *et al.*, 1999; Yang *et al.*, 2008b)
Modification: The PCL-graft-chitosan copolymers were synthesized by coupling the hydroxyl end-groups on preformed PCL chains and the amino groups present on 6-O-triphenylmethyl chitosan and by removing the protective 6-O-triphenylmethyl groups in acidic aqueous solution		
Biomedical grade chitosan	Monomethoxy poly(ethyleneglycol)	(Zhang *et al.*, 2005; Yang *et al.*, 2008b)
Modification: Chitosan was completely dissolved in formic acid by stirring and a suitable amount of mPEG was added. After 15 min, enough formaldehyde solution was added to the above mixture and was stirred for 12 h.		

Table 2. Functional molecules for modification of chitosan

3.3 Modified dextran

The development of existing materials to prepare modified dextran is the subject of numerous researches due to their surface-active properties and potential pharmaceutical, biochemical and medical applications. Modified dextran gives a large range of properties, allowing the selection of the carrier which proves the most useful for a particular drug encapsulation and release. Dextran is one of the water-soluble polysaccharides that have been modified to obtain amphiphilic polymers capable of forming micellar structures and binding organics solutes in the hydrophobic domain. Also, it is amphiphilic block copolymers that can self- assemble in selective solvents to form micelles with a core and a shell containing insoluble and soluble blocks (Lu *et al.*, 1994; Ichinose *et al.*, 2000; Lu and Tjerneld, 1997). Core-shell type nanoparticles of a poly (DL-lactide-co-glycolide) (PLGA) grafted-dextran copolymer are prepared with varying graft ratio of PLGA. The DexLG copolymer was able to form nanoparticles in water by self-aggregating process (Song *et al.*,

2006). Dextran was chemically modified by the covalent attachment of hydrocarbon groups (aliphatic or aromatic) via the formation of ether links. According to the extent of modification, either water-soluble or water-insoluble dextran derivatives were obtained. The latter exhibited solubility in organic solvents like tetrahydrofuran or dichloromethane saturated with water (Bertholon *et al.*, 2006; Durand *et al.*, 2004; Leonard *et al.*, 2003; Aumelas *et al.*, 2007; Leonard *et al.*, 2000; Osterberg *et al.*, 1995). Biodegradable hydrogel nanoparticles were prepared from glycidyl methacrylate dextran (GMD) and dimethacrylate poly(ethylene glycol) (DMP). GMD was synthesized by coupling of glycidyl methacrylate to dextran in the presence of 4-(*N*,*N*-dimethylamino)pyridine (DMAP) using dimethylsulfoxide (DMSO) as an aprotic solvent (Kim *et al.*, 2000; Vandijkwolthuis *et al.*, 1995). Dextran also was modified using click-chemistry. Each reaction step was done under aqueous conditions, including the introduction of azide functionalities to the backbone of the polysaccharide. The reaction consisted of the synthesis of 1-azido-2,3-epoxypropane, which was etherified onto the backbone of the polysaccharide using base-catalysis in water/isopropanol mixture at ambient temperature (Fringuelli *et al.*, 1999; Seppala *et al.*, 2010). Modified dextran was synthesized by conjugating the various groups to dextran such as poly (lactic-co-glycolic acid, p-hexylbenzoyl chloride. These groups are listed in Table 3.

Dextran	**Grafting agent**	**References**
Dextran (average molecular weights: 77,000)	Poly(lactic-co-glycolic acid)	(Tiera *et al.*, 2003)
Modification: The DexLG graft copolymer was synthesized by conjugating the carboxylic acid end of PLGA and the hydroxyl group of dextran using DCC as a coupling agent.		
Dextran T40 ðMw < 40; 000	P-Hexylbenzoyl chloride	(Tiera *et al.*, 2003; Bertholon *et al.*, 2006)
Modification: Dextran was dissolved under stirring in 5 ml of water containing 1.8 g of triethylamine. The resulting solution was heated at 20 8C and 1.4 g of p-hexylbenzoyl chloride was added under vigorous stirring for 1h.		
Dextran	1,2- Epoxy-3-phenoxypropan	(Durand *et al.*, 2002; Sun *et al.*, 2006; Song *et al.*, 2006)
Modification: Water-soluble amphiphilic dextran, i.e. dextran with lowsubstitution ratio – here DexP15 – was obtaiobtained after reaction with 1,2- epoxy -3-phenoxypropane in 1M NaOH as previously described.		
Dextran (Mw) 30 200	Bile acid	(Melo *et al.*, 1999; Akiyoshi *et al.*, 1993)
Modification: modified dextran were obtained by reacting dextran (*Mw*) 30 200, *Mw*/*Mn*) 1.112) with a bile acid in the presence of *N*,*N*-dicyclohexylcarbodiimide as a coupling agent and 4-(*N*,*N*-dimethylamino)pyridine as a catalyst.		
Dextran methoxypolyethylene	Glycol/poly (ϵ-caprolactone)	(Zhang *et al.*, 2008; Cao *et al.*, 2005)
Modification: A series of amphiphilic copolymers, dextran-graft-methoxypolyethylene glycol/poly (e-caprolactone) (Dex-g-mPEG/PCL) were synthesized by grafting both PCL and mPEG chains to dextran, and subsequently the micellar self-assembly behavior of resultant copolymers was investigated.		

Table 3. Functional molecules for modification of dextran

3.4 Modified pullulan

Due to their amphiphilic structure, modified pullulan has potential high surface and interfacial properties. They diffuse through the bulk phase and adsorb at the interface, inducing a sharp reduction in the surface or interfacial tension of a polymer solution (Muller *et al.,* 2003). Like other polysaccharides pullulan have been used to modify with various groups for preparation of their nanoparticles (table 4). Pullulan which is partly modified by relatively higher hydrophobic groups such as cholesteryl groups, it shows unique association behavior. Cholesterol-bearing pullulans have been studied in detail by Akiyoshi and Sunamoto. It was designed as a self-aggregate to form monodisperse and stable nanogels due to the hydrophobic moieties in an aqueous solution. The nanogels formed complexes with various drugs and proteins by hydrophobic interaction and released them upon exposure to specific proteins (Akiyoshi *et al.,* 1997; Akiyoshi *et al.,* 1993; Cheng *et al.,* 2008). Hydrophobically-modified pullulans of moderate molar mass and differing in hydrophobic modification ratio, charge ratio and the nature of the hydrophobic chains were prepared (Bataille *et al.,* 1997; Glinel *et al.,* 1999; Fischer *et al.,* 1998). Poly (DL-lactide-co-glycolide)-grafted pullulan can form self-assembling nanospheres and controll adriamycin release. Pullulan acetate (PA) is the other important hydrophobized pullulan, which can form self-aggregation nanoparticles as well as its modified materialsn (Zhang *et al.,* 2009; Na *et al.,* 2007).

Pullulan	**Grafting agent**	**References**
Pullulan	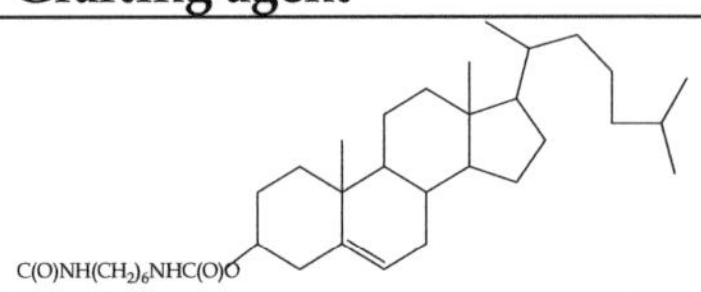	(Akiyoshi *et al.,* 1998)
Modification: Cholesterol-bearing pullulan forms a spherical andmonodisperse nanoparticle which is a self-aggregate of 10–12 CHP molecules. This nanoparticlehas several hydrophobic domains of four to five associated cholesteryl moieties.		
carboxymethylpullulan	Alkyl bromide (octyl, decyl or dodecyl)	(Bataille *et al.,* 1997; Glinel *et al.,* 1999)
Modification: Hydrophobically-modified carboxymethylpullulans (HMCMPs) were obtained by a synthetic pathway adapted from that used by Della Valle for gellan and Fischer et al. for pectin.		
Pullulan with molecular weight of 50,000–100,000 (g/mol)	Poly(DL-lactide-co-glycolide)	(Jeong *et al.,* 2006)
Modification: Pullulan (1 g) was dissolved in DMSO (15 ml) for 3 h. Various amounts of PLGA were dissolved in DMSO (5 ml) with a 1.3 equiv. amount of DCC and DMAP.		
Pullulan (*M*w = 200,000)	Acetic anhydride	(Na *et al.,* 2007; Zhang *et al.,* 2009)
Modification: 2 g of pullulan, suspended in 20ml of formamide, was dissolved by vigorous stirring at 54 °C. To this solution, 6ml pyridine and 15 ml, 10ml or 7.5ml of acetic anhydride were added to change the acetylation degree.		

Table 4. Functional molecules for modification of pullulan

3.5 Modified cyclodextrins

Modifications to the cyclodextrins(Namazi *et al.*, 2005; Namazi and Kanani, 2009) lead to a wide range of photochemistry of cyclodextrin complexes, through which the improvement of guest reactivity occurs; in addition, light harvesting molecular devices and photochemical frequency switches may be constructed. A few amphiphilic β-CD derivatives such as β-CDC_6 modified on the secondary face with 6C aliphatic esters and 6-N-CAPRO-β-CD modified on the primary face with a 6C aliphatic amide were demonstrated to give stable nanoparticles of high drug loading capacity and reduction of burst effect during the drug release process when nanoparticles are prepared directly from preformed drug/amphiphilic CD inclusion complex (Lemos-Senna *et al.*, 1998). A new nanoparticle carrier system was obtained from amphiphilic cyclodextrin bearing fatty acids (with a chain length of either 6 or 12 carbon atoms) grafted O_2 and O_3 position of the cyclodextrin. Nanoparticles with a mean diameter of several hundred nm were prepared by dispersion. Amphiphilic cyclodextrins (CDs) are obtained by the chemical per-modification of natural CDs (β-CD or γ-CD) by the selective substitution of aliphatic chains of varying length (2C to 18C), structure (linear or branched) linked with varying bonds (ester, ether, amide, thio, fluoro) of high purity. These CD derivatives were demonstrated to yield nanospheres or nanocapsules spontaneously using the nanoprecipitation technique with or without the presence of surfactants. Carboxymethyl-β-cyclodextrin modified nanoparticles were fabricated for removal of copper ions from aqueous solution by grafting CM-β-CD onto the magnetite surface via carbodiimide method. The grafted Carboxymethyl-β-cyclodextrin on the Fe_3O_4 nanoparticles contributes to an enhancement of the adsorption capacity because of the strong abilities of the multiple hydroxyl and carboxyl groups in CM-β-CD to adsorb metal ions. Double hydrophilic copolymers with one polyethylene glycol (PEG) block and one β-cyclodextrin (β-CD) flanking block (PEG-β-PCDs) were synthesized through the post-modification of macromolecules. The self-assembly of PEG-β-PCDs in aqueous solutions was studied by a fluorescence technique(Choisnard *et al.*, 2006).

3.6 Modified celullose

Modified celullose have received wide applications for the stabilization of disperse systems, in particular suspensions and emulsions (Namazi and Rad, 2004). The most important types of associating polymers are water-soluble amphiphilic polymers, notably block or graft copolymers, with hydrophobic blocks or grafts. Cellulose is the most abundant polysaccharide available worldwide and exhibits attractive structure and single properties, which are quite attractive for both academic and industrial researchers. Recently, cellulose based polymers have been widely investigated for its positive characteristics such as safety, biodegradability, biocompatibility, and protein rejecting ability, and so on(Namazi and Jafarirad, 2008). However, there have been few reports on the utilization of self-assembled micelles based on amphiphilic cellulose derivatives as delivery carriers for poorly water-soluble pharmaceutical active ingredients (Klemm *et al.*, 2005; Cheng *et al.*, 2008; Dong *et al.*, 2008). Poly (ε-caprolactone) (PCL) and poly (L-lactic acid) (PLLA) are biodegradable polymers that are potential candidates as matrixes in biocomposites. Several studies have been conducted on the PCL and PLLA modification of soluble cellulose and its derivate (Nishio and Teramoto, 2003; Nishio *et al.*, 2002; Burt and Shi, 2003). Modified cellulose was prepared with hydrophilic groups that it can be self-assemble into polymeric vesicle or as nontoxic surfactants. Sulfate was firstly introduced as hydrophilic groups, then the hydrophobic groups for cellulose derivatives. The aqueous self-assembly of the modified cellulose was investigated using transmission electron microscopy (TEM) and dynamic laser scattering (DLS). Results showed that modified cellulose were capable of forming polymeric micelles in water with an average particle diameter ranging from 20 to 67 nm (Cheng *et al.*, 2008). Novel modified cellulose derivatives were synthesized long chain alkyl groups as hydrophobic moieties and quaternary ammonium groups as hydrophilic moieties. The results of measurements (DLS, TEM) revealed that modified cellulose can be self-assembled into cationic micelles in distilled water with the average hydrodynamic radius of 320–430 nm (Zhou *et al.*, 2011).

4. Prepration methods and characterization of polysaccharide-based nanoparticle

As for polysaccharide-based nanoparticles, Alonso et al. (Alonso *et al.*, 2001) and Prabaharan et al. (Prabaharan and Mano, 2005) have made excellent reviews in 2001 and 2005, respectively, focusing on the preparation and application of chitosan nanoparticle carriers. Many studies have demonstrated that nanoparticles have a number of advantages over microparticles (Panyam and Labhasetwar, 2003). It has been reported that micro particles are less effective drug delivers than particle having size ranging in between nanometers for e.g. Nanoparticles having size range greater than 230 nm acquire in the spleen shown by body distribution studies (Kreuter, 1991). As time goes on, more polysaccharide-based nanoparticles emerge, which greatly enriches the versatility of nanoparticle carriers in terms of category and function. In this section, several mechanisms are introduced to prepare these nanoparticles, that is, emulsification solvent evaporation method, solvent diffusion method, self-assembly of hydrophobically modified, dialysis method and other methods. The select of method depends on a number of factors, such as, particle size, particle size distribution, area of application and etc. Particle size is the greatest important characteristics of nanoparticles. Some methods for the determining particle size are (Labhasetwar *et al.*, 2003)

a. Photon-correlation spectroscopy.
b. Dynamic light scattering.
c. Brownian motion and light scattering properties.
d. Scanning or transmission electron microscopy (SEM or TEM).

They determine the in vivo distribution, biological fate, toxicity and targeting ability of these delivery systems. In addition, they can influence drug loading, drug release and stability of the nanoparticles.

4.1 Polysaccharides-based nanoparticles through emulsification solvent evaporation method

Emulsification solvent evaporation is the most widely employed technique to prepare nanoparticles of polymers in the current literature on techniques using a dispersion of preformed polymers (Vanderhoff *et al.*, 1979). In the conventional methods, two main strategies are being used for the formation of emulsions: the preparation of single-emulsions, e.g., oil-in-water (o/w) or double-emulsions, e.g., (water-in-oil)-in-water, (w/o/w).

In a single emulsification solvent evaporation process, polymer dissolved in a volatile water-immiscible organic solvent such as dichloromethane, chloroform, ethyl acetate, which is also used as the solvent for dissolving the hydrophobic surfactant. This solution is emulsified in an aqueous phase containing a surfactant or stabilizer (emulsifying agent) resulting in oil-in-water (o/w) emulsion.(ODonnell and McGinity, 1997; I. *et al.*, 2004; Lee, 2001) The coalescence of the organic droplets can be avoided by continuous stirring. Emulsification can also be enhanced by using sonication or microfluidization with a homogenizer, which reduces the droplet size of the organic dispersed phase. After the formation of stable emulsion, the organic solvent is evaporated either under stirring at room temperature or by rotary evaporation under reduced pressure to transform the nano-emulsion into a nanoparticle suspension. Formed nanoparticles are harvested from the aqueous slurry by lyophilization.

For the water-soluble surfactants, a double-emulsion (water-oil-water) variation of the process is utilized. An aqueous solution of the active agent (internal water phase, w1) is emulsified into an organic solution containing the biodegradable polymer and lipophilic surfactant (oil phase, o) for resulting primary emulsification. Then, this emulsion (w1/o) is added to the large aqueous phase with emulsifier (external water phase, w2) to create w1/o/w2 double emulsion. The emulsifier amount is much higher in the first emulsion than in the second emulsion, because the droplet size of the first emulsion needs to be much smaller than in the second outer emulsion. The organic solvent is removed by evaporation or extraction and solid nanoparticles are formed. The nanoparticles are collected by centrifugation or filtration and are subsequently lyophilized.

Wouessidjewe and coworks(Lemos-Senna *et al.*, 1998) using this method for preparing nanospheres from an amphiphilic 2,3-di-O-hexanoyl-γ -cyclodextrin (γCDC_6). This preparation method involves in emulsifying an organic phase having the cyclodextrin in an aqueous phase containing Pluronic F68 as surfactant. This solution was dispersed in aqueous phase by using a high speed homogenizer. Afterward, the organic solvent was evaporated by mechanical stirring at room temperature. The influence of the process parameters, i.e. surfactant concentration and initial γCDC_6 content, on the characteristics of nanosphere preparation, as well as on the nanosphere loading of a hydrophobic drug, progesterone, was calculated. Cyclodextrin nanospheres presenting a mean diameter varying from 50 to 200 nm were obtained, even in the presence of low surfactant concentration.

Nanoparticles of dextran (Aumelas *et al.*, 2007) could be simply prepared by the o/w emulsion solvent evaporation method, with using a low modified dextran ($DexP_{15}$) as polymeric surfactant in the water phase and a highly modified dextran in the CH_2Cl_2 phase. After emulsification and solvent evaporation, core-shell particles with a dense dextran core and a dextran surface coverage are expected. Dextran segments originating from $DexP_{15}$ chains which are not embedded in the dextran core are assumed to extend freely toward the aqueous solution and to form a hydrophilic shell. The size of $DexP_{130}$ nanoparticles prepared by o/w emulsion process decreases as the amount of $DexP_{15}$ in the water phase increases. Unpredictably, dextran nanoparticles were also obtained without any polymeric surfactant in the aqueous phase. For comparison, when poly (lactic acid) was used instead of hydrophobically modified dextran, it was not possible to obtain nanoparticles without the presence of surfactant in the aqueous phase. This specific result can be explained assuming a limited solubility of highly hydrophobized dextrans in water. This solubility can be due to the presence of a fraction of low substituted dextran molecules in the final product or to partitioning of the highly substituted sample. This water-soluble fraction could act as a stabilizer for the transient oil droplets. Generally speaking, the size of bare dextran nanoparticles, i.e. prepared in the absence of $DexP_{15}$, increases with the substitution ratio of dextran, for example from 370 nm for $DexP_{65}$ nanoparticles to 850 nm for $DexP_{210}$ nanoparticles. Other dextran particles, in the size range 150–250 nm, were obtained in the presence of $DexP_{15}$. The colloidal stability of suspensions was also examined at various NaCl concentrations. For the targeted nanoparticles, surface coverage by hydrophilic loops is essential to provide a convenient colloidal stability in physiological conditions (especially with regard to the ionic strength).

In the o/w emulsion process, we showed that the size of particles is strongly related to the concentration of surfactive polymer in the aqueous phase. In generally, parameters in the emulsification solvent evaporation process that affect particle size, zeta potential, hydrophilicity, and drug loading include:

1. Homogenization intensity and duration.
2. Type and amounts of emulsifier, polymer and drug.
3. Particle hardening (solvent removal) profile (Zambaux *et al.*, 1998).

4.2 Polysaccharides-based nanoparticles through solvent diffusion method

Spontaneous emulsification or solvent diffusion method is a modified version of solvent evaporation method. The different process variants are all based upon the use of solvents which are of limited water miscibility and capable of spontaneous emulsion formation. This method thus offers the advantage of the use of pharmaceutically acceptable solvents and does not require the use of high-pressure homogenizers for the formation of the o/w emulsion as the preliminary stage of nanoparticle formation (Allemann *et al.*, 1998; Leroux *et al.*, 1995). In this method, the water-miscible solvent along with a small amount of the water-immiscible organic solvent is used as an oil phase. Due to the spontaneous diffusion of solvents an interfacial turbulence is created between the two phases leading to the formation of small particles. In this technique, the phase separation is accompanied by vigorous stirring. On the opposite with o/w, the size of nanoparticles obtained using the solvent-diffusion method is poorly affected by the concentration of polymeric surfactant added to the aqueous phase. A reduction in particle size can be gained by increasing the concentration of water miscible solvent.

Nanoparticles of dextran (Aumelas *et al.*, 2007) could be prepared by solvent-diffusion method. Dextran nanoparticles of similar size were obtained with or without using stabilizer such as $DexP_{15}$. This process avoids the use of any high energy input step. The colloidal stability of suspensions was also examined at various NaCl concentrations. The particular colloidal stability of $DexC_{1052}$ nanoparticles up to high ionic strengths without $DexP_{15}$ can be justified by assuming that the water-soluble fraction contained in that polymer is higher than in the others. Also this method was employed to prepare pullulan acetate (PA) nanoparticles.(Zhang *et al.*, 2009) This technique had some advantages compared with other methods. It is a straightforward technique and the particle size increased from 185.7 nm to

423.0 nm with the degree of acetylation increasing from 2.71 to 3.0. Briefly, PA is readily soluble in dimethyl sulfoxide (DMSO), DMF, tetrahydrofuran (THF), dichloromethane, chloroform, acetone, and pyridine. To make nanoparticles by solvent diffusion method, only water-miscible solvents were considered because the solvents could diffuse into aqueous phase. The solvent selected to dissolve the polymer, as well as the type of polymer can influence the formation of nanoparticles, due to differences in the polymer-solvent and water–solvent interactions. It was supposed that the diffusion-stranding process might be altered, thus inducing changes in the mean size. Therefore, solvents are of primary importance in the formation of nanoparticles by the solvent diffusion method. In other study, five water-miscible solvents, i.e., DMSO, DMF, acetone, THF and pyridine were used. 0.5% poly (vinyl alcohol) [PVA] or distilled water served as aqueous phase. PA2 could form nanoparticles in anyone of the five organic solvents added to water or 0.5% PVA. However, PA1 could do only in DMSO and DMF added to 0.5% PVA. Really, PA2 led to the smallest nanoparticles (185.7 nm), and the largest was PA1 nanoparticles (423.0 nm).

4.3 Polysaccharides-based nanoparticles through self-assembly method

The literature survey showed that several studies have been carried out to investigate the synthesis and the application of polysaccharide based self- aggregate nanoparticles as drug delivery systems. When hydrophilic polymeric chains are grafted with hydrophobic segments, amphiphilic copolymers are formed. Upon contact with an aqueous environment, polymeric amphiphiles spontaneously form micelles or micelle-like aggregates via undergoing intra- or intermolecular associations between hydrophobic moieties, primarily to minimize interfacial free energy. These polymeric micelles display unique characteristics, such as small hydrodynamic radius (less than microsize) with core-shell structure, unusual rheology feature, thermodynamic stability, depending on the hydrophilic/hydrophobic constituents. In specific, polymeric micelles have been recognized as a promising drug carrier, since their hydrophobic domain, surrounded by a hydrophilic outer shell, can serve as a preservatory for various hydrophobic drugs (Letchford and Burt, 2007). Usually, these hydrophobic molecules can be divided into linear, cyclic hydrophobic molecules, hydrophobic drug, polyacrylate family, etc.

4.3.1 Linear hydrophobic molecules

Poly (ε-caprolactone) (PCL) is biodegradable industrial polyester with excellent mechanical strength, non-toxicity, and biocompatibility. It has been frequently used as implantable carriers for drug delivery systems or as surgical repair materials. It is hopeful to combine chitosan with the biodegradable polyester to create amphiphilic copolymer applicable to drug delivery systems. In 2002 and 2003, (Gref *et al.*, 2002; Lemarchand *et al.*, 2003) synthesized amphiphilic dextran by coupling between carboxylic function present on preformed PCL monocarboxylic acid and the hydroxyl groups on dextran. The comb-like copolymers (dextran-PCLn) consisted of a dextran back bone on to which preformed PCL blocks were grafted. Nanoparticles of less than 200 nm were successfully prepared by using the new materials (Rodrigues *et al.*, 2003). Further, bovine serum albumin and lectin were incorporated in the nanoparticles. Lectins could also be adsorbed onto the surface of the nanoparticles. Surface-bound lectin conserved its hemagglutinating activity, suggesting the possible application of this type of surface-modified nanoparticles for targeted oral administration. Caco-2 cellular viability was higher than 70% when put in contact with the nanoparticles, even at concentrations as high as 660 mg/ml (Rodrigues *et al.*, 2003). In addition, it was found that the modification of the surface with dextran significantly reduced the cytotoxicity towards J774 macrophages. Biodegradable amphiphilic PCL-graft-chitosan copolymer was synthesized (Jing *et al.*, 2006). The copolymers could form spherical or elliptic nanoparticles in water.

Poly (ethylene glycol) has been employed extensively in pharmaceutical and biomedical fields because of its outstanding physicochemical and biological properties including hydrophilic property, solubility, non-toxicity, ease of chemical modification and absence of antigenicity and immunogenicity. Therefore, poly (ethylene glycol) is widely used as a pharmacological polymer with high hydrophilicity, biocompatibility and biodegradability. In recent years, derivative poly (ethylene glycol)-g-derivative chitosan to obtain nanoparticles has been studied by many researchers (Ouchi *et al.*, 1998; Jung *et al.*, 2006) (Park *et al.*, 2008) (Yang *et al.*, 2008b) (Opanasopit *et al.*, 2007). The grafted poly (ethylene glycol) methyl ether onto N-Phthaloyl chitosan chains, aggregated to obtain sphere-like nanoparticles (an *et al.*, 2004). When the chain length of poly (ethylene glycol) methyl ether was as high as 5×10^3 Da, the sphere size became as small as 80-100 nm. By simply adjusting the hydrophobicity/hydrophilicity of the chitosan chain, stable nanospheres could be obtained directly. Also methoxy poly (ethylene glycol)-grafted chitosan to develop polymeric micelles for the drug delivery to brain tumor was synthesized.(Jung *et al.*, 2006) Methoxy poly (ethylene glycol)-grafted-chitosan conjugates by formaldehyde linking method was synthesized(Yang *et al.*, 2008b). The conjugates formed monodisperse self-aggregated nanoparticles with a roughly spherical shape and a mean diameter of 261.9 nm. A poorly water-soluble anticancer drug, methotrexate was physically entrapped inside the nanoparticles. Other group synthesized amphiphilic grafted copolymers, N-phthaloyl chitosan- grafted poly (ethylene glycol) methyl ether (Opanasopit *et al.*, 2007). These copolymers could form micelle-like nanoparticles. The CMC of these nanoparticles in water was similar (28 μg/ml). The nanoparticles exhibited a regular spherical shape with core-shell structure with sizes in the range of 100-250 nm. Camptothecin as a model drug was loaded into the inner core of the micelles.

For modifying polysaccharides have been used some long-chain fatty acids such as hexanoic acid, decanoic acid, linoleic acid, linolenic acid, palmitic acid, stearic acid, and oleic acid. Choisnard et al. (Choisnard *et al.*, 2006) prepared decanoate β-cyclodextrin esters (DS, 2-7) and hexanoate β-cyclodextrin esters (DS, 4-8) biocatalyzed by thermolysin from native β-cyclodextrin and vinyl hexanoate or vinyl decanoate used as acyldonors. Both esters self - organized into nanoparticles by a nanoprecipitation method. Chen et al. (Chen *et al.*, 2003a) modified chitosan by coupling with linoleic acid through the 1-ethyl-3-(3-dimethylamino-propyl)-carbodiimide-mediated reaction to increase its amphipathicity for enhanced emulsification. The micelle formation of linoleic acid-modified chitosan in the 0.1 M acetic acid solution was improved by o/w emulsification with methylene chloride, an oil phase, the self-aggregation concentration from 1.0 g/L to 2.0 g/L. The addition of 1 M sodium chloride promoted the self-aggregation of linoleic acid-chitosan molecules both with and without emulsification. The micelles formed nanosize particles ranging from 200 to 600 nm. The nanoparticles encapsulated a lipid soluble model compound, retinal acetate, with 50% efficiency. The similar group modified chitosan with linolenic acid (the DS 1.8%) using the same reaction. The self-aggregated nanoparticles of linolenic acid-chitosan were also used to immobilize trypsin using glutaraldehyde as crosslinker. Results indicated that the activity of trypsin immobilized onto the nanoparticles increased with increasing concentration of glutaraldehyde up to 0.07% (v/v) and then decreased with increasing amount of glutaraldehyde. On the other side, particle size increased (from 523 to 1372 nm) with the increasing concentration of glutaraldehyde (from 0.03 to 0.1% v/v) (Liu *et al.*, 2005).

Water-soluble N-palmitoyl chitosan was prepared by swollen chitosan coupling with palmitic anhydride in dimethyl sulfoxide, which could procedure micelles in water (Jiang *et al.*, 2006). The DS of N-palmitoyl chitosan was in the range of 1.2-14.2% and the CMC of N-palmitoyl chitosan micelles was in the range of 2.0×10^{-3} to 37.2×10^{-3} mg/ml. The loading capacity of hydrophobic model drug ibuprofen in the micelles was about 10%. Also stearic acid grafted chitosan oligosaccharide by 1-ethyl-3-(3-dimethylaminopropyl) carbodiimide-mediated coupling reaction was synthesized (Hu *et al.*, 2006). The CMC of the copolymer was approximately 0.06, 0.04, 0.01 mg/ml respectively. To increase the stability of the micelle in vivo and controlled drug release, the shells of micelles were cross-linked by glutaraldehyde. Paclitaxel was used as a model drug to incorporate into the micelles, and the surfaces of the micelles were further cross-linked by glutaraldehyde to form drug loaded and shell cross-linked nanoparticles. The higher drug entrapment efficiencies (above 94%) were observed in all cases. Zhang et al. (Zhang *et al.*, 2007) developed self-assembled nanoparticles based on oleoyl-chitosan with a mean diameter of 255.3 nm. Doxorubicin was efficiently loaded into the nanoparticles with an encapsulation efficiency of 52.6%. The drug was rapidly and completely released from the nanoparticles at pH 3.8, whereas at pH 7.4 there was a sustained release after a burst release. Amylose-conjugated linoleic acid complexes were synthesized to serve as molecular nanocapsules for the protection and the delivery of linoleic acid (Shimoni *et al.*, 2005).

Pluronic tri-block copolymers collected of poly (ethylene oxide)-poly (propylene oxide) - poly (ethylene oxide) show lesser critical solution temperature behaviors over a broad temperature range depending on the composition and MW. They self-assemble to procedure a spherical micellar structure above the lower critical solution temperature by hydrophobic interaction of the poly (propylene oxide) middle block in the structure. Pluronic/heparin composite nanocapsules, which displayed a 1000-fold volume transition (ca. 336 nm at 25 °C; ca. 32 nm at 37 °C), and a reversible swelling and de-swelling behavior when the temperature was cycled between 20 and 37 °C is prepared (Choi *et al.*, 2006). Core/shell nanoparticles with the poly (lactide-co-glycolide) core and the polymeric shell made-up of pluronics and hyaluronic acid was synthesized (Yuk *et al.*, 2005).

4.3.2 Cyclic hydrophobic molecules

Cholesterol is an essential lipid in animals, which not only participates the formation of cell membranes but also works as a raw material for the synthesis of bile acids, vitamin D and steroid hormones. Conjugating hydrophobic cholesterol to hydrophilic polysaccharides may form amphiphilic copolymer which may further form self-assembly nanoparticles in aqueous solution. cholesterol-modified chitosan conjugate with succinyl linkages was synthesized (Wang *et al.*, 2007c). The conjugates formed monodisperse self-aggregated nanoparticles with a roughly spherical shape and a mean diameter of 417.2 nm by probe sonication in aqueous media. Epirubicin, as a model anticancer drug, was physically entrapped inside the nanoparticles by the remote loading technique. Epirubicin-loaded nanoparticles were almost spherical in shape and their size increased from 338.2 to 472.9 nm with the epirubicin-loading content increasing from 7.97% to 14.0%. Also was prepared self-aggregated nanoparticles of cholesterol-modified O-carboxymethyl chitosan (Wang *et al.*, 2007b).

Various cholesterol-bearing pullulans with different MWs of the parent pullulan and DS of the cholesteryl moiety was synthesized (Nishikawa *et al.*, 1996; Akiyoshi *et al.*, 1997). Irrespective of the MW of the parent pullulan and the DS, all of cholesterol-pullulans provided unimodal and mono-disperse self-aggregates in water. The size of the self-aggregate reduced with an increase in the DS of the cholesteryl moiety (hydrodynamic radius, 8.4-13.7 nm). However, the aggregation number of cholesterol-pullulans in one nanoparticle was almost independent of the DS. The polysaccharide density within the self-aggregate (0.13– 0.50 g/ml) was affected by both the MW and the DS of cholesterol-pullulans. The characteristic temperature to cause a structural change of the nanoparticles decreased with an increase in the DS and the ionic strength of the medium. Moreover, they also prepared thermo-responsive nanoparticles by self-assembly of two different hydrophobically modified polymers, namely, cholesterol-pullulan and a copolymer of N-isopropylacrylamide and N-[4-(1-pyrenyl) butyl]-N-n-octadecylacrylamide via their hydrophobic moieties (Akiyoshi *et al.*, 2000) , as well as hexadecyl group-bearing pullulan self-assembly nanoparticles (Kuroda *et al.*, 2002).

Bile acids such as deoxycholic acid and 5β-cholanic acid are known to form micelles in water as a result of their amphiphilicity, which plays an important role in the emulsification, solubilization, and absorption of cholesterol, fats, and liphophilic vitamins in human body. Therefore, it is expected that the introduction of deoxycholic acid or 5β-cholanic acid into chitosan would induce self-association to form self-aggregates. Covalently conjugated deoxycholic acid to chitosan via carbodiimide-mediated reaction to generate self-aggregated nanoparticles was prepared (Lee *et al.*, 1998; Jeong *et al.*, 1998). Adriamycin was physically entrapped inside the self- aggregates. The size of adriamycin-loaded self-aggregates increased with increasing the loading content of adriamycin (Lee *et al.*, 2000).

Chemically modified chitosan oligosaccharides with deoxycholic acid was reported (Chae *et al.*, 2005). Owing to the amphiphilic characters, the deoxycholic acid-chitosan formed self-aggregated nanoparticles in aqueous milieu. The particle size of the nanoparticles was in the range of 200-240 nm. Furthermore, deoxycholic acid-chitosan showed great potential for gene carrier with the high level of gene transfection efficiencies, even in the presence of serum. Deoxycholic acid-heparin amphiphilic conjugates with different degree of substitution of deoxycholic acid was synthesized (Park *et al.*, 2004), which provided monodispersed self-aggregates in water, with mean diameters (120-200 nm) decreasing with increasing DS. Increasing DS enhanced the hydrophobicity of the self-aggregate inner core.

However, chitosan -based self-aggregates were difficult to be widely applied for drug delivery systems because chitosan aggregates are insoluble in biological solution (pH7.4) and they are readily precipitated within a few days. Recently, water-soluble chitosan derivatives have been used to increase their stability in biological solution and decrease the cytotoxicity induced by acidic solution, where chitosan is soluble. Covalently modified glycol chitosan with deoxycholic acid self-aggregates as a new drug delivery system was prepared (Kim *et al.*, 2005) and investigated in detail the effect of deoxycholic acid attached to glycol chitosan on the formation, physicochemical characteristics, and stability of self-aggregates in aqueous media. The same group (Kwon *et al.*, 2003; Park *et al.*, 2007) covalently attached the 5β-cholanic acid to glycol chitosan through amide formation using carbodiimide as catalyzer. The 5β-cholanic acid-glycol chitosan formed self-aggregates (210-859 nm in diameter) in an aqueous phase by intra- or intermolecular association between hydrophobic 5β-cholanic acids attached to glycol chitosan.

FITC is a widely used hydrophobic fluorescein, the isothiocyanato of which can readily react with free amine to incorporate fluorescence labeling. Doxorubicin is an anti-tumor antibiotic, which can inhibit the synthesis of RNA and DNA and has a therapeutic effect on many tumors. FITC and doxorubicin themselves are hydrophobic cyclic molecules, which can be conjugated onto hydrophilic polysaccharides form amphiphilic copolymers. Hydrophobically modified glycol chitosans by chemical conjugation of FITC or doxorubicin to the backbone of glycol chitosan was prepared (Lee *et al.*, 2006; Son *et al.*, 2003). Biodistribution of self-aggregates (300 nm in diameter) was evaluated using tissues obtained from tumor-bearing mice, to which self-aggregates were systemically administered via the tail vein. Na et al. (Na *et al.*, 2003) introduced vitamin H to pullulan acetate and prepared corresponding self-assembled nanoparticles (~100 nm) in order to improve their cancer-targeting activity and internalization. Three samples of biotinylated pullulan acetate, comprising 7, 20 and 39 vitamin H groups per 100 anhydroglucose units, were synthesized. In addition, synthesized successfully N-succinyl-chitosan, which could be self-assembly of well-dispersed and stable nanospheres in distilled water with 50-100 nm in diameter (Zhu *et al.*, 2006). Experimental results indicated that a hydrophobic domain formed within these nanospheres. The assembly mechanisms were believed to be the intermolecular H-bonding of N-succinyl-chitosan and hydrophobic interaction among the hydrophobic moieties in N-succinyl-chitosan macromolecules. Park et al. (Park *et al.*, 2006) described N-acetyl histidine-conjugated glycol chitosan self-assembled nanoparticles as a promising system for intracyto-plasmic delivery of drugs.

4.3.3 Polyacrylate-based nanoparticles applicable as biomaterials

Poly (methyl methacrylate) and poly (isobutyl cyanoacrylate) (PIBCA) all belong to polyacrylate family and they were widely used for biomaterials. Containing carboxylic ester groups in their structures, they are hydrophobic. The efficient uptake of injected nanoparticles by cells of the mononuclear phagocyte system limits the development of long-circulating colloidal drug carriers. The complement system plays a major role in the opsonization and recognition processes of foreign materials. Since heparin is an inhibitor of complement activation, nanoparticles bearing heparin covalently bound to poly (methyl methacrylate) and evaluated their interactions with complement was prepared (Passirani *et al.*, 1998a). Nanoparticles bearing covalently bound dextran instead of heparin were weak activators of complement as compared with cross-linked dextran or bare poly (methyl methacrylate) nanoparticles. In addition to the specific activity of bound heparin, the protective effect of both polysaccharides is hypothesized to be due to the presence of a dense brush-like layer on the surface of the particles. Dextran nanoparticles were also eliminated very slowly over 48 h. bare poly (methyl methacrylate) nanoparticles were found to have a half-life of only 3 min. Both types of nanoparticles proved to be long-circulating. The potent capacity for opsonization of the poly (methyl methacrylate) core was hidden by the protective effect of either polysaccharide, probably due to a dense brush -like structure. In the case of heparin nanoparticles, the "stealth" effect was probably increased by its inhibiting properties against complement activation (Passirani *et al.*, 1998b).

PIBCA-chitosan nanoparticles by emulsion polymerization of IBCA in the presence of chitosan as a polymeric stabilizer at low pH were prepared (Yang *et al.*, 2000). Nimodipine as a model drug was successfully incorporated into the nanoparticles with mean particle diameter of 31.6 nm and a positive charge. Also PIBCA-chitosan, PIBCA-dextran and PIBCA-dextran sulfate core-shell nanoparticles by redox radical or anionic polymerization of IBCA in the presence of chitosan, dextran or dextran sulfate was prepared (Bertholon *et al.*, 2006). Bravo-Osuna et al. (Bravo-Osuna *et al.*, 2006; Bravo-Osuna *et al.*, 2007a; Bravo-Osuna *et al.*, 2007c; Bravo-Osuna *et al.*, 2007b) developed PIBCA-thiolated chitosan nanoparticles by radical emulsion polymerization. The nanoparticles had mean hydrodynamic diameter around 200 nm and positive zeta potential values, indicating the presence of the cationic thiolated chitosan at the nanoparticle surface. Polysaccharide-coated nanoparticles by radical emulsion polymerization of IBCA in the presence of various polysaccharides (dextran, dextran sulfate, heparin, chitosan, hyaluronic acid, pectin) was synthesized (Chauvierre *et al.*, 2003). They also measured the complement activation induced by different polysaccharide-coated nanoparticles and of the antithrombic activity of heparin. These nanoparticles maintained the heparin antithrombic properties and inhibited complement activation. This work demonstrated the hemoglobin loading on nanoparticle surface, rather than being encapsulated. With a size of 100 nm, these drug delivery systems made suitable tools in the treatment of thrombosis oxygen deprived pathologies (Chauvierre *et al.*, 2004). In addition, they investigated for the first time the mobility of dextran chains on the PIBCA nanoparticles with electronic paramagnetic resonance. This technique opens an interesting prospect of investigating surface properties of polysaccharide-coated nanoparticles by a new physicochemical approach to further correlate the mobility of the polysaccharide chains with the fate of the nanoparticles in biological systems (Vauthier *et al.*, 2004).

4.4 Polysaccharides-based nanoparticle through dialysis method

The preparation of nanoparticles was performed by a dialysis method without the use of any surfactant or emulsifiers. Dialysis offers a simple and effective method for the preparation of small, narrow-distributed polymer nanoparticle (Fessi *et al.*, 1989; Jeong *et al.*, 2001; Kostog. M *et al.*, 2010; Jeon *et al.*, 2000). Polymer is dissolved in an organic solvent and placed inside a dialysis tube with proper molecular weight cut-off. Dialysis is performed against a non-solvent miscible with the former miscible. The displacement of the solvent inside the membrane is followed by the progressive aggregation of polymer due to a loss of solubility and the formation of homogeneous suspensions of nanoparticles.

Paclitaxel-loaded HGC (PTX-HG C) nanoparticles were simply prepared by this method (Kwon *et al.*, 2006). The incorporation of PTX into the HGC nanoparticles occurred simultaneously during dialysis. The loading efficiency of PTX into HGC nanoparticles was determined by varying the feed weight ratio of PTX to HGC nanoparticles. When the feed ratio was less than 0.1, the loading efficiency was above 90%. Importantly, the PTX-HG C nanoparticles were well dispersed in an aqueous medium. However, if the feed ratio was above 0.1, the loading efficiency significantly decreased to about 42% and the excess of PTX molecules precipitated during dialysis. Thus, the maximum loading content of PT X into HGC nanoparticles was determined to 10 wt%.

To make core-shell type nanoparticles, poly (DL-lactide-co-glycolide) (PLGA) grafted-dextran (DexLG) graft copolymer was dissolved in DMSO and the core-shell type nanoparticles were prepared by dialysis method against water. The morphology of core-shell type nanoparticles of DexLG copolymer was observed by SEM and the particle size was evaluated by DLS. Core-shell type nanoparticles of DexLG copolymer has spherical shapes in their morphology and particle size was around 50-200 nm.

Starch ester nanoparticles were prepared by the dialysis method. Appropriate amount of graft polymer was dissolved in DMSO, the sample was dialysed against water using a dialysis membrane of MW 12,000 g mol–1 cut off. Starch nanoparticle formed was studied by atomic force microscopy. Nanoparticles in DMSO water solution were transferred to freshly cleaved mica sheet by drop and analyzed by tapping mode. Size of the particles was found to be in the range of 65–75 nm (diameter), and 17–19 nm (height).

5. Medical applications of polysaccharide-based nanoparticles

Polysaccharide-based nanoparticles have received considerable attention in recent years as one of the most promising nanoparticulate drug delivery systems owing to their unique potentials. Nanoparticle drug delivery systems are defined as particulate dispersions or solid particles with a size in the range of 10-1000nm and with various morphologies, including nanospheres, nanocapsules, nanomicelles, nanoliposomes, and nanodrugs, etc. The drug is dissolved, entrapped, encapsulated or attached to a nanoparticle matrix (Kommareddy *et al.*, 2005; Lee and Kim, 2005). Drug delivery systems of nanoparticles have several advantages, such as high drug encapsulation efficiency, efficient drug protection against chemical or enzymatic degradation, unique ability to create a controlled release, cell internalization as well as ability to reverse the multidrug resistance of tumor cells (Soma *et al.*, 1999). The use of starch nanoparticles is receiving a significant amount of attention due to their good hydrophilicity, biocompatibility and biodegradability. Starch nanocrystals have also been found to be excellent reinforcements (Elvira *et al.*, 2002). Hydrophobic grafted and cross-linked starch nanoparticles were used for drug delivery and Indomethacin was taken as the model drug (Abraham and Simi, 2007). Hydrophilic amylopectin was modified by grafting hydrophobic poly (lactic acid) chains (PLA) for the fabrication of polymeric micelles for drug delivery. When these spherical nano-aggregates were used as the drug carrier, it was found that they had a good loading capacity and in vitro release properties for hydrophobic indomethacin drug (Brecher *et al.*, 1997; Dufresne *et al.*, 2006).

A novel amphiphilic copolymer (dextran-g-polyethyleneglycol alkyl ether) was synthesized which resulted in polymeric micelle formation, encapsulating cyclosporine in the hydrophobic core and providing a hydrophilic corona (Na *et al.*, 2003; Francis *et al.*, 2003).

Nanoparticles of poly (DL-lactide-co-glycolide)-grafted dextran were synthesized for use as a nanoparticulate oral drug carrier. These nanoparticles were able to form nanoparticles in water by self-aggregating process, and their particle size was around 50 nm~300 nm. Core-shell type nanoparticles of DexLG copolymer can be used as a colonic drug carrier (Tiera *et al.*, 2003). Superparamagnetic chitosan–dextran sulfate hydrogels as drug carriers was synthesized. The 5- aminosalicylic acid was chosen as model drug molecule (Saboktakin *et al.*, 2010). Dextran sulphate–chitosan nanoparticles were prepared to overcome the pharmacokinetic problems and to obtain the full benefits of the drug (Anitha *et al.*, 2011). Self-assembled hydrogel nanoparticles composed of dextran and poly (ethylene glycol) was synthesized and prepared nanoparticles used for drug carrier with hydrophobic model drug in vitro (Kim *et al.*, 2000).

Hydrophobized pullulan has been used as drug delivery carrier, Specifically, cholesterol-pullulan and a copolymer of N-isopropylacrylamide and N-[4-(1-pyrenyl)butyl]-N-n-octadecylacrylamide via their hydrophobic moieties, as well as hexadecyl group-bearing pullulan self-assembly nanoparticles (Akiyoshi *et al.*, 1998; Akiyoshi *et al.*, 1993; Jung *et al.*, 2004). These hydrophobized pullulan self-associate to form colloidally stable nanoparticles with inner hydrophobic core. This hydrophobic core can only encapsulate hydrophobic substances like insoluble drugs and proteins (Gupta and Gupta, 2004). Amphiphilic polysaccharides composed of pullulan and poly (DL-lactide-coglycolide) (PLGA) were synthesized to give amphiphilicity and biodegradability as novel drug carriers. Due to its biodegradability, PLGA is commonly used for the controlled release of drugs (Jeong *et al.*, 2006). Hydrophobically modified glycol chitosan (HGC) nanoparticles showed potential as carriers for anticancer peptides and anticancer drugs because of their biocompatible in vivo (Kwon *et al.*, 2003; Yoo *et al.*, 2005). Modified chitosan derivatives, are emerging as novel carriers of drugs because of their solubility and biocompatibility in vivo (Sinha *et al.*, 2004; Jiang *et al.*, 2006; Chen *et al.*, 2003b). Nanoparticles of carboxymethyl chitosan (CM-chitosan) as carriers for the anticancer drug, were prepared by gelification with calcium ions and Doxorubicin (DOX) was chosen as a model drug.

6. Conclusions and future trends

The literature survey showed that in the last decades a lot of attention has been focused to the combination of polysaccharides based polymers with inorganic nanoparticles, to benefit from the advantages of both organic and inorganic composite components. As this chapter showed the use of polysaccharides-based nanoparticles is receiving a significant amount of interests because of the plentiful availability of natural polymer, inexpensive, renewability, biocompatibility, biodegradability and nontoxicity. Therefore, a number of formulations of such bionanocomposites exhibits some excellent characteristics such as magnetic, optical, antimicrobial functionalities, size particles, surface coverage, colloidal stability, enzyme degradability and interesting applications of the polysaccharide based nanoparticles and their derivatives for biotechnological and biomedical applications was explained. The preparation of this kind of materials strongly relies on earlier steps of their production and modification steps which emphasises the relevance of preparative strategies that take in consideration their final applications. With this respect, we introduced various methods for the preparation of polysaccharides-based nanoparticles such as: solvent evaporation method, spontaneous emulsification or solvent diffusion method, self-assembly of hydrophobically modified and dialysis method. On the other hand, the modified polysaccharides exhibit considerable potentials to utilize as stabilizers to produce stable hydrophilic nanoparticles through the o/w emulsion/evaporation technique. Modified polysaccharides were shown to exhibit surface active properties and to act as efficient emulsion stabilizers. Surface modified colloidal carriers such as nanoparticles are able to modulate the biodistribution of the loaded drug when given intravenously, but also to control the absorption of drugs administered by other routes. The amphiphilic character imparted upon polysaccharides after hydrophobic modification gives them a wide and interesting use spectrum, for instance as rheology modifiers, emulsion stabilizers, surface modifiers for liposomes and nanoparticles and as drug delivery vehicles. The recent attempts toward finding new methods for the earlier diagnosis of diseases and more effective therapies to synthesize the new generation of multifunctional nanostructured materials based on polysaccharides, modified polysaccharides and polysaccharide-based dendrimers is very fast emerging. As time goes on, more polysaccharide-based nanoparticles emerge, which greatly enriches the versatility of nanoparticle carriers agents in terms of category and function.

7. Acknowledgments

Authors are greatly acknowledging the Research Center for Pharmaceutical Nanotechnology and the University of Tabriz for their financial supports of this work.

8. References

Abraham, T. E. & Simi, C. K. (2007). Hydrophobic grafted and cross-linked starch nanoparticles for drug delivery. *Bioprocess and Biosystems Engineering* 30(3): 173-180.

Akiyoshi, K., Deguchi, S., Moriguchi, N., Yamaguchi, S. & Sunamoto, J. (1993). Self-Aggregates of Hydrophobized Polysaccharides in Water - Formation and Characteristics of Nanoparticles. *Macromolecules* 26(12): 3062-3068.

Akiyoshi, K., Deguchi, S., Tajima, H., Nishikawa, T. & Sunamoto, J. (1997). Microscopic structure and thermoresponsiveness of a hydrogel nanoparticle by self-assembly of a hydrophobized polysaccharide. *Macromolecules* 30(4): 857-861.

Akiyoshi, K., Kang, E. C., Kurumada, S., Sunamoto, J., Principi, T. & Winnik, F. M. (2000). Controlled as sociation of amphiphilic polymers in water: thermosensitive nanoparticles formed by self-assembly of hydrophobically modified pullulans and poly(N-isopropylacrylamides) *Macromolecules* 33: 3244-3249.

Akiyoshi, K., Kobayashi, S., Shichibe, S., Mix, D., Baudys, M., Kim, S. W. & Sunamoto, J. (1998). Self-assembled hydrogel nanoparticle of cholesterol-bearing pullulan as a carrier of protein drugs: complexation and stabilization of insulin. *Journal of Controlled Release* 54: 313-320.

Albertsson, A. C., Qu, X. & Wirsen, A. (1999). Synthesis and characterization of pH-sensitive hydrogels based on chitosan and D,L-lactic acid. *Journal of Applied Polymer Science* 74(13): 3193-3202.

Allemann, E., Quintanar-Guerrero, D., Fessi, H. & Doelker, E. (1998). Preparation techniques and mechanisms of formation of biodegradable nanoparticles from preformed polymers. *Drug Development and Industrial Pharmacy* 24(12): 1113-1128.

Alonso, M. J., Janes, K. A. & Calvo, P. (2001). Polysaccharide colloidal particles as delivery systems for macromolecules. *Adv Drug Deliv Rev* 47(1): 83-97.

Aminabhavi, T. M., Balundgi, R. H. & Cassidy, P. E. (1990). A Review on Biodegradable Plastics. *Polymer-Plastics Technology and Engineering* 29(3): 235-262.

an, R. Y., Matsusaki, M., Akashi, M. & Chirachanchai, S. (2004). Controlled hydrophobic/hydrophilic chitosan: colloidal phenomena and nanosphere formation. *Colloid Polym. Sci.* 282: 337-342.

Anitha, A., Deepagan, V. G., Divya Rani, V. V., Deepthy Menon, Nair, S. V. & Jayakumar, R. (2011). Preparation, characterization, in vitro drug release and biological studies of curcumin loaded dextran sulphate-chitosan nanoparticles. *Carbohydrate Polymers* 84: 1158-1164.

Aoki, N., Nishikawa, M. & Hattori, K. (2003). Synthesis of chitosan derivatives bearing cyclodextrin and adsorption of p-nonylphenol and bisphenol A. *Carbohydrate Polymers* 52(3): 219-223.

Aumelas, A., Serrero, A., Durand, A., Dellacherie, E. & Leonard, M. (2007). Nanoparticles of hydrophobically modified dextrans as potential drug carrier systems. *Colloids Surf B Biointerfaces* 59(1): 74-80.

Bataille, I., Huguet, J., Muller, G., Mocanu, G. & Carpov, A. (1997). Associative behaviour of hydrophobically modified carboxymethylpullulan derivatives. *International Journal of Biological Macromolecules* 20(3): 179-191.

Bertholon, I., Vauthier, C. & Labarre, D. (2006). Complement activation by core-shell poly(isobutylcyanoacrylate)-polysaccharide nanoparticles: influences of surface morphology, length, and type of polysaccharide. *Pharmaceutical Research* 23: 1313-1323.

Borredon, E., Aburto, J. & Alric, I. (1999). Preparation of long-chain esters of starch using fatty acid chlorides in the absence of an organic solvent. *Starch-Starke* 51(4): 132-135.

Bravo-Osuna, I., Millotti, G., Vauthier, C. & Ponchel, G. (2007a). In vitro evaluation of calcium binding capacity of chitosan and thiolated chitosan poly(isobutyl cyanoacrylate) core-shell nanoparticles. *Int J Pharm* 338(1-2): 284-290.

Bravo-Osuna, I., Ponchel, G. & Vauthier, C. (2007b). Tuning of shell and core characteristics of chitosan-decorated acrylic nanoparticles. *Eur J Pharm Sci* 30(2): 143-154.

Bravo-Osuna, I., Schmitz, T., Bernkop-Schnurch, A., Vauthier, C. & Ponchel, G. (2006). Elaboration and characterization of thiolated chitosan-coated acrylic nanoparticles. *Int J Pharm* 316(1-2): 170-175.

Bravo-Osuna, I., Vauthier, C., Farabollini, A., Palmieri, G. F. & Ponchel, G. (2007c). Mucoadhe-sion mechanism of chitosan and thiolated chitosan-poly(isobutyl cyanoacrylate) core-shell nanoparticles. *Biomaterials* 28: 2233-2243.

Brecher, M. E., Owen, H. G. & Bandarenko, N. (1997). Alternatives to albumin: Starch replacement for plasma exchange. *Journal of Clinical Apheresis* 12: 146-153.

Burt, H. M. & Shi, R. W. (2003). Synthesis and characterization of amphiphilic hydroxypropylcellulose-graft-poly(epsilon-caprolactone). *Journal of Applied Polymer Science* 89(3): 718-727.

Cao, A. I., Yang, J., Yu, Y. H., Li, Q. B. & Li, Y. (2005). Chemical synthesis of biodegradable aliphatic polyesters and polycarbonates catalyzed by novel versatile aluminum metal complexes bearing salen ligands. *Journal of Polymer Science Part a-Polymer Chemistry* 43(2): 373-384.

Chae, S. Y., Son, S., Lee, M., Jang, M. K. & Nah, J. W. (2005). Deoxycholic acid-conjugated chitosan oligosaccharide nanoparticles for ef ficient gene carrier. *Journal of Controlled Release* 109: 330-344.

Chauvierre, C., Labarre, D., Couvreur, P. & Vauthier, C. (2003). Novel polysaccharide-decorated poly(isobutyl cyanoacrylate) nanoparticles. *Pharm Res* 20(11): 1786-1793.

Chauvierre, C., Marden, M. C., Vauthier, C., Labarre, D., Couvreur, P. & Leclerc, L. (2004). Heparin coated poly(alkylcyanoacrylate) nanoparticles coupled to hemoglobin: a new oxygen carrier. *Biomaterials* 25(15): 3081-3086.

Chen, X. G., Lee, C. M. & Park, H. J. (2003a). OM emulsification for the self-aggregation and nanoparticle formation of linoleic acid-modified chitosan in the aqueous system. *Journal of Agricultural and Food Chemistry* 51: 3135-3139.

Chen, X. G., Lee, C. M. & Park, H. J. (2003b). OM emulsification for the self-aggregation and nanoparticle formation of linoleic acid-modified chitosan in the aqueous system. *Journal of Agricultural and Food Chemistry* 51: 3135-3139.

Cheng, F., Wei, Y. P., Hou, G. & Sun, S. F. (2008). Amphiphilic cellulose: Surface activity and aqueous self-assembly into nano-sized polymeric micelles. *Reactive & Functional Polymers* 68(5): 981-989.

Choi, S. H., Lee, J. H., Choi, S. M. & Park, T. G. (2006). Thermally reversible pluronic/heparin nanocapsules exhibiting 1000-fold volume transition. *Langmuir* 22(4): 1758-1762.

Choisnard, L., Geze, A., Putaux, J. L., Wong, Y. S. & Wouessidjewe, D. (2006). Nanoparticles of beta-cyclodextrin esters obtained by self-assembling of biotransesterified beta-cyclodextrins. *Biomacromolecules* 7: 515-520.

Couvreur, P. (1998). Polyalkylcyanoacrylates as colloidal drug carriers. *Crit Rev Ther Drug Carr Syst* 5: 1-20.

Couvreur, P., Lemarchand, C. & Gref, R. (2004). Polysaccharide-decorated nanoparticles. *European Journal of Pharmaceutics and Biopharmaceutics* 58(2): 327-341.

Della Valle, G., Buleon, A., Carreau, P. J., Lavoie, P. A. & Vergnes, B. (1998). Relationship between structure and viscoelastic behavior of plasticized starch. *Journal of Rheology* 42(3): 507-525.

Dong, H., Xu, Q., Li, Y., Mo, S., Cai, S. & Liu, L. (2008). The synthesis of biodegradable graft copolymer cellulose-graft-poly(L-lactide) and the study of its controlled drug release. *Colloids Surf B Biointerfaces* 66(1): 26-33.

Dufresne, A., Angellier, H., Choisnard, L., Molina-Boisseau, S. & Ozil, P. (2004). Optimization of the preparation of aqueous suspensions of waxy maize starch nanocrystals using a response surface methodology. *Biomacromolecules* 5(4): 1545-1551.

Dufresne, A., Cavaille, J. Y. & Helbert, W. (1996). New nanocomposite materials: Microcrystalline starch reinforced thermoplastic. *Macromolecules* 29(23): 7624-7626.

Dufresne, A., Thielemans, W. & Belgacem, M. N. (2006). Starch nanocrystals with large chain surface modifications. *Langmuir* 22(10): 4804-4810.

Durand, A., Marie, E., Rotureau, E., Leonard, M. & Dellacherie, E. (2004). Amphiphilic polysaccharides: Useful tools for the preparation of nanoparticles with controlled surface characteristics. *Langmuir* 20(16): 6956-6963.

Durand, A., Rouzes, C., Leonard, M. & Dellacherie, E. (2002). Surface activity and emulsification properties of hydrophobically modified dextrans. *Journal of Colloid and Interface Science* 253(1): 217-223.

Elvira, C., Mano, J. F., San Román, J. & Reis, R. L. (2002). Starch-based biodegradable hydrogels with potential biomedical applications as drug delivery systems. Biomaterials. 23: 1955-1966.

Fang, Y., Huang, M. F., Liu, L., Zhang, G. B. & Yuan, G. B. (2006). Preparation of chitosan derivative with polyethylene glycol side chains for porous structure without specific processing technique. *International Journal of Biological Macromolecules* 38(3-5): 191-196.

Fessi, H., Puisieux, F., Devissaguet, J., Ammoury, N. & Benita, S. (1989). Nanocapsule formation by interfacial polymer deposition following solvent displacement. *Int J Pharm* 55: 1-4.

Fischer, A., Houzelle, M. C., Hubert, P., Axelos, M. A. V., Geoffroy-Chapotot, C., Carre, M. C., Viriot, M. L. & Dellacherie, E. (1998). Detection of intramolecular associations in hydrophobically modified pectin derivatives using fluorescent probes. *Langmuir* 14(16): 4482-4488.

Fowler, P. A., Fang, J. M., Sayers, C. & Williams, P. A. (2004). The chemical modification of a range of starches under aqueous reaction conditions. *Carbohydrate Polymers* 55(3): 283-289.

Fowler, P. A., Fang, J. M., Tomkinson, J. & Hill, C. A. S. (2002). The preparation and characterisation of a series of chemically modified potato starches. *Carbohydrate Polymers* 47(3): 245-252.

Francis, M., Lavoie, L., Winnik, F. & Leroux, J. C. (2003). Solubilization of cyclosporin A in dextran-g-polyethyleneglycolalkylether polymeric micelles. *Eur. J. Pharm. Sci.* 56: 337-346.

French, D., Pulley, A. O., Effenberger, J. A., Rougvie, M. A. & Abdullah, M. (1965). The schardinger dextrin: molecular size and structure of the δ-, ε-, ζ-, and η-dextrin. *Arch. Biochem. Biophys.* 111: 153-160.

Fringuelli, F., Piermatti, O., Pizzo, F. & Vaccaro, L. (1999). Ring opening of epoxides with sodium azide in water. A regioselective pH-controlled reaction. *Journal of Organic Chemistry* 64(16): 6094-6096.

Fujiwara, T., Tanaka, N. & Kobayashi, S. (1990). Structure of Delta-Cyclodextrin 13.75h2o. *Chemistry Letters* (5): 739-742.

Gassan, J. & Bledzki, A. K. (1999). Composites reinforced with cellulose based fibres. *Progress in Polymer Science* 24(2): 221-274.

Glinel, K., Huguet, J. & Muller, G. (1999). Comparison of the associating behaviour between neutral and anionic alkylperfluorinated pullulan derivatives. *Polymer* 40(25): 7071-7081.

Gorochovceva, N., Naderi, A., Dedinaite, A. & Makuska, R. (2005). Chitosan-N-poly(ethylene glycol) brush copolymers: Synthesis and adsorption on silica surface. *European Polymer Journal* 41(11): 2653-2662.

Gref, R., Rodrigues, J. & Couvreur, P. (2002). Polysaccharides grafted with polyesters: Novel amphiphilic copolymers for biomedical applications. *Macromolecules* 35(27): 9861-9867.

Gref, R., Rodrigues, J. S., Santos-Magalhaes, N. S., Coelho, L. C. B. B., Couvreur, P. & Ponchel, G. (2003). Novel core (polyester)-shell(polysaccharide) nanoparticles: protein loading and surface modification with lectins. *Journal of Controlled Release* 92(1-2): 103-112.

Gupta, M. & Gupta, A. (2004). Hydrogel pullulan nanoparticles encapsulating pBUDLacZ plasmid as an efficient gene delivery carrier. *Journal of Controlled Release* 99: 157-166.

Gurruchaga, M., Silva, I. & Goni, I. (2009). Physical blends of starch graft copolymers as matrices for colon targeting drug delivery systems. *Carbohydrate Polymers* 76(4): 593-601.

Hamidi, M., Azadi, A. & Rafiei, P. (2008). Hydrogel nanoparticles in drug delivery. *Adv Drug Deliv Rev* 60(15): 1638-1649.

Hennink, W. E., van de Manakker, F., Vermonden, T. & van Nostrum, C. F. (2009). Cyclodextrin-Based Polymeric Materials: Synthesis, Properties, and Pharmaceutical/Biomedical Applications. *Biomacromolecules* 10(12): 3157-3175.

Hinrichsen, G., Mohanty, A. K. & Misra, M. (2000). Biofibres, biodegradable polymers and biocomposites: An overview. *Macromolecular Materials and Engineering* 276(3-4): 1-24.

Hong-Wei Lua & Li-Ming Zhanga (2011). Carbohydrate Preparation and properties of new micellar drug carriers based on hydrophobically modified amylopectin *Polymers* 83: 1499-1506.

Hu, F. Q., Ren, G. F., Yuan, H., Du, Y. Z. & Zeng, S. (2006). Shell cross-linked stearic acid grafted chitosan oligosaccharide self-aggregated micelles for controlled release of paclitaxel. *Colloids Surf B Biointerfaces* 50(2): 97-103.

I., B., S., H. & R., K. M. (2004). PLGA nano particles in drug delivery: the state of the art. *Crit. Rev. Ther. Drug Carrier Syst.* 21(5): 387-422.

Ichinose, K., Tomiyama, N., Nakashima, M., Ohya, Y., Ichikawa, M., Ouchi, T. & Kanematsu, T. (2000). Antitumor activity of dextran derivatives immobilizing platinum complex (II). *Anti-Cancer Drugs* 11(1): 33-38.

Jeon, H., Jeong, Y., Jang, M., Park, Y. & Nah, J. (2000). Effect of solvent on the preparation of surfactant-free poly(DL-lactide-co-glycolide) nanoparticles and norfloxacin release characteristics. *Int J Pharm* 207: 99-108.

Jeong, S. Y., Lee, K. Y., Kwon, I. C., Kim, Y. H. & Jo, W. H. (1998). Preparation of chitosan self-aggregates as a gene delivery system. *Journal of Controlled Release* 51(2-3): 213-220.

Jeong, Y., Cho, C., Kim, S., Ko, K., Kim, S., Shim, Y. & Nah, J. (2001). Preparation of poly(DL-lactide-co-glycolide) nanoparticles without surfactant. *Journal of Applied Polymer Science* 80: 2228-2236.

Jeong, Y., Na, H. S., Oh, J. S., Choi, K. C., Song, C. & Lee, H. (2006). Adriamycin release from self-assembling nanospheres of poly(DL-lactide-co-glycolide)-grafted pullulan. *International Journal of Pharmaceutics* 322: 154-160.

Jiang, G. B., Quan, D., Liao, K. & Wang, H. (2006). Novel polymer micelles prepared from chitosan grafted hydrophobic palmitoyl groups for drug delivery. *Mol Pharm* 3(2): 152-160.

Jing, X. B., Yu, H. J., Wang, W. S., Chen, X. S. & Deng, C. (2006). Synthesis and characterization of the biodegradable polycaprolactone-graft-chitosan amphiphilic copolymers. *Biopolymers* 83(3): 233-242.

Jung, S., Jeong, Y. I., Kim, S. H., Jung, T. Y., Kim, I. Y., Kang, S. S., Jin, Y. H., Ryu, H. H., Sun, H. S., Jin, S. G., Kim, K. K. & Ahn, K. Y. (2006). Polyion complex micelles composed of all-trans retinoic acid and poly (ethylene glycol)-grafted-citosan. *J Pharm Sci* 95(11): 2348-2360.

Jung, S. W., Jeong, Y. I., Kim, Y. H. & Kim, S. W. (2004). Self- assembled nanoparticles of poly (ethylene glycol) grafted pullulan acetate as a novel drug carrier. *Arch. Pharm. Res.* 27: 562-569.

Kapusniak, J. & Siemion, P. (2007). Thermal reactions of starch with long-chain unsaturated fatty acids. Part 2. Linoleic acid. *Journal of Food Engineering* 78(1): 323-332.

Kim, J. H., Kwon, H. Y., Lee, J. Y., Choi, S. W. & Jang, Y. S. (2001). Preparation of PLGA nanoparticles containing estrogen by emulsification-diffusion method. *Colloids and Surfaces a-Physicochemical and Engineering Aspects* 182(1-3): 123-130.

Kim, K., Kwon, S., Park, J. H., Chung, H., Jeong, S. Y., Kwon, I. C. & Kim, I. S. (2005). Physicochemical characterizations of self-assembled nanoparticles of glycol chitosan-deoxycholic acid conjugates. *Biomacromolecules* 6(2): 1154-1158.

Kim, S. H., Kim, I. S. & Jeong, Y. I. (2000). Self-assembled hydrogel nanoparticles composed of dextran and poly(ethylene glycol) macromer. *International Journal of Pharmaceutics* 205(1-2): 109-116.

Klemm, D., Heublein, B., Fink, H. P. & Bohn, A. (2005). Cellulose: Fascinating biopolymer and sustainable raw material. *Angewandte Chemie-International Edition* 44(22): 3358-3393.

Kommareddy, S., Tiwari, S. & Amiji, M. (2005). Long-circulating polymeric nanovectors for tumor-selective gene delivery. *Technol Cancer Res Treat* 4: 615-625.

Kostog. M, Kohler. S, Liebert. T & Heinze. T (2010). Pure cellulose nanoparticles from trimethylsilyl cellulose. *Macromol Symp* 294(2): 96-106.

Kreuter, J. (1991). Peroral administration of nanoparticles. *Adv Drug Deliv Rev* 7(1): 71-86.

Kumar, M. N. V. R. (2000). A review of chitin and chitosan applications. *Reactive & Functional Polymers* 46(1): 1-27.

Kurita, K. (2001). Controlled functionalization of the polysaccharide chitin. *Progress in Polymer Science* 26(9): 1921-1971.

Kuroda, K., Fujimoto, K., Sunamoto, J. & Akiyoshi, K. (2002). Hierarchical self-assembly of hydrophobically modified pullulan in water: gelation by networks of nanoparticles. *Langmuir* 18: 3780-3786.

Kwon, G. S. (2003). Polymeric micelles for delivery of poorly water-soluble compounds. *Critical Reviews in Therapeutic Drug Carrier Systems* 20(5): 357-403.

Kwon, I. C., Kim, J. H., Kim, Y. S., Kim, S., Park, J. H., Kim, K., Choi, K., Chung, H., Jeong, S. Y., Park, R. W. & Kim, I. S. (2006). Hydrophobically modified glycol chitosan nanoparticles as carriers for paclitaxel (Reprinted from Journal of Controlled Release, vol 109, pg 1, 2005). *Journal of Controlled Release* 111(1-2): 228-234.

Kwon, I. C., Kwon, S., Park, J. H., Chung, H., Jeong, S. Y. & Kim, I. S. (2003). Physicochemical characteristics of self-assembled nanoparticles based on glycol chitosan bearing 5 beta-cholanic acid. *Langmuir* 19(24): 10188-10193.

Kwon, I. C., Park, J. H., Kwon, S. G., Nam, J. O., Park, R. W., Chung, H., Seo, S. B., Kim, I. S. & Jeong, S. Y. (2004). Self-assembled nanoparticles based on glycol chitosan bearing 5 beta-cholanic acid for RGD peptide delivery. *Journal of Controlled Release* 95(3): 579-588.

Labhasetwar, V., Panyam, J., Dali, M. A., Sahoo, S. K., Ma, W. X., Chakravarthi, S. S., Amidon, G. L. & Levy, R. J. (2003). Polymer degradation and in vitro release of a model protein from poly(D,L-lactide-co-glycolide) nano- and microparticles. *Journal of Controlled Release* 92(1-2): 173-187.

Larsen, K. L. (2002). Large cyclodextrins. *Journal of Inclusion Phenomena and Macrocyclic Chemistry* 43(1-2): 1-13.

Larsen, K. L., Endo, T., Ueda, H. & Zimmermann, W. (1998). Inclusion complex formation constants of alpha-, beta-, gamma-, delta-, epsilon-, zeta-, eta- and theta-cyclodextrins determined with capillary zone electrophoresis. *Carbohydrate Research* 309(2): 153-159.

Lee, K. Y., Jo, W. H., Kwon, I. C., Kim, Y. H. & Jeong, S. Y. (1998). Structural determination and interior polarity of self-aggregates prepared from deoxycholic acid-modified chitosan in water. *Macromolecules* 31: 378-383.

Lee, K. Y., Kim, J. H., Kwon, I. C. & Jeong, S. Y. (2000). Self-aggregates of deoxycholic acid modified chitosan as a novel carrier of adriamycin. *Colloid and Polymer Science* 278: 1216-1219

Lee, M., Cho, Y. W., Park, J. H., Chung, H. S., Jeong, S. Y., Choi, K. W., Moon, D. H., Kim, S. Y., Kim, I. S. & Kwon, I. C. (2006). Size control of self-assembled nanoparticles by an emulsion/solvent evaporation method. *Colloid and Polymer Science* 284(5): 506-512.

Lee, M. & Kim, S. (2005). Polyethylene glycol-conjugated copolymers for plasmid DNA delivery. *Pharmaceutical Research* 22: 1-10.

Lee, V. H. L. (2001). Encyclopedia of Controlled Drug Delivery. *Journal of Controlled Release* 71(3): 353-354.

Lemarchand, C., Couvreur, P., Besnard, M., Costantini, D. & Gref, R. (2003). Novel polyester-polysaccharide nanoparticles. *Pharm Res* 20(8): 1284-1292.

Lemos-Senna, E., Wouessidjewe, D., Lesieur, S. & Duchene, D. (1998). Preparation of amphiphilic cyclodextrin nanospheres using the emulsification solvent evaporation method. Influence of the surfactant on preparation and hydrophobic drug loading. *International Journal of Pharmaceutics* 170: 119-128.

Leonard, M., Rouzes, C., Durand, A. & Dellacherie, E. (2003). Influence of polymeric surfactants on the properties of drug-loaded PLA nanospheres. *Colloids and Surfaces B-Biointerfaces* 32(2): 125-135.

Leonard, M., Rouzes, C., Gref, R., Delgado, A. D. & Dellacherie, E. (2000). Surface modification of poly(lactic acid) nanospheres using hydrophobically modified dextrans as stabilizers in an o/w emulsion/evaporation technique. *Journal of Biomedical Materials Research* 50(4): 557-565.

Leroux, J. C., Allemann, E., Doelker, E. & Gurny, R. (1995). New Approach for the Preparation of Nanoparticles by an Emulsification-Diffusion Method. *European Journal of Pharmaceutics and Biopharmaceutics* 41(1): 14-18.

Letchford, K. & Burt, H. (2007). A review of the formation and classi fication of amphiphilic block copolymer nanoparticulate structures: micelles, nanospheres, nanocap-sules and polymersomes. *Eur. J. Pharm. Biopharm.* 65: 259-269.

Lim, S. T. & Kim, J. Y. (2009). Preparation of nano-sized starch particles by complex formation with n-butanol. *Carbohydrate Polymers* 76(1): 110-116.

Liu, C. G., Desai, K. G. H., Chen, X. G. & Park, H. J. (2005). Preparation and characterization of nanoparticles containing trypsin based on hydrophobically modified chitosan. *Journal of Agricultural and Food Chemistry* 53: 1728-1733.

Lu, M., Albertsson, P. A., Johansson, G. & Tjerneld, F. (1994). Partitioning of Proteins and Thylakoid Membrane-Vesicles in Aqueous 2-Phase Systems with Hydrophobically-Modified Dextran. *Journal of Chromatography A* 668(1): 215-228.

Lu, M. & Tjerneld, F. (1997). Interaction between tryptophan residues and hydrophobically modified dextran - Effect on partitioning of peptides and proteins in aqueous two-phase systems. *Journal of Chromatography A* 766(1-2): 99-108.

Mader, K., Besheer, A., Hause, G. & Kressler, J. (2007). Hydrophobically modified hydroxyethyl starch: Synthesis, characterization, and aqueous self-assembly into nano-sized polymeric micelles and vesicles. *Biomacromolecules* 8(2): 359-367.

Martel, B., Devassine, M., Crini, G., Weltrowski, M., Bourdonneau, M. & Morcellet, M. (2001). Preparation and sorption properties of a beta-cyclodextrin-linked chitosan derivative. *Journal of Polymer Science Part a-Polymer Chemistry* 39(1): 169-176.

Melo, E., Nichifor, M., Lopes, A. & Carpov, A. (1999). Aggregation in water of dextran hydrophobically modified with bile acids. *Macromolecules* 32(21): 7078-7085.

Miyazawa, I., Ueda, H., Nagase, H., Endo, T., Kobayashi, S. & Nagai, T. (1995). Physicochemical Properties and Inclusion Complex-Formation of Delta-Cyclodextrin. *European Journal of Pharmaceutical Sciences* 3(3): 153-162.

Muller, G., Duval-Terrie, C. & Huguet, J. (2003). Self-assembly and hydrophobic clusters of amphiphilic polysaccharides. *Colloids and Surfaces a-Physicochemical and Engineering Aspects* 220(1-3): 105-115.

Na, K., Lee, T. B., Park, K. H., Shin, E. K., Lee, Y. B. & Cho, H. K. (2003). Self-assembled nanoparticles of hydrophobically-modified polysaccharide bearing vitamin H as a targeted anti-cancer drug delivery system. *Eur. J. Pharm. Sci.* 18: 165-173.

Na, K., Park, K. H., Song, H. C., Bom, H. S., Lee, K. H., Kim, S., Kang, D. & Lee, D. H. (2007). Ionic strength-sensitive pullulan acetate nanoparticles (PAN) for intratumoral administration of radioisotope: Ionic strength-dependent aggregation behavior and (99m)Technetium retention property. *Colloids and Surfaces B-Biointerfaces* 59(1): 16-23.

Namazi, H., Bahrami, S. & Entezami, A. A. (2005). Synthesis and controlled release of biocompatible prodrugs of beta-cyclodextrin linked with PEG containing ibuprofen or indomethacin. *Iranian Polymer Journal* 14(10): 921-927.

Namazi, H. & Dadkhah, A. (2008). Surface modification of starch nanocrystals through ring-opening polymerization of epsilon-caprolactone and investigation of their microstructures. *Journal of Applied Polymer Science* 110(4): 2405-2412.

Namazi, H. & Dadkhah, A. (2010). Convenient method for preparation of hydrophobically modified starch nanocrystals with using fatty acids. *Carbohydrate Polymers* 79(3): 731-737.

Namazi, H., Fathi, F. & Dadkhah, A. (2011). Hydrophobically modified starch using long-chain fatty acids for preparation of nanosized starch particles . *Scientia Iranica, Transactions C: Chemistry and Chemical Engineering* 18: 439-445.

Namazi, H. & Jafarirad, S. (2008). Preparation of the New Derivatives of Cellulose and Oligomeric Species of Cellulose Containing Magneson II Chromophore. *Journal of Applied Polymer Science* 110(6): 4034-4039.

Namazi, H. & Kanani, A. (2009). Investigation diffusion mechanism of beta-lactam conjugated telechelic polymers of PEG and beta-cyclodextrin as the new nanosized drug carrier devices. *Carbohydrate Polymers* 76(1): 46-50.

Namazi, H. & Mosadegh, M. (2011) Bio-nanocomposites based on naturally occurring common polysaccharides chitosan, cellulose and starch with their biomedical applications. In Tiwari, A. (Eds) Recent developments in bio-nanocomposites for biomedical applications (pp. 379-397)

Namazi, H. & Mosadegh, M. (2011). Preparation and Properties of Starch/Nanosilicate Layer/Polycaprolactone Composites. J Polym Environ 19: 980-987

Namazi, H., Mosadegh, M. & Dadkhah, A. (2009). New intercalated layer silicate nanocomposites based on synthesized starch-g-PCL prepared via solution intercalation and in situ polymerization methods: As a comparative study. *Carbohydrate Polymers* 75(4): 665-669.

Namazi, H. & Rad, S. J. (2004). Synthesis of block and grafted copolymers containing spacer-linked chromophore based on cellulose and polyethylene glycol. *Journal of Applied Polymer Science* 94(3): 1175-1185.

Nishikawa, T., Akiyoshi, K. & Sunamoto, J. (1996). Macromolecular complexation between bovine serum albumin and the self-assembled hydrogel nanoparticle of hydrophobized polysaccharides. *Journal of the American Chemical Society* 118(26): 6110-6115.

Nishio, Y. & Teramoto, Y. (2003). Cellulose diacetate-graft-poly(lactic acid)s: synthesis of wide-ranging compositions and their thermal and mechanical properties. *Polymer* 44(9): 2701-2709.

Nishio, Y., Teramoto, Y., Yoshioka, M. & Shiraishi, N. (2002). Plasticization of cellulose diacetate by graft copolymerization of epsilon-caprolactone and lactic acid. *Journal of Applied Polymer Science* 84(14): 2621-2628.

ODonnell, P. B. & McGinity, J. W. (1997). Preparation of microspheres by the solvent evaporation technique. *Adv Drug Deliv Rev* 28(1): 25-42.

Ogata, T. & Takahashi, Y. (1995). *Carbohydrate Research* 138: C5.

Onyuksel, H., Krishnadas, A. & Rubinstein, I. (2003). Sterically stabilized phospholipid mixed micelles: In vitro evaluation as a novel carrier for water-insoluble drugs. *Pharmaceutical Research* 20(2): 297-302.

Opanasopit, P., Ngawhirunpat, T., Chaidedgumjorn, A., Rojanarata, T., Apirakaramwong, A., Phongying, S., Choochottiros, C. & Chirachanchai, S. (2006). Incorporation of camptothecin into N-phthaloyl chitosan-g-mPEG self-assembly micellar system. *European Journal of Pharmaceutics and Biopharmaceutics* 64(3): 269-276.

Opanasopit, P., Ngawhirunpat, T., Rojanarata, T., Choochottiros, C. & Chirachanchai, S. (2007). Camptothecin-incorporating N-phthaloylchitosan-g-mPEG self-assembly micellar system: effect of degree of deacetylation. *Colloids Surf B Biointerfaces* 60(1): 117-124.

Osterberg, E., Bergstrom, K., Holmberg, K., Schuman, T. P., Riggs, J. A., Burns, N. L., Vanalstine, J. M. & Harris, J. M. (1995). Protein-Rejecting Ability of Surface-Bound Dextran in End-on and Side-on Configurations - Comparison to Peg. *Journal of Biomedical Materials Research* 29(6): 741-747.

Ouchi, T., Nishizawa, H. & Ohya, Y. (1998). Aggregation phenomenon of PEG-grafted chitosan in aqueous solution. *Polymer* 39(21): 5171-5175.

P., C. (1988). Polyalkylcyanoacrylates as colloidal drug carriers. *Crit Rev Ther Drug Carr Syst* 5: 1-20.

Panyam, J. & Labhasetwar, V. (2003). Biodegradable nanoparticles for drug and gene delivery to cells and tissue. *Adv Drug Deliv Rev* 55(3): 329-347.

Park, J. S., Han, T. H., Lee, K. Y., Han, S. S., Hwang, J. J., Moon, D. H., Kim, S. Y. & Cho, Y. W. (2006). N-acetyl histidine-conjugated glycol chitosan self-assembled nanoparticles for intracytoplasmic delivery of drugs: endocytosis, exocytosis and drug release. *Journal of Controlled Release* 115(1): 37-45.

Park, J. S., Koh, Y. S., Bang, J. Y., Jeong, Y. I. & Lee, J. J. (2008). Antitumor effect of all-trans retinoic acid-encapsulated nanoparticles of methoxy poly(ethylene glycol)-conjugated chitosan against CT-26 colon carcinoma in vitro. *J Pharm Sci* 97(9): 4011-4019.

Park, K., Kim, J. H., Nam, Y. S., Lee, S., Nam, H. Y., Kim, K., Park, J. H., Kim, I. S., Choi, K., Kim, S. Y. & Kwon, I. C. (2007). Effect of polymer molecular weight on the tumor targeting characteristics of self-assembled glycol chitosan nanoparticles. *Journal of Controlled Release* 122(3): 305-314.

Park, K., Kim, K., Kwon, I. C., Kim, S. K., Lee, S., Lee, D. Y. & Byun, Y. (2004). Preparation and characterization of self-assembled nanoparticles of heparin-deoxycholic acid conjugates. *Langmuir* 20(26): 11726-11731.

Passirani, C., Barratt, G., Devissaguet, J. P. & Labarre, D. (1998a). Interactions of nanoparticles bearing heparin or dextran covalently bound to poly(methyl methacrylate) with the complement system. *Life Sci* 62(8): 775-785.

Passirani, C., Barratt, G., Devissaguet, J. P. & Labarre, D. (1998b). Long-circulating nanoparticles bearing heparin or dextran covalently bound to poly(methyl methacrylate),. *Pharmaceutical Research* 15: 1046-1050.

Payne, G. F., Yi, H. M., Wu, L. Q., Bentley, W. E., Ghodssi, R., Rubloff, G. W. & Culver, J. N. (2005). Biofabrication with chitosan. *Biomacromolecules* 6(6): 2881-2894.

Prabaharan, M. & Mano, J. F. (2005). Chitosan-based particles as controlled drug delivery systems. *Drug Delivery* 12(1): 41-57.

Ragauskas, A. J., Zhang, J. G., Elder, T. J. & Pu, Y. Q. (2007). Facile synthesis of spherical cellulose nanoparticles. *Carbohydrate Polymers* 69(3): 607-611.

Riedel, U. & Nickel, J. (1999). Natural fibre-reinforced biopolymers as construction materials - new discoveries. *Angewandte Makromolekulare Chemie* 272: 34-40.

Rodrigues, J. S., Santos-Magalhaes, N. S., Coelho, L. C. B. B., Couvreur, P., Ponchel, G. & Gref, R. (2003). Novel core (polyester)-shell(polysaccharide) nanoparticles: protein loading and surface modification with lectins. *Journal of Controlled Release* 92: 103-112.

Saboktakin, M., Tabatabaie, R., Maharramov, A. & Ramazanov, M. (2010). Synthesis and characterization of superparamagnetic chitosan-dextran sulfate hydrogels as nano carriers for colon-specific drug delivery. *Carbohydrate Polymers* 81: 372-376.

Saenger, W. (1980). *Angewandte Chemie-International Edition* 19: 344-362.

Saenger, W. R., Jacob, J., Gessler, K., Steiner, T., Hoffmann, D., Sanbe, H., Koizumi, K., Smith, S. M. & Takaha, T. (1998). Structures of the common cyclodextrins and their larger analogues - Beyond the doughnut. *Chemical Reviews* 98(5): 1787-1802.

Sakairi, N., Tojima, T., Katsura, H., Han, S. M., Tanida, F., Nishi, N. & Tokura, S. (1998). Preparation of an alpha-cyclodextrin-linked chitosan derivative via reductive amination strategy. *Journal of Polymer Science Part a-Polymer Chemistry* 36(11): 1965-1968.

Seppala, J., Pahimanolis, N., Vesterinen, A. H. & Rich, J. (2010). Modification of dextran using click-chemistry approach in aqueous media. *Carbohydrate Polymers* 82(1): 78-82.

Shimoni, E., Lalush, I., Bar, H., Zakaria, I. & Eichler, S. (2005). Utilization of amylose-lipid complexes as molecular nanocapsules for conjugated linoleic acid. *Biomacromolecules* 6(1): 121-130.

Sinha, V. R. & Kumria, R. (2001). Polysaccharides in colon-specific drug delivery. *International Journal of Pharmaceutics* 224(1-2): 19-38.

Sinha, V. R., Singla, A. K., Wadhawan, S., Kaushik, R., Kumria, R., Nansal, K. & Dhawan, S. (2004). Chitosan microspheres as a potential carrier for drugs. *Int J Pharm* 274: 1-33.

Soma, C. E., Dubernet, C., Barratt, G., Nemati, F., Appel, M., Benita, S. & Couvreur, P. (1999). Ability of doxorubicin-loaded nanoparticles to overcome multidrug resistance of tumour cells after their capture by macrophages. *Pharmaceutical Research* 16: 1710-1716.

Son, Y. J., Jang, J. S., Cho, Y. W., Chung, H., Park, R. W., Kwon, I. C., Kim, I. S., Park, J. Y., Seo, S. B., Park, C. R. & Jeong, S. Y. (2003). Biodistribution and anti-tumor ef fi cacy of doxorubicin loaded glycol-chitosan nanoaggregates by EPR effect. *Journal of Controlled Release* 91: 135-145.

Song, C. E., Jeong, Y. I. & Choi, K. C. (2006). Doxorubicin release from core-shell type nanoparticles of poly(DL-lactide-co-glycolide)-grafted dextran. *Archives of Pharmacal Research* 29(8): 712-719.

Sun, K., Tang, M. H. & Dou, H. J. (2006). One-step synthesis of dextran-based stable nanoparticles assisted by self-assembly. *Polymer* 47(2): 728-734.

T. Nakagawa, K., Ueno, M., Kashiw, J. & Watanabe, J. (1994). Preparation of a novel cyclodextrin homologue with d.p. five. *Tetrahedron Letters* 35: 1921-1924.

Thanou, M., Kean, T. & Roth, S. (2005). Trimethylated chitosans as non-viral gene delivery vectors: Cytotoxicity and transfection efficiency. *Journal of Controlled Release* 103(3): 643-653.

Tharanathan, R. N. & Ramesh, H. P. (2003). Carbohydrates - The renewable raw materials of high biotechnological value. *Critical Reviews in Biotechnology* 23(2): 149-173.

Tiera, M. J., Vieira, N. A. B., Moscardini, M. S. & Tiera, V. A. D. (2003). Aggregation behavior of hydrophobically modified dextran in aqueous solution: a fluorescence probe study. *Carbohydrate Polymers* 53(2): 137-143.

Tukomane, T. & Varavinit, S. (2008). Influence of octenyl succinate rice starch on rheological properties of gelatinized rice starch before and after retrogradation. *Starch-Starke* 60(6): 298-304.

Ueda, H. (2002). Physicochemical properties and complex formation abilities of large-ring cyclodextrins. *Journal of Inclusion Phenomena and Macrocyclic Chemistry* 44(1-4): 53-56.

Vanderhoff, J., El Aasser, M. & Ugelstad, J. (1979).Polymer emulsification process. In *US Patent*, Vol. 4,177,177.

Vandijkwolthuis, W. N. E., Franssen, O., Talsma, H., Vansteenbergen, M. J., Vandenbosch, J. J. K. & Hennink, W. E. (1995). Synthesis, Characterization, and Polymerization of Glycidyl Methacrylate Derivatized Dextran. *Macromolecules* 28(18): 6317-6322.

Vauthier, C., Chauvierre, C., Labarre, D. & Hommel, H. (2004). Evaluation of the surface properties of dextran-coated poly(isobutylcyanoacrylate) nanoparticles by spin-labelling coupled with electron resonance spectroscopy. *Colloid and Polymer Science* 282(9): 1016-1025.

Vyas, S. P. & Sihorkar, V. (2001). Potential of polysaccharide anchored liposomes in drug delivery, targeting and immunization. *Journal of Pharmacy and Pharmaceutical Sciences* 4(2): 138-158.

Wang, P. X., Chi, H., Xu, K., Wu, X. L., Chen, Q., Xue, D. H., Song, C. & Zhang, W. (2008). Effect of acetylation on the properties of corn starch. *Food Chemistry* 106(3): 923-928.

Wang, P. X., Chi, H., Xu, K., Xue, D. H., Song, C. L. & Zhang, W. D. (2007a). Synthesis of dodecenyl succinic anhydride (DDSA) corn starch. *Food Research International* 40(2): 232-238.

Wang, Y. S., Jiang, Q., Liu, L. R. & Zhang, Q. Q. (2007b). The interaction between bovine serum albumin and the self- aggregated nanoparticles of cholesterol -modified O-carboxymethyl chitosan. *Polymer* 48: 4135-4142.

Wang, Y. S., Liu, L. R., Jiang, Q. & Zhang, Q. Q. (2007c). Self-aggregated nanoparticles of cholesterol-modified chitosan conjugate as a novel carrier of epirubicin. *European Polymer Journal* 43: 43-51.

Wu, H. Y., Gao, W. X., Lin, X. Q., Lin, X. P., Ding, J. C. & Huang, X. B. (2011). Preparation of nano-sized flake carboxymethyl cassava starch under ultrasonic irradiation. *Carbohydrate Polymers* 84(4): 1413-1418.

Yang, L. Q., Kuang, J. L., Li, Z. Q., Zhang, B. F., Cai, X. & Zhang, L. M. (2008a). Amphiphilic cholesteryl-bearing carboxymethylcellulose derivatives: self-assembly and rheological behaviour in aqueous solution. *Cellulose* 15(5): 659-669.

Yang, S. C., Ge, H. X., Hu, Y., Jiang, X. Q. & Yang, C. Z. (2000). Formation of positively charged poly (butyl cyanoacrylate) nanoparticles stabilized with chitosan. *Colloid and Polymer Science* 278: 285-292.

Yang, X. D., Zhang, Q. Q., Wang, Y. S., Chen, H., Zhang, H. Z., Gao, F. P. & Liu, L. R. (2008b). Self-aggregated nanoparticles from methoxy poly(ethylene glycol)-modified chitosan: Synthesis; characterizat ion; aggregation and methotrexate release in vitro,. *Colloids Surf.* 61: 125-131.

Yoo, H. S., Lee, J. E., Chung, H., Kwon, I. & Jeong, S. Y. (2005). Self-assembled nanoparticles containing hydrophobically modified glycol chitosan for gene delivery. *Journal of Controlled Release* 103: 235-243.

Yuan & Zhuangdong (2007). Study on the synthesis and catalyst oxidation properties of chitosan bound nickel(II) complexes. *Journal of Agricultural and Food Chemistry* 21(5): 22-24.

Yuk, S. H., Han, S. K., Lee, J. H., Kim, D. & Cho, S. H. (2005). Hydrophilized poly(lactide-co-glycolide) nanoparticles with core/shell structure for protein delivery. *Science and Technology of Advanced Materials* 6(5): 468-474.

Zambaux, M., Zambaux, X. F., Gref, R., Maincent, P., Dellacherie, E., Alonso, M., Labrude, P. & Vigneron, C. (1998). Influence of experimental parameters on the characteristics of poly(lactic acid) nanoparticles prepared by a double emulsion method. *Journal of Controlled Release* 50: 31-40.

Zhang, H. Z., Gao, F. P., Liu, L. R., Li, M. M., Zhou, Z. M., Yang, X. D. & Zhang, Q. Q. (2009). Pullulan acetate nanoparticles prepared by solvent diffusion method for epirubicin chemotherapy. *Colloids and Surfaces B-Biointerfaces* 71(1): 19-26.

Zhang, J., Chen, X. G., Li, Y. Y. & Liu, C. S. (2007). Self-assembled nanoparticles based on hydrophobically modified chitosan as carriers for doxorubicin. *Nanomed-Nanotechnol.* 3: 258-265.

Zhang, L. M., Lu, H. W., Liu, J. Y. & Chen, R. F. (2008). Synthesis of an amphiphilic polysaccharide derivative and its micellization for drug release. *Journal of Bioactive and Compatible Polymers* 23(2): 154-170.

Zhang, L. M., Lu, H. W., Wang, C. & Chen, R. F. (2011). Preparation and properties of new micellar drug carriers based on hydrophobically modified amylopectin. *Carbohydrate Polymers* 83(4): 1499-1506.

Zhang, M., Bhattarai, N. & Matsen, F. A. (2005). PEG-grafted chitosan as an injectable thermoreversible hydrogel. *Macromolecular Bioscience* 5(2): 107-111.

Zhou, J. P., Song, Y. B., Zhang, L. Z., Gan, W. P. & Zhang, L. N. (2011). Self-assembled micelles based on hydrophobically modified quaternized cellulose for drug delivery. *Colloids and Surfaces B-Biointerfaces* 83(2): 313-320.

Zhu, A. P., Chen, T., Yuan, L. H., Wu, H. & Lu, P. (2006). Synthesis and characterization of N-succinyl-chitosan and its self-assembly of nanospheres. *Carbohydrate Polymers* 66(2): 274-279.

9

Polysaccharide-Based Nanoparticles for Controlled Release Formulations

A. Martínez[1], A. Fernández[2], E. Pérez,
M. Benito[3], J.M. Teijón[2] and M.D. Blanco[2]
[1]*Departamento de Farmacología, Facultad de Farmacia, Universidad Complutense de Madrid,*
[2]*Departamento de Bioquímica y Biología Molecular, Facultad de Medicina, Universidad Complutense de Madrid,*
[3]*Centro Universitario San Rafael-Nebrija. Ciencias de la Salud, Madrid*
Spain

1. Introduction

Nanoscience is the science of the phenomena peculiar to matter on the scale from 1 to several hundred nanometers (10^{-9} m). Some unique features of matter emerge when features are on the nanoscale, and the appreciation of these new properties opens new opportunities. Ignored in the past decades due to the lack of technology, these new emerging opportunities offered by nanoscience have been one of the most important areas of researching from the middle of the twentieth century to nowadays (Tibbals, 2010).

New opportunities have been realized in a wide variety of areas of technology, ranging from intelligent nanoscale materials, faster electronics or nanomotors, to medicine and biology, where first nanotechnology applications have demonstrated an enormous potential.

While medical nanotechnology was improving a wide range of medical resources and practice, the concept of nanomedicine was taking shape. Nanomedicine has recently been referred by the National Institutes of Health as the applications of nanotechnology for treatment, diagnosis, monitoring, and control of biological systems (Moghimi et al., 2005). Although this term has been defined in the literature in many ways, nanomedicine means essentially applying nanotechnology to medicine.

In contrast with other therapies, nanomedicine attempts to use sophisticated approaches to either kill specific cells or repair them one cell at a time. This approach also offers new possibilities towards the development of personalized medicine (Gurwitz & Livshits, 2006). Because nanomedicine inherits its focus on certain diseases which are currently being investigated, its primary aims have been towards non-infectious diseases, especially cancer, and on degenerative diseases in order to characterize them in the increasingly sedentary and aging populations of the wealthiest countries that lead in medical research (Tibbals, 2010).

One of the most important and hopeful tools employed in nanomedicine for medical applications are nanoparticles. Nanoparticles are solid, colloidal particles consisting of macromolecular substances that vary in size from 10 nm to 1000 nm. However, particles >200 nm are not heavily pursued and nanomedicine often refers to devices <200 nm (i.e.,width of microcapillaries). Depending on the method of preparation nanoparticles, nanospheres, or nanocapsules can be constructed to possess different properties and release characteristics for the best delivery or encapsulation of the therapeutic agent (Barratt, 2000).

One advantage of nanovectors—nanoparticles is their ability to overcome various biological barriers and to localize into the target tissue. The nanovectors currently used and investigated can be classified into three main groups or "generations" (Sakamoto et al., 2007). The first generation comprises a passive delivery system that localizes into the target site. In case of a tumour, the system reaches the tumour through the fenestrations in the adjacent neovasculature, and is normally decorated by a "stealth" layer in order to avoid their uptake by phagocytic blood cells, thus substantially prolonging their circulation time (Romberg et al., 2008). The unique mechanism of driving systems to the tumour site is the size of particles, not specific recognition of the tumour or neovascular targets. As a case in point, particles based on albumin-paclitaxel have been recently aproved by FDA for their use in metastatic breast cancer (Kratz, 2008). The second generation of nanosystems includes additional functionalities that allow for molecular recognition of the target tissue or for active or triggered release of the payload at the disease site. These include ligands, aptamers and small peptides that bind to specific target-cell surface markers or surface markers expressed in the disease microenvironment (Kang et al., 2008). Responsive systems, such as pH-sensitive polymers, are included in this cathegory. Although the representatives of the second generation have not yet been approved by the FDA, there are numerous ongoing clinical trials involving targeted nanovectors, particularly in cancer applications. Finally, the third generation nanovectors are focused to successfully overcome the natural barriers that the vector needs to bypass to efficiently deliver the drug to the target site. This goal will only be reached by a "multistage" approach, and such a system has been recently reported (Tasciotti et al., 2008).

Polymeric nanoparticles made from natural and synthetic polymers have received the majority of attention due to their stability and ease of surface modification. Polymeric materials used for preparing nanoparticles for drug delivery must be biocompatible at least and biodegradable best. Among natural polymers, proteins or polysaccharides tend to be internalized and degraded rapidly, thus enabling a moderate intracellular release of the drug or gene (Sinha & Trehan, 2003). Polysaccharides have been especially used in the preparation of drug delivery systems.

Polysaccharides are the polymers of monosaccharides. In nature, polysaccharides have various resources from algal origin (e.g.alginate), plant origin (e.g. pectin, guar gum), microbial origin (e.g. dextran, xanthan gum), and animal origin (chitosan, chondroitin) (Sinha & Kumria, 2001). They offer a wide diversity in structure and properties due to their wide range of molecular weight and chemical composition.

Due to the presence of various reactive groups in their structure, polysaccharides can be easily modified chemically and biochemically. Moreover, the presence of hydrophilic groups in their structure, such as hydroxyl, carboxyl and amino groups, enhance bioadhesion with biological tissues, like epithelia and mucous membranes, forming non-covalent bonds, which is an useful strategy to improve bioavailability of drugs included in drug delivery systems (Lee et al., 2000).

One of the main advantages of polysaccharides as natural biomaterials is their availability in natural resources and low cost in their processing, which make them very accessible materials to be used as drug carriers. Furthermore, polysaccharides are highly stable, safe, non-toxic, hydrophilic and biodegradable (Liu et al., 2008).Thus, they have a large variety of composition and properties that cannot be easily mimicked in a chemical laboratory, and the ease of their production makes numerous polysaccharides cheaper than synthetic polymers (Coviello et al., 2007). Therefore, polysaccharides have a promising future as biomaterials.

In recent years, a large number of studies have been conducted on polysaccharides and their derivatives for their potential application as nanoparticle drug delivery systems. The number of polysaccharides that have been investigated for the preparation of nanoparticles suitable as delivery systems is extremely large. As a result, attention has been focused on the latest studies and exploitations related to such systems, including some of the most used polysaccharides, a brief description of their structural features and some of the techniques carried out to prepare polysaccharide-based nanoparticles.

2. Structural features and characteristics of polysaccharides

As the number of polysaccharides used in the preparation of drug delivery systems is very large, some of the most commonly used polymers have been collected, describing their chemical structure, chemical features, and highlighting their applications in different fields, especially, in the preparation of drug delivery systems.

2.1 Alginate

Alginate is a well known polysaccharide obtained from natural sources, such as its extraction from cell walls and intercellular spaces of marine brown algae, and its production by bacteria. It can be characterized as an anionic copolymer whose chemical structure is based on a backbone of (1-4) linked β- D-mannuronic acid (M units) and α-L-guluronic acid (G units) (Fig. 1) of widely varying composition and sequence depending on the source of the alginate, resulting in an irregular blockwise pattern of GG, MG and MM blocks. Alginate has a variable molecular weight, depending on the enzymatic control during its production and the degree of depolymerization caused by its extraction. Typically, commercial alginates have an average molecular weight of approximately 200,000 Da, but alginates with values as high as 400,000-500,000 Da are also available (Rehm, 2009).

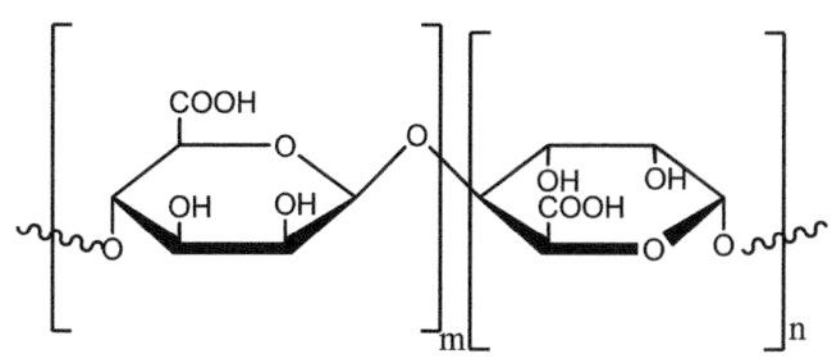

Fig. 1. Chemical structure of alginate

The physico-chemical properties of alginate have been found to be highly affected by the M/G ratio as well as by the structure of the alternating zones, which can be controlled by enzymatic pathways (Coviello et al., 2007). The alginate composition influence on the flexibility of the polysaccharide chain was first reported by Smidsrod (1973), who described that the extension of the alginate chain was dependent on its composition, with the intrinsic flexibility of the blocks decreasing in the order MG>MM>GG. M block segments show linear and flexible conformation because of the β (1→4) linkages. Besides, the guluronic acid gives rise to α (1→4) linkages, which serves to introduce a steric hindrance around the carboxylic groups, and provide folded and rigid structural conformations that are responsible for a pronounced stiffness of the molecular chains (Yang et al., 2011).

Alginate is a biopolymer and a polyelectrolyte considered to be biocompatible, non-inmunogenic, non-toxic and biodegradable, and the composition of the polymer has been reported to affect to its applications. Alginate with high content of guluronic acid block can produce, in the form of calcium salts, cross-links stabilizing the structure of the polymer in a rigid gel form. This properly enables alginate solutions to be processed into the form of films, beads and sponges (Sujata, 2002). However, high mannuronic acid alginate capsules are interesting for cell transplantation and for biohybrid organs, because of their less viscosity. In the case of cellular response, some research groups found immunostimulatory activity caused by those alginates with high mannuronic acid content, and immunosuppressive activity caused by alginates with high guluronic acid content. It was concluded that mannuronic acid oligomers would provoke cytokine release by macrophages by a receptor-mediated mechanism, whereas guluronic oligomers should inhibit this reaction (Orive et al., 2002).

Compositional modifications of natural alginates can be obtained by several mannuronan C-5 epimerases produced by alginate-producing bacteria, such as *A. vinelandii*. Recently, the combination of different epimerases has been used as a fundamental tool in order to create specific engineered alginates with any desired block length and composition (Rehm, 2009). Moreover, alginate has a large number of free hydroxyl and carboxyl groups distributed along the backbone, which are highly reactive and turn it into an ideal candidate for being appropriately modified by chemical functionalization. Thus, properties such as solubility, hydrophobicity and physicochemical and biological characteristics may be modified, having proved alginate derivatives to have a lot of potential applications. These chemical modifications of alginate have been achieved using techniques such as oxidation, sulfation, esterification, amidation, or grafting methods (Yang et al., 2011).

Due to its abundance, low price and non-toxicity, alginate has been extensively used in different industries. For instance, it has been used as food additive and thickener in salad dressings and ice-creams in the alimentary industry (Nair & Laurencin, 2007). Moreover, the biocompatibility behavior and the high functionality make alginate a favorable biopolymer material for its use in biomedical applications, such as scaffolds in tissue engineering (Barbosa et al., 2005), immobilization of cells (Lan & Starly, 2011), and controlled drug release devices (Pandey & Ahmad, 2011).

In case of its applications in nanomedicine, alginate has also been extensively investigated as a drug delivery device in which the rate of drug release can be modified by varying the drug polymer interaction, as well as by chemical immobilization of the drug in the polymer backbone using the reactive carboxylate groups (Nair & Laurencin, 2007). Apart from its easy functionalization due to its reactive structure, there are many advantages and favorable properties of alginate for its use in drug delivery. It is a natural polymer compatible with a wide variety of substances, which does not need multiple and complex drug-encapsulation process. Moreover, it is mucoadhesive and biodegradable and, consequently, it can be used in the preparation of controlled drug-delivery systems achieving an enhanced drug bioavailability (Pandey & Ahmad, 2011).

Therefore, the biocompatibility, availability and versatility of this polysaccharide make it an important and hopeful tool in the field of nanomedicine, especially in the preparation of nanoparticulate drug delivery systems.

2.2 Chitosan

Chitosan is a linear polysaccharide composed by units of glucosamine and N -acetyl-glucosamine linked by (1 → 4) β-glycosidic bonds (Fig. 2). It is a hydrophilic biopolymer obtained industrially by hydrolysing the aminoacetyl groups of chitin – which is the main component of the shells of crab, shrimp and krill – by an alkaline deacetylation treatment (Muzzarelli & Muzzarelli, 2005).

The degree of deacetylation (%DD) can be determined by NMR spectroscopy, and generally the %DD in commercial chitosan is in the range 60–100%. On average, the molecular weight of commercially produced chitosan is between 3,800 to 20,000 Da. A commonly used method for the synthesis of chitosan is the deacetylation of chitin, using sodium hydroxide in excess as a reagent and water as a solvent. This reaction pathway, when allowed to go to completion (complete deacetylation), yields up to 98% product (Yuan, 2007). So, once deacetylation happens, chitosan is consisting primarily of repeating units of β-(1,4)-2-amino-deoxy-D-glucose (D-glucosamine).

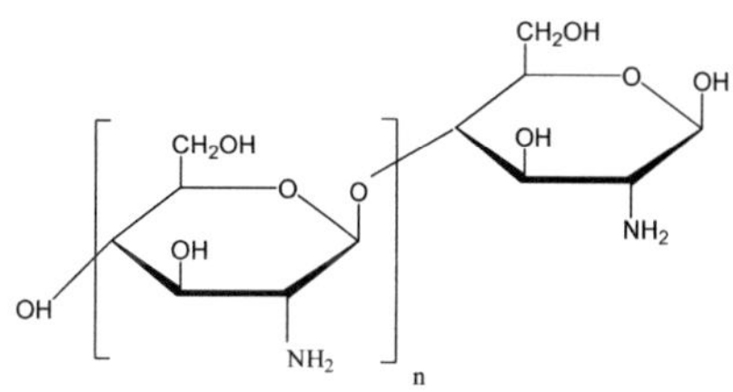

Fig. 2. Chemical structure of chitosan.

This biopolymer is accepted as a biodegradable and non toxic polymer. Despite its biocompatibility, the applications of chitosan are limited due to its insolubility above pH 6. Chitosan is a weak base and it is insoluble in water and organic solvents. However, it is soluble in diluted aqueous acidic solution (pH <6.5), which can convert the glucosamine units into a soluble form with protonated amine groups (Sinha et al., 2004). It is possible to increase the solubility of chitosan in water removing one or two hydrogen atoms from the amino groups of chitosan, and introducing some hydrophilic segments (Srinophakun & Boonmee, 2011).

The non-toxic, biodegradable and biocompatible properties of chitosan provide potential for many applications (Guerrero et al., 2010). Due to its polyelectrolyte nature, chitosan can be used as absorbent for treatment of textile industry effluents as well as for heavy metal ions uptaking from wastewater. It has been also used as template for the preparation of mesoporous metal oxides spheres (Braga et al., 2009). However, it has been more frequently proposed for applications in pharmaceutical and biomedical fields due to its biocompatibility and biodegradability. It has been assayed as biomaterial for wound healing and prosthetic material, since it can be biodegraded by enzyme action (Bernardo et al., 2003). Also it is reported to find applications as an antimicrobial compound, as a drug in the treatment of hyperbilirubinaemia and hypercholesterolaemia and, also, it has been prepared and evaluated for its antitumour activity carrying several antineoplastic agents (Blanco et al., 2000).

In the field of nanomedicine, chitosan has attracted attention as a matrix for controlled release due to its reactive functionalities, polycationic character, easily degradation by enzymes and non-toxic degradation products. Over the years, a variety of natural and synthetic polymers have been explored for the preparation of drug-loaded microparticles and chitosan has been extensively investigated (Davidenko et al., 2009; Muzzarelli & Muzzarelli, 2005). Because of its bioadhesive properties, chitosan has received substantial attention as carrier in novel bioadhesive drug delivery systems which prolong the residence time of the drugs at the site of absorption and increase the drug bioavailability (Varum et al., 2008). Thus, some drugs administered via nasal (Learoyd et al., 2008) or gastrointestinal routes have improved their treatment efficacy when they are included into chitosan-based systems (Guerrero et al., 2010).

Taking all into account, chitosan appears to be a promising matrix for the controlled release of pharmaceutical agents. Experimental *in vitro* and *in vivo* results show chitosan as an ideal carrier for a wide variety of drugs whose efficacy is increased when they are included into these systems.

2.3 Hyaluronic acid

Hyaluronic acid (HA) (also called sodium hyaluronic or hyaluronan) is a polysaccharide with a structure composed of repeating disaccharide units of D-glucuronic acid and N-acetyl D-glucosamine linked by β (1-3) and β (1-4) glycosidic bonds (Fig. 3) (Cafaggi et al., 2011). HA can be modified in many ways to alter the properties of the resulting materials, including modifications leading to hydrophobicity and biological activity. There are three functional groups that can be chemically modified: the glucuronic acid carboxylic acid, the primary and secondary hydroxyl groups, and the N-acetyl group (Burdick & Prestwich, 2011). HA has a molecular weight that can reach as high as 10^7 Da.

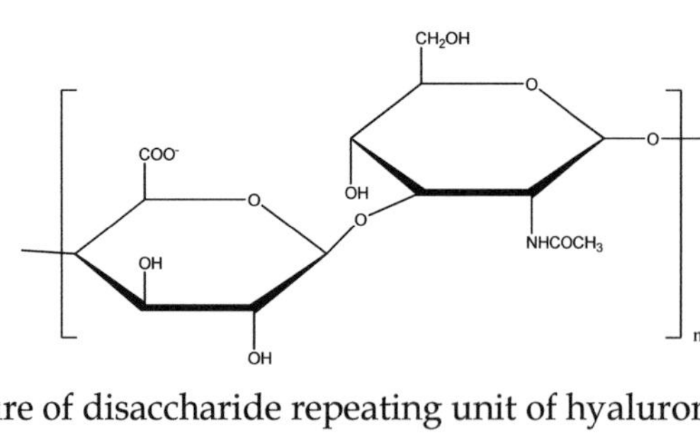

Fig. 3. Chemical structure of disaccharide repeating unit of hyaluronic acid.

It belongs to a group of substances known as glycosaminoglycans. It is the most simple among them, the only one not covalently associated with a core protein, and the only one which is non-sulfated (Kogan et al., 2007). Since hyaluronan is a physiological substance, it is widely distributed in the extracellular matrix of vertebrate tissues. It is mainly synthesized in vertebrate organisms as an essential functional component due to its viscoelastic and rheological properties. It is a major and important component of cartilage, skin and synovial fluid.

HA is usually linked to other biopolymers in the organism, and several separation procedures have to be applied in order to obtain the pure compound, such as protease digestion, HA ion-pair precipitation, membrane ultrafiltration, HA non-solvent precipitation and/or lyophilisation (Mendichi & Soltes, 2002). With these methods HA from several hundred thousand Da up to 2.5 MDa can be obtained. However some microorganisms secreted HA with a molar mass in the range of several MDa, such as attenuated strains of *Streptococcus zooepidemicus* and *S. Equi*. *Bacillus subtilis* has been recently genetically modified to culture a proprietary formula to yield hyaluronans (Mendichi & Soltes, 2002).

It is a biodegradable, bioactive, non inmunogenic, non cytotoxic and negatively charged polysaccharide (Oh et al., 2010) that has been associated with several cellular processes, including angiogenesis and the regulation of inflammation (Leach & Schmidt, 2005).

Among its applications, it is widely used as a coating for the surface modification of various biomaterials used for prosthetic cartilage, vascular graft, guided nerve regeneration and drug delivery (Li et al., 2006).

Like other glycosaminoglycans, hyaluronan can serve as a targeting vehicle for the delivery of chemotherapeutic agents to cancerous tissues, as many tumours over express the hyaluronan CD44 and RHAMM receptors (Yip et al., 2006). As a drug delivery carrier, HA has several advantages including the negligible non-especific interaction with serum components due to its polyanionic characteristics (Ito et al., 2006) and the highly efficient targeted specific delivery to the liver tissues with HA receptors (Zhou et al., 2003).

More recently, HA has become recognized as an important building block for the creation of new biomaterials with utility in tissue engineering and regenerative medicine (Allison & Grande-Allen, 2006; Prestwich, 2008). Moreover, it has been shown that HA binds to cells and effectively promotes new bone formation. Balazs classified the biomedical applications of the HA and its derivatives in areas as vicosurgery, viscoaugmentation, viscoseparation, viscosupplementation, viscoprotection (Balazs, 2004).

So, in this way, there is a wide number of usages of HA in medicine and cosmetics, such as ophthalmology, orthopaedic surgery and rheumatology, otolaryngology, wound healing, pharmacology and drug delivery (Kogan et al., 2007), which shows HA as a successful biomaterial used in different fields of biomedicine.

2.4 Dextran

Dextran is a polysaccharide made of many glucose molecules composed of chains of varying lengths. It has a substantial number of α (1→6) glucosidic linkages in its main chain (Fig. 4), and a variable amount of α (1→2), α (1→3) and α (1→4) branched linkages (Misaki et al., 1980). The degree and type of branching will be determined by the bacterial strain that synthesizes it. Its average molecular weight is as high as 10^7 - 10^8 Da (Heinze et al., 2006) but can be reduce by acidic hydrolysis obtaining molecular weight fractions that also can interest.

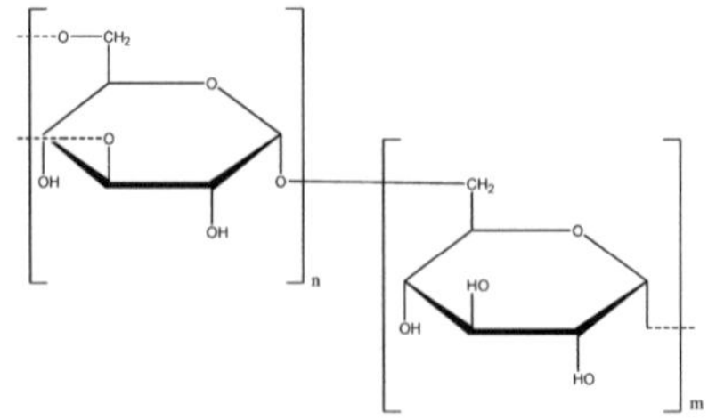

Fig. 4. Chemical structure of dextran with α (1→3) branched linkage.

The natural structure of dextran can be modified by reacting different molecules (such hydrophobic molecules) with its different hydroxyl groups (Lemarchand et al., 2003b). Many amphiphilic dextran derivates have been obtained by varying the nature of the reacting molecules (aromatic rings, aliphatic or cyclic hydrocarbons) and the number of grafted, that is the number of hydrophobic groups per 100 glucopyranose units or the degree of substitution (Rotureau et al., 2004).

Dextran is neutral, water soluble, biocompatible and biodegradable. Its features may vary depending on the molecular mass as well as the distribution, type of branches and the degree of branching, which depend on the bacterial synthesis or post-synthesis reactions to form derivatives.

Dextran is synthetized by a wide variety of bacterial strains. *Leuconostoc mesenteroides* produces dextran from sucrose and *Gluconobacter oxydans* produces dextran from maltodextrin. *Streptococcus mutans* also produces dextran from sucrose (Heinze et al., 2006). It can be also obtained enzimatically using cell-free culture supernatant (Wang et al., 2011). Apart from these methods, dextran can be also produced by chemical synthesis, developing a cationic ring opening polymerisation of levoglucosan (Heinze et al., 2006).

It has wide applications in different areas such as pharmaceutical, chemical, clinical, and food industry. Dextran is used as a drug (as blood plasma volume expander), adjuvant, emulsifier, carrier, stabilizer and thickener of jam and ice cream. Also it is widely used for the separation and purification of proteins (Naessens et al., 2005) based on size exclusion chromatography with a matrix of cross-linked dextran gel layer. Its derivatives also have multiple applications depending on the characteristics that structural modifications give them.

Both dextran and its derivatives have potential application for the preparation of modified drug delivery (Aumelas et al., 2007; Coviello et al., 2007; Chen et al., 2003). Not only has this polysaccharide been used to prepare nanoparticulate systems as a carrier, but also it has been employed to cover these systems (Gavory et al., 2011).

It seems that dextran is a very useful tool in the field of nanomedicine, showing also good availavility, biocompatibility and biodegradability, being selected by a lot of researchers as biomaterial in the preparation of nanosystems.

2.5 Other polysaccharides

2.5.1 Pullulan

Pullulan is a linear bacterial homopolysaccharide produced from starch by the fungus *Aureobasidium pullulans*. The backbone is formed by glycosidic linkages between α-(1→6) D-glucopyranose and α-(1→4) D-glucopyranose units in a 1:2 ratio (Fig. 5). The molecular weight of pullulan range from thousands to 2,000,000 Da depending on the growth conditions (Rekha & Chandra, 2007).

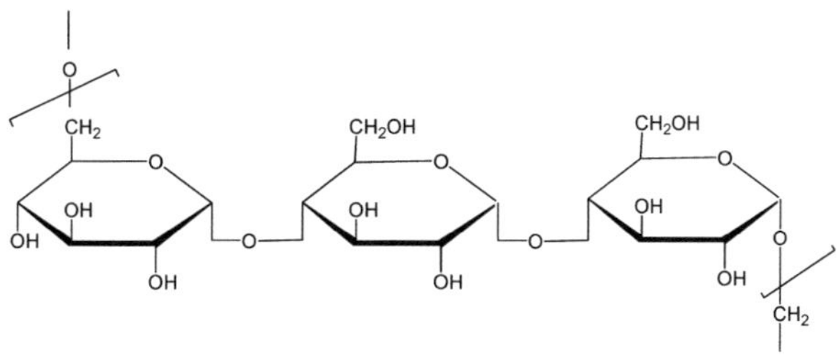

Fig. 5. Chemical structure of pullulan.

The backbone structure of pullulan tends to behave as a random expanded flexible coil in aqueous solution with modelling studies suggesting that this flexibility is imparted by the α-(1→6) linkage. This could be the reason why pullulan is biodegradable and has high adhesion, structural flexibility and solubility (Leathers, 2003). Pullulan can also be easily derivatized in order to impart new physico-chemical properties, e.g. to increase the solubility in organic solvents or to introduce reactive groups.

This polysaccharide has numerous uses: in foods and beverages as a filler; as an edible, mostly tasteless polymer, the chief commercial use of pullulan is in the manufacture of edible films that are used in various breath freshener or oral hygiene products; in pharmaceuticals as a coating agent; in manufacturing and electronics it is used because of its film- and fiber-forming properties. It is worth noting that pullulan films, formed by drying pullulan solutions, are clear and highly oxygen-impermeable and have excellent mechanical properties.

Due to it is hemocompatible, non-immunogenic, non-carcinogenic, FDA approved it for a variety of applications (Coviello et al., 2007). Recently, pullulan has been investigated for being used in various biomedical applications such as drug and gene delivery (Rekha & Chandra, 2007), tissue engineering (Thebaud et al., 2007), and wound healing (Bae et al., 2011).

Numerous papers deal with pullulan hydrogels as drug delivery systems, particularly in the form of micro and nanogels. Despite pullulan is not a natural gelling polysaccharide, an appropriate chemical derivatization of its backbone can actually lead to a polymeric system capable of forming hydrogels. The study of nanogels has been intensified over the last decade due to related potential applications in the development and implementation of new environmentally responsive or smart materials, biomimetics, biosensors, artificial muscles, drug delivery systems and chemical separations (Coviello et al., 2007).

In order to obtain nanostructures that may act as carriers of different drugs, the backbone structure of pullulan is modified with hydrophobic molecules, resulting in a molecule of hydrophobized pullulan that self-assembles in water solutions. Cholesterol, hexadecanol or vitamin H are some molecules that are attached to the structure of pullulan in order to obtain micelles in water solution (Liu et al., 2008).

2.5.2 Guar gum

Guar gum is a water soluble polysaccharide extracted from the seeds of *Cyamopsis tetragonoloba*, which belongs to Leguminosae family. Also called guaran, it is a non-ionic natural polysaccharide derived from the ground endosperm of guar beans. Its backbone consists of linear chains of (1 → 4)-β-D-mannopyranosyl units with α-D-galactopyranosyl units attached by (1 → 6) linkages (Fig. 6), forming short side-branches (Sarmah et al., 2011).

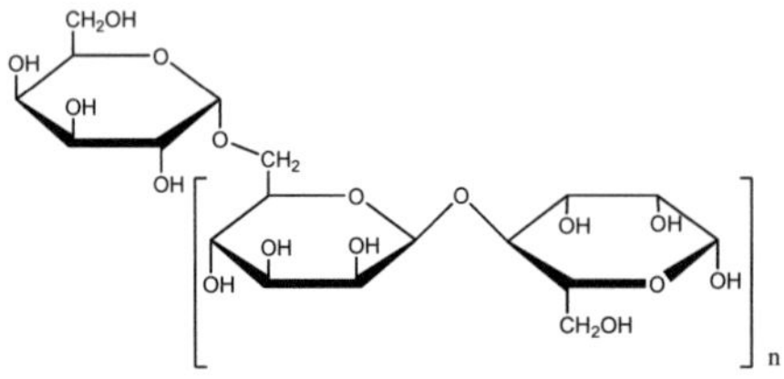

Fig. 6. Chemical structure of guar gum.

Guar gum hydrates in cold water to form a highly viscous solution in which the single polysaccharide chains interact with each other in a complex way (Barbucci et al., 2008). Its nine hydroxyl groups are available for the formation of hydrogen bonds with other molecules, but it remains neutrally charged due to the absence of dissociable functional groups. Extreme pH and high temperature conditions (e.g. pH 3 at 50°C) degrade its structure (Tiraferri et al., 2008). It remains stable in solution over pH range 5-7. Strong acids cause hydrolysis and loss of viscosity, and alkalis in strong concentration also tend to reduce viscosity. It is insoluble in most hydrocarbon solvents.

As the guar gum polymer is a low-cost, easily available and non-toxic polysaccharide, it is widely applied in many industrial fields. Thanks to its property of producing highly viscous aqueous solutions, it is commonly used as a thickening agent in cosmetics and in sauces, salad dressings and ice creams in the food industry (Barbucci et al., 2008). In pharmaceuticals, guar gum is used in solid dosage forms as a binder and disintegrant, and it has also been used as hydrophilic matrix, for designing oral controlled release dosage forms (Sarmah et al., 2011). Guar gum has been extensively used for colon delivery due to its drug release retarding property and susceptibility to microbial degradation in the large intestine (Soumya et al., 2010).

Not only the native guar-gum is used, but also chemically modified products can be used with the objective of changing its intrinsic characteristics of solubility, viscosity and rheological behaviour. For instance, hydrossilalchyl derivatives, which are often used for the formulation of cements and plasters, or carboxymethyl derivatives, which are employed as thickening agents.

In case of biomedical fields or pharmaceutical fields, such as 3D scaffolds for cell culture, fillers for tissue engineering and carriers for drugs, the physically cross-linked product is obtained through a spacer arm between the polymer chains and allows the obtainment of an insoluble compound in a wide range of pH with a good mechanical stability (Barbucci et al., 2008).

Little information is available in the literature for the possibility of using guar gum based nanosized materials as drug carriers due to its solubility in water, what makes difficult to use it as adsorbent in aqueous conditions. Some researchers have incorporated to its structure some compounds like silica, in order to obtain insoluble compounds which could act as adsorbents in aqueous media (Singh et al., 2009). Moreover, guar gum-based nanosystems have been prepared by nanoprecipitation and cross-linking methods (Soumya et al., 2010). A different application of this polysaccharide has been found as stabilizer of nanosuspensions, where the presence of guar gum during the synthesis process allows the achievement of a better stability of the nanoparticles (Tiraferri et al., 2008).

2.5.3 Pectin

Pectin is a structural polysaccharide obtained from the cell wall of all plants, where is implicated in cell adhesion. This natural polymer has a heterogeneous chemical structure based on large amounts of poly (D-galacturonic acid) bonded via α (1 → 4) glycosidic linkage (Fig. 7). Pectin has a few hundred to about one thousand building blocks per molecule, corresponding to an average molecular weight of about 50,000 to about 180,000 Da (Sinha & Kumria, 2001). The carboxyl groups are partially in the methyl ester form with different degree of esterification (DE) and amidation (DA), which determine the content of carboxylic acid in pectin chains.

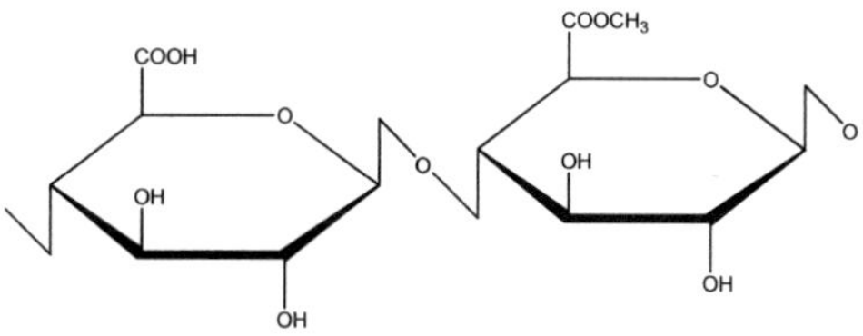

Fig. 7. Chemical structure of pectin

In the beginning, applications of pectin concentrated in food industry, as gelling or thickening agent, but lately it started being also used as an excipient for pharmaceutical purposes (Liu et al., 2003). Nowadays, some of the uses of pectin in biomedical applications include the facilitation of the delivery of specific sequences of amino acids, anti-inflammatory agents, anti-coagulants, and wound healing substances to tissue sites. Also, pectin remains intact in the physiological environment of the stomach and the small intestine, but is degraded by pectinases, which are secreted by the bacteria inhabitants of the human colon. Due to these properties it is highly possible that pectin could function as a delivery vehicle to escort protein and polypeptide drugs from the mouth to the colon (Sinha & Kumria, 2001). To be used as such, pectin based composites can be formed into membranes, microspheres, scaffolds, or injectable gels (Liu et al., 2004).

The most attractive property of pectin for industrial applications is its gelling activity. Parameters such as type and concentration of pectin (DE, DA), modification of hydroxyl groups, pH, temperature and the presence of cations, determine the gel process. For example, a high DE of pectin provides the gel formation, increasing the amounts of hydrophobic areas and reducing the solubility of pectin. In contrast, when the DE is less than 50%, pectin is highly water soluble and gel formation is only at extremely low pH solution or in the presence of divalent cations, which cross-link the galacturonic acids of the main polymer chains (Liu et al., 2003). Also, it is possible to reduce the hydrophilic property with an increasing tendency to form gels by the introduction of amide groups in low DE pectin.

With regard to its use in the preparation of drug delivery systems, pectin is not able to shield its drug load effectively during its passage through the stomach and small intestine due to its high water-solubility (Sinha & Kumria, 2001). Consequently, most of the researching groups focused on looking for water resistant pectin derivatives, which were also enzymatically degradable. For this purpose, calcium salts binding by non-covalent associations with the carbohydrate chains of pectin were investigated, which can reduce the solubility and are stable in low pH solution while resisting extensive hydration *in vivo* in the gastrointestinal tract. Thus, calcium pectinate is a potential candidate as a drug carrier for colon-specific delivery in different formulations such as microspheres, films, gels or droplets (Liu et al., 2003). Another derivative of pectin, amidated pectin cross-linked with calcium, was considered for colonic delivery, with retarding drug release and because of its biodegradability, higher tolerance to pH variations and fluctuations in calcium levels (Sinha & Kumria, 2001).

In addition, combinations of pectin with other polymers, either naturally occurring or synthetic, have been developed in order to obtain useful novel formulations. The combination of pectin and a second polymer into a composite may alter degree of swelling and change mechanical properties (Liu et al., 2003), improving in the most cases the stability of the drug and controlling the drug release. As a case in point, pectin has been combined with 4-aminothiophenol (Perera et al., 2010), chitosan (Fernandez-Hervas & Fell, 1998), hyaluronic acid (Pliszczak et al., 2011) or poly (lactide-co-glycolide) (Liu et al., 2004), showing good results as controlled drug release devices.

3. Preparation methods of polysaccharide-based nanoparticles

As for polysaccharide-based nanoparticles, it can be seen in the literature wide research carried out focusing on the preparation and application of these systems, which enhances their importance and versatility in terms of category and function.

According to the literature and the structural features of the employed polysaccharides, five mechanisms can be mainly applied in order to obtain nanoparticles, namely gelation of emulsion droplets, covalent cross-linking, ionic cross-linking, self-assembling and nanoprecipitation.

3.1 Formation of nanoparticles from an emulsion: nanoparticles obtained by gelation of emulsion droplets

Different methods to prepare emulsified systems have been significantly developed. All of them require two immiscible phases and the presence of a surface active agent, whose nature has been already evolved, replacing the commonly used pluronic or span by new amphiphilic copolymers (Qiu & Bae, 2006). These methods are two-step processes, where the first step consists of the preparation of an emulsified system while nanoparticles are formed during the second step. Generally, the principle of the second step gives its name to the method (Vauthier & Bouchemal, 2009).

Nanoparticles can be obtained from an emulsion method by gelation of the emulsion droplets where the polymer is dissolved, and that have been formed in the first step of the emulsion procedure. Polysaccharides show good gelling properties as well as good solubility in water, which make them ideal candidates to be used for preparing nanoparticles by this method (Vauthier & Couvreur, 2000). Different mechanisms of gelation can be applied depending on the gelling properties of the polymer. Changes in temperature of the emulsion system or gelation induced by covalent or ionic cross-linking are some of these mechanisms which induce the gelation of the pre-formed droplets and allow nanoparticles to be obtained.

Alginate and pectin particles have been obtained by using a modified emulsification/internal gelation method (Opanasopit et al., 2008). The preparation of two different emulsions is required: one containing the gelling polymer in the dispersed phase and the other containing the gelling agent (usually counter-ions) or the pH controlling agent in the dispersed phase. Both emulsions are mixed together under strong stirring conditions in order to achieve collisions between droplets, which are necessary to promote the gelation of the polymer and, consequently, the formation of nanoparticles (Fig. 8). In case of alginate, the size range of particles is greatly dependent on the order of addition of counter-ion to the alginate solution. Some studies show that the addition of a polyelectrolyte complexation step in this procedure shows some benefits in order to obtain a better control of size distribution. Dextran or chitosan can be used as complexing agents in the *in situ* gelation of the droplets obtained in the previous nanoemulsion (Reis et al., 2007). The resulting nanoparticles range in size from 267 nm to 2.76 μm.

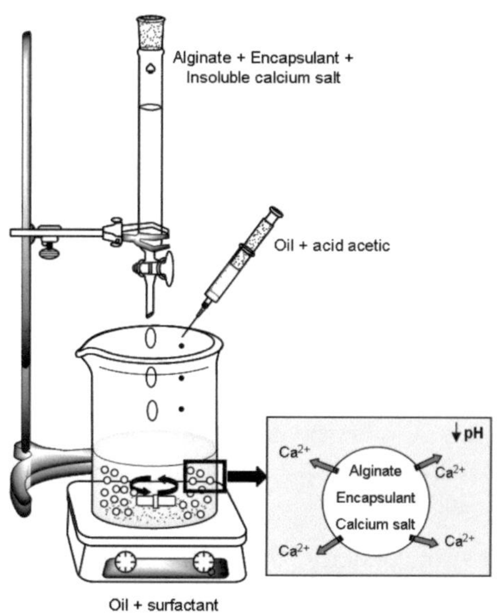

Fig. 8. Schematic representation of the emulsification-internal gelation technique using alginate.

Chitosan particles have been formed from an emulsified system by emulsion cross-linking method or by emulsion-droplet coalescence method. In the first method, the reactive functional amine group of chitosan reacts with aldehyde groups of the cross-linking agent, which usually is glutaraldehyde. A water-in-oil (w/o) emulsion is prepared by emulsifying the chitosan aqueous solution in the oil phase. Aqueous droplets are stabilized using a suitable surfactant. The stable emulsion is cross-linked by glutaraldehyde to harden the droplets. Microspheres are filtered and washed repeatedly with n-hexane, followed by alcohol and then dried (Akbuga & Durmaz, 1994). Particle size can be determined by controlling the size of aqueous droplets, but it is usually ranged in a micrometric scale (Kumbar & Aminabhavi, 2003).

Chitosan nanoparticles with a mean size of 400 nm can be obtained by emulsion-droplet coalescence method. This method, introduced by Tokumitsu and coworkers (Tokumitsu et al., 1999), utilizes the principles of both emulsion cross-linking and precipitation. A stable emulsion containing aqueous solution of chitosan and the encapsulant drug is produced in liquid paraffin. Another emulsion containing NaOH aqueous solution is produced in the same manner and is finally mixed with the other under high speed stirring. Droplets of each emulsion would collide at random and coalesce, precipitating chitosan droplets to give small solid particles.

With a polymer like agarose, gel beads can be formed by cooling down the temperature of the solution which is prepared at high temperature. Thermal gelation results from the formation of helicoidal structures responsible for a three-dimensional network in which large amounts of water can be entrapped. The hydrogel, being hydrophilic, inert, and biocompatible, forms a suitable matrix for macromolecules that can be entrapped in the gel during formation (Vauthier & Couvreur, 2000). Agarose nanoparticles are produced using an emulsion-based technology which requires the preparation of an agarose solution in corn oil emulsion at 408°C. Macromolecules to be encapsulated are initially added to the agarose solution. The small size of the dispersed aqueous nanodroplets is achieved by homogenization. Gelation of agarose is then induced by diluting the emulsion with cold corn oil under agitation at 58°C. The liquid nanodroplets then gel to macromolecule-containing agarose hydrogel nanoparticles (Wang & Wu, 1997). The mean average size of the obtained nanoparticles is 504 nm.

3.2 Polysaccharide-based nanoparticles with covalent cross-links

Among various polysaccharides, chitosan is the early one to be used to prepare nanoparticles based in covalent cross-links. Glutaraldehyde has been usually used as a cross-linker to obtain nanoparticles by emulsion cross-linking method (previously described), but its citotoxicity limits its utility in the field of drug delivery systems. However, some chitosan nanoparticles are still being produced using glutaraldehyde as cross-linker agent (Zhi et al., 2005).

To overcome the problems of toxicity that are presented by glutaraldehyde, some biocompatible cross-linkers, such as natural di- and tricarboxylic acids, including succinic acid, malic acid, tartaric acid and citric acid, are used for intermolecular cross-linking of chitosan nanoparticles (Bodnar et al., 2005). By this method, the pendant amino groups of chitosan react in aqueous media with carboxylic groups of natural acids which were previously activated by a water-soluble carbodiimide, obtaining polycations, polyanions, and polyampholyte nanoparticles with an average size in the range of 270–370 nm depending on the pH.

Hyaluronic acid is another polysaccharide used to prepare nanoparticles by using a carbodiimide method. The preparation of nano-sized particulate systems based on hyaluronic acid takes place by covalently cross-linking via carboxyl groups of the hyaluronic acid chain with a diamine in aqueous media at room temperature. Bodnar and coworkers have obtained spherical nanoparticles whose size varies less than 130 nm (Bodnár et al., 2009).

Recently, nanoparticles based on thiolated alginate and modified albumin have been synthesized and stabilized by the formation of disulphide bonds between both polymers (Martínez et al., 2011). In this case, the covalent interaction is established between the sulfhydryl groups of the albumin, obtained after a reduction process of the protein, and the sulfhydryl groups of the L-cysteine which has been attached to the polysaccharide structure using a carbodiimide reaction. Nanoparticles with a size range of 42-388 nm are obtained by this coacervation method based on a pH change that induces the disulphide bond formation between both structures (Fig. 9).

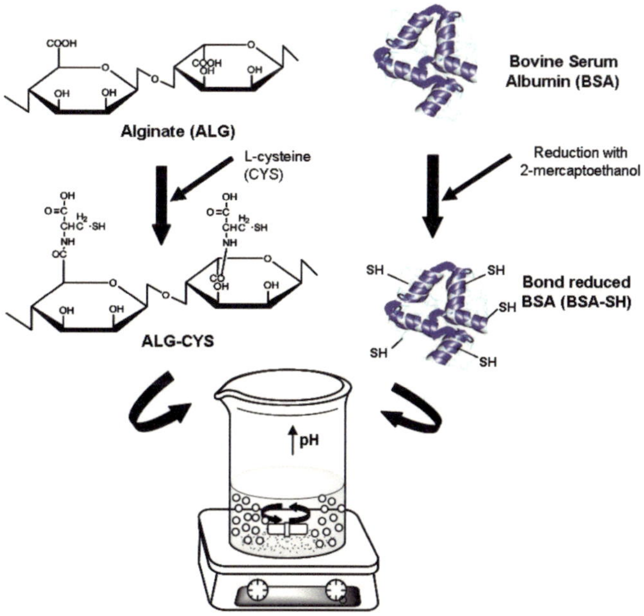

Fig. 9. Schematic representation of preparation method of nanoparticles based on thiolated alginate and modified albumin, stabilized by disulphide bond formation.

3.3 Polysaccharide-based nanoparticles with ionic cross-links

Ionic gelation procedure to obtain nanoparticles is included among the few organic solvent free methods, as nanoparticles are totally synthesized in aqueous media. Compared with covalent cross-linking, this method shows more advantages, such as simple procedures and mild preparation. Nanoparticles can be obtained from aqueous solutions of charged polysaccharides which gel in the presence of small ions of opposite charges. Thus, polyanions and polycations could act as cross-linkers with polycationic and polyanionic polysacchraides, respectively. Very dilute solutions of the polysaccharide are used to perform the gelation process, in which the chains of the polymer reacting with the gelling agent are forming small clusters. These clusters are stabilized by forming complex with opposite charged electrolytes (Vauthier & Bouchemal, 2009).

The cationic nature of chitosan when it is dissolved in an acidic aqueous solution (pH 4-6) can be exploited to form nanoparticles by adding small amounts of tri-polyphosphate (TPP) included in an alkaline phase (pH 7-9), upon mixing of the two phases through inter and intra molecular linkages are created between TPP phosphates and chitosan amino groups (Janes et al., 2001).

Depending on the pH and the ionic strength of the dispersing medium, these nanoparticles are capable of swelling and shrinking, which is used to trigger the release of a drug encapsulated in the nanoparticles upon the action of a pH or an ion concentration variation stimulus. For instance, KCl can be added to the dispersing medium to vary the ionic strength and cause the nanoparticle swelling; or glucosamin groups of chitosan can be deprotonated by raising the pH from acid to basic values causing a shrinking on the gel because the intramolecular electric repulsions inside the particle mesh are reduced. The average size of the obtained chitosan nanoparticles ranges between 20 and 400 nm (Pan et al., 2002).

Some water-soluble chitosan derivatives, like N-(2-hydroxyl) propyl-3-trimethyl ammonium chitosan chloride or N-trimethyl chitosan, have been also ionically cross-linked to prepare nanoparticles. The average size of the obtained systems is between 110 and 350 nm (Amidi et al., 2006).

A slightly modified ionotropic gelation technique was used by de la Fuente and coworkers in order to obtain nanoparticles based on chitosan and hyaluronic acid using TPP as ionic cross-linker. Their results show that hyaluronic acid/chitosan nanoparticles have a small size in the range of 110–230 nm (de la Fuente et al., 2008).

Not only TPP is used as a cross-linker to obtain chitosan naoparticles by ionic gelation method. Some researchers as Kim and co-workers have obtained chitosan-based nanoparticles by this method using the encapsulated drug itself as a cross-linker, establishing electrostatic interactions between amine group of chitosan and hydroxyl group of the drug (Kim et al., 2006).

Among negatively charged polysaccharides, alginate is one of the most used to obtain nanoparticles by ionic gelation. In this case, carboxylic groups on molecular chains of alginate structure can be cross-linked by bivalent calcium ions to form nanoparticles. Then, clusters formed in the pre-gel phase can be stabilized with polications like polylysine and chitosan (De & Robinson, 2003). Polylysine can form polyelectrolyte complexes with alginate without the previous formation of the pre-gel phase with calcium, but more compact nanoparticles are obtained when this previous step is carried out. The size of nanoparticles using polylysine as stabilizer depends, not only on the concentration of alginate, but also on the molecular weight. In fact, an optimal mass balance between sodium alginate: $CaCl_2$: cationic polymer (poly-Llysine or chitosan) has been found to obtain particles with nanometric size (Vauthier & Bouchemal, 2009). Nanoparticles with an average size ranged between 194 nm and 1.6 μm can be obtained by this method (Ahmad et al., 2006; Azizi et al., 2010).

In addition, the interaction of alginate with divalent calcium ions was used to obtain nanoparticles from water-in-oil microemulsions, as it was previously described (Reis et al., 2007).

3.4 Methods based on self assembling macromolecules

3.4.1 Polysaccharide based nanoparticles by polyelectrolyte complexation (PEC)

Polyelectrolyte complexes (PECs) are formed by the interaction between oppositely charged polymers by intramolecular electrostatic interactions. PECs are very interesting materials for different applications because some of their properties, like swelling or permeability, can be easily modified by external stimuli, such as the pH of the medium.

Positively or negatively charged nanoparticles with a core/shell structure can be obtained according to the nature of the polyelectrolyte used in excess. The hydrophobic core is composed by the complexed segments whereas the excess of component not incorporated in the polyelectrolyte complex is segregated in the outer shell ensuring the colloidal stabilization of the nanoparticles against coagulation and conferring the charge of the nanoparticle surface. This charge could affect to the interaction between cells and nanoparticles. Moreover, molecular weight of the two polyelectrolytes influences the size of the nanoparticles (Vauthier & Bouchemal, 2009).

Although any polyelectrolyte could interact with polysaccharides in order to obtain PEC nanoparticles, only water-soluble and biocompatible polymers are used as polyelectrolytes with this goal. Among the existing polyanionic and polycationic polysaccharides to form PEC nanoparticles, chitosan is widely used because it satisfies the needs of safety and solubility. It can be seen in the literature that much research has been carried out on PECs with chitosan as polycation and different negative polymers, such as negative polysaccharides, poly(acrylic acid) (PAA) or nucleic acids.

There is a wide variety of negative polysaccharides that can be attached to chitosan to form PEC nanoparticles. Carboxymethyl cellulose (Cui & Mumper, 2001), dextran-sulfate (Drogoz et al., 2007), alginate (Sarmento et al., 2006) and glucomannan (Alonso-Sande et al., 2006) are just a few examples of PEC combinations that allow the preparation of nanoparticles whose size ranges between 100 and 800 nm.

The formation of the complex between chitosan and poly(acrylic acid) (PAA) has been widely studied. The influence of molecular weight of chitosan and PAA, the ratio of the initial polyelectrolyte concentrations, dropping temperature, pH of the initial solutions and the purification process on the size, stability and morphology of the nanoparticles has been studied by different authors (Chen et al., 2005; Davidenko et al., 2009).

Nucleic acids can also be combined with chitosan to obtain nanospheres. In this case, the drug being the nucleic acid is incorporated in the nanocarrier as part of its structure (de Martimprey et al., 2009). The N/P ratio, which is defined as the ratio number of amine groups of the polycation (N) divided by the number of phosphate groups of the nucleic acid (P), has to be taken into account to obtain the desired size of nanospheres.

Apart from chitosan, polyelectrolyte complexes with nanometric size can be formed using alginate, a negatively charged polysaccharide, combined with polylysine, a positively charged peptide (George & Abraham, 2006). Although both structures could interact without the previous formation of the alginate pre-gel phase with calcium, more compact nanoparticles are obtained when this previous step is carried out. Nanoparticles with a mean size of 250–850 nm are obtained using very well defined concentrations of both electrolytes.

3.4.2 Nanoparticles obtained from self-assembling of hydrophobically modified polysaccharides

Amphiphilic copolymers are synthesized when hydrophobic segments are added to chains of hydrophilic polymers. In aqueous solutions, amphiphilic molecules orientate themselves in order to achieve a state of minimum free energy and the hydrophobic blocks are removed from the aqueous environment. Consequently, polymeric micelles with core/shell structure are formed. Thanks to their hydrophobic domain, surrounded by a hydrophilic outer shell, they can serve as reservoir for various hydrophobic drugs (Letchford & Burt, 2007).

The synthesis and application of polysaccharide-based-self-aggregate nanoparticles as drug delivery systems have been recently investigated. There are various hydrophobic molecules that can be attached to polysaccharides in order to obtain these kind of systems, such as poly(ethylene glycol) derivatives, long chain fatty acids, poly(ε-caprolactone), pluronic copolymers, cholesterol and poly(isobutilcyanoacrylate) (PIBCA).

Poly(ethylene glycol) (PEG) has been often used in pharmaceutical and biomedical fields as soluble polymeric modifier in organic synthesis, and as a pharmacological polymer with high hydrophilicity, biocompatibility and biodegradability. PEG and its derivatives can be attached to the polysaccharide structure to form micelles directly in an aqueous medium by adjusting the hydrophobicity/hydrophilicity of the polysaccharide chain. Chitosan has been grafted with different molecules of poly(ethylene glycol), obtaining nanoparticles with an average size ranged between 80-260 nm (Yang et al., 2008; Yoksan et al., 2004).

Some long-chain fatty acids like hexanoic acid, linoleic acid, linolenic acid, palmitic acid or stearic acid have been used for modifying polysaccharides and obtaining polymeric micelles. Nanoparticles based on linoleic acid-chitosan have been obtained through a carbodiimide-mediated reaction, and their size ranged between 200-600 nm. (Chen et al., 2003). Hu and coworkers employed a similar methodology in order to obtain stearic acid-chitosan nanoparticles. To increase the stability of the micelle *in vivo* and controlled drug release, the shells of micelles were cross-linked by glutaraldehyde (Hu et al., 2006). Dextran has been also employed to obtain nanoparticles by coupling lipoic acid to the structure of dextran and forming nanoparticles in water, whose size varied from 145 to 221 nm (Li et al., 2009).

Poly(ε-caprolactone) (PCL) is a well known biodegradable polyester, frequently used as implantable carrier for drug delivery systems, and a promising molecule which allows the formation of nanometric micelles when it is attached with a polysaccharide structure. The combination of the hydroxyl groups of dextran with the carboxylic function present on preformed PCL monocarboxylic acid results in the formation of nanoparticles of less than 200 nm (Lemarchand et al., 2003a).

Hyaluronic acid and pluronics form the polymeric shell of nanoparticles obtained by Han and coworkers (Han et al., 2005). Pluronics self-assemble to form a spherical micellar structure above the lower critical solution temperature by hydrophobic interaction of the poly(propylene oxide) middle block in the structure. Depending on the composition and molecular weight, they show at high concentration a sol-gel transition behaviour when raising the temperature above the lower critical solution temperature, which means a swelling or de-swelling behaviour temperature-dependent (Liu et al., 2003).

Among different cyclic hydrophobic molecules, cholesterol is one of the most used to give hydrophobic character to polysaccharide structures and form self-assembly nanoparticles in aqueous solution. Cholesterol attached to pullulans of different molecular weights, or to chitosan, allow to obtain nanoparticles of very small sizes: 20-30 nm in case of cholesterol-pullulans based nanoparticles (Akiyoshi et al., 1998) and 417 nm in case of cholesterol-chitosan systems (Wang et al., 2007). Apart from cholesterol, 5β-Cholanic acid is a non-toxic bile acid present in humans that can be chosen as the hydrophobic moiety to be attached to hyaluronic acid in order to form amphiphilic conjugates. The mean diameters of the obtained nanoparticles are in the range of 237–424 nm (Choi et al., 2010).

Poly(isobutilcyanoacrylate) (PIBCA), which belongs to polyacrylate family, can form amphiphilic copolymers in combination with polysaccharide structures due to the hydrophobic nature of this molecule that shows carboxylic esters on its structure. Among the different existing combinations of PIBCA with polysaccharides, chitosan and dextran are good examples of hydrophilic molecules that have been used to obtain nanometric micelles though an emulsion polymerization of IBCA in the presence of polysaccharides. In case of chitosan, the obtained micelles show very small sizes, lower than 35 nm (Bertholon et al., 2006).

3.5 Nanoprecipitation

Nanoprecipitation is one of the most used methods to produce nanoparticles owing to the reproducibility, the simplicity and the economy of the technique. Systems of three components are needed to perform nanoprecipitation process: the polymer, the polymer solvent and the non-solvent of the polymer. To produce nanoparticles, the polymer solution is mixed with the non-solvent. Nanoparticles are instantaneously formed during the fast diffusion of the polymer solution in the non-solvent. This technique has been up to date mainly applied for poly(lactic acid), poly(lactic-*co*-glycolic acid) (Lassalle & Ferreira, 2007) as well as for polysaccharide derivatives (Hornig & Heinze, 2008). Chitosan and amphiphilic cycodextrins are some examples of polysaccharide derivatives employed to obtain nanoparticles by this method.

In case of chitosan, sodium sulfate is commonly used as a precipitating agent to form chitosan particles. When sodium sulfate is slowly added into a solution of chitosan and polysorbate 80 under both stirring and ultrasonication, desolvated chitosan in a particulate form is obtained. The precipitated particles are at micro/nano interface (900 ± 200 nm) (Berthold et al., 1996).

Skiba and coworkers synthesised nanoparticles based on amphiphilic cyclodextrines (Skiba et al., 1996). Nanospheres are prepared by progressive dispersion of an organic solution of modified β-cyclodextrin in an aqueous phase with or without surfactant. Various physicochemical parameters were studied such as the effect of the chain length of acyl groups and type of surfactant on the size and physicochemical properties and stability of the nanospheres. Systems with a mean diameter varying between 90 and 150 nm were obtained by this method.

4. Application as drug delivery systems

In the design of drug carriers, issues of safety, toxicity and availability have to be taken into account, and the application of polysaccharides simplifies some of these issues. Thus, polysaccharides are emerging as new promising materials to be applied in the design of drug delivery systems. Some examples of drug delivery systems that have been prepared during the last years using polysaccharides in their composition have been collected below (Table 1).

4.1 Peptides and proteins included in polysaccharide-based delivery systems

Some of the most significant advances in biotechnology in recent years are related with the discovery of some therapeutic and antigenic peptides and proteins (Vila et al., 2002). Despite these great advances, some problems like low stability, short biological half-life and the need to cross biological barriers limit the use and the *in vivo* application of the most of biologically derived drugs. The inclusion of the biologically derived drug into drug delivery systems based on polysaccharides have shown good results and some of the problems of systemic administration can be overcome.

Insulin is one of the most widely used therapeutic peptides. It is a 5.8 kDa protein used exogenously to treat insulin-dependent diabetes mellitus when normal pancreatic production is insufficient. Orally administration of insulin normally shows low bioavailability due to acidic gastric pH, the enzymatic barrier of the intestinal tract and the physical barrier made up of the intestinal epithelium. The inclusion of insulin into nano- and microparticulate systems potentially provides gastric protection, controlled release and enhanced absorption by mucosal adhesion and nanoparticle direct uptake (Tiyaboonchai et al., 2003). There are various studies in which insulin has been successfully included into polysaccharide-based nanoparticles prepared by different methods, that have shown good results in loading efficacy of the drug, good release control and, in some of them, *in vivo* efficacy of the systems.

Nanoemulsion/in situ triggered gelation is one of the methods employed to prepare insulin-loaded nanoparticles. Pinto Reis and co-workers have designed nanoparticles based on alginate and alginate coated with chitosan that have been loaded with insulin obtaining an encapsulation efficiency between 76-93% due to the interaction of amino groups of insulin with carboxylic groups of alginate. They observed that insulin was strongly retained into both systems at low pH, but when the release medium was changed by one at near neutral pH, up to 40-70% of the insulin was released almost immediately. Alginate-chitosan nanospheres demonstrated the best controlled insulin profile release in simulated intestinal conditions (Reis et al., 2008). The same method has been used to prepare insulin-loaded nanoparticles based on alginate-dextran, achieving an encapsulation efficency of 82.5%. At low gastric pH, insulin was fully retained likely due to alginate polymer forming a compact acid-gel structure reducing permeability and potentially stabilizing insulin from acid attack. Up to 89% of the insulin was released almost immediately after changing the medium to near neutral pH, and full release was observed after 1h. Nanoencapsulated insulin was bioactive, which was demonstrated through both *in vivo* and *in vitro* bioassays (Reis et al., 2007).

Insulin-loaded nanoparticles have been prepared by ionic cross-link methods using chitosan as the selected polysaccharide and TPP anions as cross-linker agent. Pan and co-workers used this method in order to prepare chitosan nanoparticles which enhanced the intestinal absorption of insulin *in vivo* (Pan et al., 2002). The association between the drug and the system was up to 80%, and the drug release was very dependent on the pH of the release medium. *In vivo* studies revealed that insulin was released in its active form, and the dosages of insulin loaded into chitosan nanoparticles were found orally effective, prolonging the hypoglycemia over 15 h.

Group of drugs	Entrapped Drug	Method of synthesis	Polysaccharide	NP size (nm)	References
Peptides and proteins	Insulin	Emulsification /ionic gelation	Alginate	564	Reis et al., 2008; Reis et al., 2007
			Alginate-chitosan	1280	
			Alginate-dextran	267-2760	
		Ionic cross-link	Chitosan	300-400	Pan et al., 2002
				250-400	
		Polyelectrolyte complexation (PEC)	Chitosan-dextran	423-850	Sarmento et al., 2006
			Chitosan-alginate		
			Chitosan-glucomannan	200-700	Alonso-Sande et al., 2006
		Self-assembling	Pullulan-cholesterol	20-30	Akiyoshi et al., 1998
	Growth factors	Ionic cross-link	Chitosan	301-424	Cetin et al., 2007
	Antisenses	Ionic cross-link	Alginate-chitosan	196-430	Azizi et al., 2010
Anticancer drugs	Tamoxifen	Covalent cross-link	Alginate-albumin	42-388	Martínez et al., 2011
		Emulsification	Guar gum	200-300	Sarmah et al., 2009
	Mitoxantrone	Ionic cross-link	Chitosan	75	Lu et al., 2006
	Methotrexate	Self-assembling	PEG-chitosan	262	Yang et al., 2008
	Doxorubicin	Emulsification	Oleoyl-chitosan	255	Zhang et al., 2007
	Paclitaxel	Self-assembling	Stearic acid-chitosan	28-175	Hu et al, 2006
	Epirubicin	Self-assembling	Chitosan-cholesterol	417	Wang et al., 2007
Nucleic acids and genetic material	Nucleic acids	Ionic cross-link	Chitosan	200-700	Katas & Alpar, 2006
		Polyelectrolyte complexation (PEC)	Chitosan-alginate	323-1600	Douglas & Tabrizian, 2005
			Chitosan-CMC	180-200	Cui & Mumper, 2001

Group of drugs	Entrapped Drug	Method of synthesis	Polysaccharide	NP size (nm)	References
			Chitosan-nucleic acids	129-450	de Martimprey et al., 2009
	Antisenses	Polyelectrolyte complexation (PEC)	Alginate-poly-L-lysine	28.000-10.000	González Ferreiro et al., 2002
	Genes	Emulsification /ionic gelation	Alginate	80	You & Pen, 2005
Other drugs	Furosemide	Emulsification /ionic gelation	Chitosan	30-150	Zhi et al., 2005
	Cyclosporine A	Ionic cross-link	Chitosan	250-400	De la Fuente et al., 2010
	Rertinol	Nanoprecipitation	Chitosan	50-200	Kim et al., 2006
	Antituber-cular drugs	Ionic gelation	Alginate	236	Ahmad et al., 2006
	Anfortericin B	Polyelectrolyte complexation (PEC)	Chitosan-dextran sulfate	600-800	Tiyaboonchai & Limpeanchob, 2007

Table 1. Some groups of drugs included in polysaccharide-based nanoparticles.

Ionotropic complexation (PEC) between chitosan and different polyanions (alginate, dextran sulfate and glucomannan) has been described in different works as another method that allowed the preparation of insulin-loaded nanoparticles. Insulin association efficiency from 63 to 94% and loading capacity from 5 to 13% were obtained with chitosan/alginate and chitosan/dextran nanoparticles, providing dextran sulfate combination the highest insulin association efficiency. It might be seen that insulin release occurred very rapidly from both systems, but a significant increment of insulin retention when using dextran sulfate in the formulation compared with alginate was observed in simulated gastric conditions (Sarmento et al., 2006). Alonso-Sande and co-workers prepared insulin-loaded chitosan/glucomannan nanoparticles reaching association efficiency values of 89%. They analyzed the influence of the pH release medium and the TPP presence on the release rate, obtaining more restricted diffusion of the peptide through a highly cross-linked system with TPP and at high pH of the release medium (Alonso-Sande et al., 2006).

Finally, nanoparticles based on self-assembling methods were also obtained and loaded with insulin. In this case, insulin was easily complexed with the hydrogel nanoparticle of hydrophobized cholesterol-bearing pullulan in water. The presence of different concentrations of bovine serum albumin (BSA) in the release medium was analyzed, and approximately 15% of insulin still remained in the complex, even at concentrations of BSA comparable to its physiological value in the blood. The physiological activity of complexed insulin was preserved *in vivo* after i.v. injection, showing an excellent behaviour of the complex as a possible protein drug carrier (Akiyoshi et al., 1998).

Basic fibroblast growth factor (bFGF) is a protein with molecular mass of 18 kDa that is a potent mitogen which regulates angiogenesis during growth and development. This growth factor stimulates the proliferation of a wide variety of cells, including mesenchymal, neuroectodermal, and endothelial cells, and it is effective on protecting neurons from oxidative stress processes (Li et al., 2006). However, bFGF has minimal pharmacological effects in the central nervous system (CNS) because of the presence of bloodbrain barrier that reduces its transport to CNS. In order to enhance the amount of bFGF that could reach the CNS, Cetin and co-workers have developed bFGF-loaded chitosan nanoparticles according to a ionotropic gelation process, using TPP as cross-linker agent. Nanoparticles displayed a low bFGF-association efficiency (27.4%) leading to final bFGF-loading values as low as 0.021%, but since bFGF exerts its biological activity in the concentration range of 0.1 to 10.0 ng/mL, they concluded that the bFGF loading was acceptable. *In vitro* release studies showed that around 30% of the loaded protein was immediately released into PBS and that the highest extent of release (68%) was observed at 24h. Moreover, their results showed that integrity of the encapsulated bFGF was not affected by the entrapment procedure and release conditions (Cetin et al., 2007).

Epidermal growth factor receptor (EGFR) antisense (AS) is used to reduce the expression of EGFR, a receptor tyrosine kinase proto-oncogene that plays a central role in the initiation and development of several human malignancies, notably of the breast, brain, and lung. The problem is that the macromolecular drug can degrade during preparation of the nanoparticles, during storage, and also during *in vitro* or *in vivo* release. For this reason, the ability of nanoparticles to protect the antisense molecules from degradation in aqueous medium is necessary to be investigated. Azizi and co-workers prepared EGFR-AS-loaded chitosan/alginate nanoparticles with different composition ratios, obtaining that nanoparticles can release antisense over 45-50 hours and 67-96% loading efficacy. Thus, they concluded that the nanoparticles could retain and stabilize the content of antisense in their hydrogel matrix (Azizi et al., 2010).

4.2 Anticancer drugs included in polysaccharide-based delivery systems

One of the major problems facing cancer chemotherapy is the achievement of the required concentration of the drug at the tumour site for a desired period of time, since tumours usually present resistances to treatment, and high dosages are frequently toxic. Thus, one of the main goals of nanomedicine is to develop safe and effective drug carriers that are systemically applied but will selectively deliver cytotoxic drugs to tumour cells without harming normal cells (Gullotti & Yeo, 2009). Among the available potential drug carrier systems in this size range, polysaccharide-based nanoparticles play an important role and their use with some anticancer drugs show promising results.

Tamoxifen has been successfully used since 1970s in treatment of hormone dependent breast cancer (Ameller et al., 2004). However, tamoxifen shows low water solubility, which limits the administration of this drug only to the oral route. Furthermore, following a long-term therapy, tamoxifen has some side effects, such as endometrial cancer and development of drug resistance. To overcome the undesirable side effects of tamoxifen, and to increase the concentration at the tumour site, tamoxifen could be entrapped into polymeric nanoparticles, which may provide better means of delivery in terms of enhanced uptake by the tumour and increased local concentration of the drug at the receptor site. Tamoxifen-loaded nanoparticles were prepared by Sarmah and co-workers based on guar gum, which is commonly used for colon specific drug delivery in the pharmaceutical industry. Nanoparticles were obtained by o/w emulsification and in situ polymer cross-linking, using dichloromethane as the best solvent of the drug and glutaraldehyde as cross-linker agent during the process. An efficiency loading of 15% was obtained when dichloromethane was used as selected solvent (Sarmah et al., 2009). Recently, nanoparticles based on thiolated alginate (ALG-CYS) and disulfide bond reduced albumin (BSA-SH) have been synthesized by coacervation method and stabilized by disulfide bond formation between both polymers. *In vitro* studies revealed that total release of the drug was not achieved in any case; only the 23–61% of the drug was released. Maximum release took place between 7 and 75 h. According to the results, it was concluded that the presence of alginate in the nanoparticle composition allowed the modulation of the amount of released TMX (Martínez et al., 2011).

Mitoxantrone is often used to treat breast cancer clinically, but the prolonged treatment with this drug results in some side-effects, such as heart toxicity and myelosupression, which are often a problem. Mitoxantrone is positively charged and it can be absorbed by negatively charged polysaccharides, such as chitosan. Nanospheres can be obtained by ion gelation method using sodium TPP as gelation agent, and obtaining an encapsulation efficacy of 98%. Tests for *in vitro* release in physiological saline or physiological saline containing 0.5% (w/v) ascorbic acid by a dialysis bag showed sustained release and little burst effect (Lu et al., 2006).

Methotrexate (MTX) is a folate antimetabolite and has been used in the treatment of various malignancies, including childhood acute lymphocytic leukemia, osteosarcoma, non-Hodgkin's lymphoma, Hodgkin's disease, head and neck cancer, lung cancer, and breast cancer. However, it may cause some adverse effects such as bone marrow suppression, acute and chronic hepatotoxicity, interstitial pneumonitis and chronic interstitial obstructive pulmonary disease. Therefore, in order to reduce its toxic and side effects and to improve specificity and selectivity, it is a good candidate to be included in drug delivery systems. Yang and co-workers have designed methoxy poly(ethylene glycol)-grafted-chitosan (mPEG-g-CS) self-aggregated nanoparticles being used as a carrier of MTX . Depending on the formulation, loading efficiency varied from 21 to 95%. *In vitro* release studies showed that the MTX appeared to be released in a biphasic way, which characterized by an initial release or rapid release period followed by a step of slower release. A fast release was observed in 4 h, in which 40% of the drug was released from nanoparticles. After this initial effect, MTX was released in a continuous way for up to 48 h, reaching percentage of cumulative release close to 60%. Therefore, the self-aggregated nanoparticles delayed the drug release in the release process when these results were compared with the release of the free drug (Yang et al., 2008).

Doxorubicin (DOX) is a member of the anthracycline ring antibiotics, with a broad spectrum of antitumor activity, including a variety of human and animal solid tumors. Despite advances of this antitumor agent, it is hydrophobic and possesses inevitable, serious side effects such as nonspecific toxicity that limit the dose and use of the drug. Therefore, a lot of studies have been carried out in order to entrap this drug into different drug carriers. DOX has been successfully entrapped into oleoyl-chitosan nanoparticles prepared by o/w emulsification method. Nanoparticles had a high encapsulation efficiency of 53% in loading doxorubicin, being drug incorporation into nanoparticles generally limited by the large surface area of the latter, as well as by the solubility of the drug in water. The drug was completely released from nanoparticles in the buffer medium of Na_2HPO_4– citric acid (pH 3.8), whereas in PBS (pH 7.4) 65% of doxorubicin was released after 6 hours, followed by a sustained release until 72 hours. Approximately 72% of DOX was released for 3 days, showing the potential of the nanoparticles as a sustained drug delivery system, suggesting that the nanoparticles might act as a barrier against the release of entrapped DOX. This result indicated that the nanoparticles contributed to an extended circulation of DOX and thus an improvement in therapeutic efficacy (Zhang et al., 2007).

Paclitaxel is an anticancer drug which is used in different types of malignancies, such as ovarian and breast cancer. Despite its multiple applications, it has important problems of solubility in water and, consequently, its administration is carried out including an oily component (Cremophor) and dehydrated alcohol within its formulation. Thus, relevant side effects such as hypersensibility, hypotension or breast pain are frequently observed after its intravenous administration. In the last years, a lot of studies have included this drug as good candidate to be included in drug delivery systems to overcome its administration problems. Hu and coworkers have obtained paclitaxel-loaded nanoparticles by self-cross-linking stearic-grafted chitosan and using glutaraldehyde as cross-linker in order to stabilize the systems. These formulations showed high encapsulation efficiency which ranged from 95% to 99%. When the surfaces of micelles were cross-linked by glutaraldehyde, the burst release of the micelles at earlier stage was highly improved and the drug release time was prolonged. By controlling the amino substitution of stearic-grafted chitosan and the cross-link degree, the prolonged and controlled release could be achieved (Hu et al., 2006).

As an anthracycline anticancer agent, **epirubicin** has a wide range of antitumor activity and is used to treat various carcinomas. However, this therapy may cause some serious side effects such as allergic reactions, cardiotoxicity and blood problems. Therefore, nanoparticles being used as a carrier of epirubicin are hoped to sustain its release, prolong its circulation time, enhance its therapeutic index and decrease its toxic effects. Cholesterol-modified nanoparticles have been prepared by self-assembling method by Wang and co-workers (Wang et al., 2007). They synthesized cholesterol-modified chitosan conjugate with succinyl linkages, investigated its self-aggregation behaviour, and prepared self-aggregated nanoparticles by sonication in aqueous media. Nanoparticles loading efficiency varied from 25 to 71% depending on the weight ratio of epirubicin and the nanoparticles and exhibited the release profile relating to the pH of the release media. When the pH of the release media increased, an evident decrease of drug release rate was observed owing to the solubility of epirubicin is greatly influenced by the pH of the aqueous solution, and self-aggregated nanoparticles were also pH sensitive because of the presence of many amino groups in their molecules. Thus, the drug release from self-aggregated nanoparticles was very slow in PBS (pH 7.4) and the total release amount was about 25% in 48 h, which suggested self-aggregated nanoparticles had a potential as a sustained-release carrier of epirubicin.

4.3 Nucleic acids and genetic material included in polysaccharide-based delivery systems

Small interfering RNAs (siRNAs) have proven to be versatile agents for controlling gene expression in mammalian cells. They have been employed as a novel tool since they can block the expression of genes, such as those expressed in infectious diseases and cancers.

However, siRNA suffers particular problems including poor cellular uptake, rapid degradation as well as limited blood stability. Therefore, effective systems which can protect and transport siRNA to the cytoplasm of the targeted cells are needed to exploit the promising potential applications offered by successful delivery of siRNA. Chitosan has been used to prepare nanoparticles based on modified ionic gelation with TPP as cross-linker agent, and they have been applied as possible vector that should be able to be taken up by the cells and escape the endosomal vesicle to avoid lysosomal degradation. *In vitro* study revealed the transfection efficiency of siRNA depends on the method of siRNA association to the chitosan, and entrapping siRNA using ionic gelation has shown to yield a better biological effect than simple complexation or siRNA adsorption onto the chitosan nanoparticles (Katas & Alpar, 2006).

Ionotropic complexation (PEC) between chitosan and different polyanions has been used as an useful method of preparation of nanoparticles in order to include nucleic acids, since chitosan-**DNA** nanoparticles demonstrated low transfection efficiencies and the incorporation of secondary polymers improved the characteristics of these systems (Kaul & Amiji, 2002). Among the chitosan–polyanion complexes investigated, the combination of chitosan and alginate is considered to be among the most interesting for delivery systems. The method used to prepare the nanoparticles is a two-step method where the first step is the formation of a calcium–alginate pre-gel. Various concentrations of chitosan solutions were then added with continuous stirring. High loading efficacy was obtained (26-60%), while maximum mass loading was 60 μg DNA/mg nanoparticles and 6% and 3.5% of the adsorbed DNA was released. Therefore, the high encapsulation of DNA from alginate–chitosan nanoparticles is encouraging for application in the field of gene therapy (Douglas & Tabrizian, 2005). The complexation between chitosan and carboxymethylcellulose (CMC) is another PEC combination employed in the inclusion of nucleic acid into delivery systems. Cui and Mumper coated the surface of this pre-formed nanoparticles with plasmid-DNA, obtaining a final plasmid DNA concentration of up to 400 mg/ml. Chitosan/CMC based nanoparticles containing plasmid DNA were applied topically to the skin of shaved mice and resulted in detectable and quantifiable levels of luciferase expression in skin after 24 h (Cui & Mumper, 2001). Nucleic acids can be also combined with chitosan in order to obtain nanospheres, being part of the structure of the system at the same time they are the encapsulant of the nanospheres. Chitosan showed a very high protection efficiency for siRNA. Polyplexes prepared with this polysaccharide are able to protect a siRNA for 7 h and an efficient transfection of the cells may be obtained after the intranasal administration of siRNA/chitosan polyplexes (de Martimprey et al., 2009).

Antisense oligonucleotides are therapeutic agents known to selectively modulate gene expression. The development of non-parenteral dosage forms for these compounds is desirable. However, the high molecular weight, the hydrophilicity and multiple negative charges result in a poor absorption of antisense oligonucleotides. Moreover, the oral administration faces additional problems such as degradation in the acidic gastric environment, enzymatic metabolism in the lumen and at the gastrointestinal epithelium and first-pass hepatic clearance. To achieve a successful non-parenteral delivery of antisense therapeutics, it is necessary to solve the specific problems of the oral administration route, together with general concerns of correct time-space targeting, improved cellular uptake and nuclear localization to exert gene transfection (Akhtar et al., 2000). Using a ionotropic gelation method, microparticles made of alginate cross-linked with calcium ions and poly-L-lysine have been reported to effectively act as transfecting agents. Encapsulation efficacies of 32-95% depending on the composition of the formulation have been obtained and the *in vitro* release behaviour depended mostly on the medium composition. The importance of competitive anions and ionic strength on the mechanism of dissociation of the oligonucleotide from the polymeric matrix was observed. Thus, the presence of phosphate anions preferably displaced alginate from the structure, resulting in the release of complexed oligonucleotide. Rat *in vivo* studies showed promising oligonucleotide bioavailability for microparticles after intrajejunal administration in the presence of a mixture of permeation enhancers to achieve a successful intestinal application (Gonzalez Ferreiro et al., 2002).

Natural biopolymers are also widely used in the field of **gene delivery**. In fact, alginate nanoparticles have been prepared using w/o microemulsion as a template followed by calcium cross-linking of guluronic acid units of alginate polymer. Ca-alginate nanoparticles were loaded with GFP-encoding plasmids in order to study their potency as carriers for gene delivery. The degree of endocytosis by NIH 3T3 cells and ensuing transfection rate were investigated. Results showed that Ca-alginate nanoparticles were very efficient gene carriers (You & Peng, 2004).

4.4 Other drugs included in polysaccharide-based delivery systems

Furosemide is a loop diuretic used in the treatment of congestive heart failure and edema, and it has been recently incorporated into chitosan nanoparticles (Zhi et al., 2005). Since the chitosan molecule has strong interaction with the organic compounds, it can be applied to adsorb diuretics from the water samples. Zhi and coworkers prepared chitosan nanoparticles by a nanoemulsion system, in which the chitosan nanoparticles were prepared by adding NaOH solution or glutaraldehyde as the solidification solution. Size was smaller in case of using NaOH as cross-linker agent (30 - 150 nm). The adsorptive efficiency of furosemide on the nanoparticle was 51.9%, and the presence of less cross-linking agent could be in favour of the adsorption ability. The furosemide adsorption capacity on the chitosan nanoparticles was affected not only by the –NH2 content, but also by the molecular weight of chitosan.

Ophthalmic drug delivery, probably more than any other route of administration, may benefit from the characteristics of nanotechnology-based drug delivery systems, mainly because of their capacity to protect the encapsulated molecule while facilitating its transport to the different compartments of the eye (Raju & Goldberg, 2008). For this purpose, the common use of chitosan is justified by its mucoadhesive and penetration enhancing properties, as well as by its good biocompatibility with the ocular structures. **Cyclosporine A (CyA)** is an immunosuppressant drug widely used in post-allogeneic organ transplant to reduce the activity of the immune system, and therefore the risk of organ rejection. The entrapment of the hydrophobic polypeptide CyA was achieved by ionic gelation technique using TPP and a previous dissolution of the peptide in an acetonitrile: water mixture, and a further nanoprecipitation into the nanoparticles in the form of small nanocrystals (de la Fuente et al., 2010). Entrapment efficiencies were reported to be as high as 73.4%. The *in vivo* evaluation of this new prototype, in rabbits, evidenced the capacity of these nanoparticles providing a selective and prolonged delivery of CyA to the cornea and conjunctiva. More importantly, it was observed that CyA-loaded chitosan nanoparticles provided therapeutic levels of CyA in the conjunctiva and the cornea for up to 24 and 48 h postadministration, respectively, while reducing the access of CyA to the blood circulation. This positive behavior of chitosan nanoparticles was attributed to their improved interaction with the corneal and conjunctival and the prolonged delivery of the CyA molecules associated to them (de la Fuente et al., 2010).

Retinol and its derivatives are extensively used in the pharmaceutical and cosmetic area. Especially, retinoids are recognized as being important for modern therapy of dermatological treatment of wrinkled skin. The stability and solubility problems of these compouns make them ideal candidates to be included into carriers. Nanoparticles are reported to be useful formulation to solve the poor aqueous solubility of retinoids and are able to use it by intravenous injection. Chitosan has been used by Kim and coworkers in order to obtain retinol-encapsulated chitosan nanoparticles for application of cosmetic and pharmaceutical applications (Kim et al., 2006). Retinol-encapsulated nanoparticles were completely reconstituted into aqueous solution as same as original aqueous solution.

The requirement of multidrug administration daily or several times a week for at least 6 months is the main cause of the failure of antitubercular chemotherapy. The dose-dependent side effects with high incidence and the need of daily-dosing results in discontinuation of medication, relapse of symptoms and an alarming increase in the prevalence of multidrug-resistant strains (Pandey & Khuller, 2005). Consequently, reductions in dose and dosing frequency are the major goals of tuberculosis research and the application of drug delivery systems constitutes an important therapeutic strategy. **Isoniazid, pyrazinamide and rifampicin**, which are important antitubercular drugs, were encapsulated into alginate nanoparticles prepared by cation-induced gelification, reaching drug encapsulation efficiencies of 70–90% for isoniazid and pyrazinamide and 80–90% for rifampicin (Ahmad et al., 2006). When alginate nanoparticles were administered through nebulisation, the drugs were detected in plasma from 3 h onwards. Encapsulated drugs were observed up to 14, 10 and 14 days, respectively, in contrast with free drugs that were cleared from the circulation within 12–24 h. The levels of drugs in various organs remained above the minimum inhibitory concentration at both doses for equal periods, demonstrating their equiefficiency. Alginate nanoparticles hold great potential in reducing dosing frequency of antitubercular drugs.

Amphotericin B (AmB) is a polyene macrolide antifungal agent and the drug of choice for systemic fungal infection. Unfortunately, it is poorly absorbed from the gastrointestinal tract due to its low aqueous solubility and it must be given parenterally to treat systemic fungal infections. Currently, two types of drug formulations for AmB are available. The first one is a micellar solution of AmB and the second one are lipid-based nanoparticulate formulations. Despite these formulations reduce nephrotoxicity of the treatment, they are still quite expensive. Thus, much effort has been spent to develop cheaper delivery systems with reduced amphotericin B toxicity. Polysaccharide based nanoparticle delivery systems are one approach that has been investigated by Tiyaboonchai and coworkers (Tiyaboonchai & Limpeanchob, 2007). They prepared AmB-chitosan-dextrane sulfate nanoparticles by polyelectrolyte complexation at room temperature, showing an an association efficiency of 50–65%. They observed a fast release characteristic of AmB independent of the processing conditions, with most of AmB released from particles within 5 min. This suggested that

AmB exhibited only moderate interaction with the weakly cross-linked polymers of the nanoparticles. A reduction of nephrotoxicity was observed in an *in vivo* renal toxicity study.

5. Conclusion

As pointed out throughout this review, polysaccharides show variability and versatility, due to their complex structure, which is difficult to be reproduced with synthetic polymers. Thus, native polysaccharides and their derivatives are emerging in the last years as one of the most used biomaterials in the field of nanomedicine, especially being chosen by a lot of researchers as carriers to be used in the preparation of nanoparticulate drug delivery systems.

A wide variety of preparation methods of nanoparticles has been developed, and three aspects have marked the evolution of these methods: need for lees toxic agents, simplification of the procedures and optimization to improve yield and entrapment efficiency. Now it is possible to choose the best method of preparation and the best suitable polymer to achieve an efficient encapsulation of the drug, taking into account the drug features in this selection.

As reviewed above, so many nanoparticle drug delivery systems have been prepared using various polysaccharides as carriers combined with different groups of drugs, and they have been investigated in terms of physicochemical features, drug -loading efficiency, *in vitro* toxicity and comparative *in vivo* test. Deeper studies, such as evaluation of interaction between cells, tissues and organs, as well as how the administration of these systems can affect to the metabolism, need to be carried out. In fact, more and more nanoparticle systems are emerging nowadays, and these necessary studies will be focused on in the near future, completing the evaluation of these hopeful polysaccharide-based systems.

6. Acknowledgements

The financial support of the Ministerio de Ciencia e Innovación of Spain FIS(PS09/01513) and MAT2010-21509-C03-03, and the FPI grant from UCM to A. Martínez are gratefully acknowledged.

7. References

Ahmad, Z.; Pandey, R.; Sharma, S. & Khuller, G. K. (2006). Pharmacokinetic and pharmacodynamic behaviour of antitubercular drugs encapsulated in alginate nanoparticles at two doses. *Int J Antimicrob Agents*, 27, 5, pp. (409-416), 0924-8579 (Print) 0924-8579 (Linking).

Akbuga, J. & Durmaz, G. (1994). Preparation and evaluation of crosslinked chitosan microspheres containing furosemide. *International Journal of Pharmaceutics*, 11, pp. (217-222).

Akhtar, S.; Hughes, M. D.; Khan, A.; Bibby, M.; Hussain, M.; Nawaz, Q.; Double, J. & Sayyed, P. (2000). The delivery of antisense therapeutics. *Adv Drug Deliv Rev*, 44, 1, pp. (3-21), 0169-409X (Print) 0169-409X (Linking).

Akiyoshi, K.; Kobayashi, S.; Shichibe, S.; Mix, D.; Baudys, M.; Kim, S. W. & Sunamoto, J. (1998). Self-assembled hydrogel nanoparticle of cholesterol-bearing pullulan as a carrier of protein drugs: complexation and stabilization of insulin. *J Control Release*, 54, 3, pp. (313-320), 0168-3659 (Print) 0168-3659 (Linking).

Alonso-Sande, M.; Cuna, M. & Remunan-Lopez, C. (2006). Formation of new glucomannan-chitosan nanoparticles and study of their ability to associate and deliver proteins. *Macromolecules*, 39, pp. (4152-4158).

Allison, D. D. & Grande-Allen, K. J. (2006). Review. Hyaluronan: a powerful tissue engineering tool. *Tissue Eng*, 12, 8, pp. (2131-2140), 1076-3279 (Print) 1076-3279 (Linking).

Ameller, T.; Legrand, P.; Marsaud, V. & Renoir, J. M. (2004). Drug delivery systems for oestrogenic hormones and antagonists: the need for selective targeting in estradiol-dependent cancers. *J Steroid Biochem Mol Biol*, 92, 1-2, pp. (1-18), 0960-0760 (Print) 0960-0760 (Linking).

Amidi, M.; Romeijn, S. G.; Borchard, G.; Junginger, H. E.; Hennink, W. E. & Jiskoot, W. (2006). Preparation and characterization of protein-loaded N-trimethyl chitosan nanoparticles as nasal delivery system. *J Control Release*, 111, 1-2, pp. (107-116), 0168-3659 (Print) 0168-3659 (Linking).

Aumelas, A.; Serrero, A.; Durand, A.; Dellacherie, E. & Leonard, M. (2007). Nanoparticles of hydrophobically modified dextrans as potential drug carrier systems. *Colloids Surf B Biointerfaces*, 59, 1, pp. (74-80), 0927-7765 (Print) 0927-7765 (Linking).

Azizi, E.; Namazi, A.; Haririan, I.; Fouladdel, S.; Khoshayand, M. R.; Shotorbani, P. Y.; Nomani, A. & Gazori, T. (2010). Release profile and stability evaluation of optimized chitosan/alginate nanoparticles as EGFR antisense vector. *Int J Nanomedicine*, 5, pp. (455-461), 1178-2013 (Electronic) 1176-9114 (Linking).

Bae, H.; Ahari, A. F.; Shin, H.; Nichol, J. W.; Hutson, C. B.; Masaeli, M.; Kim, S. H.; Aubin, H.; Yamanlar, S. & Khademhosseini, A. (2011). Cell-laden microengineered pullulan methacrylate hydrogels promote cell proliferation and 3D cluster formation. *Soft Matter*, 7, 5, pp. (1903-1911), 1744-6848 (Electronic) 1744-683X (Linking).

Balazs, E. A. (2004). *Viscoelastic properties of hyaluronan and its therapeutics use*, Elsevier, Amsterdam

Barbosa, M.; Granja, P.; Barrias, C. & Amaral, I. (2005). Polysaccharides as scaffolds for bone regeneration. *ITBM-RBM*, 26, pp. (212-217).

Barbucci, R.; Pasqui, D.; Favaloro, R. & Panariello, G. (2008). A thixotropic hydrogel from chemically cross-linked guar gum: synthesis, characterization and rheological behaviour. *Carbohydr Res*, 343, 18, pp. (3058-3065), 1873-426X (Electronic) 0008-6215 (Linking).

Barratt, G. M. (2000). Therapeutic applications of colloidal drug carriers. *Pharm Sci Technolo Today*, 3, 5, pp. (163-171), 1461-5347 (Electronic) 1461-5347 (Linking).

Bernardo, M. V.; Blanco, M. D.; Sastre, R. L.; Teijon, C. & Teijon, J. M. (2003). Sustained release of bupivacaine from devices based on chitosan. *Farmaco*, 58, 11, pp. (1187-1191), 0014-827X (Print) 0014-827X (Linking).

Berthold, A.; Cremer, K. & Kreuter, J. (1996). Preparation and characterization of chitosan microspheres as drug carrier for prednisolone sodium phosphate as model for anti-inflammatory drugs. *journal of Controlled Release*, 39, pp. (17-25).

Bertholon, I.; Lesieur, S.; Labarre, D.; Besnard, M. & C., V. (2006). Characterization of Dextran-Poly(isobutylcyanoacrylate) Copolymers Obtained by Redox Radical and Anionic Emulsion Polymerization. *Macromolecules*, 39, pp. (3559-3567).

Blanco, M. D.; Gomez, C.; Olmo, R.; Muniz, E. & Teijon, J. M. (2000). Chitosan microspheres in PLG films as devices for cytarabine release. *Int J Pharm*, 202, 1-2, pp. (29-39), 0378-5173 (Print) 0378-5173 (Linking).

Bodnár, M.; Daróczi, L.; Batta, G.; Bakó, J.; Hartmann, J. F. & Borbély, J. (2009). Preparation and characterization of cross-linked hyaluronan nanoparticles. *Colloid & Polymer Science*, 287, pp. (991-1000).

Bodnar, M.; Hartmann, J. F. & Borbely, J. (2005). Preparation and characterization of chitosan-based nanoparticles. *Biomacromolecules*, 6, 5, pp. (2521-2527), 1525-7797 (Print) 1525-7797 (Linking).

Braga, T. P.; Chagas, E. C.; Freitas de Sousa, A.; Villarreal, N. L.; Longhinotti, N. & Valentini, A. (2009). Synthesis of hybrid mesoporous spheres using the chitosan as template. *Journal of Non-Crystalline Solids*, 355, pp. (860-866).

Burdick, J. A. & Prestwich, G. D. (2011). Hyaluronic acid hydrogels for biomedical applications. *Adv Mater*, 23, 12, pp. (H41-56), 1521-4095 (Electronic) 0935-9648 (Linking).

Cafaggi, S.; Russo, E.; Stefani, R.; Parodi, B.; Caviglioli, G.; Sillo, G.; Bisio, A.; Aiello, C. & Viale, M. (2011). Preparation, characterisation and preliminary antitumour activity evaluation of a novel nanoparticulate system based on a cisplatin-hyaluronate complex and N-trimethyl chitosan. *Invest New Drugs*, 29, 3, pp. (443-455), 1573-0646 (Electronic) 0167-6997 (Linking).

Cetin, M.; Aktas, Y.; Vural, I.; Capan, Y.; Dogan, L. A.; Duman, M. & Dalkara, T. (2007). Preparation and in vitro evaluation of bFGF-loaded chitosan nanoparticles. *Drug Deliv*, 14, 8, pp. (525-529), 1071-7544 (Print) 1071-7544 (Linking).

Coviello, T.; Matricardi, P.; Marianecci, C. & Alhaique, F. (2007). Polysaccharide hydrogels for modified release formulations. *J Control Release*, 119, 1, pp. (5-24), 1873-4995 (Electronic) 0168-3659 (Linking).

Cui, Z. & Mumper, R. J. (2001). Chitosan-based nanoparticles for topical genetic immunization. *J Control Release*, 75, 3, pp. (409-419), 0168-3659 (Print) 0168-3659 (Linking).

Chen, Q.; Hu, Y.; Chen, Y.; Jiang, X. & Yang, Y. (2005). Microstructure formation and property of chitosan-poly(acrylic acid) nanoparticles prepared by macromolecular complex. *Macromol Biosci*, 5, 10, pp. (993-1000), 1616-5187 (Print) 1616-5187 (Linking).

Chen, X. G.; Lee, C. M. & Park, H. J. (2003). O/W emulsification for the self-aggregation and nanoparticle formation of linoleic acid-modified chitosan in the aqueous system. *J Agric Food Chem*, 51, 10, pp. (3135-3139), 0021-8561 (Print) 0021-8561 (Linking).

Choi, K. Y.; Chung, H.; Min, K. H.; Yoon, H. Y.; Kim, K.; Park, J. H.; Kwon, I. C. & Jeong, S. Y. (2010). Self-assembled hyaluronic acid nanoparticles for active tumor targeting. *Biomaterials*, 31, 1, pp. (106-114), 1878-5905 (Electronic) 0142-9612 (Linking).

Davidenko, N.; Blanco, M. D.; Peniche, C.; Becherán, L.; Guerrero, S. & Teijón, J. M. (2009). Effects of different parameters on characteristics of chitosan-poly(acrilic acid) nanoparticles obtained by the method of coacervation. *Journal of Applied Polymer Science*, 111, pp. (2362-2371).

de la Fuente, M.; Ravina, M.; Paolicelli, P.; Sanchez, A.; Seijo, B. & Alonso, M. J. (2010). Chitosan-based nanostructures: a delivery platform for ocular therapeutics. *Adv Drug Deliv Rev*, 62, 1, pp. (100-117), 1872-8294 (Electronic) 0169-409X (Linking).

de la Fuente, M.; Seijo, B. & Alonso, M. J. (2008). Design of novel polysaccharidic nanostructures for gene delivery. *Nanotechnology*, 19, 7, pp. (075105), 0957-4484 (Print) 0957-4484 (Linking).

de Martimprey, H.; Vauthier, C.; Malvy, C. & Couvreur, P. (2009). Polymer nanocarriers for the delivery of small fragments of nucleic acids: oligonucleotides and siRNA. *Eur J Pharm Biopharm*, 71, 3, pp. (490-504), 1873-3441 (Electronic) 0939-6411 (Linking).

De, S. & Robinson, D. (2003). Polymer relationships during preparation of chitosan-alginate and poly-l-lysine-alginate nanospheres. *J Control Release*, 89, 1, pp. (101-112), 0168-3659 (Print) 0168-3659 (Linking).

Douglas, K. L. & Tabrizian, M. (2005). Effect of experimental parameters on the formation of alginate-chitosan nanoparticles and evaluation of their potential application as DNA carrier. *J Biomater Sci Polym Ed*, 16, 1, pp. (43-56), 0920-5063 (Print) 0920-5063 (Linking).

Drogoz, A.; David, L.; Rochas, C.; Domard, A. & Delair, T. (2007). Polyelectrolyte complexes from polysaccharides: formation and stoichiometry monitoring. *Langmuir*, 23, 22, pp. (10950-10958), 0743-7463 (Print) 0743-7463 (Linking).

Fernandez-Hervas, M. & Fell, J. (1998). Pectin/chitosan mixtures as coatings for colon-specific drug delivery: an in vitro evaluation. *International Journal of Pharmaceutics*, 169, pp. (115-119).

Gavory, C.; Durand, A.; Six, J. L.; Nouvel, C.; Marie, E. & Leonard, M. (2011). Polysaccharide-covered nanoparticles prepared by nanoprecipitation. *Carbohydrate Polymers*, 84, pp. (133-140).

George, M. & Abraham, T. E. (2006). Polyionic hydrocolloids for the intestinal delivery of protein drugs: alginate and chitosan--a review. *J Control Release*, 114, 1, pp. (1-14), 0168-3659 (Print) 0168-3659 (Linking).

Gonzalez Ferreiro, M.; Tillman, L.; Hardee, G. & Bodmeier, R. (2002). Characterization of alginate/poly-L-lysine particles as antisense oligonucleotide carriers. *Int J Pharm*, 239, 1-2, pp. (47-59), 0378-5173 (Print) 0378-5173 (Linking).

Guerrero, S.; Teijón, C.; Muñiz, E.; Teijón, J. M. & Blanco, M. D. (2010). Characterization and in vivo evaluation of ketotifen-loaded chitosan microspheres. *Carbohydrate Polymers*, 79, pp. (1006-1013).

Gullotti, E. & Yeo, Y. (2009). Extracellularly activated nanocarriers: a new paradigm of tumor targeted drug delivery. *Mol Pharm*, 6, 4, pp. (1041-1051), 1543-8384 (Print) 1543-8384 (Linking).

Gurwitz, D. & Livshits, G. (2006). Personalized medicine Europe: health, genes and society: Tel-Aviv University, Tel-Aviv, Israel, June 19-21, 2005. *Eur J Hum Genet*, 14, 3, pp. (376-380), 1018-4813 (Print) 1018-4813 (Linking).

Han, S. K.; Lee, J. H.; Kim, D.; Cho, S. H. & Yuk, S. H. (2005). Hydrophilized poly(lactide-coglycolide) nanoparticles with core/shell structure for protein delivery. *Science and Technology of Advanced Materials*, 6, pp. (468-474).

Heinze, T.; Liebert, T.; Heublein, B. & Hornig, S. (2006). Functional Polymers Based on Dextran. *Advances in Polymer Science*, 205/2006, pp. (199-291).

Hornig, S. & Heinze, T. (2008). Efficient approach to design stable water-dispersible nanoparticles of hydrophobic cellulose esters. *Biomacromolecules*, 9, 5, pp. (1487-1492), 1526-4602 (Electronic) 1525-7797 (Linking).

Hu, F. Q.; Ren, G. F.; Yuan, H.; Du, Y. Z. & Zeng, S. (2006). Shell cross-linked stearic acid grafted chitosan oligosaccharide self-aggregated micelles for controlled release of paclitaxel. *Colloids Surf B Biointerfaces*, 50, 2, pp. (97-103), 0927-7765 (Print) 0927-7765 (Linking).

Ito, T.; Iida-Tanaka, N.; Niidome, T.; Kawano, T.; Kubo, K.; Yoshikawa, K.; Sato, T.; Yang, Z. & Koyama, Y. (2006). Hyaluronic acid and its derivative as a multi-functional gene expression enhancer: protection from non-specific interactions, adhesion to targeted cells, and transcriptional activation. *J Control Release*, 112, 3, pp. (382-388), 0168-3659 (Print) 0168-3659 (Linking).

Janes, K. A.; Calvo, P. & Alonso, M. J. (2001). Polysaccharide colloidal particles as delivery systems for macromolecules. *Adv Drug Deliv Rev*, 47, 1, pp. (83-97), 0169-409X (Print) 0169-409X (Linking).

Kang, J.; Lee, M. S.; Copland, J. A., 3rd; Luxon, B. A. & Gorenstein, D. G. (2008). Combinatorial selection of a single stranded DNA thioaptamer targeting TGF-beta1 protein. *Bioorg Med Chem Lett*, 18, 6, pp. (1835-1839), 1464-3405 (Electronic) 0960-894X (Linking).

Katas, H. & Alpar, H. O. (2006). Development and characterisation of chitosan nanoparticles for siRNA delivery. *J Control Release*, 115, 2, pp. (216-225), 0168-3659 (Print) 0168-3659 (Linking).

Kaul, G. & Amiji, M. (2002). Long-circulating poly(ethylene glycol)-modified gelatin nanoparticles for intracellular delivery. *Pharm Res*, 19, 7, pp. (1061-1067), 0724-8741 (Print) 0724-8741 (Linking).

Kim, D. G.; Jeong, Y. I.; Choi, C.; Roh, S. H.; Kang, S. K.; Jang, M. K. & Nah, J. W. (2006). Retinol-encapsulated low molecular water-soluble chitosan nanoparticles. *Int J Pharm*, 319, 1-2, pp. (130-138), 0378-5173 (Print) 0378-5173 (Linking).

Kogan, G.; Soltes, L.; Stern, R. & Gemeiner, P. (2007). Hyaluronic acid: a natural biopolymer with a broad range of biomedical and industrial applications. *Biotechnol Lett*, 29, 1, pp. (17-25), 1573-6776 (Electronic) 0141-5492 (Linking).

Kratz, F. (2008). Albumin as a drug carrier: design of prodrugs, drug conjugates and nanoparticles. *J Control Release*, 132, 3, pp. (171-183), 1873-4995 (Electronic) 0168-3659 (Linking).

Kumbar, S. G. & Aminabhavi, T. M. (2003). Synthesis and characterization of modified chitosan microspheres: effect of the grafting ratio on the controlled release of nifedipine through microspheres. *Journal of Applied Polymer Science*, 89, pp. (2940-2949).

Lan, S. F. & Starly, B. (2011). Alginate based 3D hydrogels as an in vitro co-culture model platform for the toxicity screening of new chemical entities. *Toxicol Appl Pharmacol*, pp.), 1096-0333 (Electronic) 0041-008X (Linking).

Lassalle, V. & Ferreira, M. L. (2007). PLA nano- and microparticles for drug delivery: an overview of the methods of preparation. *Macromol Biosci*, 7, 6, pp. (767-783), 1616-5187 (Print) 1616-5187 (Linking).

Leach, J. B. & Schmidt, C. E. (2005). Characterization of protein release from photocrosslinkable hyaluronic acid-polyethylene glycol hydrogel tissue engineering scaffolds. *Biomaterials*, 26, 2, pp. (125-135), 0142-9612 (Print) 0142-9612 (Linking).

Learoyd, T. P.; Burrows, J. L.; French, E. & Seville, P. C. (2008). Chitosan-based spray-dried respirable powders for sustained delivery of terbutaline sulfate. *Eur J Pharm Biopharm*, 68, 2, pp. (224-234), 0939-6411 (Print) 0939-6411 (Linking).

Leathers, T. D. (2003). Biotechnological production and applications of pullulan. *Appl Microbiol Biotechnol*, 62, 5-6, pp. (468-473), 0175-7598 (Print) 0175-7598 (Linking).

Lee, J. W.; Park, J. H. & Robinson, J. R. (2000). Bioadhesive-based dosage forms: the next generation. *J Pharm Sci*, 89, 7, pp. (850-866), 0022-3549 (Print) 0022-3549 (Linking).

Lemarchand, C.; Couvreur, P.; Besnard, M.; Costantini, D. & Gref, R. (2003a). Novel polyester-polysaccharide nanoparticles. *Pharm Res*, 20, 8, pp. (1284-1292), 0724-8741 (Print) 0724-8741 (Linking).

Lemarchand, C.; Couvreur, P.; Vauthier, C.; Costantini, D. & Gref, R. (2003b). Study of emulsion stabilization by graft copolymers using the optical analyzer Turbiscan. *Int J Pharm*, 254, 1, pp. (77-82), 0378-5173 (Print) 0378-5173 (Linking).

Letchford, K. & Burt, H. (2007). A review of the formation and classification of amphiphilic block copolymer nanoparticulate structures: micelles, nanospheres, nanocapsules and polymersomes. *Eur J Pharm Biopharm*, 65, 3, pp. (259-269), 0939-6411 (Print) 0939-6411 (Linking).

Li, Y.; Nagira, T. & Tsuchiya, T. (2006). The effect of hyaluronic acid on insulin secretion in HIT-T15 cells through the enhancement of gap-junctional intercellular communications. *Biomaterials*, 27, 8, pp. (1437-1443), 0142-9612 (Print) 0142-9612 (Linking).

Li, Y. L.; Zhu, L.; Liu, Z.; Cheng, R.; Meng, F.; Cui, J. H.; Ji, S. J. & Zhong, Z. (2009). Reversibly stabilized multifunctional dextran nanoparticles efficiently deliver doxorubicin into the nuclei of cancer cells. *Angew Chem Int Ed Engl*, 48, 52, pp. (9914-9918), 1521-3773 (Electronic) 1433-7851 (Linking).

Liu, L.; Fishman, M. L.; Kost, J. & Hicks, K. B. (2003). Pectin-based systems for colon-specific drug delivery via oral route. *Biomaterials*, 24, 19, pp. (3333-3343), 0142-9612 (Print) 0142-9612 (Linking).

Liu, L.; Won, Y. J.; Cooke, P. H.; Coffin, D. R.; Fishman, M. L.; Hicks, K. B. & Ma, P. X. (2004). Pectin/poly(lactide-co-glycolide) composite matrices for biomedical applications. *Biomaterials*, 25, 16, pp. (3201-3210), 0142-9612 (Print) 0142-9612 (Linking).

Liu, Z.; Jiao, Y.; Wang, Y.; Zhou, C. & Zhang, Z. (2008). Polysaccharides-based nanoparticles as drug delivery systems. *Adv Drug Deliv Rev*, 60, 15, pp. (1650-1662), 1872-8294 (Electronic) 0169-409X (Linking).

Lu, B.; Xiong, S. B.; Yang, H.; Yin, X. D. & Zhao, R. B. (2006). Mitoxantrone-loaded BSA nanospheres and chitosan nanospheres for local injection against breast cancer and its lymph node metastases. I: Formulation and in vitro characterization. *Int J Pharm*, 307, 2, pp. (168-174), 0378-5173 (Print) 0378-5173 (Linking).

Martínez, A.; Iglesias, I.; Lozano, R.; Teijón, J. M. & Blanco, M. D. (2011). Synthesis and characterization of thiolated alginate-albumin nanoparticles stabilized by disulfide bonds. Evaluation as drug delivery systems. *Carbohydrate Polymers*, 83, 3, pp. (1311-1321).

Mendichi, R. & Soltes, L. (2002). Hyaluronan molecular weight and polydispersity in some commercial intra-articular injectable preparations and in synovial fluid. *Inflamm Res*, 51, 3, pp. (115-116), 1023-3830 (Print) 1023-3830 (Linking).

Misaki, A.; Torii, M.; Sawai, T. & Goldstein, I. J. (1980). Structure of the dextran of Leuconostoc mesenteroides B-1355. *Carbohydrate Research*, 84, pp. (273-285).

Moghimi, S. M.; Hunter, A. C. & Murray, J. C. (2005). Nanomedicine: current status and future prospects. *FASEB J*, 19, 3, pp. (311-330), 1530-6860 (Electronic) 0892-6638 (Linking).

Muzzarelli, R. A. A. & Muzzarelli, C. (2005). Chitosan chemistry: Relevance to the biomedical sciences Polysaccharides 1: Structure, characterization and use. *Advances in Polymer Science*, 186, pp. (151-209).

Naessens, M.; Cerdobbel, A.; Soetaert, W. & Vandamme, E. J. (2005). Leuconostoc dextransucrase and dextran: production, properties and applications. *Journal of Chemical Technology & Biotechnology*, 80, pp. (845-860).

Nair, L. S. & Laurencin, C. T. (2007). Biodegradable polymers as biomaterial. *Progress in Polymer Science*, 6, pp. (762-798).

Oh, E. J.; Park, K.; Kim, K. S.; Kim, J.; Yang, J. A.; Kong, J. H.; Lee, M. Y.; Hoffman, A. S. & Hahn, S. K. (2010). Target specific and long-acting delivery of protein, peptide, and nucleotide therapeutics using hyaluronic acid derivatives. *J Control Release*, 141, 1, pp. (2-12), 1873-4995 (Electronic) 0168-3659 (Linking).

Opanasopit, P.; Apirakaramwong, A.; Ngawhirunpat, T.; Rojanarata, T. & Ruktanonchai, U. (2008). Development and characterization of pectinate micro/nanoparticles for gene delivery. *AAPS PharmSciTech*, 9, 1, pp. (67-74), 1530-9932 (Electronic) 1530-9932 (Linking).

Orive, G.; Ponce, S.; Hernandez, R. M.; Gascon, A. R.; Igartua, M. & Pedraz, J. L. (2002). Biocompatibility of microcapsules for cell immobilization elaborated with different type of alginates. *Biomaterials*, 23, 18, pp. (3825-3831), 0142-9612 (Print) 0142-9612 (Linking).

Pan, Y.; Li, Y. J.; Zhao, H. Y.; Zheng, J. M.; Xu, H.; Wei, G.; Hao, J. S. & Cui, F. D. (2002). Bioadhesive polysaccharide in protein delivery system: chitosan nanoparticles improve the intestinal absorption of insulin in vivo. *Int J Pharm*, 249, 1-2, pp. (139-147), 0378-5173 (Print) 0378-5173 (Linking).

Pandey, R. & Ahmad, Z. (2011). Nanomedicine and experimental tuberculosis: facts, flaws, and future. *Nanomedicine*, 7, 3, pp. (259-272), 1549-9642 (Electronic) 1549-9634 (Linking).

Pandey, R. & Khuller, G. K. (2005). Antitubercular inhaled therapy: opportunities, progress and challenges. *J Antimicrob Chemother*, 55, 4, pp. (430-435), 0305-7453 (Print) 0305-7453 (Linking).

Perera, G.; Barthelmes, J. & Bernkop-Schnurch, A. (2010). Novel pectin-4-aminothiophenole conjugate microparticles for colon-specific drug delivery. *J Control Release*, 145, 3, pp. (240-246), 1873-4995 (Electronic) 0168-3659 (Linking).

Pliszczak, D.; Bourgeois, S.; Bordes, C.; Valour, J. P.; Mazoyer, M. A.; Orecchioni, A. M.; Nakache, E. & Lanteri, P. (2011). Improvement of an encapsulation process for the preparation of pro- and prebiotics-loaded bioadhesive microparticles by using experimental design. *Eur J Pharm Sci*, 44, 1-2, pp. (83-92), 1879-0720 (Electronic) 0928-0987 (Linking).

Prestwich, G. D. (2008). Engineering a clinically-useful matrix for cell therapy. *Organogenesis*, 4, 1, pp. (42-47), 1547-6278 (Print) 1547-6278 (Linking).

Qiu, L. Y. & Bae, Y. H. (2006). Polymer architecture and drug delivery. *Pharm Res*, 23, 1, pp. (1-30), 0724-8741 (Print) 0724-8741 (Linking).

Raju, H. B. & Goldberg, J. L. (2008). Nanotechnology for ocular therapeutics and tissue repair. *Expert Review of Ophthalmology*, 3, pp. (431-436).

Rehm, B. H. A. (Ed). (2009). *Alginates: Biology and Applications*, Springer, 978-3-540-92678.

Reis, C. P.; Ribeiro, A. J.; Houng, S.; Veiga, F. & Neufeld, R. J. (2007). Nanoparticulate delivery system for insulin: design, characterization and in vitro/in vivo bioactivity. *Eur J Pharm Sci*, 30, 5, pp. (392-397), 0928-0987 (Print) 0928-0987 (Linking).

Reis, C. P.; Ribeiro, A. J.; Veiga, F.; Neufeld, R. J. & Damge, C. (2008). Polyelectrolyte biomaterial interactions provide nanoparticulate carrier for oral insulin delivery. *Drug Deliv*, 15, 2, pp. (127-139), 1071-7544 (Print) 1071-7544 (Linking).

Rekha, M. R. & Chandra, P. S. (2007). Pullulan as a promising biomaterial for biomedical applications: A perspective. *Trends in Biomaterials & Artificial Organs*, 20, pp. (116-121).

Romberg, B.; Hennink, W. E. & Storm, G. (2008). Sheddable coatings for long-circulating nanoparticles. *Pharm Res*, 25, 1, pp. (55-71), 0724-8741 (Print) 0724-8741 (Linking).

Rotureau, E.; Leonard, M.; Dellacherie, E. & Durand, A. (2004). Amphiphilic derivatives of dextran: adsorption at air/water and oil/water interfaces. *J Colloid Interface Sci*, 279, 1, pp. (68-77), 0021-9797 (Print) 0021-9797 (Linking).

Sakamoto, J.; Annapragada, A.; Decuzzi, P. & Ferrari, M. (2007). Antibiological barrier nanovector technology for cancer applications. *Expert Opin Drug Deliv*, 4, 4, pp. (359-369), 1742-5247 (Print) 1742-5247 (Linking).

Sarmah, J. K.; Bhattacharjee, S. K.; Mahanta, R. & Mahanta, R. (2009). Preparation of cross-linked guar gum nanospheres containing tamoxifen citrate by single step emulsion in situ polymer cross-linking method. *Journal of Inclusion Phenomena and Macrocyclic Chemistry*, 65, pp. (329-334).

Sarmah, J. K.; Mahanta, R.; Bhattacharjee, S. K. & Biswas, A. (2011). Controlled release of tamoxifen citrate encapsulated in cross-linked guar gum nanoparticles. *Int J Biol Macromol*, 49, 3, pp. (390-396), 1879-0003 (Electronic) 0141-8130 (Linking).

Sarmento, B.; Martins, S.; Ribeiro, A.; Veiga, F.; Neufeld, R. & Ferreira, D. (2006). Development and comparison of different nanoparticulate polyelectrolyte complexes as insulin carriers. *International Journal of Peptide Research and Therapeutics*, 12, pp. (131-138).

Singh, V.; , P., S.; Singh, S. K. & Sanghi, R. (2009). Removal of cadmium from aqueous solutions by adsorption using poly(acrylamide) modified guar gum–silica nanocomposites. *Separation and Purification Technology*, 67, pp. (251-261).

Sinha, V. R. & Kumria, R. (2001). Polysaccharides in colon-specific drug delivery. *Int J Pharm*, 224, 1-2, pp. (19-38), 0378-5173 (Print) 0378-5173 (Linking).

Sinha, V. R.; Singla, A. K.; Wadhawan, S.; Kaushik, R.; Kumria, R.; Bansal, K. & Dhawan, S. (2004). Chitosan microspheres as a potential carrier for drugs. *Int J Pharm*, 274, 1-2, pp. (1-33), 0378-5173 (Print) 0378-5173 (Linking).

Sinha, V. R. & Trehan, A. (2003). Biodegradable microspheres for protein delivery. *J Control Release*, 90, 3, pp. (261-280), 0168-3659 (Print) 0168-3659 (Linking).

Skiba, M.; Wouessidjewe, D.; Puisieux, F.; Duchène, D. & Gulik, A. (1996). Characterization of amphiphilic fl-cyclodextrin nanospheres. *International Journal of Pharmaceutics*, 142, pp. (121-124).

Soumya, R. S.; Ghosh, S. & Abraham, E. T. (2010). Preparation and characterization of guar gum nanoparticles. *Int J Biol Macromol*, 46, 2, pp. (267-269), 1879-0003 (Electronic) 0141-8130 (Linking).

Srinophakun, T. & Boonmee, J. (2011). Preliminary Study of Conformation and Drug Release Mechanism of Doxorubicin-Conjugated Glycol Chitosan, via cis-Aconityl Linkage, by Molecular Modeling. *Int J Mol Sci*, 12, 3, pp. (1672-1683), 1422-0067 (Electronic) 1422-0067 (Linking).

Sujata, V. B. (2002). *Biomaterials*, Springer, 0-7923-7058-9, India.

Tasciotti, E.; Liu, X.; Bhavane, R.; Plant, K.; Leonard, A. D.; Price, B. K.; Cheng, M. M.; Decuzzi, P.; Tour, J. M.; Robertson, F. & Ferrari, M. (2008). Mesoporous silicon particles as a multistage delivery system for imaging and therapeutic applications. *Nat Nanotechnol*, 3, 3, pp. (151-157), 1748-3395 (Electronic) 1748-3387 (Linking).

Thebaud, N. B.; Pierron, D.; Bareille, R.; Le Visage, C.; Letourneur, D. & Bordenave, L. (2007). Human endothelial progenitor cell attachment to polysaccharide-based hydrogels: a pre-requisite for vascular tissue engineering. *J Mater Sci Mater Med*, 18, 2, pp. (339-345), 0957-4530 (Print) 0957-4530 (Linking).

Tibbals, H. F. (2010). *Medical Nanotechnology and Nanomedicine*, CRC Press, 978-1-4398-0876-4, USA.

Tiraferri, A.; Chen, K. L.; Sethi, R. & Elimelech, M. (2008). Reduced aggregation and sedimentation of zero-valent iron nanoparticles in the presence of guar gum. *J Colloid Interface Sci*, 324, 1-2, pp. (71-79), 1095-7103 (Electronic) 0021-9797 (Linking).

Tiyaboonchai, W. & Limpeanchob, N. (2007). Formulation and characterization of amphotericin B-chitosan-dextran sulfate nanoparticles. *Int J Pharm*, 329, 1-2, pp. (142-149), 0378-5173 (Print) 0378-5173 (Linking).

Tiyaboonchai, W.; Woiszwillo, J.; Sims, R. C. & Middaugh, C. R. (2003). Insulin containing polyethylenimine-dextran sulfate nanoparticles. *Int J Pharm*, 255, 1-2, pp. (139-151), 0378-5173 (Print) 0378-5173 (Linking).

Tokumitsu, H.; Ichikawa, H. & Fukumori, Y. (1999). Chitosan-gadopentetic acid complex nanoparticles for gadolinium neutron-capture therapy of cancer: preparation by novel emulsion-droplet coalescence technique and characterization. *Pharm Res*, 16, 12, pp. (1830-1835), 0724-8741 (Print) 0724-8741 (Linking).

Varum, F. J.; McConnell, E. L.; Sousa, J. J.; Veiga, F. & Basit, A. W. (2008). Mucoadhesion and the gastrointestinal tract. *Crit Rev Ther Drug Carrier Syst*, 25, 3, pp. (207-258), 0743-4863 (Print) 0743-4863 (Linking).

Vauthier, C. & Bouchemal, K. (2009). Methods for the preparation and manufacture of polymeric nanoparticles. *Pharm Res*, 26, 5, pp. (1025-1058), 1573-904X (Electronic) 0724-8741 (Linking).

Vauthier, C. & Couvreur, P. (2000). Development of nanoparticles made of polysaccharides as novel drug carrier systems, In: *Handbook of Pharmaceutical Controlled Release Technology*, D. L. Wise, (Ed), pp. (413-429), Marcel Dekker, New York.

Vila, A.; Sanchez, A.; Tobio, M.; Calvo, P. & Alonso, M. J. (2002). Design of biodegradable particles for protein delivery. *J Control Release*, 78, 1-3, pp. (15-24), 0168-3659 (Print) 0168-3659 (Linking).

Wang, N. & Wu, X. S. (1997). Preparation and characterization of agarose hydrogel nanoparticles for protein and peptide drug delivery. *Pharm Dev Technol*, 2, 2, pp. (135-142), 1083-7450 (Print) 1083-7450 (Linking).

Wang, S.; Mao, X.; Wang, H.; Lin, J.; Li, F. & Wei, D. (2011). Characterization of a novel dextran produced by Gluconobacter oxydans DSM 2003. *Appl Microbiol Biotechnol*, 91, 2, pp. (287-294), 1432-0614 (Electronic) 0175-7598 (Linking).

Wang, Y. S.; Liu, L. R.; Jiang, Q. & Zhang, Q. Q. (2007). Self-aggregated nanoparticles of cholesterol-modified chitosan conjugate as a novel carrier of epirubicin. *European Polymer Journal*, 43, pp. (43-51).

Yang, J. S.; Xie, Y. J. & He, W. (2011). Research progress on chemical modification of alginate: A review. *Carbohydrate Polymers*, 84, pp. (33–39).

Yang, X.; Zhang, Q.; Wang, Y.; Chen, H.; Zhang, H.; Gao, F. & Liu, L. (2008). Self-aggregated nanoparticles from methoxy poly(ethylene glycol)-modified chitosan: synthesis; characterization; aggregation and methotrexate release in vitro. *Colloids Surf B Biointerfaces*, 61, 2, pp. (125-131), 0927-7765 (Print) 0927-7765 (Linking).

Yip, G. W.; Smollich, M. & Gotte, M. (2006). Therapeutic value of glycosaminoglycans in cancer. *Mol Cancer Ther*, 5, 9, pp. (2139-2148), 1535-7163 (Print) 1535-7163 (Linking).

Yoksan, R.; Matsusaki, M.; Akashi, M. & Chirachanchai, S. (2004). Controlled hydrophobic/hydrophilic chitosan: colloidal phenomena and nanosphere formation. *Colloid & Polymer Science*, 282, pp. (337-342).

You, J. O. & Peng, C. A. (2004). Calcium-alginate nanoparticles formed by reverse microemulsion as gene carriers. *Macromolecular Symposia*, 219, pp. (147-153).

Yuan, Z. (2007). Study on the synthesis and catalyst oxidation properties of chitosan bound nickel(II) complexes. *Journal of Agricultural and Food Chemistry*, 21, 5, pp. (22-24).

Zhang, J.; Chen, X. G.; Li, Y. Y. & Liu, C. S. (2007). Self-assembled nanoparticles based on hydrophobically modified chitosan as carriers for doxorubicin. *Nanomedicine*, 3, 4, pp. (258-265), 1549-9642 (Electronic) 1549-9634 (Linking).

Zhi, J.; Wang, Y. J. & Luo, G. S. (2005). Adsorption of diuretic furosemide onto chitosan nanoparticles prepared with a water-in-oil nanoemulsion system. *Reactive and Functional Polymers*, 249-257, pp.),

Zhou, B.; McGary, C. T.; Weigel, J. A.; Saxena, A. & Weigel, P. H. (2003). Purification and molecular identification of the human hyaluronan receptor for endocytosis. *Glycobiology*, 13, 5, pp. (339-349), 0959-6658 (Print) 0959-6658 (Linking).

10

Electrospray Production of Nanoparticles for Drug/Nucleic Acid Delivery

Yun Wu[1], Anthony Duong[1,2], L. James Lee[1,2] and Barbara E. Wyslouzil[1,2,3]
[1]*NSF Nanoscale Science and Engineering Center, The Ohio State University, Columbus, Ohio*
[2]*William G. Lowrie Department of Chemical and Biomolecular Engineering, The Ohio State University, Columbus, Ohio*
[3]*Department of Chemistry, The Ohio State University, Columbus, Ohio*
USA

1. Introduction

Nanomedicine - the application of nanotechnology to medicine - shows great potential to positively impact healthcare. Nanoparticles, including solid lipid nanoparticles, lipoplexes and polyplexes, can act as carriers to deliver the drugs or nucleic acid-based therapeutics that are particularly promising for advancing molecular and genetic medicine.

Many techniques have been developed to produce nanoparticles. Among them, electrospray has attracted recent research interest because it is an elegant and versatile way to make a broad array of nanoparticles. Electrospray is a technique that uses an electric field to disperse or break up a liquid. Compared with current technologies, such as bulk mixing, high pressure homogenization and double emulsion techniques, electrospray has three potential advantages. First, electrospray can generate monodisperse droplets whose size can vary from tens of nanometer to hundreds of micrometers, depending on the processing parameters. Secondly, it is a very gentle method. The free charge, induced by the electric field, only concentrates at the surface of the liquid, and does not significantly affect sensitive biomolecules such as DNA. Finally electrospray has the ability to generate structured micro/nanoparticles in a more controlled way with high drug/nucleic acid encapsulation efficiency.

Electrospray is a critical element of electrospray ionization mass-spectrometry, an analytical technique used to detect macromolecules that was developed by the 2002 chemistry Nobel Prize winner, Dr. John B. Fenn. (Fenn et al. 1989) Since then research has focused both on developing a fundamental understanding of the process as well as exploring potential applications of electrospray in fields ranging from the semiconductor industry to life science. During the past two decades, electrospray has been used to assist pyrolysis reactions and chemical vapor deposition processes, to produce inorganic particles including fine metal powder (Sn, Ag, Au, etc.), metal oxide particles (ZrO2, TiO2, etc.), ceramic particles (Si, SiO2), and semiconductor quantum dots (CdSe, GaAs, etc.) (Jaworek 2007; Salata 2005). Various structured particles have been produced via electrospray, including poly(methyl methacrylate)-pigment nanoparticles (Widiyandari et al. 2007), cocoa butter microcapsules containing a sugar solution or an oil-in-water emulsion (Loscertales et al. 2002; Bocanegra et al. 2005) and microbubble suspensions (Farook et al. 2007), to name a few. Electrospray has been used to deposit particle suspensions to form thin film (Jaworek & Sobczyk 2008, Jaworek 2010) or on-demand patterns, such as silica particle coatings on a quartz glass (Jaworek 2007; Salata 2005). Electrospray has been successfully applied in tissue engineering, for example, polymer materials including poly(lactide-co-glycolide) or poly(ethylene glycol) were electrospray-coated on biomedical implants (Kumbar et al. 2007). Combined with electrospinning, electrospray was used to fabricate smooth muscle cell integrated blood vessel constructs. (Stankusa et al. 2007) In the area of drug/nucleic acid delivery, many biological materials, such as DNA, proteins, and lipids have been electrosprayed without changing their biological activity (Pareta et al. 2005; Jayasinghe et al. 2005; Davies et al. 2005; Wu et al. 2009a, 2009b, 2010, 2011). Proteins, such as bovine serum albumin, have been encapsulated in biodegradable polymeric microcapsules (Pareta and Edirisinghe, 2006; Xie & Wang, 2006), and small molecule drugs, such as taxol and griseofulvin, have been encapsulated in polymeric microparticles for systemic or oral delivery. (Xie et al. 2006; Zhang et al, 2011)

2. Principles of electrospray

As illustrated in Figure 1a, in the standard electrospray configuration, a conducting liquid is slowly injected through a needle by a syringe pump. An electrical potential is applied to the needle to introduce free charge at the liquid surface. The free charge generates electric stress that causes the liquid to accelerate away from the needle. When the electrical potential rises to several kilovolts, the liquid meniscus at the needle opening develops into a conical shape, commonly called the Taylor cone. At the cone apex, where the free charge is highly concentrated, a liquid jet with high charge density is observed. Monodisperse particles are formed when the jet breaks into fine particles due to varicose or kink instabilities. (Cloupeau et al. 1994; Jaworek 2007; Salata 2005; Loscertales et al. 2002). Typically, the initial micron size droplets contain both solvent and a non-volatile solute and nanoparticles are produced as solvent evaporates from the high surface area aerosol.

Other electrospray configurations are also possible. Figure 1(b) shows a coaxial electrospray configuration, where two liquids are fed through the inner needle and the outer needle, respectively. This configuration is widely used to produce structured nanoparticles. Finally, since the flow rate in a single electrospray setup is always low (uL/hr or mL/hr), multiplexed electrospray configurations have been developed to scale up production. (Deng et al, 2006) As shown in Figure 1(c), the single needle is replaced by a micro-nozzle array. All nozzles work simultaneously and produce nanoparticles with the liquid flow increased by orders of magnitude.

The electrospray modes and resulting droplet sizes d_D are controlled by the process parameters that include the applied voltage V, liquid flow rate Q, and liquid properties including electrical conductivity γ, surface tension σ, and liquid density ρ. (Cloupeau et al. 1994; Jaworek 2007; Salata 2005; Loscertales et al. 2002, Ganan-Calvo AM. 2004; Hartman et al. 1999; Basak et al. 2007) As illustrated in Figure 2, different electrospray modes are obtained as the applied voltage increases. (Cloupeau et al. 1994; Jaworek and Krupa 1999; Chen et al. 2005; Yurteri et al. 2010) These include dripping, micro-dripping, spindle,

Taylor cone-jet and multi-jet mode. Taylor cone-jet mode is the most common used electrospray mode, because it can produce highly monodisperse particles in a stable manner.

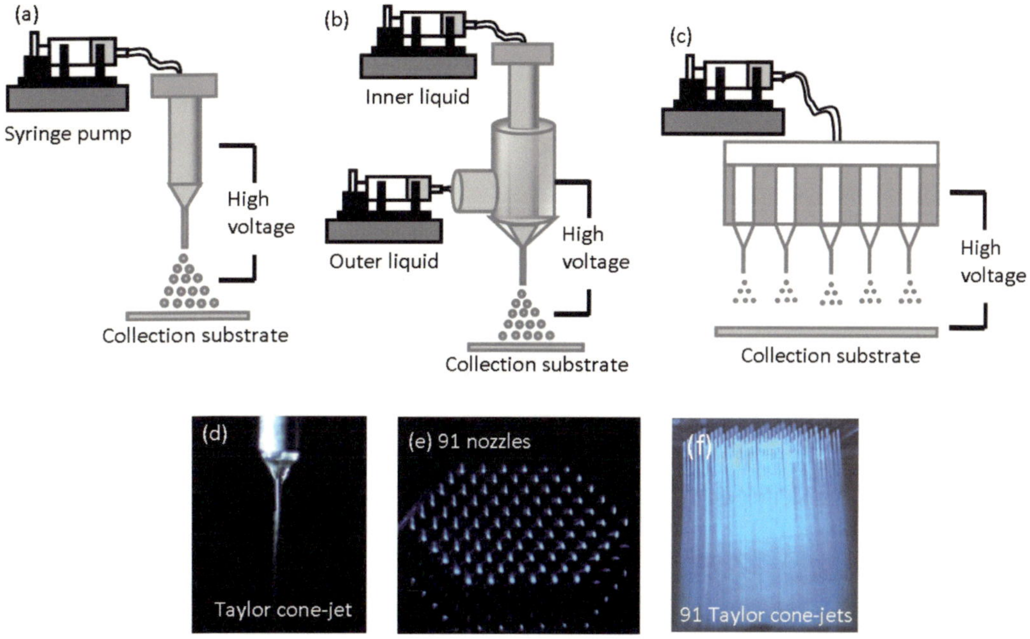

Fig. 1. Schematic diagrams of **(a)** standard electrospray **(b)** coaxial electrospray and **(c)** multiplexed electrospray. **(d)** Taylor cone-jet from standard electrospray **(e)** 91 nozzles used in multiplexed electrospray and **(f)** 91 Taylor cone-jets formed simultaneously. (Figures (e) and (f) are courtesy of Dr. Weiwei Deng).

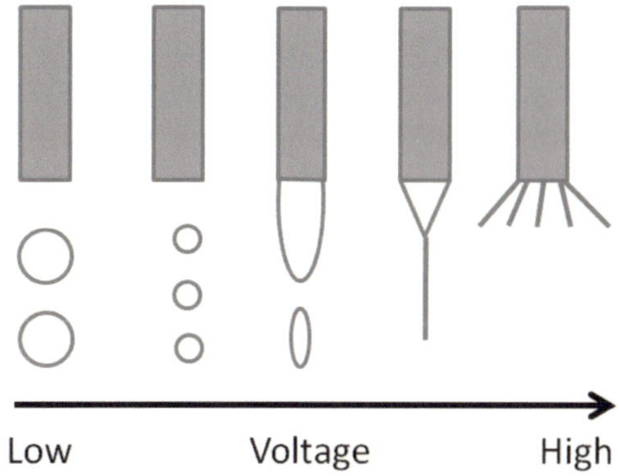

Fig. 2. The electrospray modes change as the voltage increases. From left to right: dripping, microdripping, spindle, Taylor cone-jet and multi-jet.

In cone-jet mode, the droplet size scales with the liquid flow rate, and is inversely proportional to the liquid conductivity. Theoretically droplet sizes can be determined from the following scaling law, confirmed by many experiments (Jaworek 2007).

$$d_D = \alpha \left(\frac{Q^3 \varepsilon_0 \rho}{\pi^4 \sigma \gamma} \right)^{1/6} \quad (\alpha{=}2.9)$$

where ε_0 is the vacuum permittivity. In general, droplet size decreases with decreasing flow rate and increasing electrical conductivity of the liquid. By adjusting the flow rate and the liquid properties, droplets with the desired size can be produced in a well-controlled manner. Frequently, the initial solution contains both solvent and solute, and the final particle size (d_p) is related to the initial droplet size d_D by the following equation:

$$d_p = d_D \left(\frac{c\rho}{\rho_p} \right)^{1/3}$$

where c is the concentration of solute and ρ_p is the density of the particle.

3. Electrospray production of nanoparticles for drug/nucleic acid delivery

Our research group has been the first to explore electrospray as a means to produce solid lipid nanoparticles, lipoplexes and polyplexes for drug/nucleic acid delivery. Our results demonstrate the great potential of electrospray to produce nanoparticles in a well-controlled manner, and opens new avenues for the development of nanomedicine.

3.1 Solid lipid nanoparticles for hydrophobic drug delivery

3.1.1 Introduction

Solid lipid nanoparticles (SLNs) have been proposed as an alternative drug delivery system to traditional colloidal carriers, such as emulsions, liposomes and polymeric nanoparticles. As drug delivery vehicles, emulsions and liposomes are often limited by physical instability and low drug loading capacity. (Müller et al. 2000, 2004) Polymeric nanoparticles can exhibit cytotoxicity, and there are no industrialized production methods. SLNs may provide a drug delivery system that avoids many of these problems, and high pressure homogenization (HPH), the most popular preparation technique, is a reasonable method for large-scale production. Unfortunately, HPH requires molten lipids and high energy input. Thus, it is often unsuitable for use with temperature and shear sensitive biomolecules. Furthermore, as the mixtures cool, some drugs do not incorporate into the lipids but stick on the surface, resulting in burst drug release kinetics and low drug loading capacity. (Müller et al. 2000, 2004)

In our work we used electrospray to produce solid lipid nanoparticles. (Wu et al. 2009a, 2011) Cholesterol was chosen as a model since it is not only an important lipid but also a good model for lipophilic drugs. Similar to most lipophilic drugs, cholesterol has very low solubility in aqueous solutions, on the order of 1.8μg/mL or 4.7μM. (Haberland & Reynolds, 1973) Cholesterol has similar structure as the well-known anticancer drug Taxol®, and cholesterol (molecular weight = 387.6) is also a small molecule. In addition, the solubility of lipophilic drugs in lipids is much higher than in aqueous solutions, and, therefore, lipid nanoparticles, like the cholesterol nanoparticles discussed here, may also be useful as carriers to deliver lipophilic drugs.

3.1.2 Production of cholesterol nanoparticles

The standard electrospray configuration (Figure 1a) was used to produce cholesterol nanoparticles dispersed in aqueous media. Cholesterol powder (Sigma C3045) was dissolved in ethanol (Pharmco-Aaper E200GP) at a concentration of 2 mg/mL and then delivered to a 27 gauge needle by a syringe pump at a flow rate of 2 mL/h. A voltage of approximate 2.5 kV was applied to the needle and the observation of a stable Taylor cone-jet mode confirmed the proper conditions for generating fine droplets. The ethanol evaporated from the droplets, aided by heat input from a lamp, and the residual cholesterol particles were captured in 1X PBS supplemented with 1% Pluronic F68 placed in a grounded aluminum dish 5.5 cm below the needle tip. Figure 3 illustrates a typical size distribution measured for the cholesterol nanoparticle suspension. Highly monodisperse cholesterol nanoparticles were produced with sizes ~150 nm. The cholesterol concentration in the final solution was 180±10 μg/mL prior to sterile filtration and 150±8 μg/mL afterwards, i.e. up to 100 times higher than the solubility limit for cholesterol.

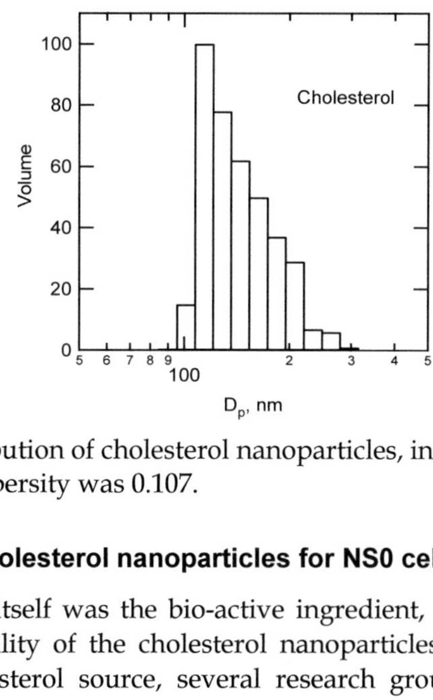

Fig. 3. A typical size distribution of cholesterol nanoparticles, in which Dp, mean, v was 140 nm and polydispersity was 0.107.

3.1.3 Bioavailability of cholesterol nanoparticles for NS0 cells

In this study, cholesterol itself was the bio-active ingredient, and NS0 cells were used to investigate the bioavailability of the cholesterol nanoparticles. While NS0 cells normally require serum as a cholesterol source, several research groups have demonstrated the growth of NS0 cells in protein free and chemically defined medium, using chemically defined cholesterol supplements. (Zhang & Robinson, 2005; Talley et al. 2005; Ojito et al. 2001) The auxotrophic nature of NS0 cells with respect to cholesterol makes this cell line a natural choice for characterizing the bioavailability of cholesterol nanoparticles produced by electrospray.

NS0 cells were continuously cultured in 250 mL shake flasks with 50 mL working volume in the basal medium supplemented with cholesterol nanoparticles or SyntheChol NS0 supplement (Sigma S5442, SyntheChol for short) at a cholesterol concentration level of 3.5 μg/mL. SyntheChol is a proprietary cholesterol supplement in liquid form and acts as the positive control. Figure 4 illustrates the viable cell density and cell viability as a function of time. NS0 cells grew better with cholesterol nanoparticles than with SyntheChol reaching much higher peak cell densities: 5.28e6 cells/mL vs. 3.35e6 cells/mL.

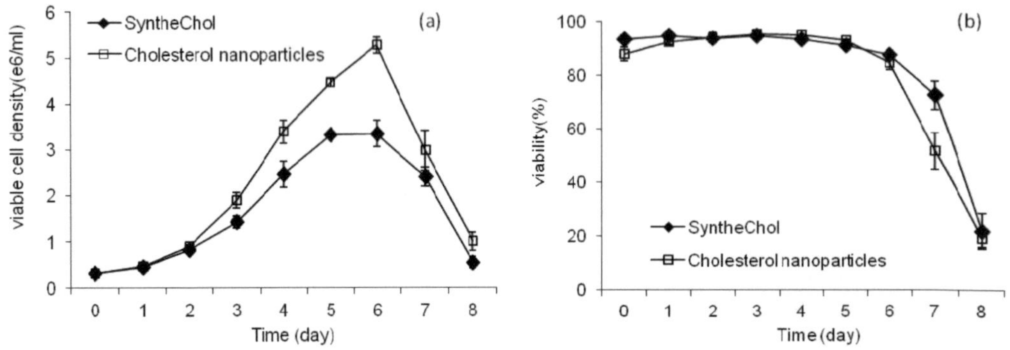

Fig. 4. Growth of NS0 cells in media supplemented with 3.5 μg/mL of cholesterol nanoparticles or SyntheChol: viable cell density **(a)** and cell viability **(b)**. The results are averages of triplicate runs.

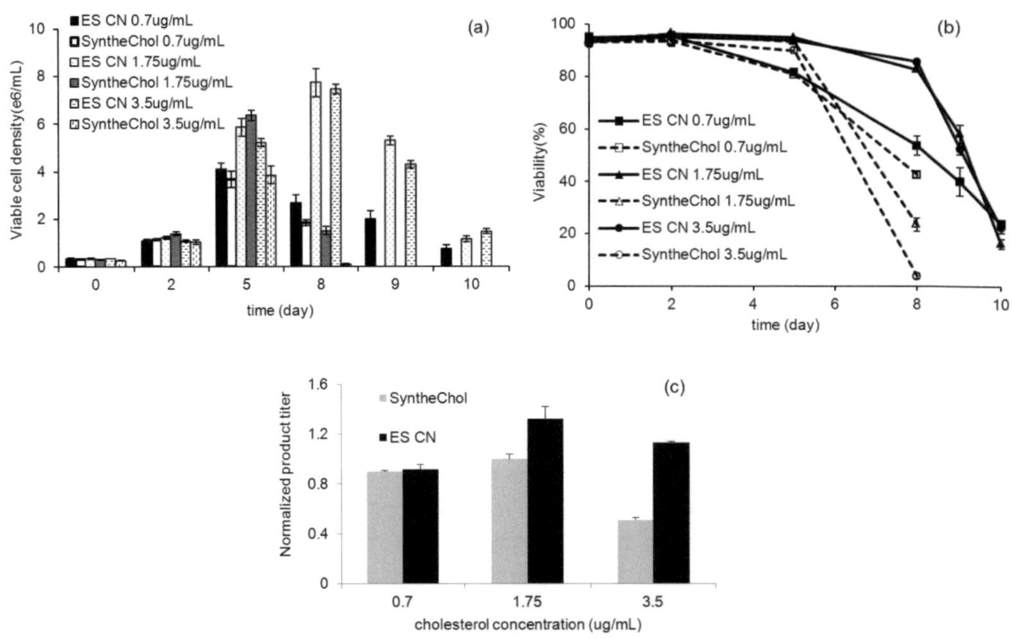

Fig. 5. The effect of cholesterol concentration on **(a)** the growth, **(b)** the viability, and **(c)** the product titer of NS0 cells during fed-batch culture. Cholesterol nanoparticles or SyntheChol supplement were fed on day 0 and daily from day 2 till the end of culture at cholesterol concentration levels of 0.7 μg/mL, 1.75 μg/mL or 3.5 μg/mL. Error bars indicate one standard deviation of the results of triplicate runs. (ES CN: electrosprayed cholesterol nanoparticles)

The NS0 cell line is one of the important cell lines widely used by the pharmaceutical industry to produce therapeutic antibodies. We therefore investigated the effect of the cholesterol source on product titer by conducting a fed-batch culture of NS0 cells using electrosprayed cholesterol nanoparticles and the SyntheChol supplement at feed rates of 0.7, 1.75 and 3.5 μg/mL/day. Figure 5 summarizes the viable cell density, cell viability and normalized product titer (relative to the 1.75 μg/mL/day SyntheChol supplement). At the lowest feed rate of 0.7 μg/mL/day, neither cholesterol nanoparticles nor SyntheChol supplement provided sufficient cholesterol to the cells and so cell growth was limited and the corresponding product titer was low (Figure 5c). When the feed rate of SyntheChol was increased, cell growth improved up to a feed rate of 1.75 μg/mL/day but was compromised at feed rate of 3.5 μg/mL/day. The inhibitory effect of 3.5 μg/mL/day feeding was presumably due to the compounds in SyntheChol that solubilize the cholesterol rather than to the presence of cholesterol. For cholesterol nanoparticles, both cell growth and viability improved at feed rates higher than 0.7 μg/mL/day. Compared to SyntheChol, culturing was extended from 8 days to 10 days. In general, the performance of cholesterol nanoparticles was better than that of SyntheChol supplementation at the same cholesterol feed rates. For both cholesterol nanoparticles and SyntheChol supplements, the highest product titer was achieved at a feed rate of 1.75 μg/mL/day and the product titer with cholesterol nanoparticles was ~32% higher than that with SyntheChol.

3.1.4 Conclusions

Compared with conventional technologies, we have demonstrated that electrospray is an efficient way to produce solid lipid nanoparticles for drug delivery. In our application we used electrospray to produce highly monodisperse cholesterol nanoparticles, and investigated the bioavailability of the nanoparticles using the cholesterol auxotrophic NS0 cell line with SyntheChol as the positive control. We found that the cholesterol nanoparticles not only supported NS0 cell growth but also improved the titer of therapeutic antibody by 32% when compared with SyntheChol.

3.2 DNA/polycation polyplexes for gene delivery

3.2.1 Introduction

Viral and non-viral vectors have been developed to deliver nucleic acids to treat genetic and acquired disease. (Pack et al., 2005; Mastrobattista et al., 2006) Compared to viral vectors, non-vrial vectors show lower immunogenicity, lower toxicity, better stability and lower cost. (Laporte et al., 2006; Kircheis, et al., 2001; Gebhart and Kabanov, 2001). Among the nonviral vectors, polyplexes and lipoplexes represent two major carrier systems. In this section, we focus on polyplexes for gene delivery, while in the following section we discuss the use of lipoplexes for oligonucleotides delivery.

Polyethylenimines (PEI) is a cationic polymer widely used to condense DNA and form polyplexes. Although PEI shows great promise for gene delivery both *in vitro* and *in vivo* (Godbey, et al., 1999; Neu et al., 2005), the delivery efficiency highly depends on the N/P ratio (the molar ratio of nitrogen in PEI to phosphate in DNA) and the preparation method. Bulk mixing (BM), the most commonly used method, is a simple and straightforward way to prepare DNA/PEI polyplexes. Unfortunately, the formation of the polyplexes is not well controlled in bulk mixing, significantly affecting the particle size, structure, and, thus, the delivery efficiency. (Kircheis, et al., 2001; Gebhart and Kabanov, 2001) Alternative methods are needed to overcome this challenge.

Because electrospray is a gentle method that does not damage biomolecules it has attracted research interest in the realm of gene delivery. In some of the earliest work, enhanced green fluorescent protein (eGFP) plasmid was successfully delivered into African Green Monkey fibroblast cells (COS-1) using coaxial electrospray. (Chen et al. 2000) Since electrospray generates an aerosol, Davies et al. (2005) investigated gene delivery by exposing mice to naked pCIKLux aerosol generated via electrospray. Unfortunately, the *in vivo* delivery efficiency, 0.075%, was much lower than the 0.2% delivery efficiency achieved by Koshkina et al. (2003) when they nebulized pDNA/PEI polyplexes for pulmonary gene delivery. One possible reason for the low delivery efficiency of electrosprayed pCIKLux might be that the naked pDNA used is not stable in biological fluids and does not interact with cell membrane due to its negative charges. Our hypothesis was that pDNA/PEI polyplexes produced via electrospray might overcome limitations in the earlier studies. Thus, we used coaxial electrospray to achieve better control over the mixing of the DNA solution and the PEI solution, and thus to produce DNA/PEI polyplexes in a well-controlled manner. (Wu et al. 2010)

3.2.2 Coaxial electrospray DNA/PEI polyplexes

Coaxial electrospray was explored as a method to produce pDNA/PEI polyplexes using pGFP and pSEAP for qualitative and quantitative studies, respectively. The pDNAs were dispersed in OPTI-MEM medium at a concentration of 20 μg/mL. Branched 25 kDa PEI was dissolved in OPTI-MEM medium at concentrations of 17.2 μg/mL. In the coaxial electrospray setup, the pDNA solution flowed through the inner 27 gauge needle and the PEI solution flowed through the outer 20 gauge needle. The flow rates of the pDNA and PEI solutions were both set at 6mL/hr to yield an N/P ratio of 6.7. A positive voltage, typically 5~6 kV, was applied between the inner needle and a grounded copper ring electrode. The polyplexes were produced in the stable Taylor cone jet mode, captured in a grounded aluminum dish 5 cm below the needle tip, and used within 10 minutes.

The conventional bulk mixing method was also used to prepare pDNA/PEI polyplexes. Here PEI solution (17.2 μg/mL) was added into equal volume pDNA (20 μg/mL) solution to achieve an N/P ratio of 6.7. The resulting mixture was vortexed for a few seconds, incubated at room temperature for 10 minutes and used immediately.

3.2.3 pDNA damage analysis

To investigate the effects of coaxial electrospray on pDNA integrity, pGFP and pSEAP solutions were coaxial electrosprayed with OPTI-MEM medium containing no PEI. The flow rate for both pDNAs was set at 6mL/h, and the voltages were set at 5.0 kV for pGFP and 5.6 kV for pSEAP. Agarose gel electrophoresis was used to detect if the pDNA was damaged by coaxial electrospray. As illustrated in Figure 6, compared to the non-electrosprayed pDNA controls, little degradation of pDNA is observed. In addition, the ratio of supercoiled to open circular pDNA of the sprayed samples appears unchanged relative to the controls. Our observations are consistent with those reported by Davies et al. (2005) and confirm that coaxial electrospray is a gentle but effective way to spray delicate bio-molecules.

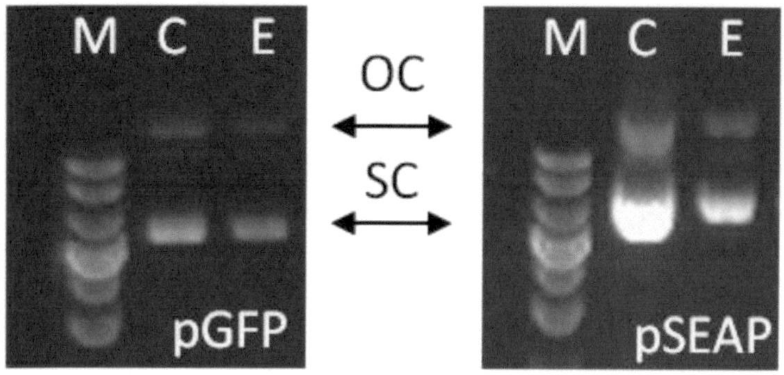

Fig. 6. Gel electrophoresis shows that the pDNA is not damaged during the electrospray process. M: marker lane. C: control. E: electrosprayed pDNAs. Electrospray conditions: pGFP: flowrate of 6 mL/h and voltage of 5.0 kV; pSEAP: flowrate of 6 mL/h and voltage of 5.6 kV. OC: open circular. SC: supercoiled plasmid.

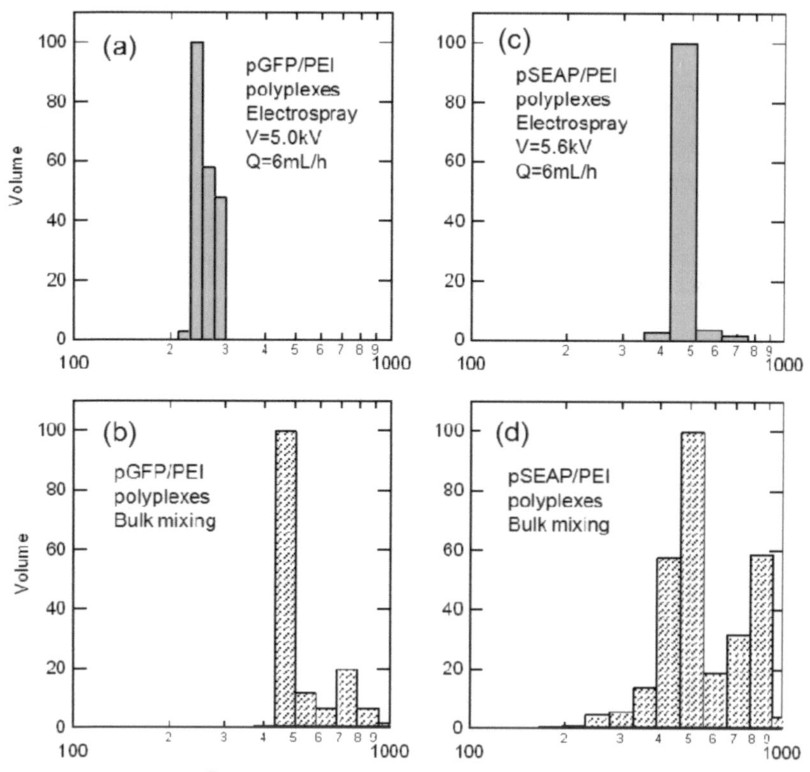

Fig. 7. Typical size distributions of the pDNA/PEI polyplexes prepared by coaxial electrospray and bulk mixing. The flow rate was 6 mL/h and the voltage was 5.0 kV for pGFP/PEI polyplexes and 5.6 kV for pSEAP/PEI polyplexes.

3.2.4 Size distribution of pDNA/PEI polyplexes

Dynamic light scattering was used to measure the size distribution of pGFP/PEI and pSEAP/PEI polyplexes prepared by coaxial electrospray, and bulk mixing. (Figure 7) The pDNA/PEI polyplexes prepared by either method are relatively monodisperse. The pGFP/PEI polyplexes prepared by coaxial electrospray had smaller particle size compared to those prepared by bulk mixing, while the sizes of the pSEAP/PEI polyplexes prepared by coaxial electrospray and bulk mixing were comparable. We note that the polyplexes prepared in this work were almost always larger than 400 nm because OPTI-MEM medium rather than NaCl solutions were used. OPTI-MEM medium was chosen to prepare polyplexes because it better mimics the *in vivo* situation than NaCl or PBS, and because its physical properties (electric conductivity and surface tension) are appropriate for the electrospray process. The downside is that OPTI-MEM medium contains reduced serum, and the adsorption of serum albumin and other negatively charged proteins enhances aggregation of polyplexes (Pack et al., 2005). Although the size distribution of each sample was measured within 10 minutes after preparation, the polyplexes were always larger than 300 nm suggesting that aggregates already formed. The adsorption of proteins to the polyplexes was also demonstrated by zeta potential measurements. The zeta potentials of all polyplexes samples were either close to zero or only slightly positive. Approaches such as grafting hydrophilic polymers, including polyethylene glycol (PEG) or polysaccharides, onto the PEI may help improve the stability of the polyplexes in serum (Pack et al., 2005; Laporte et al., 2006; Neu et al., 2001) and is one approach that could reduce aggregation.

3.2.5 pDNA/PEI polyplex transfection, DNA expression, and cell viability

The delivery efficiency of pGFP/PEI polyplexes produced by coaxial electrospray and bulk mixing was evaluated in NIH 3T3 cells and compared to Lipofectamine™ 2000, the positive control. As illustrated in Figure 8(a), at N/P ratio of 6.7, cells transfected with polyplexes produced by coaxial electrospray showed similar GFP expression as those transfected with Lipofectamine™ 2000, while cells transfected with polyplexes produced by bulk mixing showed much less GFP expression. No significant difference in cell viability was found among coaxial electrospray, BM and Lipofectamine™ 2000 ($p<0.05$).

Since the evaluation of GFP expression is only qualitative, pSEAP was used to compare the delivery efficiency of polyplexes in a more quantitative way. As shown in Figure 9 (a), at N/P ratio of 6.7, cells transfected with the pSEAP/PEI polyplexes produced by coaxial electrospray gave 2.6 times higher SEAP expression than those produced by bulk mixing. Although cells transfected with Lipofectamine™ 2000 showed the highest SEAP expression, 3.3 times higher than bulk mixing, the cell viability was low because of the toxicity of this material. (Figure 9(b))

3.2.6 Conclusions

Both pGFP/PEI and pSEAP/PEI polyplexes were successfully produced using coaxial electrospray. The delivery efficiency of pDNA/PEI polyplexes was evalulated in NIH 3T3 cells. At N/P ratio of 6.7, polyplexes produced by coaxial electrospray were more effective at delivering genes to NIH 3T3 cells than those prepared by bulk mixing. Since coaxial electrospray is an aerosol technique, it should be a useful way to deliver pDNA/PEI polyplexes directly to the lungs and achieve higher gene delivery efficiency than either electrospraying naked pDNA or nebulizing pDNA/PEI polyplexes produced by conventional methods.

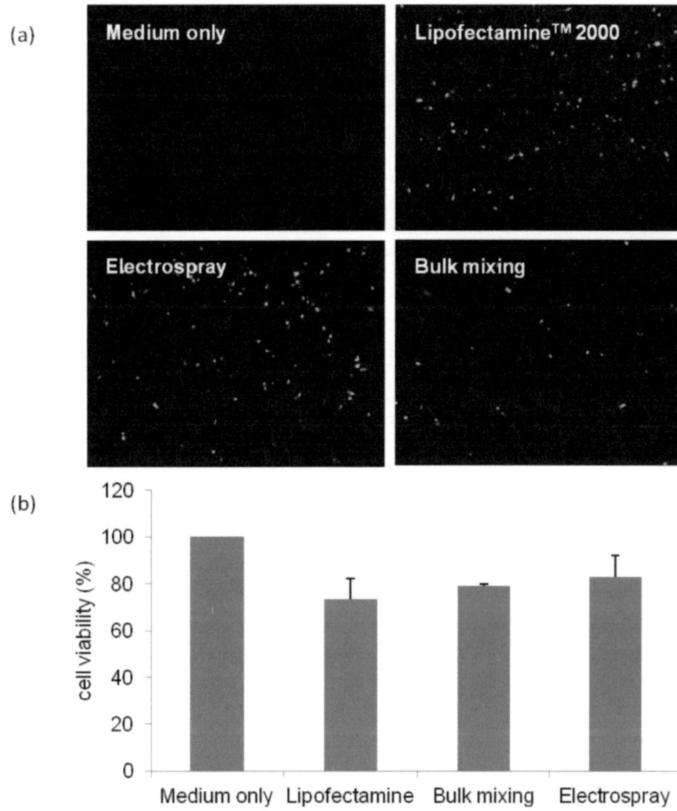

Fig. 8. **(a)** The GFP expression in NIH 3T3 cells and **(b)** the cell viability measured 2 days post transfection. All polyplexes were produced at the N/P ratio of 6.7. For coaxial electrospray, the flow rate was 6mL/h and voltage was 5.0kV. (n=4)

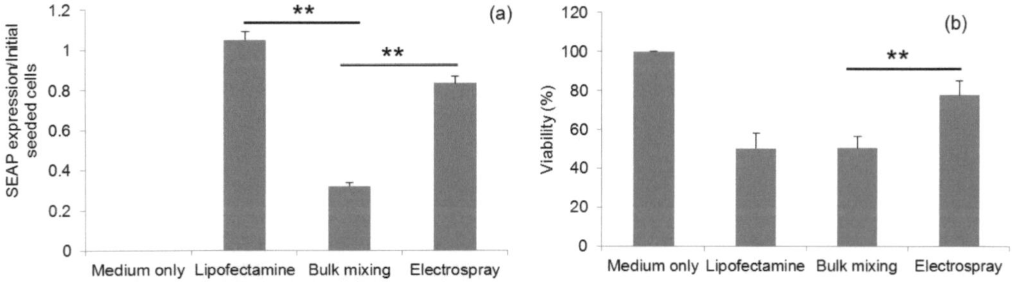

Fig. 9. The pSEAP expression in NIH 3T3 **(a)** and the cell viability **(b)** measured 2 days post transfection. All polyplexes were produced at the N/P ratio of 6.7. For coaxial electrospray, the flow rate was 6 mL/h and voltage was 5.6 kV. (n=5, **: $p<0.005$)

3.3 Coaxial electrospray lipoplexes for oligodeoxynucleotides (ODN) delivery

3.3.1 Introduction

Lipoplexes, a major non-viral carrier system, have attracted a lot of research interest in the drug and nucleic acid delivery field. The phospholipid bilayer structure of lipoplexes provides great flexibility in encapsulating various drugs or nucleic acids, either within the hydrophobic bilayers or in the hydrophilic core. Many techniques, such as bulk mixing (Chrai et al. 2001; Elouahabi et al. 2005), the film method (Bangham method) (Fan et al. 2007; Otake et al. 2006) and alcohol dilution (Stano et al. 2004), have been developed to produce lipoplexes. However, non-uniform particle size, low encapsulation efficiency and multistep production process remain as major challenges. (Müller et al. 2000, 2004)

Antisense oligodeoxynucleotides (ODN) are short pieces of specially designed DNA. ODN has been widely investigated as a potential therapeutic agent against viral infections, cardiovascular inflammation, hematological malignancies, pulmonary diseases and cancer. (Chiu et al. 2006; Yang et al. 2004) ODN inhibits gene expression by hybridizing to specific mRNA sequences, which interferes with the target protein expression. For the preparation of ODN encapsulated lipoplexes, the ethanol dilution method is very popular (Stano et al. 2004; Jeffs et al. 2005). In this approach, the lipid mixture is first dissolved in ethanol and then mixed with the aqueous ODN solution. The final mixture is dialyzed against buffer solutions to remove ethanol and as the ethanol is removed ODN encapsulated lipoplexes are formed. The ethanol dilution method is clearly a batch production method, and batch to batch variation of the product can be an issue. In addition, multiple steps that may take days to complete are involved, and contamination of the final product can be an issue.

Our alternative approach was to use coaxial electrospray to produce ODN encapsulated lipoplexes, where the lipids/ethanol mixture initially surrounded the aqueous-ODN core. (Wu et al. 2009b) As the liquids leave the needles, the electrical field breaks the compound liquid stream into fine droplets. The large surface area of the droplets results in rapid ethanol evaporation and the ODN encapsulated lipoplexes are formed immediately. Compared to ethanol dilution, coaxial electrospray is a simple, one step and continuous production process. In addition, lipoplexes produced by coaxial electrospray can be either collected in buffer solutions for intravenous injection or used directly in pulmonary delivery.

3.3.2 Electrospray G3139 encapsulated lipoplexes

In our study, G3139 (Genasense or oblimerson sodium) was chosen as the model ODN. G3139 is an 18-mer ODN (5'-TCT CCC AGC GTG CGC CAT-3'). It is specially designed to bind the first six codons of the human Bcl-2 mRNA and thus inhibit Bcl-2 expression, and may provide a way to decrease the resistance of tumor cells to chemotherapy. (Chiu et al. 2006) To prepare G3139 encapsulated lipoplexes, G3139 was dispersed in 1X PBS solution fed through the inner 27 gauge needle. The lipid mixture (DC-Choleserol: EggPC:DSPE-PEG=30:68:2 molar ratio) was dissolved in ethanol and fed through the outer 20G needle. A positive voltage, typically ~3 kV, was applied between the inner needle and a grounded copper ring electrode to break the liquid into fine droplets. As the ethanol evaporates the lipoplexes are formed and then captured in a grounded aluminum dish 10 cm below the needle tip containing 15 mL PBS solution.

3.3.3 Size distribution and zeta potential of G3139 encapsulated lipoplexes

The size distribution and surface charge of G3139 encapsulated lipoplexes prepared by coaxial electrospray depend on the concentrations and flow rates of G3139 solution and lipid mixtures. By adjusting operation parameters, the size of lipoplexes can vary between 100 nm and 2500 nm and the surface charge of lipoplexes can vary between -16 mV and +20 mV. (Wu et al. 2009b) The best balance between lipoplex productivity, size and surface charge, was achieved by setting the flowrates of the G3139 solution and the lipid mixture to 1.2 mL/h, the concentration of the G3139 solution to 0.5 mg/mL and the concentration of the lipid mixture to 10 mg/mL. Under these conditions Figure 10 shows that the lipoplex average mean diameter by volume was 190 ± 39 nm. The corresponding zeta potential was +4.5 ± 0.43 mV and the final lipid/G3139 ratio was 20.

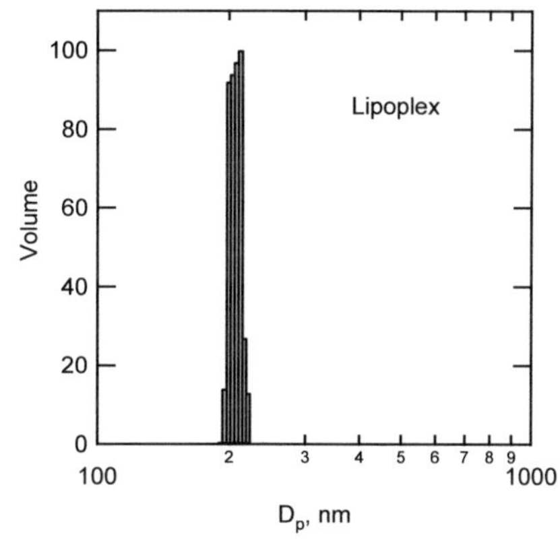

Fig. 10. A typical size distribution of G3139 encapsulated lipoplexes. The average mean diameter by volume was 190 ± 39 nm.

3.3.4 Structure of G3139 encapsulated lipoplexes

Small angle neutron scattering (SANS) and cryo-transmission electron microscopy (cryo-TEM) were used to characterize the structures of G3139 encapsulated lipoplexes. Figure 11 illustrates typical SANS spectra for lipoplexes made by coaxial electrospray and by an ethanol dilution method.

For the sample prepared by the ethanol dilution method, a Bragg Peak was observed, indicating the multi-lamellar structure of the lipoplexes, while the Bragg Peak was absent in lipoplexes produced by coaxial electrospray. From the position of the Bragg Peak we determined the inter-lamellar spacing d using the relation $d = 2\pi/q$, where $q = (4\pi/\lambda)\sin(\theta/2)$ is the momentum transfer vector, λ is the neutron wavelength, and θ is the scattering angle. When we fit the peak for the lipoplexes using a Gaussian function, we found a center-to-center lamellar spacing of 6.6 +/- 0.2 nm. The cryo-TEM images support the SANS results by showing that most electrosprayed lipoplexes have a uni-lamellar structure, while lipoplexes prepared by ethanol dilution have onion like multi-lamellar structure with center to center inter-lamella distance measured from the picture of about 7-10 nm. The difference in the structure of lipoplexes produced by these two methods may reflect the removal rate of ethanol from the solution. In coaxial electrospray, the liquids break into tiny droplets, and the ethanol evaporates in less than 1 s due to the large surface area of the fine droplets. Thus, the lipids and G3139 may be "locked" and do not have time to rearrange and form the multi-lamellar structure. In contrast, the dialysis step in the ethanol dilution method is a very slow process, usually requiring about 24 h to remove ethanol from the solution, and therefore the lipids and G3139 have enough time to form the more complex multi-lamellar structure.

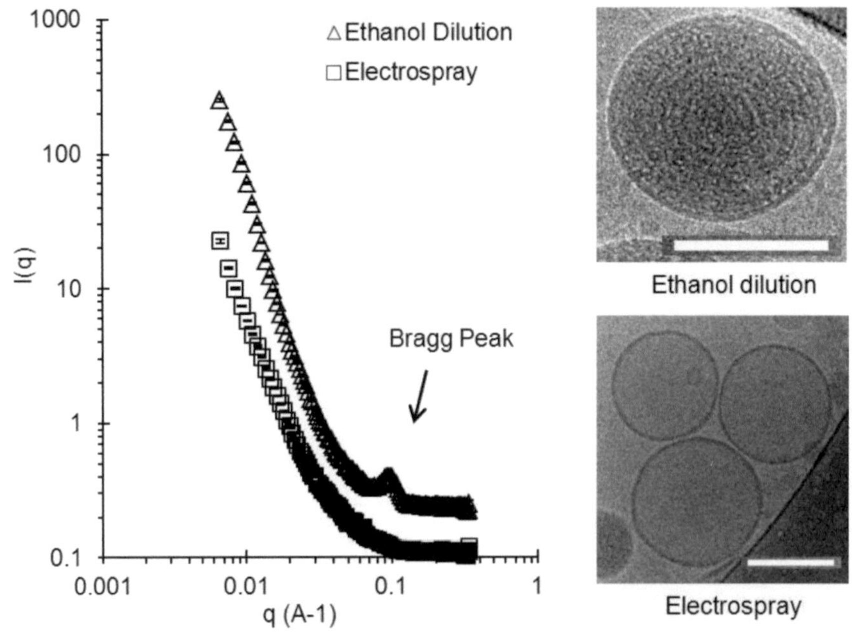

Fig. 11. Typical SANS spectra and cryo-TEM images show that the electrosprayed lipoplexes have unilamellar structure, while those prepared by ethanol dilution have a multi-lamellar structure. The error bars on the SANS data represent ± one standard deviation and are generally smaller than the symbol size. (Scale bar: 100nm)

3.3.5 Bcl-2 down regulation

The bioactivity of G3139 encapsulated lipoplexes was evaluated in K562 cells (chronic myelogenous leukemia cell line). In addition to non-targeted G3139 encapsulated lipoplexes, transferrin (Tf) conjugated G3139 encapsulated lipoplexes were also used in this study. Chiu et al. (2006) reported that transferrin (Tf) conjugated lipoplexes had targeting ability because they could binding to the transferrin receptor (TfR), a transmembrane glycoprotein over expressed on cancer and leukemia cells. Thus, transferrin conjugated lipoplexes provided better down regulation of Bcl-2 in K562 cells and were more effective for *in vivo* applications. K562 cells were treated with both non-targeted and Tf-targeted G3139 encapsulated lipoplexes at G3139 concentration of 1 μM. Bcl-2 expression were measured by western blotting 48 h post transfection. Figure 12 shows that the Bcl-2 expression was decreased by up to ~55% when cells were transfected with Tf-targeted G3139 encapsulated lipoplexes, compared to ~40% by non-targeted G3139 encapsulated lipoplexes and ~15% by free G3139.

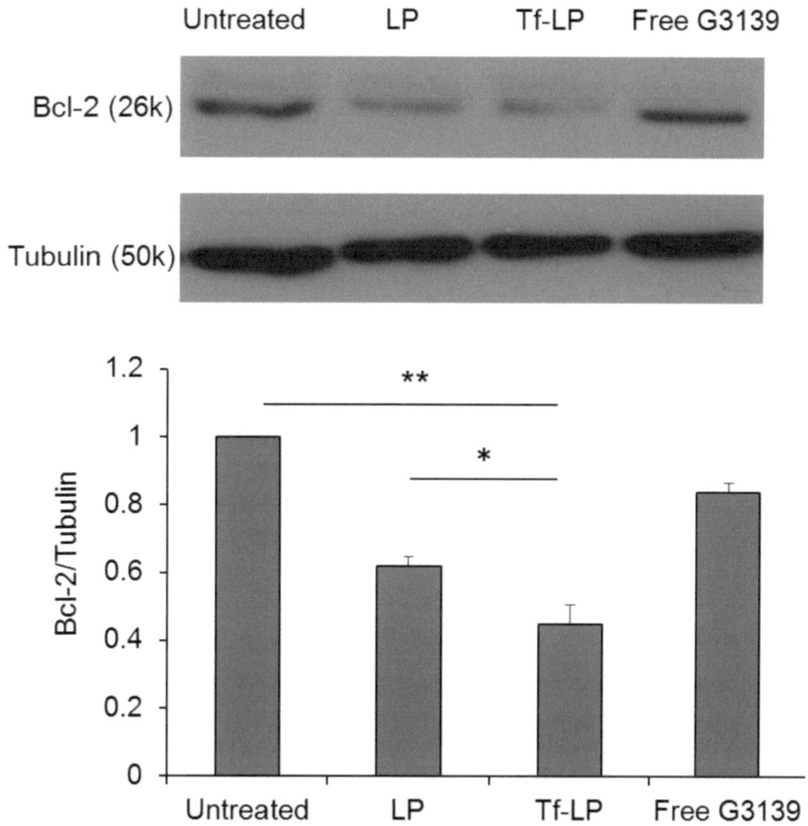

Fig. 12. A representative western blotting of Bcl-2 expression in K562 cells. LP: non-targeted G3139 encapsulated lipoplexes; Tf-LP: Tf-targeted G3139 encapsulated lipoplexes. The concentration of G3139 was 1 μM. (n=3, *: $p < 0.05$, **: $p < 0.01$.)

3.3.6 Cellular uptake of ODN encapsulated lipoplexes

K562 cells were transfected by free FAM-ODN, non-targeted FAM-ODN encapsulated lipoplexes and Tf-targeted FAM-ODN encapsulated lipoplexes at FAM-ODN concentration level of 1μM. Flow cytometry and confocal microscopy were used to investigate the cellular uptake of FAM-ODN encapsulated lipoplexes 4 h post transfection. As shown on Figure 13 the cellular uptake of FAM-ODN delivered by lipoplexes, particularly Tf-targeted lipoplexes, was much more efficient compared to cells treated with free FAM-ODN. Compared to cells transfected with non-targeted FAM-ODN encapsulated lipoplexes, those treated with Tf-targeted FAM-ODN encapsulated lipoplexes had ~40% higher fluorescence signal, indicating that transferrin improved the interaction between the lipoplexes and the K562 cells and facilitated the cellular uptake of lipoplexes, and thus more efficient Bcl-2 down-regulation was observed. For future *in vivo* applications Tf-targeted lipoplexes might also be more effective targeting cancer cells and minimize the rapid clearance by the reticuloendothelial system, or side effects, such as nonspecific cytokine production.

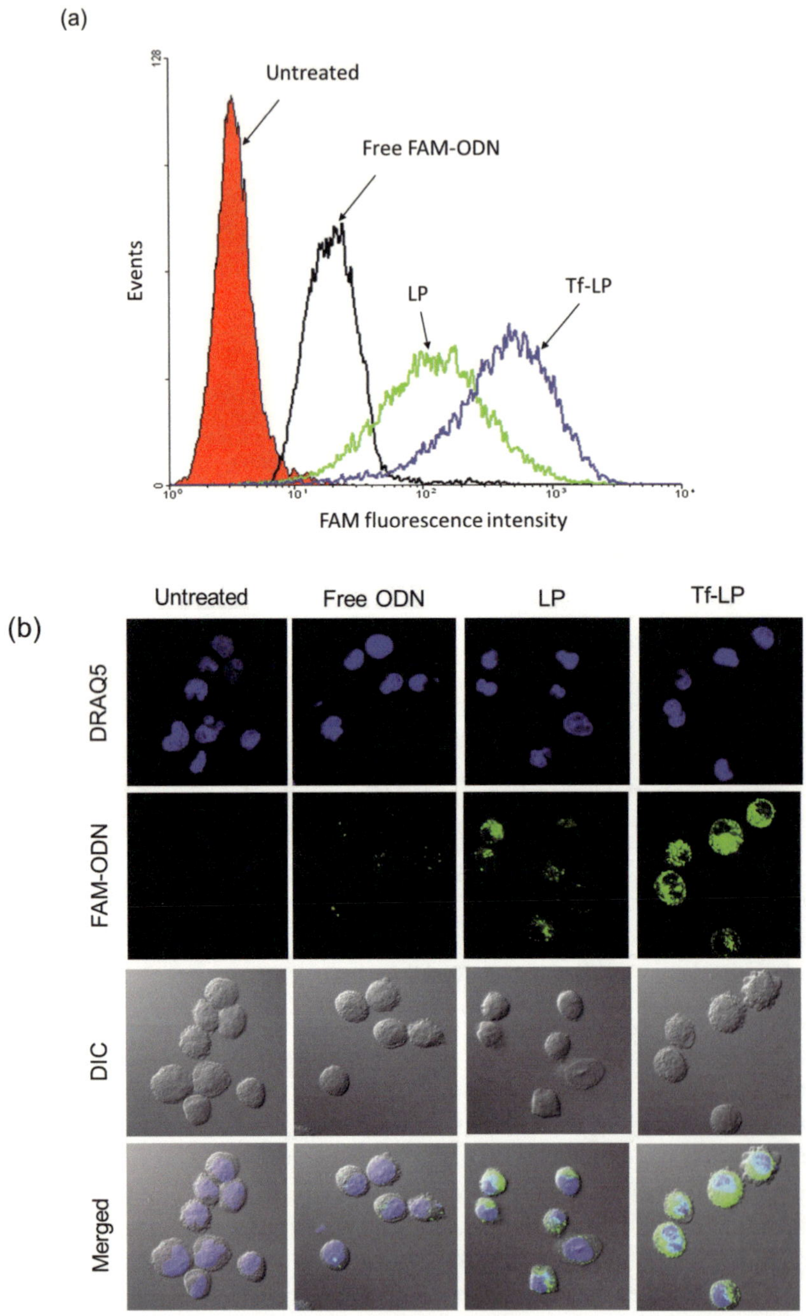

Fig. 13. **(a)** Flow cytometery and **(b)** confocal microscopy images showed the cellular uptake of FAM-ODN by K562 cells. K562 cells were treated with free FAM-ODN, non-targeted FAM-ODN encapsulated lipoplexes (LP) and Tf-targeted FAM-ODN encapsulated lipoplexes (Tf-LP) at FAM-ODN concentration of 1 μM. Control: cells cultured in medium with no transfection. DIC: differential interference contrast.

3.3.7 Conclusions

We developed a coaxial electrospray process to produce oligonucleotide encapsulated lipoplexes for nucleic acid delivery. The lipoplexes produced by coaxial electrospray can either be collected for intravenous injection or delivered as aerosol for inhalation therapy. This method allows for better control over the way in which the lipid and aqueous phases are mixed. Compared with the standard ethanol dilution technique, coaxial electrospray is a simple, one step, continuous process that significantly reduces the time and effort required to produce the lipoplexes. By adjusting operating parameters, such as flow rates and liquid concentrations, monodisperse lipoplexes with different size and surface charge can be easily produced to meet various application needs. In this work, G3139 encapsulated lipoplexes were successfully produced via coaxial electrospray with diameter of ~190nm and zeta potential of ~+4.5 mV. Due to fast ethanol removal, the lipoplexes produced by coaxial electrospray showed unilamellar structure compared to the multi-lammellar structure of lipopelxes produced by ethanol dilution method. Tranferrin was successfully conjugated to the G3139 encapsulated lipoplexes. Flow cytometry and confocal microscopy analysis showed that transferrin provided targeting ability for the lipoplexes, which greatly improved the cellular uptake of lipoplexes. Compared to ~40% Bcl-2 protein down-regulation observed by non-targeted G3139 encapsulated lipoplexes, Tf-targeted lipoplexes was more efficiently delivered to K562 cells and down regulated the Bcl-2 protein expression by ~55%.

4. Summary

To summarize, we have demonstrated the great potential of electrospray to produce nanoparticles for a variety of drug/nucleic acid delivery applications, including solid lipid nanoparticles for hydrophobic drug delivery, as well as polyplexes and lipoplexes for nucleic acid delivery. We hope this review can stimulate further development and utilization of electrospray in nanobiotechnology.

5. Acknowledgement

This work was supported by the National Science Foundation under Grant No. EEC-0425626. We acknowledge and thank Dr. Weiwei Deng at the University of Central Florida for providing the multiplexed electrospray pictures.

6. References

Bocanegra, R.; Gaonkar, A. G.; Barrero, A.; Loscertales, I. G.; Pechack, D. & Marquez, M. (2005). Production of Cocoa Butter Microcapsules Using an Electrospray Process. *Journal of Food Science*, Vol. 70, No. 8, pp. E492-E497, ISSN 1750-3841

Chen, D. R.; Wendt, C. H. & Pui, D. Y. H. (2000). A Novel Approach for Introducing Bio-materials into Cells. *Journal of Nanoparticle Research*, Vol. 2, No. 2, pp. 133-139, ISSN 1388-0764

Chen, X.; Jia, L.; Yin, X.; Cheng, J. & Lu, J. (2005). Spraying Modes in Coaxial Jet Electrospray with Outer Driving Liquid. *Physics of Fluids*, Vol. 17, No. 3, pp. 032101. ISSN 1070-6631

Chiu, S.; Liu, S.; Perrotti, D.; Marcucci, G. & Lee, R. J. (2006). Efficient delivery of a Bcl-2-Specific Antisense Oligodeoxyribonucleotide (G3139) via Transferrin Receptor-Targeted Liposomes. *Journal of Controlled Release*, Vol. 112, No. 2, pp. 199–207, ISSN 0168-3659

Chrai, S. S.; Murari, R. & Ahmad, I. (2001). Liposomes (a Review) Part One: Manufacturing Issues. *Biopharm International*, Vol. 14, No. 11, pp. 10-14, ISSN 1542-166X

Cloupeau, M. & Prunet-Foch, B. (1994) Electrohydrodynamic Spraying Functioning Modes: a Critical Review. *Journal of Aerosol Science*, Vol. 25, No. 6, pp. 1021-1036, ISSN 0021-8502

Davies, L. A.; Hannavy, K.; Davies, N.; Pirrie, A.; Coffee, R. A.; Hyde, S. C. & Gill, D. R. (2005). Electrohydrodynamic Comminution: a Novel Technique for the Aerosolisation of Plasmid DNA. *Pharmaceutical Research*, Vol. 22, No. 8, pp. 1294-1304, ISSN 0724-8741

Deng, W.; Klemic, J. F.; Li, X.; Reed, M. A. & Gomez, A. (2006). Increase of Electrospray Throughput Using Multiplexed Microfabricated Sources for the Scalable Generation of Monodisperse Droplets. *Journal of Aerosol Science*, Vol. 37, pp. 696–714, ISSN 0021-8502

Elouahabi, A. & Ruysschaert, J. (2005). Formation and Intracellular Trafficking of Lipoplexes and Polyplexes. *Molecular Therapy*, Vol. 11, pp. 336-347, ISSN 1525-0016

Fan, M.; Xu, S.; Xia, S. & Zhang, X. (2007). Effect of Different Preparation Methods on Physicochemical Properties of Salidroside Liposomes. *Journal of Agricultural and Food Chemistry*, Vol. 55, No. 8, pp. 3089-3095, ISSN 0021-8561

Farook, U.; Zhang, H.B.; Edirisinghe, M. J.; Stride, E. & Saffari, N. (2007). Preparation of Microbubble Suspensions by Co-axial Electrohydrodynamic Atomization. *Medical Engineering and Physics*. Vol. 29, No. 7, pp. 749–754, ISSN 1350-4533

Fenn, J. B.; Mann, M.; Meng, C. K.; Wong, S. F. & Whitehouse, C. M. (1989). Electrospray Ionization for Mass Spectrometry of Large Biomolecules. *Science*, Vol. 246, No. 4926, pp. 64-71, ISSN 0036-8075

Gebhart, C. L. & Kabanov, A. V. (2001). Evaluation of Polyplexes as Gene Transfer Agents. *Journal of Controlled Release*, Vol. 73, No. 2-3, pp. 401-416, ISSN 0168-3659

Godbey, W. T.; Wu, K. K. & Mikos, A. G. (1999). Poly(ethylenimine) and Its Role in Gene Delivery. *Journal of Controlled Release*, Vol. 60, No. 2-3, pp. 149-160, ISSN 0168-3659

Haberland, M. E. & Reynolds, J. A. (1973). Self-association of Cholesterol in Aqueous Solution. *Proceedings of the National Academy of Sciences*, Vol. 70, No. 8, pp. 2313-2316, ISSN 1091-6490

Jaworek, A. (2010). *Electrospray Technology for Thin-Film Deposition*, Nova Science Publishers Inc, ISBN 9781617612015, Hauppaug, New York, USA

Jaworek, A. (2007). Micro- and Nanoparticle Production by Electrospraying. *Powder Technology*, Vol. 176, No. 1, pp. 18–35, ISSN 0032-5910

Jaworek, A. & Krupa, A. (1999). Classificantion of the Modes of EHD Spraying. *Journal of Aerosol Science*, Vol. 30, No. 7, pp. 873-893, ISSN 0021-8502

Jaworek, A. & Sobczyk, A. T. (2008). Electrospraying Route to Nanotechnology: an Overview. *Journal of Electrostatics*. Vol. 66, No. 3-4, pp. 197–219, ISSN 0304-3886

Jayasinghe, S. N.; Qureshi, A. N. & Eagles, P. A. M. (2006). Electrohydrodynamic Jet Processing: an Advanced Electric-Field-Driven Jetting Phenomenon for Processing Living Cells. *Small*, Vol. 2, No. 2, pp. 216-219, ISSN 1613-6829

Jeffs, L. B.; Palmer, L. R.; Ambegia, E. G.; Giesbrecht, C.; Ewanick, S. & MacLachlan, I. (2005) A Scalable Extrusion-Free Method for Efficient Liposomal Encapsulation of Plasmid DNA. *Pharmaceutical Research*. Vol. 22, No. 3, pp. 363-372, ISSN 0724-8741

Kircheis, R.; Wightman, L. & Wagner E. (2001). Design and Gene delivery Activity of Modified Polyethylenimines. *Advanced Drug Delivery Reviews*, Vol. 53, No. 3, pp. 341-358, 0169-409X

Koshkina, N. V.; Agoulink, I. Y.; Melton, S. L.; Densmore, C. L. & Knight V. (2003). Biodistribution and Pharmacokinetics of Aerosol and Intravenously Administered DNA-Polyethylenimine Complexes: Optimization of Pulmonary Delivery and Retention. *Molecular Therapy*, Vol. 8, No. 2, pp. 249-254, ISSN 1525-0016

Kumbar, S. G.; Bhattacharyya, S.; Sethuraman, S. & Laurencin, C. T. (2007). A Preliminary Report on a Novel Electrospray Technique for Nanoparticle Based Biomedical Implants Coating: Precision Electrospraying. *Journal of Biomedical Materials Research Part B: Applied Biomaterials*, Vol. 81B, No. 1, pp. 91-103, ISSN 1552-4973

Laporte, L. D.; Rea, J. C. & Shea, L. D. (2006). Design of Modular Non-Viral Gene Therapy Vectors. Biomaterials, Vol. 27, No. 7, pp. 947-954, ISSN 0142-9612

Loscertales, I. G.; Barrero, A.; Guerrero, I.; Cortijo, R.; Marquez, M. & Ganan-Calvo, A. M. (2002). Micro/Nano Encapsulation Via Electrified Coaxial Liquid Jets. *Science*, Vol. 295, No. 5560, pp. 1695-1698, ISSN 0036-8075

Mastrobattista, E.; Van Der Aa, M. A. E. M.; Hennink, W. E. & Crommelin, D. J. A. (2006). Artificial Viruses: a Nanotechnological Approach to Gene Delivery. *Nature Reviews Drug Discovery*, Vol. 5, No. 2, pp. 115-121, ISSN 1474-1776

Müller, R. H. & Keck, C. M. (2004). Challenges and Solutions for the Delivery of Biotech Drugs - a Review of Drug Nanocrystal Technology and Lipid Nanoparticles. *Journal of Biotechnology*. Vol. 113, No. 1-3, pp. 151-170, ISSN 1684-5315

Müller, R. H.; Mader, K. & Gohla, S. (2000). Solid Lipid Nanoparticles (SLN) for Controlled Drug Delivery - a Review of the State of the Art. *European Journal of Pharmaccutics and Biopharmaceutics*. Vol. 50, No. 1, pp. 161-177, ISSN 0939-6411

Neu, M.; Fisher, D. & Kissel T. (2005). Recent Advances in Rational Gene Transfer Vector Design Based on Poly(ethyleneimine) and Its Derivatives. *Journal of Gene Medicine*, Vol. 7, No. 8, pp. 992-1009, ISSN 1521- 2254

Ojito, E.; Labrada, G.; Garcia, Z.; Garcia, N. & Chico, E. (2001). Study of NS0 Cell Line Metabolism in Lipid Supplemented Protein Free Media, *Animal Cell Technology: From Target to Market: Proceedings of the 17th ESACT Meeting*, pp. 179-182, ISBN 978-1-4020-0264-9, Tylösand, Sweden, June 10-14, 2001.

Otake, K.; Shimomura, T.; Goto, T.; Imura, T.; Furuya, T.; Yoda, S.; Takebayashi, Y.; Sakai, H. & Abe, M. (2006). Preparation of Liposomes Using an Improved Supercritical Reverse Phase Evaporation Method. *Langmuir*, Vol. 22, No. 6, pp. 2543-2550, ISSN 0743-7463

Pack, D. W.; Hoffman, A. S.; Pun, S. & Stayton, P. S. (2005). Design and Development of Polymers for Gene Delivery. *Nature Reviews Drug Discovery*, Vol. 4, No. 7, pp. 581-593, ISSN 1474-1776

Pareta, R. & Edirisinghe, M. J. (2006). A Novel Method for the Preparation of Biodegradable Microspheres for Protein Drug Delivery. *Journal of the Royal Society Interface*, Vol. 3, No. 9, pp. 573–582, ISSN 1742-5689

Pareta, R.; Brindley, A.; Edirisinghe, M. J.; Jayasinghe, S. N. & Lukinska, Z. B. (2005). Electrohydrodynamic Atomization of Protein (Bovine Serum Albumin). *Journal of Materials Science: Materials in Medicine*, Vol. 16, No. 10, pp. 919– 925, ISSN 0957-4530

Salata, O. V. (2005). Tools of Nanotechnology: Electrospray. *Current Nanoscience*, Vol. 1, No.1, pp. 25-33, ISSN 1573-4137

Stankusa, J. J.; Solettib, L.; Fujimoto, K.; Hong, Y.; Vorp, D. A. & Wagnera W. R. (2007). Fabrication of Cell Microintegrated Blood Vessel Constructs Through Electrohydrodynamic Atomization. *Biomaterials*, Vol. 28, No. 17, pp. 2738–2746, ISSN 0142-9612

Stano, P.; Bufali, S.; Pisano, C.; Bucci, F.; Barbarino, M.; Santaniello, M.; Carminati, P. & Luisi, P. L. (2004). Novel Camptothecin Analogue (Gimatecan)-Containing

Liposomes Prepared by the Ethanol Injection Method. *Journal of Liposome Research.* Vol. 14, No. 1-2, pp. 87–109, ISSN 0898-2104

Talley, D.; Cutak, B.; Rathbone, E.; Al-Kolla, T.; Allison, D.; Blasberg, J.; Kao, K. & Caple, M. (2003). SyntheChol™ Synthetic Cholesterol for Cholesterol Dependent Cell Culture-Development of Non-Animal Derived Chemically Defined NS0 Medium, *Animal Cell Technology Meets Genomics: Proceedings of the 18th ESACT Meeting,* pp. 577-580, ISBN 978-4020-2791-5, Granada, Spain, May 11-14, 2003.

Widiyandari, H.; Hogan, C. J. J.; Yun, K. M.; Iskandar, F.; Biswas, P. & Okuyama, K. (2007). Production of Narrow-Size-Distribution Polymer-Pigment-Nanoparticle Composites via Electrohydrodynamic Atomization. *Macromolecular Materials and Engineering,* Vol. 292, No. 4, pp. 495–502, ISSN 1438-7492

Wu, Y.; Chalmers, J. J.; Wyslouzil, B. E. (2009). The Use of Electrohydrodynamic Spraying to Disperse Hydrophobic Compounds in Aqueous Media. *Aerosol Science and Technology,* Vol. 43, pp. 902-910, ISSN 0278-6826

Wu, Y.; Yu, B.; Jackson, A.; Zha, W.; Lee L. J. & Wyslouzil, B. E. (2009). Coaxial Electrohydrodynamic Spraying: a Novel One-Step Technique to Prepare Oligodeoxynucleotide Encapsulated Lipoplex Nanoparticles. *Molecular Pharmaceutics,* Vol. 6, No. 5, pp. 1371–1379, ISSN 1543-8384

Wu. Y.; Fei, Z.; Lee, L. J. & Wyslouzil, B. E. (2010). Electrohydrodynamic Spraying of DNA/Polyethylenimine Polyplexes for Nonviral Gene Delivery. *Biotechnology and Bioengineering,* Vol. 105, No. 4, pp. 834-841, ISSN 0006-3592

Wu, Y.; Ma, N.; Chalmers, J.; Wyslouzil, B. E.; McCormick, E. L. & Casnocha, S. (2011). Enhanced Productivity of NS0 Cells in Fed-Batch Culture with Cholesterol Nanoparticle Supplementation. *Biotechnology Progress,* Vol. 27, No. 3, pp. 796–802, ISSN 1520-6033

Xie, J.; Lim, L. K.; Phua, Y.; Hua, J. & Wang, C. (2006). Electrohydrodynamic Atomization for Biodegradable Polymeric Particle Production. *Journal of Colloid and Interface Science,* Vol. 302, No. 1, pp. 103–112, ISSN 0021-9797

Xie, J. & Wang, C. (2007). Encapsulation of Proteins in Biodegradable Polymeric Microparticles Using Electrospray in the Taylor Cone-Jet Mode. *Biotechnology and Bioengineering,* Vol. 97, No. 5, pp. 1278–1290, ISSN 0006-3592

Yamane, I.; Kan, M.; Minamoto, Y. & Amatsuji, Y. (1981). α-Cyclodextin, a Novel Substitute for Bovine Albumin in Serum-Free Culture of Mammalian Cells. *Proceedings of the Japan Academy, Ser. B, Physical and Biological Sciences,* Vol. 57, No. 10, pp. 385-389 ISSN 0386-2208

Yang, L.; Li, J.; Zhou, W.; Yuan, X. & Li, S. (2004). Targeted delivery of antisense oligodeoxynucleotides to folate receptor-overexpressing tumor cells. *Journal of Controlled Release,* Vol. 95, No. 2, pp. 321– 331, ISSN 0168-3659

Yurteri, C. U.; Hartman, R. P. A. & Marijnissen J. C. M. (2010). Producing Pharmaceutical Particles via Electrospraying with an Emphasis on Nano and Nano Structured Particles - A Review. *KONA Powder and Particle Journal* No.28, pp. 91-115, ISSN 0288-4534

Zhang, J. & Robinson, D. (2005). Development of Aninal-free, Protein-free and Chemically-defined Media for NS0 Cell Culture. *Cytotechnology,* Vol. 48, No. 1-3, pp. 59-74, ISSN 0920-9069

Zhang, S.; Kawakami, K.; Yamamoto, M.; Masaoka, Y.; Kataoka, M.; Yamashita, S. & Sakuma, S. (2011). Coaxial Electrospray Formulations for Improving Oral Absorption. *Molecular Pharmaceutics,* Vol. 8, No. 3, pp. 907-813, ISSN 1543-8384

11

Microbubble Therapies

Ajay Sud[1,2,3] and Shiva Dindyal[1,2]
[1]*Whipps Cross University Hospital, London,*
[2]*The Royal London Hospital, London,*
[3]*The Royal Liverpool University Hospital, Liverpool,*
United Kingdom

1. Introduction

Microbubble technologies have recently gained acceptance as radiological molecular imaging or perfusion agents. These intravenous ultrasound contrast agents consist of gas-filled microspheres stabilized by a 'shell' layer which interfaces with a surrounding solvent pool. Development in the form of encapsulation with complex biocompatible polymer shell consisting of proteins, lipids and surfactants has improved their stability and half-life significantly. These shell modifications have optimised the mechanical index of microbubbles, making them more resistant to compression rather than expansion.

The relatively inert nature of microbubbles contributes to a favourable safety profile and 'stealth' delivery profile. These indolent properties of microspheres are fundamental to their ability to function as targeted delivery vehicles for potential therapeutic modalities. Their small size (1-8 micrometre) enables unobstructed diffusion between the vascular and capillary compartments. Complete isolation of the active therapeutic components from normal native tissues and liberal exposure to the target disease tissue is the panacea of drug delivery.

Incorporation of lipopolysaccharide ligands adhering to selective cell surface epitopes, to the microbubble surface interface, is the principal method by which these vehicles can be precisely targeted. Selective bioactive material delivery into the target cell significantly improves therapeutic efficacy and can completely negate the adverse side-effects.

Externally applied focused ultrasound directed at the ligand bound microbubbles and target cells complex is the crucial final step ensuring optimal delivery. Application of sonication results in the oscillation followed by destruction of the bubbles and liberation of the bioactive compound; the high intensity focused acoustic energy disrupts the microbubbles. Passage of the active product into the intracellular space is facilitated through cellular sonoporation, at least in part as a result of acoustic cavitation of the target cell's membrane.

It has been demonstrated that microbubble technology can accommodation active constituents, hence be 'drug loaded'. Microbubble delivery technology has been utilised as a feasible drug delivery modality for gene therapies, cytotoxics, drugs and dyes. Co-administration of the microbubbles with the active product, without drug loading, has also been utilised to improve targeted drug delivery.

The premise of the clinically applicable microbubble delivered therapeutic agent delivery is an attractive proposition, but the clinical experience is still limited. Establishing the clinical benefits of microbubble technologies in the field of diagnosis has opened the door to a precise and accurate drug delivery vector. The proliferation in interest surrounding microbubbles has yet to meet its climax; the scientific, translational and clinical challenges of this exciting technology are still to be fully elucidated.

In this chapter we will consider the current understanding of the elucidated theory underlying this technology, and the potential clinical applications with drug and gene delivery. We also endeavour to summaries uncertainties and controversies.

2. The evolution of microbubble

High-intensity focused ultrasound is used routinely for the lithotripsy of the renal calculi. The ablation of the solid tumours and regions of tissue with high-intensity focused ultrasound is a less proven application. Since the inception of the microbubbles their application has grown exponentially with incremental advances in their formulations (Tinkov et al., 2009).

The dawn of microbubble technologies began with an innocuous *in vitro* observation that simple agitated saline enhanced ultrasound signal. This observation was demonstrated in cardiological and aortic ultrasound echocardiogram signal within the aortic root and chambers of the heart (Gramiak & Shah, 1968). The first generation microbubble contrast ultrasound agents were largely only applicable to diagnostic situations where additional opacification of the myocardium and coronary vessels is was necessary. The limited spectrum of clinical application was the result of the inherent characteristics of these first generation microbubbles agents (Kaul. 1997).

The fragile *in vitro* composition of these microbubbles is the result of unstable gas fluid interface and large bubble size. First generation microbubble agents consist of simple air bubbles dispersed in aqueous solution with an absence of a stabilising shell. The large bubble diameter ensured these agents do not to pass through the small capillaries, including the intrapulmonary circulation. This with combined with the exceptionally short life-span beds means that these first generation agents will never become detectable in within the left ventricle after intravenous injection (Kaul. 1997).

Later generations of microbubble agents have increasingly sophisticated colloidal gaseous/aqueous shells, which results in enhanced stability characteristics. The improved modalities of the later generations of microbubbles contrast dyes are now are recognised part imaging adjunct. The smaller size distribution of second generation enabled transpulmonary passage. Following intravenous administration of this generation of microbubble agents with stabilised solvent-gaseous interface they can become detectable within the left heart chambers. This generation is limited by unstable gaseous cores. Despite the composite shells the air cores have a tendency to dissolve in the blood within minutes. Subsequent generations replaced these highly soluble gaseous air nucleuses to overcome this inherent characteristic which limited their half-life (Voci et al, 1994).

Third generation microbubble compounds contained physiologically inert gas cores propagated their stability almost tripling half-life to at least 15 minutes. This generation has extended the application of microbubble agents to any ultrasound amenable organs. Further manipulation of the agents will ensure further broadening of their scope of application (Tinkov et al., 2009).

The trends in the developments in later generations of microbubbles agents have moved away from the diagnostic efficacy and towards therapeutic modalities with drug carrying capacities. Precise formulation and engineering is necessary to ensure the efficacious delivery of specific active products (Pitt et al., 2004).

3. Drug targeting

The essential facet of effective targeted drug delivery is the ability to improve the therapeutic efficacy and minimise the quantity of circulating application of unbind active drug. The implications of this include reduction of the adverse effects and a significant minimisation of the effective required dosage necessary. The second property is paradoxically essential for the premise of the drug-loaded microbubbles to be efficacious because of the shell drug-loading capacity is limited. Because of this, only extremely potent agents with a low toxicity threshold would be amenable to this form of delivery. Poly/oligonucleotide gene therapy, cytotoxic, steroid, antimicrobials and active protein compounds may benefit from microbubble compounds.

It is necessary to expose the systemic circulation to the active compound bound to microbubble to ensure close enough proximity to the target organ. It is undeterminable and unpredictable to accurately the proportion of the active product liberated by focused ultrasound destruction at the desired target organ site. Of the remaining non-utilised microbubble complexes some will be eliminated or be dispersed remotely from the target organ. Well perfused organs, such as the renal (Koike et al 2005) and myocardial (Vannan et al 2002; Shohet et al 2000) systems will witness higher microbubble agent concentrations, making an ultrasound liberation of the active products more effective. Organs with inferior blood perfusion levels may not witness the necessary concentrations to make the targeted ultrasound microbubble destruction efficacious. Hence, the extent of blood supply alone may significantly ameliorate any possible projections of the efficacy of targeted therapeutic microbubbles (Tinkov et al., 2009).

The bodies' innate mechanisms can be harnessed to preferentially channel microbubbles to the desire target tissues. This approach of passive targeting relies on the innate non-specific macro- or microcellular characteristic. Reliance on these non-specific in vitro mechanisms means that these agents will be amenable to the influences of pharmacodynamics forces. Two relevant mechanisms include the particle clearance, aspects of the innate and even adaptive immune systems. An example of such a modality includes the effects of immuno-receptive microbubbles typically include those with phospholipid microbubbles containing phosphatidylserine and those which possess polymer-shells. Phagocytic immune white blood cells consume these antigenic microbubbles and convey these microbubbles to sites of active inflammation. Spontaneous non-macrophage mediate passive accumulation of ligand-containing microbubble agents via pathological or physiological enhancement in endothelial permeability and then retention has been found not to be possible (Bloch et al., 2004).

Active molecular targeting is essentially the incorporation of intentionally designed microbubble manipulations to ensure preferential affinity to the desire target tissues. Cell targeting can be achieved with a series of complementary immunoliposome (Siwak et al., 2002) and nanoparticle ligands (Shi et al., 2007) to specific cell surface receptors attached to the microbubble shells. Isolated or preferential expression of particular diseased cell-specific receptors on their surface provides a means by which targeting discrimination can be achieved. Complex formation between the complementary cell surface and attached microbubble shell protein ensure both selectivity and persistence of the vector to the target.

A major area of research has been into the active molecular targeting of microbubble agents are that they wholly reside within the intravascular compartment and will only come into contact with potential target markers situated on the endothelial wall. The shear force of vessel Blood-flow does significantly weaken microbubble ligand complexes-receptor binding complexes. Increasing the ligand density present on the microbubble surface has been found to be a reliable method by which to target binding can be enhanced. No further improvement in receptor binding is exhibited above a certain ligand concentration (Tinkov et al 2009). Supple ligand-microbubble spacer arms are a promising method by which endothelial wall receptor binding can be stabilised against shear blood-vessel forces. It has been theorised that the formation of the ligand endothelial wall receptor complexes induces microbubble core pressurization. The resultant deformation in the shape of the microbubble shell produces limited gaseous leakage and wrinkling microbubbles with outward protrusions, which enhance adhesion characteristics (Kilbanov et al 1999).

The laminar flow patterns present in larger vessels may ameliorate the possibility for significant target binding. The application of ultrasound beams can deflect circulating microbubbles on to the endothelial vessel walls enabling improved contact targeting. This property can help overcome the problem of sclerotic plaque targeting in larger vessels and the microbubble targeting inflammatory deep venous areas (Dayton et al 1999).

4. Therapeutic microbubble vectors

Specificity of drug delivery and targeting is the panacea pharmaceutical technologies. The biological barrier forms a worthy barrier to entry of the therapeutic compounds. These biological barriers include the cell membrane, capillary endothelium, and blood - brain barrier and vessel walls. These barriers can selectively transport molecules of varying molecular sizes. Ultrasound targeted microbubble destruction will enable the transport of 2-3 MDa, 6-8 μm, ~ 9 nm,~100 nm through the cell membrane, capillary endothelium (vessel diameter <7 μm), vessel wall diameter ~55 μm) and blood brain barrier respectively (Schlicher et al., 2006; Skyba et al., 1998).

A dichotomy exists between the desired characteristics necessary for the delivery of compounds to intracellular target sites. A simple innately lipophilic natured compound will preferentially access the intracellular compartment but, lacks the ability to freely transit in the aqueous haematological system to the cellular compartment. Uncomplicated hydrophilic agents can easily around the circulation, but without the benefit of selective specialised transport proteins are unable to enter intracellular compartments. Interiorisation of active compounds may be augmented through the application of ultrasound to cell-membrane barrier. Small molecular compounds, including polynucleotides and proteins can pass into the intracellular compartment through the process of cavitation when facilitated with focused ultrasound. High energy ultrasound cellular damage is an undesirable effect of focused high-energy ultrasound. The acoustic cellular damage induced at these energies required to induce significant gases cavitation in vitro make this prospect clinically unreasonable (O'Brien 2007).

The application of ultrasound resonance field to microbubbles induces their frequency dependent oscillation. Microbubbles destabilisation and destruction can be induced with targeted high-energy ultrasound. Local delivery can be achieved through the liberation of the active product at the target site with the microbubbles acting cavitation nuclei. Significantly lower levels of the ultrasound energy are necessary when used in combination with microbubble agents.

If insonated microbubbles pass in the near vicinity of cellular membranes, the characteristics of the cell membranes may be altered and demonstrate the presence of sonopores. Their appearance may be the result of the proposed phenomena specific to microbubbles, these include microstream swirling, micro-jetting, the impact of enhanced ion-channel conductance, the formation of hydrodynamic shock waves or controversially free radical formation (12-24)(Miller & Quddus 2000; Miller & Pislaru 2002; Marmottant & Hilgenfeldt 2003; Barnett 1998; Guzman et al., 2003; Wang et al., 1999; Wei et al., 2004; Juffermans et al., 2006; Miller & Thomas 1993). There remain dissenters as to the whole validity of the cavitation theory, whilst others propose alternative ionic channel conductance mechanisms (Lawrie et al., 2003; Bouakaz A et al., 2006).

Bioactive substances may be able to infiltrate into the intracellular compartment through these sonate induced pores. These induced cellular membrane transit pores are transient (seconds to minutes) in their nature because of an endogenous vesicle-based healing responses resulting in these channels resealing. These healing processes are dependent calcium and ATP based processes. Their estimated sizes range from 30-100nm in size up to a maximum of a couple of millimetres (Pan et al., 2005).

It is questionable whether the formation membrane pores results in proportional changes in the nuclear membrane. Some studies have found that the viscous cytoplastic nature meant that the formation of these pores do not affect the nucleus. In contradiction to this one in vivo study has demonstrated that targeted ultrasound microbubble destruction resulted in nuclear uptake of rhodamine-labelled (Duvshani-Eshet et al., 2006).

It has been proposed that the ultrasound focused destruction of microbubbles increases the capillary permeability at the cellular microvascular level. The implication is that permeability can be enhanced active compounds can be enhanced. The reciprocal undesirable effect is that the application of focused-ultrasound microbubble destruction can result in local haemorrhage (Tinkov et al., 2009)

5. Drug loading

Manipulation to the microbubble formulation alters their physiochemical characteristics significantly, particularly the shell volumes and widths. There are four classes of microbubble structures with the capacity as gene and drug carriers. Potentially higher drug loads can be loaded into acoustically-active lipospheres (shell thickness [triacetin layer 500-

1000 nm], [soya-bean oil layer 300-700nm]), microcapsules(usually emulsification method, shell thickness 50-200 nm) and protein-shelled microbubbles (HSA-shelled microbubbles, probe-type sonication method, thickness 200-300 nm) possess larger shell volumes to which active products can be embedded. Whereas, Phospholipid-microbubbles (shell thickness 2-3 nm), because they have smaller shell volumes have better acoustic properties (Tinkov et al., 2009).

The most common approach for drug-loading in phospholipid and protein microbubbles is for shell surface loading. The loading process of the active agents within the entire volume of the shell is appropriate for microcapsules and on occasion protein composite microbubbles.

Microbubble drug coupling can be achieved through a number of the modalities. In essence microbubbles are gas-filled colloidal particle, which comprises a surfactant wrapping a flexible protein and polymer shell.

The outer shell surface can be used as a foundation by which active products can be attached. Drug compound attachment of the microbubble composite polymer shell can be utilise van-der-waal, hydrophobic or electrostatic forces. The association of active agents can be achieved through the association of smaller secondary shell surface anchor/carrier particles proteins or complexation to adhesive human serum albumin molecules. The nucleic acids, DNA and RNA can be attached to the outer microbubble shells with charge coupling.

Physical encapsulation in biodegradable polymeric shells offers the advantage of facilitation of the appropriate elimination from the systemic circulation. These biocompatible shells can consist of compounds such as gelatin, and may reduce immunogenicity and unfortunately ligand affinity.

Hydrophobic drugs molecules incorporated within can be submerged within the oily layer of lipospheres or intercalated between the monolayer of phospholipids.

A 'phase-shift' colloid emulsion vehicle is an elaborate system consisting of a perfluoropentane microemulsion system stabilised by biodegradable surfactants. This colloid emulsion is then loaded with the active compound possessing the appropriate physiochemical properties. The droplet-colloid composition is phase-shifted to bubble form through the application of heated sonication (Rapport et al., 2007).

6. Co-administration drug targeting

The drug loading of microbubbles is not essential to glean the benefits of their potential drug targeting properties. Administration of microbubble agents can be accompanied by the active product. Ultrasonic beam application at the target sites may facilitate the entrance of otherwise circulating active product into the interstitial compartment and into the intracellular compartment. Further study is necessary to study the therapeutic modality of this approach (Tinkov et al., 2009).

7. Targeted microbubble production

The process production and formulation begins with the identification of an amenable disease and correlated target disease specific receptor. A therapeutic mechanism and active compound are prerequisites to ensure the efficacy of any potential enhancement with microbubble targeting. Comprehensive understanding of pharmacological, pharmacodynamics, physiochemical and dosages are essential to ensure realistic and cost-effective production of clinically applicable microbubble compounds. The active compounds must be amenable to the microbubble formulation process, which in many instances may denature and degrade biological compounds (e.g peptide, nucleic acid and monoclonal antibodies) and many synthetic compounds. Highly thermodynamic and immunogenical components be they active compounds or ligands are unsuitable for incorporation into microbubble formulations. For several molecular therapeutic compounds the impact of the formulation and manufacture process makes targeted microbubble modalities non-feasible.

The production process must incorporate uncompromising quality control based on sound reproducible manufacturing practices and in-depth understanding of both components and the required end-product.

Complexation of the active products to the shell utilises the relative physiochemical characteristics of the shell and the active product. Non-covalent bonding to the shell ensures appropriate strength to their loading and for ultrasonic liberation. Electrostatic charged albumin and phospholipids can be coupled to charged active products.

7.1 Phospholipids

Semi-synthetic flexible phospholipid thin monolayer possesses suitable acoustic properties. When present in solution the microbubble the hydrophobic portion orientated internally towards the gaseous centre and the electrostatic hydrophilic elements coming in continuity with the aqueous solvent. The condensation of the saturated fatty acid chains portions of the phospholipids structure imparts integral stability to the gel micelle shells.

There is a wide spectrum of the phospholipid compositions, including unsaturated fatty acid chains. Amalgamation of certain phospholipids and excipients can drastically alter the microbubble stability. An example, includes the integration of unsaturated chained phospholipids into the shell significantly reduces the stability of the microbubble shells. Alternatively, the steric stabilisation can be achieved with the inclusion of PEGylated phospholipid.

Other phospholipid compounds or phospholipid adjuvants impart an extended range of characteristics to these microbubble vectors. The inclusion of PEGylated or non-bilayer phospholipids can prolong circulation and gene payload transfection, respectively. Alteration of the condensed and disordered phase domains within the phospholipid layers changes the miscibility of components within the shell, with a result in its characteristics.

The heterogeneity of constituent components changes the degree of phase separation, which itself alters the capacity for drug/ligand loading, stability, and acoustic characteristics. Heterogeneity of phase separation can confer desirable characteristics to the microbubble formulations. One such advantage imparted by the incorporation of brush moieties is through the steric protection they can provide. The major disadvantage is the threat to formulation quality, which may be the result of phase separation in the lateral orientation. It has been suggested that the degree lateral phase separation may be the limited through adjustment of the fabrication process and through manipulation of the shells composition.

The condensation state is a crucial in determining the *in vivo* stability and is reliant upon on the temperature. To ensure that an agent has sufficient in vivo stability the physiological body temperature must be below the threshold phase-transition temperature (Borden et al., 2005).

When considering phospholipid constructs the stability is very dependent upon the hydrophobic carbon chain lengths. Longer carbon chained microbubble phospholipid shells possess less surface tension and an increased tendency to resist gaseous permeation. Increased viscosity with enhanced durability, but reduced acoustic characteristics with diminished echogenicity (Duncan P.B. & Needham D. 2004).

The ideal compromise to ensure optimal phospholipid microbubble stability and the existence of the minimal phase-shift characteristics, which are achieved by ensuring the equivalence of the various constituents phase-transition temperatures and phospholipid fatty-acid residue lengths. Avoidance of unsaturated phospholipid components possessing a large degree of conformational liberty would also improve stability. These 'loose' phospholipids destabilise the order of the otherwise densely packed monolayer. Microbubble monolayer destabilising and disorganising abundant surface electrostatic charges should also be avoided because they induce electrostatic repulsions within a layer orientation (Borden et al., 2006).

Non-conventional production of the phospholipid can be undertaken using a flow-focused method, which encompasses the passage of the inert gaseous core and shell material through a fine nozzled device and into a water reservoir.

7.2 Polymeric

In terms of polymeric microcapsule formulations characteristics are dependent upon material biocompatibility polymeric component weight, which possesses a direct relation to the shell thickness. Variability to the polymer constituents influences the stability, echogenic, time to destruction and acoustic characteristics (Forsberg et al., 2004).

Inappropriate selection of the incorporated of Proteinaceous excipients into the microbubble shell can also embarrass the functional stability of therapeutic microbubble formulations. Exposural factors, which may disrupt the self-association forces between shell components, which must be mitigated against, include heat and free-radicals. Adhesive shell building factors include thiol-rich proteins including HSA. These forces are imparted through the sonication of the shell constituents' results in intensive cavitation and free-radical production, which result in the production of HSA linking thiol bridges. These bridges can also be formed with the addition of glutaraldehyde and formaldehyde during the shell formation process (Tinkov et al., 2009).

7.3 Mechanical agitation

The most frequently utilised method for the production of acoustically active lipsomes and phospholipid-shelled microbubbles. The initial phase of the process involves the formation of a thin hydrated phospholipid sheet, which is then either infiltrated with an alcoholic component or phase inverted by an alternate approach. A gaseous component is then pressed in the space above the liposomal dispersion and then agitated at least several thousand oscillations every second. Loading of the active agents can take place at initial microbubble production or after this second gaseous core forming stage. To ensure a predictable desire loading pattern it essential to take steps to ensure loading of the active molecular compounds are bound to the liposomal shell instead of encapsulated within the water soluble liposomal centre (Tinkov et al., 2009).

The key to the production of the acoustically active liposomes revolves around the production of the appropriate micro-emulsion. This can be undertaken through the combination of the perpetual aqueous phase and injectable oil phase. Suitable constituents of the injectable oil phase include castor or triacetin oil. Successful emulsion formation requires the incorporation of co-solubilizers, such as synthetic block copolymers, because of the defoaming characteristics of the oil phase. The adequately lipophilic agent can then be dissolved and remain persistently encapsulated within the oily phase and acoustically active liposomes (Fang et al 2007)

This versatile modus operandi for the production of therapeutic microbubbles is advantageous when integrating delicate targeting ligands and drugs molecules. Successful utilisation of this drug-loading approach requires meticulous control of important variable production parameters. Components concentrations, viscosity, Agitation time and temperature must be all thoroughly controlled to ensure predictable characteristics.

7.4 Emulsification

Some phospholipid-shelled microbubble and many polymer shelled microcapsule formulations are produced with this oil-in-water freeze-drying emulsification production method. Some of the commonly used later generations of contrast enhanced ultrasound agents have been manufactured in this manner. The inner emulsion phase consists of a volatile solid compound (such as camphor), an organic solvent phase, which has the characteristic of being lyophilizable water-immiscibility and polymeric shell material. The water-water-insoluble benzene compound based solvents, such as toluene or p-xylene. The polymeric shell material consists of compounds similar to the highly biodegradable and biocompatibile co-polymer poly-lactide-co-glycolide (PLGA). The emulsion matrix is the residual after the removal of the volatile aqueous and organic phases remaining after the freeze-drying process. The gaseous cores are formed after reconstitution of the gaseous vial and emulsion matrix with the injection solvent medium. Coverage of these microbubbles with gelatin or albumin compounds included in the aqueous phase of this emulsification production method may significantly improve their biocompatibility profile (Schneider et al., 1991; El-Sherif et al, 2003; Short 2005).

Lipophilic active products can be easily accommodated within organic emulsion phase of polymeric shelled microbubbles compounds. Water-soluble drugs and biological agents can be accommodated in these formulations through encapsulation. Incorporation of hydrophilic compounds within the mixed-phase emulsion is undertaken through their dispersion within the shell containing the organic phase. The next step of the process involves the production of a double emulsion, which consists of water in organic in water emulsion. The lypophilization of the double emulsion is the final synthetic mechanism resulting in hollow shell drug-loaded microbubbles. The important factors that need to be controlled include the emulsion droplet sizes; quantity and molecular weight of the shell material monomers (Tinkov et al, 2009).

7.5 Probe-type sonication

Low frequency and high intensity ultrasound is the commonest technique for the production of protein-shelled microbubbles and less commonly phospholipid microbubbles. Ultrasound induces exuberant convection streams to disperse the inert gaseous phase into the aqueous phase and produces the microbubble gaseous core. The process generates free radicals and temperature rises in excess of seventy degrees centigrade, which itself produces insoluble-shell microbubble shells. The proteins can be denatured with unveiling of their thiol-groups with the subsequent formation of covalent and non-covalent bridging bonds. Phospholipid microbubbles do not benefit from higher temperatures as their exceeding their phase-transition temperature can be deleterious to the formulation.

The most commonly utilised shell building component was HSA because of its exceptional affinity to a wide spectrum of active agents, including nucleic acid products. Drug loading to only the outer shell can be undertaken with the cell surface loading approach. This involves the incubation of the active consignment compounds to composed microbubbles. The disadvantage of this approach is a relatively low concentration of active product can be achieved through this process of surface absorption. This limitation can be overcome by the inclusion of the compound throughout the depth of the shell. This can be achieved through the application of sonication to the shell active product mixture prior to the microbubble formulation. Fragile active compounds may not be amenable to the entire shell loading approach as they will be damaged by the harsh cavitation and chemical and thermal stresses.

7.6 Spray-drying method

This versatile approach is essentially the formation of hollow particles through a spray-drying process. Encapsulation with the spraying technique is commonly utilised for the production conventional microbubbles. It provides an opportunity for a reliable method by which drug-loaded polymeric- and protein-microbubbles can be manufactured. This spray-drying method provides drug volume loading, a stable dried product and a final composition produced in favourable conditions.

Phospholipid-, protein- and polymer- microbubble formation utilises volatile ammonium derivatives and organic liquids (e.g. halogenated hydrocarbons) to induce pores and cavities in spray-dried particles. The commonly utilised HSA components of microbubble shells can be thermally stabilised and undergo chemical, but not free-radical cross- processes. The premise of this approach is the application of an evaporation process to a solvent mixture containing the dissolved shell material. The resultant saturation of the air-water interface with solvent-mixture forms an elastic film layer. On the droplets surface the shell components within the film solidifies, whilst the associated solvent completely evaporates. The particle formation is determined by the pressure induced fluid conversion into solid. The optimal acoustic characteristics are exhibited by microbubbles with the largest possible core cavity volumes. This spray-drying technique does not result in a uniform population of microbubble particle characteristics. The core cavity sizes can vary dramatically from exceptionally large cavities with fragile shells to multiple small voids (Porter 1997; Lentacker et al., 2007).

8. Stability

The shell is fundamental for microbubble stability is maintains the integrity of the gaseous core by preventing the leaching of the gaseous core and minimising interface tensions. There is a complex gaseous balance reached between the solvent saturated gases and the gaseous core content. The passage of gas occurs from the core into the plasma solvent and simultaneously from the solvent into the gaseous core. This continues until the gaseous partial pressure equilibrium has been achieved. When the net gaseous movement is static the equilibrium osmotic gradient is zero. The aqueous insoluble Perfluorocarbon microbubbles vivo have negative osmotic gradient, so initially tend to increase in size because the net passage of aqueous dissolved gases into the gaseous core. Formulation techniques to improve stability include the incorporation of mixed gaseous cores, such core have been termed osmotically stabilized. Specifically engineered shells may reduce the degree of gaseous passage, but will be unlikely completely ameliorate it. Alteration of the gaseous core constituents will prove a more robust modification (Tinkov et al., 2009).

Air-filled microbubbles consisting of proteineous shells possess inferior echogenicity characteristics than fluorinated gas core microbubbles. The water insoluble perfluorobutane and sulphur hexafluoride consisting gaseous core microbubbles confer some superior properties to more advanced microbubble generations. Perfluorocarbon containing third generation microbubbles are disadvantaged by their tendency to absorb blood containing gases resulting in an inconsistent size profile and potentially reduced stability due to swelling. Perfluorocarbon and nitrogen mixtures may fare better and possess inherently superior pharmacodynamics properties (Tinkov et al 2009).

9. Quality control

Quality assurance requires a detailed understanding of microbubble properties. The formulation characteristics, which complicate analysis, revolve around the dynamic nature of microbubbles and their delicate quiescent phases. Their complicated characteristics include their unpredictable buoyancy and excessive sensitivity to shear-stress, pressure, temperature. These drug-loaded microbubbles are relatively complex composite mixture of molecular interacting structures, which can behave in a broad manner at any one time. Despite many of these unique properties many conventional assays can be utilised in an efficacious manner. Specific analytic tests are necessary to characterise therapeutic microbubble destructibility and echogenicity (Tinkov et al 2009).

9.1 Size

The terminological dogma underlying therapeutic microbubbles dictates size limitations of between 500 nm and 1μm in size. This size ranges enables optimal compromise between drug-loading in the gaseous core and echo signal. Nanobubble preparations have a greater predilection to take advantage of the permeability and retention effect. Nanoparticles more easily cross leaky capillary walls. Nanobubble units can have a tendency to coalescence into a microbubble conformation, thus ameliorating this advantageous property of the smaller bubbles (Tinkov et al 2009).

Acoustic activity, pharmacodynamics and pharmacokinetics of microbubble preparations are highly dependent upon their size distrubtion. A Broad, mult- or bimodal distribution rather than a more conventional normal distributions have been found. The upper size limit should be between 5 to 10μm to enable safe clearance through the pulmonary vasculture. The size distribution can be sized and calculated by using electromagnetic impedance field calculation. As the particles pass through the 'coulter counter's' miniscule apertures located between two the complementary electrode the reading are produced. The detection and measurement of microbubble shadows cast as microbubble particles passage before a narrow uniform light source, is termed the 'light obscuration' method. The size range of this technique is has a maximum size tolerance of 400μm and requires low concentrations to be accurate. Laser diffraction approach can measure in the presence of high concentrations, is unaffected by buoyancy and can accurately assess any population distribution. Its particle size tolerance is 40-2 mm and requires calibration/optimisation for individual compounds. The dynamic light scattering method relies upon the 'Brownian motion of particles' is seldom utilised due its unreliability (Tinkov et al 2009).

9.2 Zeta potential

The zeta potential has important ramifications to both the intrinsic physiochemical characteristics of microbubble formulations and molecular pharmacokinetics. The active-product loading capacity, stability of the colloidal dispersion and the interaction with the endothelial cell wall properties can be usefully represented by the zeta potential behaviour. Preferable zeta potentials may significantly extend capillary retention time duration (Fisher et al., 2002).

The electrophoretic mobility can be determined by the degree of laser light scattering, which results from particle oscillation induced by an alternating electrical field. Using mathematical modelling the electrophoretic potential be translated into the zeta potential. The accuracy and precision of this presumptive calculation can be compromised by the inherent characteristics of microbubbles. The convection associated buoyancy compromises the reproducibility of original electrophoretic observations. Several environmental, physiochemical and electrochemical can affect the calculated zeta potential measurements (Tinkov et al 2009).

9.3 Mechanical properties

Critical to the modality of microbubbles is their characteristic acoustic behaviour on the application of an ultrasounic field. This behaviour is dictated by the shell characteristics. Advanced in the field of dynamic imaging has enabled the robust characterisation of ultrasound manipulated microbubbles.

The shell hardness plays a disproportionally large role in the determination of the ultrasonic threshold energy microbubble cracking threshold. Upon reaching their reaching their cracking threshold the encapsulating shell wall ruptures and the gaseous bubble is released instantaneously leaving a partially maintained shell. Harder shells require in the order of twice to twenty-fold the ultrasonic energy to rupture and liberate is payload, in comparison to softer shelled formulations. These necessarily high energy levels and more pronounced harder-shell microbubble oscillations may have a deleterious effect upon the adjacent cells (Postma et al., 2004; Tinkov et al., 2009).

Soft more pliable shells, such as phospholipid shell components oscillate through the process of expansion and contraction. With the expansions lipid aggregates are expelled and then reseals. Significantly lower ultrasonic energies are necessary to rupture softer shells (Sboros et al., 2006; Tinkov et al., 2009)

9.4 Lateral phase separation

Microbubble shells can be significantly compromised by the phenomenon of phase separation. The molecular findings demonstrated within the shell during lateral phase-shift are an uneven distribution of constituents. The differentials in inherent forces include heterogeneous electrostatic binding and steric-shielding binding forces. Quality assurance methods are essential to ensure microbubble stability. Fluorescent dye tagged microbubble laser scanning confocal microscopy can employed to characterise the quality of dispersion and drug-loading (Borden et al., 2006).

9.5 Chemical integrity

Novel and broad characterisation techniques are necessary to ensure quality control of payload compounds incorporated within microbubble formulations.

Reversed-phase and normal-phase liquid chromatography are useful tools for quality assurance analysis of smaller molecular therapeutic compounds and small phospholipid compounds respectively (Lentacker et al., 2006; Hvattum et al., 2006).

10. Safety

There is relatively little clinical experience of the microbubble agents as a result the extent and spectrum of adverse events has not yet completely revealed themselves. Thus far, the clinical use of these agents as largely contrast ultrasound agents has proven to be safe. The complicity of the therapeutic adjuncts will probably introduce further adverse effects.

Myocardial exposure to microbubble agents may induce a physiological response of premature cardiac contractions. Microvascular damage, pectechial haemorrhage, free radical damage and single strand DNA fractures are also postulated side-effect of the focused ultrasound microbubble destruction process. In clinical practice these complications have been seldom encountered.

In regards to the reasonable ultrasound energies feasible for focused therapeutic ultrasound microbubble destruction, 'the principle of as low as reasonably achievable' has been deemed appropriate by radiological committee guidelines. For clinical ultrasound of microbubble agents the accepted safe mechanical index parameters are 0.05 to 0.5. For harmonic and non-harmonic clinical imaging modalities exceeding the mechanical index of 0.5 would result in microbubble destruction. The American Food and Drug

Administration (FDA) have set the maximum accepted mechanical index tolerance of 1.9 (Tinkov et al., 2009).

10.1 Immunogenicity

Although microbubbles themselves are not immunogenic, the incorporation of targeted ligands to their shells does render them immunogenic. Immunoreactive components are desirable to target delivery, but, undesirable for ability to attract unwanted immunological attention whilst in transit to the target site.

Conventional exposed spacer-linked or directly linked ligands loaded upon the gaseous microbubble cores may be inappropriately degraded or trigger an aggressive detrimental immunology response. Buried ligands within an overbrushed grafted polymeric-PEG layer may enable the camouflage of these immunoreactive elements. This stealth layer can then be unveiled and the immunoreactive layer underneath revealed with the application of focused ultrasound beam. The ultrasonic waves push the microbubbles towards the vessel walls and the polymer sheath is unbrushed and enables ligand binding to the specific receptor targeting sites. The caveat of these stealth particle modifications is reduced target receptor site affinity (Kilbanov et al., 1999).

11. Gene-loading microbubbles

Gene-therapies have proven an intriguing, but nonetheless frustrating proposition. Their potentially enduring therapeutic promise has been slighted by some critical inherent deficiencies. The absence of any effective conventional carriers has hindered the progression of more widely clinically applicable nucleic acid therapeutic options. The essential criteria an advanced gene carrier must possess an exceptional safety profile. The carriage of their genetic payload must be hostile and protect it from the otherwise deleterious effects of native enzymes, immunological elements and pharmacokinetic elimination. The aim is the controlled and timely liberation of the active agent followed by appropriate transfection of the target cells. Transfection is mediated by attenuated viruses, thus rendered incapable of replication and instigating disease. Amongst critics of gene-therapy options typical concerns over safety exist as with non-microbubble preparations. These safety concerns pertain to the risks of genetic mutations, pathogenesis and immunological reactions. Work utilising non-viral vectors aimed at excluding these viron-vector related risks has not yielded practical or efficacious alternatives. Alternative delivery systems investigated include microbubble, lipoplexes, electroporation and microinjection.

Genetic therapy delivery mediated through microbubble formulations is widely considered to be a safe prospect. The enhancement in transfected genetic material quantities is significant. In vitro models have demonstrated transfection magnitiude enhancement rates of 10-3,000 fold in comparison to naked plasmid DNA constructs alone. Despite the observation under in vivo conditions, which have demonstrated 1,000 fold transfection improvement, true proven in vivo transfection enhancement remains elusive. Serum nucleases are pose a significant risk upon naked and viron loaded nucleic acid compounds. The shielding effect of microbubbles has been proposed in microbubble constructs loaded with adenovirus-associated microbubbles (Tinkov et al., 2009).

Unlike other therapeutic opportunities the response to therapy rarely correlates with the quantity delivered. It is essential however, to deliver a threshold quantity to the appropriate area. The limited surface absorption binding capacity of plasmid DNA to produce maximally loaded gene loaded microbubble is in the order of approximately 0.001-0.005 pg/μm^2. Adjuvants can be utilised to increase the gene binding and loading properties. These include the use dipalmitoylphosphatidylethanolamine (DPPE) containing phospholipid microbubbles, which have a higher binding capacity to plasmid DNA. Layer-by-layer technique with poly-L-lysine may enable pDNA packing to decimal points higher than conventional microbubble loading. The absorbed loading of genes on to albumin microbubbles has been found not to enhance pDNA loading. Utilising a formulation approach a greater than 200-fold loading to entire shell volume has been reported. More recent developments are secondary-carriers associated microbubbles and the active nucleic acid components. These constitute nanoparticles, polyplexes, liposomes and lipoplexes. They have multiple functions, including enhancement of the transfection and microbubble strong shell attachment to the microbubble shell with improved loading capacity. They function through the condensation of large DNA molecules, thus protecting them from the action of serum nucleases. Entrance into the intracellular compartment involves both endocytosis and pinocytosis (Frenkel et al., 2002; Tinkov et al., 2009).

In vivo microbubble transfection delivery studies have been undertaken in organ systems already validated for diagnostic contrast ultrasonography. Vascular paradigms are particular relevant to study therapeutic microbubble compounds. Numerous pathological vascular paradigms have indicated that the duration of action varies from condensed intensive therapeutic windows to the extremes of a month. No meaningful conclusions can be drawn from study of these gene-therapy microbubble therapies for atherosclerotic or intimal disease. Ultrasound induced sonoporation and of the microbubble containing genetic-material enables delivery into the cytoplastic compartment. Nanobubble delivery does not seem to enhance entry into the nuclear compartment resulting in its deactivation/destruction prior to activity (Newman & Bettinger 2007; Liu et al., 2006; Unger et al., 2006).

Elucidation of formulation, pharmacodynamics and pharmacokinetic principles is necessary to definitively ascertain the modality of this novel disease-specific therapeutic option. Establishing such studies are complicated by the absence of standardised models with comparable measurable parameters. Also more modern genetic therapy incorporated microbubble formulations have not been studied to the same extent as co-administration preparations. This leaves the door ajar for more modern formulations fulfilling their yet unrealised potential (Tinkov et al., 2009)

12. Conclusion

Safe *in vivo* utilisation of microbubble contrast enhanced ultrasound agents is well established and validated. The idea of directly translating microbubble into routine used therapeutic treatments is oversimplification of the challenge ahead. The premise of diagnostic microbubble is with low administrated doses to produce a sustained and clearly visible echogenic signal. Whereas, with therapeutic microbubbles the all the associated paradigm variables still remain too uncertain. The quantities of both ultrasonographic energy and microbubble dosage are many magnitudes greater, which imparts numerous complications. The most serious of these is significant unintentional tissue damage related to cavitation-inducing ultrasound beams. A pattern of cellular damage has been demonstrated on myocardial tissue between the upper safe ultrasonic energy limit and 2.5 times of this limit. Such a degree of irreversible cellular destruction will prove detrimental to organ function and subsequently result in significant morbidity. Despite these concerns, it has been proven that efficacious microbubble therapeutics used at the effective dosages with safely ultrasonography energies.

Further studies on all aspects of formulation, production and trials of therapeutic microbubbles are necessary as a precursor to establishing these agents as realistic therapeutic options. The promise in their therapeutic role may lie as co-administration agents to optimise selective uptake of the active agents. Therapeutic microbubbles may prove the 'missing-link' in fixing the inherent shortfalls of promising therapies, particularly in reference to gene therapy.

13. References

Barnett S. (1998). Nonthermal Issues: Cavitation-Its nature, detection and measurement. *Ultrasound in Medicine and Biology*, Vol 24, Suppl 1 pp 11-21. PMID:9841460

Bloch S.H., Dayton P.A., Ferrara K.W. (September-October 2004). Targeted Imaging using Ultrasound Contrast Agents. *IEEE Engineering in Medicine and Biology Magazine*. Vol 23, No. 5, pp 18-29. PMID:15565796

Borden M.A., Kruse D.E., Caskey C.F., Zhao S., Dayton P.A. & Ferrara K.W. (November 2005). Influence of lipid shell physicochemical properties on ultrasound-induced microbubble destruction. *IEEE Transaction in Ultrasonics, Ferroelectrics, and Frequency Control*, Vol 52, No. 11, 1992-2002. PMID:16422411

Borden M.A., Martinez G.V., Ricker J., Tsvetkova N., Longo M., Gillies R.J., Dayton P.A. & Ferrara K.W. (April 2006). Lateral Phase Separation in Lipid-coated Microbubbles. *Langmuir. the ACS Journal of Surfaces and Colloids*, Vol. 22, No. 9, pp 4291-4297. PMID:16618177

Bouakaz A., Tran T.A., Roger S., Leguennec J.Y. & Tranquart F (2006). On the Mechanisms of Cell Membrane Permeabilization with Ultrasound an Microbubbles. *Ultrasound in Medicine and Biology*, Vol 32, pp 90.

Dayton P., Kilbanov A., Brandenburger G. & Ferrara K. (October 1999). Acoustic Radiation Force In-vivo: A Mechanism to Assist Targeting of Microbubbles. *Ultrasound in Medicine and Biology*. Vol 25, No. 8, pp 1195-1201. PMID:10576262

Duncan P.B. & Needham D. (March 2004). Test of the Epstein-Plesset Model for Gas Microparticle Dissolution in Aqueous Media: Effect of Surface Tension and Gas Undersaturation in Solution. *Langmuir: the ACS Journal of Surfaces and Colloids*, Vol 20, No. 7, pp 2567-2578. PMID:15835125

Duvshani-Eshet M., Baruch L., Kesselman E., Shimoni E. & Machluf M. (January 2006). Therapeutic Ultrasound-mediated DNA to Cell and Nucleus: Bioeffects Revealed by Confocal and Atomic-force Microscopy. *Gene Therapy*, Vol 13, No. 2, pp 163-172. PMID16177822

El-Sherif D.M. & Wheatley M.A. (August 2003). Development of a Novel Method for Synthesis of a polymeric Ultrasound Contrast Agent. *Journal of Biomedical Materials Research. Part A*, Vol 66, No. 2, pp 347-355.

Fang J.Y., Hung C.F., Liao M.H & Chien C.C. (August 2007). A Study of the Formulation Design of Acoustically Active Lipospheres as Carriers for Drug Delivery. *European Journal of Pharmaceutical and Biopharmaceutics*, Vol 67, No. 1, pp 67-75. PMID:17320362

Fisher N.G., Christiansen J.P., Kilbanov A.L., Taylor R.P., Kaul S., Linder J.R. (August 2002). Influence of Microbubble Surface Charge on Capillary Transit and Myocardial Contrast Enhancement. *Journal of the American College of Cardiology*, Vol 40, No. 4, pp 811-819. PMID:12204515

Forsberg F., Lathia J.D., Merton D.A., Liu J.B., Le N.T. & Goldberg B.B., Wheatley M.A. (October 2004). Effect of Shell Type on the In-vivo Backscatter from Polymer-encapsulated Microbubbles. *Ultrasound in Medicine and Biology*, Vol 30, No. 10. pp 1281-1287. PMID:15582227

Frenkel P.A., Chen S., Thai T., Shohet R.V. & Grayburn P.A. (June 2002). DNA-loaded Albumin Microbubbles Enhance Ultrasound-mediated Transfection In-vitro. *Ultrasound in Medicine and Biology*, Vol 28, No. 6, pp 817-822.

Gramiak R., & Shah P.M. (September-October 1968). Echocardiography of the aortic root. *Investigative Radiology*, Vol 3, No. 3, pp 356-366. PMID:5688346

Guzman H.R., McNamara A.J., Nguyen D.X. & Prausnitz M.R. (August 2003). Bioeffects caused by changes in Acoustic Cavitation Bubble Density and Cell Concentration: A unified Explanation Based on Cell-to-bubble Ratio and Blast Radius. *Ultrasound in Medicine and Biology*, Vol 29, No. 8, pp 1211-1222. PMID:12946524

Hvattum E., Uran S., Sandeek A.G., Karlsson A.A. & Skotland T. (October 2006). Quantification of phosphatidylerine, phosphatidic acid and free fatty acids in an Ultrasound Contrast Agent by Normal-phase High-performance Liquid Chromatography with Evaporation Light Scattering Detection. *Journal of Pharmaceutical and Biomedical Analysis*, Vol 42, No. 4, pp 506-512.

Kaul S. (October 2010). New developments in ultrasound systems for contrast echocardiography. *Clinical Cardiology*, Vol 20. No. 10 Suppl 1 pp I27-30. PMID:9383599

Kilbanov A.L., Gu H., Wokdyla J.K., Wible J.H., Kim D.H., Needham D., Villanueva F.S. & Brandenburger G.H. (1999). Attachment of Ligands to Gas-filled microbubbles via PEG-spacer and Lipid Residues Anchored at the Interface. *Proc International Symposium Control Rel Bioact Mater*: 26.

Koike H., Tomita N., Azuma H., Taniyama Y., Yamasaki K., Kunugiza Y., Tachibana K., Ogihara T. & Morishita R. (January 2005). An Efficient Gene Transfer Method Mediated by Ultrasound and Microbubbles into the Kidney. *The journal of Gene Medicine*. Vol 7, No. 1, pp108-116. PMID:15515148

Pan H., Zhou Y., Izadnegahdar O., Cui J., Deng C.X. (June 2005). Study of Sonoporation Dynmaics Affected by Ultrasound Duty Cycle. *Ultrasound in Medicine and Biology*. Vol 31, No. 6, pp 849-856. PMID:15936500

Lawrie A., Brisken A.F., Francis S.E., Wyllie D., Kiss-Toth E., Qwarnstrom E.E., Dower S.K., Crossman D.C. & Newman C.M. (October 2003). Ultrasound-enhanced Transgene Expression in Vascular Cells is not Dependent upon Cavitation-induced Free Radicals. *Ultrasound in Medicine and Biology*. Vol 29, No. 10, pp 1453-1461. PMID:14597342

Lentacker I., De Smedt S.C., Demeester J., Van Marck V., Bracke M., Sanders N.N. (2007). Lipoplex-loaded Microbubbles for Gene Delivery: A Trojan Horse Controlled by Ultrasound. *Advanced Functional Material*, Vol. 17, pp 1910-1916.

Liu Y., Miyoshi H. & Nakamura M. (August 2006). Encapsulated Ultrasound Microbubles: Therapeutic Application in Drug /Gene Delivery. Journal of Control Release, Vol 114, No. 1, pp 89-99. PMID:16824637

Marmottant P., Hilgenfeldt S. (2003). Controlled Vesicle Deformation and Lysis by Single Oscillating Bubbles. *Nature*, Vol 423, No. 6936, pp152-156. PMID:12736680

Miller D.L. & Quddus J. (May 2000). Sonoporation of Monolayer Cells by Diagnostic Ultrasound Activation of Constrast-agent Gas Bodies. *Ultrasound in Medicine and Biology*, Vol 26, No. 4, pp 661-667 (7). PMID:10856630

Miller D.L., Pislaru S.V. & Greenleaf J.F. (November 2002). Mechanical DNA Delivery by Ultrasonic Cavitation. *Somatic and Cell Molecular Genetics*, Vol 27 No. 1-6, pp 115-134. PMID:12774945

Miller D.L. & Thomas R.M. (1993). A comparison of haemolytic and Sonochemical Activity of Ultrasonic Cavitation in a Rotating Tube. *Ultrasonic in Medicine and Biology*, Vol 19, No. 1, pp 83-90. PMID:8456532

Newman C.M. & Bettinger T. (March 2007). Gene Therapy Progress and Prospects: Ultrasound for Gene Transfer. Gene Therapy. Vol 14, No. 6, pp 465-475. PMID:17339881

O'Brien W.D. (January-April 2007). Ultrasound-Biophysics Mechanisms. Progress in *Biophysics and Molecular Biology*. Vol 93, No. 1-3, pp 823-829. PMID:16934858

Pitt W.G., Husseini G.A. & Staples B.J. (November 2004). Expert Opinion on Ultrasound Drug Delivery _ A General Review. *Expert Opinion on Drug Delivery*, Vol 1, No. 1, pp 37-56. PMID:16296719

Porter T.R. (1997). Perfluorobutane Ultrasound Contrast Agents and Methods for its Manufacture and Use. Board of Regents of the University of Nebraska. *United States of America Patents*, US 5,567,415.

Postma M., Van Wamel A., Lancee C.T. & De Jong N (June 2004). Ultrasound-induced Encapsulation Microbubble Phenomena. *Ultrasound in Medicine and Biology*, Vol 30, No. 6, pp 827-840. PMID:15219962

Rapoport N., Gao Z., Kennedy A. (July 2007). Multifunctional Nanoparticles for Combining Ultrasonic Tumor Imaging and Targeted Chemotherapy. *Journal of the National Cancer Institution*, Vol 99, No. 14 pp 1095-1106.

Sboros V., Glynos E., Pye S.D., Moran C.M., Butler M., Ross J., Short R., McDicken W.N. & Koutsos V. (April 2006). Nanointerrogation of Ultrasonic Contrast Agent Microbubbles Using Atomic Force Microscopy. *Ultrasound in Medicine and Biology*, Vol 32, No. 4, pp 579-585.

Schlicher R.K., Radhakrishna H., Tolentino T.P., Apkarian R.P., Zarnitsyn V. & Prausnitz M.R. (June 2006). Mechanism of Intracellular Delivery by Acoustic Cavitation. *Ultrasound Medicine and Biology*, Vol 32, No. 6, pp 915-924. PMID:16785013

Schneider M., Bichon D., Dussat P. & Puginier J. & Hybl E. (1991). Ultrasound Contrast agents and Methods of Making and Using them. *Patent* WO 91/15244.

Shi M., Wosnik J.H., Ho K., Keating A. & Shoichet M.S. (2007). Immuno-polymeric nanoparticles by Diels-Alder Chemistry. *Angewandte Chemie International Edition*, Vol 46,No. 32, pp 6126-6131

Short R.E. (2005). Method of Preparing Gas Filled Polymer Matrix Microparticles Useful for Echographic Imaging. *United States of America Patents Office*. Patent No. US 6,919,068 B2.

Siwak D.R., Tary A.M. & Lopez-Berestein G. (April 2002). The potential of drug-carrying immunoliposomes as anticancer agents. Commentary re: J. W. Park et al., Anti-HER2 immunoliposomes: enhanced efficacy due to targeted delivery. *Clinical Cancer Research.*, 8: 1172-1181, 2002. Clinical Cancer Research Vol 8, No. 4, pp 955-956.

Shohet R.V., Chen S., Zhou Y.T., Wang Z., Meidell R.S., Unger R.H. & Grayburn P.A. (June 2000). Echocardiographic Destruction of Albumin Microbubbles Directs Gene Delivery to The Myocardium. *Circulation*, Vol 101, pp 2554-2556 PMID:10840004

Skyba D., Price R., Linka A., Skalak T., & Kaul S. (July 1998) .Direct In-vivo Visualisation of Intravascular Destruction of Microbubbles by Ultrasound and Its Local Effects on Tissue. *Circulation*, Vol 98, Nov 4, pp 290-293. PMID:9711932

Tinkov S., Berkeredjian R., Winter G., & Coester C. (June 2009). Microbubbles as Ultrasound Triggered Drug Carriers. *Journal of Pharmaceutical Sciences*, Vol 98, No. 6, pp 1935-1961. PMID:18979536

Unger E.C., McCreey T.P., Sweitzer R.H., Shen D. & Wu G. (June 1998). In-Vitro Studies of a new Thrombus-specific Ultrasound Contrast Agent. *American Journal of Cardiology*, Vol 81, No. 12A, pp 58-61G. PMID:9662229

Vannan M., McCreery T., Li T., Han Z., Unger E., Kuersten B., Nabel E. & Rajagopalan S. (March 2002). Ultrasound-mediated Transfection of Canine Myocardium by Intravenous Administration of Cationic Microbubble-linked Plasmid DNA. *Journal of the American Society of Echocardiography*. Vol 15, pp 214-218. PMID:11875383

Voci P., Bilotta F., Merialdo P. & Agati L. (July-August 1994). Myocardial contrast enhancement after intravenous injection of sonicated albumin microbubbles: a transesophageal echocardiography dipyridamole study. *Journal of the American Society of Echocardiography*. Vol 7, No. 4. Pp 337-46. PMID:7917341

Wang Z.Q., Pecha R., Gompf B. & Eisenmenger W. (1999). Single Bubble Sonoluminescence: Investigations of the Emitted Pressure Wave with Fiberoptic Probe Hydrophone. Physical Review E 59:1777-1780.

Wei W., Zheng-Zhong B., Yong-Jie W., Qing-Wu Z., Ya-Lin M. (December 2004). Journal of Ultrasonic Gene Delivery and Safety on Cell Membrane Permeability Control. *Journal of Ultrasound in Medicine*. Vol 23, No. 12, pp 1569-1582.

12

Advanced Core-Shell Composite Nanoparticles Through Pickering Emulsion Polymerization

Lenore L. Dai
Arizona State University
USA

1. Introduction

Solid particles have been identified as a new type of emulsifying agent in addition to surfactants and amphiphilic polymers since the pioneer studies by Ramsden in 1903 (Ramsden, 1903) and Pickering in 1907 (Pickering, 1907). Such emulsions are later on named as Pickering emulsions. In Pickering emulsions, solid particles of intermediate wettability in the size range from several nanometers to several micrometers attach to liquid-liquid interfaces and provide emulsion stability. Recently, there has been growing interest in Pickering emulsions because they open new avenues of emulsion stabilization and have numerous practical applications. For instance, we have studied the fundamentals of particle assembly in Pickering emulsions (Dai et al., 2005; Tarimala and Dai, 2004) , utilized them as templates to investigate the dynamics of particles (Tarimala et al., 2004; Tarimala et al., 2006), and developed microrheology at liquid-liquid interfaces (Wu and Dai, 2006; Wu and Dai, 2007; Wu et al., 2009). In this chapter, we further apply the concept of Pickering emulsions to synthesize core-shell composite nanoparticles.

Organic-inorganic composites are vital in biological, medical, and chemical applications. Among them, core-shell composite nanoparticles are a unique class of materials which are attractive for wide applications. It is worthwhile to note that the composite nanoparticle structure in this study is opposite to the often reported core-shell structure in which inorganic particles serve as the core and polymer serves as the shell; here the polymer serves as the core and the inorganic particles serve as the shell. Such materials provide a new class of supramolecular building blocks and can "exhibit unusual, possibly unique, properties which cannot be obtained simply by co-mixing polymer and inorganic particles."(Barthet et al., 1999) In comparison with the recently reported methods to synthesize core-shell composite particles, for example, post-surface-reactio (Ding et al., 2004; Lynch et al., 2005), electrostatic deposition (Dokoutchaev et al., 1999), and layer-by-layer self-assembly (Caruso, 2001; Caruso et al., 1999; Caruso et al., 2001), we synthesize core-shell composite nanoparticles through a novel Pickering emulsion polymerization route (Ma and Dai, 2009; Ma et al., 2010). Figure 1 illustrates the polymerization route and its comparison with the conventional emulsion polymerization.

Pickering emulsion polymerization is superior in several aspects: (1) no sophisticated instrumentation is needed; (2) a commercialized nanoparticle powder or solution can be used without further treatment; (3) the synthesis can be completed in one-step; and (4) the produced particle dispersion is surfactant-free. Despite of these advantages, efforts to explore and utilize this approach are limited, although pioneer explorations have been initiated some approaches including miniemulsion polymerization (Bon and Colver, 2007; Cauvin et al., 2005), dispersion polymerization (Schmid et al., 2006; Schmid et al., 2007; Yang et al., 2008) inverse suspension polymerization (Duan et al., 2009; Gao et al., 2009), and inverse emulsion polymerization (Voorn et al., 2006) stabilized by fine solid particles.

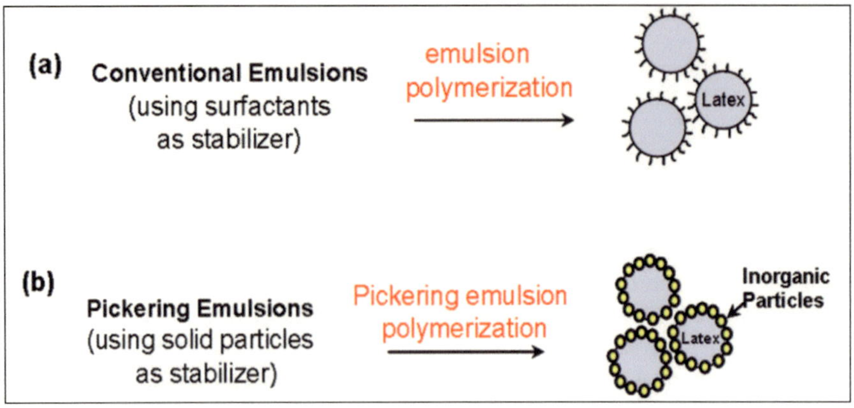

Fig. 1. Comparison between (a) conventional emulsion polymerization vs. (b) Pickering emulsion polymerization.

2. Pickering emulsion polymerization and its possible mechanisms

The general scope of Pickering emulsion polymerization is similar to that of emulsion polymerization, with the exception of using solid nanoparticles as emulsion stabilizers. Here we employ polystyrene (PS)-silica as a model system to illustrate its synthesis and explore possible mechanisms.

2.1 Synthesis and characterization of polystyrene-silica core-shell composite particles through Pickering emulsion polymerization

2.1.1 Materials and synthesis

IPA-ST silica solution, obtained from Nissan Chemicals, is 10–15 nm silica nanoparticles dispersed in 2-isopropanol. The silica concentration is 30–31% by weight. Nonionic azo initiator VA-086 (98%, 2,2-azobis(2-methyl-*N*-(2-hydroxyethyl)propionamide)), styrene monomer (99.9%), and HPLC grade water were purchased from Wako Chemicals, Fisher Scientific, and Acro Organics, respectively. All materials were used as received.

First, water, IPA-ST and styrene were agitated mechanically at 600 rpm for 8 min using Arrow 6000 (Arrow Engineering) in an ice bath to emulsify. Second, the emulsion was degassed with nitrogen and kept in nitrogen atmosphere under magnetic stir. When the temperature was raised to 70 °C, the initiator solution was added to start the polymerization. The composite particles were sampled at different time intervals ranging from 3 h to 24 h. Before characterization, samples were washed twice by centrifuging - redispersing cycles using an Eppendorf 5810R centrifuge. In each cycle, the sample was centrifuged at 7000 rpm for 5 min, the supernatant was replaced with water and the sediment was redispersed by shaking manually.

2.1.2 Characterization of the polystyrene-silica core-shell composite particles

The synthesized polystyrene-silica core-shell composite particles through Pickering emulsion polymerization were characterized by various techniques. Figure 2(a) is a representative scanning electron microscope (SEM) image of the composite particles sampled at 5 hour reaction time. The roughness of the composite particle surfaces suggests that the particles are covered by silica nanoparticles. The core-shell structure can be clearly observed in the transmission electron miscroscope (TEM) image presented in Figure 2 (b). In many regions, the thickness of the shell is close to the size of one silica nanoparticle (10-15 nm), which may suggest a monolayer coverage. Furthermore, we employed hydrofluoric acid (HF) to dissolve the silica shell which led to the smooth PS core in Figure 2(c). The PS-silica composparticle size and its distribution agree well with the dynamic light scattering (DLS) measurement, as shown in Figure 2(d). Finally, we performed energy dispersive x-ray (EDX) spectrum to confirm that a substantial amount of Si and O exist, as shown in Figure 2(e). Note that the relative intensity of the peak does not necessarily correspond to the true atom ratio in the sample. One of the main reasons is that the penetration depth of the electron beam is unknown. The penetration depth depends on various factors, such as the electron beam voltage, the nature of the sample, and the Au/Pd coating thickness during the sample peparation. The EDX result only provides qualitative information regarding the existence of silica, which composes of Si and O, in the composite particles.

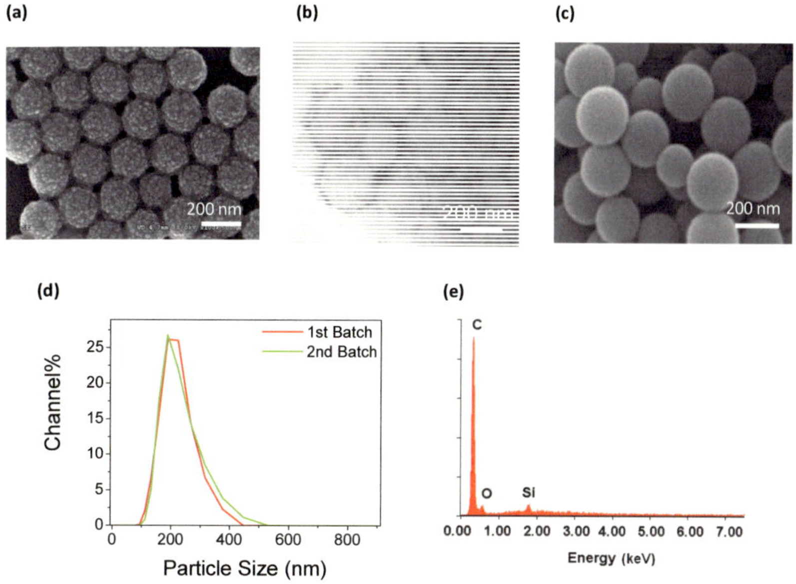

Fig. 2. Polystyrene-silica core-shell composite particles synthesized by Pickering emulsion polymerization: (a) a SEM image; (b) a TEM image of the cross-sectioned composite particles; (c) a SEM image of HF etched composite particles; (c) an overlay of DLS measurements of two batches; (d) an EDX spectrum.

It is worthwhile to note that we carefully selected VA-086 as the initiator. VA-086 is a water-soluble nonionic initiator and no success has been reported in surfactant-free emulsion polymerization of styrene (Tauer et al., 1999). In order to verify the sole stabilizing effect of silica nanoparticles, emulsifier-free emulsion polymerization using VA-086 as the initiator in the absence of nanoparticles was performed. No polystyrene particle formation was observed in the product, evidenced by SEM experiments. These experiments show that the initiator VA-086 has little effect on stabilizing the system in emulsion polymerization and therefore silica nanoparticles are the only source of stabilizer when present. In addition, VA-086 is neutral in charge thus is expected to minimize any electrostatic interactions with the negatively-charged silica nanoparticle surfaces which may complicate identifying silica nanoparticles as the sole stabilizer in emulsion polymerization.

The silica content is quantitatively determined by thermalgravimetric analysis (TGA), as shown in Figure 3. Two samples were measured: the composite particles (solid line) and the composite particles after removal of the silica component by hydrofluoric acid etching, which is essentially polystyrene cores (dashed line). The polystyrene cores show a residual weight of approximately zero at 800 °C. Thus it is reasonable to assume that the major weight loss during heating is associated with the thermo-oxidative degradation of polystyrene and the residue close to 800 °C is solely silica. The silica content of the composite particles is approximately 20 wt%. Although some silica nanoparticles remain in the continuous phase and are washed off by centrifuging-redispersing cycles, the silica content of particles prepared via solid-stabilized emulsion polymerization using nonionic initiator VA-086 is significantly higher than that of particles (1.1 wt%) prepared via dispersion polymerization using nonionic initiator AIBN (Schmid et al., 2007). The improvement is likely due to the distinct polymerization mechanisms. In contrast to the dispersion polymerization in which the polystyrene monomers are dissolved in alcohols, the emulsion polymerization here contains distinguished liquid-liquid interfaces due to the immiscibility between the monomers and the aqueous continuous phase. Therefore the nanoparticles, even in the absence of electrostatic interactions, are thermodynamically favorable to self-assemble and remain at the liquid-liquid interfaces, following the same argument in Pickering emulsions (Dai et al., 2005; Tarimala and Dai, 2004). At the initial

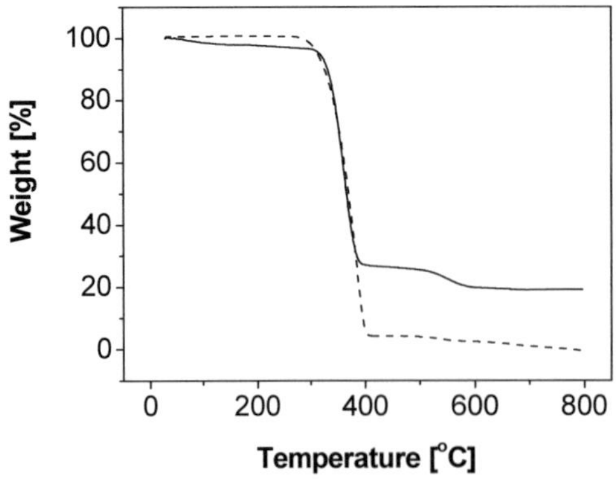

Fig. 3. Thermogravimetric analysis of the PS-silica core-shell composite particle prepared usinVA-086 as the initiator before (solid line) and after (dashed line) HF etching treatment.

stage of polymerization, the nanoparticles provide stability to the monomer droplets. During the nucleation stage, silica nanoparticles are at the interfaces between the monomer phase and continuous phase. It is worthwhile to note that the role of silica nanoparticles described here is not the same as that in the polymerization involving oppositely charged initiator and nanoparticles (Schmid et al., 2007). In the latter case, the initiator molecules or residues adsorb onto the silica nanoparticle surfaces after initiation (Schmid et al., 2007) thus the silica nanoparticles function as the surface-active initiator residue. The mechanism of the core-shell structure formation in Pickering emulsion polymerization will be detailed later on.

2.2 Possible mechanisms of Pickering emulsion polymerization

The mechanism of conventional emulsion polymerization stabilized by surfactants has been under active discussion for over half a century and some consensus has been reached. Harkins proposed three loci of particle nucleation in 1947 (Harkins, 1947), which are later developed into at least three different nucleation mechanisms (Chern, 2006): the micellar nucleation, the homogeneous coagulative nucleation, and the droplet nucleation. Upon initiator addition and decomposition, free radicals form in the aqueous phase. The micellar nucleation (Chern, 2006; Tauer et al., 2008) begins with the capture of free radicals by micelles, proceeds with the continuous swelling and polymerization of monomers in the monomer-swollen particles, and finally terminates with the exhaustion of monomers. While some researchers believe that the micellar nucleation mechanism dominates at a surfactant concentration above the critical micelle concentration, doubts have also been raised (Tauer et al., 2008). In the absence of micelles, the homogeneous coagulative nucleation mechanism is likely dominant. In homogeneous coagulative nucleation (Chern, 2006; Feeney et al., 1984, 1987; Yamamoto et al., 2004), monomers dissolve in water and undergo radical polymerization to form oligomers. The oligomers coagulate to form embryos, nuclei, and primary particles sequentially. These primary particles, stabilized by the adsorption of surfactant molecules, could grow either *via* swelling of particles by monomers or deposition of oligomers onto their surfaces (Yamamoto et al., 2006). Finally, droplet nucleation is another possible mechanism in conventional emulsion polymerization. Here the monomer droplets may be subjected to the oligomeric radical entry and solidify into particles, following the droplet nucleation mechanism. Droplet nucleation is usually minor in emulsion polymerization, except in miniemulsion polymerization when hydrophobic initiators are used.

Based on the fundamental understandings in conventional emulsion polymerization, we propose possible Pickering emulsion polymerization mechanisms, taking into account the differences between fine solid particles and surfactant molecules. Since the nanoparticles do not form micelles like surfactant molecules, micellar nucleation is excluded. Thus, there are two possible nucleation mechanisms involved in the initial stage of Pickering emulsion polymerization. Homogeneous coagulative nucleation is likely to be the dominating mechanism here, which yields the sub-micron-sized particles. The droplet nucleation might also occur, which yields micron-sized particles. The two mechanisms are illustrated in Figure 4. Upon initiator addition, monomers dissolved in the aqueous phase react with decomposed initiators and form oligomers with radicals. In homogeneous coagulative nucleation, the oligomers coagulate into nuclei, which subsequently become monomer swollen particles. Nanoparticles self-assemble at the interfaces between monomer and the continuous phase to provide stability. With the continuous supply of monomer molecules from the monomer droplets through diffusion, the particle size growth is mainly achieved by monomer swelling followed by polymerization within the core. In contrast, in droplet nucleation, initiated oligomers with radicals enter monomer droplets and subsequently polymerize into solid cores without significant size growth.

We use the hypothesized mechanisms to interpret the formation of polystyrene-silica nanocomposite particles prepared using VA-086 as the initiator. Figure 5 shows the dependence of particle size and surface coverage on reaction time and initiator concentration [Ma et al., 2010]. The composite particles are sampled from 3 h to 24 h reaction time and the initiator concentration relative to monomer is selected to be 0.83, 2.5, and 4.2 wt % respectively. At 3 h reaction time, well after the nucleation stage, composite particles with dense silica coverage are obtained. Since VA-086 initiator residues cannot provide sufficient stabilization to the monomer-swollen particles, silica nanoparticles would self-assemble at interfaces to provide stabilization and thus lead to high silica coverage. At initiator concentration 0.83 wt %, the silica coverage decreases with the particle size growth and the silica nanoparticles form patches on the nanocomposite particle surface with a low coverage. This might be an indication that the surface area of the polystyrene core increases with the particle growth without a significant increase of silica continuously attaching onto the polystyrene core. The particle growth mechanism is likely due to swelling of particles by monomers in the continuous phase. The same mechanism explains the surface coverage decrease in the system containing 2.5 wt % of initiator (images not shown) and from 3 h to 11 h in the system containing 4.2 wt % of initiator. These observations suggest that the Pickering emulsion polymerization using VA-086 as the initiator mainly follows the homogeneous coagulative nucleation mechanism.

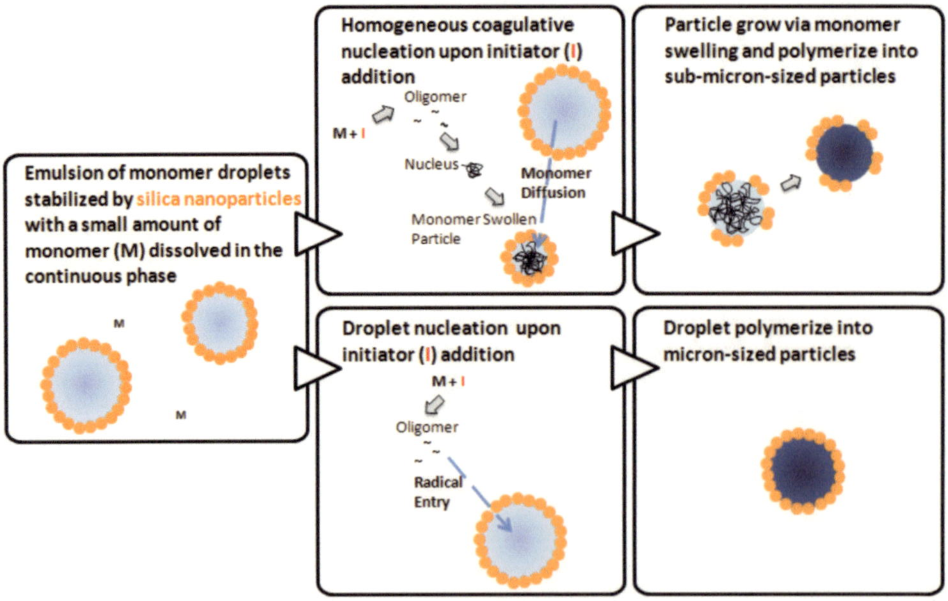

Fig. 4. Schematic illustration for possible mechanisms of Pickering emulsion polymerization.

One remaining mystery is the unexpected silica coverage from 11 h to 24 h in the system with 4.2 wt % initiator. Although the origin is unclear, we tentatively attribute the unusual silica coverage increase to the deposition of oligomers on the polystyrene core [Yamamoto et al., 2006], which adsorbed onto silica nanoparticles in the continuous phase [Yamamoto et al., 2006]. Excess initiator molecules might generate a large number of oligomers in the continuous phase, which could possibly adsorb onto silica nanoparticles. Thus when the oligomers on silica nanoparticles attach to preformed polystyrene surfaces, the silica nanoparticles are anchored there. It is also possible that the surface coverage increase might be due to the adsorption of depleted or close to depleted monomer droplets with a size below that of particles. It is worthwhile to note that the continuous phase contains approximately 21% isopropanol. The existence of isopropanol might increase the solubility of the monomer and the degree of polymerization required for an oligomer to be insoluble in the continuous phase, however, the solubility of monomer in the continuous phase is still low enough to enable emulsification and subsequent emulsion polymerization.

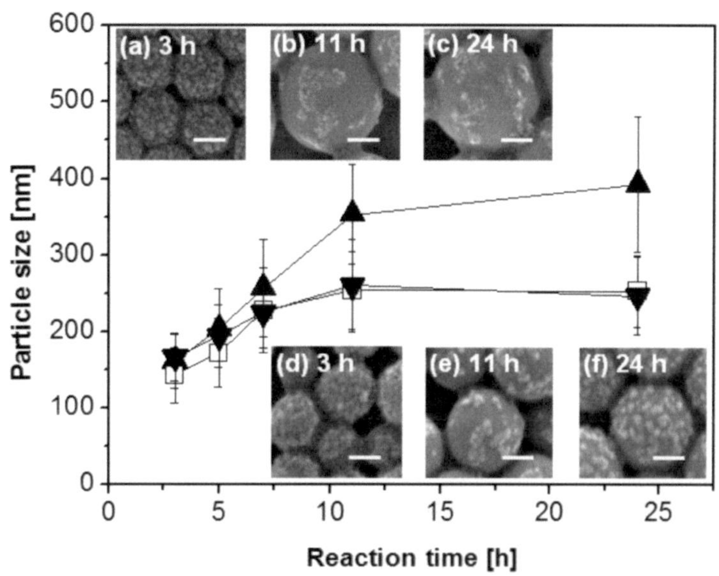

Fig. 5. Plot of particle size versus reaction time and representative SEM images with different initiator VA-086 concentrations: 0.83 wt % (▲, inset images a, b and c), 2.5 wt % (□) and 4.2 wt % (▼, inset images d, e and f). The error bars indicate the width of the particle size distribution and the scale bars represent 100 nm.

3. Environmentally responsive core-shell composite nanoparticles from Pickering emulsion polymerization

The Pickering emulsion polymerization opens a new and convenient way to synthesize core-shell composite nanoparticles. The simplicity enables further design and development advanced functional core-shell composite nanoparticles. One particular example to be illustrated here is the development of environmentally responsive core-shell composite nanoparticles from Pickering emulsion polymerization. By encapsulating drugs or chemicals, such nanoparticles enable controlled release upon environmental changes, as shown in Figure 6. For example, other than the styrene monomer, we have incorporated NIPAAm (N-isopropylacrylamide) as co-monomer into the Pickering polymerization and synthesized temperature responsive PS/PNIPAAm-silica composite nanopariticles. PNIPAAm is a well-understood temperature sensitive gel, which undergoes volume shrinkage at a transition temperature of approximately 32 °C in pure water (Schild, 1992). The mechanism of this change is based on different solubility below and above the lower critical solution temperature (LCST) in aqueous media. Below the LCST, the polymer chain is hydrophilic as the hydrogen bonding between the hydrophilic groups and water molecules dominates; above the LCST, the polymer chain becomes hydrophobic due to the weakened hydrogen bonding at elevated temperature and the hydrophobic interactions among hydrophobic groups (Qiu and Park, 2001). Figure 7(a) is a representative SEM image of the composite particles sampled at 5-hour reaction time which shows that the particles tend to be spherical. The roughness of the composite nanoparticle surfaces suggests that the nanoparticles are covered by silica nanoparticles; this is contrasted by the smooth surface of the hydrofluoric acid (HF)-treated particles in Figure 7(b). HF dissolves the silica layer and leaves behind the smooth polymer surface. It is also evidenced by the blue line in the Fourier transform infrared (FTIR) spectrum in Figure 7(c) which shows that the composite nanoparticles have a characteristic strong peak at 1104 cm^{-1}, corresponding to the asymmetrical vibration of the Si-O-Si bond. Such a peak is absent in the red line in Figure 7(c) which represents the HF-treated composite particles. FTIR is a strong analytical tool which gives information about specific chemical bonds simply by interpreting the infrared absorption spectrum; here it is used to identify the presence of silica.

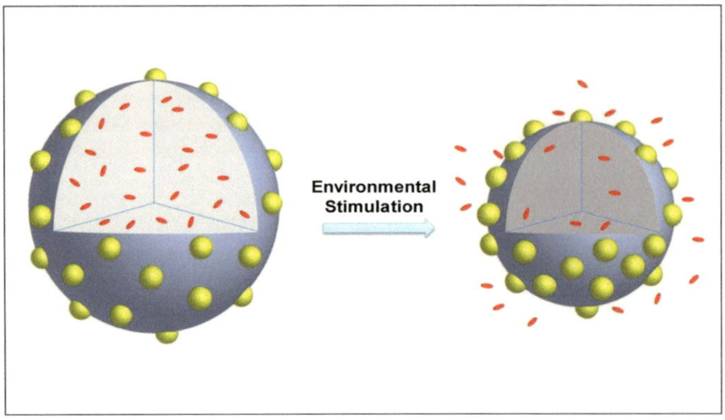

Fig. 6. Schematic illustration of environmentally responsive composite nanoparticles responding to an environmental change and releasing encapsulated materials such as a drug.

During our experiments, we also incorporated different ratios of monomer/comonomer in the formulation of the PS/PNIPAAm-silica core-shell composite nanoparticles. It is found that when the concentration of the NIPAAm monomer is high, the volume change of the nanoparticles is significantly greater with change in temperature, as shown in Figure 8. Control experiments of polystyrene-silica nanoparticles did not show a size transition over a temperature range of 25-45°C (data not shown). The transition temperature is not shifted by the silica nanoparticle encapsulation. This is consistent with the recently reported composite microspheres with a PNIPAAm core and a silica shell which also show a volume transition starting at 32°C (Qiu and Park, 2001). It is likely due to the fact that silica particles are physically adsorbed on the surfaces of PNIPAAm microspheres thus no chemical bond formation with silica occurs which might change the transition temperature. Moreover, the copolymerization with styrene has no significant effect on the transition temperature. One hypothesis is the relative phase separation of PNIPAAm and polystyrene within the core. Duracher et al. studied PNIPAAm-polystyrene particles and suggested a PNIPAAm-rich shell and a polystyrene-rich core structure (Duracher et al., 1998). Such phase separation may also occur in the core of the composite particles here although detailed morphology is unknown.

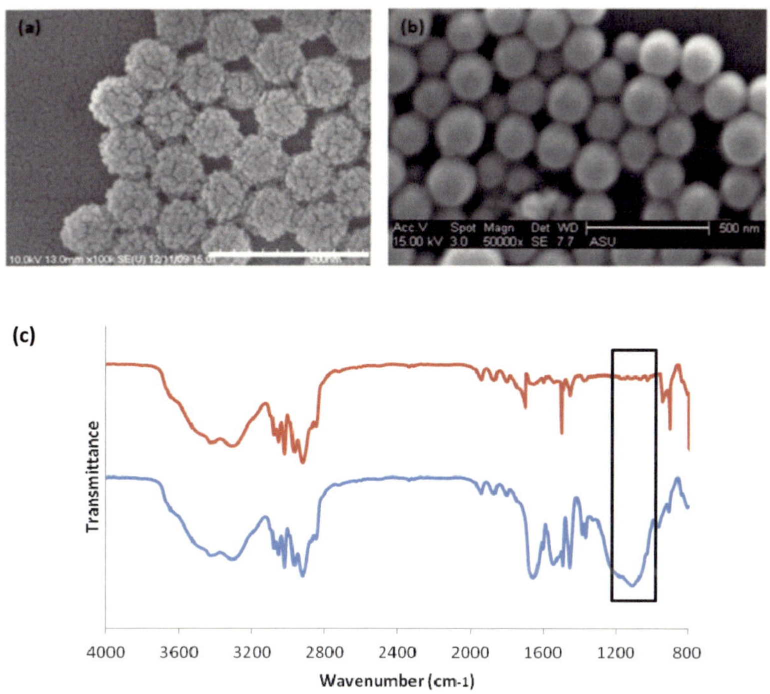

Fig. 7. (a) An SEM image of the composite particles; (b) SEM image taken after HF etching process (the scale bar represents 500 nm); (c) An FTIR spectrum of the composite nanoparticles where the blue line represents the composite particles and the red line is a sample of composite particles treated with HF. The box highlights the difference between the two spectra near 1104 cm^{-1}, which corresponding to the asymmetrical vibration of the Si-O-Si bond the two spectra near 1104 cm^{-1} which corresponding to the asymmetrical vibration of the Si-O-Si bond.

Figure 9(a) shows the dependence of average diameter of the composite particles on temperature with 15% NIPAAm. The average particle size at 25°C is approximately 92 nm. The size decreases sharply as the temperature reaches 32°C, around the LCST for homopolymer PNIPAAm and size change is nearly reversible upon cooling. In addition, we have encapsulated a drug, 17-(Allylamino)-17-demethoxygeldanamycin (17AAG), during the Pickering emulsion polymerization and performed cumulative drug release measurements. Figure 9(b) depicts the cumulative fractional drug release at 25 °C and 40 °C of 17-AAG from the drug-loaded nanoparticles. No significant release of the drug was observed at room temperature (25°C). However, at a higher temperature of 40°C, the drug

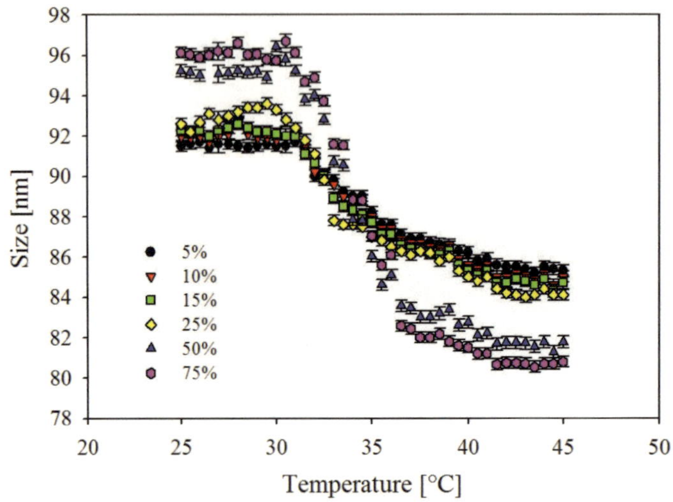

Fig. 8. Hydrodynamic diameters of PS/PNIPAAm-silica composite core-shell nanoparticles measured by DLS, decrease significantly near the transition temperature of PNIPAAm (32°C). The legend shows the various concentrations of NIPAAm, it is observed for higher concentrations of NIPAAm there is greater size change of nanoparticles.

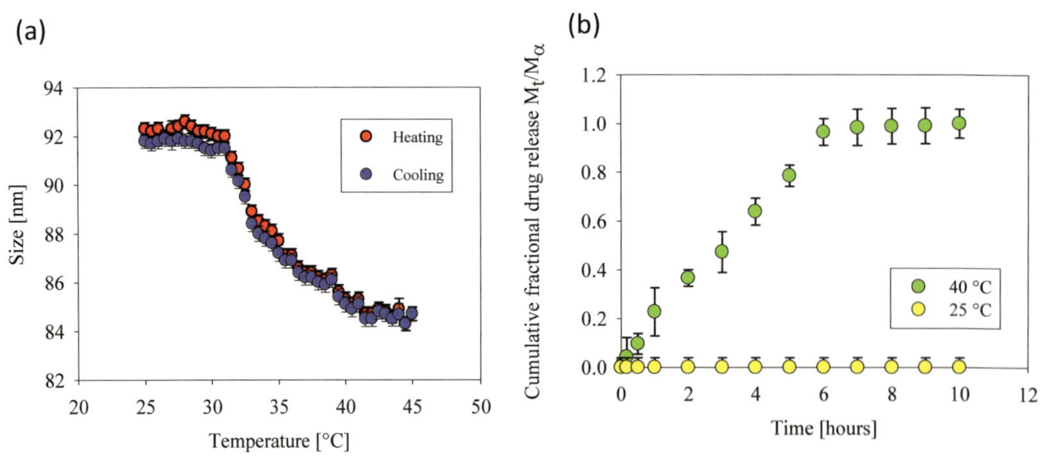

Fig. 9. (a) The dependence of average diameter of composite nanoparticles on temperature. The error bars show standard deviations of particles made in three different batches. The transition temperature is around 32°C. There size transition is nearly reversible; (b) Cumulative fractional drug release versus time curve indicating release at room temperature and at 40 °C. There is minimum release at 25 °C which is below the transition temperature of the nanoparticles.

releases from the nanoparticles reached a maximum after 7 h. The cumulative fractional drug release is calculated as M_t/M_∞, where t is the release time, M_t is the amount of drug released at a time t and M_∞ is the amount of drug released at time infinity. Infinity is taken to be when the maximum amount of drug gets released and there is no subsequent release after infinity. The concentration of the drug in the sample solution was read from the calibration curve as the concentration corresponding to the absorbance of the solution. To determine the release mechanisms of the composite nanoparticle system an equation proposed by Yasuda et al. was used (Yasuda et al., 1968), which analyses the release behavior of a solute from a polymer matrix, $\frac{M(t)}{M(\infty)} = kt^n$ where k is a constant related to the physical properties of the system, and the index, n, is the diffusional component that depends on the release mechanism. When $n<0.5$, the solute is released by Fickian diffusion; when $0.5<n<1.0$, the solute is released by non-Fickian diffusion and when $n=1$, there is zero order release [Yasuda et al., 1968]. The calculated n value is 0.73 which indicates the non-Fickian diffusion. The mathematical model indicates that the drug diffusion behavior is non-Fickian and the rate of drug release is due to the combined effect of drug diffusion and polymer response due to increase in temperature.

4. Conclusion

In summary, polystyrene-silica core-shell composite particles were successfully synthesized by a novel one-step Pickering emulsion polymerization. The sole stabilizing effect of silica nanoparticles in the emulsion polymerization was verified. In addition, possible mechanisms of Pickering emulsion polymerization were explored and suggested that homogeneous coagulative nucleation is likely the dominating mechanism here. Finally, the temperature responsiveness of core-shell composite nanoparticles and drug release were validated by incorporating NIPAAM as a co-monomer into the Pickering emulsion polymerization.

5. Acknowledgments

We gratefully acknowledge Professor Paul Westerhoff at Arizona State University for the DLS usage and the Imaging Center at Texas Tech University/Solid State Center at Arizona State University for electron microscope usage. We are also grateful for the financial support provided by the National Science Foundation (CBET-0918282, CBET-0922277, and CBET-1050045) and the Texas Higher Education Coordinating Board.

6. References

Barthet, C.; Hickey, A. J.; Cairns, D. B.; Armes, S. P. (1999). Synthesis of novel polymer-silica colloidal nanocomposites via free-radical polymerization of vinyl monomers. *Adv. Mater.*, *11*, 408-410.

Bon, S. A. F.; Colver, P. J. (2007). Pickering miniemulsion polymerization using laponite clay as a stabilizer. *Langmuir*, *23*, 8316-8322.

Caruso, F. (2001). Nanoengineering of particle surfaces. *Adv. Mater.*, *13*, 11-22.

Caruso, F.; Susha, A. S.; Giersig, M.; Möhwald, H. (1999). Magnetic core-shell particles: Preparation of magnetite multilayers on polymer latex microspheres. *Adv. Mater., 11*, 950-953.

Caruso, R. A.; Susha, A.; Caruso, F. (2001). Multilayered titania, silica, and laponite nanoparticle coatings on polystyrene colloidal templates and resulting inorganic hollow spheres. *Chem. Mater., 13*, 400-409.

Cauvin, S.; Colver, P. J.; Bon, S. A. F. (2005). Pickering stabilized miniemulsion polymerization: Preparation of clay armored latexes. *Macromolecules, 38*, 7887-7889.

Chern, C. S. (2006). Emulsion polymerization mechanisms and kinetics. *Prog. Polym. Sci., 31*, 443-486.

Dai, L. L.; Sharma, R.; Wu, C. Y. (2005). Self-assembled structure of nanoparticles at a liquid-liquid interface. *Langmuir, 21*, 2641-2643.

Ding, X. F.; Jiang, Y. Q.; Yu, K. F.; Hari-Bala; Tao, N. N.; Zhao, J. Z.; Wang, Z. C. (2004). Silicon dioxide as coating on polystyrene nanoparticles in situ emulsion polymerization. *Mater. Lett., 58*, 1722-1725.

Dokoutchaev, A.; James, J. T.; Koene, S. C.; Pathak, S.; Prakash, G. K. S.; Thompson, M. E. (1999). Colloidal metal deposition onto functionalized polystyrene microspheres. *Chem. Mater., 11*, 2389-2399.

Duan, L.; Chen, M.; Zhou, S.; Wu, L. (2009). Synthesis and characterization of poly(N-isopropylacrylamide)/Silica composite microspheres via inverse pickering suspension polymerization. *Langmuir, 25*, 3467-3472.

Duracher, D.; Sauzedde, F.; Elaissari, A.; Perrin, C. (1998). Pichot, *Colloid Polym. Sci., 276*, 219.

Feeney, P. J.; Napper, D. H.; Gilbert, R. G. (1987). Surfactant-Free Emulsion Polymerizations: Predictions of the Coagulative Nucleation Theory. *Macromolecules*, 20, 2922-2930.

Feeney, P. J.; Napper, D. H.; Gilbert, R. G. (1984). Coagulative Nucleation and Particle Size Distributions in Emulsion Polymerization. *Macromolecules*, 17, 2520-2529.

Gao, Q.; Wang, C.; Liu, H.; Wang, C.; Liu, X.; Tong, Z. (2009). Suspension polymerization based on inverse pickering emulsion droplets for thermo-sensitive hybrid microcapsules with tunable supracolloidal structures. *Polymer, 50*, 2587-2594.

Harkins, W. D. (1947). A general theory of the mechanism of emulsion Polymerization 1. *J. Am. Chem. Soc., 69*, 1428-1444.

Lynch, D. E.; Nawaz, Y.; Bostrom, T. (2005). Preparation of sub-micrometer silica shells using poly(1-methylpyrrol-2-ylsquaraine). *Langmuir, 21*, 6572-6575.

Ma, H.; Dai, L. L. (2009). Synthesis of polystyrene-silica composite particles via one-step nanoparticle-stabilized emulsion polymerization. *Journal of Colloid and Interface Science. 333*, 807-811.

Ma, H.; Luo, M. X.; Sanyal, S.; Rege, K.; Dai, L. L. (2010). The one-step Pickering emulsion polymerization route for synthesizing organic-inorganic nanocomposite particles. *Materials, 3*, 1186-1202.

Pickering, S. U. (1907). Emulsions. *J. Chem. Soc., 91*, 2001-2021.

Qiu, Y.; Park, K. (2001). Environment-sensitive hydrogels for drug delivery. *Adv. Drug Deliv. Rev., 53*, 321-339.

Ramsden, W. (1903). Separation of solids in the surface-layers of solutions and "suspensions" (observations on surface-membranes, bubbles, emulsions, and mechanical coagulation). preliminary account.W. *Proc. R. Soc. London, 72*, 156-164.

Schild, H. G. (1992). Poly(N-isopropylacrylamide): Experiment, theory and application. *Prog. Polym. Sci., 17*, 163-249.

Schmid, A.; Fujii, S.; Armes, S. P. (2006). Polystyrene-silica nanocomposite particles via alcoholic dispersion polymerization using a cationic azo initiator. *Langmuir, 22*, 4923-4927.

Schmid, A.; Fujii, S.; Armes, S. P.; Leite, C. A. P.; Galembeck, F.; Minami, H.; Saito, N.; Okubo, M. (2007). Polystyrene-silica colloidal nanocomposite particles prepared by alcoholic dispersion polymerization. *Chem. Mater., 19*, 2435-2445.

Tarimala, S.; Dai, L. L. (2004). Structure of microparticles in solid-stabilized emulsions. *Langmuir, 20*, 3492-3494.

Tarimala, S.; Ranabothu, S. R.; Vernetti, J. P.; Dai, L. L. (2004). Mobility and in situ aggregation of charged microparticles at oil-water interfaces. *Langmuir, 20*, 5171-5173.

Tarimala, S.; Wu, C.; Dai, L. L. (2006). Dynamics and collapse of two-dimensional colloidal lattices. *Langmuir, 22*, 7458-7461.

Tauer, K.; Deckwer, R.; Kuhn, I.; Schellenberg, C. (1999). A comprehensive experimental study of surfactant-free emulsion polymerization of styrene. *Colloid Polym. Sci., 277*, 607-626.

Tauer, K.; Hernandez, H.; Kozempel, S.; Lazareva, O.; Nazaran, P. (2008). Towards a consistent mechanism of emulsion polymerization - new experimental details. *Colloid Polym. Sci., 286*, 499-515.

Voorn, D. J.; Ming, W.; van Herk, A. M. (2006). Polymer-clay nanocomposite latex particles by inverse pickering emulsion polymerization stabilized with hydrophobic montmorillonite platelets. *Macromolecules, 39*, 2137-2143.

Wu, J.; Dai, L. L. (2006). One-particle microrheology at liquid-liquid interfaces. *Appl. Phys. Lett., 89*, 094107.

Wu, J.; Dai, L. L. (2007). Apparent microrheology of oil-water interfaces by single-particle tracking. *Langmuir, 23*, 4324-4331.

Wu, C.; Song, Y.; Dai, L. L. (2009). Two-particle microrheology at oil-water interfaces. *Appl. Phys. Lett., 95*, 144104.

Yamamoto, T.; Kanda, Y.; Higashitani, K. (2006). Initial growth process of polystyrene particle investigated by AFM. *J. Colloid Interface Sci., 299*, 493-496.

Yamamoto, T.; Kanda, Y.; Higashitani, K. (2004). Molecular-scale observation of formation of nuclei in soap-free polymerization of styrene. *Langmuir, 20*, 4400-4405.

Yamamoto, T.; Nakayama, M.; Kanda, Y.; Higashitani, K. (2006). Growth mechanism of soap-free polymerization of styrene investigated by AFM. *J. Colloid Interface Sci., 297*, 112-121.

Yang, J.; Hasell, T.; Wang, W. X.; Li, J.; Brown, P. D.; Poliakoff, M.; Lester, E.; Howdle, S. M. (2008). Preparation of hybrid polymer nanocomposite microparticles by a nanoparticle stabilised dispersion polymerisation. *J. Mater. Chem., 18,* 998-1001.

Yasuda, H.; Lamaze, C. E.; Ikenberry, L. D. (1968). *Die Makromolekulare Chemie, 118,* 19.

Section 3

Biological Activities

13

Biological Activities of Carbon Nanotubes

Anurag Mishra[1,2], Yon Rojanasakul[2] and Liying Wang[1]
[1]*National Institute for Occupational Safety and Health, HELD/PPRB, Morgantown, WV,*
[2]*West Virginia University, Department of Pharmaceutical Sciences, Morgantown, WV, USA*

1. Introduction

During the past several years, nanotechnology based on novel nanomaterials has gained considerable attention in various scientific disciplines such as biotechnology, medicine and material engineering (McCarthy and Weissleder 2008). According to British Standards Institute Report (2007), nanoparticles are those particles with at least one dimension of less than or equal to 100 nm (1 nm = 1 x 10^{-6} m) in size. Since particle size is directly related to surface area and associated surface energy, nanoscaled materials relatively exhibit unique physicochemical, optical and electrical properties than micron-sized particles. Nanomaterials have exceptional properties and are beneficial in a wide range of applications. Nanotechnology based on these novel nanomaterials is fueling the modern industrial revolution which is already a multi-billion dollar market capitalization. Among the different types of nanomaterials, carbon nanoparticles have gained much attention in recent years due to their exceptional physicochemical properties. Some of the most popular carbon-based nanomaterials are fullerene (C_{60}), carbon nanohorn, single wall carbon nanotubes (SWCNT), and multi wall carbon nanotubes (MWCNT).

Carbon nanotubes (CNT) are one of the most commonly used nanomaterials possessing unique physicochemical properties such as high aspect ratio and a diameter of less than 100 nm (Iijima 1991). Due to their exceptional characteristics, CNT, if incorporated will enhance the efficiency of a number of applications including electronics (Bandaru 2007), biosensors (Le Goff, Holzinger et al. 2011), drug and biomolecule carriers (Prato, Kostarelos et al. 2007). Other potential biomedical applications of CNT include bone scaffold, dental tissue support, and neuronal cell growth scaffold (Li, Fan et al. 2010). Increasing evidence has shown that certain CNT properties such as nano-sized dimension, high surface energy, and large reactive surface area are directly correlated to their biological activities (Oberdorster, Ferin et al. 1994; Oberdorster, Oberdorster et al. 2005). The bioactivity of nanoparticles differs from micron-size particles of the same material. Although the underlying mechanism remains to be understood, small size, high surface area and chemical composition of nanomaterials play an important role. Recent studies have shown that CNT could be harmful to human health. Fiber morphology and high surface energy of CNT raise health concerns among scientists due to their structural similarities with asbestos fibers (Donaldson, Murphy et al. 2010). The biological properties of nanoparticles are currently under intense investigations and are the subject of this review.

2. Routes of nanoparticle exposure and associated pathologies

Nanomaterials such as CNT have very low specific weight and can be easily aerosolized and come in contact with humans during manufacturing, transportation or usage of the CNT-based products. Apart from unintentional exposures, for certain biomedical applications such as drug delivery, artificial tissues and diagnostic agents, CNT need to be introduced into the human body. Therefore, it is important to consider the potential adverse effects of nanoparticles. Most recent studies have focused on the adverse effects of nanoparticle exposure on pulmonary or dermal tissues.

2.1 Pulmonary exposure

Lung is the major target organ for nanoparticle exposure. Because of their low density and small size, CNT can be aerosolized and inadvertently inhaled during their manufacturing or handling. Therefore, respiratory exposure of nanomaterials including CNT has been the focus of intense research. Lung exposure to solid particles has been linked to asthma (Bonner 2010), fibrosis (Shvedova, Kisin et al. 2005), mesothelioma (Sakamoto, Nakae et al. 2009), and other inflammatory diseases (Li, Muralikrishnan et al. 2010). The region of lung affected by accidently inhaled materials depends on the shape (fibrous, spherical), size (aerodynamic diameter), and other physical and chemical properties of the particles. Nano-sized particles are deposited deep inside the lung compared to micron-sized particles of similar chemical composition (Oberdorster, Ferin et al. 1994). As shown in Figure 1, a large fraction of inhaled particles with the size of less than 100 nm deposits mainly in the terminal alveolar region of the lung, whereas most of the micron-sized inhaled particles remain in the upper respiratory tract. Recent studies have shown that 80% of pulmonary exposed SWCNT reach the alveolar space of the lung in mice (Shvedova, Kisin et al. 2005; Mercer, Scabilloni et al. 2008). Pulmonary exposure of SWCNT and MWCNT has been shown to induce rapid interstitial lung fibrosis with non-persistent inflammatory response in rodents (Shvedova, Kisin et al. 2005; Porter, Hubbs et al. 2010). Inhaled CNT can also translocate to the surrounding regions of the lung such as pleural space (Mercer, Scabilloni et al. 2008; Wilson and Wynn 2009). In several *in vivo* studies, pulmonary exposure to CNT has been shown to induce granuloma formation in the terminal alveolar region of the lung (Lam, James et al. 2004; Warheit, Laurence et al. 2004; Shvedova, Kisin et al. 2005). Inflammatory granulomas are accumulation of epithelioidal macrophages engulfing persistent and non-biodegradable particles. Neutrophil infiltration was also observed in these studies.

Usually inhalation mimics pulmonary exposure of non-soluble particles in animal models, while intratracheal instillation or pharyngeal aspiration have shown similar results (Li, Li et al. 2007; Shvedova, Kisin et al. 2008). Upon alveolar deposition, an unexpectedly rapid translocation of dispersed SWCNT into the alveolar interstitium (1 day post-exposure) has been observed (Shvedova 2005). Subsequent development of lung fibrosis occurs as early as 1 week post-exposure and progresses through 60 days post-exposure without persistent inflammation. The mechanism of nanoparticle-induced lung fibrosis is still under investigation. Data suggest oxidative stress, cytotoxicity and apoptosis induction via DNA damage (Nam, Kang et al. 2011; Ravichandran, Baluchamy et al. 2011) or direct stimulation of lung fibroblasts (Wang, Mercer et al. 2010). *In vitro* studies provide detailed information on the mechanism of the unusual CNT-induced lung fibrosis. Several key lung cells have been selected to determine the specific CNT-lung cell interactions. For example, SWCNT directly stimulate fibroblasts to produce collagen (Wang, Mercer et al. 2010) or induce oxidative stress through ROS production in macrophages (Migliore, Saracino et al. 2010; Palomäki, Välimäki et al. 2011). A study using lung epithelial cells has shown that chronic exposure of low-dose SWCNT can induce cancer-like cell transformation (Wang, Luanpitpong et al. 2011). Several other studies have shown that CNT can induce toxicity of alveolar epithelial cells through a suppression of immune response or oxidative stress (Simon-Deckers, Gouget et al. 2008; Herzog, Byrne et al. 2009). Due to the large surface area and proximity to the circulatory system, inhaled nanoparticles can potentially translocate to other parts of the body and can cause toxic effects such as cardiovascular abnormality (Li, Hulderman et al. 2006).

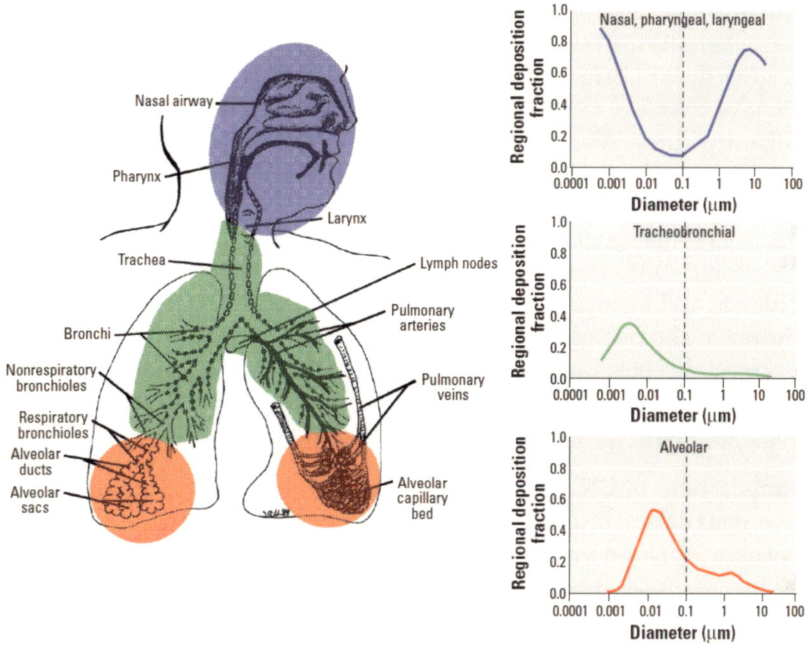

Fig. 1. Predicted fractional deposition of inhaled particles in the nasopharyngeal, tracheobronchial, and alveolar region of the human respiratory tract during nose breathing (Oberdorster, et al., 2005)

2.2 Skin exposure

Skin is the primary barrier preventing the entry of foreign particles into the body. Exposure of skin to nanoparticles can occur in the form of cosmetic formulations containing nanomaterials such as nano-sized titanium dioxide and accidental exposure of nanoparticles during manufacturing or handling. Recent *in vivo* and *in vitro* studies have shown that skin-exposed CNT can translocate to the deeper layers of skin such as dermis or subcutaneous layer. *In vivo* studies with dermal exposure of CNT have shown inflammation and inflammatory granuloma formation in dermal and subcutaneous tissues (Yokoyama, Sato et al. 2004; Koyama, Endo et al. 2006). Other studies have also shown that exposure to CNT can induce dermal granuloma formation, thickness of skin folding and neutrophil and macrophage-mediated inflammation (Sato, Yokoyama et al. 2005; Murray, Kisin et al. 2009).

A number of *in vitro* dermal studies has shown that SWCNT exposure can cause inflammation and oxidative stress in dermis and keratinocyte cells (Li, Hulderman et al. 2006; Poland, Duffin et al. 2008; Msiska, Pacurari et al. 2009). The inflammation and toxicity are less severe in skin exposure than lung exposure due to the stiffness and thick layer of the skin cutaneous tissue (Sato, Yokoyama et al. 2005).

2.3 Oral exposure

Oral exposure of nanoparticles could be through food packaging, contaminated food or water. In addition, ingestion of nano-drugs or nano-delivery systems can lead to gastrointestinal track (GIT) exposure to nanoparticles. Absorption from GIT depends on the physiochemical properties of nanomaterials. For example, Kolosnjaj-Tabi et al. (2010) reported that when SWCNT were administered orally (1000 mg/kg body weight), neither death nor growth or behavioral troubles were observed. However, intraperitoneal administration of SWCNT (50-1000 mg/kg body weight) can coalesce inside the body to form fiber-like structures. When SWCNT length exceeded 10 μm, they irremediably induced granuloma formation compared to smaller aggregates which did not induce granuloma but persisted inside the cells for up to 5 months post-administration. Individualized SWCNT (< 300 nm) can escape the reticuloendothelial system to be excreted through the kidney and bile ducts (Kolosnjaj-Tabi, Hartman et al. 2010). Oral exposure of corn oil-suspended CNT induced oxidative and genotoxic changes in the lung and liver of rats (Folkmann, Risom et al. 2009). However, the risk of CNT ingestion is not clearly understood because of the short history of usage of this new nanomaterial and the limited knowledge/studies in the field.

2.4 Cardiovascular exposure

One of the applications of CNT is as a carrier for drugs and biomolecules into the body. This requires these materials to be able to get into the systemic circulation. Therefore, implication of CNT exposure in blood vessels and heart muscles should be thoroughly investigated. Apart from directly injected into the blood for medical applications, nanoparticles can also enter blood via translocation from the lung or skin. Intravenous exposure in mice to SWCNT induces inflammatory reactions and an up-regulation of pro-inflammatory cytokines such as tumor necrosis factor-α (Li, Hulderman et al. 2007; Yang, Wang et al. 2008). Another study reported that CNT exposure to the heart tissue induces cellular mitochondrial damage, oxidative stress and apoptosis (Li, Hulderman et al. 2007). Additionally, vascular thrombosis and platelet aggregation was observed in both *in vivo* and *in vitro* exposures to CNT (Radomski, Jurasz et al. 2005). These studies suggest that cardiovascular exposure of nanomaterials could be hazardous to normal cardiac function. However, studies on the dose, time, physicochemical properties and pharmacokinetic parameters of the administrated CNT are limited.

2.5 Neuronal exposure

There are only a few studies on the effect of nanomaterial exposure on the nervous system. When exposed to neuroblast-glioma cells *in vitro*, SWCNT induced DNA damage and cytotoxicity (Belyanskaya, Weigel et al. 2009). Also SWCNT induced oxidative stress, cell membrane damage and DNA damage in neuronal pc12 cells (Zhang, Ali et al. 2010). *In vivo* systemic exposure studies did not find distribution or translocation of CNT to neuronal tissues or brain (Wang, Wang et al. 2004). However, another *in vivo* study showed the translocation of CNT across blood brain barrier (Yang, Guo et al. 2007)

3. Absorption, distribution and metabolism of carbon nanotubes *in vivo*

Biodegradation of nanoparticles after exposure depends on the nature of nanomaterial and its chemical constituents. It also depends on the bio-distribution and persistence of nanoparticles in the body. Much of the absorption of CNT depends on the translocation of CNT from the site of exposure (lung, skin, etc.) to the systemic circulation (Yang, Wang et al. 2008). Only a few studies have shown the translocation and distribution of systemically exposed CNT into organs such as liver, lung, spleen and kidney where a high level of CNT accumulation was observed (Wang, Wang et al. 2004; Rotoli, Bussolati et al. 2008). Usually biodegradable nanomaterials are broken down into constituent molecules. However, some nanomaterials including SWCNT and MWCNT are bio-persistent and difficult to excrete, which become a safety concern and deter their biomedical applications. Studies have shown that these CNT can stay in the lung tissue for months after exposure and lead to granuloma or progressive interstitial fibrosis (Shvedova, Kisin et al. 2005; Mercer, Hubbs et al. 2011). Bio-persistency of CNT increases their interaction with body's cells which can lead to harmful effects to specific tissues and organs. Contaminants, mainly metal ions which are used as catalysts of CNT synthesis, can induce reactive oxygen species (ROS) generation which can cause tissue inflammation, cell damage and even carcinogenesis.

4. Physicochemical properties of nanomaterials affecting their biological activities

Physicochemical properties of nanomaterials can have a great influence on the biological activities of the materials. Some key physicochemical properties affecting their bioactivities include size, shape, surface activity, dispersion status, and metal contaminants.

4.1 Size

Studies with micron- and nano-sized particles of the same material, e.g., silica and CNT, have shown that the nano-sized particles have a deeper lung deposition and are more toxic than large particles. As the particle size reduces, the surface atom increases. As the particles size reduces to less than 100 nm, the surface atom increases exponentially (Figure 2). Properties such as surface energy and electrical force also change accordingly (Garg and Sinnott 1998; Folkmann, Risom et al. 2009). These properties have been shown to influence the translocation and distribution of nanoparticles in the body.

4.2 Shape

High aspect-ratio fibers like asbestos are historically known to be cytotoxic (Jaurand, Renier et al. 2009). Compared to asbestos fibers, the aspect ratio of MWCNT is similar but larger. Recent animal studies have shown that the toxicity of CNT is several times higher than that of asbestos. Pulmonary exposed long fibers like CNT (> 20 μm) are difficult to be cleared by macrophages (Lam, James et al. 2004). Other studies have also reported that short CNT evade phagocytosis by macrophages when compared to larger micron-sized particles of similar composition (Muller, Huaux et al. 2005; Mercer, Scabilloni et al. 2008; Zhang, Bai et al. 2010; Palomäki, Välimäki et al. 2011)

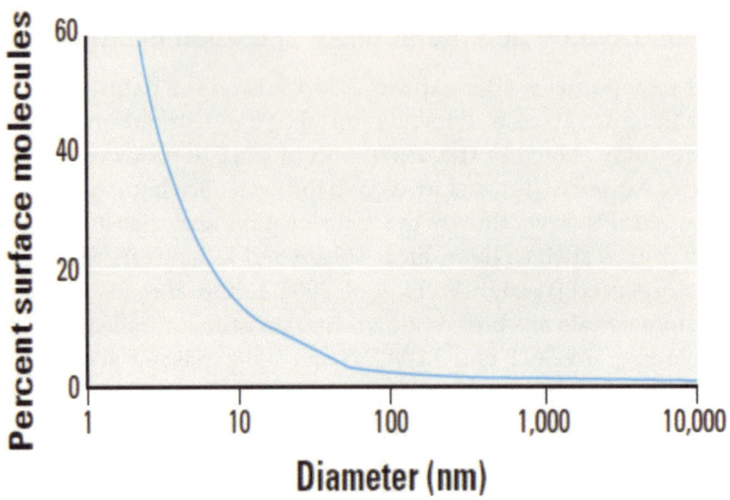

Fig. 2. Surface molecules as a function of particle size. Surface molecules increase exponentially when particle size decreases to less than 100 nm (Oberdorster et al. 2005).

4.3 Surface activity

The biological activity of CNT is directly related to its surface activity. CNT have a large surface area and high surface energy. Studies have shown that surface modifications can lead to changes in biological activity and toxicity. Surface modification with different chemical functionalities can affect the overall biological activity of nanoparticles. For example, acid treatment of CNT introduces carboxyl groups onto the surface of particles leading to an increase in dispersion and water solubility. Functionalization of CNT with COOH reduces the attractive electrical force between CNT surfaces and affects the biological response. COOH-functionalized MWCNT are more water soluble and dispersed than non-functionalized MWCNT (Upadhyayula and Gadhamshetty; Cao, Chen et al. 2011; Jacobs, Vickrey et al. 2011)

4.4 Dispersion status

Particle agglomeration and aggregation is a common phenomenon for CNT. High surface energy and surface area of nanoparticles result in Van der Waals interactions leading to agglomeration. For full exploitation of CNT, they should be used as well dispersed particles such as those observed with aerosolized particles. Dispersion results in an increase in particle number per unit mass and an associated increase in contact surface area with exposed cells, thus affecting their biological activities. Published data have shown that dispersed CNT exhibit more pronounced effects on cell proliferation and cytotoxicity than their non-dispersed counterpart. Several methods of nanoparticle dispersion have been investigated including the use of natural lung surfactants such as Survanta® (Wang et al. 2010), phospholipids such as dipalmitoylphosphatidylcholine (Sager, Porter et al. 2007), organic solvents such as acetone and dimethyl sulphoxide (Soto, Carrasco et al. 2005) and biomolecules such as single-stranded DNA, albumin, and cell culture serum (Cherukuri, Bachilo et al. 2004; Jia, Wang et al. 2005; Muller, Huaux et al. 2005)

4.4.1 Dispersion of CNT using natural lung surfactants

Survanta® is a surfactant replacement used in clinic which provides some advantages as a dispersing agent for CNT, i.e., it is commercially available, biocompatible and safe. Previous study has also shown that it is effective in dispersing CNT into fine particles comparable in size to that of aerosolized CNT (Figure 3). It has no apparent cytotoxic effect and does not mask the biological effect of CNT (Wang et al. 2010).

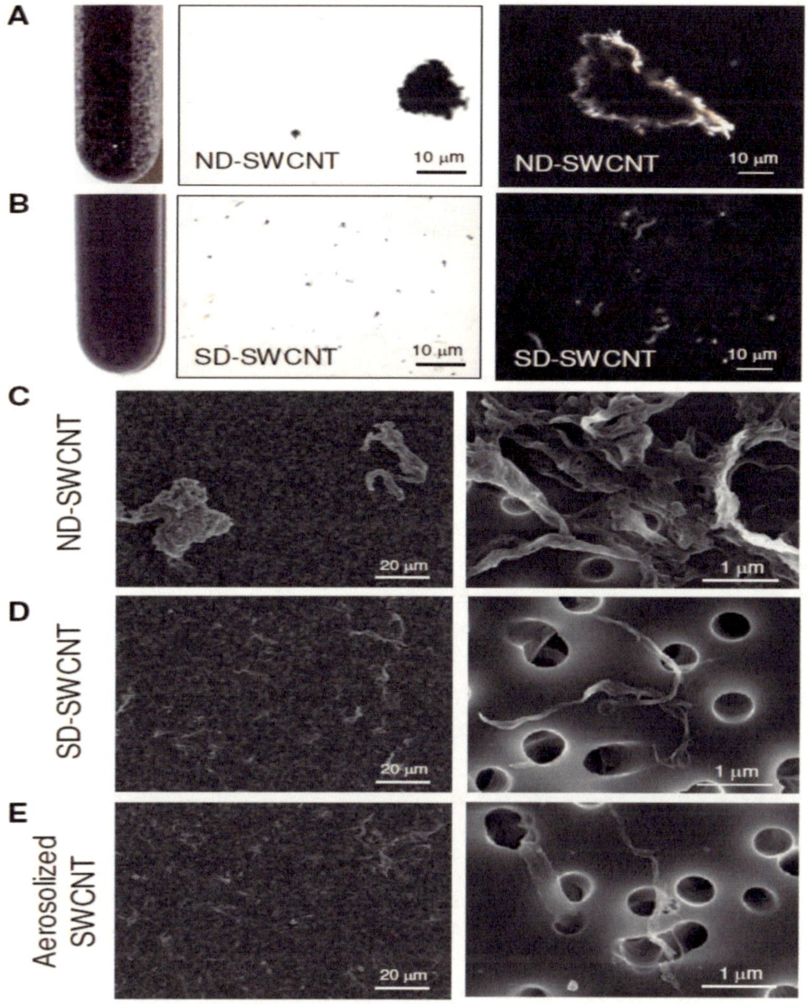

Fig. 3. Comparison of Survanta®-dispersed and non-dispersed SWCNT. (A) Non-dispersed SWCNT suspension showing visible clumping of SWCNT (left panel) with corresponding light microscopy (middle panel, 100x) and hyperspectral imagining of an individual clump (right panel, 400x). (B) Survanta®-dispersed SWCNT suspension showing much improved

dispersion with no visible large clumps (left panel) observed in the corresponding light microscopy and hyperspectral imaging. (C-E) Scanning electron microscopy of non-dispersed, Survanta®-dispersed and aerosolized SWCNT at low magnification (400x, left panel) and high magnification (30,000x, right panel)(Wang, Castranova et al. 2010)

4.4.2 Effect of dispersed CNT on fibrogenesis

Well dispersed SWCNT and MWCNT can deposit deep in the interstitial area of the alveoli, where it can enhance the fibrotic process by directly stimulating interstitial lung fibroblasts (Mercer, Scabilloni et al. 2008; Monteiro-Riviere, Inman et al. 2009). Dispersed SWCNT have been shown to evade engulfment by lung macrophages where 80% of pulmonary exposed SWCNT reach the alveolar interstitial space (Shvedova, Kisin et al. 2005; Mercer, Scabilloni et al. 2008). In case of exposure to micron-sized particles, most of these particles are engulfed by macrophages which induce a robust inflammatory response. Dispersed SWCNT bypass the inflammatory response which might be due to evading macrophage engulfment. Dispersed SWCNT can stimulate resident fibroblasts in the interstitial space to produce collagen. Ongoing research suggests that dispersed nano-sized SWCNT can induce lung fibrosis without persistent inflammation. Usually micron-sized particles induce robust inflammation followed by slow developing fibrosis in contrast to nano-sized particles like CNT which induce rapid fibrosis. The mechanism of this unusual fibrosis is unclear and is under investigation.

4.5 Metal contaminants

Metals are routinely used as catalysts in nanomaterial production. These metal ions are usually incorporated into the nanomaterials during the manufacturing process and contribute towards metal contaminants. Generally metal ions present in CNT as impurities are Fe, Ni, Co, etc. During the purification process, manufactured CNT are thoroughly washed in acid to remove these metal ions or any amorphous carbon. The metal ion concentration depends on the method used to manufacture CNT and the purification process. In some cases, metal ion contamination can reach 30% by weight in unpurified CNT. Metal impurities such as Fe in CNT are known to induce oxidative stress in cells (Warheit, Laurence et al. 2004; Le Goff, Holzinger et al. 2011). These ions can influence redox reactions by inducing ROS or inhibiting antioxidant enzymes. Formation of ROS leads to oxidative stress, inflammation and apoptosis.

5. Mechanisms of CNT toxicity

5.1 Oxidative stress

ROS are generated by distressed cells leading to oxidative stress and apoptosis. It is an imbalance between antioxidant proteins and ROS produced in cells and induced by toxic substances such as asbestos (Liu, Ernst et al. 2000). CNT have been found to induce oxidative stress and ROS in the lung (Manna, Sarkar et al. 2005; Shvedova, Kisin et al. 2005; Pacurari, Yin et al. 2008), skin (Murray, Kisin et al. 2009), and heart (Li, Hulderman et al. 2007). Metal contaminants such as Fe are the major source of ROS generation induced by CNT. Studies have shown that SWCNT containing up to 30% of Fe (%wt) are more toxic than SWCNT with 0.23% iron content (Shvedova, Castranova et al. 2003). Iron ion can initiate Fenton reaction with hydrogen peroxide in cells to produce highly oxidative species such as hydroxyl radicals.

$$Fe^{2+} + H_2O_2 \rightarrow Fe^{3+} + OH\cdot + OH^-$$

As mentioned above, metal ions are key contributor to nanomaterial-induced ROS generation. Since many nanoparticles are prepared by using metal ions, exposure to raw or as manufactured materials in occupational workers can exacerbate the toxic response due to these impurities. Oxidative stress further leads to inflammation and toxicity which ultimately results in cell death. Other studies have also suggested that oxidative stress plays an important role in nanoparticle-induced toxicities (Warheit, Laurence et al. 2004; Li, Muralikrishnan et al. 2010). The damage caused by nanoparticle-induced oxidative stress can be reduced by pretreatment with antioxidants (Shvedova, Kisin et al. 2007).

5.2 Inflammation

CNT have been shown to induce inflammatory response in a number of *in vivo* studies (Shvedova, Kisin et al. 2005; Poland, Duffin et al. 2008). Macrophages engulf inhaled particles and produce various inflammatory cytokines and chemokines, which attract and amplify inflammatory responses in the body. Pulmonary exposed CNT induce transient inflammation whereas micro-sized particles usually induce persistent inflammation. Previous study has shown that alveolar microphages ignore certain small-sized CNT (Shvedova, Kisin et al. 2005), which may explain the less inflammation induced by the nanoparticles. Some studies suggest that CNT suppress the immune response by reducing the inflammatory signal and preventing macrophage activation (Mitchell, Gao et al. 2007). *In vitro* studies using keratinocytes and macrophages showed that these cells secrete pro-inflammatory cytokines such as interlukin-8 and tumor necrosis factor-α in response to CNT stimulation (Monteiro-Riviere, Inman et al. 2005; Brown, Kinloch et al. 2007).

5.3 Fibrosis

Unlike micron-sized particles, nanoparticles induce an unusual rapid fibrosis. For example, inhaled CNT can quickly penetrate the alveolar epithelial barrier into the interstitial tissue to form a matrix which stimulates resident fibroblasts to produce collagen (Mercer, Scabilloni et al. 2008; Monteiro-Riviere, Inman et al. 2009). Persistent stimulation of fibroblasts has been shown to result in interstitial lung fibrosis *in vivo* and collagen production by lung fibroblasts *in vitro* (Wang, Mercer et al. 2010). Data have also shown that CNT induce fibrogenic cytokines and growth factors such as transforming growth factor-β1, matrix metalloproteinase-9, and fibroblast growth factor-2 in human lung cells both *in vitro* and *in vivo* (Shvedova, Kisin et al. 2008; Wang, Mercer et al. 2010).

5.4 DNA damage

CNT can interfere with the genetic constituents of the cells such as DNA and RNA (Zhu, Chang et al. 2007; Pacurari, Yin et al. 2008; Bonner 2010). These changes in nucleic acid structures can affect cell survival and genomic integrity. MWCNT induce clastogenic (DNA break) and aneuogenic response (chromosomal loss) raising the possibility of mutational changes in the genetic materials of the cell (Muller, Decordier et al. 2008). Mutational changes in K-ras have been observed in the lung of mice after CNT exposure (Shvedova, Kisin et al. 2008). A recent study showed that chronic exposure of SWCNT to lung epithelial cells causes malignant transformation of the cells and tumorigenesis in nude mice (Wang, Luanpitpong et al. 2011). Changes in p53 protein phosphorylation in embryonic stem cells after SWCNT exposure further support the potential genotoxicity and tumorigenicity of the nanomaterial (Zhu, Chang et al. 2007). Genotoxic damage due to CNT exposure has also reported in lung fibroblasts (Kisin, Murray et al. 2007).

6. Conclusion

The unique characteristics of nanomaterials offer potential novel applications as well as potential toxicities. A number of factors including the route of exposure and physico-chemical properties of nanoparticles can affect the biological activities of CNT and their toxic responses. Among the various exposure routes, the pulmonary route is the most common route of exposure to airborne nanoparticles, which have been shown to induce fibrotic and toxicological responses. Properties of nanoparticles such as dispersion status, size and shape, chemical composition and surface functionalization play an important role in the biological activities of nanoparticles. Mechanistic understanding of nanoparticle interactions with cells and tissues is still lacking. Most of the reported toxic effects of nanoparticles are caused by tissue penetration and induction of oxidative stress, DNA damage, inflammation and fibrosis. Careful evaluations of the toxic and fibrogenic effects of nanomaterials are critically needed for the safe and effective use of nanomaterials.

7. Disclaimer

The findings and conclusions in this report are those of the authors and do not necessarily represent the views of the National Institute for Occupational Safety and Health.

8. References

Bandaru, P. R. (2007). "Electrical Properties and Applications of Carbon Nanotube Structures." *Journal of Nanoscience and Nanotechnology* 7(4-5): 1239-1267.

Belyanskaya, L., S. Weigel, et al. (2009). "Effects of carbon nanotubes on primary neurons and glial cells." *Neurotoxicology* 30(4): 702-711.

Bonner, J. C. (2010). "Nanoparticles as a Potential Cause of Pleural and Interstitial Lung Disease." *Proc Am Thorac Soc* 7(2): 138-141.

Brown, D. M., I. A. Kinloch, et al. (2007). "An in vitro study of the potential of carbon nanotubes and nanofibres to induce inflammatory mediators and frustrated phagocytosis." *Carbon* 45(9): 1743-1756.

Cao, X., J. Chen, et al. (2011). "Effect of surface charge of polyethyleneimine-modified multiwalled carbon nanotubes on the improvement of polymerase chain reaction." *Nanoscale* 3(4): 1741-1747.

Cherukuri, P., S. M. Bachilo, et al. (2004). "Near-Infrared Fluorescence Microscopy of Single-Walled Carbon Nanotubes in Phagocytic Cells." *Journal of the American Chemical Society* 126(48): 15638-15639.

Donaldson, K., F. Murphy, et al. (2010). "Asbestos, carbon nanotubes and the pleural mesothelium: a review of the hypothesis regarding the role of long fibre retention in the parietal pleura, inflammation and mesothelioma." *Particle and Fibre Toxicology* 7(1): 5.

Folkmann, J. K., L. Risom, et al. (2009). "Oxidatively Damaged DNA in Rats Exposed by Oral Gavage to C(60) Fullerenes and Single-Walled Carbon Nanotubes." *Environmental Health Perspectives* 117(5): 703-708.

Garg, A. and S. B. Sinnott (1998). "Effect of chemical functionalization on the mechanical properties of carbon nanotubes." *Chemical Physics Letters* 295(4): 273-278.

Herzog, E., H. Byrne, et al. (2009). "Swcnt suppress inflammatory mediator responses in human lung epithelium in vitro." *Toxicol Appl Pharmacol* 234: 378 - 390.

Iijima, S. (1991). "Helical microtubules of graphitic carbon." *Nature* 354(6348): 56-58.

Jacobs, C. B., T. L. Vickrey, et al. (2011). "Functional groups modulate the sensitivity and electron transfer kinetics of neurochemicals at carbon nanotube modified microelectrodes." *Analyst* 136(17): 3557-3565.

Jaurand, M.-C., A. Renier, et al. (2009). "Mesothelioma: Do asbestos and carbon nanotubes pose the same health risk?" *Particle and Fibre Toxicology* 6(1): 16.

Jia, G., H. Wang, et al. (2005). "Cytotoxicity of Carbon Nanomaterials: Single-Wall Nanotube, Multi-Wall Nanotube, and Fullerene." *Environmental Science & Technology* 39(5): 1378-1383.

Kisin, E. R., A. R. Murray, et al. (2007). "Single-walled carbon nanotubes: Geno- and cytotoxic effects in lung fibroblast V79 cells." *Journal of Toxicology and Environmental Health-Part a-Current Issues* 70(24): 2071-2079.

Kolosnjaj-Tabi, J., K. B. Hartman, et al. (2010). "In Vivo Behavior of Large Doses of Ultrashort and Full-Length Single-Walled Carbon Nanotubes after Oral and Intraperitoneal Administration to Swiss Mice." *ACS Nano* 4(3): 1481-1492.

Koyama, S., M. Endo, et al. (2006). "Role of systemic T-cells and histopathological aspects after subcutaneous implantation of various carbon nanotubes in mice." *Carbon* 44(6): 1079-1092.

Lam, C. W., J. T. James, et al. (2004). "Pulmonary toxicity of single-wall carbon nanotubes in mice 7 and 90 days after intratracheal instillation." *Toxicol Sci* 77(1): 126 - 134.

Le Goff, A., M. Holzinger, et al. (2011). "Enzymatic biosensors based on SWCNT-conducting polymer electrodes." *Analyst* 136(7): 1279-1287.

Li, J.-G., W.-X. Li, et al. (2007). "Comparative study of pathological lesions induced by multiwalled carbon nanotubes in lungs of mice by intratracheal instillation and inhalation." *Environmental Toxicology* 22(4): 415-421.

Li, J. J. e., S. Muralikrishnan, et al. (2010). "Nanoparticle-induced pulmonary toxicity." *Experimental Biology and Medicine* 235(9): 1025-1033.

Li, X. and et al. (2010). "Current investigations into carbon nanotubes for biomedical application." *Biomedical Materials* 5(2): 022001.

Li, Z., T. Hulderman, et al. (2007). "Cardiovascular effects of pulmonary exposure to single-wall carbon nanotubes." *Environmental Health Perspectives* 115(3): 377-382.

Liu, W., J. Ernst, et al. (2000). "Phagocytosis of crocidolite asbestos induces oxidative stress, DNA damage, and apoptosis in mesothelial cells." *Am J Respir Cell Mol Biol* 23: 371 - 378.

Manna, S. K., S. Sarkar, et al. (2005). "Single-Walled Carbon Nanotube Induces Oxidative Stress and Activates Nuclear Transcription Factor-κB in Human Keratinocytes." *Nano Letters* 5(9): 1676-1684.

McCarthy, J. R. and R. Weissleder (2008). "Multifunctional magnetic nanoparticles for targeted imaging and therapy." *Advanced Drug Delivery Reviews* 60(11): 1241-1251.

Mercer, R. R., A. Hubbs, et al. (2011). "Pulmonary fibrotic resposnse to aspiration of multi-walled carbon nanotubes." *Particle and Fibre Toxicology*.

Mercer, R. R., J. Scabilloni, et al. (2008). "Alteration of deposition pattern and pulmonary response as a result of improved dispersion of aspirated single-walled carbon nanotubes in a mouse model." *Am J Physiol Lung Cell Mol Physiol* 294(1): L87-97.

Migliore, L., D. Saracino, et al. (2010). "Carbon nanotubes induce oxidative DNA damage in RAW 264.7 cells." *Environmental and Molecular Mutagenesis* 51(4): 294-303.

Mitchell, L. A., J. Gao, et al. (2007). "Pulmonary and Systemic Immune Response to Inhaled Multiwalled Carbon Nanotubes." *Toxicol. Sci.* 100(1): 203-214.

Monteiro-Riviere, N. A., A. O. Inman, et al. (2005). "Surfactant effects on carbon nanotube interactions with human keratinocytes." *Nanomedicine: Nanotechnology, Biology and Medicine* 1(4): 293-299.

Monteiro-Riviere, N. A., A. O. Inman, et al. (2009). "Limitations and relative utility of screening assays to assess engineered nanoparticle toxicity in a human cell line." *Toxicology and Applied Pharmacology* 234(2): 222-235.

Msiska, Z., M. Pacurari, et al. (2009). "DNA Double Strand Breaks by Asbestos, Silica and Titanium dioxide: Possible Biomarker of Carcinogenic Potential?" *Am. J. Respir. Cell Mol. Biol.*: 2009-0062OC.

Muller, J., I. Decordier, et al. (2008). "Clastogenic and aneugenic effects of multi-wall carbon nanotubes in epithelial cells." *Carcinogenesis* 29(2): 427-433.

Muller, J., F. Huaux, et al. (2005). "Respiratory toxicity of multi-wall carbon nanotubes." *Toxicology and Applied Pharmacology* 207(3): 221-231.

Murray, A. R., E. Kisin, et al. (2009). "Oxidative stress and inflammatory response in dermal toxicity of single-walled carbon nanotubes." *Toxicology* 257(3): 161-171.

Nam, C.-W., S.-J. Kang, et al. (2011). "Cell growth inhibition and apoptosis by SDS-solubilized single-walled carbon nanotubes in normal rat kidney epithelial cells." *Archives of Pharmacal Research* 34(4): 661-669.

Oberdorster, G., J. Ferin, et al. (1994). "Correlation between particle size, in vivo particle persistence, and lung injury." *Environ Health Perspect* 102(Suppl 5): 173 - 179.

Oberdorster, G., E. Oberdorster, et al. (2005). "Nanotoxicology: an emerging discipline evolving from studies of ultrafine particles." *Environ Health Perspect* 113: 823 - 839.

Pacurari, M., X. Yin, et al. (2008). "Raw single-wall carbon nanotubes induce oxidative stress and activate mapks, ap-1, nf-kappab, and akt in normal and malignant human mesothelial cells." *Environ Health Perspect* 116: 1211 - 1217.

Palomäki, J., E. Välimäki, et al. (2011). "Long, Needle-like Carbon Nanotubes and Asbestos Activate the NLRP3 Inflammasome through a Similar Mechanism." *ACS Nano* 5(9): 6861-6870.

Poland, C. A., R. Duffin, et al. (2008). "Carbon nanotubes introduced into the abdominal cavity of mice show asbestos-like pathogenicity in a pilot study." *Nature Nanotechnology* 3(7): 423-428.

Porter, D. W., A. F. Hubbs, et al. (2010). "Mouse pulmonary dose- and time course-responses induced by exposure to multi-walled carbon nanotubes." *Toxicology* 269(2-3): 136-147.

Prato, M., K. Kostarelos, et al. (2007). "Functionalized Carbon Nanotubes in Drug Design and Discovery." *Accounts of Chemical Research* 41(1): 60-68.

Radomski, A., P. Jurasz, et al. (2005). "Nanoparticle-induced platelet aggregation and vascular thrombosis." *British Journal of Pharmacology* 146(6): 882-893.

Ravichandran, P., S. Baluchamy, et al. (2011). "Pulmonary Biocompatibility Assessment of Inhaled Single-wall and Multiwall Carbon Nanotubes in BALB/c Mice." *Journal of Biological Chemistry* 286(34): 29725-29733.

Rotoli, B. M., O. Bussolati, et al. (2008). "Non-functionalized multi-walled carbon nanotubes alter the paracellular permeability of human airway epithelial cells." *Toxicology Letters* 178(2): 95-102.

Sager, T. M., D. W. Porter, et al. (2007). "Improved method to disperse nanoparticles for in vitro and in vivo investigation of toxicity." *Nanotoxicology* 1(2): 118-129.

Sakamoto, Y., D. Nakae, et al. (2009). "Induction of mesothelioma by a single intrascrotal administration of multi-wall carbon nanotube in intact male fischer 344 rats." *J Toxicol Sci* 34: 65 - 76.

Sato, Y., A. Yokoyama, et al. (2005). "Influence of length on cytotoxicity of multi-walled carbon nanotubes against human acute monocytic leukemia cell line THP-1 in vitro and subcutaneous tissue of rats in vivo." *Molecular BioSystems* 1(2): 176-182.

Shvedova, A., E. Kisin, et al. (2008). "Inhalation vs. Aspiration of single-walled carbon nanotubes in c57bl/6 mice: Inflammation, fibrosis, oxidative stress, and mutagenesis." *Am J Physiol Lung Cell Mol Physiol* 295: L552 - 565.

Shvedova, A. A., V. Castranova, et al. (2003). "Exposure to carbon nanotube material: Assessment of nanotube cytotoxicity using human keratinocyte cells. "*Journal of Toxicology and Environmental Health-Part A* 66(20): 1909-1926.

Shvedova, A. A., E. R. Kisin, et al. (2005). "Unusual inflammatory and fibrogenic pulmonary responses to single-walled carbon nanotubes in mice." *Am J Physiol Lung Cell Mol Physiol* 289(5): L698-708.

Shvedova, A. A., E. R. Kisin, et al. (2007). "Vitamin E deficiency enhances pulmonary inflammatory response and oxidative stress induced by single-walled carbon nanotubes in C57BL/6 mice." *Toxicology and Applied Pharmacology* 221(3): 339-348.

Simon-Deckers, A., B. Gouget, et al. (2008). "In vitro investigation of oxide nanoparticle and carbon nanotube toxicity and intracellular accumulation in A549 human pneumocytes." *Toxicology* 253(1-3): 137-146.

Soto, K. F., A. Carrasco, et al. (2005). "Comparative <i>in vitro</i> cytotoxicity assessment of some manufacturednanoparticulate materials characterized by transmissionelectron microscopy." *Journal of Nanoparticle Research* 7(2): 145-169.

Upadhyayula, V. K. K. and V. Gadhamshetty "Appreciating the role of carbon nanotube composites in preventing biofouling and promoting biofilms on material surfaces in environmental engineering: A review." *Biotechnology Advances* 28(6): 802-816.

Wang, H., J. Wang, et al. (2004). "Biodistribution of carbon single-wall carbon nanotubes in mice." *J Nanosci Nanotechnol* 4: 1019 - 1024.

Wang, L., V. Castranova, et al. (2010). "Dispersion of single-walled carbon nanotubes by a natural lung surfactant for pulmonary in vitro and in vivo toxicity studies." *Particle and Fibre Toxicology* 7(1): 31.

Wang, L., S. Luanpitpong, et al. (2011). "Carbon Nanotubes Induce Malignant Transformation and Tumorigenesis of Human Lung Epithelial Cells." *Nano Letters* 11(7): 2796-2803.

Wang, L., R. R. Mercer, et al. (2010). "Direct Fibrogenic Effects of Dispersed Single-Walled Carbon Nanotubes on Human Lung Fibroblasts." *Journal of Toxicology and Environmental Health, Part A* 73(5-6): 410-422.

Warheit, D. B., B. R. Laurence, et al. (2004). "Comparative pulmonary toxicity assessment of single-wall carbon nanotubes in rats." *Toxicological Sciences* 77(1): 117-125.

Wilson, M. S. and T. A. Wynn (2009). "Pulmonary fibrosis: pathogenesis, etiology and regulation." *Mucosal Immunol* 2(2): 103-121.

Yang, S.-t., W. Guo, et al. (2007). "Biodistribution of Pristine Single-Walled Carbon Nanotubes In Vivo†." *The Journal of Physical Chemistry C* 111(48): 17761-17764.

Yang, S.-T., X. Wang, et al. (2008). "Long-term accumulation and low toxicity of single-walled carbon nanotubes in intravenously exposed mice." *Toxicology Letters* 181(3): 182-189.

Yokoyama, A., Y. Sato, et al. (2004). "Biological Behavior of Hat-Stacked Carbon Nanofibers in the Subcutaneous Tissue in Rats." *Nano Letters* 5(1): 157-161.

Zhang, Y., S. F. Ali, et al. (2010). "Cytotoxicity Effects of Graphene and Single-Wall Carbon Nanotubes in Neural Phaeochromocytoma-Derived PC12 Cells." *ACS Nano* 4(6): 3181-3186.

Zhang, Y., Y. Bai, et al. (2010). "Functionalized carbon nanotubes for potential medicinal applications." *Drug Discovery Today* 15(11-12): 428-435.

Zhu, L., D. W. Chang, et al. (2007). "DNA Damage Induced by Multiwalled Carbon Nanotubes in Mouse Embryonic Stem Cells." *Nano Letters* 7(12): 3592-3597.

14

Nanoparticles and Nanostructures for Biophotonic Applications

Enzo Di Fabrizio[1,2] et al.*
[1]*Nanostructures Department, Italian Institute of Technology, Genova,*
[2]*BioNEM lab., Departement of Clinical and Experimental Medicine, Magna Graecia University, Viale Europa, Catanzaro, Italy*

1. Introduction

The aim of this chapter is to expound on the theoretical analysis and experimental assessment of NanoParticles (NPs) for imaging, early detection and therapeutic applications. NPs are extremely small particulates with dimensions ranging from few micrometers down to few tens of nanometers. Their characteristics, including size, shape, physical and chemical properties, can be tailored during the fabrication/synthesis process and, on account of these, they would feature certain aspects that may be exploited for applications ranging from drug delivery to the enhancement of the local electric field, and thus the detection of few molecules. In particular, intravascularly injectable NPs (that are sometimes called nanovectors or nanocarriers) are probably the major class of nanotechnological devices of interest for use in cancer or, in general, for the treatment of diseases. On the other hand, aggregates of metallic NPs, either of silver or gold, represent extremely efficient SERS (Surface Enhanced Raman Scattering) substrates. In the following, after a brief description of NPs as a whole, a number of different applications will be discussed.

2. Nanoporous silicon nanoparticles: A drug delivery system

Intravascularly injectable NPs can be conveniently designed or engineered to release drug molecules or imaging tracers with a superior performance with respect to freely administrated agents (Ferrari, 2005; Whitesides, 2003; La Van et al., 2003); to this extent, they represent smart Drug Delivery Systems (DDS).

---

* Francesco Gentile[1,2], Michela Perrone Donnorso[1,3], Manohar Chirumamilla Chowdary[1], Ermanno Miele[1], Maria Laura Coluccio[1,2], Rosanna La Rocca[1], Rosaria Brescia[4], Roman Krahne[1], Gobind Das[1], Francesco De Angelis[1], Carlo Liberale[1], Andrea Toma[1], Luca Razzari[1], Liberato Manna[4] and Remo Proietti Zaccaria[1]
[1]*Nanostructures Department, Italian Institute of Technology, Genova, Italy*
[2]*BioNEM lab., Department of Clinical and Experimental Medicine, Magna Graecia University, Viale Europa, Catanzaro, Italy*
[3]*Nanophysics Department, Italian Institute of Technology, Genova, Italy*
[4]*Nanochemistry Department, Italian Institute of Technology, Genova, Italy*

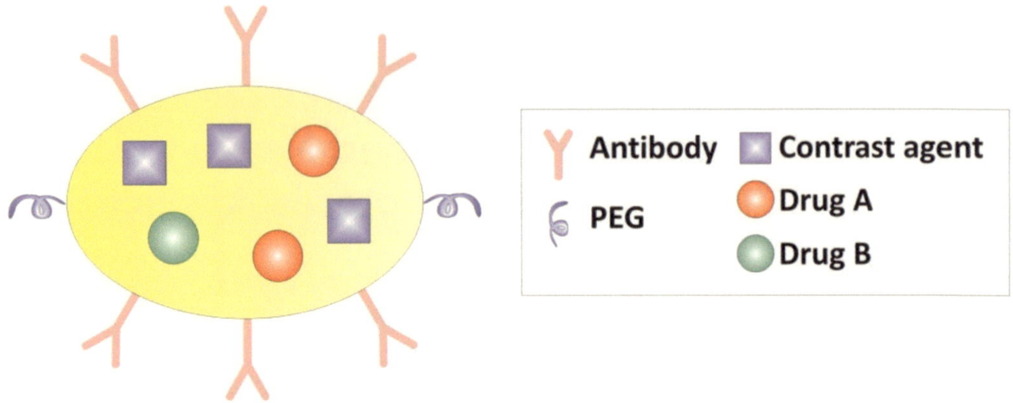

Fig. 1. Cartoon representing a nanoparticle featuring different coverings and payloads.

In other terms, on account on their geometrical, physical and chemical properties (**Fig.1**), the therapeutic or contrast agents would be targeted directly towards the site of interest (malignant cells) with a significant reduction of side effects, and a concurrent increase in the efficiency of delivery.

Liposomes are the simplest form of NPs (and, accordingly, they are sometimes referred to as first generation nanovectors) (Klibanov et al., 1991; Park, 2002; Crommelin & Schreier, 1994). Established for the treatment of Kaposi's sarcoma more than 10 years ago, nowadays certain liposomes based NPs are still being used for cancer treatment. Literature records a massive number of second generation NPs, (which, differently from liposomes, can be artificially produced using nanofabrication techniques), including polymer-based nanovectors, silicon and silica NPs and metal-based nanovectors like nanoshells (Kircher et al., 2003; Schellenberger et al., 2002; Zhang & Shang, 2004; Cohen et al., 2003; Hirsch et al., 2003; Langer, 1998; Duncan, 2003; Gilles & Frechet, 2002).

Despite this, and notwithstanding the merits that such an abundance of devices provides in terms of novel technological foundations and opportunities, only a small amount of these nanovectors would be really effective in delivering drugs or contrast agents (Ferrari, 2005).

A number of third generation NPs is currently under development which feature advanced properties and thus more efficacious delivery modalities, including nanoporous NPs with/without reduced silver for SERS analysis, mesoporous silicon particles as a multistage delivery systems, nanoporous NPs enhancing, via geometrical confinement of gadolinium-based contrast agents, T_1 contrast (Tasciotti et al., 2008; Jeyarama et al., 2010).

In general, an ideal nanoparticle should be able (i) to navigate into the circulatory system avoiding the immune system and recognizing the diseased cells (biological target) with high selectivity; (ii) to adhere firmly to them and (iii) to allow for endocytosis process. Targeting methods have been largely investigated and range from specific (covalently linked ligands decorate nanovectors and may recognize antibodies over-expressed on the cells of interest) to non specific (that are, mechanisms based on the size, shape and physical properties of the nanovector including density, porosity, surface charge) (Decuzzi, 2006a, 2006b, 2007).

Regardless the particular mechanisms of transport and adhesion that NPs can experience, it is clear that a thorough understanding of the physics behind these phenomena plays a fundamental role. Realistically, mathematical models provide an unprecedented tool for predicting the behaviour of NPs within the macro/microcirculation, thus also supplying a rationale for the best design of NPs (Gentile, 2007, 2008). In DDS the choice of bulk material constituent the nanocarrier is a key point since it must fulfill a tailored biological behavior (bioactivity, biocompatibility, biodegradability), it must improve payload capacity and be harmlessly eliminated from the body in a reasonable period of time after releasing the cargo and having carried out possible diagnostic function (Park et al., 2009). While many proposed nanocarriers do not meet these requirements, porous silicon (PSi), considered as silicon crystal having a series of voids, is a promising materials for its biocompatibility and biodegradability (Granitzer & Rumpf, 2010; Canham, 1997; O Farrel et al., 2006) which results in decomposition products not harmful for biological system. Porous silicon dissolves in body fluids into orthosilicic acid, commonly found in everyday foods and efficiently excreted from the body through the urine (Park et al., 2009). Thanks to the porous structures, silicon nanoporous nanoparticles show a great surface/volume ratio (200-800 m^2/cm^3) (Halimaoui, 1995), a very attractive characteristic since surface can be used to load drugs by physisoprtion process and modified with molecules that promote cell adhesion. Pore width, porosity and nanoparticles size can be tuned by adjustment of the parameters during the fabrication process. When the pore size is in the range of 2-5 nm, physisorbed drugs are efficiently entrapped and, due to the effect of the quantum confinement of the silicon structure (Godefroo et al., 2008), particles show emission at 620 nm (red-orange) at room temperature under UV illumination (wavelength of 365 nm), a very attractive feature for the realization of theranostic nanoparticles (Janib, 2010). Pores population can be divided in two different types: open pores that are connected to each other and to the external surface and closed pores isolated from the outside, that are less useful for drug delivery purpose.

Porosity is defined as the fraction of void in the porous structure. High porosity means great surface/volume ratio and it is preferable for drug delivery device, however, due to the surface tension, too high porosity silicon layer can undergo to collapse during the fabrication process and the obtained nanoparticles can disintegrate in water solution. Size and shape are two important features of nanoparticles used in drug delivery systems. Nanoporous silicon nanoparticles with size of 30-100 nm are suitable carriers for many anticancer agents (Petros & DeSimone, 2010; Mitragotri & Lahan, 2009).

2.1 Fabrication and characterization of NanoPorous Silicon Particles

The most widely used method to fabricate Nanoporous Silicon Nanoparticles (NPNPs) is a process that includes several steps, starting from the production of nanoporous silicon film by an electrochemical etching in ethanol/HF solution of Si wafer, in which silicon acts as an anode and a platinum electrode acts as a cathode. Since Si wafer surface is hydrophobic, ethanol increases the wettability of the substrate allowing the electrolyte penetrating into the pores and also helps in removing the H_2 bubble from the sample surface formed during the anodization process (Canham, 1997). A constant current density is applied to allow the formation of homogeneous porous layer. Samples are rinsed in de-ionized water, then in ethanol and pentane, and sonicated in water for a time required to remove all the nanoporous silicon film from the crystalline silicon substrate. The film is then fractured by ultrasonication and filtered through filtration membrane and centrifuged in order to select the desired particles size. It is possible to vary NPNPs morphology (pore size and distribution, pores interconnectivity, porosity and particles diameter) changing the fabrication parameters such as etching time, anodization current density, HF concentration, wafer type and ultrasonication time. Tab.1 shows the effect of some fabrication parameters on the formation of NPNPs as produced by the authors. Freshly anodized PSi has a remarkable surface hydrophobicity (contact angle ~110°, depending on porosity and surface roughness), a very attractive property that could be used in biomedical field to design nanostructure that enable plasma protein harvesting and concentration (Pujia et al., 2010). However hydrophobicity is a limitation in case of hydrophilic drugs loading, in this case porous silicon hydrophobic behavior has to be changed in hydrophilic one by thermal treatment just before porous silicon film sonication. To prevent particles aggregation and dissolution fabricated NPNPs are stored at 4°C until using.

Sample	**Etching conditions (HF 25%/ethanol 1:2 v/v)**	**Ultrasonication time**	**Particles Size**	**Pore size**	**Gravimetric Porosity**
p-type	J =10 mA/cm^2 for 300 sec	10 min	101± 34 nm	< 20 nm	80%
p-type	J =10 mA/cm^2 for 720 sec	10 min	75 ±24nm	< 10 nm	87%
p-type	J =10 mA/cm^2 for 720 sec	20 min	30 ±15nm	< 10 nm	87%
p++	J =10 mA/cm^2 for 720 sec	10 min	65±24 nm	< 5 nm	65%

Table 1. Characterization of p-type silicon nanoparticles varying the fabrication parameters.

The porosity of the porous silicon layer, is determined by gravimetric measurements, weighing the silicon substrate both before and after anodization (m1 and m2 respectively) and again after the complete dissolution of the porous layer by 25% KOH solution (m3). Porosity(P) is calculated by the relation P = (m1-m2)/(m1-m3). Particles Dynamic Light Scattering (DLS), Transmission Electron Microscopy (TEM) and Scanning TEM (STEM) analyses are commonly used to determine hydrodynamic size and pores width of NPNPs. Porous silicon nanoparticles can be investigated using BET – BJH theory as well. BET (Brunauer - Emmet - Teller) adsorption isotherms allow for the calculation of surface / volume ratio (m^2/cm^3) and Barrett-Joyner-Halenda (BJH) method can be used in pores width (nm) and volume (cm^3/g) determination. **Fig.2** shows TEM, STEM and DLS analysis of nanoparticles as developed in the authors' laboratory. Nanoparticles chemical analysis, performed by X-ray energy dispersive spectroscopy (EDS), indicates that fluoride is a common contaminant that is residual from the electrolyte solution used during the fabrication and disappears during annealing at 300°C as shown in **Fig.3**.

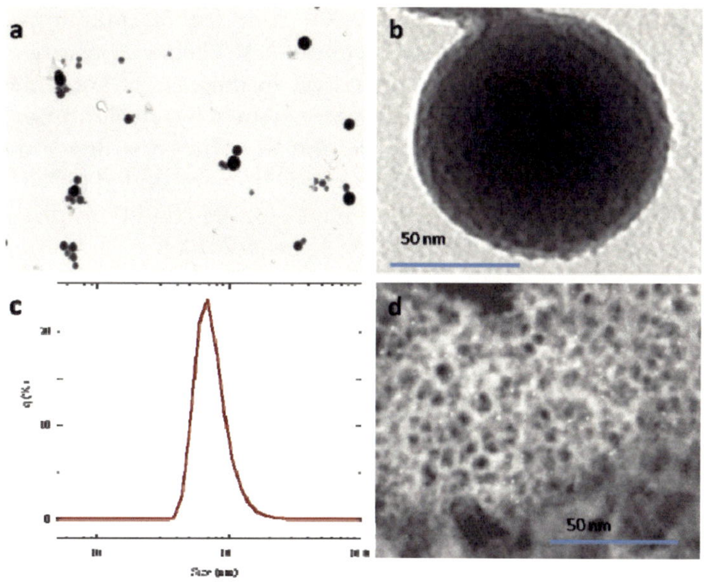

Fig. 2. TEM images (a, b), DLS size distribution (c) and STEM-HAADF image (d) of 75 nm porous silicon nanoparticles

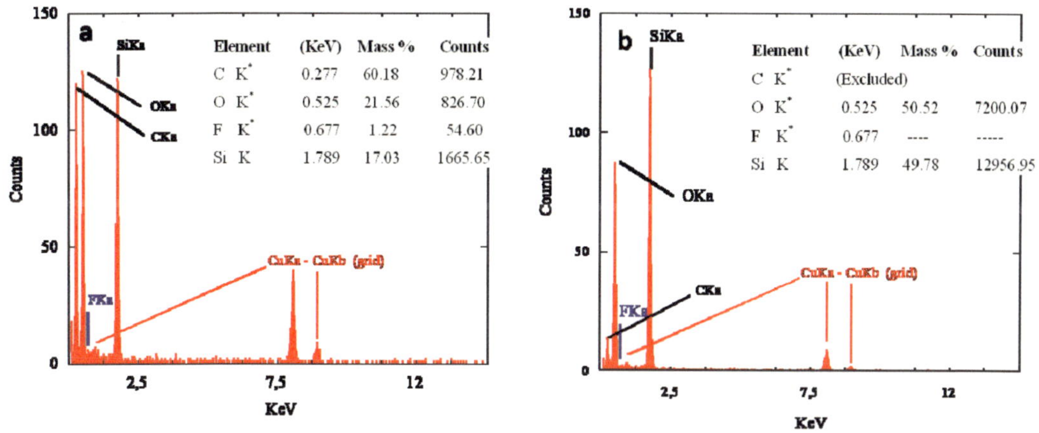

Fig. 3. Chemical analysis with EDS of nanoporous silicon NPs before (a) and after (b) annealing at 300°C.

In order to evaluate the cytotoxicity of NPNPs, apoptosis test has been performed by using iodure propidium agent and cytofluorimetric analysis. Human colon carcinoma cells (HCT116) and health monocytes (THP1) were incubated with 100 µg of two different types of NPNPs (before and after annealing) for 48 h in RPMI 1640 medium at 37°C at 5% CO_2 and cell viability detected and compared with control samples treated with medium only (**Fig.4**). No significant toxic effect was observed in the two cell lines incubated with NPNPs compared with controls. In order to evaluate the really usefulness for therapeutic applications, drug loading can be carried out by physisorption process, incubating NPNPs with drug in water at room temperature overnight. Loaded nanoparticles can be recovered from the solution by centrifugation at 12,000 rpm for 30 minutes. The amount of incorporated drug can be quantified through UV-Visible spectroscopy analysis of supernatant compared with drug standard curve. In the case of anti cancer Doxorubicin, loading test performed in the authors' laboratory, shows 5% (weight) payload for NPNPs obtained from silicon film annealed at 260°C for 4h, while a stronger thermal treatment (500°C for 12 h) can enhance physisorption process. Drug embedded NPNPs could be stored in DI water at 4°C for long time while under physiological conditions dissolve in 120 hours releasing the loaded drug (Pujia et al., 2010; De Angelis, 2010a).

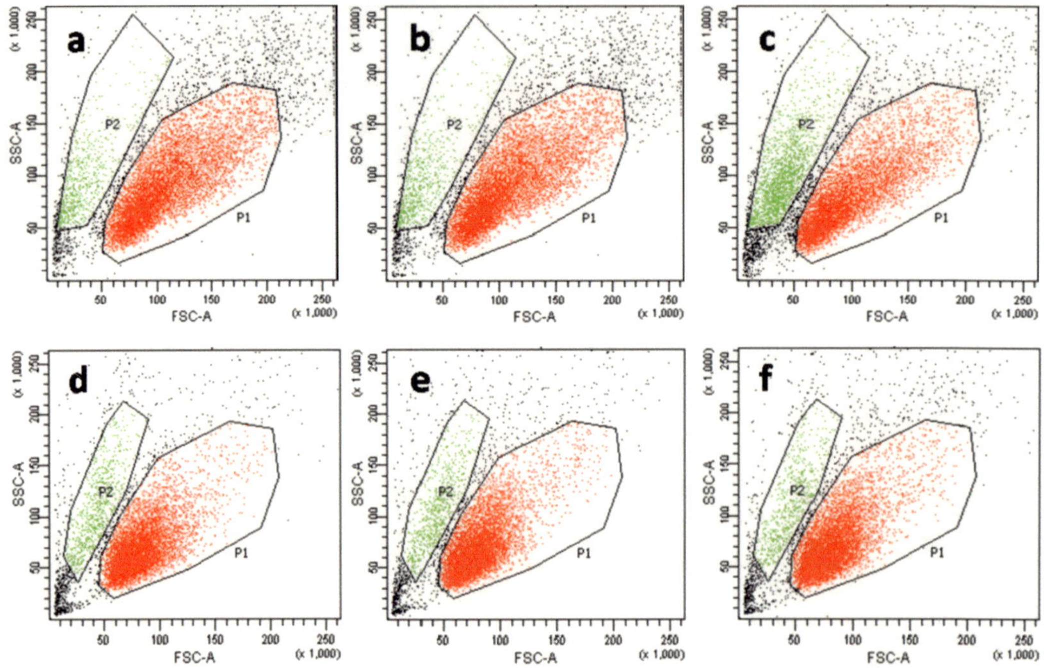

Fig. 4. Density plot of HCT116 (a, b, c) and THP1 (d, e, f) cell lines. Gate P1 shows cell viability %. a) HCT116 cells control without NPNPs, (P1 79%). b) HCT116 cells incubated with not annealed NPNPs (P1 78%). c) HCT116 cells incubated with annealed NPNPs (P1 68%). d) THP1 cells control without NPNPs (P1 83%). e) THP1 cells incubated with not annealed NPNPs (P1 82%). f) THP1 cells incubated with annealed NPNPs (P1 82%).

In conclusion, porous silicon nanoparticles represent a powerful and versatile tool in developing new drug delivery strategies, with high loading capacity, biocompatibility, cheap and scalable fabrication process. Porous silicon-based nanocarriers properties (size, shape and surface chemistry) can be easily tuned operating on the fabrication parameters to achieve a tailored biological behavior and improved bioavailability of transported drug.

3. Core-shell nanorods for light emitting applications

Inorganic semiconductor nanocrystals hold a great promise for emerging nanotechnologies, since they feature physical and chemical properties unique to the nanometer length scale, which are useful for diverse applications such as microelectronic and optical devices, as well as for sensing. These properties depend not only on the nanoparticle's composition, but also on its *size, shape,* and *mode of organization.* For this reason, a major effort has been dedicated in recent years to the development of synthetic methods providing control over morphological parameters of the nanocrystals and their assembly into organized structures. Colloidal synthesis, where judicious choice of protective ligands is employed to morphologically control crystal growth, has proven to be the most versatile approach for the synthesis of non-spherical nanoparticles, providing a variety of shapes, such as rods, prisms, cubes, disks, and others. In this section we will focus on *rod-shaped* nanocrystals, whose shape anisotropy is translated into polarized physical properties (Hu et al., 2001), and which can be harnessed for the creation of polarized light emitting materials (Kazes et al., 2002, 2004; Rizzo et al., 2009). For an extensive review on the physical properties of rod-shaped nanocrystals we refer the reader to reference (Krahne et al., 2011a). In particular, we will discuss a novel core-shell architecture for nanorods, in which a spherical core of a smaller band gap material is embedded in a rod-shaped shell of a high band gap semiconductor (Talapin et al., 2003; Carbone et al., 2007) as sketched in **Fig.5a**.

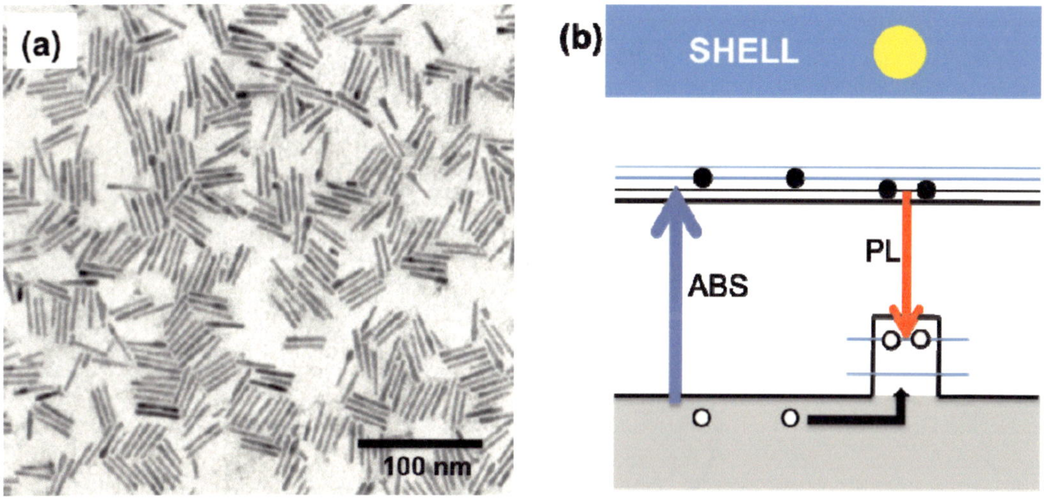

Fig. 5. (a) Transmission electron microscopy image of core shell nanorods. Scheme of the CdSe/CdS core-shell nanorod architecture and of the related band structure.

Such a system is often referred as "dot-in-a-rod", where the charge carriers can be excited in the UV-blue spectral region and then quickly (on the picosecond timescale) relax into the more long living core states from which the light emission occurs (Lupo et al., 2008). The core-shell nanorods have numerous advantages concerning their optical emission properties: (i) the stronger confinement in the core leads well defined energy levels for the optical transitions. (ii) The shell passivates the surface states of the emitting low band gap material. (iii) Due to their rod shape they have an enhanced absorption cross section with respect to spherical particles. In the specific case of CdSe/CdS core shell nanorods the lower energy levels of the holes are localized in the core, whereas the electrons are mostly delocalized over the rod volume, as illustrated in **Fig.5**. This particular electronic level structure proved to reduce non-radiative Auger recombination significantly (Zavelani-Rossi, 2010a) and therefore provides great advantages for lasing devices. Furthermore, the energy of the emitting light is determined by the quantum confinement in the core, which allows to tune their emission wavelength independently of the rod length, at least to a certain extent (Krahne, 2011b), as can be seen in **Fig.6**, which shows emission spectra of core shell nanorods with different core diameter and rod length.

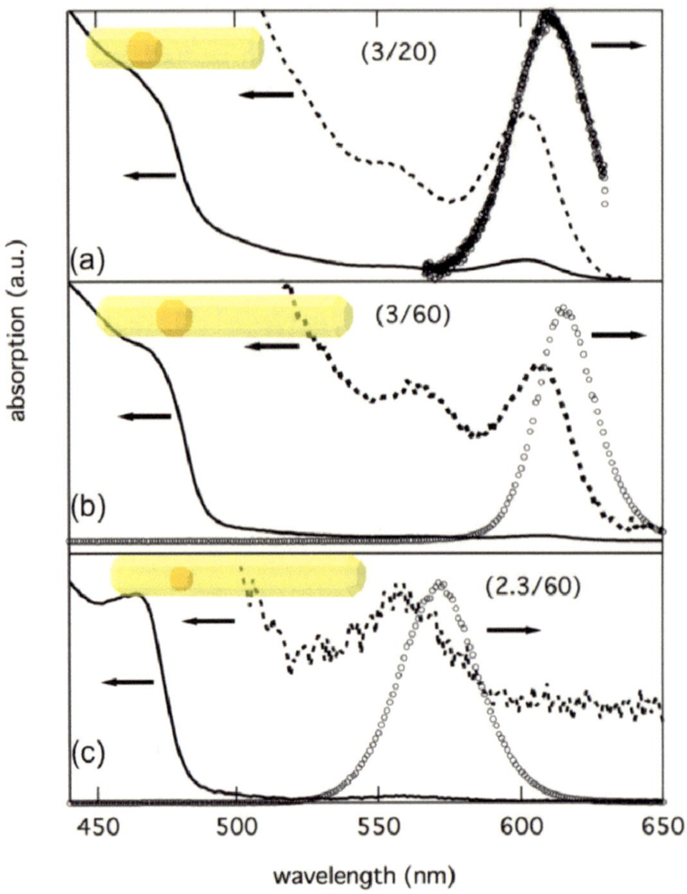

Fig. 6. Absorption and emission spectra of core-shell CdSe/CdS nanorods with different core sizes and different lengths. The core diameter and the rod length are specified by the numbers in brackets in units of nm. The emission wavelength is dominated by the core size. Taken with permission from (Krhane et al., 2011a).

Light emission from the nanorods can be obtained either by optical or via electrical pumping. In the latter case the charge carriers are injected via external electrodes that are in contact with the nanorods. Metal electrodes have the disadvantage that a Schottky barrier is formed at the interface with semiconductor material which significantly hinders the charge injection. Furthermore, the nanorod luminescence gets quenched by a direct contact of the nanorods with a metal, for example with gold. Instead, the implementation of a nanorod layer in a sandwich-like geometry, in between hole- and electron-injection organic layers, has proven to be a successful approach to fabricate light emitting diodes (LED) based on semiconductor nanorods as the active material (Rizzo et al., 2009). **Fig.7** illustrates this fabrication scheme where an ITO (indium-tin-oxide) substrate is used as a back electrode onto which *a* *N*,*N*=-bis(naphthalen-1-yl)-*N*,*N*=-bis(phenyl)benzidine (-NPD) hole injection layer (HIL) doped with 2,3,5,6-tetrafluoro-7,7,8,8-tetracyanoquinodimethane (F4-TCNQ) and a CBP hole transporting layer (HTL) were thermally evaporated

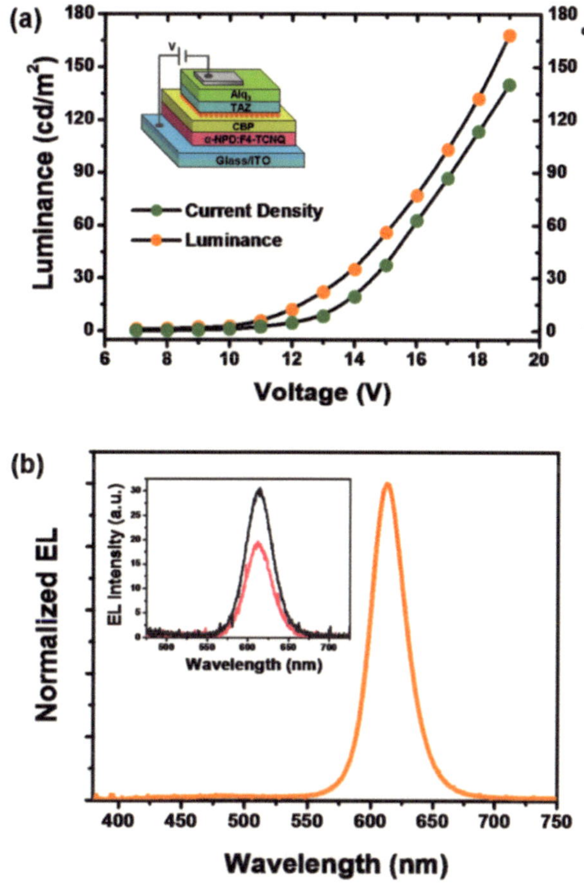

Fig. 7. (a) Current density and luminance from a LED based on an oriented layer of laterally aligned core-shell nanorods as illustrated by the scheme in the inset.
(b) Electroluminescence spectrum. The inset shows two spectra for orthogonal polarization directions. Taken with permission from (Rizzo et al., 2009).

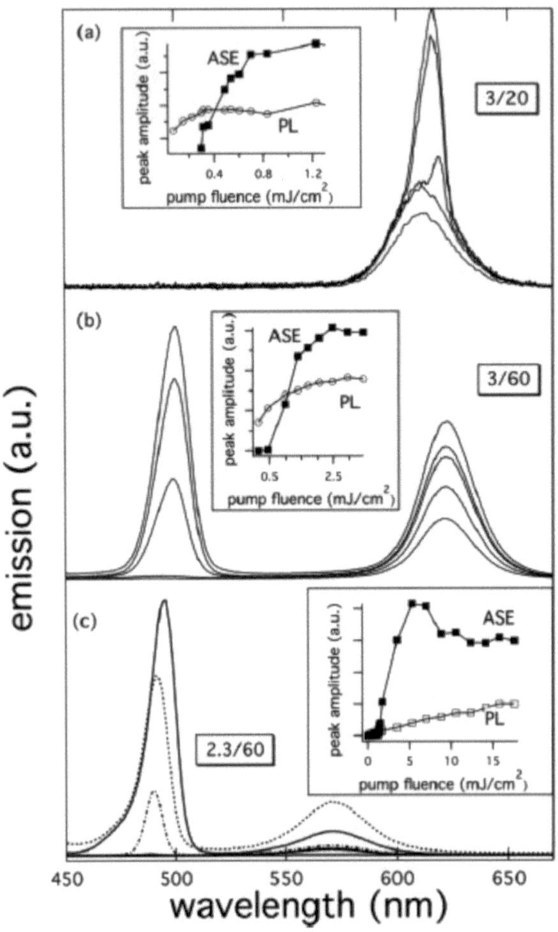

Fig. 8. Amplified spontaneous emission from a dense layers of core-shell nanorods with different core and shell size recorded at different values of pump fluence. Taken with permission from (Krhane et al., 2011a).

Then, a self-assembled oriented layer of laterally aligned nanords was transferred onto the CBP via a stamping technique. After the nanorod deposition the structure was over-coated with a 3-(4-biphenylyl)-4-phenyl-5-*t*-butylphenyl-1,2,4-triazole (TAZ) hole blocking layer (HBL), the tris(8-(hydroxyl-quinoline) aluminum (Alq3) electron transporting layer (ETL), and LiF/Al electrodes. In this type of device structure the excitation formation in the nanorod layer can occur either *via* charge trapping or *via* Forster energy transfer process from the organic material (Li et al, 2005; Anikeeva, 2007).

Amplified spontaneous emission (ASE) can be observed from dense aggregates of nanorods, for example in the form of nanorod layers fabricated by drop deposition from highly concentrated nanorod solutions onto planar surfaces. **Fig.8a** shows ASE from core states recorded from relatively short nanorods with large CdSe core. For rods with a length significantly larger than 25 nm ASE from the shell states, at 490 nm, was observed, while the photoluminescence from the core transitions was maintained, as can be seen in **Fig.8b-c**.

Lasing devices based on colloidal semiconductor nanorods have, to the best of our knowledge, only been obtained by optical pumping so far. In order to obtain lasing, an optical gain medium has to be positioned into a resonant cavity that provides sufficient feedback. Core-shell nanorods have demonstrated optical gain both from the core (Zavelani-Rossi, 2010b), and recently also from the shell emission (Krahne et al., 2011b). A conventional approach to obtain optical feedback is to embed the optical gain medium into an external resonator, for example a physical cavity consisting of a series of Bragg mirrors. However, this approach is not straightforward for self assembled layers of nanocrystals as gain medium because the roughness and thickness of this layer cannot be well controlled. An innovative solution to this problem was demonstrated by Zavelani et al. who used the nanorod layer itself as a resonant cavity. In this work ordered assemblies of nanorods were obtained via the coffee stain effect, i.e. the fluid dynamics in an evaporating droplet (Zavelani-Rossi, 2010b). Here the nanorods self assembled in large-scale ordered superstructures that are reminiscent of nematic/smectic liquid crystal phases. In particular, a dense and highly ordered region was obtained at the edge of the film that formed the outer ring (see **Fig.9a**). Within this outer ring the rods were well aligned and on average the long axis of the nanorods was oriented parallel to the ring edge. From such ordered regions of nanorods polarized emission (Carbone et al., 2007) and directionally dependent photoconductivity (Persano et al., 2010) have been observed. **Fig.9b** shows lasing spectra recorded from regions of the outer ring demonstrating that the lateral facets of the ring can function as a Fabry-Perot resonator.

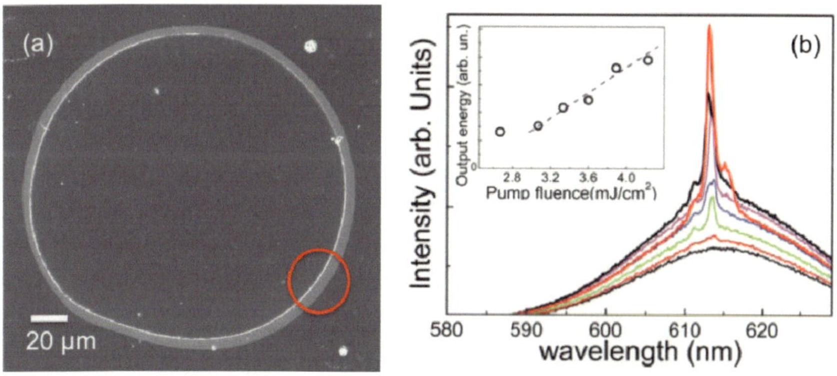

Fig. 9. Scanning electron microscopy image of a coffee stain ring that showed lasing. The red circle illustrates a spot from which lasing spectra were recorded. (b) Emission spectra above and below lasing threshold. The inset shows the characteristic input-output curve of a lasing device.

Such self-assembled micro-lasers provide new possibilities for the integration of narrow band emitters into device architectures such as lab-on-chip for point of care diagnostics, or for optical components in local area network datacom structures. Although the technology regarding light emitting devices based on colloidal nanorods is still in its infancy, the results described in this section are very encouraging and this bottom -up technology can be expected to take larger impact in the lighting industry in the near future.

4. Elongated nanoparticle arrays

The label-free ultrasensitive detection of biological molecules such as proteins, nucleic acids etc. is of utmost importance in the field of clinical medicine, especially with regards to the early diagnosis of diseases (Rosi & Mirkin, 2005). Vibrational spectroscopies (i.e. Raman Scattering and IR absorption) allow for direct detection of biological species (Colthup et al., 2010), but their cross-sections are extremely low in common experimental conditions. The possibility to excite localized surface plasmon resonances (LSPRs) in metallic nanoparticles has indicated a feasible and efficient way for enhancing the electromagnetic field on the local scale (Bohren & Huffman, 1998; Nie & Emory, 1997).

Therefore, the combination of plasmonic nanostructures with vibrational spectroscopies can be used to manipulate light-matter interactions, giving rise to fascinating perspective towards the production of novel biosensor devices (Anker et al., 2008; Das et al., 2009; De Angelis, 2008, 2010b). There is thus intensive activity oriented at the fabrication of tailored nanostructures endowed with the desired plasmonic properties. Metal nanoparticles present specific optical properties that depend on their size and geometry. While a symmetrical shape allows for polarization-independent plasmonic excitation, a dichroic absorption is expected in elongated NPs (Toma et al., 2008; Fazio et al., 2011). Beside this, the opto-plasmonic response can be additionally tailored by acting on the nanoparticle distribution and mutual coupling (Fischer & Martin, 2008).

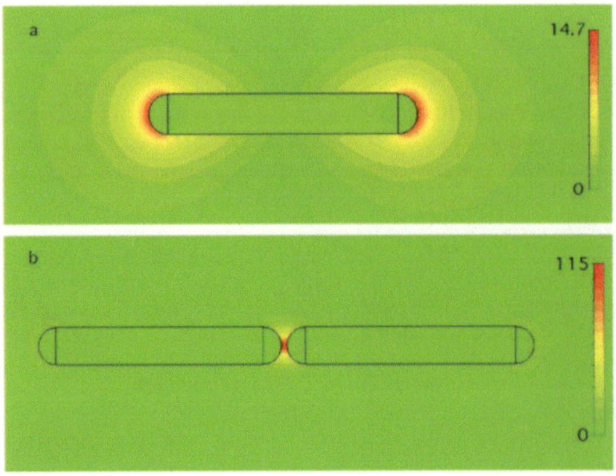

Fig. 10. (a) Calculated absolute value of the electric field distribution around an isolated nanostructure. The illuminating plane wave impinges perpendicularly on the array, and is polarized along the long axis of the elongated nanoparticle. The metallic structure length is 410 nm, while its width and height are set to 60 nm. The excitation wavelength is ≅ 1.9 μm (b) Field distribution in the case of coupled nanostructures (gap width: 10 nm).

In order to elucidate this peculiar behavior, we have performed 3D numerical simulations using a commercial software based on a finite integration technique (CST, Computer Simulation Technology, Darmstadt, Germany). **Fig.10** shows the absolute value of the electric field around the nanostructures, on a plane that is perpendicular to the direction of the illuminating wave and cuts the nanostructure exactly at its half height. While an isolated particle concentrates the radiation at its extremities **(Fig.10a)**, a dimer (i.e. a couple of closely spaced particles) creates a "hot spot" (Fischer & Martin, 2008; Stockman et al., 1994) in correspondence of the inter-particle nanocavity (**Fig.10b**).

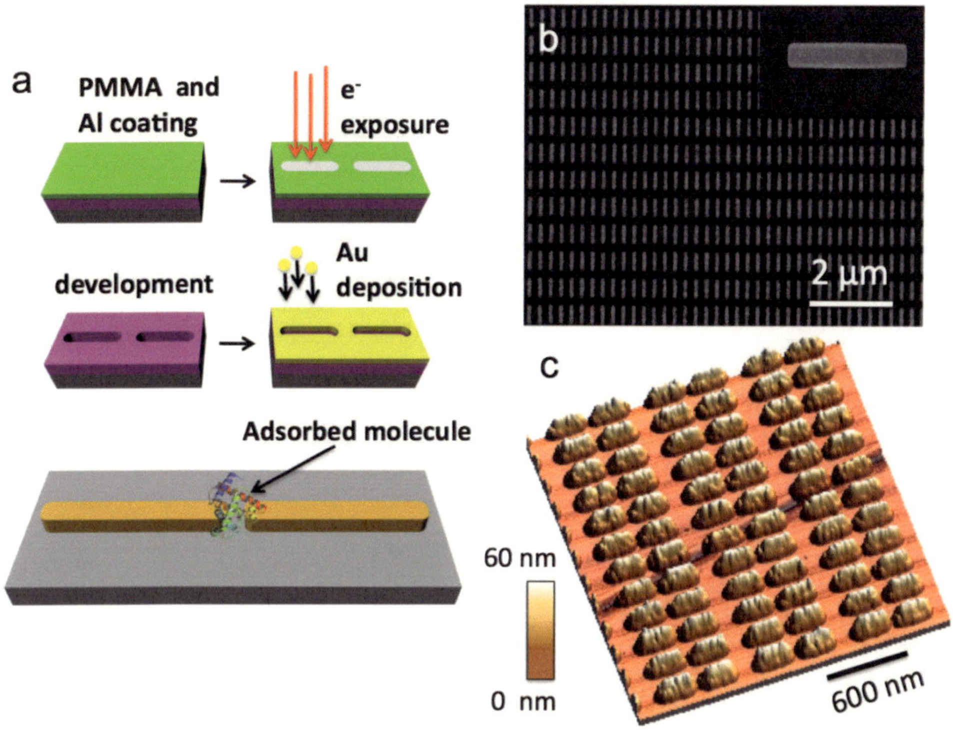

Fig. 11. (a) Schematic block diagram of the fabrication process. The last cartoon highlights the nano-biosensor concept: fabrication of a highly sensitive and specific device based on plasmonic signal enhancement. (b, c) Representative SEM and AFM images of two different nanoparticle arrays fabricated by EBL technique. The NPs are 410 nm (b) and 200 nm (c) long, while width and height are both set at 60 nm for all the structures.

Here we investigate these two specific examples, i.e. uncoupled and dimer nanoparticle arrays, with the aim to prove that their field enhancement and localization capabilities can be used for high-sensitivity Raman spectroscopy. A schematic diagram, summarized in **Fig.11a**, elucidates the main steps involved in the fabrication process. A 120 nm thick layer of PMMA (950K) was spin-coated on a CaF_2 (100) substrate. To prevent charging effects during the electron exposure, a 10 nm thick Al layer was thermally evaporated on the PMMA surface. Electron beam direct-writing of the nanoparticle patterns was carried out using a high resolution Raith150-Two e-beam writer at 15 keV beam energy and 25 pA beam current. After the Al removal in a KOH solution, the exposed resist was developed in a conventional solution of MIBK:IPA (1:3) for 30 s. Then, a 5 nm adhesion layer of Ti and a 60 nm Au film were evaporated, using a 0.3 Å/s deposition rate in a 10^{-7} mbar vacuum chamber (Kurt J Lesker PVD75). Finally, the unexposed resist was removed with acetone and rinsed out in IPA. Large scan overviews of different nanostructure arrangements are reported in Fig. 2b,c. The sample topography has been characterized recurring to scanning electron microscopy (SEM) and atomic force microscopy (AFM, Veeco MultiMode with NanoScope V controller) equipped with ultra-sharp Si probes (ACLA-SS, AppNano) and operating in tapping mode. The resulting arrays present, as shown in **Fig.11b,c**, a high degree of reproducibility with a surface RMS roughness value of around 1 nm.

The optical properties of the nanoparticle array were investigated by means of spectroscopic transmission of polarized light in the range between 450 and 900 nm (see **Fig.12a,b**). For this purpose, we used a fiber-optic spectrometer (AvaSpec-256, Avantes) while the light source was a combined deuterium-halogen lamp (AvaLight-DHc, Avantes). The polarization of the incident light was varied from parallel to perpendicular orientation with respect to the long axis of the nanostructure. **Fig.12a** shows the results concerning isolated nanoparticles while in **Fig.12b** we report transmission spectra for coupled nanostructures. In both cases the optical transmittance spectra present evidence of a clear anisotropic behavior; for perpendicular polarization a localized minimum around 620 nm is found. This is the typical behavior exhibited by sub-wavelength metal nanoparticles sustaining a localized surface plasmon resonance (Bohren & Huffman, 1998). For parallel polarization a red shifted extinction peak, centered around 700 nm, is observed. This can be attributed to a higher order plasmonic resonance (Aizpurua et al., 2005).

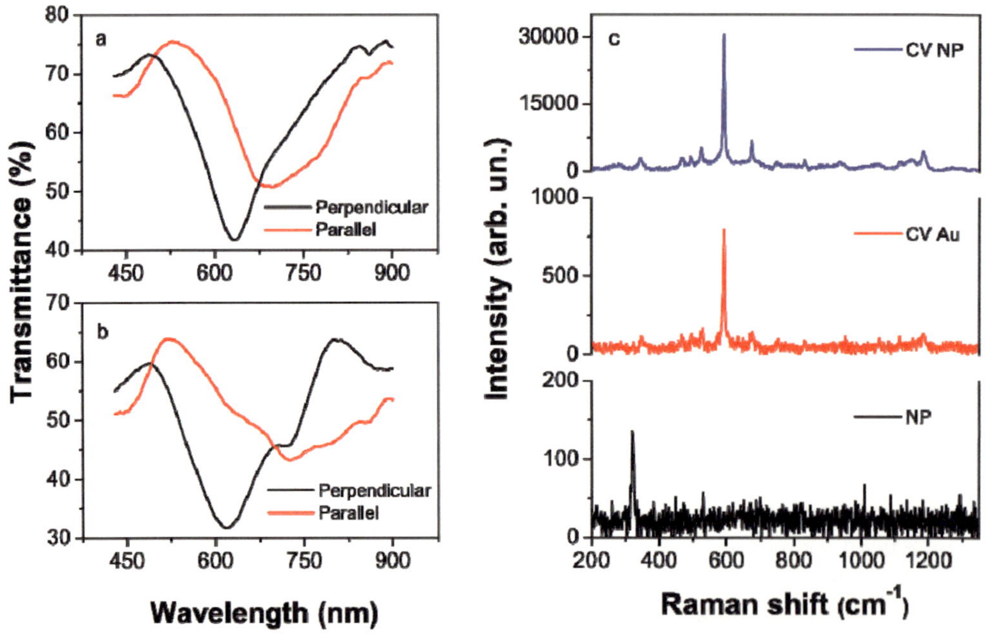

Fig. 12. Transmission optical spectra of single (a) and coupled (b) nanoparticle arrays. Red/black lines are for perpendicular and parallel polarization respectively. (c) SERS spectra of cresyl violet molecules deposited on NP sample (blue line) and flat Au film (red line). Background measurement performed on bare NP structure is also reported (black line).

Moving from the isolated nanoparticle array (**Fig.12a**) to the dimer case (**Fig.12b**) we can notice a broadening and a slightly shift of the resonance peaks due to near field coupling (Fischer & Martin, 2008).

In order to probe the enhancement behavior in Raman spectroscopy related to the plasmon excitation, the sample of **Fig.11b** was immersed in a cresyl violet (CV) solution (3.46 μM in H_2O) for 15 min. The sample was gently rinsed in DI-water to remove molecules in excess not chemisorbed on the metal surface and then dried in nitrogen flow. SERS measurements were carried out by means of inVia microspectroscopy (Renishaw). The samples were excited by 633 nm laser wavelength (laser power = 0.14 mW and accumulation time = 50 s) through a 150X objective. The measurements were performed on bare nanoparticle arrays (sample NP), on CV molecules chemisorbed on a flat Au film (sample CV Au) and on CV deposited on the nanostructure arrays (sample CV NP). As shown in **Fig.12c** (NP line) no characteristic Raman band, except at around 320 cm^{-1} related to the Ca-F vibration from the substrate, is observed. The characteristic vibrational bands of CV are observed in the SERS spectrum (CV NP trace). Intense Raman bands centered at around 591, 882, 927 and 1189 cm^{-1} can be attributed to the N-H_2 rocking vibration, two benzene group bending, out-of phase N- H_2 rocking vibration, and combination of N-H_2 rocking and C-H_X rocking, respectively (Vogel, 2000; Sackmann et al., 2007). CV molecules chemisorbed over a flat Au film are also shown in **Fig.12c**. We can clearly observe that there is a giant enhancement in Raman signal for CV NP sample with respect to the CV Au one. The evaluated SERS enhancement for the fabricated device is around 10^7.

5. NPs based plasmonics devices for SERS applications

Metallic NPs are characterized from a great mobility of theirs electrons, resulting in a characteristic ability of sustaining coherent electronic oscillations, when light is impinged on them. The phenomenon, interesting generally the metallic surfaces, consists of the localized surface plasmons (LSP) formation (Raether, 1988; McCall et al., 1980; Haynes et al., 2005; Das et al., 2009; Kneipp et al., 2002) associated to the collective oscillation of the electrons moving in the small metal particles volume. The local electromagnetic field near the nanostructures surface results enormously enhanced. This property makes metallic NPs particularly interesting in spectroscopy. In Raman spectroscopy, this phenomenon is known as Surface Enhanced Raman Scattering (SERS) effect; it allows the detection or very diluted solutions, where very few molecules are present, overcoming the normal limit of Raman due to the great fluorescence which, in these cases, may cover molecules signals (Kneipp et al., 1997; De Angelis, 2008, 2010b). In fact, when a molecule is very close to a metallic nanostructure the LSP formation, under the laser effect, generates a giant enhancement of Raman signal (Das et al., 2009; Nie & Emory, 1997; Creighton et al., 1979).

It is clear that LSPR and, consequently, SERS intensity is strongly influenced from size, shape, inter-particle spacing of the nanoparticles and the dielectric environment of material. For particles diameters d « λ (λ = wave length of excitation laser), electrons move in phase on the excitation wave plan with the consequently formation of polarization charges and, then, of a dipolar field on particles surface. The estimated enhancement for Au and Ag isolated particles is around to 10^6 –10^7 (Otto, 1984; Kneipp & Kneipp, 2006).

For this reason the roughness of a surface obtained from a NPs deposition is of fundamental importance and overall the fabrication of specific metal nanostructures, in which size, shape, inter-particle spacing of the nanoparticles may be chosen carefully, has become an important factor for research on plasmonic devices (Gunnarsson et al., 2001). Traditionally, as already mentioned above, NPs for SERS applications were prepared in chemical way, producing Ag colloids to whom attacking bio-molecules for the detection (Xu et al., 1999; Kneipp et al., 2004; Hao & Schatz, 2004) with good results regarding SERS effect, but with some limits in the reproducibility of the systems. Afterwards, other kinds of SERS substrates were realized, as metallic films (Constantino et al., 2002; Garoff et al., 1983) and nanostructures (NPs) with characteristic length scale in the nanometer range, as nanovoids, nanoshells, nanorods, nanorings and nanocubes (Le Ru et al., 2008; Grzelczak et al., 2010; Tao et al., 2008). Recently development of nanofabrication techniques, as electron beam lithography (EBL) or focalized ion beam (FIB), has improved plasmonic nanostructures control and consequently has improved the efficiency of these nano-devices, focalizing the attention on substrate-bound nanostructure fabrication (Zhang et al., 2006).

Here, the fabrication of silver/gold nano-aggregates into well-defined lithographic structures is described. Micro- and nano- structured patterns are realized by EBL. The procedure of fabrication consists in three steps: 1)spin-coating of the silicon wafer by a resist, ZEP for microstructures and PMMA-A2 for the nano-structures; 2)the spinned resist is exposed to high energy electrons to design the pattern; 3)the exposed resist is removed by the appropriate solvent (developer). The exposition phase is regulated by means of typical EBL parameters, as electron beam current and exposition time, regulated on the basis of resist characteristics, on its thickness and on pattern size.

The metallic nanoparticles (especially gold and silver) may be fabricated into the lithographic pattern by physical techniques (evaporation), chemical techniques (reduction from metallic ions) or self-assembly techniques (Jensen et al., 1999; Felidj et al., 2002).

Among these, electroless deposition is a very fast, simple and economic technique, based on the auto-catalytic reduction of metal salts. The redox reaction consists in an electrons exchange between the metallic ions of an opportune solution where the substrate is immersed, and a reducing agent, which may be in the same solution or may be the substrate itself. In literature, several studies report the deposition of metals, as silver, gold, copper, nickel or their alloy, by this technique, obtaining thin films with higher roughness, dendritic structures, submicrometric metallic structures or NPs (Qiu & Chu, 2008; Gao et al., 2005; Yang et al., 2008; Peng & Zhu, 2004; Goia & Matijevic, 1998).

Here, the attention is centered on deposition of silver and gold nanoparticles for SERS application by electroless technique in which the reducing agent is the substrate itself (Coluccio et al., 2009). After the resist developing (the 3° steps of lithographic process) the pattern on Si wafer consists in holes where silicon remains uncovered. The patterned substrate is then dept into a solution of silver nitrate (AgNO3) or chloroauric acid ($HAuCl_4$) prepared in fluoridric acid. The standard solution concentration is around 1 mM, but it is possible to modulate it according to the morphology that we want to obtain. The solution enters the cavity and the reaction between silver/gold ions and the silicon wafer occurs. The metal deposition increases with the increase of temperature and reaction time.

The electroless process may be divided in two steps:

1. the patterned silicon reacts with fluoridric acid (HF) determines the removing of the superficial silicon oxide (SiO_2) layer and the formation of a hydrogenated surface, inert

to reactions with O_2, CO_2, CO, etc., while presenting a good reactivity with silver ions (Palermo & Jones, 2001);

2. the metal nano grains growth into the holes obtained onto the substrate, according the following reactions:

$$4\ Ag^+ + Si(s) + 6\ HF \rightarrow 4Ag^0 + H_2SiF_6 + 4H^+ \quad (1)$$

which can be separated into two half-cell reactions (2) and (3):

Anode

$$Si + 2H_2O \rightarrow SiO_2 + 4H^+ + 4e^- \text{(silicium oxidation)} \quad (2)$$

Cathode

$$Ag^+ + e^- \rightarrow Ag^0 \text{ (silver reduction).} \quad (3)$$

The oxidation of silicon, produced the electrons, is necessary for the silver reduction (Goia & Matijevic, 1998; Coluccio et al., 2009; Palermo & Jones, 2001). The mechanism of formation of the nanoparticles begins with few silver ions that react directly with the substrate forming metallic nuclei. These Ag nuclei are strongly electronegative, thereby, attracting other electrons from the silicon bulk and consequently getting negatively charged; then they attract silver ions which reduced to Ag^0 thus inducing the growth of the original Ag nuclei (Qiu & Chu, 2008). The gold reactions are very similar, the ion reacting obviously is Au^{3+}. Gold forms smaller NPs then silver and the reaction kinetic is slower.

The morphology of nano-aggregates depends on the lithographic pattern and, in particular, on the size: in nano sized structures, the grains grow compact and more mild reaction conditions are necessary for avoiding an excessive nano grains formation, in particular, concentrations under 1 mM and temperatures under 50°C may be used. In micro sized structures, for obtaining a uniform surface covering is important to use concentration up to 1mM and temperature around 50°C. For every SERS structures, the optimal reaction conditions must be searched, modulating the parameter for obtaining the correct particles sizes and density.

Both gold and silver nano aggregates give a good SERS effect, even if silver has the disadvantage that oxidizes with time, giving a decreasing of SERS efficiency. To overcome this problem bimetallic substrates may be fabricated depositing in a first step silver nano grains and over them gold nano grains. Gold assures the SERS device protection from time depending oxidation and maintains a good efficiency of the substrate. SEM images of silver and gold SERS devices are shown in **Fig.13**.

As mentioned above the fabrication of regular array of metallic NPs, may increase the efficiency and the reproducibility of the SERS device, and their optical characteristics may be selected varying size and mutual distance between particles. NPs arrays examples are reported in **Fig.14**.

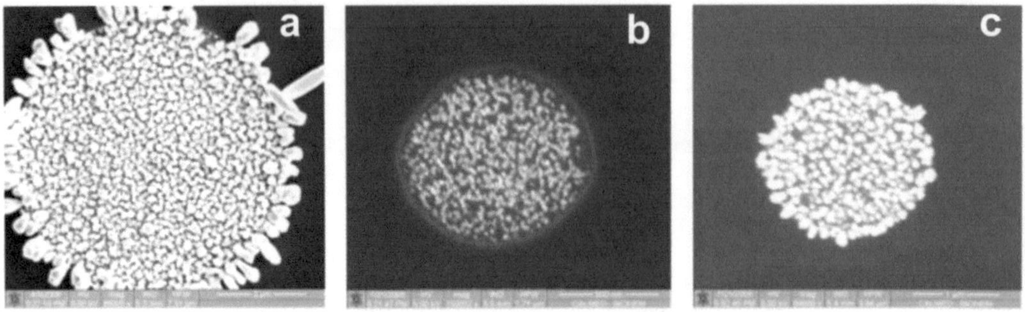

Fig. 13. SEM images of nano grains assembled on microstructures, fabricated using e-beam lithography and electroless deposition: (a) silver nano grains, (b) gold nano grains, (c) silver/gold nano grains.

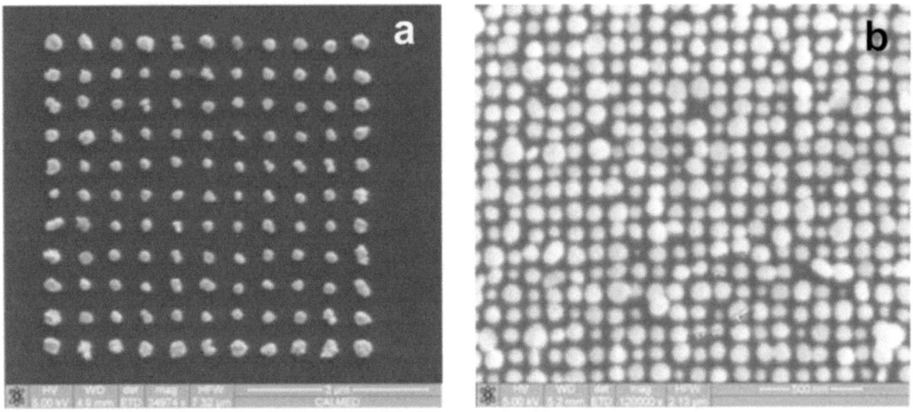

Fig. 14. SEM images of nano grains assembled on nanostructures, fabricated using e-beam lithography and electroless deposition.

The SERS effect of gold, silver or of bimetallic (silver/gold) microstructures is investigated using rhodamine-6G as probe molecules. The substrates are dipped into rhodamine-6G water solution with different concentrations, rinsed with water dried in N_2 and then used for the spectroscopic investigation. SERS experiments show spectra with well-defined peaks at very low molecular concentration, for all the metallic/bimetallic substrates (**Fig.15**). The highest SERS efficiency is found for SERS device based on gold over silver surface.

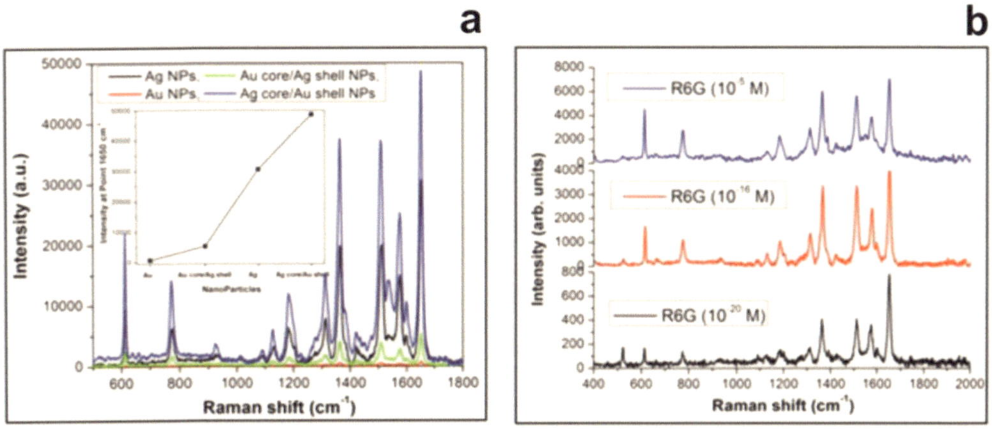

Fig. 15. a) R6G [10^{-12}M], absorbed on the silver, gold, Ag/Au and Au/Ag structures. Raman signal increases from Au, Au core/Ag shell, Ag and Ag core/Au shell; b) 3. SERS spectra acquired from 10^{-5}, 10^{-16} and 10^{-20} M R6G, absorbed on the silver nanostructures.

6. References

Aizpurua, J.; Bryant, G. W.; Richter, L. J.; Garcia de Abajo, F. J.; Kelley, B. K. & Mallouk, T. (2005). Optical properties of coupled metallic nanorods for field-enhanced spectroscopy. *Phys. Rev. B,* Vol. 71, pp. 235420.

Anikeeva, P.O., et al., Electroluminescence from a mixed red-green-blue colloidal quantum dot monolayer. Nano Letters, 2007. 7(8): p. 2196-2200.

Anker, J. N.; Hall, W. P.; Lyandres, O.; Shah, N. C.; Zhao, J. & Van Duyne, R. P. (2008). Biosensing with plasmonic nanosensors. *Nature Materials,* Vol. 7, pp. 442-453.

Bohren, C. F. & Huffman, D. R., (1998). *Absorption and Scattering of light by small particles* (Wiley, New York).

Canham L., Properties of Porous Silicon. INSPEC - The Institution of Electrical Engineers London - United Kingdom, 1997.

Carbone, L., et al., Synthesis and micrometer-scale assembly of colloidal CdSe/CdS nanorods prepared by a seeded growth approach. Nano Letters, 2007. 7(10): p. 2942-2950.

Cohen M. H., Melnik K., Boiasrki A., Ferrari M. & Martin F. J., Microfabrication of silicon-based nanoporous particulates for medical applications, *Biomed. Microdevices* (5), 253–259, 2003.

Colthup, N. B., Daly, L. H. & Wiberley, S. E., (2010). *Introduction to Infrared and Raman Spectroscopy* (Academic Press, USA).

Coluccio M.L., Das G., Mecarini F., Gentile F., Pujia A., Bava L., Tallerico R., Candeloro P., Liberale C., De Angelis F., Di Fabrizio E., Silver-based surface enhanced Raman scattering (SERS) substrate fabrication using nanolithography and site selective electroless deposition, Microelectronic Engineering 86 (2009) 1085–1088.

Constantino C.J.L., Lemma T., Antunes P.A., Aroca R., Single molecular detection of a perylene dye dispersed in a Langmuir–Blodgett fatty acid monolayer using surface-enhanced resonance Raman scattering, Spectro chimica Acta Part A 58 (2002) 403–409.

Creighton J. A., Blatchford C. G., Albrecht M. G., Plasma Resonance Enhancement of Raman-Scattering by Pyridine Adsorbed on Silver or Gold Sol Particles of Size Comparable to the Excitation Wavelength, J. Chem. Soc., Faraday Trans II 75 (1979) 790-798.

Crommelin D.J.A., Schreier H., Liposomes, in: *Colloidal Drug Delivery Systems*, J. Kreuter, editor. Marcel Dekker, Inc. New York, 1994.

Das G., Mecarini F., Gentile F., De Angelis F., Kumar M., Candeloro P., Liberale C., Cuda G., Di Fabrizio E., Nano-patterned SERS substrate: Application for proteinanalysis vs. temperature, Biosens. Bioelectron. 24 (2009) 1693–1699.

De Angelis, F.; Patrini, M.; Das, G.; Maksymov, I.; Galli, M.; Businaro, L.; Andreani, L. C. & Di Fabrizio, E. (2008). A Hybrid Plasmonic Photonic Nanodevice for Label-Free Detection of a Few Molecules. *Nano Letters*, Vol. 8, No. 8, pp. 2321–2327

De Angelis F., Pujia A., Falcone C., Iaccino E., Palmieri C., Liberale C., Mecarini F., Candeloro P., Luberto L., de Laurentiis A., Das G., Scala G., and Di Fabrizio E., Water soluble nanoporous nanoparticles for in vivo targeted drug delivery and controlled release in b cells tumor context. Nanoscale, 2:2230-2236, 2010.

De Angelis, F.; Das, G.; Candeloro, P.; Patrini, M.; Galli, M.; Bek. A.; Lazzarino, M.; Maksymov, I.; Liberale, C.; Andreani, L. C. & Di Fabrizio, E. (2010). Nanoscale chemical mapping using three-dimensional adiabatic compression of surface plasmon polaritons. *Nature Nanotech.*, Vol. 5, pp. 67-72

Decuzzi P. and Ferrari M., The Adhesive Strength of Non-Spherical Particles Mediated by Specific Interactions, *Biomaterials*, 27(30):5307-1534, 2006.

Decuzzi P., Causa F., Ferrari M. and Netti P. A., The Effective Dispersion of Nanovectors Within the Tumor Microvasculature, *Ann Biomed Eng.*, 34(4):633-641, 2006.

Decuzzi P. and Ferrari M., The Role of Specific and Non-Specific Interactions in Receptor-Mediated Endocytosis of Nanoparticles, *Biomaterials*, 28(18):2915-2922, 2007.

Duncan R., The dawning era of polymer therapeutics, *Nature Rev. Drug Discov.* (2), 347–360, 2003.

Fazio B.; D'Andrea, C.; Bonaccorso, F.; Irrera, A.; Calogero, G.; Vasi, C.; Gucciardi, P. G.; Allegrini, M.; Toma, A.; Chiappe, D.; Martella, C. & Buatier De Mongeot F. (2011). Re-radiation Enhancement in Polarized Surface-Enhanced Resonant Raman Scattering of Randomly Oriented Molecules on Self-Organized Gold Nanowires. *ACS Nano*, Vol. 5 No. 7, pp. 5945-5956.

Felidj N., Aubard J., Levi G., Krenn J.R., Salerno, Schider G., Lamprecht B., Leitner A., Aussenegg F.R., Controlling the optical response of regular arrays of gold particles for surface enhanced Raman scattering, Phys. Rev. B Condens. Matter Mater. Phys. 65 (2002) 075419/1–075419/9.

Ferrari M., Cancer Nanotechnology: opportunities and challenges, *Nature Reviews Cancer*, 2005; (5):161-171.

Fischer, H. & Martin, O. J. F. (2008). Engineering the optical response of plasmonic nanoantennas. *Optics Express,* Vol. 16 No.12, pp. 9144-9154.

Gao J., Tang F., Ren J., Electroless nickel deposition on amino-functionalized silica spheres, Surface & Coatings Technology 200 (2005) 2249–2252.

Garoff S., Weitz D.A., Alvarez M.S. and Chung J.C., Electromagnetically Induced Changes in Intensities, Spectra and Temporal Behavior of Light Scattering from Molecules on Silver Island Films, J. Phys. Colloq. 44, C10 (1983) 345-348.

Gentile F., Ferrari M., Decuzzi P. (2007), Transient Diffusion of Nanovectors in Permeable Capillaries, J. Serbian Soc. Comput. Mech., 1(1), 1-19.

Gentile F., Ferrari M. and Decuzzi P., `The Transport of Nanoparticles in Blood Vessels: The Effect of Vessel Permeability and Blood Rheology`, Annals of Biomedical Engineering, 2008:2(36); 254–261.

Gilles E. M. & Frechet J. M. J., Designing macromolecules for therapeutic applications: Polyester dendrimerpolyethylene oxide 'bow-tie' hybrids with tunable molecular weights and architecture, *J. Am. Chem. Soc.* (124), 14137–14146, 2002.

Godefroo S., Hayne M., Jivanescu M., Stesmans A., Zacharias M., Lebedev O.I., van Tendeloo G., and Moschchalkov V. V., Classification and control of the origin of photoluminescence from si nanocrystals. Nature Nanotechnology, 3:174-178, 2008.

Granitzer P. and Rumpf K., Porous silicon - a versatile host material. Materials, 3:943-998, 2010.

Grzelczak M., Vermant J., Furst E. M., Liz-Marza´n L. M., Directed Self-Assembly of Nanoparticles, ACSNano VOL. 4 ▪ NO. 7 ▪ 3591–3605 ▪ 2010;

Gunnarsson L., Bjerneld E.J., Xu H., Petronis S., Kasemo B. and Käll M., Interparticle-coupling effects in Nanofabricated Substrates for Surface Enhanced Raman Scattering, Applied Physics Letters 78, 802-804 (2001).

Halimaoui A., Porous silicon: material processing, properties and applications, in JC Vial and J. Derrien (editors), Porous silicon science and technology, Springer-Verlag (1995).

Hao E. and Schatz G. C., Electromagnetic fields around silver nanoparticles and dimmers, J. Chem. Phys. 120 (2004) 357-367.

Haynes C.L., McFarland A.D., VanDuyne R.P., Surface-Enhanced Raman Spectroscopy, Anal. Chem. 77 (2005) 338A-346A.

Hirsch, L. R., Halas, N. J. & West, J. L. Nanoshell-mediated near-infrared thermal therapy of tumors under magnetic resonance guidance, *Proc. Natl Acad. Sci. USA* (100), 13549–13554, 2003.

Hu, J.T., et al., Linearly polarized emission from colloidal semiconductor quantum rods. Science, 2001. 292(5524): p. 2060-2063.

Jensen T.R., Schatz G.C., Van Duyne R.P., Nanosphere Lithography: Surface plasmon resonance spectrum of a periodic array of silver nanoparticles by UV-vis extinction spectroscopy and electrodynamic modeling, J. Phys. Chem. B 103 (1999) 2394–2401 .

Jeyarama S. Ananta, Biana Godin, Richa Sethi, Loick Moriggi, Xuewu Liu, Rita E. Serda, Ramkumar Krishnamurthy, Raja Muthupillai, Robert D. Bolskar, Lothar Helm, Mauro Ferrari, Lon J. Wilson, Paolo Decuzzi, Geometrical confinement of gadolinium-based contrast agents in nanoporous particles enhances T_1 contrast, *Nature Nanotechnology* 5, 815–821 (2010).

Ji-Ho Park, Luo Gu, Geo rey von Maltzahn, Erkki Ruoslahti, Sangeeta N. Bhatia, and Michael J. Sailor. Biodegradable luminescent porous silicon nanoparticles for in vivo applications. Nature Materials, 8:331, 336, 2009.

Kazes, M., et al., Lasing from semiconductor quantum rods in a cylindrical microcavity. Advanced Materials, 2002. 14(4): p. 317-321.

Kazes, M., et al., Lasing from CdSe/ZnS quantum rods in a cylindrical microcavity. Quantum Dots, Nanoparticles and Nanowires, 2004. 789: p. 11-16429.

Kircher M. F., Mahmood U., King, R. S., Weissleder R. & Josephson L., A multimodal nanoparticle for preoperative magnetic resonance imaging and intraoperative optical brain tumor delineation, *Cancer Res.* (63), 8122–8125, 2003.

Klibanov A. L. et al., Activity of amphipathic PEG 5000 toprolong the circulation time of liposomes depends on the liposome size and is unfavourable for immunoliposome binding to target, *Biochem. Biophys. Acta* (1062), 142–148, 1991.

Kneipp K., Wang Y., Kneipp H., Perelman L.T., Itzkan I., Dasari R. R., and Feld M. S., Single Molecule Detection using Surface-Enhanced Raman Scattering, Phys. Rev. Lett. 78 (1997) 1667-1670.

Kneipp K., Kneipp H., Itzkan I., Dasari R. R., Feld M.S., Kneipp H., Itzkan I., Dasari R. R., and Feld M. S., Surface-enhanced Raman scattering and biophysics, J. Phys. Condens. Matter 14 (2002) R597-R624.

Kneipp, K., Kneipp, H., Abdali, S. et al., Single Molecule Raman Detection of Enkephalin on Silver Colloidal Particles. Spectroscopy 18 (2004) 433–440.

Kneipp K., Kneipp H., Kneipp J., Surface-Enhanced Raman Scattering in Local Optical Fields of Silver and Gold Nanoaggregatess From Single-Molecule Raman Spectroscopy to Ultrasensitive Probing in Live Cells, Acc. Chem. Res. 39 (2006) 443-450.

Krahne, R., et al., Amplified Spontaneous Emission from Core and Shell Transitions in CdSe/CdS Nanorods fabricated by Seeded Growth. Applied Physics Letters, 2011. 98(6): p. 063105.

Krahne, R., et al., Physical properties of elongated inorganic nanoparticles. Physics Reports, 2011. 501(3-5): p. 75-221.

La Van D. A., McGuire T. & Langer R., Small-scale systems for *in vivo* drug delivery, *Nature Biotechnol.*, (21), 1184–1191, 2003.

Langer R., Drug delivery and targeting, *Nature* (392), 5–10, 1998.

Le Ru E.C., Etchegoin P.G., Grand J., Félidj N., Aubard J., Lévi G., Hohenau A., Krenn J.R., Surface enhanced Raman spectroscopy on nanolithography-prepared substrates, Current Applied Physics 8 (2008) 467–470.

Li, Y.Q., et al., White organic light-emitting devices with CdSe/ZnS quantum dots as a red emitter. Journal Of Applied Physics, 2005. 97(11): p. art. n. 113501.

Lupo, M.G., et al., Ultrafast Electron-Hole Dynamics in Core/Shell CdSe/CdS Dot/Rod Nanocrystals. Nano Letters, 2008. 8(12): p. 4582-4587.

McCall S.L., Platzman P.M., Wolf P.A., Surface enhanced Raman scattering, Phys. Lett. 77A (1980) 381-383.

Mitragotri Samir and Lahan Joerg, Physical approaches to biomaterial design. Nature Materials, 8:15-23, 2009.

Nie S. and Emory S. R., Probing Single Molecules and Single Nanoparticles by Surface-Enhanced Raman Scattering, Science 275 (1997) 1102-1106.

Nobile, C., et al., Self-assembly of highly fluorescent semiconductor nanorods into large scale smectic liquid crystal structures by coffee stain evaporation dynamics. Journal Of Physics-Condensed Matter, 2009. 21(26): p. 264013.

O Farrel N., Houlton A., B. R. Horrocks, Silicon nanoparticle application in cell biology and medicine. International Journal of Nanomedicine, 1(4):451-472, 2006.

Otto A., In Light scattering in solids IV. Electronic scattering, spin effects, SERS and morphic effects; M. Cardona, G. Guntherodt, Eds.; Springer-Verlag: Berlin, Germany, 1984; Vol. 1984, pp 289- 418.

Palermo, D. Jones, Morphological changes of the Si [100] surface after treatment with concentrated and diluted HF, Materials Science in Semiconductor Processing, 4 (2001) 437–441.

Park J. W., Liposome-based drug delivery in breast cancer treatment, *Breast Cancer Res.* (4), 95–99, 2002.

Peng K., Zhu J., Morphological selection of electroless metal deposits on silicon in aqueous fluoride solution, ElectrochimicaActa 49 (2004) 2563–2568.

Persano, A., et al., Photoconduction Properties in Aligned Assemblies of Colloidal CdSe/CdS Nanorods. Acs Nano, 2010. 4(3): p. 1646-1652.

Petros Robby A. and DeSimone Joseph M., Strategies in the design of nanoparticles for therapeutic applications. Nature Reviews Drug Discovery, 9:615-627, 2010.

Pujia A., De Angelis F., Scumaci D., Gaspari M., Liberale C., Candeloro P., Cuda G., Di Fabrizio E., Highly efficient human serum filtration with water-soluble nanoparticles. International Journal of Nanomedicine, 5:1005-1015, 2010.

Qiu T., Chu P.K., Self-selective electroless plating: An approach for fabrication of functional 1D nanomaterials, Mater. Sci. Eng. R 61 (2008) 59-77.

Raether H., Surface Plasmonson Smooth and Rough Surfaces and on Gratings (Springer, Berlin, 1988).

Rizzo, A., et al., Polarized Light Emitting Diode by Long-Range Nanorod Self-Assembling on a Water Surface. Acs Nano, 2009. 3(6): p. 1506-1512.

Rosi, N. L. & Mirkin, C. A., (2005). Nanostructures in Biodiagnostics. *Chem. Rev.,* Vol. 105, pp. 1547-1562.

Sackmann, M.; Bom, S.; Balster, T. & Materny, A. (2007). Nanostructured gold surfaces as reproducible substrates for surface-enhanced Raman spectroscopy. *J. Raman Spect.,* Vol. 38, pp. 277-282.

Schellenberger E. A. et al., Annexin V-CLIO: a nanoparticle for detecting apoptosis by MRI, *Mol. Imaging* (1), 102–107, 2002.

Siti M. Janib, Ara S. Moses, and J. Andrew MacKay. Imaging and drug delivery using theranostic nanoparticles. Advanced Drug Delivery Reviews, 62:1052-1063, 2010.

Stockman, M. I.; Pandey, L. N.; Muratov, L. S. & George, T. F. (1994). Giant fluctuations of local optical fields in fractal clusters. *Phys. Rev. Lett.,* Vol. 72, pp. 2486-2489.

Talapin, D.V., et al., Highly emissive colloidal CdSe/CdS heterostructures of mixed dimensionality. Nano Letters, 2003. 3(12): p. 1677-1681.

Tao A. R., Habas S., Yang P., Shape Control of Colloidal Metal Nanocrystals , Small 4 No 3 (2008), 310 - 325.

Tasciotti E, Liu X, Bhavane R, Plant K, Leonard AD, Price BK, Cheng MM, Decuzzi P, Tour JM, Robertson F, Ferrari M. Mesoporous silicon particles as a multistage delivery system for imaging and therapeutic applications, Nature Nanotechnol. 2008 3(3):151-157.

Toma, A.; Chiappe, D.; Massabò, D.; Boragno, C. & Buatier de Mongeot, F. (2008). Self-organized metal nanowire arrays with tunable optical anisotropy. *Appl. Phys. Lett.,* Vol. 93, pp. 163104.

V. Dan G., Matijevic E., Preparation of monodispersed metal particles, New J. Chem. 22 (1998) 1203–1215.

Vogel, E.; Gbureck, A. & Kiefer, W. (2000). Vibrational spectroscopic studies on the dyes cresyl violet and coumarin 152*, *J. Mol. Struct.*, Vol. 550, pp. 177-190.

Whitesides G. M., The 'right' size in nanotechnology, *Nature Biotechnol.*, (21), 1161–1165, 2003.

Xu H., Bjerneld E.J., Kall M., et al..Spectroscopy of Single Hemoglobin Molecules by Surface Enhanced Raman Scattering. Phys. Rev. Lett. 83 (1999) 4357–4360.

Yang Y., Shi J., Kawamura G., Nogami M., Preparation of Au-Ag, Ag-Au core-shell bimetallic nanoparticles for surface-enhanced Raman scattering, Scr. Mater. 58 (2008) 862-865.

Zavelani-Rossi, M., et al., Lasing in self-assembled microcavities of CdSe/CdS core/shell colloidal quantum rods. Nanoscale, 2010. 2(6): p. 931-935.

Zavelani-Rossi, M., et al., Suppression of Biexciton Auger Recombination in CdSe/CdS Dot/Rods: Role of the Electronic Structure in the Carrier Dynamics. Nano Letters, 2010. 10(8): p. 3142-3150.

Zhang Y. & Shang M., Self-assembled coatings on individual monodisperse magnetite nanoparticles for efficient intracellular uptake, *Biomed. Microdevices* (6), 33–40, 2004.

Zhang, X., Yonzon, C.R., Duyne, R.P.V. 2006. Nanosphere Lithography Fabricated Plasmonic Materials and Their Applications. *J. Mater. Res.* 21: 1083–1092.

15

Water-Soluble Single-Nano Carbon Particles: Fullerenol and Its Derivatives

Ken Kokubo
Division of Applied Chemistry, Graduate School of Engineering, Osaka University
Japan

1. Introduction

Since its discovery in 1985, fullerene has been extensively investigated as a unique, "dissolvable," and "modifiable" nanocarbon material. The most representative fullerene, C_{60}, is a perfectly spherical molecule with a diameter of ca. 1 nm (0.7 nm when the distance between the furthest C–C bond is considered, and 1 nm when the π-orbitals are included). It has many interesting electronic and biological properties owing to its spherical π-conjugation. While fullerenes satisfactorily dissolve in aromatic solvents such as toluene and *o*-dichlorobenzene as well as in carbon disulfide, they dissolve poorly in most common solvents such as hexane, chloroform, diethyl ether, ethyl acetate, tetrahydrofuran (THF), acetone, acetonitrile, ethanol, and even in benzene. This limitation has been one of the important issues hindering their practical application, especially in the field of life sciences.

Although the single-crystal X-ray structural analysis of C_{60} has been successful, it is generally difficult to grow the crystal of fullerene derivatives. Such a poor crystallinity is because of the lack of molecular orientation and intermolecular interaction that are crucial for determining the molecular alignment in a crystal. The lack of molecular orientation and interaction is attributed to the unidirectional spherical shape of these derivatives, which is different from the shapes of other organic molecules such as cubic- or plate-shaped ones. For the same reason, the solid form of C_{60} is known to easily afford its nanoparticles with a top-down approach, in which the solid is reduced to small particles (as small as 20 nm) by applying mechanical forces; such particles can even be obtained by hand-grinding (Deguchi et al., 2006; Deguchi et al., 2010). These small nanocarbon particles, so-called nC_{60} (Oberdörster, 2004; Brant et al., 2005) or nano-C_{60} (Fortner et al., 2005), can be dispersed even in neutral water, and they remain dispersed for a long time, especially in the presence of humic acid (Chen & Elimelech, 2007; Isaacson & Bouchard, 2010). On the other hand, the aggregate of fullerene can be easily obtained by a bottom-up approach in many solvents such as benzene (Ying et al., 1994), benzonitrile (Nath et al., 1998), *N*-methylpyrrolidone (Yevlampieva et al., 2002; Kyzyma et al., 2010), and *o*-dichlorobenzene (Gun'kin & Loginova, 2006). The nanoparticle formation and aggregation behavior are among the outstanding features of fullerene in terms of both its practical application and the safe use of nanomaterials.

Polyhydroxylated fullerenes, so-called fullerenols or fullerols, are a class of fullerenes that has many hydroxyl groups, formed by the chemical modification of covalent C–O bonds, on their spherical surfaces. The chemical formula $C_{60}(OH)_n$ represents an average structure that consists of a mixture of fullerenols having different number of hydroxyl groups, each with its own regioisomer. The solubility of a fullerene molecule is dependent on the number of hydroxyl groups introduced; i.e., the low-degree hydroxylated fullerenols $C_{60}(OH)_{10-12}$ (Chiang et al., 1994) can dissolve in some polar solvents, e.g., THF, dimethylformamide (DMF), and dimethyl sulfoxide (DMSO), and the medium-degree fullerenols $C_{60}(OH)_{16}$ (Wang et al., 2005) and $C_{60}(OH)_{20-24}$ (Li et al., 1993) are reported to dissolve even in water. However, these later fullerenols may be contaminated with Na salt because of the reagents used in synthesis, resulting in compositions with the formula $Na^{+}_{n}[C_{60}Ox(OH)_y]^{n-}$ and exhibiting high water solubility in spite of their small number of hydroxyl groups (Husebo et al., 2004). In contrast, the high-degree hydroxylated fullerenols $C_{60}(OH)_{36}$ and $C_{60}(OH)_{44}$, which are synthesized without using any Na salt, are completely water soluble by as much as 17 and 65 mg/mL, respectively (Kokubo et al., 2008; Kokubo et al., 2010). Particle size analysis revealed that the high-degree fullerenols exhibited high dispersion properties at a molecular level (ca. 1 nm, which is as large as the molecular diameter). The behavior of water-soluble carbon particles in the single-nano region (1–10 nm) is less well understood in terms of their chemical and physical properties.

This chapter focuses on the methods of synthesizing fullerenols, provides examples of applications, and describes the particle-size measurements of the high-degree fullerenols as water-soluble single-nano carbon particles.

2. Synthesis of fullerenols

2.1 Hydroxylation of fullerene

Various types of fullerenols having different number of hydroxyl groups have been synthesized so far because of their promising water solubility and the expected bioactivities. In general, the structure of fullerenols is qualitatively identified by infrared spectroscopy as having a characteristic broad νO–H band, along with three broad peaks assigned for νC = C, δ_sC–O–H, and νC–O. The number of hydroxyl groups introduced is quantitatively determined by either elemental analysis or X-ray photoelectron spectroscopy (XPS). Such a quantitative analysis is founded on the hypothesis that the addend of a fullerene is composed of only hydroxyl group.

One of the most well-known fullerenols, $C_{60}(OH)_{12}$, was synthesized by L. Y. Chiang using oleum ($H_2SO_4 \cdot SO_3$), followed by the hydrolysis of the intermediate cyclosulfated fullerene (Scheme 1a) (Chiang et al., 1994). The compound is soluble in an alkaline solution and some polar solvents such as THF and DMSO, but it is not soluble in neutral water because it has few hydroxyl groups. L. Y. Chiang also studied its antioxidant activity toward a superoxide, a reactive oxygen species (Chiang et al., 1995). Some other related methods of accessing the lower-degree fullerenols have been reported (Zhang et al., 2010).

The most frequently used method for synthesizing the medium-degree fullerenol $C_{60}(OH)_{22-26}$ is the one reported by J. Li et al. (Scheme 1b) (Li et al., 1993). This method employs sodium hydroxide (NaOH) as a hydroxylation reagent, and molecular oxygen is required to neutralize the intermediate fullerenyl anion, which is formed by the attack of -

OH to C_{60}, in order to induce the successive attack of ^{-}OH. The fullerenols $C_{60}(OH)_{22\text{-}26}$, as well as $C_{60}(OH)_{16}$ synthesized by a similar method using NaOH and H_2O_2 (Scheme 1c) (Wang et al., 2005), exhibits sufficient solubility in neutral water. The fullerenol has also been known to have many bioactivities, including antioxidant activity similar to pristine C_{60} (Bosi et al., 2003; Bakry et al., 2007; Partha & Conyers, 2009). However, the practical use of these types of fullerenols might be restricted because of their unfavorable contamination by Na^+ ions, which are inevitably introduced during treatment with NaOH, and the purification is rather difficult except when done by repeated gel column chromatography (Husebo et al., 2004). The relatively higher water solubility than that expected, given the number of hydroxyl groups, is reasonably explained by the corrected chemical formula $Na^+_n[C_{60}O_x(OH)_y]^{n-}$. Such a Na salt form is attributed to the weak acidity of the phenolic O–H of the low-degree fullerenols, in contrast to the alcoholic O–H of the high-degree fullerenols. Therefore, there is a demand for the development of new, facile, and scalable methods for synthesizing highly water-soluble and pure fullerenols.

$$C_{60} \xrightarrow[\text{2. } H_2O]{\text{1. } H_2SO_4\bullet SO_3} C_{60}(OH)_{12} \qquad \text{(1-a)}$$

$$C_{60} \xrightarrow[\text{r.t. grinding}]{\text{NaOH, 30\% } H_2O_2} C_{60}(OH)_{16} \qquad \text{(1-b)}$$

$$C_{60} \xrightarrow[\text{benzene, r.t.}]{\text{NaOH, TBAH}} C_{60}(OH)_{22\text{–}26} \qquad \text{(1-c)}$$

Scheme 1. Synthesis of low- and medium-degree fullerenols

2.2 Highly polyhydroxylated fullerenols

To avoid contamination by Na^+ ions, we examined the use of hydrogen peroxide (H_2O_2) as a hydroxylation reagent instead of NaOH. Thus, we have found a new and facile approach for synthesizing high-degree fullerenols that have high water solubility without using any Na salts (Kokubo et al., 2008). The reddish brown suspension of fullerenol $C_{60}(OH)_{12}$ in 30% aqueous H_2O_2 was stirred vigorously at 60 °C under air until it turned to a transparent yellow solution, which occurred within 2–4 days (Scheme 2a). To this solution, a mixture of 2-propanol, diethyl ether, and hexane as an antisolvent was added to afford a yellowish brown to milky white precipitation of the desirable high-degree fullerenol $C_{60}(OH)_{36}\cdot 8H_2O$. A longer reaction time of up to two weeks gave the similar but more water-soluble fullerenol $C_{60}(OH)_{40}\cdot 9H_2O$.

This new approach using H_2O_2 to synthesize high-degree fullerenols was useful; however, the starting material was limited to $C_{60}(OH)_{12}$ and was not applicable to pristine C_{60}. We then improved the method in order to provide a facile, one-step method for synthesizing fullerenol from pristine C_{60}; we added an NH_3 aqueous solution to the H_2O_2 aqueous solution to give similar water-soluble fullerenols, although they contained some undesirable nitrogen-containing groups, along with hydroxyl groups (Matsubayashi et al., 2009). We further improved the method, synthesizing pure $C_{60}(OH)_{44} \cdot 8H_2O$ fullerenol with no nitrogen in one step from pristine C_{60}. To the best of our knowledge, this fullerenol has the largest number of hydroxyl groups per C_{60} among the fullerenols reported so far; this fullerenol was obtained by a two-phase synthesis in the presence of tetrabutylammonium hydroxide (TBAH) as a phase transfer catalyst (PTC) (Scheme 2b) (Kokubo et al., 2010). The fullerenol exhibits a very high water solubility of up to 64.9 mg/mL.

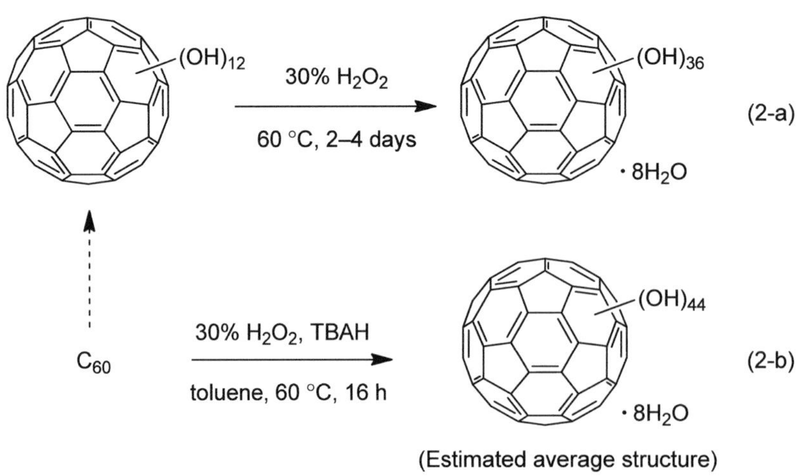

Scheme 2. Synthesis of high-degree fullerenols

2.3 Structural characterization

The structural characterization of the high-degree fullerenols was conducted by infrared spectroscopy (Fig. 1). The spectra of the fullerenols closely resembled each other, although their relative peak intensities differed somewhat, suggesting a difference in the number of introduced hydroxyl groups. Four characteristic broad bands were observed at 1080, 1370, 1620, and 3400 cm^{-1} and were assigned to νC–O, δ_sC–O–H, νC=C, and νO–H, respectively. A small shoulder peak at 1720 cm^{-1} may imply the existence of a carboxylic acid group, O=C–OH, which might have formed by the further oxidation of a hydroxyl group associated with C-C bond cleavage of the fullerene nucleus. However, such partial oxidation must not be crucial because the generally strong C=O absorption is much smaller than the other generally weak or medium C=C or C–O absorptions, which is consistent with the results of the elemental analysis.

The quantitative analysis to determine the number of hydroxyl groups was conducted by elemental analysis. As shown in Table 1, the average structure of the high-degree fullerenol $C_{60}(OH)_{44} \cdot 8H_2O$ would be deduced to be $C_{60}(OH)_{52}$ if just the results of elemental analysis was used. However, the largest reported number of substituents in one C_{60} moiety is $C_{60}F_{48}$ (Tuinman et al., 1992; Troyanov et al., 2010), and thus $C_{60}(OH)_{52}$ is unlikely to be formed due to the enormous strain energy. On the other hand, it is known that the tightly entrapped water molecules, the secondary bound water, in highly hydroxylated fullerenols cannot be dissociated by the usual method of heating the fullerenols to about 120–150 °C. Therefore, water content measurements using thermogravimetric analysis was conducted (Figure 2). With a water content of 9.4 wt%, as shown in Table 1, the average structure of fullerenol was deduced to be $C_{60}(OH)_{44} \cdot 8H_2O$, using both elemental analysis and water content measurements.

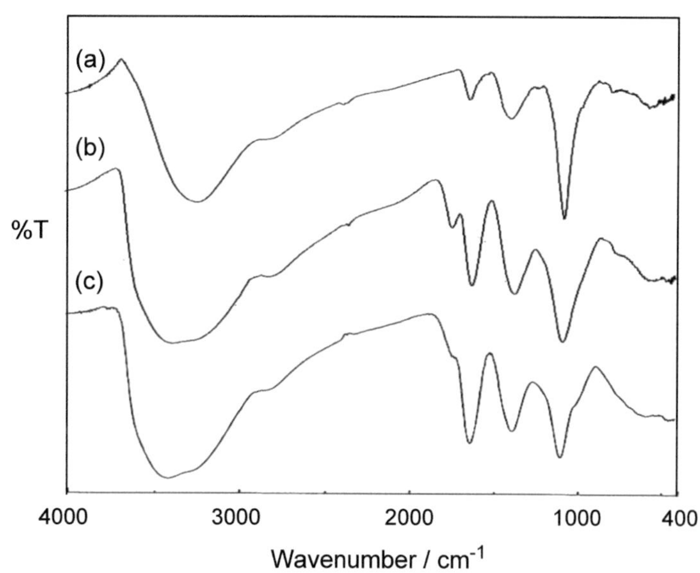

Fig. 1. IR spectra of (a) $C_{60}(OH)_{12}$, (b) $C_{60}(OH)_{36} \bullet 8H_2O$, and (c) purified $C_{60}(OH)_{44} \bullet 8H_2O$.

Average structure	Elemental analysis (%) Found (Calcd)[a]	Water content (wt%)[a,b]	Solubility (mg/mL)[c]
$C_{60}(OH)_{36} \cdot 8H_2O$	C: 48.06, H:3.61 (C: 48.79, H:3.54)	8.9 (9.7)	17.5
($C_{60}(OH)_{44}$)	(C: 49.06, H:3.02)	(0)	
$C_{60}(OH)_{44} \cdot 8H_2O$	C: 44.68, H:3.56 (C: 44.70, H:3.75)	9.4 (8.9)	64.9
($C_{60}(OH)_{52}$)	(C: 44.90, H:3.27)	(0)	

[a]Values in parentheses are calculated data. [b]Determined by TGA. [c]At 25 °C in neutral water (pH = 7).

Table 1. Average structure and water solubility of high-degree fullerenols

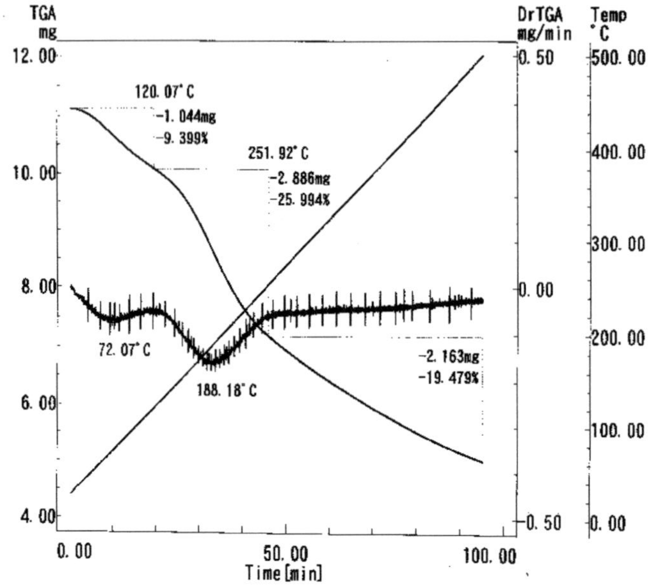

Fig. 2. Thermogravimetric analysis of fullerenol $C_{60}(OH)_{44} \cdot 8H_2O$ under N_2 flow.

2.4 Reaction mechanism

Possible reaction mechanisms for the formation of fullerenols are illustrated in Schemes 3 and 4. For the reaction using NaOH, the attack of ^-OH to C_{60} followed by the oxidation with molecular oxygen via one electron transfer from the C_{60} anion gives the hydroxylated C_{60}. The successive attack of ^-OH and repeated oxidation finally gives the medium-degree fullerenol (Scheme 3) (Husebo et al., 2004).

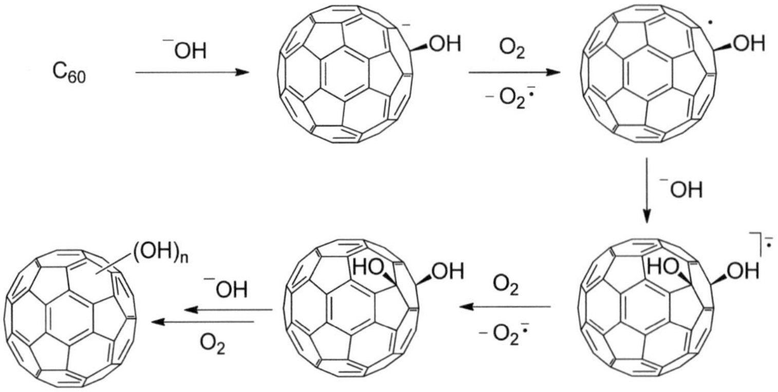

Scheme 3. A possible reaction mechanism for medium-degree fullerenols using the NaOH method

In contrast, for the reaction using H_2O_2, the basic hydroxide ion from TBAH initially induces the hydroperoxide ion ^-OOH because of the slightly higher acidity of H_2O_2 than that of H_2O (Scheme 4) (Wang et al., 2005; Kokubo et al., 2011). The ^-OOH thus formed attacks C_{60} to give fullerene oxide $C_{60}O$, followed by the attack of ^{-}OH and protonation. The epoxidation process may be repeated to give $C_{60}O_2$, $C_{60}O_3$, and so on (Tajima & Takeuchi, 2002), which are more susceptible than $C_{60}O$ to the subsequent nucleophilic attack of ^{-}OH (or ^{-}OOH) because of the higher strain. These fullerene oxide intermediates were detected in the reaction mixture by liquid chromatography-mass spectrometry (LC-MS) (APCI; m/z = 736, 752, and 768) and were proven to be the intermediates by their kinetic behavior. The role of the quaternary ammonium salt TBAH is that of the promotion of ^{-}OOH formation and its transfer from the hydrophilic aqueous phase to the hydrophobic fullerenyl sites in the organic phase as PTC.

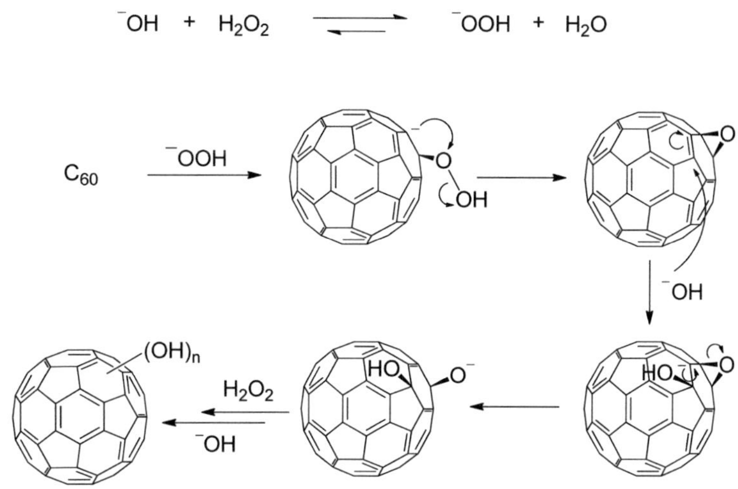

Scheme 4. Proposed reaction mechanism for high-degree fullerenols using the H_2O_2 method

3. Measurement of particle size distribution

3.1 Dynamic light scattering method

Although a fullerenol seems to completely dissolve in a solution, it may be aggregated in the nano-size region, as seen for many fullerene derivatives and some fullerenols (Mohan et al., 1998; Husebo et al., 2004; Brant et al., 2007; Chae et al., 2009; Su et al., 2010). The particle size of medium-degree fullerenol $C_{60}(OH)_{24}$ in aqueous solution is reported as between ca. 20 and 450 nm depending on the measurement conditions. Even such a high number of hydroxyl groups results in the formation of aggregation due to the large hydrophobic and π- π interactions between fullerenyl cores.

In order to investigate the dispersant behavior, the particle size measurement of high-degree fullerenols in the 0.1 wt% aqueous solution was carried out using the common dynamic light scattering (DLS) method (Berne & Pecora, 1976). The narrow particle-size distributions around 1–2 nm in terms of the number of $C_{60}(OH)_{36}$ and $C_{60}(OH)_{44}$ molecules are essentially the same, indicating the highly dispersed nature of the fullerenols at a molecular level (Figure 3) (Kokubo et al., 2008; Kokubo et al., 2011). The average particle size of $C_{60}(OH)_{44}$ was determined to be 1.46 ± 0.38 nm (N = 8). The particle size by intensity was not applicable to the fullerenol solution because of the interference from the intensity of contamination in the range of 60–130 nm.

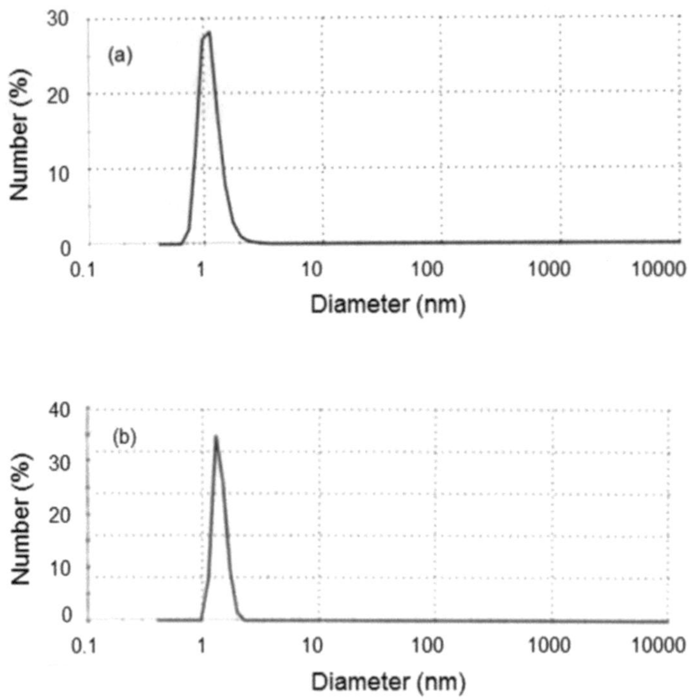

Fig. 3. DLS analysis of (a) $C_{60}(OH)_{36}$ and (b) $C_{60}(OH)_{44}$ in water (0.1 wt%) expressed by size distribution in number.

3.2 Induced grating method

Recently, the induced grating (IG) method was developed to improve the reproducibility of particle size measurements, especially in the single-nano region (Wada et al., 2006). DLS measurements provide the diffusion coefficient D, which is converted to the diameter by monitoring the fluctuations in scattering intensity due to the Brownian motion of particles. However, because the efficiency of DLS is proportional to the sixth power of the particle diameter (Kerker, 1969), the detection sensitivity strongly depends on the particle size and thus the presence of impurities. The IG method also determines the diffusion coefficient D given by the following Einstein–Stokes equation:

$$D = \frac{k_B T}{3\pi \eta d}$$

where k_B is the Boltzmann's constant, T is the temperature in Kelvin, η is the viscosity of the medium, and d is the diameter of the particles. When the radio frequency voltage is turned off, the diffraction light intensity I begins to decrease according to the following equation:

$$I = I_0 exp(-2Dq^2 t)$$

where I_0 is the initial intensity and q is the value of 2π divided by the pitch of the grating. However, the measurement also includes an activation procedure induced by dielectrophoresis to form a particle grating (Pohl, 1978). Dielectrophoresis first generates a periodic density modulation of the particles, which are then relaxed to a diffuse state until they reach a steady state. Thus, the diffraction light is less affected by the presence of impurities as compared with DLS.

We therefore also conducted particle size measurements of high-degree fullerenols using the IG method (Kokubo et al., 2011). In the diffusion region, sufficient photointensity was observed for the $C_{60}(OH)_{44}$ aqueous solution (0.1 wt%) and the logarithmic value of the relative photointensity correlated well linearly with the time scale (Fig. 4).

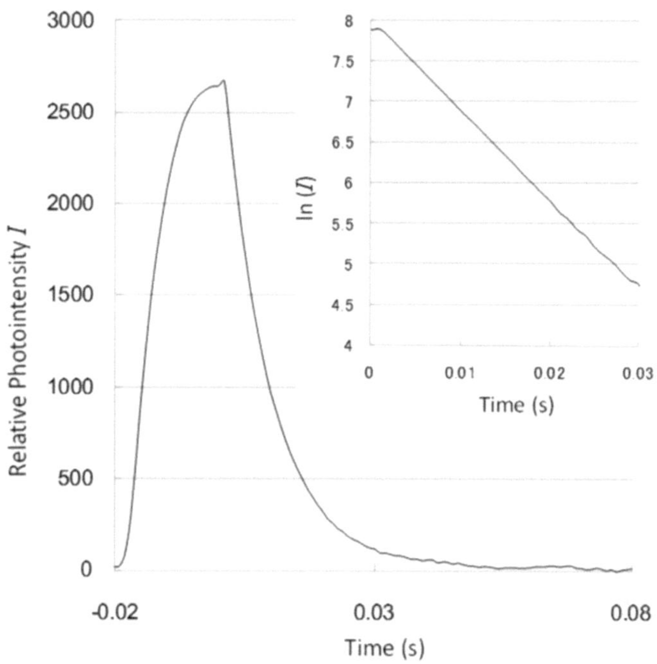

Fig. 4. Time course of relative photointensity I of $C_{60}(OH)_{44}$ aqueous solution (0.1 wt%) measured by the IG method.

The particle size distribution was narrow, in the range of 0.7–1 nm, and the average particle size was determined to be 0.806 ± 0.022 nm ($N = 8$), which was fairly consistent with the DLS results (Fig. 5). Therefore, it was confirmed that the high-degree fullerenols have high dispersion properties in water on a molecular level around their diameter of ca. 1 nm. It is remarkable that the reproducibility of the data measured over eight runs was 10 times higher for the IG method than for the DLS method.

3.3 Other methods

The particle size measurements by the DLS and IG methods were verified by a scanning probe microscope (SPM, Kokubo et al., 2011). We directly measured the particle size of the fullerenol as a function of protrusion height observed on a mica plate on which a highly diluted aqueous solution of fullerenol was applied and dried. As shown in Fig. 6, the protrusions were clearly observed as scattered spots, whereas a mica plate without the fullerenol treatment, used as a control, did not show any spots at all. From the height of eight spots, the average particle size was determined to be 1.03 ± 0. 28 nm (N = 8).

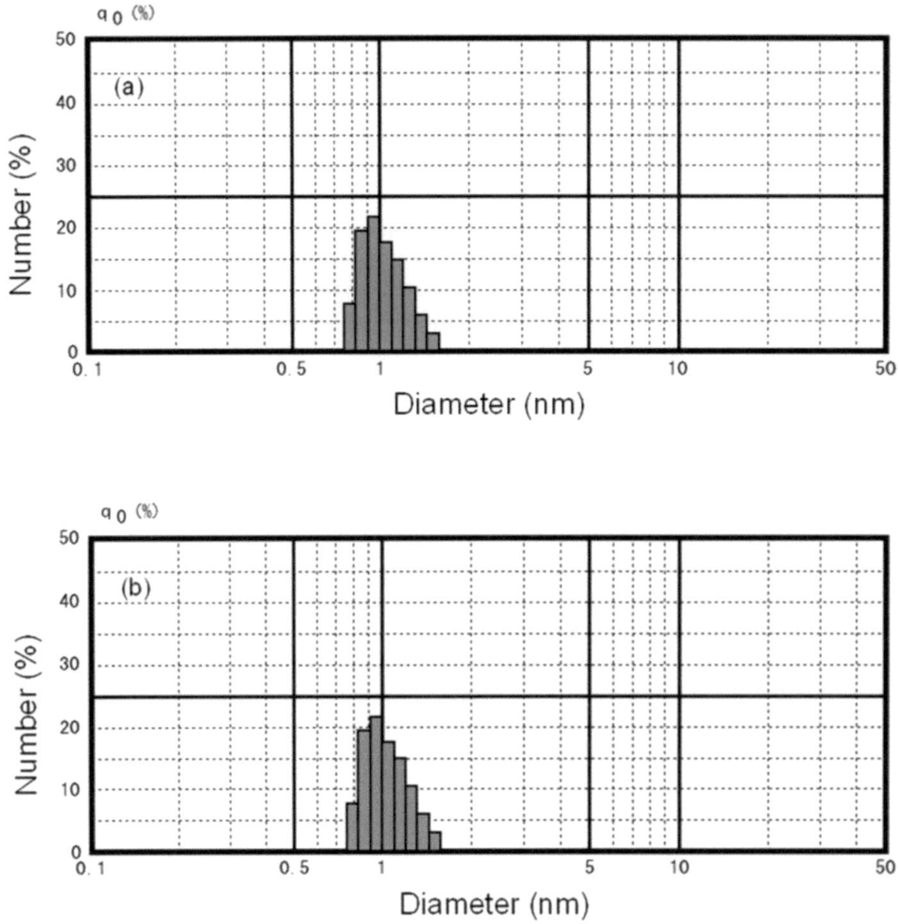

Fig. 5. IG analysis of (a) $C_{60}(OH)_{36}$ and (b) $C_{60}(OH)_{44}$ in water (0.1 wt%) expressed by size distribution in number.

The size distribution of fullerenols has also been investigated using flow field-flow fractionation (FFF) technique (Assemi et al., 2010) and transmission electron microscopy (TEM) (Wang et al., 2010). The FFF is an elution technique that analyzes ensembles of the sample that have a similar property and produces a size distribution rather than an average size. They found that the size of medium-degree fullerenol $C_{60}(OH)_{24}$ nanoparticles was ranging from about 1.8 nm (0.001 M NaCl) to 6.7 nm (0.1 M NaCl). However, this result is in contrast to some DLS data that reports sizes on the order of 100 nanometers for fullerenol nanoparticles. This is because of the fact that the impurities and the large aggregation are separated from the monodispersed fraction eluting from the FFF channel. The TEM observation revealed that the aggregation form of the low-degree fullerenol $C_{60}(OH)_{12}(ONa)_2$ in the solution with a particle size of 50–250 nm.

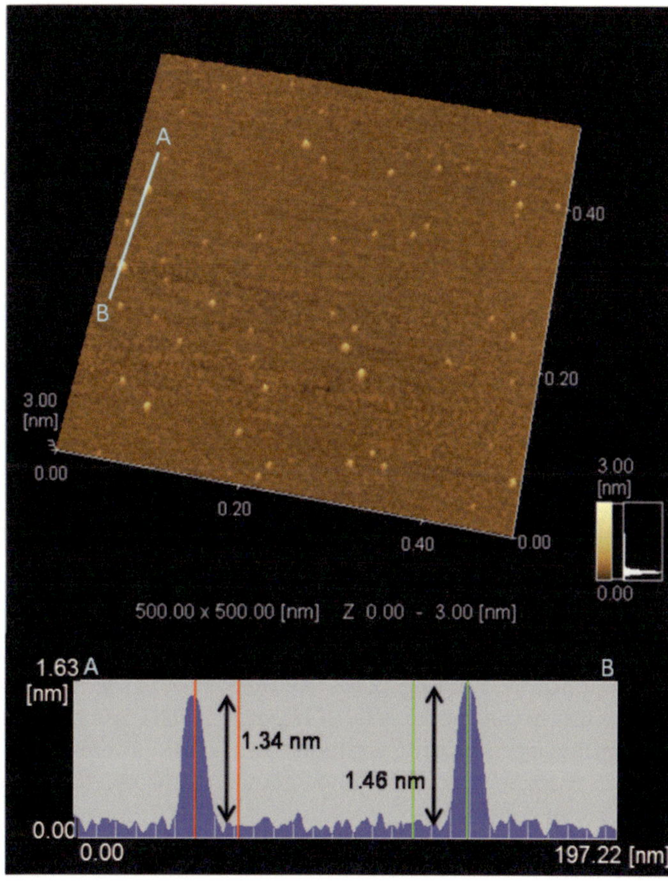

Fig. 6. SPM analysis of $C_{60}(OH)_{36}$ applied as a diluted solution and dried on a mica plate.

4. Application of fullerenols

4.1 Biochemical application

The water-soluble medium-degree fullerenols have been demonstrated to be useful as free-radical scavengers for the absorption of superoxide radicals generated by in vitro xanthine and xanthine oxidase in aqueous solution, suggesting potential use of fullerenols in biochemical or pharmaceutical applications (Chiang et al., 1995). Since then, much research has been devoted to studies on the antioxidant (Dugan et al., 1996; Djordjevic et al., 2004; Bogdanovic et al., 2008), antimicrobial (Aoshima et al., 2009), anti-cancer (Chaudhuri et al., 2009; Krishna et al., 2010), antitumor, and antimetastatic activities of medium-degree fullerenols (Jiao et al., 2010).

Recently, N. Miwa et al. reported the antioxidant activity (Kato et al., 2009) and related bioactivities (Saitoh et al., 2010; Saitoh et al., 2011) of high-degree fullerenols. Some reviews relevant to the biochemical application of fullerenes and fullerenols have also been reported (Nielsen et al., 2008; Rade et al., 2008; Partha & Conyers, 2009). Fullerenes are not considered to have highly significant acute toxicity and genotoxicity, although some toxicological results have also been reported. In contrast, fullerenols are considered to be less toxic than C_{60} due to the introduction of hydrophilic groups that reduce their cytotoxicity.

4.2 Industrial application

Because of a 1 nm grain size, high water dispersibility on a molecular level and metal-free material, the high-degree fullerenol has been proposed as a chemical mechanical polishing (CMP) slurry for use during planarization in the Cu damascene process for the fabrication of next-generation semiconductors. Y. Takaya et al. found that the Cu-surface roughness was improved from 20 to 0.6 nm root mean square (RMS) by using $C_{60}(OH)_{36}$ as functional molecular abrasive grains to achieve better polishing performance than could be achieved using conventional processes (Takaya et al., 2009). The etching ability of $C_{60}(OH)_{36}$ for a Cu surface evaluated in static etch was also found to be high in relation to the achievement of a highly planar surface by polishing experiment. Very recently, further XPS analysis and SEM observation revealed that the chemical effect of fullerenol plays a key role in high polishing performance; i.e., the fullerenol chemically reacted with the copper surface to form a complex brittle layer that was fragile enough to be removed by rubbing with a polishing pad (Takaya et al., 2011).

4.3 Other applications

Other examples of applications using low- to medium-degree fullerenols have been reported, such as polymer-based solar cells (Cao et al., 2001; Rincón et al., 2005), drug delivery and MRI contrast agents using endohedral metallofullerenol (Sitharaman et al., 2004), macromolecular materials and polymer nanocomposites (Goswami et al., 2003; Ouyang et al., 2004), proton conductors (Hinokuma et al., 2001; Maruyama et al., 2002), and electrodeposited films (Wang et al., 2010). Following these applications, water-soluble high-degree fullerenols will open the new pathways for the new fullerenol chemistry.

5. Summary

Fullerenols are one of the most important and promising fullerene derivatives that can be easily synthesized with tunable properties by varying the number of hydroxyl groups introduced. Their water solubility, high dispersing nature as single-nano carbon particles, and varied biochemical properties are extremely attractive from the viewpoint of materials chemistry as well as life science applications. The analytical methods for particle size measurement of single-nano particles will continue to be improved by the further development of these kinds of nanocarbon materials.

6. References

Aoshima, H., Kokubo, K., Shirakawa, S., Ito, M., Yamana, S. & Oshima, T. (2009). Antimicrobial activity of fullerenes and their hydroxylated derivatives, *Biocont. Sci.* 14: 69–72.

Assemi, S., Tadjiki, S., Donose, B. C., Nguyen, A. V. & Millere, J. D. (2010). Aggregation of fullerenol $C_{60}(OH)_{24}$ nanoparticles as revealed using flow field-flow fractionation and atomic force microscopy, *Langmuir* 26: 16063–16070.

Bakry, R., Vallant, R. M., Najam-ul-Haq, M., Rainer, M., Szabo, Z., Huck, C. W. & Bonn, G. K. (2007). Medicinal applications of fullerenes, *Int. J. Nanomed.* 2: 639–649.

Berne, B. J. & Pecora, R. (1976). *Dynamic light scattering with application to chemistry, biology, and physics*, General Publishing Company, Tronto.

Bogdanovic, V., Stankov, K., Icevic, I., Zikic, D., Nikolic, A., Solajic, S., Djordjevic, A. & Bogdanovic, G. (2008). Fullerenol $C_{60}(OH)_{24}$ effects on antioxidative enzymes activity in irradiated human erythroleukemia cell line, *J. Radiat. Res.* 49: 321–327.

Bosi, S., Da Ros, T., Spalluto, G. & Prato, M. (2003). Fullerene derivatives: an attractive tool for biological applications, *Eur. J. Med. Chem.* 38: 913–923.

Brant, J. A., Labille, J., Robichaud, C. O. & Wiesner, M. (2007). Fullerenol clulster formation in aqueous solutions: implications for environmental release, *J. Colloid Interf. Sci.* 314: 281–288.

Brant, J., Lecoanet, H. & Wiesner, M. R. (2005). Aggregation and deposition characteristics of fullerene nanoparticles in aqueous systems, *J. Nanopart. Res.* 7: 545–553.

Cao, T. B., Yang, S. M., Yang, Y. L., Huang, C. H. & Cao, W. X. (2001). Photoelectric conversion property of covalent-attached multilayer self-assembled films fabricated from diazoresin and fullerol, *Langmuir* 17: 6034–6036.

Chae, S.-R., Hotze, E. M. & Wiesner, M. R. (2009). Evaluation of the oxidation of organic compounds by aqueous suspensions of photosensitized hydroxylated-C_{60} fullerene aggregates, *Environ. Sci. Technol.* 43: 6208–6213.

Chaudhuri, P., Paraskar, A., Soni, S., Mashelkar, R. A. & Sengupta, S. (2009). Fullerenol-cytotoxic conjugates for cancer chemotherapy, *ACS Nano* 3: 2505–2514.

Chen, K. L. & Elimelech, M. (2007). Influence of humic acid on the aggregation kinetics of fullerene (C_{60}) nanoparticles in monovalent and divalent electrolyte solutions, *J. Colloid Interf. Sci.* 309: 126–134.

Chiang, L. Y., Wang, L.-Y., Swirczewski, J. W., Soled, S. & Cameron, S. (1994). Efficient synthesis of polyhydroxylated fullerene derivatives via hydrolysis of polycyclosulfated precursors, *J. Org. Chem.* 59: 3960–3968.

Chiang, L. Y., Lu, F.-J. & Lin, J.-T. (1995). Free radical scavenging activity of water-soluble fullerenols, *J. Chem. Soc., Chem. Commun.* 1283–1284.

Deguchi, S., Mukai, S., Tsudome, M. & Horikoshi, K. (2006). Facile generation of fullerene nanoparticles by hand-grinding, *Adv. Mater.* 18: 729–732.

Deguchi, S., Mukai, S., Yamazaki, T., Tsudome, M. & Horikoshi, K. (2010). Nanoparticles of fullerene C_{60} from engineering of antiquity, *J. Phys. Chem. C* 114: 849–856.

Djordjevic, A., Canadanovic-Brunet, J. M., Vojinovic-Miloradov, M. & Bogdanovic, G. (2004). Antioxidant properties and hypothetic radical mechanism of fullerenol $C_{60}(OH)_{24}$, *Oxid. Commun.* 27: 806–812.

Dugan, L. L., Gabrielsen, J. K., Yu, S. P., Lin, T. S. & Choi, D. W. (1996). Buckminsterfullerenol free radical scavengers reduce excitotoxic and apoptotic death of cultured cortical neurons, *Neurobiol. Dis.* 3: 129–135.

Fortner, J. D., Lyon, D. Y., Sayes, C. M., Boyd, A. M., Falkner, J. C., Hotze, E. M., Alemany, L. B., Tao, Y. J., Guo, W., Ausman, K. D., Colvin, V. L. & Hughes, J. B. (2005). C_{60} in water: nanocrystal formation and microbial response, *Environ. Sci. Technol.* 39: 4307–4316.

Goswami, T. H., Nandan, B., Alam, S. & Mathur, G. N. (2003). A selective reaction of polyhydroxy fullerene with cycloaliphatic epoxy resin in designing ether connected epoxy star utilizing fullerene as a molecular core, *Polymer* 44: 3209–3214.

Gun'kin, I. F. & Loginova, N. Y. (2006). Aggregation of fullerene C_{60} in *o*-dichlorobenzene, *Russ. J. General Chem.* 76: 1914–1915.

Hinokuma, K. & Ata, M. (2001). Fullerene proton conductors, *Chem. Phys. Lett.* 341: 442–446.

Husebo, L. O., Sitharaman, B., Furukawa, K., Kato, T. & Wilson, L. J. (2004). Fullerenols revisited as stable radical anions, *J. Am. Chem. Soc.* 126: 12055–12064.

Isaacson, C. W. & Bouchard, D. C. (2010). Effects of humic acid and sunlight on the generation and aggregation state of aqu/C_{60} nanoparticles, *Environ. Sci. Technol.* 44: 8971–8976.

Jiao, F., Liu, Y., Qu, Y., Li, W., Zhou, G., Ge, C., Li, Y., Sun, B. & Chen, C. (2010). Studies on anti-tumor and antimetastatic activities of fullerenol in a mouse breast cancer model, *Carbon* 48: 2231–2243.

Kato, S., Aoshima, H., Saitoh, Y., Miwa, N. (2009). Highly hydroxylated or γ-cyclodextrin-bicapped water-soluble derivative of fullerene: the antioxidant ability assessed by electron spin resonance method and β-carotene bleaching assay, *Bioorg. Med. Chem. Lett.* 19: 5293–5296.

Kerker, K. (1969). *The scattering of light*, Academic, New York, pp. 31–39.

Kokubo, K., Matsubayashi, K., Tategaki, H., Takada, H. & Oshima, T. (2008). Facile synthesis of highly water-soluble fullerenes more than half-covered by hydroxyl groups, ACS *Nano* 2: 327–333.

Kokubo, K., Shirakawa, S., Kobayashi, N., Aoshima, H. & Oshima, T. (2011). Facile and scalable synthesis of a highly hydroxylated water-soluble fullerenol as a single nanoparticle, *Nano Res.* 4: 204–215.

Krishna, V., Singh, A., Sharma, P., Iwakuma, N., Wang, Q., Zhang, Q., Knapik, J., Jiang, H., Grobmyer, S. R., Koopman, B. & Moudgil, B. (2010). Polyhydroxy fullerenes for non-invasive cancer imaging and therapy, *Small* 6: 2236–2241.

Kyzyma, O. A., Korobov, M. V., Avdeev, M. V., Garamus, V. M., Snegir, S. V., Petrenko, V. I., Aksenov, V. L. & Bulavin, L. A. (2010). Aggregate development in C_{60}/*N*-methyl-2-pyrrolidone solution and its mixture with water as revealed by extraction and mass spectroscopy, *Chem. Phys. Lett.* 493: 103–106.

Li, J., Takeuchi, A., Ozawa, M., Li, X., Saigo, K. & Kitazawa, K. (1993). Fullerol formation catalyzed by quaternary ammonium hydroxides, *J. Chem. Soc., Chem. Commun.* 1784–1785.

Maruyama, R., Shiraishi, M., Hinokuma, K., Yamada, A. & Ata, M. (2002). Electrolysis of water vapor using a fullerene-based electrolyte, *Electrochem. Solid-State Lett.* 5: A74–A76.

Matsubayashi, K., Kokubo, K., Tategaki, H., Kawahama, S. & Oshima, T. (2009). One-step synthesis of water-soluble fullerenols bearing nitrogen-containing substituents, *Fuller. Nanotub. Carbon Nanostruct.* 17: 440–456.

Mohan, H., Palit, D. K., Mittal, J. P., Chiang, L. Y., Asmus, K.-D. & Guldi, D. M. (1998). Excited states and electron transfer reactions of $C_{60}(OH)_{18}$ in aqueous solution, *J. Chem. Soc., Faraday Trans.* 94: 359–363.

Nath, S., Pal, H., Palit, D. K., Sapre, A. V. & Mittal, J. P. (1998). Aggregation of fullerene, C_{60}, in benzonitrile, *J. Phys. Chem. B* 102: 10158–10164.

Nielsen, G. D., Roursgaard, M., Jensen, K. A., Poulsen, S. S. & Larsen, S. T. (2008). In vivo biology and toxicology of fullerenes and their derivatives, *Basic Clin. Pharmacol. Toxicol.* 103: 197–208.

Oberdörster, E. (2004). Manufactured nanomaterials (fullerenes, C_{60}) induce oxidative stress in the brain of juvenile largemouth bass, *Environ. Health. Perspect.* 112: 1058–1062.

Ouyang, J., Zhou, S., Wang, F. & Goh, S. H. (2004). Structures and properties of supramolecular assembled fullerenol/poly(dimethylsiloxane) nanocomposites, *J. Phys. Chem. B* 108: 5937–5943.

Partha, R. & Conyers, J. L. (2009). Biomedical applications of functionalized fullerene-based nanomaterials, *Int. J. Nanomed.* 4: 261–275.

Pohl, H. A. (1978). *Dielectrophoresis*, Cambridge University Press.

Rade, I., Natasa, R., Biljana, G., Aleksandar, D. & Borut, S. (2008). Bioapplication and activity of fullerenol $C_{60}(OH)_{24}$, *Afr. J. Biotechnol.* 7: 4940–4950.

Rincón, M. E., Guirado-López, R. A., Rodríguez-Zavala, J. G. & Arenas-Arrocena, M. C. (2005). Molecular films based on polythiophene and fullerol: theoretical and experimental studies, *Sol. Energy Mater. Sol. Cells* 87: 33–47.

Saitoh, Y., Miyanishi, A., Mizuno, H., Kato, S., Aoshima, H., Kokubo, K. & Miwa, N. (2011). Super-highly hydroxylated fullerene derivative protects human keratinocytes from UV-induced cell injuries together with the decreases in intracellular ROS generation and DNA damages, *J. Photochem. Photobiol. B* 102: 69–76.

Saitoh, Y., Xiao, L., Mizuno, H., Kato, S., Aoshima, H., Taira, H., Kokubo, K. & Miwa, N. (2010). Novel polyhydroxylated fullerene suppresses intracellular oxidative stress together with repression of intracellular lipid accumulation during the differentiation of OP9 preadipocytes into adipocytes, *Free Radic. Res.* 44: 1072–1081.

Sitharaman, B., Bolskar, R. D., Rusakova, I. & Wilson, L. J. (2004). $Gd@C_{60}[(COOH)_2]_{10}$ and $Gd@C_{60}(OH)_x$: nanoscale aggregation studies of two metallofullerene MRI contrast agents in aqueous solution, *Nano Lett.* 4: 2373–2378.

Su, Y., Xu, J., Shen, P., Li, J., Wang, L., Li, Q., Li, W., Xu, G., Fan, C. & Huang, Q. (2010). Cellular uptake and cytotoxic evaluation of fullerenol in different cell lines, *Toxicology* 269: 155–159.

Tajima, Y. & Takeuchi, K. (2002). Discovery of $C_{60}O_3$ isomer having C_{3v} symmetry, *J. Org. Chem.* 67: 1696–1698.

Takaya, Y., Kishida, H., Hayashi, T., Michihata, M. & Kokubo, K. (2011). Chemical mechanical polishing of patterned copper wafer surface using water-soluble fullerenol slurry, *CIRP Ann. - Manuf. Techn.* 60: 567–570.

Takaya, Y., Tachika, H., Hayashi, T., Kokubo, K. & Suzuki, K. (2009). Performance of water-soluble fullerenol as novel functional molecular abrasive grain for polishing nanosurfaces, *CIRP Ann. – Manuf. Techn.* 58: 495–498.

Troyanov, S. I., Troshin, P. A., Boltalina, O. V., Ioffe, I. N., Sidorov, L. N. & Kemnitz, E. (2010). Two isomers of $C_{60}F_{48}$: an indented fullerene, *Angew. Chem. Int. Ed.* 40: 2285–2287.

Tuinman, A. A., Mukherjee, P., Adcock, J. L., Hettich, R. L. & Compton, R. N. (1992). Characterization and stability of highly fluorinated fullerene, *J. Phys. Chem.* 96: 7584–7589.

Wang, F. F., Li, N., Tian, D., Xia, G. F. & Xiao, N. (2010). Efficient synthesis of fullerenol in anion form for the preparation of electrodeposited films, *ACS Nano* 4: 5565–5572.

Wada, Y., Totoki, S., Watanabe, M., Moriya, N., Tsunazawa, Y. & Shimaoka, H. (2006). Nanoparticle size analysis with relaxation of induced grating by dielectrophoresis, *Opt. Express* 14: 5755–5764.

Wang, S., He, P., Zhang, J.-M., Jiang, H. & Zhu, S.-Z. (2005). Novel and efficient synthesis of water-soluble [60]fullerenol by solvent-free reaction, *Synth. Commun.* 35: 1803–1807.

Yevlampieva, N. P., Biryulin, Y. F., Melenevskaja, E. Y., Zgonnik, V. N. & Rjumtsev, E. I. (2002). Aggregation of fullerene C_{60} in N-methylpyrrolidone, *Colloid. Surface. A* 209: 167–171.

Ying, Q., Marecek, J. & Chu, B. (1994). Slow aggregation of buckminsterfullerene (C_{60}) in benzene solution, *Chem. Phys. Lett.* 219: 214–218.

Zhang, G., Liu, Y., Liang, D., Gan, L. & Li, Y. (2010). Facile synthesis of isomerically pure fullerenols and formation of spherical aggregates from $C_{60}(OH)_8$, *Angew. Chem. Int. Ed.* 49: 5293–5295.

16

Self-Organization and Morphological Characteristics of the Selenium Containing Nanostructures on the Base of Strong Polyacids

S.V. Valueva and L.N. Borovikova
The Institution of the Russian Academy of Science,
The Institute of High-Molecular Compounds, Saint-Petersburg,
Russia

1. Introduction

This article represents the results of the study made by methods of molecular optics of nanostructures formed in process of reduction of ionic selenium in selenite-ascorbate redox system in water solutions of high molecular polymeric stabilizers of anion type: synthetic polyacid - poly-2-acrylamide-2-methylpropansulfacid with MM $M_w = 3\times10^6$ and biopolyanion - deoxyribonucleic acid with $M_w = 20\times10^6$. It was shown that polyanion - nanoparticle complex obtained under conditions of total saturation of adsorption capacity of selenium nanoparticles (mass ratio v of the components selenium : polymer is equal to 0.1) is close to its thermodynamic stability boundary: the second virial coefficient made up $A^*_2 = -0.07\times10^{-4}$ cm^{3}mol/g^2 for system of deoxyribonucleic acid - nano - Se^0- H_2O and $A^*_2 = 0.2\times10^{-4}$ cm^{3}mol/g^2 for system of poly-2-acrylamide-2- methylpropansulfacid - nano - Se^0- H_2O. In the field of formation of stable dispersions the values of free energy ΔG^* of interaction of macromolecule - nanoparticle of selenium have been calculated for anion type nanostructures. It was found that in both cases high-molecular structures with $M^*_w = 200\times10^6$ (deoxyribonucleic acid as stabilizer) and with $M^*_w = 75\times10^6$ (poly-2-acrylamide-2-methylpropansulfacid as stabilizer) were formed with close dimensions (R^*_g - statistic dimensions of the nanoformation, R^*_h - hydrodynamic dimensions of the nanoformation) and average densities Φ^*. The values of conformational parameters ρ^* and p^* testify a form of nanostructures approximating to spherical form: $\rho^* = R^*_g / R^*_h = 1$ (for both systems), $p^* = 1.1$ for a system of deoxyribonucleic acid - nano - Se^0- H_2O and $p^* = 1.5$ for a system of poly-2-acrylamide-2- methylpropansulfacid - nano - Se^0- H_2O. On the base of the experimental data related to the values of M^*_w for spherical nanostructures assuming their mono-nuclear morphology we estimated a radius of selenium nucleus that made up $R_{\mathbf{nucl}}$ = 12 nm (deoxyribonucleic acid as stabilizer) and 9 nm (poly-2-acrylamide-2-methylpropansulfacid as stabilizer). The following conclusion has been made: under adsorption of macromolecules of strong acids on nanoparticles of selenium the spherical nanoparticles with similar types of morphology with similar dimensions, densities and width of polymer shell are formed.

At present many original articles and monographs are devoted the issues of synthesis of nanoparticles and nanosystems and studying of their properties. (see e.g. [1]). Nano-dimensional particles (NDP) in absence of stabilizer represent typical liophobic colloids characterized by very low stability. The most various substances, from which high-molecular compounds are the most important, are applied for increase of the stability of NDP.

In connection with development of biological nanotechnology the above mentioned particles represent interest as active ingredients of medicines [2, 3] as well as transport systems and adsorption matrixes for bioactive substances [4].

It should be pointed out that nanoparticles of nonmetals stabilized by polymer have not been practically studied. In this respect nanoparticles of amorphous selenium (nano- a-Se^0) that are used as high-sensitive bio-sensors for immunoassay technology [5] and chromatography mobile affine reagents [6]. Nano-particles of amorphous selenium are characterized by the exclusive spectrum of bioactivity: even at very low concentrations in water (0.005 – 0.1%) they can adsorb antigens antibodies at their surface [7].

Previously (e.g. in [18 – 13]) it was demonstrated that under reduction of selenium ions in solutions of polymers two processes are going on: formation of selenium nanoparticles with narrow unimodal distribution by dimensions and adsorption of macromolecules on them. In the result self-organization of polymeric molecules on surface of nanoparticles and formation of nanostructures with high molecular masses of significant density and various morphologies occur.

This work is focused on studying of process of self-organization and structural-morphological characteristics of selenium containing nano-structures on the basis of strong polyacids: synthetic polyacid - poly-2-acrylamide-2- methylpropansulfacid (PAMS) and biopolyanion - deoxyribonucleic acid (DNA).

2. The objects and methods of research

Selenium represents an exclusive interest as a chemical element with unique semiconductor, photoelectric and X-ray sensitivity properties as well as bioactive substance with anti-oxidant, anti-inflammatory, anticarcinogenic, antimutagenic and detoxicant activities.

In this study nanoparticles of a-Se^0 by reduction of selenitic acid by means of ascorbic acid were selected as subject of inquiry. Compared to nanoparticles of metals a-Se^0 represents an inorganic polymer including fragments of cycles of Se_6 and Se_8 linked by covalent bonds Se-Se [14] in its backbone.

We used the following compounds as polymeric stabilizers: poly-2-acrylamide-2-methylpropansulfacid and bioplyanion - deoxyribonucleic acid (DNA).

-[-CH_2-CH-]$_n$-

| **PAMS**

O=C-NH-C(CH_3)$_2$-CH_2- SO_3Na

Molecular Masses (MM) of polymeric matrix made up: $M_w = 3\times10^6$ for PAMS (characteristic viscosity 1M $NaNO_3$ at 20^{0}C made up $[\eta] = 5.2$ dl/g) and $M_w = 20\times10^6$ for DNA (see table).

Reduction reaction of ionic selenium was carried out at concentration of polymer in water solution equal to 0.1% and mass ratio of selenium : polymer $v = 0.1$ i.e. under conditions of total saturation of adsorption capacity of nanoparticles as it was found for selenium - polymeric nanostructures [8].

Study of kinetics of selenium reduction (IV) was carried out at initial concentration of selenitic acid equal to 1.0 mmole/l in the regime with constant temperature (20^0 C) by means of spectrophotometer «Specord M-40» by means of registering changes of optical density of the solutions at wavelength equal to 320 nm. pH values of the solutions of the reaction mass made up 3.5. The values of the constant of the rate of reaction of formation of nanocomposites k^* calculated by well known method of Guggenheim [15] are given in the table. Calculation of k^* was carried out by the following formula:

$$k^* = \ln(D_k/(D_k - D_i))/t_i , \quad (1)$$

where D_k - optical density characterizing the end of process, D_i - optical density at given moment of time t_i. Conditions of conducting of reaction of formation of nano - a-Se^0 polymer solution were kept constant: value of k^* was influenced only by structure of polymeric matrix stabilizing nano - a-Se^0 being formed.

MM M_w^* and root-mean-square radiuses of inertia R_g^* of nanostructures were determined by means of elastic (static) scattering of light in solutions in water [16] and their affinity to solutions was determined also by the values of the second virial coefficient A_2^*. Quantity of N^* of adsorbed macromolecules on the surface of nano-a-Se^0 was calculated by means of ratio MM for polyacids and nanostructures formed by these polyacids. The values of M_w^*, N^*, R_g^*, and A_2^* are shown in the table. Wy used photogoniodiffusiometer «Fica» for determination of the reduced intensity of scattering of solutions R_θ. Wavelength of vertical incident polarized light made up $\lambda = 546.1$nm. Measurements were carried out at scattering angles range $\theta = 30^0 - 150^0$. Cleaning of cautions was carried out through millipore (Millex-HV) with diameter 0.45 µm. The values of increment of the refractive index dn/dc were obtained from refractometric measurements by means of instrument IRF-23. Processing of experimental data of light scattering for solutions of nanostructures was carried out by means of Zimm method (see figures 1 and 2) using double extrapolation (to $c = 0$ and $\theta = 0$) of dependence of Kc/R_θ on $\sin^2(\theta/2) + k'c$ (K - calibration constant, k' - numeric constant).

Basing on data for M_w^* and root-mean-square radiuses of inertia by formula (2) we evaluated the values of average density of nanostructures

$$\Phi^* = 3\, M_w^*/4\pi N_a R^3_{sph}, \quad (2)$$

where $R_{sph} = 1.29\, R_g^*$ [17]. The values of Φ^* are given in the table.

Average hydrodynamic dimensions of nanostructures R_h^* (see table) were determined by the method of quasi-elastic (dynamic) scattering [18]. The radiuses of the equivalent hydrodynamic spheres R_h^* were calculated from the values of diffusion coefficients (D^*) by Einstein-Stokes equation $R_h^* = kT/6\pi\eta_0 D^*$ (η_0 - viscosity of solvent). The value of conformation-structural parameter ρ^* (see table) was found by means of ratio of experimental values of R_g^* and R_h^* [19 - 22].

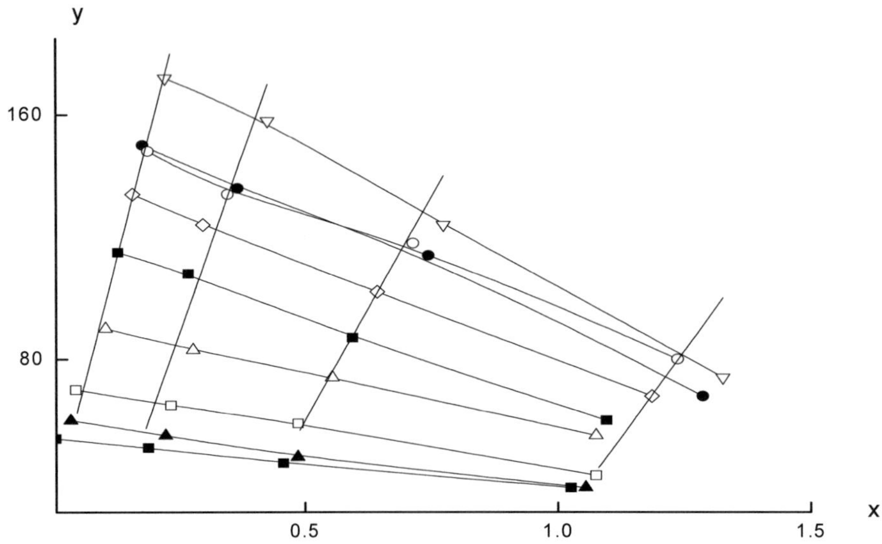

Fig. 1. Zimm Diagram for a system of DNA-Se^0- water:
x – $\sin^2(\theta/2)$ +k'c,
y- $Kc/R_\theta \times 10^8$

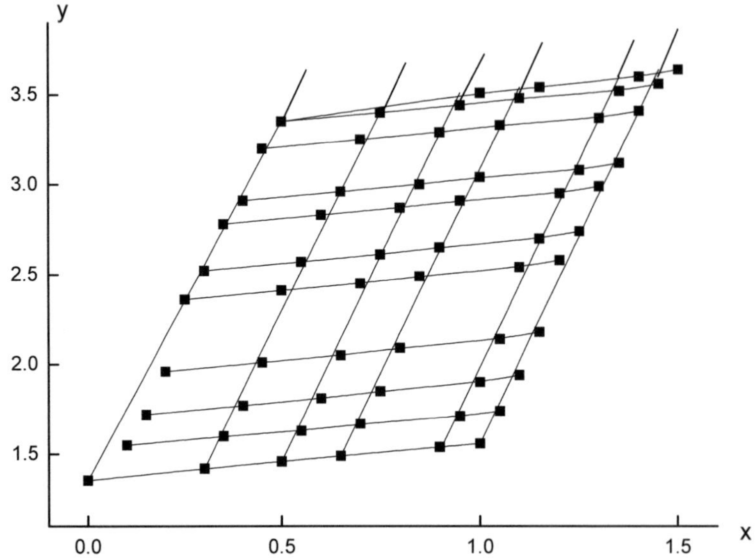

Fig. 2. Zimm diagram for a system of PAMS-Se^0- water:
x – $\sin^2(\theta/2)$ +k'c,
y- $Kc/R_\theta \times 10^8$

Optical part of the installation for measurement of quasi-elastic (dynamic) scattering was equipped with goniometer ALV-SP (Germany) with the following source of light: helium – neon (He – ne) laser Spectra-Physics with wavelength λ = 632.8 nm. Correlation function of intensity of the scattered light was obtained by means of correlator Photo Cor-FC with 288.

Analysis of correlation function was carried out by means of quasi-elastic (dynamic) scattering data processing program Dynals (Helios firm, Russia).

Molecular dispersion of the solutions of the forming nanostructures was estimated by means of birefringence (double refraction) method [23] using a character of gradient dependence of the value of double refraction Δn. The value of Δn was determined subject to rate of rotation of the rotor "g" and concentration of the solution "c" provided constant ratio selenium : polymer. Titanium dynamo-optimeter with internal rotor with 4 cm height and 0.03 cm gap between rotor and stator was used for those measurements. All measurements of double refraction Δn were carried out using thermostating at 21^{0}C to avoid changes of viscosity of solutions and optical distortions caused by temperature gradient. Phenethyl alcohol that has significant double refraction value ($\Delta n/g$ = 17 × 10^{-12}) and system of polystyrene - bromoform were used for calibration of the installation. Inaccuracy of determination of characteristic value of догиду refraction $[n] = \lim_{g\to 0,\, c\to 0} (\Delta n/gc\eta_0)$ did not exceed 10%. The measurements were carried out at $g < g_k$, where g_k is a gradient of velocity at which flow turbulence occurs.

In general case when $dn/dc \neq 0$ experimental value [n] is formed from three effects: $[n] = [n]_e + [n]_{fs} + [n]_f$ where $[n]_e$ - intrinsic anisotropy, , $[n]_{fs}$- microform effect, $[n]_f$ - macroform effect [23]. In that a value of total segment anisotropy $[n]_{fs} + [n]_e$ is determined by equilibrium stiffness of the polymer chain and structure of elementary unit of the polymer, and value $[n]_f$ is connected with asymmetry of form of the particle by the following relation:

$$[n]_f = ((n^2_s + 2)/3)^2 \times (M_w^*(dn/dc)^2 f(p))/(30\pi RTn_s) =$$

$$\text{const } M_w^*(dn/dc)^2 f(p), \quad (3)$$

where n_s - solvent refractory index, T - absolute temperature, R - universal gas constant, f(p) - tabulated function of ratio of axes of stiff ellipsoid approximating the particle [23].

Time variations of effective viscosity of PAMS solution during reduction of selenitic acid we registered by means of rheoviscosimeter «Brookfield» at rate of rotation of the rotor equal to 12 rpm.

Value of relative viscosity η_r was determined by means of Ostwald capillary viscosimeter with water outflow time equal to 120±0.2 sec at 21^{0}C.

In this study we used for the first time a method of polarized light scattering for studying structural features of nanocomposite on the base of PAMS. The studies were carried out by means of ФПС-3М instrument with photoelectric system of registration of light intensity scattered in the range of angles θ = 40^0-140^0 (wavelength of incident light = 578 nm). Calibration of the instrument was done by benzol: Rayleigh ratio for benzol at given wavelength of incident light made up 13.1×10^{-6} cm^{-1}. Angle dependencies of vertical (V_V) and horizontal (H_V) polarized components of the scattered light were measured by method described in [16]. Calculation of the parameters of scattering media was carried out within the frames of Debye-Bekey [24]. Isotropic parameters of structure such as average square of scattering micro-volume polarizability fluctuation $<\eta>^2$ and radius of polarizability fluctuation correlations α_V were determined from angle dependence V_V; the following isotropic parameters such as square of optical anisotropy average density $<\delta>^2$ and radius of correlation of optical axes of scattering elements of volume α_H. The parameters of isotropic structure are connected with dimensions α_V of macro-molecules or associates and with micro-heterogeneity $<\eta>^2$ as a function of density fluctuation [25, 26]. Nature of dependence of the parameters of anisotropic structure on concentration allowed determining of system order and statistic dimensions of oriented regions.

3. The results and discussion

When reducing selenitic acid by ascorbate in presence of polymeric stabilizers the rate constants were equal to $k^* = 0.5\times10^{-3}$ sec^{-1} for DNA and 0.4×10^{-3}sec^{-1} for PAMS correspondingly that significantly differs from the value of k^* in absence of stabilizers when $k^*=1.6\times10^{-3}$ sec^{-1}. In addition, the values of the rate constants illustrate comparable influence of selected polymeric matrixes on process of self-organization of nanostructures.

In that during the first 10 minutes of reaction efficient viscosity of nanocomposite solutions on the base of PAMS has been changed from 165 centipoise to 55 centipoise, and after that during 2 days it reduced to 18 centipoise that verified reduction of total number of the particles in the volume of solution due to adsorption of macromolecules on the surface of Se nanoparticles being formed. Studying of characteristic viscosity of water solutions of original PAMS and formed nanostructure of PAMS –nano- Se^0 has demonstrated that estimated value [η]was reduced significantly from 70 to 12 dl/g (Figure 3). However, in that according to the data за static light scattering the value of molecular mass M_w^* of nanostructure made up 75×10^6 (table) i.e. it has been increased by 25 times ($N^* = 25$) compared to free macromolecules of PAMS.

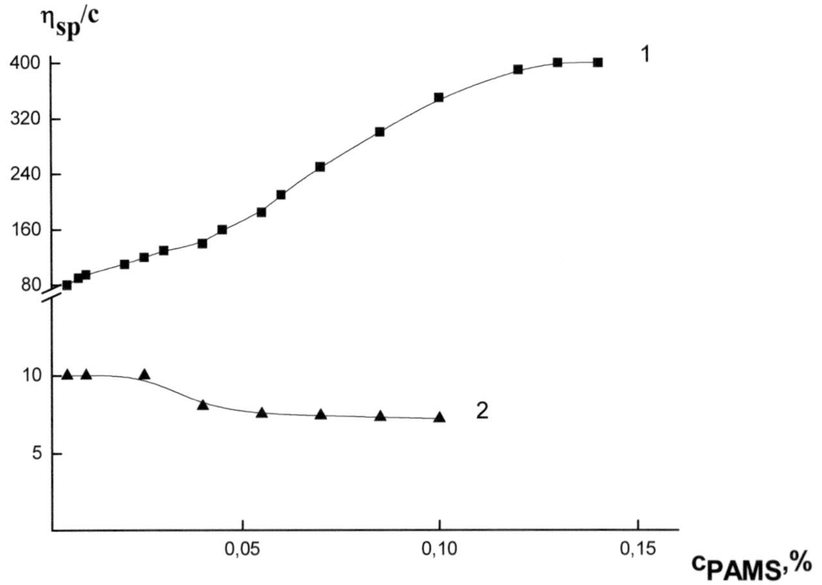

Fig. 3. Concentration dependence of reduced viscosity ηsp/c for systems PAMS-water (1) and PAMS-Se^0- water (2).

Estimation of hydrodynamic radius R_h of Se nanoparticles by the method of dynamic light scattering demonstrated that in absence of polymer big particles of selenium with radius ~ 100 nm (the first day) and ~ 180 nm (the second day) were formed with initial narrow unimodal distribution by dimensions. However, after ~ 50 hours this distribution by dimensions became bimodal and wide. Visible aggregation of selenium occurred after two days expiration. When using PAMS as polymeric stabilizer of nanoparticles of selenium in water solutions at v=0.1 the nanostructures with dimensions < 100 nm (table) are formed with narrow unimodal dimension distribution without a tendency to aggregation during a week period at least.

As it follows from the data related to optical properties of PAMS polyanion in water solutions represents an non-penetrable asymmetric swelled ball with asymmetry p> 2 [27]. For system PAMS-nano- Se^0-H_2O evaluation of the value of parameter p^* in approximation $[n] \approx [n]_f$ has demonstrated that nanostructure has conformation approximating to spherical one: $p^* = 1.5$. It is compatible with the data related to parameter ρ^*: $\rho^*=1.0$ that corresponds to spherical conformation [19, 22].

Assuming spherical conformation of nanostructure on the base of PAMS we determined its packing factor k [28]:

$$k=(N^* M^*_w \Sigma_I \Delta V_I)/(v^* M_0), \quad (4)$$

where $\Sigma_I \Delta V_I$ - intrinsic (Van der Waals) volume of the repetitive link if polymer formed from increments of Van der Waals volumes of separate atoms included into this link:

M^*_w - MM of the particle;
N^* – number of molecules of polymers in adsorbed state
M_0 – MM of the repetitive link;
v^*- volume of nanostructure equal to $(4\pi/3)\times(R_{sph})^3$ where R_{sph}=1.29 R_g^*.

Value of k made up 0.2 that is approximating to the value for globular protein (k = 0.6 ÷ 0.8) by the order of magnitude.

Polarized light scattering method was used for quantitative estimation of order and micro-heterogeneity of the system PAMS – nano - Se^0- H_2O.

Compared to the system PAMS - H_2O, for which parameter $<\delta>^2$ is not determined in general (water solutions do not reveal anisotropic scattering), this parameter for the system PAMS - Se^0- H_2O reaches a value ~ 1.5×10^{-7} that is more typical for solid bodies (Figure 4).

Sharp increase of parameter $<\eta^2>$ is observed for the same system in the range of concentrations of polymer c< 0.05 % (Figure 4), that verifies increase of micro-heterogeneity of the solution due to increase of scattering centers per unit volume.

In that statistic dimensions of optical dense regions α_v remain invariable in the whole range of the studied concentrations (Figure 5).

A distinctive feature of the system PAMS-Se^0-H_2O in the range of concentrations for polymers equal to 0.05 – 0.10 mass % is that inverse (abnormal) slope of angle dependence H_v scattering component is observed. It may be a result of nonrandom fluctuations of

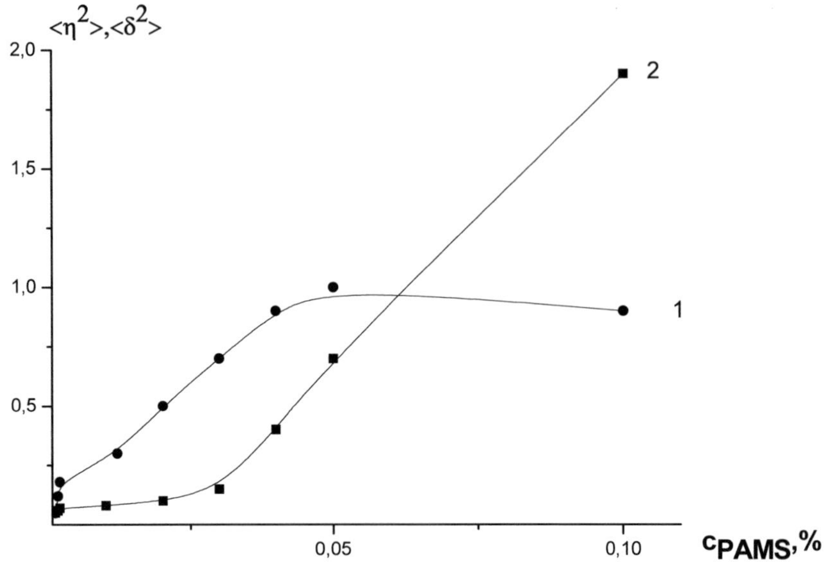

Fig. 4. Concentration dependences of mean square of polarizability fluctuations $<\eta^2>$ (1) and mean square of density of optical anisotropy $<\beta^2>$ (2) for a system PAMS-Se^0- water.

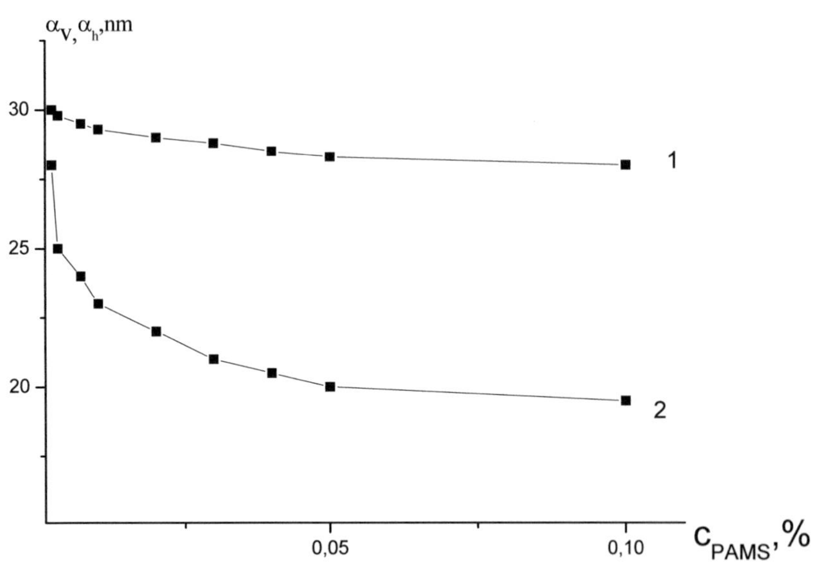

Fig. 5. Concentration dependences of correlation radius of polarizabaility fluctuations αv (1) and correlation radius of correlation of optical axes of scattering elements α_h (2) for a system of PAMS-Se^0- water.

anisotropy density for this system. Statistics processing of the data for anisotropic scattering within the angles 40-80^0 has provided a possibility to determine dimensions of the oriented regions (α_H ~ 20 nm) inside this system (Figure 5). Additional verification of high level of order of nanostructure is represented by the fact that value of local order factor F_v

determined by relation (5) reaches the value equal to 0.7 that characterizes intermediate state of total disordering of macromolecules F_v= 0.33) and crystal structure state of the macromolecules (F_v=1).

$$F_v = 1/3\,(1 + 2\exp(-\alpha_v/\alpha_h)), \tag{5}$$

Since PAMS from structural point of view is N-substituted analog of polyacrylamide that is not adsorbed on nanoparticles of selenium [8] the main input into binding of PAMZZS with nano- a-Se^0 can be made by hydrophobic fragments as well as sulfonate groups of side links of the polymer.

According to the data of static light scattering the value of molecular mass M_w^* for nano-structure DNA-nano- Se^0 made up 200×10^6 (Table) that is it was increased by 10 times (N^* = 10) compared to free macromolecules of DNA. Statistics (R_g^*) and hydrodynamic (R_h^*) dimensions of the nanostructure DNA-nano- Se^0 coincide with each other ($R_g^*=R_h^*$ =100 нм) and correspond to nanodimensional level (Table).

System	$k^*\times 10^{-3}$ s^{-1}	$M_w\times 10^{-6}$	$M^*_w\times 10^{-6}$	N^*	$A_2^*\times10^4$, $cm^3 mole/g^2$	R_g^*, nm	R_h^*, nm	$R_я$, nm	ΔR, nm	Φ^* g/cm^3	p^*	$-G^*\times10^5$, J/m^2
DNA	0.50	20.00	200.00	10.00	-0.07	100.00	100.00	12.00	88.00	0.04	1.10	0.02
PAMS	0.4	3	75	25	0.2	90	90	9	81	0.02	1,5	0.01

Table 1. The constants of the rate of reaction of formation of nanococmposites and structural-conformation parameters of the corresponding nanostructures.

It is known that in the field of acid pH (pH<4.5) oxygen destabilization/denaturation of the DNA macromolecules takes place. Within the range of pH from 3.0 to 4.5 partial destabilization of twin helix is observed, further decrease of pH results in denaturation of DNA already that is verified by studies of UV absorption of DNA, circular dichroism spectra of DNA, viscosimetry data, and double refraction (Δn) data in flow [29 - 32].

For the system DNA-nano- Se^0–H_2O at pH = 3.5 destabilization of twin helix in process of self-organization of the structure is observed: the given value of double refraction (Δn) has a positive sign compared to corresponding value for negative DNA (Figure 6). If for isolated DNA macromolecule in native condition $[n] \approx [n]_e$ [23] then for the system DNA-nano-Se^0–H_2O approximation $[n] \approx [n]_f$ is fulfilled. Estimation of the asymmetry level of the form of the nanostructure resulted in the value p^*= 1.1. It agrees to the light scattering data ($\rho^* = R_g^*/R_h^*$ =1.0).

It seems that formation of adsorbates of polymers on nanoparticles of nonmetals with the form close to spherical one is a universal phenomenon. For nanostructures, which are different from morphology point of view in the range of saturation of adsorption capacity v = 0.1, the particles with a form close to spherical one were found experimentally [8-13, 33, 34].

Calculation of average density of nanostructures at v = 0.1 by formula (2) demonstrated that though the nanostructures with density Φ^*=0.02 and 0.04 g/cm^3 are formed (Table) that exceed a density of polymer ball [23] but less than for selenium containing spherical nanostructures on the base of nonionic rigid chain molecules (Φ^*=(0.12 -0.14) g/cm^3) [12].

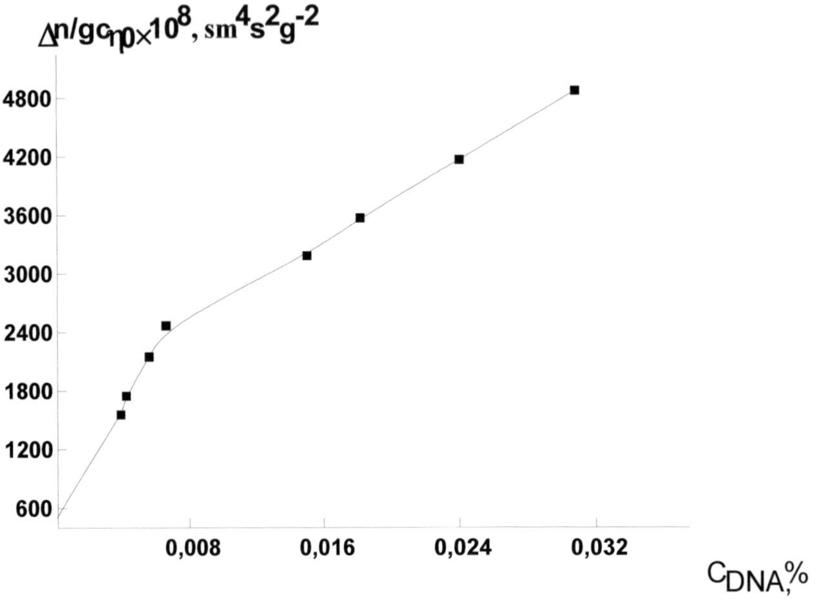

Fig. 6. Concentration dependence of the reduced optical anisotropy $\Delta n/gc\eta_0 \times 10^8$, $sm^4s^2g^{-2}$ for a system of DNA-Se^0- water.

Thus it was revealed that in both cases super high molecular nanostructures are formed (M^*_w = 200×10^6 /DNA as stabilizer/ and M^*_w = 75×10^6 /PAMS as stabilizer/) with close dimensions (R^*_g – statistic dimensions of nanostructure, R^*_h – hydrodynamic dimensions of nanostructure) and average densities Φ^* (Table). The values of conformation parameters ρ^* and p^* testify a form of nanostructures approaching to spherical one: $\rho^* = R^*_g / R^*_h = 1$ (for both systems); $p^* = 1.1$ for system DNA-nano- Se^0- H_2O and $p^* = 1.5$ for system PAMS-nano-Se^0- H_2O (see Table).

Basing on experimental data for values of M^*_w for spherical nanostructures and assuming their mononuclear morphology we evaluated a radius of selenium nucleus that made up R_{nucl} = 12 nm (DNA as stabilizer) and 9 nm (PAMS as stabilizer). Thus in water solution a width of polymer shell ΔR on the particles of nano- Se^0 stabilized by strong polyacids has small differences and makes up ~ 80 – 90 nm (see Table).

With the purpose of experimental determination of the dimensions of nuclei in nanostructures PAMS-Se^0 and DNA-Se^0 their water solutions were passed through cool dehumidification, and the samples obtained in the form of pellets were studied by X-ray diffraction method. It was found that amorphous particles of nano- Se^0 of spherical form for both nanocomposites had unimodal distribution by dimensions and average radius of selenium nucleus made up 10 – 15 nm. These values have good correlation with design quantity of R_{nucl} for nanocomposites.

It was shown that a complex polyanion – nanoparticle obtained under conditions of total saturation of adsorption capacity of selenium nanoparticles (mass ratio v of the components of the selenium: polymer complex was equal to 0.1) was close to its thermodynamic limit:

the second virial coefficient made up $A^*_2 = -0.07\times10^{-4}$ cm^3mole/g^2 for DNA-nano- Se^0- H_2O system and $A^*_2 = 0.2\times10^{-4}$ cm^3mole/g^2 for PAMS-nano- Se^0- H_2O system. It agrees with the data obtained at v =0.1 for selenium containing nanostructures on the base of the system of nonionic polymers and polybases [8, 12, 35].

Basing in relation [36] we have:

$$\Delta G^* = kTlnC_e/\pi D_{c\phi}^2, \quad (6)$$

that is applicable in the field of stable dispersions for spherical nanostructures of arbitrary morphologies, we obtained the values of free energy ΔG^* of interaction of macromolecule - nanoparticle per unit area of the surface of the particle (see Table). In equation (6) C_e - equilibrium concentration of polymer in molar fractions of monomeric units, $D_{sph} = 2R_{sph}$, ($R_{sph} = 1.29\ R_g^*$). You may see from the Table that quantity of free energy at v =0.1does not depend practically on structure of monomeric unit and nature of polyacids. It seems that in both cases the same mechanism of adsorption of macromolecules on selenium nanoparticles is realized. So, for ionogenic polymeric matrix (at pH = 3.5 strong polyacids are in ionized state) it is observed electrosteric stabilization stipulated for by electrostatic effects from one side and entropic and osmotic effects from another side [37] appearing due to hydrophobic interactions in aqueous medium between hydrophobic fragments of organic macromolecules and energy saturated surface if selenium nanoparticles.

From totality of the presented data we can make conclusion that at adsorption of macromolecules of strong acids on selenium nanoparticles spherical nanostructures of similar morphology types with close dimensions, densities and width of polymer shell are formed.

Thus, selenium nanoparticles represent adsorption matrixes where high local concentrations of polyanions containing hydrophobic fragment can be reached.

4. Conclusions

1. It has been determined by means of molecular optics that macromolecules of PAMS and DNA adsorbed on nano-Se^0 form high molecular nanostructures (with MM $M_w^*=75\times10^6$ и 200×10^6) with dimensions equal to 90 - 100 nm including 10/25 macromolecules densely packed into its composition.
2. It was shown that polyanion - nanoparticle complex obtained under conditions of total saturation of adsorption capacity of selenium nanoparticles (mass ratio v of the components of selenium: polymer complex was equal to 0.1) is close to the limit of its thermodynamic stability. It agrees with the data obtained at v =0.1 for selenium containing nanostructures on the base of the systems of non-iongenic polymers and polybases.
3. In saturation range of adsorption capacity v = 0.1 universality of formation selenium - polymeric nanostructures of spherical form was experimentally determined.

5. References

[1] Pomogailo A.D., Rosenberg A.S., Uflyand I.E. (2000). *Metal nanoparticles in polymers.* M. Chemistry.

[2] Kopeikin V.V., Panarin E.F. (2001). Water soluble nanocomposites of zero-valent silver with increased antiinfection activity, *J.Reports of the Academy of Science*. Vol. 380, No. 4, 497 -500.

[3] Balogh L., Swanson D. R., Tomalia D. A., Hagnauer G. L., McManus A. T. (2001). Antimicrobial Dendrimer Nanocomposites and a Method of Treating Wounds, US 6 224 898 B1.

[4] Connelly S., Fitzmaurice D. (1999). Programmed Assembly of Gold Nanocrystals in Aqueous Solution, *Adv.Mater.*, Vol. 11, No. 14, 1202-1205.

[5] Yost D. A., Russell J.C., Yang H. (1990). Non-Metal Colloidal Particle Immunoassay, US Patent 4 954 452.

[6] Ching S., Gordon J., Billing P.A. (1988). Process for Immunochromatography with Colloidal Particles, EP App. 0 299428 A2.

[7] Zhang J.S., Gao X.Y., Zhang L.D., Bao Y.P. (2001). Biological effects of a nano red elemental selenium, *Biofactors*. Vol. 15, No. 1, 27-38.

[8] Kopeikin V.V., Valueva S.V., Kipper A.I., Borovikova L.N., Filippov A.P. (2003). Synthesis of selenium nanoparticles in water solutions of polyvinylpyrrolidone and morphologic characateristcs of the nanocomposites being formed, *High-molecular compounds*, Vol. 45 A, No. 4, 615 - 619.

[9] Kopeikin V.V., Valueva S.V., Kipper A.I., Kalinina N.A., Silinskaya I.G., Khlebosolova E.N., Shishkina G.V., Borovikova L.N. (2003) Study of processes of self-organization of macromolecules of poly-2-acrylamido-2-methylpropan of sulfacids and sodium dodecyl sulfate on nanoparticles of zero valent selenium, *High molecular compounds*, Vol. 45 A, No. 6, 963-967.

[10] Valueva S.V., Kopeikin V.V., Kipper A.I., Filippov A.P., Shishkina G.V., Khlebosolova E.N., Rumyantseva N.V., Nazarkina Y.I., Borovikova L.N. (2005). Formation of nanoparticles of zero valent selenium in water solutions of polyampholit in presence of different redox systems, *High-molecular compounds*, Vol. 47Б, No. 5, 857-860.

[11] Valueva S.V., Kipper A.I., Kopeikin V.V., Borovikova L.N., Ivanov D.A., Filippov A.P. (2005). Influence of molecular mass of polymeric matrix on morphologic characteristics of the selenium containing nanostructures and their resistance to impact of hydrodynamic field, *High-molecular compounds*, Vol. 47A, No. 3, 438-443.

[12] Valueva S.V., Kipper A.I., Kopeiki V.V., Borovikova L.N., Lavrentyev V.K., Ivanov D.A., Filippov A.P. (2006). Studying of process of formation of selenium containing nanostructures and their characteristics on the base of the rigid molecules of cellulose derivatives, *High molecular compounds*, Vol. 48A, No. 8, 1403 - 1409

[13] Valueva S.V., Borovikova L.N., Koreneva V.V, Nazarkina Y.I., Kipper A.I., Kopeikin V.V. (2007). Structural-morphologic and biologic properties of nanoparticles of selenium stabilized by bovine serum albumin, *Journal of physical chemistry*, Vol.81, No. 7, 1329 - 1333.

[14] Handbook of Chemistry and Physics (2001). Ed. Lide R.D., 81th Edition, Chapman & Hill CRC.

[15] Berezin I.V., Klesov A.A. (1976). *Practical course of chemical and enzyme kinetics*, M.: Moscow State University.

[16] Eskin V.E. (1986). *Light scattering by polymer solutions and properties of macromolecules*, L.: Nauka, Science.

[17] Pogodina N.V., Tsvetkov N.V. (1997). Structure and Dynamics of the Polyelectrolyte Complex Formation, *Macromolecules*. Vol. 30, No. 17, 4897-4904.

[18] Brown W. (1993). *Dynamic Light Scattering: the Method and Some Application.* Oxford: Clarondon Press.

[19] Meewes M., Ricka J., De Silva M., Nuffengger R., Binkert Th. (1991). Coll-globule transition of poly (N-isopropylacrylamide). A study of surfactant effects by light scattering, *Macromolecules*. Vol. 24, No. 21, 5811-5816.

[20] Nishio I., Shao Thang Sun, Swislow G., Tanaka T. (1979). First observation of the coll-globule transition in a single polymer chain, *Nature*. Vol. 281, No. 5728, 208-209.

[21] Konishi T., Yoshizaki T., Yamakawa H. (1991). On the "universal constants" p and Φ of flexible polymers, *Macromolecules*. Vol. 24, No. 20, 5614-5622.

[22] Burchard B.W. (1992). *Static and dynamic light scattering approaches to structure determination of biopolymers// Laser Light Scattering in Biochemistry* Eds. by Harding S.E., Satelle D.B., Bloomfild V.A. Cambridge : Royal Soc. Chem. Information Services.

[23] Tsvetkov V.N., Eskin V.E., Frenkel S.Y. (1964). *Structure of macromolecules in solutions.* M.: Nauka, Science.

[24] Stein P. (1978). *Polymer Blends.* Ed. by Poul D.P., Newman S. New York; San Francisko; London: Acad. Press.

[25] Kalinina N.A., Kallistov O.V., Kuznetsov N.P., Batrakova T.V., Romashkova K.A., Gusinskaya V.A., Sidorovich A.V. (1990). Light scattering and structure of moderate concentrated solutions and films of polyamides, *High molecular compounds*, Vol. 32 A, No. 4, 695 -700.

[26] Kallistov O.V., Krivibokov V.V., Kalinina N.A., Silinskaya I.G., Kutuzov Y.I., Sidorovich A.V. (1985). Structural peculiarities of the moderate concentrated solutions of polymers with different rigidity of molecular chain, *High molecular compounds*, Vol. 27 A, No. 5, 968 - 971.

[27] Valuev S.V., Kipper S.V. (2001). Influence of ionic force on asymmetry of p form of macromolecule of poly-2-acrylamido-2-methylpropan sulfacids in diluted water-salt solutions, *Journal of applied chemistry*, Vol. 74, No. 9, 1513 - 1517.

[28] Askadsk A.A., Matveyev Y.I. (1983). *Chemical structure and physical properties of polymers.* M.: Chemistry.

[29] Dove W.F., Wallace T., Davidson N. (1959). Spectrofotometric stady of the protonation of undenatured DNA, *Biochem. Biophys. Res. Com.*, Vol. 1, No. 6. 312-317.

[30] Luck G., Zimmer Ch., Snatzke G. (1968). Circular dichroism of protonated DNA, *Biophys. Biochem. Acta*, Vol. 169, 548-549.

[31] Struts A.V., Slonitsky S.V. (1983). Study of conformation of DNA molecule in water solutions of acrylamide and semicarbazide, *Bulletin of the Leningrad State University, Physics, Chemistry*, Vol. 4, No. 22, 33 - 39.

[32] Kasyanenko N.A., Bartoshevich S.F., Frisman E.V. (1985). Study of influence of pH media on conformation of DNA molecule, *Molecular biology*, Vol. 19, No. 5, 1386 - 1393.

[33] Storhoff J.J., Lazaorides A.A., Mucic R.C., Mirkin C.A., Letsinger R.L., Schatz G.C. (2000). DNA-linked gold nanoparticle assemblies, *J.Amer.Chem.Soc.* Vol. 122, 4640 - 4651.

[34] Bronstein L.M., Valetsky P.M., Antonietti M. (1997). Formation of nanoparticles of metals in organized polymeric stsructures, *High molecular compounds*, Vol. 39 A, No. 11, 1847 - 1855.

[35] Kopeikin V.V., Valueva S.V., Kipper A.I., Filippov A.P., Khlebosolova E.N., Borovikova L.N., Lavrentyev V.K. (2003). Study of formation of the particles of nano- Se^0 in water solution of cation polyelectrolyte, *Journal of applied chemistry*, Vol. 76, No. 5, 847 – 851.

[36] Litmanovich O.E., Papisov I.M. (1999). Influence of length of macromolecule on dimension of the particle of metal reduced in polymeric solution, *High molecular compounds*, Vol. 41 A, No. 11, 1824 – 1830.

[37] Mayer A.B.R. (2001). Colloidal Metall Nanoparticles Dispersed in Amphiphilic Polymers, *Polym. Adv. Technol.* Vol. 12, 96 - 102.

17

Rapid Nanoparticle Characterization

Rajasekhar Anumolu[1] and Leonard F. Pease III[1,2,3]
[1]*Department of Chemical Engineering, University of Utah, Salt Lake City, UT*
[2]*Department of Internal Medicine, Division of Gastroenterology, Hepatology, and Nutrition, University of Utah, Salt Lake City, UT*
[3]*Department of Pharmaceutics and Pharmaceutical Chemistry, University of Utah, Salt Lake City, UT*
USA

1. Introduction

Nanoparticles are the focus of intense scientific and engineering attention, due to their unique properties and wide array of potential biomedical, optical, and electronic applications. Their unique properties bridge those of bulk materials and atomic or molecular structures. Whereas bulk materials display constant physical properties regardless of size, nanomaterial properties may be size-dependent, such as quantum confinement in semiconductor particles, surface plasmon resonance in many metal particles and superparamagnetism in magnetic materials. However, design, synthesis, and fabrication of nanoparticles for specific applications or fundamental inquiry remains immature without accurate and well resolved characterization of nanoparticle size and structure. To date, nanoparticles may be characterized using a variety of traditional techniques including electron microscopy (EM), atomic force microscopy (AFM), dynamic and static light scattering (DLS and SLS), x-ray photoelectron spectroscopy (XPS), powder X-ray diffraction (XRD), Fourier transform infrared spectroscopy (FTIR), size exclusion chromatography (SEC), asymmetric flow field flow fractionation (AFFFF), X-ray crystallography, small angle neutron scattering (SANS), matrix-assisted laser desorption/ionization time-of-flight mass spectrometry (MALDI-TOF), ultraviolet-visible spectroscopy, dual polarization interferometry, and nuclear magnetic resonance (NMR) [Attri & Minton, 2005; Bondos, 2006; Casper & Clug, 1962; Chang et al., 1992; Colter & Ellem, 1961; Dai et al., 2008; Kim et al., 2010; Knapman et al., 2010; Pease et al., 2009; Russel et al., 1989; Shoemaker et al., 2010; Siuzdak et al., 1996; Swann et al., 2004; Umbach et al., 1998; Wang, 2005]. However, techniques such as electron microscopy and X-ray crystallography are expensive, time consuming, and require extensive computational resources. Other techniques such as SANS also suffer from limited availability [Kuzmanovic et al., 2008]. Despite the wide array of available techniques, there remains a need for rapid, label free, and statistically powerful characterization techniques to resolve dynamic multimodal distributions within approximately an hour or less.

Electrospray differential mobility analysis (ES-DMA) is a rapid technique (analysis time scales on the order of 1-100 min) with sub-nanometer resolution. ES-DMA can detect particles from 0.7 nm - 700 nm [Hollertz et al., 2011; Tsai et al., 2008]. At its smallest, ES-DMA has sub-angstrom resolution on the molecular diameter. ES-DMA can be used to determine the concentration of nanoparticles and also deposit the nanoparticles for additional characterization. ES-DMA is particularly attractive because it operates at atmospheric pressure and requires neither fluorescent tagging nor calibration curves. It separates particles based on their charge-to-size ratio, similar to mass spectrometry or capillary electrophoresis but without the expensive equipment such as turbo pumps [Pease et al., 2008]. While ES-DMA does not possess the atomic scale resolution of X-ray crystallography, EM, or SANS, its simplicity gives it decided advantages in speed, cost, and statistical significance [Pease et al., 2011]. The "coarse-grain" structures it resolves provide significant information regarding the structure of several biological particles such as assembling viruses, virus-like particles, and vaccines for biomedical applications and the structure of nanoparticle clusters and aggregates [Cole et al., 2009; Hogan et al., 2006; Lute et al., 2008; Thomas et al., 2004; Wick et al., 2005].

Here we review this label-free, quantitative, and rapid technique that provides full multimodal size distributions with sub-nanometer resolution. We first describe the operation and physics underlying this instrument. We then describe exemplary applications of this instrument to nanoparticle characterization. We finally conclude by comparing ES-DMA to several of the techniques listed above and provide an outlook for future growth of the technique.

1.1 ES-DMA – operation and physics

In ES-DMA, also referred to as a scanning mobility particle sizing (SMPS) or gas-phase electrophoretic mobility molecular analysis (GEMMA) [Bacher et al., 2001; Saucy et al., 2004], a particle suspension is first conveyed into the gas phase (Figure 1 shows ES-DMA system components). This is achieved by electrospray ionization, which produces a narrow distribution of droplets, typically 150-400 nm in initial diameter.

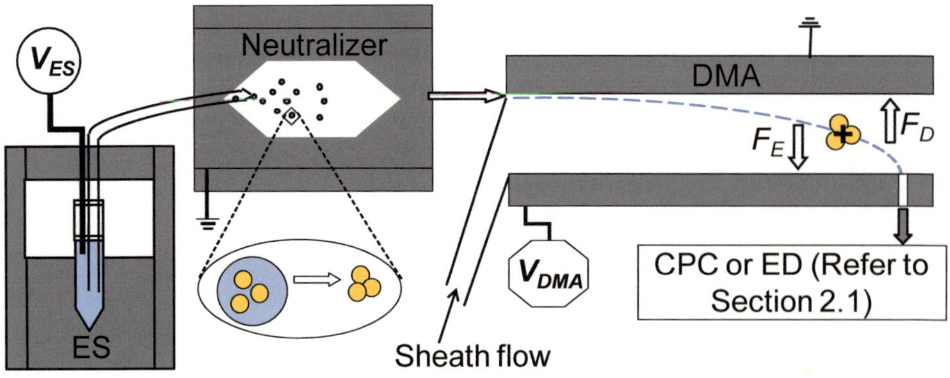

Fig. 1. Schematic of ES-DMA nanoparticle characterization system. ES-DMA produces either a size distribution using an ultrafine condensation particle counter (CPC) or deposits nanoparticles using an electrostatic deposition (ED) chamber as discussed in Section 2.1. Here, F_E is the electrostatic force and F_D is the radial component of drag force acting on the nanoparticle system.

The electrospray system uses pressure (3-5 psig) to drive flow through a 25-40 micron inner diameter capillary. At the capillary tip, a Taylor cone forms due to an applied electric field (1-3.7 kV across ~0.3 cm). Droplets emitted from the cone encapsulate one or multiple discrete nanoparticles (e.g. DNA coated gold nanoparticles or recombinant polymer strands) and are entrained in a mixed stream of air (1.0 L/min) and carbon dioxide (0.2 L/min) at atmospheric pressure. The droplets quickly evaporate leaving a dry nanoparticle. For example, Figure 1 shows three gold nanoparticles forming into a trimeric cluster. As these drying particles pass through a neutralizing chamber, collisions with charged ions reduce the charge on the nanoparticle received in the electrospray to a modified Boltzmann distribution [Kaddis et al., 2007; Loo et al., 2005; Wiedensohler, 1988]. Consequently, the positively charged particles analyzed in the DMA carry predominantly a single net positive charge. Within the annular DMA analysis chamber, a potentiated (≤-10 kV) center electrode attracts charged particles dragged towards the exit of the DMA by a carrier gas (nitrogen). The electrical force acting on the particles carrying n_e electron charges of magnitude e (=1.602 x 10^{-19} C) through an electric field, E_s, is given by $F_E = n_e e E_s$. The mobility of the particle, Z_p, is then defined as the ratio of the particle's velocity v to the force giving rise to that velocity [Anumolu et al., 2011; Knutson & Whitby, 1975]. This force gives rise to a resulting radial drag force acting on the particle, $F_D = 3\pi\mu_g v d_p / C_c$, where d_p is the spherically equivalent particle mobility diameter and μ_g is the gas viscosity. This form of Stokes' Law corrects for moderate to high Knudsen number flow using the Cunningham slip correction factor, C_c, because many particle sizes are equal to or less than the mean free path of air, λ (=66 nm) [Radar, 1990]. This correction factor is given by $C_c = 1 + Kn\ [a + \beta\ \mathrm{Exp}(-\gamma/Kn)]$, where $Kn = 2\lambda/d_p$, $a = 1.257$, $\beta = 0.40$, and $\gamma = 1.110$ [Allen & Raabe, 1985; Radar, 1990]. When the particles are not spherical, several authors show that the drag force in the free molecular regime (where Kn>1) depends on the projected area [Epstein, 1924; Hollertz et al., 2011; Pease et al., 2011]. Particles quickly reach their terminal velocity in a small fraction of the particle's DMA residence time allowing us to equate the radial component of the two forces to obtain the particle's electrical mobility as

$$Z_p = \frac{n_e e C_c}{3\pi\mu_g d_p}. \quad (1)$$

The instrument also has a unique mobility, which holds for any laminar velocity profile, given by [Knutson & Whitby, 1975],

$$Z_p^* = \frac{(q_a - q_s)/2 + q_{sh}}{2\pi V L} \ln\frac{r_2}{r_1}, \quad (2)$$

where, V is the average voltage on the center electrode, q_{sh} is the sheath flow, q_s is the sampling or monodispersed flow out of the DMA, q_a is the aerosol flow out of the ES into the DMA, L is the length between polydisperse aerosol inlet and exit slit (4.987 cm), r_1 is the inner radius of annular space of the DMA (0.937 cm), and r_2 is the outer radius of annular space of the DMA (1.905 cm). Here the values of L, r_1, and r_2 represent the dimensions of nano-DMA. When the particle and instrument mobilities are equal, the particle passes through a collection slit at the distal end of the center electrode to be counted or further analyzed. Combining particle and instrument mobilities determines the mobility diameter of the particle to be

$$\frac{d_p}{C_c} = \frac{2n_e eVL}{3\mu_g q_{sh} \ln\left(\frac{r_2}{r_1}\right)}. \quad (3)$$

Here C_c is grouped with the diameter d_p, because C_c is diameter dependent. Stepping through a series of voltages (or sizes) while counting with a condensation particle counter (CPC), yields a complete multimodal distribution. The continuous and high throughput (millions of particles per run) design of the instrument ensures statistical significance of both mean and tails of ES-DMA size distributions. The distributions are also highly repeatable with a standard deviation on the number-average diameter of only ± 0.1 nm for nominally 10 nm gold nanoparticles [Pease et al., 2007].

By collecting data for several seconds and averaging over time the mean or number–average diameter may be calculated with $d_{ave} = \sum_i N_i d_i / \sum_i N_i$, where N_i is the number of particles counted by the CPC of size d_i [Fissan et al., 1983]. Because we apply a negative bias to ions within the DMA, only particles that acquire a positive charge are detected. A modified expression for the Boltzmann distribution [Weidensohler, 1988] is used to correct for this effect, transforming the distribution of positively charged particles into the complete distribution of all particles regardless of charge. The fraction of singly and doubly charged particles was determined by Weidensohler,

$$f(j) = 10^{\sum_{l=0}^{5} a_l^j [\log(d_p / d_o)]^l}, \quad (4)$$

where d_p represents the mobility diameter of the particle, d_o = 1 nm, j is the number of charges on a particle, and a_o^1 through a_5^1 are -2.3484, 0.6044, 0.4800, 0.0013, -0.1553, and 0.0320, while a_o^2 through a_5^2 are -44.4756, 79.3772, -62.8900, 26.4492, -5.7480, and 0.5049, respectively, based on the probability of charging provided by Fuchs [Fuchs, 1963]. To remove the influence of doubly counted particles from the size distribution, the charge corrected count is multiplied by the overlap factor, f_{no}, which becomes unity for $d_p < 5$ nm and $10.94/d_p - 29.94/d_p^2$ for $d_p > 5$ nm as described by Pease, et al. [Pease et al., 2010a].

Alternatively, monodispersed nanoparticles or nanoparticle clusters from the DMA (or polydispersed nanoparticles bypassing the DMA) may be directed into an electrostatic deposition (ED) chamber. Within the chamber an electrode beneath a collecting substrate exerts an electrostatic field (≤-10 kV over ~2.5 cm) to attract entering particles. Electrostatic deposition is particularly effective at assembling nanoparticles onto the substrates, in contrast to impaction, because nanoparticle electrical mobilities are high (Eq. 1) and their inertia is relatively low (i.e., the Stokes number of nanoparticles remains much less than unity) such that they follow gas streamlines in the absence of an external field [Hinds, 1999]. Several factors affect the collection efficiency with smaller particles, higher particle charging, higher applied voltage, and lower gas flow rates increasing deposition. Figure 2 shows that smaller nanoparticles indeed have higher deposition efficiency (~100% when $d < 30$ nm) [Dixkens & Fissan, 1999]. The electrode features also affect the deposition profiles. For example, smaller electrodes lead to more focused deposition.

Fabrication of nanoparticle-based devices requires addressing nanoparticles to specific locations, which may be accomplished using electrostatic forces. For instance, several authors have demonstrated that charged nanoparticles may be directed to specific substrate locations by tuning electric fields near surfaces using charge patterning [Barry et al., 2003a, 2003b; Fissan et al., 2003; Jacobs et al., 2002; Krinke et al., 2002, 2003]. Similarly, Tsai, et al., used planar p-n junction patterned substrates to generate an array of tunable electric fields [Park & Phaneuf, 2003; Tsai et al., 2005]. Several other substrates are used for electrostatic deposition such as TEM grids (~3 mm in diameter) with thin holey carbon films [Anumolu et al., 2011], silicon wafers, and glass substrates pretreated in KOH and UVO-cleaner [Kang et al., 2011, 2012]. The optimum deposition time onto holey carbon TEM grids requires the product of the aerosol number density and time to exceed 3000 particle·hr/cm^3, while ~100 particle·hr/cm^3 is optimal to prepare glass rounds for precision optical measurements. Pease, et al., used freshly cleaved mica surfaces for deposition of carbon nanotubes for AFM observation [Pease et al., 2009b]. More recently, Saffari, et al., deposited particles into live DU145 prostate cancer cells in culture stored in Petri dishes [Hedieh et al., 2012].

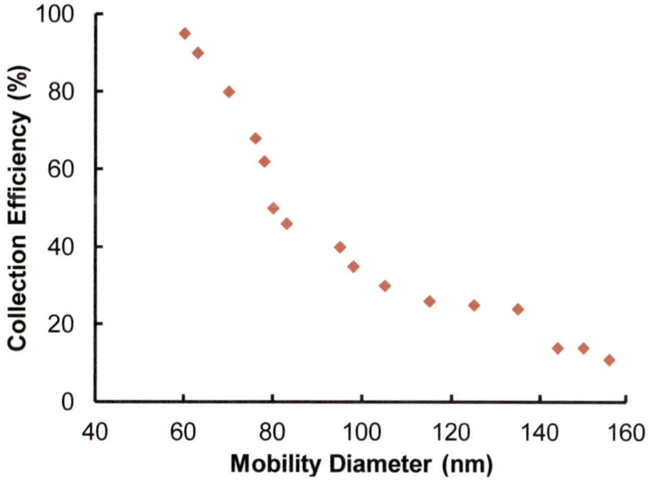

Fig. 2. Nanoparticle collection efficiency via ED at 1 L/min of inlet flow and 10 kV [Adapted from Dixkens & Fissan, 1999].

1.1.1 Predicting molecular size and structure

Bacher, et al., correlated empirically the mobility diameter (d_p in nm) of a variety of large biomolecules including proteins, antibodies, and viral proteins with their molecular weight, M_w (in kDa), assuming proteins to be globular spheres of constant density. Their composite empirical expression, $M_w = -22.033 + 9.830d_P - 1.247d_P^2 + 0.228\, d_P^3$ (which when inverted for d_P becomes $d_P = 1.832M_W^{0.3256} \approx (6M_w/\pi\rho)^{1/3}$ [Pease et al., 2001], where ρ is the density of protein), predicted individual protein and protein aggregates sizes [Bacher et al., 2001]. Pease, et al., demonstrated how to convert the projected areas, A_i, of DNA coated gold particles into d_p, accounting for Brownian motion in the DMA [Pease et al., 2007] with

$$d_p = \left(\frac{\pi^{1/2}}{6} \sum_{i=1}^{3} A_i^{-1/2} \right)^{-1} . \tag{5}$$

This equation may be applied to a wide variety of biomolecules using coordinates from the protein databank to obtain the projected areas. For example, Pease, et al., used protein databank coordinates to predict the ES-DMA measured size of IgG antibodies and insulin oligomers within 1 nm [Pease et al., 2008, 2010c]. Additionally, this formula may be used to characterize the icosahedral structure of viruses by converting the ES-DMA measured size into the edge-to-edge lengths of regular icosahedra [Pease et al., 2011]. The equation also accommodates more challenging geometries such as nanorods, clusters composed of heterogeneously sized nanospheres, and clusters composed of both nanospheres and nanorods [Pease et al., 2010a, 2010b]. Pease further extended this formulation to determine the selectivity of specific cluster compositions (dimers, trimers, etc.), as will be discussed subsequently [Pease, 2011a].

2. Materials characterized by ES-DMA

A wide variety of materials have been characterized by ES-DMA including gold nanoparticles, nanotubes (e.g. single wall carbon nanotubes), nanorods (e.g. gold nanorods), bionanoparticles (polymers, proteins, viruses, etc.), functionalized nanoparticles (e.g. with DNA), quantum dots, aggregated and conjugated nanoparticles, etc. Here we review the contributions of ES-DMA to the characterization of these materials.

2.1 Gold and metallic nanoparticles

Gold nanoparticles remain among the most extensively analyzed materials by ES-DMA because they are very stable, inert and (by many definitions) biocompatible. Figure 3 shows the size distribution of nominally 15 nm gold nanoparticles and the TEM image of size-separated gold nanoparticles using ES-DMA. The National Institute of Standards and Technology (NIST) and the Nanotechnology Characterization Lab (NCL) issued a Joint Assay Protocol, "PCC-10: Analysis of Gold Nanoparticles by Electrospray Differential Mobility Analysis (ES-DMA)," which details a protocol for size analysis of liquid borne gold nanoparticles via ES-DMA [Pease et al., 2010d]. The use of ES-DMA to measure the size distribution of colloidal nanoparticles was first demonstrated well over a decade ago [Juan & Fernandez de la Mora, 1996]. Since then gold and silver nanoparticles have been analyzed using ES-DMA [Elzey & Grassian, 2010; Lenggoro et al., 2002, 2007] by comparing with colloids of known sizes, such as polystyrene particles. However, polystyrene particles are not monodispersed. In 2007, NIST issued 10 nm, 30 nm, and 60 nm gold nanoparticles as official NIST reference materials, which are very monodispersed and can replace polystyrene particles for DMA calibration. Various techniques including DLS, EM, and SAXS were used to characterize these gold nanoparticles, and the diameter of gold nanoparticles measured by each of the different techniques was in reasonable agreement with ES-DMA (see Section 3). These reference materials were issued at the behest of the National Cancer Institute (NCI) to evaluate and qualify the methodology and instrument performance related to the physical characterization of nanoparticle systems.

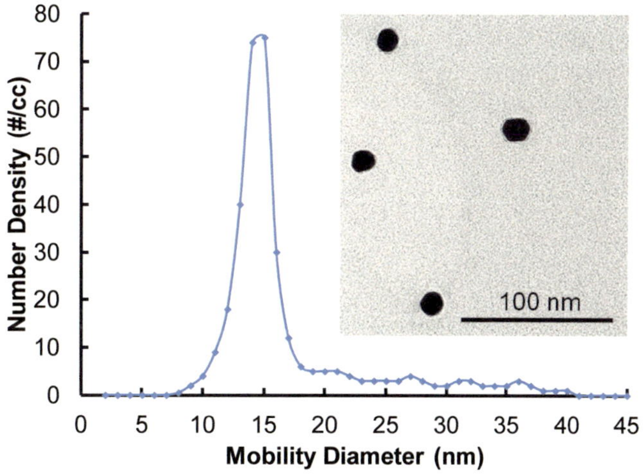

Fig. 3. ES-DMA size distribution and the TEM image (inset) following electrostatic deposition of gold nanoparticles nominally 15 nm in diameter. Number density is the number of particles per cubic centimeter of gas flow through the CPC at a rate of 1.5 L/min.

2.2 Nanotubes and nanorods

Characterization of non-spherical particles is of immense interest, as the shape of a particle along with its size greatly influences particle properties (optical, mechanical, etc.). Though several techniques are available to quantify the length distributions of nanorods and nanotubes, ES-DMA remains competitive (see Section 3 for a detailed comparison). Several authors have reported the characterization of rod- or tube-like particles, such as, gold nanorods and multi-walled carbon nanotubes [Baron et al., 1994; Chen et al., 1993; Deye et al., 1999; Kim & Zachariah, 2005, 2006, 2007a; Moisala et al., 2005; Song et al., 2005]. Song, et al., used a shape factor analysis to convert the average mobility diameter into the average length of gold nanorods. Kim, et al., used Dahneke's theory to convert ES-DMA mobility size distributions into length distributions [Kim et al., 2007b]. Their adaptation includes the orientation of the nanowire as a key factor. Below 70 nm Brownian motion randomizes the orientation of the particles, whereas longer particles align with the electric field in DMA. Pease, et al., used ES-DMA to characterize rapidly the length distribution of single-walled carbon nanotubes from liquid suspensions [Pease et al., 2009b]. Their model, also based on Dahneke's theory, converts the mobility diameter distribution to a length distribution but also accounts for thin salt layers present on the nanotubes that form during electrospray. They found d_{nt}, the diameter of an individual nanotube, and d_s, the diameter of a spherical salt particle, to be key parameters in their model. Figure 4 shows the conversion of mobility size into length distribution based on nanotube diameter and salt layer thickness. Figure 4 also suggests that neglecting to correct for the salt layer overestimates the nanotube length by 42 nm to 56 nm, whereas a small increase in d_{nt} (from 1.4 nm to 2 nm) for a 20 nm mobility diameter results in a decrease in carbon nanotube length of 71 nm [Pease et al., 2009b].

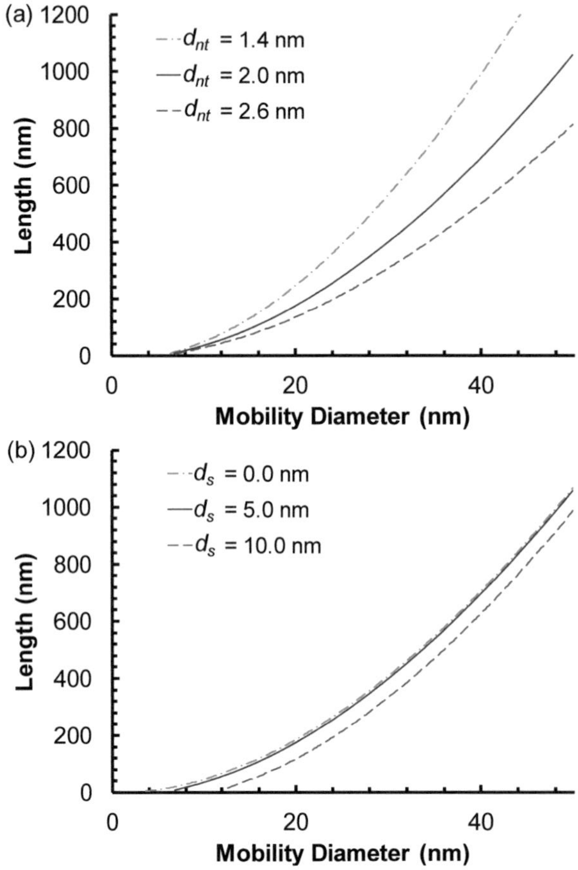

Fig. 4. Length of carbon nanotubes versus mobility diameter as a function of nanotube diameter, d_{nt}, and salt particle size, d_s, for (a) d_{nt} values of 1.4 nm (dash dot), 2.0 nm (solid), and 2.6 nm (long dash) with d_s = 5.0 nm; (b) for d_s = 0.0 nm (dash dot), 5.0 nm (solid), and 10.0 nm (long dash) with d_{nt} = 2.0 nm. Reprinted with permission from [Pease et al., 2009b]. Copyright (2009) Wiley-Interscience.

2.3 Bionanoparticles

Biological systems are of immense interest and ES-DMA is perfectly suited to analyze their soft components. Several biological particles such as viruses, proteins, and protein polymers have soft structures for which ES-DMA is sufficiently gentle to preserve their structure.

2.3.1 Viral nanoparticles

Despite the importance of viral structure, very few methods quantify or validate it. ES-DMA has been demonstrated to be gentle for both enveloped viruses and protein complexes [Wick et al., 2005]. For instance, Wick, et al., used ES-DMA to measure the enveloped alpha virus to be 70 nm ± 3 nm in good agreement with that reported in the structural databases of viruses (~70 nm), indicating electrospray and neutralizer to be sufficiently gentle to preserve the lipid envelop despite shear forces present in the Taylor cone at exit of the electrospray capillary. Furthermore, Hogan, et al., and Thomas, et al., have shown that icosahedral viruses remain infectious following electrospray [Hogan et al., 2006; Thomas et al., 2004].

Pease, et al., demonstrated that the ES-DMA technique can be used to quantify the dimensions of icosahedral viruses [Pease et al., 2011]. A recent review highlights the use of ES-DMA to analyze virus particles [Pease, 2012]. Previous ES-DMA studies of viruses report only the mobility diameter, neglecting the inherent geometry of the virus [Bacher et al., 2001; Hogan et al., 2006; Lute et al., 2008; Thomas et al., 2004]. This left the connection between mobility and actual dimensions of the virus unclear and poorly defined. Pease, et al., converted the mobility diameter, d_p, into the icosahedral geometry expected of *Tectiviridae* viruses using the projected area formulation introduced previously [Pease et al., 2010a]. Figure 5 shows the three orthogonal projections for icosahedra and also gives the corresponding projected areas, where a represents the length of an edge between neighboring vertices.

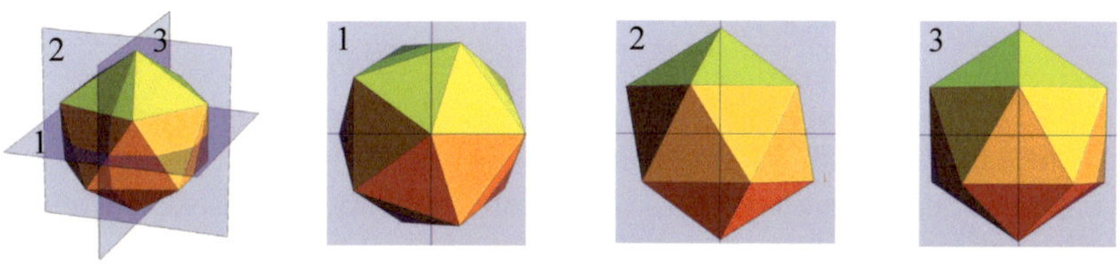

Fig. 5. Three orthogonal projections and projected areas for an icosahedron. Here, $A_1 = 5a^2/(2(5-\sqrt{5}))^{1/2}$, $A_2 = (2+\sqrt{5})a^2/2$, $A_3 = a^2((25+11\sqrt{5})/10)^{1/2}$. Reprinted with permission from [Pease et al., 2011]. Copyright (2011) American Chemical Society.

Figure 6 shows serial ES-DMA size distributions of a PR772 virus sample tracking the temporal disintegration of the capsid. The primary capsid peak in the size distribution was identified at 61.4 nm, allowing determination of the symmetry and the number of major capsid proteins per capsid. Capsomers and capsomer assemblies were identified by estimating their mobility diameter via Eq. 5 using coordinates from the protein data bank [Pease et al., 2010a]. Figure 6 also provides insight into the mechanism of degradation. Initial degradation products take the form of mostly individual capsomers though some larger assemblies (peak at 18.6 nm) are observed. Continued loss of smaller pieces leaves partially degraded capsids (peak at 51.6 nm). The small but nonzero number density between 20 and 40 nm suggests a continuum of degradation products. However, only after many weeks do the individual capsomers begin to breakdown as seen by the appearance of peaks <7 nm at t = 20 weeks.

2.3.2 Polymeric nanoparticles

Nanoparticles hold potential for a variety of biomedical applications including targeted gene and drug delivery. However, most nanoparticles with refined size (*i.e.*, where the ratio of standard deviation in the diameter to the mean diameter, also called coefficient of variation, is <0.15) are metallic with potential *in vivo* toxicity issues (e.g. quantum dots, silver particles, *etc.*) [Elzey & Grassian, 2010; Pease, 2011a]. Anumolu, et al., fabricated nanoparticles using ES-DMA from recombinant silk elastin-like protein polymers (SELPs) by encapsulating multiple polymer strands in evaporating electrospray droplets. SELPs were selected because they showed promise in clinical trials for localized gene delivery (i.e. by direct injection into tumor containing tissue) [Gustafson & Ghandehari, 2010].

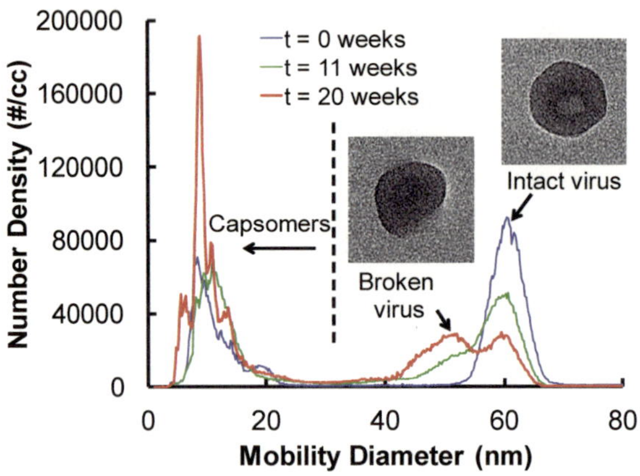

Fig. 6. Serial ES-DMA size distributions of a single PR772 sample showing the disintegration of the capsid (61.4 nm) at $t = 0$ weeks (short dash), 11 weeks (solid), and 20 weeks (long dash) after storage at room temperature. The number density was corrected with a modified Boltzmann distribution [Wiedensohler, 1988] to reconstruct the full size distribution. Insets are TEM images of representative capsids electrostatically collected at (a) 61.4 nm and (b) 50.0 nm without fixing or stain. Reprinted with permission from (Pease et al., 2011). Copyright (2011) American Chemical Society.

A key feature of their work is the use of the DMA to purify the particles. Figure 7a shows that prior to separation in the DMA, the particles are heterogeneous in size, but after separation the distributions narrow dramatically. Indeed, Figure 7b shows two histograms of SELP particles nominally 24.0 nm and 36.0 nm in diameter, each assembled from nearly 200 nanoparticle TEM images. Statistical compilation shows the standard deviation on the diameter of these purified particles to be 1.2 nm and 1.4 nm for the two sizes, respectively, leading to coefficients of variation of <5% [Anumolu et al., 2011]. This manufacturing precision meets or exceeds that of metallic nanoparticles and rivals that of biologically assembled particles such as viruses [Cole et al., 2009; Lute et al., 2008; Pease, 2011a; Pease et al., 2007]. These results provide the first compelling evidence that ES-DMA can both *generate* and *purify* polymeric nanoparticles with high dimensional uniformity without the addition of hazardous solvents.

These highly uniform nanoparticles may be developed into carriers of therapeutic agents [Anumolu et al., 2011]. Simply including the therapeutic agent in the polymer solution to be electrosprayed, leads to incorporation within the nanoparticle. For example, SELP-815K was mixed with plasmid DNA and fluorescein isothiocyanate (FITC). In both cases new peaks arise 7-8 nm from the primary peak and the distribution of all particles is wider, confirming incorporation of these model agents of gene and drug delivery.

Combining precise control over nanoparticle size with precise control over polymer structure enabled by recombinant techniques presents a unique opportunity to precisely tune the payload and rate of release of the therapeutic agents as well as their biological fate.

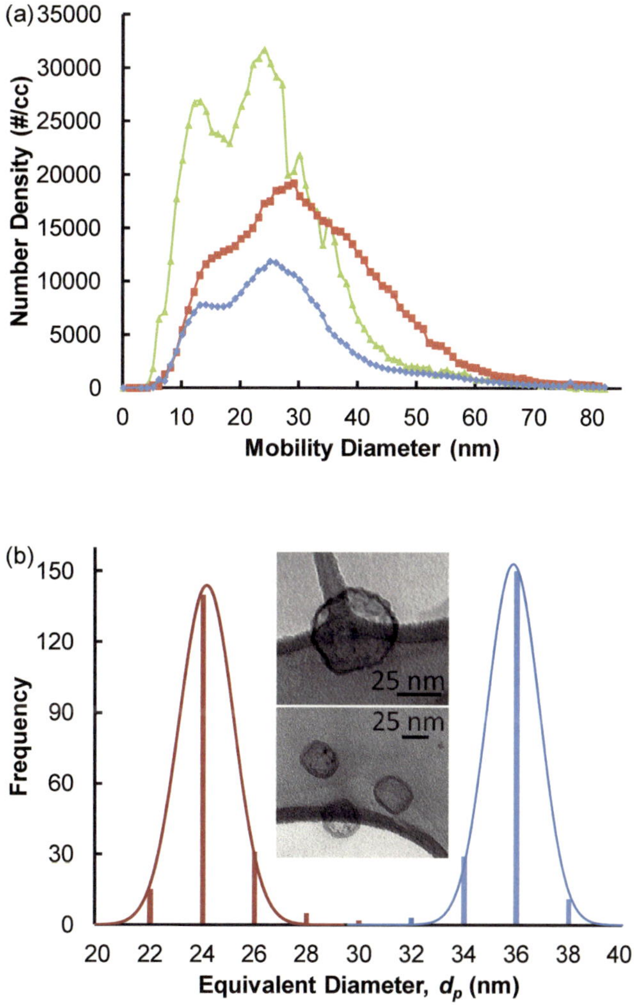

Fig. 7. (a) Size distributions of nanoparticles fabricated from polymers SELP-815K (▲), SELP-415K (■), and SELP-47K (♦) at a polymer weight fractions and buffer concentrations of 0.00133 and 2 mM, respectively. The designation 815K indicates that the strands are assembled from multiple consecutive repeats of 8 silk units, 15 elastin units, and one lysine modified elastin. (b) Histograms representing the diameter of SELP-815K nanoparticles as determined from TEM following electrostatic deposition of nominally 24.0 nm and 36.0 nm. The mean and standard deviation of the size distribution of these particles are 24.2 ± 1.2 nm and 35.8 ± 1.4 nm, respectively. The insets show micrographs of SELP-415K nanoparticles electrostatically collected on TEM grids. Reprinted with permission from [Anumolu et al., 2011]. Copyright (2011) American Chemical Society.

2.4 Surface-functionalized nanoparticles

Determining the surface density of ligands attached to nanoparticle surfaces is a challenging problem in nanoscience and nanotechnology and a major barrier to commercial development. Pease, et al., used ES-DMA to determine the surface density of thiol terminated single-stranded DNA (ssDNA) tethered to 20 nm gold nanoparticles. Comparing the diameter of coated and bare particles (Figure 8a) shows that ES-DMA measured sizes are sensitive to *both* hard and soft components of complex nanoparticles. Pease, et al., investigated the dependence of the coating thickness on the number of deoxythymine (dT) nucleotides or bases, N_b, within a ssDNA strand. The dependence of the coating thickness on N_b is related to the spatial configuration of the bases within the strand in the dry state. If the strands pack together tightly in a brush structure, similar to alkanethiol self-assembled monolayers, the coating thickness should scale linearly on the length of the ssDNA backbone [Tsai et al., 2008]. However, if packing allows for sufficient space between the strands, the bases may adopt a random coil configuration to maximize entropy (appropriate for dried strands) such that the coating thickness is proportional to the linear end-to-end distance to the ½ power. Other exponents are available for hydrated or collapsed configurations due to interactions with solvent, if present. Figure 8b shows that the data follow square root curve fits, indicating that the strands adopt a random coil configuration on the nanoparticle surface [Adamuţi-Trache et al., 1996; Netz & Andelman, 2003; Russel et al., 1989]. Knowing the configuration of the strands enables estimation of the surface density because the drag force experienced by the coated particle in the DMA depends on the diameter of the particle, the projected area of the coiled strands, and the number of those strands on the surface. The "lumpy sphere" model combines these variables by approximating each strand as a hemispherical cap, enabling direct determination of the surface density [Mansfield, 2007]. The reported densities are in reasonably agreement with those for "brushes" prepared under similar conditions [Demers et al., 2000; Liao & Roberts, 2006; Petrovykh et al., 2006; Xu & Craig, 2007]. Tsai, et al., used a similar strategy to determine the size, thermal stability, and surface density of alkanethiol self-assembled monolayers (SAMs) on gold nanoparticles. They measured the coating thickness and binding energy of SAMs with excellent precision [Tsai et al., 2008]. These results indicate the potential of ES-DMA to quantify the surface density, configuration and binding energy of biological molecules and organic coatings on nanoparticles.

2.5 Quantum dots and nanoparticle conjugates

Quantum dots (QDs) are semiconducting nanoparticles or nanocrystals that exhibit quantum confinement and are useful in nanophotonics, advanced lighting and displays, as the next generation of photovoltaics (i.e. solar cells), and as dye replacements for molecular biology. Incident photons elevate electrons from the valence to the conduction band, leaving an excited electron-hole pair called an exciton. When the exciton is confined within a nanoparticle that is smaller than its Bohr radius, quantum confinement leads to increased separation between the valance and conduction bands with corresponding increases in the band gap and energy of the photon emitted upon recombination. Therefore, QD size is an essential feature because small changes in QD diameter lead to large changes in the energy of the emitted photon.

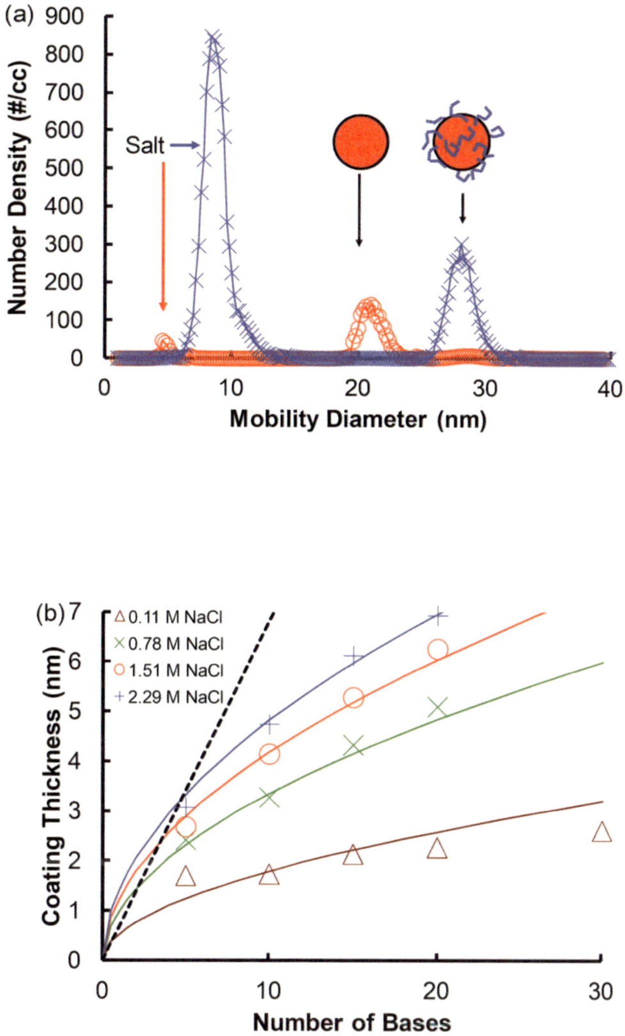

Fig. 8. (a) Size distributions of nominally 20 nm Au nanoparticles, one bare (O) and the other coated (x) with tethered single stranded DNA $(dT)_{20}$-SH. The difference between the two particle size distributions determines the apparent coating thickness. (b) Apparent coating thickness, H, versus number of dT nucleotides per strand, N_b, for a variety of salt concentrations, n_s. The dashed and solid lines respectively represent fits for a contour length model for fully stretched out DNA ($H \sim N_b$) versus that of a square root dependence ($H \sim N_b^{1/2}$) characteristic of strands coiled into low-grafting density layers (see text above). Reprinted with permission from [Pease et al., 2007]. Copyright (2011) American Chemical Society.

Figure 9a shows a typical ES-DMA size distribution of carboxylic acid terminated QDs, with peaks at 15.0 and 18.0 nm representing individual QDs and dimeric QD clusters. This size varies from that anticipated from optical measurements between 5 nm and 7 nm for several reasons. First, these rod-shaped particles (see inset of Figure 9a) are coated with 40 kDa polyethylene glycol (PEG) and salt layers; the equivalent external size of an individual quantum dot without salt and polymer coatings is ~12.4 nm. Second, optical measurements (e.g., Figure 9b-d) reflect the size of the core for QDs with type I band alignment rather than the external size measured by ES-DMA. These differences indicate that optical, TEM, and ES-DMA measurements are mutually complementary for core-shell QDs: optical measurements ascertain the core, TEM gives the combined core and shell measurements, and ES-DMA provides core, shell and coating thicknesses.

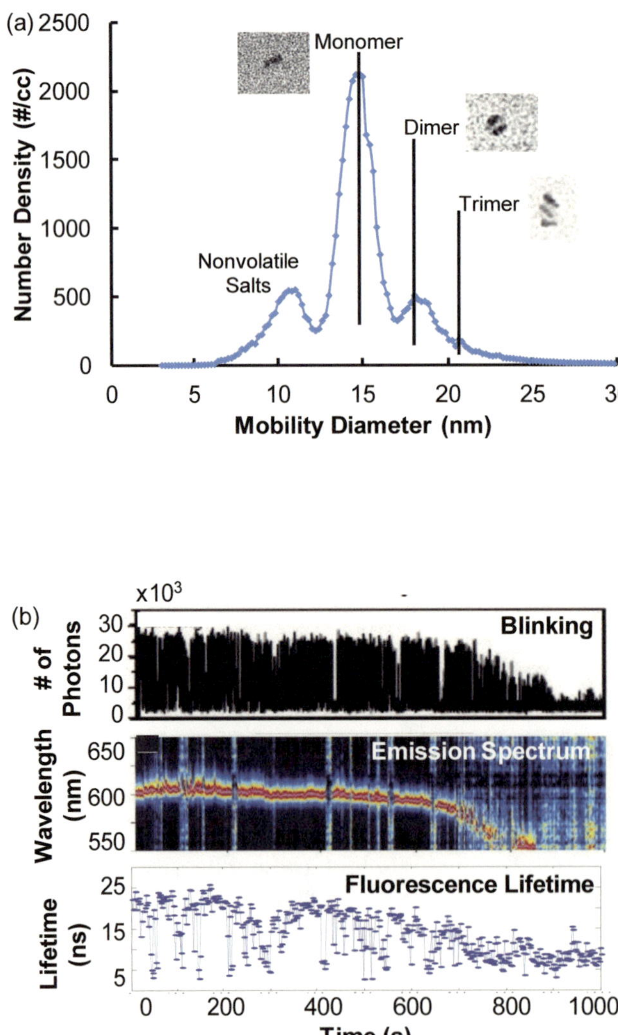

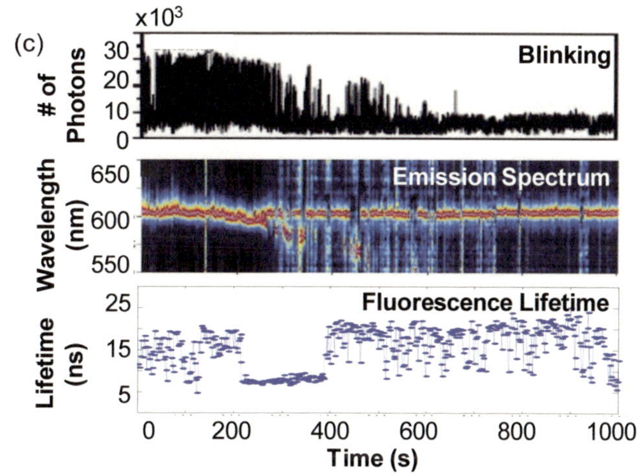

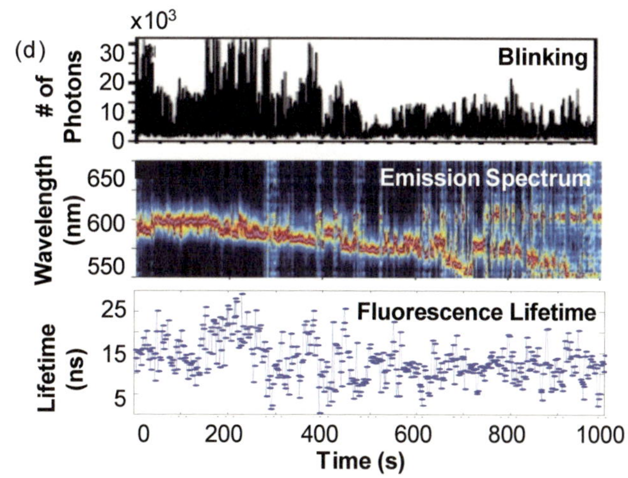

Fig. 9. (a) Raw ES-DMA size distribution of quantum dots at 0.4 µmol/L ($2.4 \cdot 10^{14}$ particles/mL) in 11 mmol/L acetic acid with insets representing TEM images of monomer, dimer and trimer. The first peak at ~11 nm is due to nonvolatile salts present in the original quantum dot solution. The intensity, spectrum, and lifetime of QD (b) monomeric, (c) dimeric, and (d) trimeric clusters

The ability to distinguish the soft and hard components indicates the potential to determine macromolecular conjugation. For example, Figure 10a compares QDs having a ZnS shell coated with mercaptoethanol (ME), ssDNA and double stranded DNA (dsDNA). ES-DMA reported an increase in the mobility diameter of the QDs complexes in the order of $d_{dsDNA} > d_{ssDNA} > d_{ME}$. This order is expected because mercaptoethanol is a shorter molecule than ssDNA, and dsDNA (persistence length ~50 nm) has a more extended conformation than ssDNA (persistence length ~1-3 nm). The extended conformation provides additional drag to the nanoparticle complex resulting in the larger apparent size measured. Similar detection of macromolecular binding is seen in Figure 10b where dextran coated particles are conjugated with the lectins concanavalin (ConA) and wheat germ agglutinin (WGA). The conjugation clearly increased the apparent size of the nanoparticle complex.

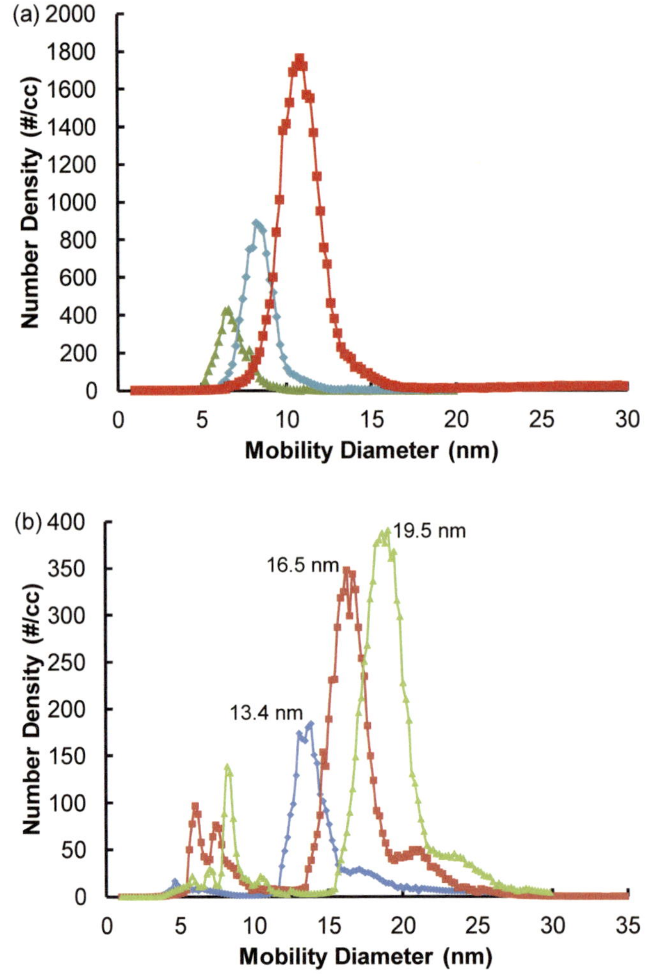

Fig. 10. ES-DMA size distributions of (a) QDs coated with mercaptoethanol (▲), ssDNA (♦), and dsDNA (■) and (b) dextran decorated bare nanoparticles bare (♦) or conjugated with WGA (■) or ConA (▲).

A further highlight of the use of ES-DMA to evaluate macromolecular conjugation is the effort of Yim, et al., to determine the number of QDs attached to genetically modified lambda phage [Yim et al., 2009]. Their biotinylated lambda phage present a rapid means of detecting harmful E. Coli bacteria because bacteriophages replicate much more rapidly than bacteria. Streptavidin coated QDs then bind to the biotinylated phage to provide colorimetric detection. The amount of bacteria is proportional to the amount of phage produced, and the signal from the assay is proportional to the average number of QDs bound to each phage. By comparing the conjugate size in ES-DMA size distributions with that of individual QDs and the lambda phage, the number of QDs attached to the bacteriophage was determined using Eq. 5 by modeling the QD-phage conjugates as clusters of spheres (i.e., the phage) and rods (the streptaviding coated QDs) [Pease, 2011a]. More recently Tsai, et al., studied viral neutralization by quantifying antibody-virus binding [Tsai et al., 2011]. They compared the size of bare MS2 virus particles to those conjugated with anti-MS2 antibodies and reported an increase in the apparent size. They found that antibodies were responsible for formation of multiple virus complexes by acting as connectors between two virus particles (no dimerization was observed with the initial virus sample), indicating that ES-DMA is able to quantify viral aggregation as a form of virus neutralization. They then used their model to determine the stoichiometric ratio of bound antibodies to virus particles to find that infectivity strongly depends on binding stoichiometry not the type of antibody. Monoclonal and polyclonal antibodies both fell on a single infectivity versus stoichiometry curve.

2.6 Nanoparticles aggregates – Packing structure and kinetics

The development of structured materials for nanotechnology and nanobiotechnology requires readily available, rapid analytical techniques to determine the composition of nanostructured particles and clusters. ES-DMA is uniquely qualified to determine the packing and aerodynamic size of colloidal clusters in the nanometer range. For instance, aggregation of metallic nanoparticles may be detected via coupled plasmon resonances using UV-vis, however, this approach does not provide direct information on aggregate size and structure [Weisbecker et al., 1996]. Figure 11a shows an ES-DMA size distribution of multimers of gold nanoparticles with each multimer (e.g. dimers or trimers) appearing as a distinct peak separated by 2-3 nm. These multimers were prepared at the limit of colloidal stability by adding ammonium acetate buffer to a solution of gold nanoparticles to suppress electrostatic screening forces between like charged particles. Figure 11b shows how the ammonium acetate concentration induces aggregation in solution. Tsai, et al., followed the temporal kinetics of the nanoparticle aggregation in the liquid-phase using ES-DMA, as shown in Figure 11c [Tsai et al., 2008]. The monomer concentration decreases with the appearance of each higher order cluster (dimer, trimer, etc.) as anticipated from simple Brownian flocculation/aggregation. By fitting the data to a first principles kinetic model of the aggregation process, the surface charge on particles was extracted from the kinetic rate constants and found to be in good agreement with values reported in the literature (see Figure 11d).

ES-DMA can also be used to monitor protein aggregation. For example, Pease, et al., demonstrated that protein aggregates can be individually resolved and determined an equilibrium constant for the formation of dimers. This equilibrium constant was evaluated based on the number density, which can be obtained from the ES-DMA distribution [Pease et al., 2008]. Pease, et al., also demonstrated that ES-DMA can be used to characterize the size and packing structure of small clusters of identical nanoparticles in colloidal suspension [Pease et al., 2010a]. Essential to their analysis was the projected area model (Eq. 5) to calculate the cluster size, distinguishing collinear from close packed sphere arrangements without resorting to TEM. Several others used ES-DMA to analyze aggregates. For example, Lall, et al., used ES-DMA to determine the ultrafine aggregate surface area and volume distribution of silver aggregates [Lall et al., 2006a, 2006b]. This model is particularly apt for very large aggregates comprising >10 particles/aggregate. Agglomeration of silver nanoparticles was also reported by Elzey & Grassian using ES-DMA [Elzey & Grassian, 2010]. They investigated agglomeration as a function of pH, which determines the dissolution of silver ions in aqueous environments.

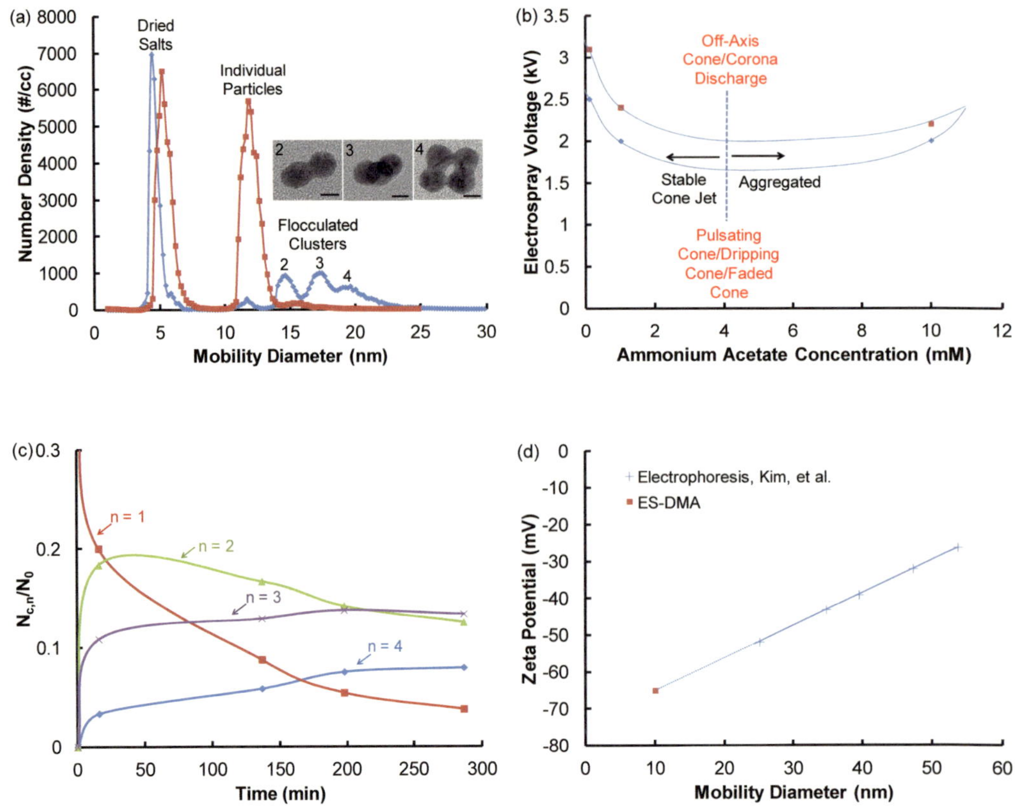

Fig. 11. (a) Size distribution of multimers of gold nanoparticles. The inset shows the TEM images of dimers, trimers, etc. (b) Onset of gold nanoparticle aggregation as a function of ammonium acetate buffer concentration. (c) Normalized concentration of aggregates vs reaction time at an ammonium acetate concentration of 7.89 mmol/L. (d) Relation between surface potential and particle diameter for gold nanoparticles as measured by ES-DMA (■) and reported in the literature (+). The absolute value from ES-DMA was 65 mV. Panel c reprinted with permission from [Tsai et al., 2008]. Copyright (2008) American Chemical Society.

2.7 Optimizing cluster production and purity

Though most ES-DMA research centers on analytical scale characterization, this technique also holds potential as a preparative scale purification tool. For example, ES-DMA separates two particle clusters (dimers) from three particle clusters (trimers), and Kang, et al., have shown that cluster composition may affect cluster photonic properties such as the wavelength evolution, intensity, blinking and radiative lifetime [Kang et al., 2012] (see Figures 9b, 9c, and 9d). To enhance the yield and purity (i.e., how many nominally three-particle clusters actually contain three particles), Pease modeled the ES-DMA technique as a function of nanoparticle properties (size, concentration, etc.) using Monte Carlo simulations [Pease, 2011a]. Figure 12a shows that tuning the initial solution concentration of nanoparticles optimizes the yield of each kind of cluster. The concentration of dimers and subsequently trimers begins to increase as the fraction of monomers declines. Figure 12a also shows that at a concentration of ~7.0·10^{14} particles/mL, only 5% of the particles/strands in the size distribution are not monomers providing a sensitive metric for the onset of aggregation. Around 10^{15} particles/mL the concentrations of monomer, dimer, trimer, and higher order clusters all approach parity.

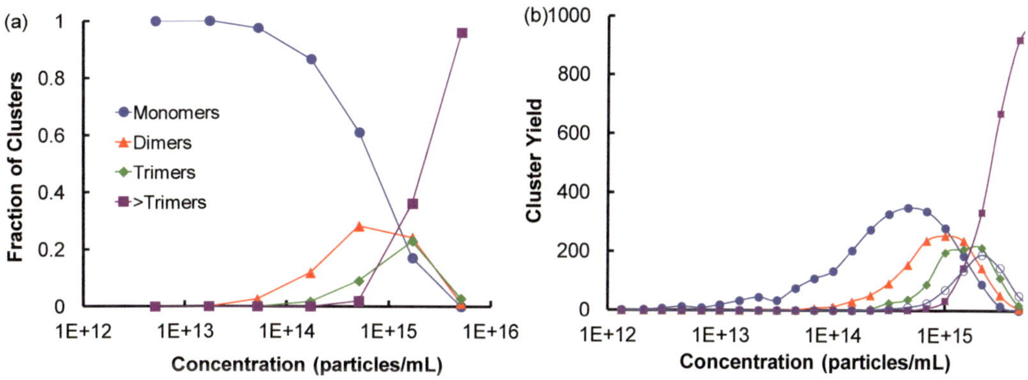

Fig. 12. (a) Monte Carlo simulation using 1000 droplets each exactly 150 nm in diameter for solution concentrations ranging from 10^{12} to 10^{16} particles/mL. (b) The yield for monomers (filled circles), dimers (filled triangles), trimers (filled diamonds), tetramers (open circles), and clusters composed of five or more particles, as a function of initial solution concentration.

To determine the cluster purity, the size distribution of heterogeneously sized particles must be calculated. There are two options for determining the average mobility diameter, d_{ave}. The first option as outlined in Pease, et al., calculates the three projected areas, A_i, orthogonal to the cluster's principle axes (see Figure 13) as given in Eqs. 5 and 6 [Pease et al., 2007]. The area of each projection may be determining from the center-to-center distances, h_i, between the circles in that projection as given by

$$A_1 = \pi/4(d_1^2 + d_2^2 + d_3^2)$$
$$A_2 = \pi/4(d_1^2 + d_3^2) - A_{ov}(h_1)$$
$$A_3 = A_1 - A_{ov}(h_2) - A_{ov}(h_3), \tag{6}$$

where d_1, d_2, and d_3 are the diameters of 3 clusters and $d_1 \geq d_2 \geq d_3$ [Pease, 2011a].

The second option is to employ the proportionality constants determined previously by Pease, et al., for homogeneous clusters [Pease et al., 2010a]. They show that $d_{ave} = kd_p$, where k is the proportionality with values of 1.526, 1.724, and 1.875 for close packed trimers, tetramers, and pentamers, respectively. For heterogeneously sized particles, $d_{ave} = \sum_{i=1}^{n} d_i / n$. Overlaps in the size distribution (see Figure 12b) increase the probability that particles of neighboring size will contaminate each other, decreasing the purity of selected clusters. Pease found that and when the coefficient of variation is <13%, this overlap in the size distributions is not significant. In more recent work, Li, et al., also investigated the formation of clusters during the electrospray process [Li et al., 2011]. They developed a statistical model to determine the extent of ES droplet induced aggregation and how it alters the results.

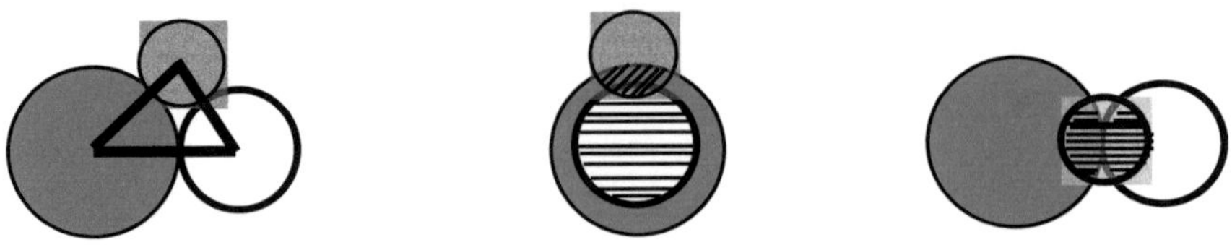

Fig. 13. Depiction of the three projected areas, A_i, used to determine the mobility of a close packed trimeric cluster. Crosshatching in the second and third geometries represent overlap area, A_{ov}.

3. Comparison of ES-DMA to similar techniques

Each of the above examples highlights the potential of ES-DMA to analyze nanoparticles. The remainder of this section compares this powerful technique to generally available classes of techniques commonly used to characterize nanomaterials such as electron microscopy and light scattering.

3.1 Electron microscopy

EM techniques such as transmission electron microscopy (TEM) and scanning electron microscopy (SEM) are direct methods to characterize nanoparticles with subnanometer resolution. EM methods provide compelling images that provide unbiased insight into the *actual* geometry and structure of electro-opaque particles, especially for the hard portions of bionanoparticles as described above. Crystallinity of the particles can also be investigated by generating diffraction patterns using EM. Despite providing very high resolution and beautiful images, EM methods remain expensive to install and operate. To rigorously determine the average nanoparticle size, several hundred to thousands of nanoparticles must be measured [Tsai et al., 2008a, 2008b], which is time consuming (of the order of days to weeks) without automation that is not widely available. For less homogeneous systems or soft materials, multimodal distributions and electron beam damage present further challenges to rigorous TEM measurement. Pease, et al., compared TEM with ES-DMA for quantitative characterization of virus particles and concluded that ES-DMA has a greater rapidity (millions of particles per hr compared to thousands of particles per hour) and statistical significance than TEM [Pease et al., 2009a, 2010b, 2011].

3.2 Light scattering techniques

Light scattering techniques such as multiangle light scattering (MALS) and dynamic light scattering (DLS) measure the intensity fluctuations of scattered light in solution. As light is focused using a laser on the particles moving due to Brownian motion, measured photon intensity fluctuates, from which the z-average hydrodynamic diameter may be estimated. The z-average molecular weight typically exceeds the mass average molecular weight for the most common distributions. In contrast, ES-DMA measures the number average diameter. Otherwise, light scattering and ES-DMA techniques are similar in that they are rapid, inexpensive, and produce statistical significant averages. Though the suspending medium of the sample influences the DLS measurements, it may be preferable for dirty samples minimizing the preparation time required for sample clean up. When characterizing structures with high aspect ratios (e.g., carbon nanotubes), MALS and DLS produce ensemble-average lengths of broad distributions using root-mean square length metrics assuming the nanotubes to have monomodal length distributions [Bauer et al., 2008; Chun et al., 2008; Hinds, 1999]. In contrast ES-DMA measures the size of each nanotube individually to assemble the full multimodal length distributions [Pease et al., 2009b]. Also, DLS does not directly correlate the size or number of individual particles with the size of an aggregate and cannot resolve individual aggregate concentrations within multimodal distributions. Clearly distinguishing clusters differing by only a single particle remains challenging, especially for particles less than 30 nm [Nguyent & Flagan, 1991; Nie et al., 2006].

3.3 Statistical comparison with different techniques

Perhaps the most comprehensive comparison of the primary techniques available to analyze nanoparticle size was performed at NIST on gold nanoparticles. Table 1 shows the statistical data comparison of mean diameters of gold nanoparticles using at least 7 different techniques. Though all the values are in reasonable agreement, some uncertainty is to be expected even from well calibrated systems. Several other authors have compared techniques including MALS, DLS, AFFFF, EM, and ES-DMA using organic (e.g., viruses and virus like particles) and inorganic (carbon nanotubes) nanomaterials [Lute et al., 2008; Pease et al., 2008, 2009a, 2009b]. For example, Lute, et al., and Pease, et al., both found ES-DMA measurements of viruses and virus like particles to be smaller than those of AFFFF-MALS. Reasons for the difference include the following. ES-DMA measures the external size of the virus particle, whereas AFFFF-MALS reports the radius of gyration that depends on the mass distribution of the particle. Also, ES-DMA measures the dry particle size in contrast to the light scattering techniques and AFFFF that measure particles in solution [Pease et al., 2008]. Organic particles are more likely to experience favorable enthaplic interactions in water rich environments that increase their hydrodynamic radius. In dry environments TEM and ES-DMA report similar averages for sufficiently large sample sizes [Lute et al., 2008; Pease, 2012]. Pease, et al., also report analysis of the length distribution of carbon nanotubes using several techniques including AFM, MALS, DLS, and FFF. Here hydrophilic interactions exert little to no influence on the size of the particle and each of these techniques reports similar results [Pease et al., 2009b]. Most recently, Kapelios, et al., characterized the size and molecular mass of proteins and complexes, comparing ES-DMA with light scattering techniques, multi-angle laser light scattering (MALLS) and quasi-elastic light scattering (QELS) [Kapelios et al., 2011]. Similar to Pease, et al., they found that ES-DMA measurements are smaller than those of MALLS and QELS, as expected for proteinacious materials.

Technique	Analyte form	Size of nominally 10 nm particles (nm)	Size of nominally 30 nm particles (nm)	Size of nominally 60 nm particles (nm)
ES-DMA	Dry, aerosol	11.3±0.1	28.4±1.1	56.3±1.5
TEM	Dry, on substrate	8.9±0.1	27.6±2.1	56.0±0.5
DLS (a)	Diluted liquid suspension	13.5±0.1	28.6±0.9	56.6±1.4
DLS (b)	Diluted liquid suspension	13.5±0.1	26.5±3.6	55.3±8.3
AFM	Dry, on substrate	8.5±0.3	24.9±1.1	55.4±0.3
SEM	Dry, on substrate	9.9±0.1	26.9±0.1	54.9±0.4
SAXS	Native liquid suspension	9.1±1.8	24.9±1.2	53.2±5.3

Table 1. Mean diameter and expanded uncertainty average particle diameter of gold nanoparticles [adapted from, NIST certificate, 2007]. DLS results: (a) 173° scattering angle, (b) 90° scattering angle. The uncertainties represent repeatability not the width of the particle size distributions.

4. Future of ES-DMA

In summary, ES-DMA is a powerful tool to characterize and fabricate nanoparticles. Several nanoparticle systems have been characterized using ES-DMA, including quantum dots, linear polymers, and complex colloidal nanomaterials. However, a major growth area for this technique is likely to be in the analysis of soft complex nanomaterials. For example. ES-DMA has significant potential to characterize biochemical reactions including those that assay for nano-therapeutics and bio-pharmaceuticals. Indeed, real-time detection of biological reactions is one of the most promising emerging applications of this instrument. ES-DMA is well suited for this future given its dynamic operating range from 150 Da up to 10^{11} Da. The lower limit of detection is still under refinement with significant improvements underway using high precision home-built DMAs that employ funnel shaped flow injectors to delay the onset of the turbulence. These systems boast sub-angtrom precision on the molecule's mobility diameter [Eichler et al., 1998; Fernandez de la Mora, 2011; Hollertz et al., 2011]. ES-DMA is also well suited to evaluate the role of aggregation, which may be particularly advantageous for detecting protein aggregates in biomanufacturing environments. However, to realize its full potential, further exploration and excellent coordination between biologists, chemists and engineers is required to expand the boundaries of ES-DMA.

5. Acknowledgements

We express appreciation for helpful conversations with De-Hao Tsai, Michael R. Zachariah and Michael J. Tarlov. We sincerely appreciate Hedieh Saffari, Joshua Gustafson, and Hamid Ghandehari for their help with biological particle systems. We thank Silvia De Paoli Lacerda, HyeongGon Kang, Peter Yim, Jeremy I. Feldblyum, Jeeseong Hwang, Matthew L. Clarke, and Phillip Deshong for the nanoparticle samples and conjugate data.

6. References

Adamuti-Trache, M.; McMullen, W. E. & Douglas, J. F. (1996). Segmental Concentration Profiles of End-Tethered Polymers with Excluded-Volume and Surface Interactions. *J. Chem. Phys.* Vol. 105, No. 11 (September 1996), pp. 4798-4811.

Allen, M. D. & Raabe, O. G. (1985). Slip Correction Measurements of Spherical Solid Aerosol Particles in an Improved Millikan Apparatus. *Aerosol Sci. Technol.*. Vol. 4, No. 3 (1985), pp. 269-286.

Anumolu, R.; Gustafson, J. A.; Magda, J. J.; Cappello, J.; Ghandehari, H. & Pease III, L. F. (2011) Fabrication of Highly Uniform Nanoparticles from Recombinant Silk-Elastinlike Protein Polymers for Therapeutic Agent Delivery. *ACS Nano*. Vol. 5, No. 7, (July 2011), pp. 5374-5382.

Attri, A. K. & Minton, A. P. (2005). New Methods for Measuring Macromolecular Interactions in Solution via Static Light Scattering: Basic Methodology, and Application to Non-Associating and Self-Associating Proteins. *Anal. Biochem.* Vol. 337, No. 1 (February 2005), pp. 103–110.

Bacher, G.; Szymanski, W. W.; Kaufman, S. L.; Zollner, P.; Blaas, D. & Allmaier, G. (2001). Charge-Reduced Nano Electrospray Ionization Combined with Differential Mobility Analysis of Peptides, Proteins, Glycoproteins, Noncovalent Protein Complexes and Viruses. *J. Mass Spectrom.* Vol. 36, No. 9, (September 2001), pp. 1038-1052.

Baron, P.A.; Deye, G. J. & Fernback, J. (1994). Length Separation of Fibers. *Aerosol Sci. Technol.* Vol. 21, No. 2, (1994), pp. 179–192.

Barry, C. R.; Steward, M. G.; Lwin, N. Z. & Jacobs, H. O. (2003). Printing Nanoparticles from the Liquid and Gas Phases Using Nanoxerography. *Nanotechnology*. Vol. 14, No. 10, (August 2003), pp. 1057-1063.

Barry, C. R.; Lwin, N. Z.; Zheng, W. & Jacobs, H. O. (2003). Printing Nanoparticle Building Blocks from the Gas Phase Using Nanoxerography. *Appl. Phys. Lett.* Vol. 83, No. 26, (October 2003), pp. 5527-5529.

Bauer, B. J.; Fagan, J. A.; Hobbie, E. K.; Chun, J. & Bajpai, V. (2008). Chromatographic Fractionation of SWNT/DNA Dispersions with On-Line Multi-Angle Light Scattering. *J. Phys. Chem. C.* Vol. 112, No. 6, (January 2008), pp. 1842-1850.

Benson, S. D.; Bamford, J. K.; Bamford, D. H. & Burnett., R. M. (2002). The X-ray Crystal Structure of P3, the Major Coat Protein of the Lipid-Containing Bacteriophage PRD1, at 1.65 A° Resolution. *Acta Crystallogr., Sect D: Biol. Crystallogr.* Vol. 58, No. 1, (January 2002), pp. 39–59.

Bondos, S. E. (2006). Methods for Measuring Protein Aggregation. *Curr. Anal. Chem.* Vol. 2, No. 2, (April 2006), pp. 157–170, ISSN: 1573-4110.

Casper, L. D. & Klug, A. (1962) Physical Principles in the Construction of Regular Viruses. *Cold Spring Harbor Symp. Quant. Biol.* Vol. 27, (1962), pp. 1–24.

Chen, B. T.; Yeh, H. C. & Hobbs, C. H. (1993). Size Classification of Carbon Fiber Aerosols. *Aerosol Sci. Technol.*, Vol. 19, No. 2, (August 1993), pp. 109–120.

Cole, K. D.; Pease III, L. F.; Tsai, D. H.; Singh, T.; Lute, S.; Brorson, K. A. & Wang, L. (2009). Particle Concentration Measurement of Virus Samples Using Electrospray Differential Mobility Analysis and Quantitative Amino Acid Analysis. *J. Chromatogr. A.* Vol. 1216, No. 30, (July 2009), pp. 5715–5722.

Chang, T. H.; Cheng, C. P. & Yeh, C. T. (1992). Deuterium Nuclear Magnetic Resonance Characterization of Particle Size Effect in Supported Rhodium Catalysts. *J. Catal.* Vol. 138, No. 2, (December 1992), pp. 457-462.

Chun, J.; Fagan, J. A.; Hobbie, E. K. & Bauer, B. (2008). Size Separation of Single-Wall Carbon Nanotubes by Flow-Field Flow Fractionation. *J. Anal. Chem.* Vol. 80, No. 7, (February 2008), pp. 2514-2523.

Colter, J. S. & Ellem, K. A. O. (1961). Structure of Viruses. *Annu. Rev. Microbiol.* Vol. 15, (October 1961), pp. 219–244.

Dai, Q.; Liu, X.; Coutts, J.; Austin, L. & Huo, Q. (2008). A One-Step Highly Sensitive Method for DNA Detection Using Dynamic Light Scattering. *J. Am. Chem. Soc.* Vol. 130, No. 26, (June 2008), pp. 8138-8139.

Deye, G. J.; Gao, P.; Baron, P. A. & Fernback, J. (1999). Performance Evaluation of a Fiber Length Classifier. *Aerosol Sci. Technol.* Vol. 30, No. 5, (1999), pp. 420–437.

Dixkens, J. & Fissan, H. (1999). Development of an Electrostatic Precipitator for Off-Line Particle Analysis. *Aerosol Sci. Technol.* Vol. 30, No. 5, (1999), pp. 438-453.

Eichler, T.; Juan, L. D. & Fernandez de la Mora, J. (1998). Improvement of the Resolution of TSI's 3071 DMA via Redesigned Sheath air and aerosol inlets. *Aerosol Sci. Technol.* Vol. 29, No. 1, (1998), pp. 39–49.

Elzey, S. & Grassian, V. H. (2010). Agglomeration, Isolation and Dissolution of Commercially Manufactured Silver Nanoparticles in Aqueous Environments. *J. Nanopart. Res.* Vol. 12, No. 5, (2010), pp. 1945-1958.

Epstein, P. S. (1924). On the Resistance Experienced by Spheres in their Motion through Gases. *Phys. Rev.* Vol. 23, No. 6, (1924), pp. 710-733.

Fernandez de la Mora, J. (2011). Electrical Classification and Condensation Detection of Sub-3 nm Aerosols, In: *Aerosol Measurement: Principles, Techniques, and Applications* (3rd edition), Baron, P.A.; Kulkarni, P. & Willeke, K., pp. 607-721, John Wiley & Sons, Inc., ISBN 978-0-470-38741-2, New York, USA.

Fissan, H. J.; Helsper, C. & Thielen, H. J. (1983). Determination of Particle-Size Distributions by Means of an Electrostatic Classifier. *J. Aerosol Sci.* Vol. 14, No. 3, (1983), pp. 354-357.

Fissan, H.; Kennedy, M. K.; Krinke, T. J. & Kruis, F. E. (2003). Nanoparticles from the Gas phase as Building Blocks for Electrical Devices. *J. Nanopart Res.* Vol. 5, No. 3-4, (August 2003), pp. 299-310.

Fuchs, N. A. (1963). On the Stationary Charge Distribution on Aerosol Particles in a Bipolar Ionic Atmosphere. *Pure Appl. Geophys.* Vol. 56, No. 1, (1963), pp. 185-193.

Hinds, W. C. (January 1999). *Aerosol Technology: Properties, Behavior, and Measurement of Airborne Particles* (2nd edition), John-Wiley & Sons, Inc., ISBN 978-0-471-19410-1, New York, USA.

Hogan, C. J.; Kettleson, E. M.; Ramaswami, B.; Chen, D. R. & Biswas, P. (2006). Charge Reduced Electrospray Size Spectrometry of Mega- and Gigadalton Complexes: Whole Viruses and Virus Fragments. *Anal. Chem.* Vol. 78, No. 3, (February 2006), pp. 844–852.

Hogan, C. J.; Yun, K. M.; Chen, D. R.; Lenggoro, I. W.; Biswas, P.; Okuyama, K. (2007) Controlled Size Polymer Particle Production via Electrohydrodynamic Atomization. *Colloids Surf., A: Physicochem. Eng. Aspects.* Vol. 311, No. 1-3, (December 2007), pp. 67-76.

Hogan, C. J. & Biswas, P. (2008). Monte Carlo Simulation of Macromolecular Ionization by Nanoelectrospray. *J. Am. Soc. Mass Spectrom.* Vol. 19, No. 8, (August 2008), pp. 1098–1107.

Hollertz, M.; Elliott, J. T.; Lewis, J.; Mansfield, E. R.; Whetten, W. D.; Knotts IV, T. A.; Tarlov, M. J.; Zachariah, M. R. & Pease III, L. F. (2011). Structural Differentiation of Oxytocin in Cyclical and Linear Conformations Using High Resolution Differential Mobility Analysis. *Anal. Chem.* (2011), submitted.

Hung, L. H. & Lee, A. P. (2007). Microfluidic Devices for the Synthesis of Nanoparticles and Biomaterials. *J. Med. Biol. Eng.* Vol. 27, No. 1, (2007), pp. 1-6.

Jacobs, H. O.; Campbell, S. A. & Steward, M. G. (2002). Approaching Nanoxerography: The Use of Electrostatic Forces to Position Nanoparticles with 100 nm Scale Resolution. *Adv. Mater.* 2002, Vol. 14, No. 21, (November 2002), pp. 1553-1557.

Johnson, B. K. & Prud'homme, R. K. (2003). Mechanism for Rapid Self-Assembly of Block Copolymer Nanoparticles. *Phys. Rev. Lett.* Vol. 91, No. 11, (2003), pp. 118302-1-118302-4.

Juan, L. D. & Fernandez de la Mora, J. (March 1996). On-line Sizing of Colloidal Nanoparticles via Electrospray and Aerosol Techniques, In: *Nanotechnology*: Molecularly Designed Materials, (Vol. 622), Chow, G. M. & Gonsalves, K. E., pp. 20-41, ACS Publications, ISBN13: 9780841233928

Kang, H.; Clarke, M. L.; Pease III, L. F.; DePaoli Lacaerda, S. H.; Karim, A. & Hwang, J. (2011). Analysis of the Optical Properties of Clustered Colloidal Quantum Dots by the Chi-Square Distribution of the Fluorescence Lifetime Curves, *ASC Nano* (2011) manuscript.

Kang, H.; Clarke, M. L.; DePaoli Lacaerda, S. H.; Pease III, L. F. & Hwang, J. (2012). Multimodal Optical Studies of Single and Clustered Colloidal Quantum Dots Towards the Long-term Performance Evaluation of Optical Properties of Quantum Dot-included Molecular Imaging Phantoms. *Biomed. Opt. Express* (2012), submitted.

Kapellios, S.; Karamanou, M. F.; Sardis, M.; Aivaliotis, A.; Economou, S. & Pergantis, A. (2011). Using Nanoelectrospray Ion Mobility Spectrometry (GEMMA) to Determine the Size and Relative Molecular Mass of Proteins and Protein Assemblies: A

Comparison with MALLS and QELS. *Anal. Bioanal. Chem.* Vol. 399, No. 7, (March 2011), pp. 2421-2433.

Kim, S. H. & Zachariah, M. R. (2005). In-Flight Size Classification of Carbon Nanotubes by Gas Phase Electrophoresis. *Nanotechnology.* Vol. 16, No. 10, (August 2005), pp. 2149–2152.

Kim, S. H. & Zachariah, M. R. (2006). In-Flight Kinetic Measurements of the Aerosol Growth of Carbon Nanotubes by Electrical Mobility Classification. *J. Phys. Chem. B.* Vol. 110, No. 10, (February 2006), pp. 4555–4562.

Kim, S. H. & Zachariah, M. R. (2007a). Gas-Phase Growth of Diameter-Controlled Carbon Nanotubes. *Mater. Lett.* Vol. 61, No. 10, (April 2007), pp. 2079-2083.

Kim, S. H. & Zachariah, M. R. (2007b). Understanding Ion-Mobility and Transport Properties of Aerosol Nanowires. *J. Aerosol Sci.* Vol. 38, No. 8, (August 2007), pp. 823-842.

Kim, S. K.; Ha, T. & Schermann, J. P. (2010). Advances in Mass Spectrometry for Biological Science. *Phys. Chem. Chem. Phys.* Vol. 12, No. 41, (November 2010), pp. 13366–13367.

Knapman, T. W.; Morton, V. L.; Stonehouse, N. J.; Stockley, P. G. & Ashcroft., A. E. (2010). Determining the Topology of Virus Assembly Intermediates Using Ion Mobility Spectrometry-Mass Spectrometry. *Rapid Commun. Mass Spectrom.* Vol. 24, No. 20, (October 2010), pp. 3033–42.

Knutson, E. O. & Whitby, K. T. (1975). Aerosol Classification by Electric Mobility: Apparatus, Theory, and Applications. *J. Aerosol Sci.* Vol. 6, No.6, (November 1975), pp. 443-451.

Krinke, T. J.; Deppert, K.; Magnusson, M. H. & Fissan, H. (2002). Nanostructured Deposition of Nanoparticles from the Gas Phase. *Part. Part. Syst. Char.* Vol. 19, No. 5, (November 2002), pp. 321-326.

Krinke, T. J.; Fissan, H. & Deppert, K. (2003). Deposition of Aerosol Nanoparticles on Flat Substrate Surfaces. *Phase Transitions.* Vol. 76, No. 4-5, (May 2003), pp. 333-345.

Kuzmanovic, D. A.; Elashvili, I.; O'Connell, C. & Krueger, S. (2008). A Novel Application of Small-Angle Scattering Techniques: Quality Assurance Testing of Virus Quantification Technology. *Radiat. Phys. Chem.* Vol. 77, No. 3, (March 2008), pp. 215–224.

Lall, A. A. & Friedlander, S. K. (2006). On-Line Measurement of Ultrafine Aggregate Surface Area and Volume Distributions by Electrical Mobility Analysis: I. Theoretical Analysis. *J. Aerosol Sci.* Vol. 37, No. 3, (March 2006), pp. 260-271.

Lall, A. A. & Friedlander, S. K. (2006). On-Line Measurement of Ultrafine Aggregate Surface Area and Volume Distributions by Electrical Mobility Analysis: II. Comparison of Measurements and Theory. *J. Aerosol Sci.* Vol. 37, No. 3, (March 2006), pp. 272-282.

Lenggoro, I. W.; Xia, B. & Okuyama, K. (2002). Sizing of Colloidal Nanoparticles by Electrospray and Differential Mobility Analyzer Methods. *Langmuir.* Vol. 18, No. 12, (May 2002), pp. 4584–4591.

Lenggoro, I. W.; Widiyandari, H.; Hogan Jr., C. J.; Biswas, P. & Okuyama, K. (2007). Colloidal Nanoparticle Analysis by Nanoelectrospray Size Spectrometry with a Heated Flow. *Anal. Chim. Acta.* Vol. 585, No. 2, (March 2007), pp. 193-201.

Li, M.; Guha, S.; Zangmeister, R. A.; Tarlov, M. J. & Zachariah, M. R. (2011). Quantification and Compensation of Nonspecific Analyte Aggregation in Electrospray Sampling. *Aerosol Sci. Technol.*. Vol. 45, No.7, (March 2011), pp. 849-860.

Loo, J. A.; Berhane, B.; Kaddis, C. S.; Wooding, K. M.; Xie, Y. M.; Kaufman, S. L. & Chernushevich, I. V. (2005). Electrospray Ionization Mass Spectrometry and Ion Mobility Analysis of the 20S Proteasome Complex. *J. Am. Soc. Mass Spectrom.* Vol. 16, No. 7, (July 2005), pp. 998-1008.

Lute, S.; Riordan, W.; Pease III, L. F.; Tsai, D. H.; Levy, R.; Haque, M.; Martin, J.; Moroe, I.; Sato, T.; Morgan, M.; Krishnan, M.; Campbell, J.; Genest, P.; Dolan, S.; Tarrach, K.; Meyer, A.; the PDA Virus Filter Task Force; Zachariah, M. R.; Tarlov, M. J.; Etzel, M. & Brorson, K. (2008). A Consensus Rating Method for Small Virus-Retentive Filters. I. Method Development. *PDA J. Pharm. Sci. Technol.* Vol. 62, No. 5, (October 2008), pp. 318–333.

Moisala, A.; Nasibulin, A. G.; Shandakov, S. D.; Jiang, H. & Kauppinen, E. I. (2005). On-line Detection of Single-Walled Carbon Nanotube Formation During Aerosol Synthesis Methods. *Carbon.* Vol. 43, No. 10, (August 2005), pp. 2066-2074.

Netz, R. R. & Andelman, D. (2003). Neutral and Charged Polymers at Interfaces. *Phys. Rep.* Vol. 380, No. 1-2, (June 2003), pp. 1-95.

Nguyent, H. V. & Flagan, R. C. (1991). Particle Formation and Growth in Single-Stage Aerosol Reactors. *Langmuir.* Vol. 7, No. 8, (August 1991), pp. 1807-1814.

Nie, Z.; Xu, S.; Seo, M.; Lewis, P. C. & Kumacheva, E. (2005). Polymer Particles with Various Shapes and Morphologies Produced in Continuous Microfluidic Reactors. *J. Am. Chem. Soc.* Vol. 127, No. 22, (May 2005), pp. 8058-8063.

Nie, Z. X.; Tzeng, Y. K.; Chang, H. C.; Chiu, C. C.; Chang, C. Y.; Chang, C. M. & Tao, M. H. (2006). Microscopy-Based Mass Measurement of a Single Whole Virus in a Cylindrical Ion Trap. *Angew. Chem. Int. Ed. Engl.* 2006, Vol. 45, No. 48, (December 2006), pp. 8131-8134.

NIST Certificate (2007). *Nanoparticle Standards at NIST: Gold Nanoparticle Reference Materials*.http://www.mel.nist.gov/tripdf/NIST/05_Hackley%20TNW%20AuRM %20talk%2002.05.08.VAH_final.pdf (2007).

Park, J. Y. & Phaneuf, R. J. (2003) Investigation of the Direct Electromigration Term for Al Nanodots within the Depletion Zone of a pn Junction. *J. Appl. Phys.* Vol. 94, No. 10, (2003), pp. 6883-6886.

Pease III, L. F.; Tsai, D. H.; Zangmeister, R. A.; Zachariah, M. R. & Tarlov, M. J. (2007). Quantifying the Surface Coverage of Conjugated Molecules on Functionalized Nanoparticles. *J. Phys. Chem. C.* Vol. 111, No. 46, (November 2007), pp. 17155-17157.

Pease III, L.F.; Elliott, J. T.; Tsai, D. H.; Zachariah, M. R. & Tarlov, M. J. (2008) Determination of Protein Aggregation with Differential Mobility Analysis: Application to IgG Antibody. *Biotechnol. Bioeng.* Vol. 101, No. 6, (December 2008), pp. 1214–1222.

Pease III, L. F.; Lipin, D. I.; Tsai, D. H.; Zachariah, M. R.; Lua, L. H. L.; Tarlov, M. J. & Middelberg, A. P. J. (2009a). Quantitative Characterization of Virus-like Particles by Asymmetrical Flow Field Flow Fractionation, Electrospray Differential Mobility Analysis, and Transmission Electron Microscopy. *Biotechnol. Bioeng.* Vol. 102, No. 3, (February 2009), pp. 845-855.

Pease III, L.F.; Tsai, D. H.; Fagan, J. A.; Bauer, B. J.; Zangmeister, R. A.; Tarlov, M. J. & Zachariah, M. R. (2009b). Length Distributions of Single Wall Carbon Nanotubes in Aqueous Suspensions Measured by Electrospray-Differential Mobility Analysis. *Small*. Vol. 5, No. 24, (December 2009), pp. 2894-2901.

Pease III, L. F.; Tsai, D.-H.; Zangmeister, R. A.; Hertz, J. L.; Zachariah, M. R. & Tarlov, M. J. (2010a). Packing and Size Determination of Colloidal Nanoclusters. *Langmuir*. Vol. 26, No. 13, (May 2010), pp. 11384–11390.

Pease III, L. F.; Feldblyum, J. I.; DePaoli Lacerda, S. H.; Liu, Y.; Hight-Walker, A.; Anumolu, R.; Yim, P. B.; Clarke, M. L.; Kang, H. G. & Hwang, J. (2010b). Structural Analysis of Soft Multicomponent Nanoparticle Clusters. *ACS Nano*. Vol. 4, No. 11, (November 2010), pp. 6982-6988.

Pease III, L. F.; Sorci, M.; Guha, S.; Tsai, D. H.; Zachariah, M. R.; Tarlov, M. J. & Belfort, G. (2010c). Probing the Nucleus Model for Oligomer Formation During Insulin Amyloid Fibrillogenesis. *Biophys. J.* Vol. 99, No. 12, (December 2010), pp. 3979-85.

Pease III, L. F.; Tsai, D. H.; Zangmeister, R. A.; Zachariah, M. R. & Tarlov, M. J. (2010d). Analysis of Gold Nanoparticles by Electrospray Differential Mobility Analysis (ES-DMA). NIST-NCL Joint Assay Protocol, PCC-10, Version 1.1.

Pease III, L. F.; Tsai, D. H.; Brorson, K. A.; Guha, S.; Zachariah, M. R.; & Tarlov, M. J. (2011). Physical Characterization of Viral Ultra Structure, Stability, and Integrity. *Anal. Chem.* Vol. 83, No. 5, (February 2011), pp. 1753-1759.

Pease III, L. F. (2011a). Optimizing the Yield and Selectivity of High Purity Nanoparticle Clusters. *J. Nanopart. Res.* Vol. 13, No. 5, (2011), 2157-2172.

Pease III, L.F. (2012). Physical Analysis of Virus Particles. *Trends Biotechnol.* (2012), in press, DOI: 10.1016/j.tibtech.2011.11.004.

Rader, D. J. (1990). Momentum Slip Correction Factor for Small Particles in Nine Common Gases. *J. Aerosol Sci.* Vol. 21, No. 2, (1990), pp. 161-168.

Regenmortel, M. H. V.; Fauquet, C. M. & Bishop, D. H. L. (2000). Virus Taxonomy: Classification and Nomenclature of Viruses: Seventh Report of the International Committee on Taxonomy of Viruses. (October 2000), Academic Press, San Diego, USA.

Russel, W. B.; Saville, D. A. & Schowalter, W. R. (1989). *Colloidal Dispersions* (1st edition). Cambridge University Press, ISBN 9780521426008, New York, USA.

Saffari, H.; Malugin, A.; Ghandehari, H. & Pease III, L.F. (2012). Electrostatic Deposition of Nanoparticles into Live Cell Culture Using an Electrospray Differential Mobility Analyzer (ES-DMA). *J. Aerosol Sci.* (2012), in press.

Saucy, D. A.; Ude, S.; Lenggoro, I. W. & Fernandez de la Mora, J. (2004). Mass Analysis of Water-Soluble Polymers by Mobility Measurement of Charge-Reduced Ions Generated by Electrosprays. *Anal. Chem.* Vol. 76, No. 4, (January 2004), pp. 1045-1053.

Shoemaker, G. K.; Duijn, E. V.; Crawford, S. E.; Uetrecht, C.; Baclayon, M.; Roos, W. H.; Wuite, G. J. L.; Estes, M. K.; Prasad, B. V. & Heck, A. J. R. (2010). Norwalk Virus Assembly and Stability Monitored by Mass Spectrometry. *J. Mol. Cell Proteomics.* Vol. 9, No. 8, (April 2010), pp. 1742–51.

Siuzdak, G.; Bothner, B.; Yeager, M.; Brugidou, C.; Fauquet, C. M.; Hoey, K. & Change, C.-M. (1996). Mass Spectrometry and Viral Analysis. *Chem. Biol.* Vol. 3, No. 1, (January 1996), pp. 45–48.

Song, D. K.; Lenggoro, I. W.; Hayashi, Y.; Okuyama, K. & Kim, S. S. (2005). Changes in the Shape and Mobility of Colloidal Gold Nanorods with Electrospray and Differential Mobility Analyzer Methods. *Langmuir.* Vol. 21, No. 23, (November 2005), pp. 10375-10382.

Swann, M. J.; Peel, L. L.; Carrington, S. & Freeman, N. (2004). Dual-Polarization Interferometry: An Analytical Technique to Measure Changes in Protein Structure in Real Time, to Determine the Stoichiometry of Binding Events, and to Differentiate between Specific and Nonspecific Interactions. *J. Anal. Biochem.* Vol. 329, No. 2, (June 2004), pp. 190-198.

Thomas, J. J.; Bothner, B.; Traina, J.; Benner, W. H. & Siuzdak, G. (2004). Electrospray Ion Mobility Spectrometry of Intact Viruses. *Spectroscopy.* Vol. 18, No. 1, (January 2004), pp. 31–36.

Tsai, D. H.; Kim, S. H.; Corrigan, T. D.; Phaneuf, R. J. & Zachariah, M. R. (2005). Electrostatic-Directed Deposition of Nanoparticles on a Field Generating Substrate. *Nanotechnology.* Vol. 16, No. 9, (July 2005), pp. 1856-1862.

Tsai, D. H.; Zangmeister, R. A.; Pease III, L. F.; Zachariah, M. R. & Tarlov, M. J. (2008). Gas-phase Ion-mobility Characterization of SAM functionalized Au Nanoparticles. *Langmuir.* Vol. 24, No. 16, (July 2008), pp. 8483-8490.

Tsai, D. H.; Pease III, L. F.; Zachariah, M. R. & Tarlov, M. J. Aggregation Kinetics of Colloidal Particles Measured by Gas-phase Differential Mobility Analysis. *Langmuir.* Vol. 25, No. 1, (December 2008), pp. 140-146.

Tsai, D.H.; Lipin, D. I.; Guha, S.; Feldblyum, J. I.; Cole, K. D.; Brorson, K. A.; Zachariah, M. R.; Tarlov, M. J.; Middelberg, A. P. J. & Pease III, L. F. (2011). Process Analytical Technology for Recombinant Pandemic Flu Vaccines: Viral Ultrastructure, Aggregation, and Binding. *AIChE Annual Meeting.* CD-ROM. (2011), New York.

Umbach, P.; Georgalis, Y. & Saenger, W. (1998). Time-Resolved Small-Angle Static Light Scattering on Lysozyme During Nucleation and Growth. *J. Am. Chem. Soc.* Vol. 120, No. 10, (March 1998), pp. 2382-2390.

Wang, W. (2005). Protein Aggregation and its Inhibition in Biopharmaceutics. *Int. J. Pharm.* Vol. 289, No. 1–2, (January 2005), pp. 1–30.

Weisbecker, C. S.; Merritt, M. V. & Whitesides, G. M. (1996). Molecular Self-Assembly of Aliphatic Thiols on Gold Colloids. *Langmuir.* Vol. 12, No. 16, (August 1996), pp. 3763–3772.

Wick, C. H.; McCubbin, P. E. & Birenzvige, A. (2006). Detection and Identification of Viruses using the Integrated Virus Detection System (IVDS); ECBC Technical Report: 2006, ECBC-TR-463.

Wiedensohler, A. (1988). An Approximation of the Bipolar Charge-Distribution for Particles in the Sub-Micron Size Range. *J. Aerosol Sci.* Vol. 19, No. 3, (1988), pp. 387–389.

Yim, P. B.; Clarke, M. L.; McKinstry, M.; De Paoli Lacerda, S. H.; Pease III, L. F.; Dobrovolskaia, M. A.; Kang, H. G.; Read, T. D.; Sozhamannan, S. & Hwang, J. (2009). Quantitative Characterization of Quantum Dot-Labeled Lambda Phage for

Escherichia Coli Detection. *Biotechnol. Bioeng.* Vol. 104, No. 6, (December 2009), pp. 1059-1067.

Zhu, Z.; Anacker, J. L.; Ji, S.; Hoye, T. R.; Macosko, C. W. & Prud'homme, R. K. (2007). Formation of Block Copolymer-Protected Nanoparticles via Reactive Impingement Mixing. *Langmuir.* Vol. 23, No. 21, (September 2007), pp. 10499-10504.

The Reactivity of Colloidal Inorganic Nanoparticles

Neus G. Bastús[1], Eudald Casals[1], Isaac Ojea[1],
Miriam Varon[1] and Victor Puntes[1,2]
[1]*Institut Català de Nanotecnologia, Barcelona,*
[2]*Institut Català de Recerca i Estudis Avançats (ICREA) Barcelona,*
Spain

1. Introduction

The development of functional colloidal inorganic nanoparticles (INPs) has increased exponentially over the past decades offering a "toolbox" ready to be used in a wide range of applications such as materials science, catalysis, biology and medicine (Freitas, 1999; Alivisatos, 2001). This applicability relies on their unique size, morphology and structure, which determines not only their properties but also its reactivity. At the nanometer scale, the confinement of electrons, phonons and photons leads to a new generation of materials which have improved or new physico-chemical properties in comparison with bulk materials. Well-known examples are the size and shape dependence absorption and scattering in noble metal INPs (e.g. Au or Ag), the enhanced fluorescence in semiconductor quantum dots (QDs) (e.g. CdSe or PbS) and the superparamagnetic moment in magnetic NPs (e.g. iron oxide or cobalt) (Burda et al., 2005). Additionally, as the size of the material is reduced and percentage of atoms at the surface becomes significant, the entire particles become more reactive (Bastús et al., 2008). Thus, in very small crystals, both the thermodynamics and kinetics of reactions can change with size. For instance, a large surface-to-volume ratio can be accompanied by a lowering of phase transition temperatures (Goldstein et al., 1992). Additionally, this high fraction of unsaturated atoms at INP's surface may lead to some instabilities which can further lead to degradation and corrosion processes. Although this secondary processes are often uncontrollable, the feature is extremely useful for catalysis applications (e.g. Pt NPs) allowing reactions at the active sites of their surfaces (Li & Somorjai, 2010).

All in all, the reactivity of an INP is not only determined by the size, composition and structure of its core. In fact, an INP is the combination of an inorganic core, which determines its physico-chemical properties, and an organic/inorganic shell, which dictates the interfacial interactions by the chemical nature of the organic layer. Therefore, the surface coating play a dominant role when controlling and tuning the reactivity of the particle, determining its solubility and selectivity against a desired target.

In this context, we identify some critical factors that one should take into account when studying the reactivity of INPs **(Figure 1)**. Thus, by reducing the size (or tuning the shape) of a nanostructured material, the atomic surface-to-volume ratio increases. In this process, most of the atoms are preferentially located at the edges and kinks of the particle, which dramatically increases its reactivity. Another consequence of the high surface-to-volume ratio of INPs is their high ability to suffer chemical transformations, degradation and corrosion, either caused by the removal of ions which they are in equilibrium with, or by the addition of other ions, leading to its growth. In all cases, these chemical reactions lead to accidental or intended morphological transformations of INPs. Independent of the particular reactivity of the inorganic core, INP's surface coating, its substitution and degradation drastically determine the final interaction between the particle and the surrounding medium. Thus, same core material can interact distinctly depending on the surface nature of the coating layer or its time-dependent evolution. Finally, it is important to consider the interactions between INPs in a colloidal solution. INPs are not isolated entities. On the contrary, they are in a constant Brownian movement which forces their collisions towards an aggregated state. Although this reactivity is often undesirable (physico-chemical properties of INPs are restricted to individual particles in non-physical contact) it can be used as driving force for the formation of more complex structures either by an oriented attached mechanism or by their use as catalysts.

An important prerequisite for boosting the applicability of INPs is developing an in-depth study of INP's reactivity. This condition is especially important in biological and medical systems where the interactions and interferences of INPs with cells and tissues determine not only the potential toxicity of engineered materials but also its biodistribution, degradation and biocompatibility (Casals et al., 2008). The response of nanostructured materials in biological fluids is extremely complex and diverse and depends on a variety of

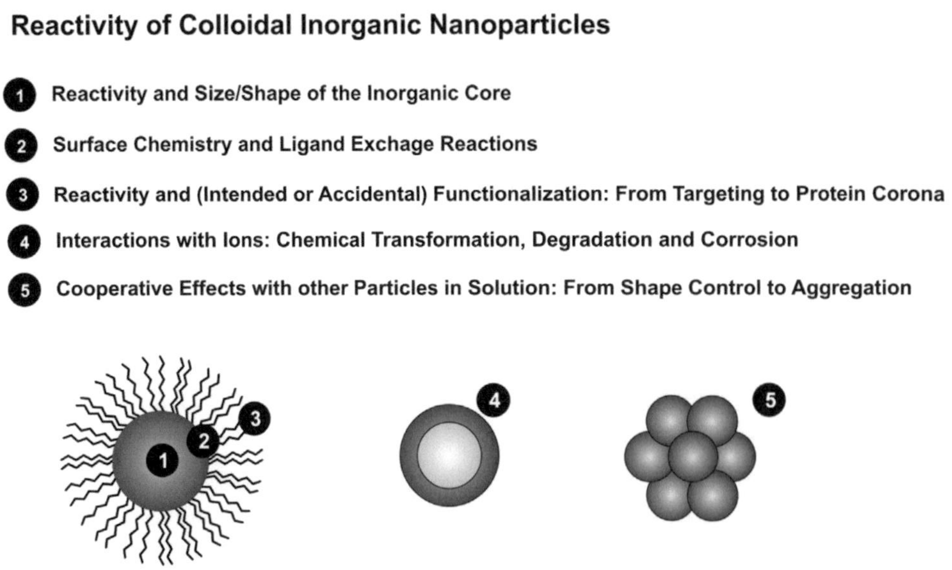

Fig. 1. Reactivity of Colloidal Inorganic Nanoparticles. The critical factors to consider on the study of INP's reactivity are i) Effects of inorganic core's size and shape, ii) the surface molecule substitution and ligand exchange, iii) the particular reactivity of the coating molecule, iv) the interactions with ions present in the colloidal solution and v) cooperative effects with other NPs present in solution.

involved parameters. INPs can aggregate into microscopic structures or may be encapsulated into surrounding material. NP's surface, which determine its final activity, may experience constant modifications such as corrosion or degradation.

In this context, this chapter explores the reactivity of colloidal INPs. It focuses on the particular physical-chemical properties of INPs and in the understanding of the factors that determine its chemical reactivity and therefore its applicability, in nanobiomedicine and catalysis among others. We summarize the potential reactivity of INPs attributed either to their size, shape, chemical composition and structure, focussing of the special features that make them unique and potentially more toxic and risky. Thus, interactions of INPs with surrounding environments will determine their stability while (intended or accidental) surface coating defines much of their bioactivity. Consequences of this are the different final bioactivity of the same element depending on their surface charge (INPs positively charged predominantly attach negative biological surfaces as cell membranes, leading to cell death) and the nature of the non-specific coating that evolves as time progresses leading to different responses at different exposure times. Additionally, toxicological aspects associated to the reactivity of INPs (size, shape, chemical composition and structure) will be widely discussed. The aim of this chapter is not to provide in-depth insights into INP's features, but to tackle the study from a holistic perspective, in the framework of INP's full life cycle.

2. Reactivity and morphology

2.1 Reactivity and size

The size of nanostructured materials determines, among others, its reactivity. The reasons are clear and easily defined. Whereas in a bulk material, surface atoms contribute only a relatively small fraction (only a few per cent) of the total number of atoms, in INPs the curvature radii is so high that all the atoms of the particle lay close or at the surface of the kinks, steps and edges. Therefore, this small particles contain all (or almost all) surface atoms, and those atoms have lower coordination numbers than in the bulk material. Since the reactivity of a single atom tend to increase as its coordination decreases, it follows that the reactivity of nanostructured materials increases as particle size decreases. Thus, in the case of a nanocrystal (NC) with a face-centered cubic (f.c.c.) structure, when the size of the crystal is small enough (1 nm) all the atoms of the particle are located at the surface (and edges) while it decreases to 50% when the size is 2 nm. For larger diameters, the number of surface atoms is extremely low (< 5% for 20 nm) and the reactivity of the crystal is determined by the quality of the surface, i.e., the number steps, kinks and terraces of its surface **(Figure 2-A)**. Thus, there is a combined effect in INPs. First, greater accessibility to all constituent (surface) atoms and secondly, enhanced reactivity because of the low coordination number they exhibit.

Among other features, the size of INPs has been extensively investigated for their implications in catalysis applications since it deeply impacts on both surface structure and electronic properties. Rather than by the exact number of atoms in a cluster (particle size and size distribution), the catalytic reactivity is determined by the arrangement of the surface atoms (number of edges and corners) and the exposure of certain crystallographic planes. Especially interesting is the case of Au NPs which, although considered chemically inert, has been found to be a very effective catalyst for sizes below 5 nm (Hvolbaek et al., 2007). Catalysts based on Au NPs allow significant lower reaction temperatures than used in existing cases which is promising for the development of energy efficient processes. Thus, Xu et al. (Xu et al., 2006) showed that unsupported nanoporous Au, with pores sizes less than 6 nm, made by selective dissolution (de-alloying) of Ag from Ag/Au alloy, are active catalysts for CO oxidation. Further evidences for the scaling of the catalytic activity with the number of corner atoms were provided by Overbury et al. (Overbury et al., 2006) by correlating the CO oxidation activity of a Au/TiO_2 catalyst with Au particle size.

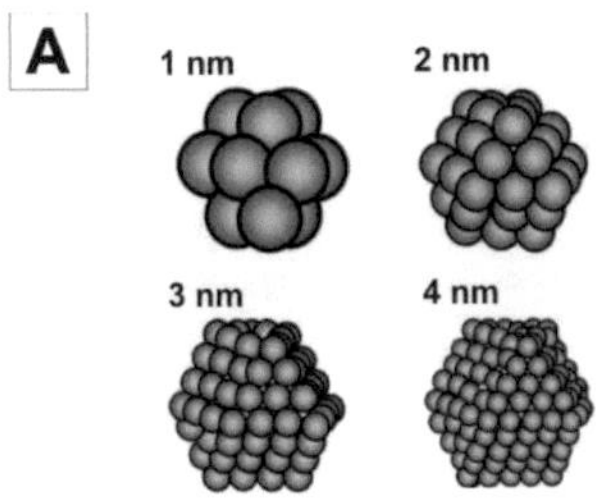

Particle diameter (nm)	Total number atoms	Total number surface atoms	Surface atoms / Total atoms (%)	Surface area (nm²)
1	76	76	100	3.1
2	610	303	50.2	12.5
5	9541	1897	19.7	78.5
10	76332	7591	9.9	314
20	610658	30364	4.9	1256
50	9541533	189779	1.9	7859
100	76103500	757600	1.0	31416

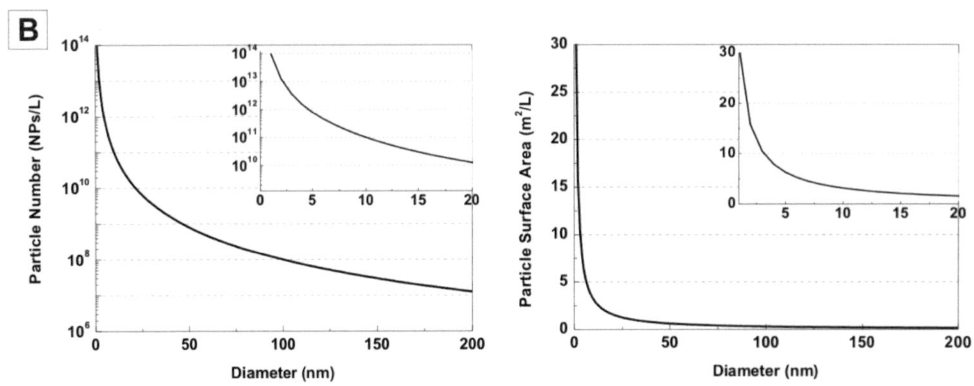

Fig. 2. Core,surface and edge distribution of atoms in inorganic nanoparticles. (A) Atomic distribution in icosahedral nanoparticles from 1 to 4 nm in diameter (left) and total number of atoms, surface atoms, surface-to-volume ratio and total surface area for Au nanoparticles of different sizes. (B) Particle number (left) and total particle surface area (right) for a colloidal solution which contains 1g/1 L of Au atoms.

Another effect of reducing the size of the nanocrystal is the increase of the particle number for a given mass. Thus, considering a colloidal solution of Au NPs with a fixed mass (1 g/1 L) and density (19g/cm³), as the particle size decreases 10 times, the concentration increases by a factor 1000 (**Figure 2-B**). This phenomenon has been a source of controversy when studying the toxicity of INPs as a function of their size, being often shown that smaller particles exhibit more acute toxicity and had higher inflammatory effects that small ones. Toxicity and INP's dose are directly interrelated. Almost any type of nanomaterial has been found to be toxic at high concentrations, and *vice versa,* nothing is toxic at enough small doses. It is often observed that doses used in *in vitro* studies are very high while there is little consideration or discussion about the realistic *in vivo* exposures (Oberdorster et al., 2005). For instance, 100 µg of INPs/ml in cell culture media, labelled as a low dose, is high and unlikely to be encountered *in vivo*. Considering the concentration of as-synthesized engineered INPs, where colloidal synthesis represents the major technique to get individual size and shape controlled NPs, the standard concentrations are ranging from 10^{12} to 10^{16} INPs/ml, depending on the material, that corresponds to an upper limit for very dense materials of few mg/ml. Alternatively, total surface area has been found to be a better dose-parameter than particle mass or particle number in toxicity studies (Oberdörster, 2000; Oberdörster et al., 2000), especially when comparing particles of different sizes. Furthermore, it has been hypothesized that phenomena that operate at low realistic doses are likely to be different from those operating at very high doses when organism or cell defences are overwhelmed (Oberdorster et al., 2005). In addition to that, it is important pointing out that INP's concentration may change in different parts of the body, either by aggregation or accumulation (trapping) in special organs as the liver or kidneys (Bastús et al., 2008; Casals et al., 2008).

The increase of INP's concentration as decreasing INP's size withstands with an increase of the total surface area, and hence the reactivity of the system. This enhancement on reactivity could be either beneficial (antioxidant, carrier capacity for drugs, increase uptake and interaction with biological tissues) or disadvantageous (toxicity, instability, induction of oxidative stress) depending on the intended use and the material used (Bastús et al., 2008). For instance, small CeO_2 INPs are known to exhibit superior antioxidant properties due to the presence of oxygen vacancies on its surface and a cycle of dual oxidation states (Ce^{3+} and Ce^{4+}), which makes them extremely useful as a quenchers of Reactive Oxygen Species (ROS) in biological systems (Karakoti et al., 2008; Garcia et al., 2010). On the other hand, INP's degradation increases as INP's size decreases.

2.2 Reactivity and shape

Similarly, the reactivity of a particle also depends on its shape. As explained above, INP's reactivity is directly associated to surface atoms, whose number is determined not only by particle's size but also by the distribution of atoms within the particle. Therefore, particle's shape is as important as size when studying INP's reactivity. This dependence can be easily evaluated by calculating the number of surface atoms for different particle's geometry while keeping constant the total number of atoms in the particle **(Figure 3)**. Thus, in the case of 10 nm spherical Au NPs, only ~ 10% of atoms are located at the surface of the particle while this number increases up to ~ 29% when same number of atoms are distributed forming a disk. Although obtaining much more smaller values, same tendency was observed in larger particles. Thus, ~ 0.47% are surface atoms in a 200 nm spherical Au NP whereas this value increases up to ~ 0.53% or to ~ 1.46% when moving to anisotropic shapes such as cubes and disks. Beyond the number of surface atoms, INP's reactivity is also directly determined by the quality of the surface. Therefore, atoms in flat surfaces, which have a higher degree of coordination than spheres, have a lower reactivity compared to the atoms on vertices and edges present in the anisotropic shapes. This allows a greater portion of atoms or molecules to be oriented on the surface rather than within the interior of the material, hence allowing adjacent atoms and molecules to interact more readily. As further discussed, this fact has important consequences on the functionalization of INPs, in particular when studying how its surface curvature affects the organization of molecules (as alkanethiols) attached at their surfaces. Thus, the degree of packaging of molecules attached to relatively small Au NPs (less than 20 nm) is significantly lower than their large particle counterparts, allowing for less density of molecules to be attached to the highly curved surfaces than to the larger flatter surfaces, when equal areas are compared.

Moreover, anisotropic shapes expose different crystal faces. In fact, shape control is usually achieved by the selective interaction of surfactant molecules with a particular crystallographic facet, which leads to a preferential growth along a certain direction. Well known examples are the synthesis of anisotropic semiconductor NCs (Kumar & Nann, 2006). For semiconductor systems, where different crystal planes are chemically distinct, this has been demonstrated by using mixtures of surfactants. Thus, Alivisatos' group induced anisotropy in CdSe spherical NCs by using a tri-n-octylphosphine oxide (TOPO) and phosphonic acids, instead of TOPO alone (Peng et al., 2000). Resultant CdSe nanorods displayed many important property modifications compared to spherical NCs, such as polarized light emission (Hu et al., 2001). In continuation of this success, other complex morphologies such as arrows and tetrapods, were successfully synthesized (Manna et al., 2000), all of them explained on the basis of the selective adsorption of surfactants on different crystallographic faces. For the fcc noble metals, the low-energy {111} and {100} facets possess similar surface energies and, therefore, chemical reactivities, which limit the number of candidate molecules for directing shape control by selective binding preference for different crystal planes. Specially interesting is the case of the polymer poly(vinyl pyrrolidone) (PVP) since it has been demonstrated to stabilize the lowest-energy crystal facets of these fcc crystal systems to give colloidal structures bounded by {111}, {100}, and {110} planes (Tao et al., 2008).

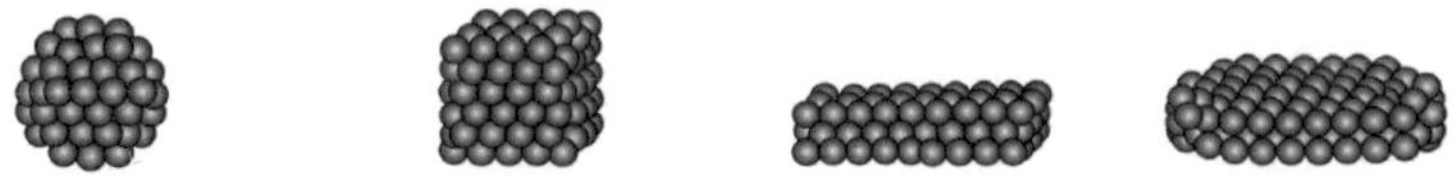

		Sphere		Cube		Rod		Disk	
Particle diameter (nm)	Total number atoms	Surface atoms / Total atoms (%)	Surface area (nm^2)	Surface atoms / Total atoms (%)	Surface area (nm^2)	Surface atoms / Total atoms (%)	Surface area (nm^2)	Surface atoms / Total atoms (%)	Surface area (nm^2)
10	76080	9.4	314	10.6	354	13.5	451	29.2	971
20	608642	4.7	1256	5.3	1416	6.8	1804	14.5	3886
30	2054169	3.1	2826	3.5	3187	4.5	4059	9.7	8743
50	9512937	1.9	7850	2.1	8854	2.7	11274	5.8	24287
70	26095556	1.35	15394	1.5	17364	1.9	22109	4.2	47629
100	76080340	0.94	31400	1.06	35419	1.35	45097	2.9	97151
200	608828006	0.47	125600	0.53	141676	0.67	180386	1.46	388686

Number of total atoms for each size was kept constant while ratio and total surface area was calculated for each geometry

Fig. 3. Shape dependent surface atomic distribution of nanoparticles . Total number of atoms was kept constant (considering Au spheres of different diameters) while the number of surface atoms and the surface area was calculated by different geometries. Rod particles were considered to have a cubic section with a long length 10 times larger than the short length and disk particles have a diameter 10 times larger than their height.

The possibility of controlling the exposition of different crystal faces by tuning NC's morphology has been widely exploited for its applicability in catalysis, where the yield and selectivity of the catalyzed reaction depends on the nature of the catalyst. This dependence has been observed in the Pt-catalyzed hydrogenation of benzene where the {100} surface yields only cyclohexane while the {111} surface gives both cyclohexene and cyclohexane (Bratlie et al., 2007). Similarly, electron-transfer reaction between hexacyannoferrate and thiosulfate has been catalyzed by Pt nanocrystals with different shapes obtaining that tetrahedral NCs, completely bound by {111} crystal facets, exhibit the highest catalytic activity whereas cubic nanocrystals exhibit the lowest activity (Tao et al., 2008). Although promising results have been obtained, the largest obstacles facing shape-controlled nanocatalysis are compatible surface chemistry and shape retention. Catalytic studies need naked surfaces, which is never the case of colloidal NCs, protected by a layer of organic material. Although removing these stabilizing agents may be required to create accessible active sites, it is difficult to do it without inducing significant morphological changes via surface reconstruction, particle ripening, melting, or oxidation (Barghorn et al., 2005).

The correlation between reactivity and INP's shape has also relevant consequences on the study of the potential toxicity of nanomaterials. A typical example is the common synthesis of Au NPs in the presence of hexadecyltrimethylammonium bromide (CTAB). CTAB is a toxic surfactant that strongly binds to the surface of Au NPs confining its growth and conferring stability. Whereas at relative low CTAB concentrations spherical Au NPs are obtained, at high concentrations (beyond the second critical micellar concentration threshold), CTAB forces crystal growth to anisotropic shapes leading to rod-like structures (Nikoobakht & El-Sayed, 2003). This correlation between CTAB concentration and morphology can be understood by a stronger interaction between the atoms at the surface of the rod particle and CTAB molecules. In fact, these hypothesis correlates with shape-dependent toxicity results. Thus, after observation that CTAB-stabilized Au NPs were toxic (rods and spheres), similar cleaning processes lead to non-toxic spheres but still toxic rods (Connor et al., 2005).

3. Corrosion and interactions with ionic species

3.1 Degradation and corrosion

Another consequence of the high surface-to-volume ratio and low atomic surface coordination of nanostructured materials is its degradation, much more relevant at the nanoscale than in bulk materials. INPs are constantly exposed to degradation processes, which reduce them to their constituent atoms either by chemical reactions with the surrounding media or easily because the degradated state is thermodynamically more favorable.

All chemical reactions follow the laws of the chemical equilibrium between reactants and products even though this equilibrium may be displaced to one of the components. This simple idea can be applied to an INP's colloidal solution. A colloidal INP's solution is a chemical reaction where the reagent is the precursor compound and the product is the synthesized colloidal solution. In this regard, it has been observed when synthesizing INPs that only a percentage of initial precursor used is finally converted to INPs. There is therefore an equilibrium between INPs and precursor ion **(Figure 4)**. This fact has important implications not only on the final INP's morphology but also when studying its toxicity. First, because it can lead to the morphological evolution of colloidal INPs (such as a broadening of the size distribution), following an Ostwald ripening process. Additionally, it is especially important when calculating INP's concentration from the initial amount of precursor weighted. As previously mentioned, toxicity and INP's concentration are directly interrelated and hence concentration value must be always corrected by the precursor to INPs conversion rate. Moreover, this chemical equilibrium is also important when purifying INPs (by means of precipitation or other washing methods) since this process displace the equilibrium towards colloidal species, which leads to the new release of ions at the expense of INPs' dissolution and further modifications on INPs' morphology. This decrease of ion concentration in solution is, in fact, a general phenomenon when studying INP's full life cycle since scavengers (as serum proteins), or the simple sample dilution can deplete ionic concentration, forcing the partial or complete INP's degradation. Although degradation process is difficult to prevent, it is possible to protect INPs by designing engineered coatings. Thus, several studies show how the encapsulation of INPs in micelles or their functionalization with self-assembled monolayers (SAM) of molecules could protect them from corrosion.

3.2 Toxicity and ion release

A critical parameter when studying the potential toxicity of INPs is the composition and reactivity of the materials used in their synthesis. Due to the abundance and high reactivity of metals in the presence of oxygen, metal oxide INPs are especially interesting. Metal oxide surfaces are known to be very biocompatible due to their charge organization at the surface (Weissleder et al., 1989). However, not all the oxide INPs display the same toxicity (Gojova et al., 2007). Different stochiometries give different crystal structures with different properties and different surfaces, leading to different potential toxicities. This is the case of magnetic conductors (magnetite), magnetic insulators (maghemite) or non-magnetic insulators (hematite) where the reactivity of conducting and non-conducting materials and the degree of aggregation of magnetic and non-magnetic particles can be significantly different.

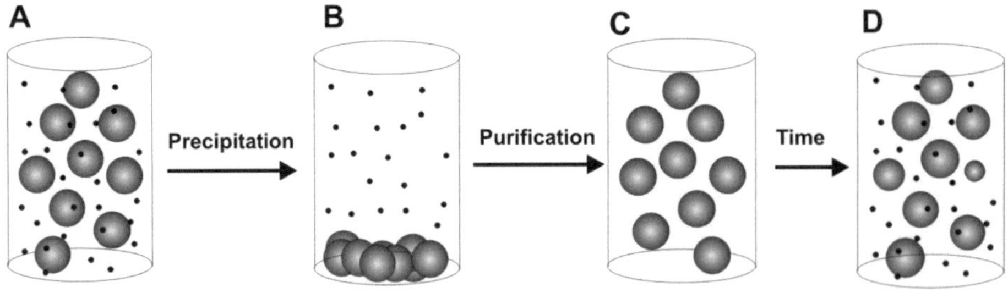

Fig. 4. **Degradation of inorganic nanoparticles**. As-synthesized INPs in equilibrium with precursor species (A). Purification and washing processes displace the equilibrium towards colloidal species (B, C), which leads to the new release of ions at the expense of NP's dissolution. On this stage, INP's morphology is modified due to an Ostwald ripening process (D).

Toxicity is frequently related to the ability of nanostructured materials to release ions. In nanostructured metallic implants, were wear-corrosion greatly contributes to ion release processes, it often correlated with health problems (Ito et al., 2001). Even fairly stable oxides, as in the case of magnetite (Fe_3O_4), may continue oxidizing when exposed to biological environments (Lazaro et al., 2005). In these cases, released ions may undergo chemical transformations which may lead to different chemical species, with altered isotopic composition, electronic or oxidation state and molecular structure (Auffan et al., 2009), which may modify its biological impact (Gupta & Gupta, 2005). In other cases, the corrosion process can be toxic by itself, modifying the redox biochemistry activity of the surrounding biological media and leading to an oxidative stress both *in vivo* and *in vitro* (Oberdorster et al., 2005). For example, Derfus et al. (Derfus et al., 2004) reported the intracellular oxidation and toxicity of CdSe QDs associated to the release of Cd^{2+} cations (Cd^{2+} binds to mitochondria proteins leading to hepatic injuries) while Franklin et al. (Franklin et al., 2007) revealed comparable toxicity of ZnO INPs (30 nm) and $ZnCl_2$ salts, therefore presuming that effects may be attributed to Zn ions released from ZnO INPs. All in all, the environmental and health effects of INPs have contributed to study corrosion processes on nanomaterials.

3.3 Reactivity and ionic/cationic Interactions

Relevant consequences of INP's reactivity are their chemical transformations *via* insertion or exchange of atoms. In extended solids, reactions involving chemical transformations are in general very slow because of high activation energies for the diffusion of atoms and ions in the solid. However, it has been proved that cation exchange reaction can occur completely and reversibly in ionic NCs at room temperature with unusually fast reactions rates. A clear example is the interaction of semiconductor nanocrystals with metal cations, CdSe reacts with Ag^+ ions to yield Ag_2Se NCs by the forward cation exchange reaction, and *vice versa*, Ag_2Se reacts with Cd^{2+} ions to yield CdSe, for the reverse cation exchange reaction (**Figure 5-A**) (Son et al., 2004). While this reaction is extremely fast (< 1s) for NCs, similar experiments carried out with micrometer-sized powders of CdSe found the cation exchange to be virtually prohibited. The speed of the reaction can be explained on the basis of simple scaling of the size in diffusion-controlled reaction schemes, when the reaction time is roughly proportional to the square of the size. This cation exchange reaction, initially investigated with Ag^+ ions, was easily extended not only to other cations but also to the formation of complex hybrid structures with precise structural and compositional tailoring (Zhang et al., 2010). Thus, CdSe NCs can be transformed into CuSe and PbSe though the cations exchange reaction with Cu^{2+} and Pb^{2+} ions while Au-CdS-CdSe core-shell-shell hybrid structures can be obtained from Au-Ag core-shell INPs.

Another chemical transformation extremely interesting is the galvanic corrosion between two different metals and/or alloys having electrical contact with each other. This effect is a coupled reaction where the more active metal corrodes at an accelerated rate and the more noble metal corrodes at a retarded rate. Galvanic replacement has been widely exploited for the obtention of hollow nanostructures of metals by reacting metals solutions (such as Au^{3+}, Pt^{2+}, and Pd^{2+} salts) with pre-formed NCs of a more reactive metal (Sun et al., 2002). An example is the synthesis of Au nanoboxes from Ag nanocubes (**Figure 5-B**). Because the standard reduction potential of $AuCl^{4-}$/Au pair is higher than the one of Ag^+/Ag pair, Ag nanocubes suspended in solution can be oxidized by $HAuCl_4$. The elemental Au produced in this reaction is confined to the vicinity of the template surface, nucleating into very small particles and growing, forming a thin shell around the Ag template. This process continues until the Ag template has been completely consumed. At the last stages of the reaction, Au shells reconstruct their walls into highly crystalline structures via processes such as Ostwald ripening. These Au shells have a morphology similar to that of the Ag templates, with their void sizes mainly determined by the dimensions of the templates. These chemically transformed particles exhibit plasmonic properties completely different from those solid (even made of the same metal).

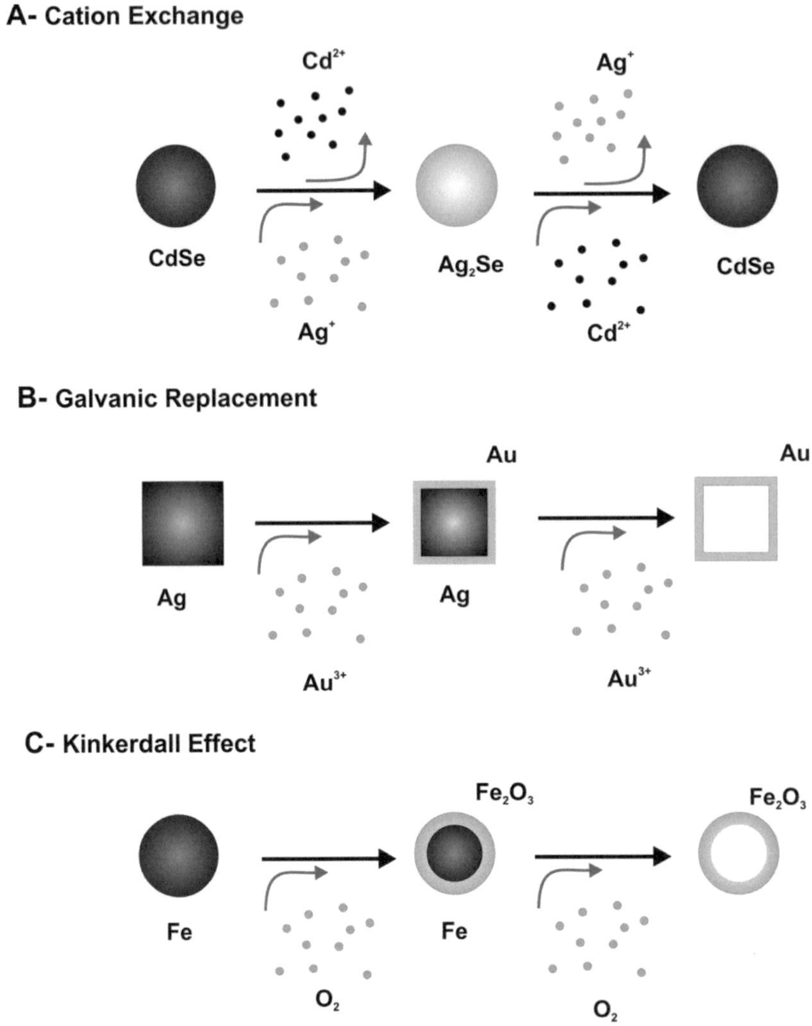

Fig. 5. **Chemical transformations of inorganic nanoparticles via insertion or exchange of atoms.** (A) Cation exchange reactions in semiconductor nanocrystals. (B) Galvanic replacement in metal nanoparticles, (C) Kinkerdall effect in magnetic nanoparticles.

Solid diffusion processes are also a synthetic strategy rather used to prepare hollow solid materials. This phenomenon, called Kirkendall effect, results from the difference of the solid-state diffusion rates of the reactants in an alloying or oxidation reaction (**Figure 5-C**). A significant difference in the diffusion coefficients produce an accumulation of vacancies at the interface of the two components that can lead to the formation of cavities. Thus, the atomic diffusion occurs through vacancy exchange and not by interchange of atoms. The first experimental proof of this phenomenon, originally identified in metal alloys, was reported by Smigelkas and Kirkendall in 1947 (Smigelskas & Kirkendall, 1947). Subsequently, this process was extended into other systems such as oxides, sulfides and nitrides, in which the metallic atoms diffuse faster. To mention some examples at the nanoscale, Alivisatos and co-workers described the formation of hollow oxides (CoO) and chalcogenides (Co_3S_4) nanostructures through the Kirkendall effect employing Co NCs as starting material and reacting them with oxygen and sulfur, respectively, at 182 °C (Yin et al., 2004). As well, Cabot et al. studied the vacancy coalescence during the oxidation process of iron INPs in solution with a controlled oxygen flow and temperature, which leads to the formation of hollow iron oxide INPs. (Cabot et al., 2007). Also has to be considered that the properties of the INPs exposed to transformation processes can be affected. Thus, INPs can lose the applicability in where they are interesting for. Worthy to mention is the work reported by Cabot et al. who studied the magnetic properties and surface effects in hollow maghemite INPs where the single dipole cannot stay at the core of the INPs (Cabot et al., 2009).

3.4 Morphology control and seeded growth methods

The reactivity of colloidal INPs against ionic precursor molecules can be used to precisely control INP's morphology following seeded-growth methods in which pre-synthesized INP's are used as catalysts for the growth of larger particles. The catalyst is usually a colloidal INP that is either injected into a solution together with the precursor molecules needed to growth the desired crystal, or is nucleated *in situ*. Well known examples are the synthesis of nanowires of CdSe, InP, Si and Ge (Trentler et al., 1995; Ahrenkiel et al., 2003; Grebinski et al., 2004) from semiconductor seeds and the size and shape control of Au NPs (Bastús et al.; Pérez-Juste et al., 2005). In the later case, the reduction of the metal salt, such as $HAuCl_4$, was typically carried out in an aqueous system in the presence of ligands such as cetyltrimethylammonium bromide (CTAB) or sodium citrate. **(Figure 6)**. Even more complex structures can be synthesized by taking advantage of NP's reactivity. This is the case of hybrid NCs (HNCs), composed by the combination of different materials assembled together in a single nanostructure. A clear example is the combination of fluorescence and magnetism into a single nanostructure which represents the possibility of bio-detect and bio-separate biological entities. Specially interesting are HNCs based on semiconductor materials, mainly for their utility on optoelectronic, photovoltaic and catalysis applications, where the combination of metal and semiconductor domains allows photo-generated charge carriers performing redox reactions with high efficiency (Kamat, 2007). In the case of semiconductor materials with anisotropic shapes (such as rods, tetrapods or even pyramids), the higher reactivity of the polar facets opens up the possibility of nucleating a second material exclusively at these locations. This idea has been clearly demonstrated for the growth of Au domains onto CdSe NCs (Mokari et al., 2004; Meyns et al., 2010).

3.5 Adsorption, alloying and environmental remediation

Taking advantage from their reduced size, large surface area and highly reactive surface sites, nanomaterials have become particularly attractive as separation media for environmental remediation (Diallo & Savage, 2005; Savage & Diallo, 2005; Karn et al., 2009). For certain reduced diameters, INPs are sustained by Brownian dispersion avoiding sedimentation, which means that they can be maintained exploring the volume without additional agitation. This capacity to be dispersed in groundwater allows them to travel farther than larger macro-sized particles, thus achieving a wider distribution and permitting the whole volume to be quickly scanned with a relatively low number of particles. For example, in a rough estimation, a 10 nm metallic INP in water at room temperature (RT) will experience Brownian relaxation in the order of the nanosecond, and after each Brownian step in solution it will move about 10 to 20 nm. Therefore, a typical INP concentration of few nM will explore the total volume in the time order of the cent-second (assuming a 10 % efficiency, i.e. the INP would repeat a previous position up to 10 times before visiting a new one).

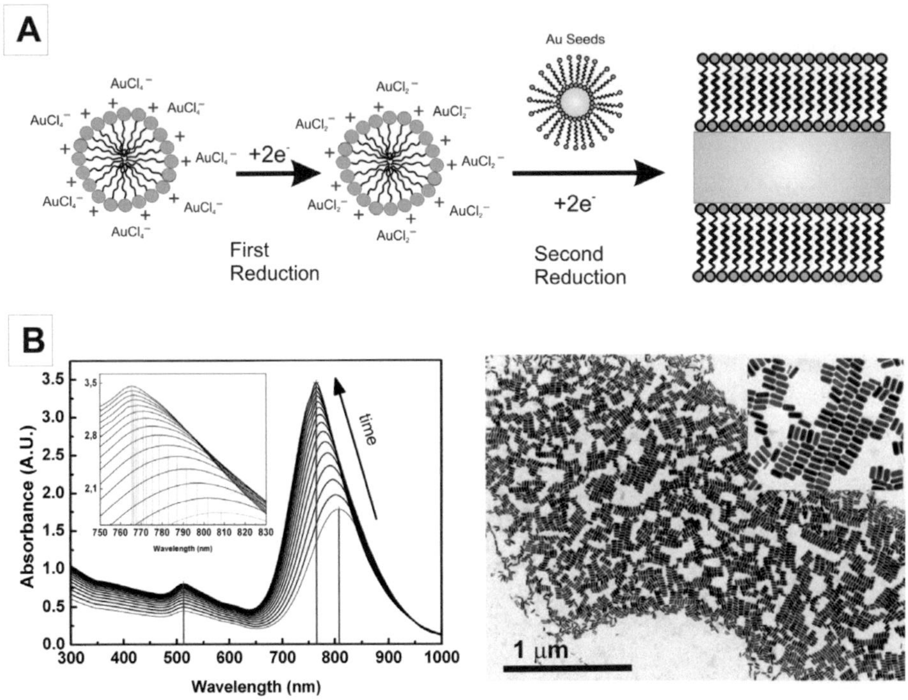

Fig. 6. Reactivity and shape control. Growth of Au nanorods in the presence of a seed NCs used as catalyst. **(A)** The transport of Au ions bound to the CTAB micelles to the growing seed particles is controlled by the double layer interaction. The first reduction is confined in metallomicelles while the second reduction only begins after the seed solution is added. **(B)** Vis-NIR absorption spectra of a representative Au NRs solution growing from Au seeds as a function of time from 20 to 60 min after the addition of seed to the growth solution and morphological characterization of resultant particles by transmission electron microscopy.

The high adsorption capacity of nanomaterials for certain pollutants has been demonstrated in many cases. For example, heavy metals have been effectively removed from contaminated water, as is the case of Cr^{+6} and Cr^{+3} employing maghemite (Hu et al., 2005) and silica NPs (López & Castaño, 2008) respectively, or Hg^{+2} with alumina INPs (Pacheco et al., 2006). In addition, the magnetic properties of some nanomaterials have been employed for magnetic separation after the absorption process, which allowed mixtures of different-sized INPs to be separated by the application of different magnetic fields (Yavuz et al.,

2006). With this methodology As^{+3} and As^{+5} could be eliminated from drinking water supplies employing iron oxide INPs, which involved the formation of weak arsenic-iron oxide complexes at INP's surface. An overview on the use of nano-size iron particles for environmental remediation has been reported by Zhang et al. (Zhang, 2003) Furthermore, the incorporation of some other metal ions, such as TiO_2, into the lattice structure of iron oxide enhances the material properties namely surface adsorption and photo-induced catalysis (Gupta & Ghosh, 2009). The surface functionalization is also a very important factor in environmental remediation since it can enhance the affinity towards specific target molecules. A well known example is the case of magnetite INPs functionalized with di-mercaptosuccinic acid, where the thiolated ligand acts as an effective sorbent for toxic soft metals such as Hg, Ag, Pb, Cd and Ti while As binds to the iron oxide lattices (Yantasee et al., 2007). Alternatively, specific functionalization can also be used on organic–inorganic nanocomposite semiconductor systems to trap organic molecules and then promote their complete degradation making use of the photocatalytic properties (Beydoun et al., 1999). A comprehensive overview of the different manufactured nanomaterials along with the pollutants they could potentially remediate has been reviewed elsewhere (Karn et al., 2009).

Apart from the adsorption capacity, the chemistry of metal alloying has also been used for sequestration of mercury. Thus, in the presence of Au (Henglein & Giersig, 2000; Lisha et al., 2009) or Ag NPs (Katsikas et al., 1996), the Hg^{2+} atoms present in solution can easily be reduced followed by adsorption on the INP's surface and further amalgam formation. As a consequence of the mercury uptake, remarkable morphological transformations can be observed leading to non-spherical shapes and coalescence of particles to an increasing extent. The optical absorption spectra of the mercury containing Au particles is shifted and damped. Similar chemical and optical transformations of spherical CdTe and CdSe NCs into anisotropic alloyed CdHgTe and CdHgSe particles respectively have been observed in the presence of mercury, which have been attributed to the positive redox potential of mercury species (Taniguchi et al., 2011).

4. Reactivity and surface chemistry

Independent of particle size and core composition and structure, the surface coating of the nanostructured material play a dominant role when controlling and tuning the reactivity of the particle and defining its bioactivity. Although as-synthesized INPs present excellent physical and chemical properties, they do not always possess suitable surface properties for specific applications. Then, it is necessary to modify their surface for several reasons such as solubility and stabilization against aggregation, selectivity against a desired target or for surfactant-mediated self-organization. INPs are necessary coated by an intended or accidental coating which contributes to the final reactivity. Thus, the same core material could exhibit different behavior depending on the nature of its surface coating and its aggregation state.

4.1 Surface chemistry and phase transfer

Although there have been many significant developments in the synthesis of INPs, maintaining their stability for long times avoiding its aggregation it is an important issue yet to be deeply explored. Pure metals, such as Fe, Co, Ni and their metal alloys as well as quantum dots (QDs) and Au NPs are usually synthesized in organic solvents and then stabilized using hydrophobic surfactants which provide steric stabilization of the particles (Peng et al., 2000; Puntes et al., 2001; Talapin et al., 2001). The general structure of the surfactant consists of a head group that "sticks" (adsorbed or chemically linked) to the INP's surface while the other end points towards the solution and provides colloidal stability. For biological applications, surface coating should be polar to provide high aqueous solubility and prevent INPs' aggregation and hence the hydrophobic surfactant has to be either replaced by others hydrophilic surfactant molecules in a ligand exchange reaction (Aldana et al., 2001; Gerion et al., 2001) or, alternatively, encapsulated within a polymeric sphere (Pöselt et al., 2009). In this way, by choosing the surfactant molecules, it is possible not only to adjust the surface properties of the particles but also tune its final reactivity. All of these surface modifications alter the chemical composition of INP's surface and therefore, their integration in different environments. An interesting example was the possibility to tune the toxicity of C_{60} by modifying the chemical nature of their surface (Sayes et al., 2004). C_{60} is an hydrophobic organic nanomaterial that forms aggregates when, pristine, gets in contact with water. These aggregates are cytotoxic at low concentration (20 parts per billion) causing cell death via disruption of cell membranes and oxidation of lipids. However, the increase of the solubility of fullerene derivatives decreases the toxic pronounced effects, obtaining a toxicity seven orders of magnitude lower in the case of $C_{60}(OH)_{24}$. Other similar examples are the coating of carbon nanotubes (CNTs) with heparin (Murugesan et al., 2006) or poly(L-lactide) (Zhang et al., 2006) to yield biocompatible structures. Although this increase of solubility may lower the toxicity of nanomaterials, the biopersistence of these coatings is as essential as the knowledge about the bioavailability of the core material, which could exhibit intrinsic toxic properties.

4.2 Surface chemistry and functionalization

One of the most challenging ligand exchange reactions is the INPs functionalization for biomedical applications. In biomedicine, INPs have to be functionalized in order to selectively interact with biological targets. For this purpose, they are usually coated with biological molecules such as antibodies (Wang et al., 2002; El-Sayed et al., 2005), aptamers (Farokhzad et al., 2006), peptides (Kogan et al., 2006; Bastús et al., 2009), and oligonucleotides (Rosi et al., 2006). Some relevant examples are the functionalization of INPs with uptake ligands that specifically bind to receptors of the cell membrane (ex. transferrin (Wagner et al., 1994)), vector ligands that specifically interact with receptors which are over-expressed on the cell surface (Dixit et al., 2006) and peptides that facilitates crossing the cell membrane, either using cell penetrating peptides (Tkachenko et al., 2003; Pujals et al., 2009) or achieving the release of the INP to the cytosol with membrane-disruptive peptides (Han et al., 2006).

However, coatings used for INPs may not be persistent and could be metabolized exposing INP's core. Vector and targeting agents bind usually to INPs' surface by reactive (terminal) moieties. Thus, thiols, amines and carboxilic acids strongly bind to metal, semiconductor and magnetic cores with different and specific affinities. As a consequence of these different activities it is necessary to be aware of the surfactant exchange possibilities: If INP's media is rich in new species that bind stronger than those already present at the INP's surface, a ligand exchange process would take place. This is the case of Au NPs coated with amine or carboxilic groups and further exposed to thiol molecules (Woehrle et al., 2005). In fact, it has even been observed that Au NPs coated with thiol molecules may undergo a ligand exchange process if other thiols are present in solution at similar or higher concentrations. Additionally, this process could lead to a decrease of INP's solubility and a consequent non-reversible aggregation. In more sophisticated situations, as in the case of alloys presenting different types of surface atoms, they may spontaneously segregate forming a core-shell structure (atoms with lower surface tension migrates outwards at room temperature conditions (Margeat et al., 2007)), specially for smaller INPs. This phenomenon directly represents important modification of the INP's surface coating. Other studies have been shown that surface coating can be weathered either by its exposition to oxigen-rich environments or by ultraviolet (UV) irradiation (Rancan et al., 2002; Derfus et al., 2004).

4.3 Surface chemistry and protein corona

As previously mentioned, surface coating modifications are not always intended. They can eventually be accidental. One interesting case is the modification of INP's surface when exposed to biological fluids and their non-specific binding of media proteins (Bastús et al., 2009). This is the case of loosely coated INPs, such as citrate-stabilized Au NPs, where citrate molecules are immediately replaced by media proteins. This process, known as INP's protein corona (INP-PC), is one of the most significant alterations and may, in turn, strongly influence the biocompatibility and biodistribution of these INPs (Casals et al., 2010). INP-PC is a process governed by the competitive time-dependent adsorption of complex mixture of proteins (as serum) by a finite number of surface sites (Vroman, 1962). Thus, highest mobility proteins arrive first and are later (may take several hours) replaced by less motile proteins that have a higher affinity for the surface (Slack & Horbett, 1995). Therefore, after exposition to cell culture media, INPs coating evolves in a time-dependent manner from a soft PC, a transient agglomeration of proteins onto INP's surface, to a hard PC, a dense albumin-rich proteic coating (Casals et al., 2010). This modifications of INP's coating alter INP's reactivity and strongly determine INP-biology interactions, since they are expected to happen at different time-scales. Whereas the removal of particles from the bloodstream takes place in a few minutes, interactions with cells of distant organs may be days after the exposure. Although this PC coating may increase the biocompatibility of the material, it may also alter their toxicity profile, something that has to be carefully considered before using INPs in biomedical applications. All in all, the increasing number of recently publications (Dobrovolskaia & McNeil, 2007), which include the study of different materials such as Au (Casals et al., 2010), Ag, Fe_2O_3, CoO, CeO_2 (Casals et al., 2011) FePt and CdSe/ZnS (Rocker et al., 2009) and the biological implications of the process, indicates the increasing importance of the topic.

5. Reactivity between Inorganic nanoparticles in a colloidal solution

INPs are in a constant Brownian movement which force their collisions towards an aggregated state. Although this reactivity is often undesirable (some physico-chemical properties of INPs are restricted to individual particles in non-physical contact) it can be used as driving force for the formation of more complex structures by an oriented attached mechanism.

5.1 Reactivity and agglomeration

The revolutionary control of physico-chemical properties of nanostructured materials is restricted to individual particles in non-physical contact. Therefore, INP's surface needs to be stabilized in order to avoid irreversible agglomeration and, consequently, there is a significant effort to obtain isolated INPs during long periods of time, which is usually achieved by electrostatic, steric or electro-steric means (Roucoux et al., 2002). In fact, this may be one of the critical differences with the previously existing INPs, since natural and un-intentional occurring INPs tend to agglomerate readily. The degree of agglomeration has relevant consequences on the final behavior of the nanometric material. For instance, INPs used for sunscreens are usually composed of large agglomerates of nanometric domains (Bastús et al., 2008). Although these agglomerates are still efficient sun blockers, their optical properties are less intense than the individual INPs. Additionally, their large size avoids the penetration of such particles beneath the external skin.

INPs' aggregation can be exploited to be used as test to probe either the success of its functionalization or the presence of biological molecules (**Figure 7**). Metal particles

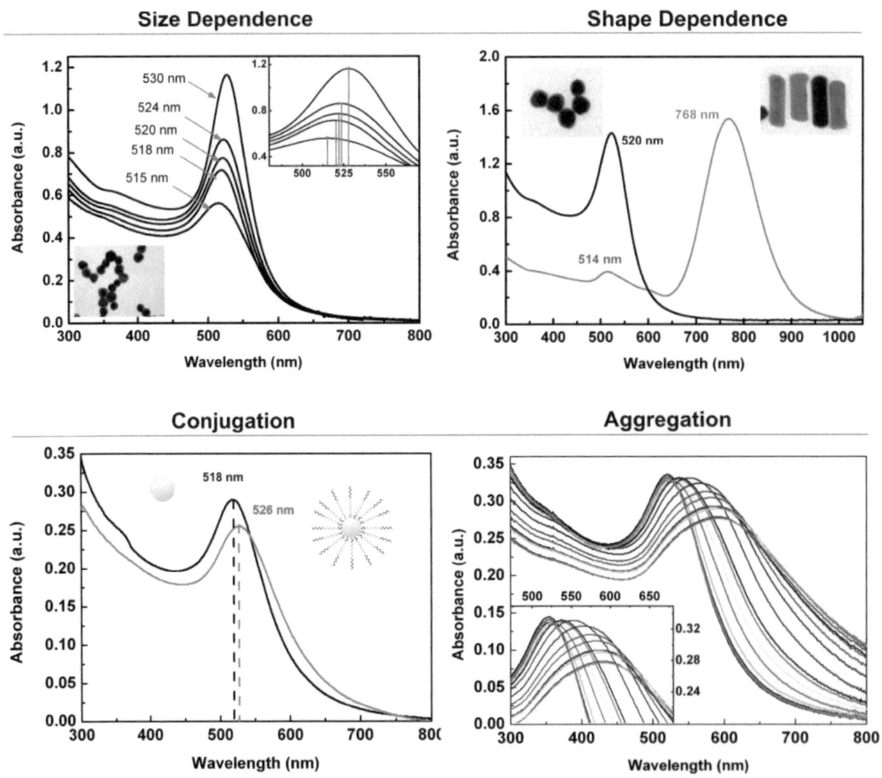

Fig. 7. **Optical properties of gold nanoparticles.** The position of the surface plasmon resonant band depends on the size (A) and shape (B) of Au NPs. This peak position depends on the chemical composition of particle surfactant shell (by changing the refractive index of the surrounding environment), being therefore sensible to functionalization processes (C). The aggregation of Au NPs correlates with a red-shift of the SPR band position along with an increase in the absorbance from 600 to 800 nm (D).

(especially Au and Ag) interact resonantly with visible light absorbing a certain fraction of the visible wavelengths while another fraction is reflected. Small particles absorbs resonantly light in the blue region (~ 500 nm) while as Au NP's size increases, surface plasmon band (SPR) shifts to longer wavelengths, yielding Au NPs a pale-blue colour. SPR band is not only influenced by INP's size, it depends as well on its shape, the chemical composition of the INP's surfactant shell and its agglomeration state (due to dipole coupling between the plasmons of neighbouring particles). This property allows the use of Au NPs as sensors. Thus, when particles aggregate absorption wavelengths red-shifts and the colloidal solution changes its color from red to purple. One of the most typical examples that takes advantage of SPR changes is the colorimetric test to detect single-stranded oligonucleotide targets by using DNA-modified Au NP probes. Thus, two species are present in solution such that each is functionalized with a DNA–oligonucleotide complementary to one half of a given target oligonucleotide. Mixing the two probes with the target results in the formation of a polymeric network of DNA–Au-NPs with a concomitant red-to-purple colour change (Mirkin et al., 1996). Similarly, INP's aggregation can be use as a test to evaluate the INP's stabilization mechanism which allow, under certain conditions, determine the success of INP's functionalization process. Thus, electrostatically-stabilized citrate-coated Au NPs easily aggregate when changing the ionic strength and/or the pH of the medium in which they are solved, while peptidic conjugates with significant steric repulsion (stabilized by the interactions of molecules side chains or domains) withstand much higher changes without compromising stability.

5.2 Reactivity and shape control

The reactivity of colloidal INPs against other INPs is the driving force for the formation of complex structures by an oriented attachment mechanism. Thus, nearly isotropic NCs previously formed in solution coalesce and fuse along some preferential crystallographic directions leading to wires (Tang et al., 2002), rings (Cho et al., 2005) or even sheets (Schliehe et al., 2010). Although this mechanism is not completely understood, one possible explanation is the NC's coalescence along certain directions, which might reduce the overall surface energy by eliminating some high energy facets. Additionally, this process may be facilitated when the organic shell is partially removed from the surface of the initial seed particle or when weakly coordinating molecules are used as stabilizers, so that dipole-dipole inter-particle interactions are enhanced and one-directional NC attachment is spontaneously promoted (Cozzoli et al., 2006).

6. Conclusion

This chapter overviews the particular physical-chemical properties of INPs in order to understand what factors determine its chemical reactivity and thus its applicability in nanobiomedicine and catalysis among others. According to this, the critical factors that determine the interactions of INPs with surrounding environment are: i) Effects of inorganic core's size and shape, ii) the particular reactivity of the coating molecule, iii) the surface molecule substitution and ligand exchange, iv) the interactions with ions present in the colloidal solution and v) the cooperative effects with other NPs present in solution. Along with this study, toxicological aspects associated to the reactivity of INPs (size, shape, chemical composition and structure) have been widely discussed.

7. Acknowledgment

The authors acknowledge financial support from the grants "Plan Nacional" from the Spanish Ministry of Science and Innovation (MAT2009-14734-C02-01) and "Nanotecnologías en biomedicina (NANOBIOMED)" from the Consolider-Ingenio 2010 Program (CSD2006-00012).

8. References

Ahrenkiel, S. P., Micic, O. I., Miedaner, A., Curtis, C. J., Nedeljkovic, J. M. & Nozik, A. J. (2003). Synthesis and Characterization of Colloidal InP Quantum Rods. *Nano Letters*, Vol.3, No. 6, pp. 833-837, ISNN 1530-6984

Aldana, J., Wang, Y. A. & Peng, X. G. (2001). Photochemical instability of CdSe nanocrystals coated by hydrophilic thiols. *Journal of the American Chemical Society*, Vol.123, No. 36, pp. 8844-8850, ISNN 0002-7863

Alivisatos, P. (2001). Less is more in medicine. *Scientific American*, Vol.285, No. 3, pp. 66-73, ISNN 0036-8733

Auffan, M., Rose, J., Bottero, J.-Y., Lowry, G. V., Jolivet, J.-P. & Wiesner, M. R. (2009). Towards a definition of inorganic nanoparticles from an environmental, health and safety perspective. *Nat Nano*, Vol.4, No. 10, pp. 634-641, ISNN 1748-3387

Barghorn, S., Nimmrich, V., Striebinger, A., Krantz, C., Keller, P., Janson, B., Bahr, M., Schmidt, M., Bitner, R. S., Harlan, J., Barlow, E., Ebert, U. & Hillen, H. (2005). Globular amyloid-peptide 1-42 oligome ;a homogenous and stable neuropathological protein in Alzheimer's disease. *Journal of Neurochemistry*, Vol.95, No. 3, pp. 834-847, ISNN

Bastús, N. G., Casals, E., Vázquez-Campos, S. & Puntes, V. (2008). Reactivity of engineered inorganic nanoparticles and carbon nanostructures in biological media. *Nanotoxicology*, Vol.2, No. 3, pp. 99 - 112, ISNN 1743-5390

Bastús, N. G., Comenge, J. & Puntes, V. Kinetically Controlled Seeded Growth Synthesis of Citrate-Stabilized Gold Nanoparticles of up to 200 nm: Size Focusing versus Ostwald Ripening. *Langmuir*, Vol.27, No. 17, pp. 11098-11105, ISNN 0743-7463

Bastús, N. G., Sanchez-Tillo, E., Pujals, S., Farrera, C., Lopez, C., Giralt, E., Celada, A., Lloberas, J. & Puntes, V. (2009). Homogeneous Conjugation of Peptides onto Gold Nanoparticles Enhances Macrophage Response. *ACS Nano*, Vol.3, No. 6, pp. 1335-1344, ISNN 1936-0851

Bratlie, K. M., Lee, H., Komvopoulos, K., Yang, P. & Somorjai, G. A. (2007). Platinum Nanoparticle Shape Effects on Benzene Hydrogenation Selectivity. *Nano Letters*, Vol.7, No. 10, pp. 3097-3101, ISNN 1530-6984

Burda, C., Chen, X., Narayanan, R. & El-Sayed, M. A. (2005). Chemistry and Properties of Nanocrystals of Different Shapes. *Chemical Reviews*, Vol.105, No. 4, pp. 1025-1102, ISNN 0009-2665

Cabot, A., Alivisatos, A. P., Puntes, V. F., Balcells, L., Iglesias, O. & Labarta, A. (2009). Magnetic domains and surface effects in hollow maghemite nanoparticles. *Physical Review B*, Vol.79, No. 9, pp. 094419

Cabot, A., Puntes, V. F., Shevchenko, E., Yin, Y., Balcells, L. s., Marcus, M. A., Hughes, S. M. & Alivisatos, A. P. (2007). Vacancy Coalescence during Oxidation of Iron

Nanoparticles. *Journal of the American Chemical Society*, Vol.129, No. 34, pp. 10358-10360, ISNN 0002-7863

Casals, E., Pfaller, T., Duschl, A., Oostingh, G. J. & Puntes, V. (2010). Time Evolution of the Nanoparticle Protein Corona. *ACS Nano*, Vol.4, No. 7, pp. 3623-3632, ISNN 1936-0851

Casals, E., Pfaller, T., Duschl, A., Oostingh, G. J. & Puntes, V. (2011). Hardening of the Nanoparticle Protein Corona in Metal (Au, Ag) and Oxide (Fe_3O_4, CoO and CeO_2) Nanoparticles. *Small*, Vol.7, No. 24, pp. 3479-3486

Casals, E., Vázquez-Campos, S., Bastús, N. G. & Puntes, V. (2008). Distribution and potential toxicity of engineered inorganic nanoparticles and carbon nanostructures in biological systems. *TrAC Trends in Analytical Chemistry*, Vol.27, No. 8, pp. 672-683, ISNN 0165-9936

Cho, K.-S., Talapin, D. V., Gaschler, W. & Murray, C. B. (2005). Designing PbSe Nanowires and Nanorings through Oriented Attachment of Nanoparticles. *Journal of the American Chemical Society*, Vol.127, No. 19, pp. 7140-7147, ISNN 0002-7863

Connor, E. E., Mwamuka, J., Gole, A., Murphy, C. J. & Wyatt, M. D. (2005). Gold nanoparticles are taken up by human cells but do not cause acute cytotoxicity. *Small*, Vol.1, No. 3, pp. 325-327, ISNN 1613-6810

Cozzoli, P. D., Pellegrino, T. & Manna, L. (2006). Synthesis, properties and perspectives of hybrid nanocrystal structures. *Chemical Society Reviews*, Vol.35, No. 11, pp. 1195-1208, ISNN 0306-0012

Derfus, A. M., Chan, W. C. W. & Bhatia, S. N. (2004). Probing the cytotoxicity of semiconductor quantum dots. *Nano Letters*, Vol.4, No. 1, pp. 11-18, ISNN 1530-6984

Diallo, M. S. & Savage, N. (2005). Nanoparticles and Water Quality. *Journal of Nanoparticle Research*, Vol.7, No. 4, pp. 325-330, ISNN 1388-0764

Dixit, V., VandenBossche, J., Sherman, D. M., Thompson, D. H. & Andres, R. P. (2006). Synthesis and Grafting of Thioctic Acid-PEG-Folate Conjugates onto Au Nanoparticles for Selective Targeting of Folate Receptor-Positive Tumor Cells. *Bioconjugate Chem.*, Vol.17, No. 3, pp. 603-609, ISNN 1043-1802

Dobrovolskaia, M. A. & McNeil, S. E. (2007). Immunological properties of engineered nanomaterials. *Nat Nano*, Vol.2, No. 8, pp. 469-478, ISNN 1748-3387

El-Sayed, I. H., Huang, X. & El-Sayed, M. A. (2005). Surface Plasmon Resonance Scattering and Absorption of anti-EGFR Antibody Conjugated Gold Nanoparticles in Cancer Diagnostics: Applications in Oral Cancer. *Nano Lett.*, Vol.5, No. 5, pp. 829-834, ISNN 1530-6984

Farokhzad, O. C., Cheng, J. J., Teply, B. A., Sherifi, I., Jon, S., Kantoff, P. W., Richie, J. P. & Langer, R. (2006). Targeted nanoparticle-aptamer bioconjugates for cancer chemotherapy in vivo. *Proceedings of the National Academy of Sciences of the United States of America*, Vol.103, No. 16, pp. 6315-6320, ISNN 0027-8424

Franklin, N. M., Rogers, N. J., Apte, S. C., Batley, G. E., Gadd, G. E. & Casey, P. S. (2007). Comparative Toxicity of Nanoparticulate ZnO, Bulk ZnO, and ZnCl2 to a Freshwater Microalga (Pseudokirchneriella subcapitata): The Importance of Particle Solubility. *Environmental Science & Technology*, Vol.41, No. 24, pp. 8484-8490, ISNN 0013-936X

Freitas, R. A. J. (1999). Nanomedicine, Volume I: Basic Capabilities. Georgetown, Landes Bioscience., ISNN 978-1-57059-680-3

Garcia, A., Espinosa, R., Delgado, L., Casals, E., Gonzalez, E., Puntes, V., Barata, C., Font, X. & Sanchez, A. (2010). Acute toxicity of cerium oxide, titanium oxide and iron oxide nanoparticles using standardized tests. *Desalination*, Vol.269, No. 1-3, pp. 136-141, ISNN 0011-9164

Gerion, D., Pinaud, F., Williams, S. C., Parak, W. J., Zanchet, D., Weiss, S. & Alivisatos, A. P. (2001). Synthesis and Properties of Biocompatible Water-Soluble Silica-Coated CdSe/ZnS Semiconductor Quantum Dots. *J. Phys. Chem. B*, Vol.105, No. 37, pp. 8861-8871, ISNN 1520-6106

Gojova, A., Guo, B., Kota, R. S., Rutledge, J. C., Kennedy, I. M. & Barakat, A. I. (2007). Induction of inflammation in vascular endothelial cells by metal oxide nanoparticles: Effect of particle composition. *Environmental Health Perspectives*, Vol.115, No. 3, pp. 403-409, ISNN 0091-6765

Goldstein, A. N., Echer, C. M. & Alivisatos, A. P. (1992). Melting in Semiconductor Nanocrystals. *Science*, Vol.256, No. 5062, pp. 1425-1427,

Grebinski, J. W., Hull, K. L., Zhang, J., Kosel, T. H. & Kuno, M. (2004). Solution-Based Straight and Branched CdSe Nanowires. *Chemistry of Materials*, Vol.16, No. 25, pp. 5260-5272, ISNN 0897-4756

Gupta, A. K. & Gupta, M. (2005). Cytotoxicity suppression and cellular uptake enhancement of surface modified magnetic nanoparticles. *Biomaterials*, Vol.26, No. 13, pp. 1565-1573, ISNN 0142-9612

Gupta, K. & Ghosh, U. C. (2009). Arsenic removal using hydrous nanostructure iron(III)-titanium(IV) binary mixed oxide from aqueous solution. *Journal of Hazardous Materials*, Vol.161, No. 2-3, pp. 884-892, ISNN 0304-3894

Han, G., You, C.-C., Kim, B.-j., Turingan, R. S., Forbes, N. S., Martin, C. T. & Rotello, V. M. (2006). Light-Regulated Release of DNA and Its Delivery to Nuclei by Means of Photolabile Gold Nanoparticles. *Angewandte Chemie International Edition*, Vol.45, No. 19, pp. 3165-3169, ISNN 1521-3773

Henglein, A. & Giersig, M. (2000). Optical and Chemical Observations on Gold-Mercury Nanoparticles in Aqueous Solution. *The Journal of Physical Chemistry B*, Vol.104, No. 21, pp. 5056-5060, ISNN 1520-6106

Hu, J., Chen, G. & Lo, I. M. C. (2005). Removal and recovery of Cr(VI) from waste water by maghemite nanoparticles. *Water Research*, Vol.39, No. 18, pp. 4528-4536, ISNN 0043-1354

Hu, J., Li, L.-s., Yang, W., Manna, L., Wang, L.-w. & Alivisatos, A. P. (2001). Linearly Polarized Emission from Colloidal Semiconductor Quantum Rods. *Science*, Vol.292, No. 5524, pp. 2060-2063

Hvolbaek, B., Janssens, T. V. W., Clausen, B. S., Falsig, H., Christensen, C. H. & Norskov, J. K. (2007). Catalytic activity of Au nanoparticles. *Nano Today*, Vol.2, No. 4, pp. 14-18, ISNN 1748-0132

Ito, A., Sun, X. & Tateishi, T. (2001). In-vitro analysis of metallic particles, colloidal nanoparticles and ions in wear-corrosion products of SUS317L stainless steel. *Materials Science and Engineering: C*, Vol.17, No. 1-2, pp. 161-166, ISNN 0928-4931

Kamat, P. V. (2007). Meeting the Clean Energy Demand: A Nanostructure Architectures for Solar Energy Conversion. *The Journal of Physical Chemistry C*, Vol.111, No. 7, pp. 2834-2860, ISNN 1932-7447

Karakoti, A., Monteiro-Riviere, N., Aggarwal, R., Davis, J., Narayan, R., Self, W., McGinnis, J. & Seal, S. (2008). Nanoceria as antioxidant: Synthesis and biomedical applications. *JOM Journal of the Minerals, Metals and Materials Society*, Vol.60, No. 3, pp. 33-37, ISNN 1047-4838

Karn, B., Kuiken, T. & Otto, M. (2009). Nanotechnology and *in Situ* Remediation: A Review of the Benefits and Potential Risks. *Environ Health Perspect*, Vol.117, No. 12, pp. 1813-1831

Katsikas, L., Gutierrez, M. & Henglein, A. (1996). Bimetallic Colloids: Silver and Mercury. *The Journal of Physical Chemistry*, Vol.100, No. 27, pp. 11203-11206, ISNN 0022-3654

Kogan, M. J., Bastus, N. G., Amigo, R., Grillo-Bosch, D., Araya, E., Turiel, A., Labarta, A., Giralt, E. & Puntes, V. F. (2006). Nanoparticle-mediated local and remote manipulation of protein aggregation. *Nano Letters*, Vol.6, No. 1, pp. 110-115, ISNN 1530-6984

Kumar, S. & Nann, T. (2006). Shape Control of II–VI Semiconductor Nanomaterials. *Small*, Vol.2, No. 3, pp. 316-329, ISNN 1613-6829

Lazaro, F. J., Abadia, A. R., Romero, M. S., Gutierrez, L., Lazaro, J. & Morales, M. P. (2005). Magnetic characterisation of rat muscle tissues after subcutaneous iron dextran injection. *Biochimica Et Biophysica Acta-Molecular Basis of Disease*, Vol.1740, No. 3, pp. 434-445, ISNN 0925-4439

Li, Y. & Somorjai, G. A. (2010). Nanoscale Advances in Catalysis and Energy Applications. *Nano Letters*, Vol.10, No. 7, pp. 2289-2295, ISNN 1530-6984

Lisha, K., Anshup, A. & Pradeep, T. (2009). Towards a practical solution for removing inorganic mercury from drinking water using gold nanoparticles. *Gold Bulletin*, Vol.42, No. 2, pp. 144-152, ISNN 0017-1557

López, X. & Castaño, V. M. (2008). Chromium Removal from Industrial Water Through Functionallized Nanoparticles *J. Nanosci. Nanotechnol*, Vol.8, No., pp. 5733–5738 ISNN

Manna, L., Scher, E. C. & Alivisatos, A. P. (2000). Synthesis of Soluble and Processable Rod-, Arrow-, Teardrop-, and Tetrapod-Shaped CdSe Nanocrystals. *Journal of the American Chemical Society*, Vol.122, No. 51, pp. 12700-12706, ISNN 0002-7863

Margeat, O., Ciuculescu, D., Lecante, P., Respaud, M., Amiens, C. & Chaudret, B. (2007). NiFe nanoparticles: A soft magnetic material? *Small*, Vol.3, No. 3, pp. 451-458, ISNN 1613-6810

Meyns, M., Bastus, N. G., Cai, Y., Kornowski, A., Juarez, B. H., Weller, H. & Klinke, C. (2010). Growth and reductive transformation of a gold shell around pyramidal cadmium selenide nanocrystals. *Journal of Materials Chemistry*, Vol. 20, pp. 10602-10605, ISNN 0959-9428

Mirkin, C. A., Letsinger, R. L., Mucic, R. C. & Storhoff, J. J. (1996). A DNA-based method for rationally assembling nanoparticles into macroscopic materials. *Nature*, Vol.382, No. 6592, pp. 607-609, ISNN 0028-0836

Mokari, T., Rothenberg, E., Popov, I., Costi, R. & Banin, U. (2004). Selective growth of metal tips onto semiconductor quantum rods and tetrapods. *Science*, Vol.304, No. 5678, pp. 1787-1790, ISNN 0036-8075

Murugesan, S., Park, T. J., Yang, H. C., Mousa, S. & Linhardt, R. J. (2006). Blood compatible carbon nanotubes - Nano-based neoproteoglycans. *Langmuir*, Vol.22, No. 8, pp. 3461-3463, ISNN 0743-7463

Nikoobakht, B. & El-Sayed, M. A. (2003). Preparation and growth mechanism of gold nanorods using seed-mediated growth method. *Chemistry of Materials*, Vol.15, No. 10, pp. 1957-1962, ISNN 0897-4756

Oberdörster, G. (2000). Toxicology of ultrafine particles: in vivo studies. *Philosophical Transactions of the Royal Society of London Series a-Mathematical Physical and Engineering Sciences*, Vol.358, No. 1775, pp. 2719-2739, ISNN 1364-503X

Oberdörster, G., Finkelstein, J., Johnston, C., Gelein, R., Cox, C., Baggs, R. & Elder, A. (2000). Acute Pulmonary Effects of Ultrafine Particles in Rats and Mice. *Res Rep Health Eff Inst*, Vol.96, No., pp. 5-74

Oberdorster, G., Oberdorster, E. & Oberdorster, J. (2005). Nanotoxicology: an emerging discipline evolving from studies of ultrafine particles. *Environ Health Perspect*, Vol.113, No. 7, pp. 823-39, ISNN 0091-6765 (Print)

Overbury, S. H., Schwartz, V., Mullins, D. R., Yan, W. & Dai, S. (2006). Evaluation of the Au size effect: CO oxidation catalyzed by Au/TiO_2. *Journal of Catalysis*, Vol.241, No. 1, pp. 56-65, ISNN 0021-9517

Pacheco, S., Medina, M., Valencia, F. & Tapia, J. (2006). Removal of Inorganic Mercury from Polluted Water Using Structured Nanoparticles, ASCE.

Peng, X., Manna, L., Yang, W., Wickham, J., Scher, E., Kadavanich, A. & Alivisatos, A. P. (2000). Shape control of CdSe nanocrystals. *Nature*, Vol.404, No. 6773, pp. 59-61, ISNN 0028-0836

Peng, X. G., Manna, L., Yang, W. D., Wickham, J., Scher, E., Kadavanich, A. & Alivisatos, A. P. (2000). Shape control of CdSe nanocrystals. *Nature*, Vol.404, No. 6773, pp. 59-61, ISNN 0028-0836

Pérez-Juste, J., Pastoriza-Santos, I., Liz-Marzán, L. M. & Mulvaney, P. (2005). Gold nanorods: Synthesis, characterization and applications. *Coordination Chemistry Reviews*, Vol.249, No. 17-18, pp. 1870-1901, ISNN 0010-8545

Pöselt, E., Fischer, S., Foerster, S. & Weller, H. (2009). Highly Stable Biocompatible Inorganic Nanoparticles by Self-Assembly of Triblock-Copolymer Ligands *Langmuir*, Vol.25, No. 24, pp. 13906-13913, ISNN 0743-7463

Pujals, S., Bastus, N. G., Pereiro, E., Lopez-Iglesias, C., Punte, V. F., Kogan, M. J. & Giralt, E. (2009). Shuttling Gold Nanoparticles into Tumoral Cells with an Amphipathic Proline-Rich Peptide. *Chembiochem*, Vol.10, No. 6, pp. 1025-1031, ISNN 1439-4227

Puntes, V. F., Krishnan, K. M. & Alivisatos, A. P. (2001). Colloidal nanocrystal shape and size control: The case of cobalt. *Science*, Vol.291, No. 5511, pp. 2115-2117, ISNN 0036-8075

Rancan, F., Rosan, S., Boehm, F., Cantrell, A., Brellreich, M., Schoenberger, H., Hirsch, A. & Moussa, F. (2002). Cytotoxicity and photocytotoxicity of a dendritic C-60 mono-adduct and a malonic acid C-60 tris-adduct on Jurkat cells. *Journal of Photochemistry and Photobiology B-Biology*, Vol.67, No. 3, pp. 157-162, ISNN 1011-1344

Rocker, C., Potzl, M., Zhang, F., Parak, W. J. & Nienhaus, G. U. (2009). A quantitative fluorescence study of protein monolayer formation on colloidal nanoparticles. *Nat Nano*, Vol.4, No. 9, pp. 577-580, ISNN 1748-3387

Rosi, N. L., Giljohann, D. A., Thaxton, C. S., Lytton-Jean, A. K. R., Han, M. S. & Mirkin, C. A. (2006). Oligonucleotide-modified gold nanoparticles for intracellular gene regulation. *Science*, Vol.312, No. 5776, pp. 1027-1030, ISNN 0036-8075

Roucoux, A., Schulz, J. r. & Patin, H. (2002). Reduced Transition Metal Colloids: Novel Family of Reusable Catalysts? *Chemical Reviews*, Vol.102, No. 10, pp. 3757-3778, ISNN 0009-2665

Savage, N. & Diallo, M. S. (2005). Nanomaterials and Water Purification: Opportunities and Challenges. *Journal of Nanoparticle Research*, Vol.7, No. 4, pp. 331-342, ISNN 1388-0764

Sayes, C. M., Fortner, J. D., Guo, W., Lyon, D., Boyd, A. M., Ausman, K. D., Tao, Y. J., Sitharaman, B., Wilson, L. J., Hughes, J. B., West, J. L. & Colvin, V. L. (2004). The differential cytotoxicity of water-soluble fullerenes. *Nano Letters*, Vol.4, No. 10, pp. 1881-1887, ISNN 1530-6984

Schliehe, C., Juarez, B. H., Pelletier, M., Jander, S., Greshnykh, D., Nagel, M., Meyer, A., Foerster, S., Kornowski, A., Klinke, C. & Weller, H. (2010). Ultrathin PbS Sheets by Two-Dimensional Oriented Attachment. *Science*, Vol.329, No. 5991, pp. 550-553

Slack, S. M. & Horbett, T. A. (1995). The Vroman Effect. Proteins at Interfaces II, American Chemical Society. 602: 112-128.

Smigelskas, A. D. & Kirkendall, E. O. (1947). Zinc Diffusion in Alpha Brass. *Transactions of the American Institute of Mining and Metallurgical Engineers*, Vol.171, No. 130-142

Son, D. H., Hughes, S. M., Yin, Y. & Paul Alivisatos, A. (2004). Cation Exchange Reactions in Ionic Nanocrystals. *Science*, Vol.306, No. 5698, pp. 1009-1012

Sun, Y., Mayers, B. T. & Xia, Y. (2002). Template-Engaged Replacement Reaction: A One-Step Approach to the Large-Scale Synthesis of Metal Nanostructures with Hollow Interiors. *Nano Letters*, Vol.2, No. 5, pp. 481-485, ISNN 1530-6984

Talapin, D. V., Rogach, A. L., Kornowski, A., Haase, M. & Weller, H. (2001). Highly luminescent monodisperse CdSe and CdSe/ZnS nanocrystals synthesized in a hexadecylamine-trioctylphosphine oxide-trioctylphospine mixture. *Nano Letters*, Vol.1, No. 4, pp. 207-211, ISNN 1530-6984

Tang, Z., Kotov, N. A. & Giersig, M. (2002). Spontaneous Organization of Single CdTe Nanoparticles into Luminescent Nanowires. *Science*, Vol.297, No. 5579, pp. 237-240

Taniguchi, S., Green, M. & Lim, T. (2011). The Room-Temperature Synthesis of Anisotropic CdHgTe Quantum Dot Alloys: A Molecular Welding Effect. *Journal of the American Chemical Society*, Vol.133, No. 10, pp. 3328-3331, ISNN 0002-7863

Tao, A. R., Habas, S. & Yang, P. (2008). Shape Control of Colloidal Metal Nanocrystals. *Small*, Vol.4, No. 3, pp. 310-325, ISNN 1613-6829

Tkachenko, A. G., Xie, H., Coleman, D., Glomm, W., Ryan, J., Anderson, M. F., Franzen, S. & Feldheim, D. L. (2003). Multifunctional Gold Nanoparticle-Peptide Complexes for Nuclear Targeting. *J. Am. Chem. Soc.*, Vol.125, No. 16, pp. 4700-4701, ISNN 0002-7863

Trentler, T. J., Hickman, K. M., Goel, S. C., Viano, A. M., Gibbons, P. C. & Buhro, W. E. (1995). Solution-Liquid-Solid Growth of Crystalline III-V Semiconductors: An Analogy to Vapor-Liquid-Solid Growth. *Science*, Vol.270, No. 5243, pp. 1791-1794

Vroman, L. (1962). Effect of Adsorbed Proteins on the Wettability of Hydrophilic and Hydrophobic Solids. *Nature*, Vol.196, No. 4853, pp. 476-477

Wagner, E., Curiel, D. & Cotten, M. (1994). Delivery of drugs, proteins and genes into cells using transferrin as a ligand for receptor-mediated endocytosis. *Advanced Drug Delivery Reviews*, Vol.14, No. 1, pp. 113-135

Wang, S., Mamedova, N., Kotov, N. A., Chen, W. & Studer, J. (2002). Antigen/Antibody Immunocomplex from CdTe Nanoparticle Bioconjugates. *Nano Lett.*, Vol.2, No. 8, pp. 817-822, ISNN 1530-6984

Weissleder, R., Stark, D. D., Engelstad, B. L., Bacon, B. R., Compton, C. C., White, D. L., Jacobs, P. & Lewis, J. (1989). Superparamagnetic Iron-Oxide - Pharmacokinetics and Toxicity. *American Journal of Roentgenology*, Vol.152, No. 1, pp. 167-173, ISNN 0361-803X

Woehrle, G. H., Brown, L. O. & Hutchison, J. E. (2005). Thiol-functionalized, 1.5-nm gold nanoparticles through ligand exchange reactions: Scope and mechanism of ligand exchange. *Journal of the American Chemical Society*, Vol.127, No. 7, pp. 2172-2183, ISNN 0002-7863

Xu, C., Su, J., Xu, X., Liu, P., Zhao, H., Tian, F. & Ding, Y. (2006). Low Temperature CO Oxidation over Unsupported Nanoporous Gold. *Journal of the American Chemical Society*, Vol.129, No. 1, pp. 42-43, ISNN 0002-7863

Yavuz, C. T., Mayo, J. T., Yu, W. W., Prakash, A., Falkner, J. C., Yean, S., Cong, L., Shipley, H. J., Kan, A., Tomson, M., Natelson, D. & Colvin, V. L. (2006). Low-Field Magnetic Separation of Monodisperse Fe_3O_4 Nanocrystals. *Science*, Vol.314, No. 5801, pp. 964-967

Yin, Y., Rioux, R. M., Erdonmez, C. K., Hughes, S., Somorjai, G. A. & Alivisatos, A. P. (2004). Formation of Hollow Nanocrystals Through the Nanoscale Kirkendall Effect. *Science*, Vol.304, No. 5671, pp. 711-714

Zhang, D. H., Kandadai, M. A., Cech, J., Roth, S. & Curran, S. A. (2006). Poly(L-lactide) (PLLA)/multiwalled carbon nanotube (MWCNT) composite: Characterization and biocompatibility evaluation. *Journal of Physical Chemistry B*, Vol.110, No. 26, pp. 12910-12915, ISNN 1520-6106

Zhang, J., Tang, Y., Lee, K. & Ouyang, M. (2010). Nonepitaxial Growth of Hybrid Core-Shell Nanostructures with Large Lattice Mismatches. *Science*, Vol.327, No. 5973, pp. 1634-1638

Zhang, W. X. (2003). Nanoscale Iron Particles for Environmental Remediation: An Overview. *Journal of Nanoparticle Research*, Vol.5, No. 3, pp. 323-332, ISNN 1388-0764

Section 4

Nano-Technology

19

Platinum Fuel Cell Nanoparticle Syntheses: Effect on Morphology, Structure and Electrocatalytic Behavior

C. Coutanceau, S. Baranton and T.W. Napporn
e-Lyse, Laboratoire de Catalyse en Chimie Organique, UMR CNRS-Université de Poitiers, France

1. Introduction

Because of high cost and low availability of platinum, which yet remains unavoidable as catalyst in proton exchange membrane fuel cells for achieving acceptable electric performances, numerous synthesis methods of Pt nanoparticles were developed. The decrease of the particle size in a certain extent leads to a decrease of the noble metal content in the fuel cell electrodes and also to an increase of their real surface area. Physical methods such as plasma sputtering of metals [Brault et al., 2004; Caillard et al., 2007, Cho et al., 2008], laser ablation [Perrière et al., 2001; Boulmer et al., 2006], metal organic chemical vapor deposition [Billy et al., 2010], etc., have been and are still developed for preparing low loaded platinum based electrodes for Proton Exchange Membrane Fuel Cell (PEMFC). Although such methods present important interests due to the high control of metal deposition parameters, they involve also important material losses in the deposition chamber and on the electrode mask. Metal electrodeposition can also be classified as a physical method of electrode preparation [Chen et al., 2003; Ayyadurai et al., 2007]. This method consists in the electrochemical reduction of a metallic salt on an electron conductive support. It is then possible by varying the current or potential sequence applied to the electrode to control or at least to change the structure of the catalyst. However, the faradic yield is often very low because hydrogen evolution can occur as soon as platinum is deposited on the substrate [Coutanceau et al., 2004].

Chemical methods for platinum nanoparticle synthesis allow obtaining platinum nanoparticles deposited and well disseminated on a carbon powder in a simplest and more versatile manner. Amongst the chemical methods for nanostructured catalyst preparation, the impregnation-reduction method is often used in the field of heterogeneous catalysis. The principle of such method consists in the reduction of metallic salts impregnating a carbon powder with either a reducing gas or by thermal treatment under inert atmosphere [Vigier et al., 2004; Coutanceau et al., 2008]. Different possibilities exist. For example, the cationic exchange method [Richard & Gallezot, 1987] consists in activating a carbon support with sodium hypochlorite to form surface carboxylic acid groups, which are transformed into ammonium salts after treatment with ammonia. The ammonium groups are exchanged by contact with a $Pt(NH_3)_4OH_2$ alkaline complex, and the catalyst precursor is reduced at 300°C under pure hydrogen flow to form metallic particles. Pt/C catalyst prepared by this method Were well disseminated on the support, and displayed a mean particle size close to 2 nm. But, the total loading of metal was very limited and did not exceed 10 wt%, which is a too low value for PEMFC application. Higher metal loadings could be obtained by slightly modifying the synthesis protocol [Pieck et al., 1996; Roman-Martinez et al. 2000, Vigier et al., 2004]. The anionic exchange method consists in the modification of the carbon surface with aqua regia as reactant to form surface hydroxyl groups. After acidification of the OH groups into OH_2^+ groups, an exchange reaction is performed with H_2PtCl_6 acid complex. The catalyst precursor is then reduced at 300°C under pure hydrogen flow to form metallic particles. By this way, Pt/C catalysts with loadings up to 30 wt%, which is suitable for PEMFC application, could be obtained, but the size distribution became multimodal [Coutanceau et al., 2008].

In the present book chapter, we will focus on the synthesis of platinum supported nanocatalysts by colloidal routes. Colloidal methods have been extensively developed in the literature for the synthesis of platinum nanoparticles. Most of them are based on the use of surfactants for stabilizing the colloidal suspension and controlling the particle size. Amongst these methods, nano-encapsulation methods [Hwang et al., 2007] are based on the use of reducing agent which will act as surfactant after metal salt reduction reaction. The "Bönnemann method" [Bönnemann & Brijoux, 1995] of metal colloid synthesis belongs to this family. The methods called "water in oil" microemulsion are based on the formation of water nanodroplets as microreactors, dispersed in a continuous oil phase, protected by a surfactant - cetyltrimethyl ammonium bromide (CTAB), sodium bis(2-ethylhexyl)sulphosuccinate (AOT), poly(ethylene glycol) dodecyl ethers such as Brij®30, etc. [Boutonnet et al., 1982; Ingelsten et al., 2001; Solla-Gullon et al., 2000], in which metal salts are dissolved. After addition of a reducing agent ($NaBH_4$ [Brimaud et al., 2007], N_2H_2 [Solla-Gullon et al., 2000], H_2 [Erikson et al., 2004]), platinum nanoparticles are formed and the surfactant is fixed on the metal surface, forming the colloidal solution. The first method involves a thermal treatment at 300°C under air to recover the Pt/C catalyst, whereas the second is carried out at room temperature. Other synthesis methods, such as polyol method [Liu et al., 2005; Oh et al. 2007 ; Lebègue et al., 2011] and instant method [Reetz & Lopez, 2004; Reetz et al., 2004; Devadas et al., 2011], activated by thermal treatment or microwave irradiation, do not necessitate the presence of an organic surfactant. In the case of polyol methods, the reducing and acid-base properties of the solvent allow obtaining a colloidal solution of metallic nanoparticles protected by glycolate species [Fievet et al., 1989; Oh et al., 2007; Grolleau et al;, 2010]. In the case of the instant method, metal oxide nanoparticles are formed and reduced using H_2 or $NaBH_4$ as reducing agents.

The metallic nanoparticle synthesis starts with the nucleation step [Lamer and Dinegar, 1950]. The nucleation step occurs out of thermodynamic equilibrium and involves the co-existence of the formed species with the reactant ones. The formation of the stable phase occurs via the local creation of solid seeds. Under these conditions, the seed surface area is very high in comparison with their volume, and surface tension is also very high. Therefore, to obtain stable platinum seeds, the energetic saving has to be higher than the energy involved for the solid-liquid interface formation; in other words, the seeds have to reach a critical size where the ratio volume/surface is high enough to stabilize them. Now, under thermodynamic equilibrium conditions, the equilibrium shape of the particle is given by the Wulf theorem [Henry, 2005]. In the case of a fcc structure material, such as platinum, the equilibrium shape is a truncated octahedron. Numerical simulations have shown that the transition between a non-crystal icosahedron structure and a fcc structure occurred for about 200 platinum atoms [Mottet et al., 2004]. Once platinum seeds are formed, the growth step can occur. Growth step can involve the collision of seeds due to Brownian agitation, and their fusion (decreasing the surface tension by increasing the size [Wang et al., 2000]); then little seeds can fusion with bigger particles participating to the so-called Otswald ripening [Wilson et al., 2006]. It is also possible that platinum ions are reduced on the surface of existing platinum particle leading to the particle growth trough a homoepitaxial mechanism [Park et al., 2007]. The growth step is then stopped due to ionic metal species depletion in the reaction mixture or to the adsorption of surfactant on the crystal surface. The growth rate, the growth mechanism, and the growth stop have obviously an influence on the particle morphology while the growth step leads to define the size, size distribution, the shape, the shape distribution and internal constraints of the particles [Shevchenko et al., 2003]. Reactions occurring in PEMFCs are known to be structure sensitive, therefore the preparation method of Pt/C catalysts, i.e. the history of the Pt/C catalysts, is expected to have a great importance on the catalyst activity, selectivity and stability towards oxygen reduction reaction and on the tolerance toward poison species contained in reformate gas, such as carbon monoxide.

2. Catalysts preparation

Different colloidal methods have been proposed in the present chapter for the synthesis of Pt/C catalysts. Some of them ("Bônnemann method" and "water in oil" microemulsion) were performed in the presence of organic surfactant, the first one involving a thermal treatment at 300°C, whereas the second one is carried out at room temperature. Two other synthesis methods presented here were carried out without addition of an organic surfactant, the polyol method involving an activation step by either temperature (ca. 200°C) or microwave irradiation, and the instant method involving low temperature activation (from 40 to 80°C).

2.1 Synthesis of Pt/C by the Bönnemann method

This method consists in the chemical reduction of metal ions using a molecule which after the reduction step will act as surfactant to protect metal particles [Bönnemann et al., 1991; Bönnemann et al., 1994]. In the present case, the tetraalkylammonium triethylborohydride reducing agent $(N(alk)_4)^+(Bet_3H)^-$ was first prepared by mixing tetraalkylbromide $(N(alk)_4)^+Br^-$ and triethylborohydrure $(Bet_3H)^-K^+$ in anhydrous tetrahydrofuran, according to the following equation:

$$[N(alk)_4)^+]Br^- + [Bet_3H^-]K^+ \rightarrow [N(alk)_4^+](Bet_3H)^- + KBr \quad (1)$$

then, after KBr crystallization, filtration and removal, the reducing agent is conserved in THF at 0°C before use for the reduction of anhydrous $PtCl_2$ platinum salt. A given volume of $PtCl_2$ in anhydrous THF (1 g_{Pt} L^{-1}) [Grolleau et al., 2008; Sellin et al., 2009], are placed in a balloon and the reducing agent is dropwisely added to the mixture; the Pt^{2+} reduction reaction occurs according to the following equation:

$$PtCl_2 + 2\,[N(C_nH_{2n+1})_4^+][B(C_2H_5\,)_3H^-] \rightarrow Pt([\,N(C_nH_{2n+1})_4^+]Cl^-)_2 + 2B(C_2H_5)_3 + H_2 \quad (2)$$

A colloidal solution of platinum nanoparticles protected by a tetraalkyl chloride surfactant and dispersed in anhydrous THF is then obtained. Under these conditions, the particle growth is controlled by the presence of the surfactant which adsorbs on the platinum surface and prevents any growth of their size, as it is shown in Fig. 1a. Because the reducing agent acts as surfactant after the reduction process, the particle size appears very homogeneous and the size distribution is relatively narrow as it can be seen in the TEM image of a $Pt[N(octyl)_4Cl]_2$ colloid precursor in Fig. 1b. Once the reduction step is finished, a very stable colloid solution is obtained and can be kept for several months without any decantation of the platinum particles. This colloidal solution will be used for the impregnation of a carbon powder in order to obtain carbon supported platinum nanoparticles which will be further used as PEMFC catalysts. A given amount of carbon powder (Vulcan XC72 from Cabot inc.) is added to the platinum colloid solution (in our case in order to obtain a platinum loading of 40 wt%) and THF is completely removed by evaporation under ultrasonication. Then, the surfactant is removed by calcination of the remaining powder at 300°C under air atmosphere for 90 minutes [Grolleau et al., 2008; Sellin et al., 2009]. By this way, anchored platinum particles on the carbon support are obtained. A last step of catalytic powder washing with ultrapure water is carried out in order to remove chlorides and bromides coming from the synthesis procedure.

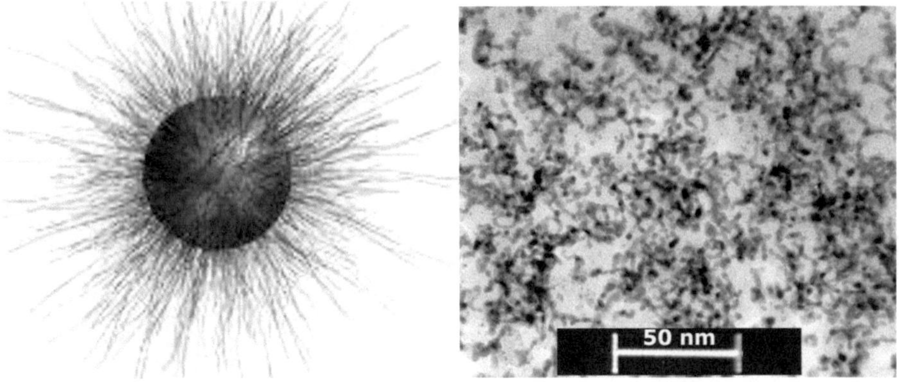

a b

Fig. 1. **Synthesis by the Bönnemann method.** (a) scheme of a platinum particle protected by tetraalkyl chloride surfactant chains; (b) transmission microscopy image of a $Pt[N(octyl)_4Cl]_2$ colloid precursor.

The platinum loading is determined using thermogravimetric analysis under air atmosphere, and was evaluated to be ca. 37.5 wt%; such value is conform to the expected one (40 wt%) [Sellin et al., 2010]. The structural and morphological parameters of the Pt/C catalyst were determined by transmission electron microscopy (TEM), high resolution transmission electron microscopy (HRTEM), X-ray diffraction (XRD) and X-ray photoelectron spectroscopy (XPS). Both the mean particle size as determined by TEM (Fig. 2a) and the mean crystallite size as determined by XRD (Fig. 2c) were estimated to be ca. 2.5 nm, with a standard dispersion of 0.9 nm in the case of the crystallite size. The TEM size d_{TEM} was determined using the following equation,

$$d_{TEM} = \frac{\sum_{i=1}^{n} n_i d_i}{n} \tag{3}$$

where n is the total number of measured particles, n_i is the number of particles with a size d_i, whereas the crystallite size L_v was determined using the simple Sherrer equation:

$$L_v = \Phi \frac{\lambda}{FMWH \cos\theta} \tag{4}$$

Where L_v is the Sherrer length, φ is the shape factor (0.89 for spherical crystallite), λ the radiation wavelength (1.5406 Å), FWHM the full width at half-maximum, and θ the angle at the maximum intensity. Apparently L_v is extracted from the diffraction peak located close to $2\theta = 40°$, which corresponds to the (111) crystallographic plane of platinum. The diffraction pattern was analyzed by the method of Levenberg-Marquardt, using a Pearson VII fit by means of a computer refinement program (Fityk free software).

A calculation of the exhibited surface area has been performed and led to a value of 110 m^2 g^{-1}, assuming that Pt/C nanoparticles have a spherical shape (HRTEM image in Fig. 2b) and that the whole Pt surface is available. However, agglomerated platinum particles are visible on the TEM image in Fig. 2a. After counting on several tenths of images of different regions of the TEM grid, it has been estimated that 15% of the platinum entities were particle agglomerates with a size higher than 14 nm. This will certainly decrease the available platinum surface for electrochemical reaction. XRD characterization in Fig. 2c allowed estimating the value of the cell parameter (ca. 0.392 nm). Lattice defects like stacking faults and microstrains are common for metal nanoparticles, and could add substantially to the width of diffraction peaks [Vogel et al., 1998; Habrioux et al., 2009]. But in the present case, certainly due to the synthesis method of Pt/C catalysts, which involves a thermal treatment at 300°C for 90 min, no stacking fault was detected from HRTEM observations as shown in Fig. 2b as an example. Moreover, the lattice parameter evaluated from XRD measurements is ca. 0.392 mm, i.e., close to the value for platinum bulk. This seems to indicate that the contribution of microstrains is low. Therefore, by correlating the TEM, HRTEM and XRD data, it can be proposed that isolated Pt nanoparticles prepared via the Bönnemann method are *à priori* round shaped single crystals without or with few defects. XPS (X-ray photoelectron microscopy) measurements carried out on a Pt/C powder (Fig. 2d) indicated that platinum was only present under 0 and +II oxidation state. The peaks corresponding to Pt 4f7/2 are located at binding energies of 71.65 eV and 72.65 eV, corresponding to metallic Pt^0 and to $Pt(OH)_2$, whereas the Pt 4f7/2 peak related to PtO species should be located at binding energy higher than 73.5 eV.

2.2 Synthesis of Pt/C by the "water in oil" microemulsion method

The "water in oil" microemulsion is a ternary system currently composed of an aqueous phase dispersed in a continuous n-heptane phase. In order to obtain a translucent mixture thermodynamically stable of both these phases (aqueous and organic), the presence of a surfactant (Brij®30) is necessary. The surfactant will act as a separator between both phases leading to the formation of nano-droplets of the minority phase in the majority one. In the

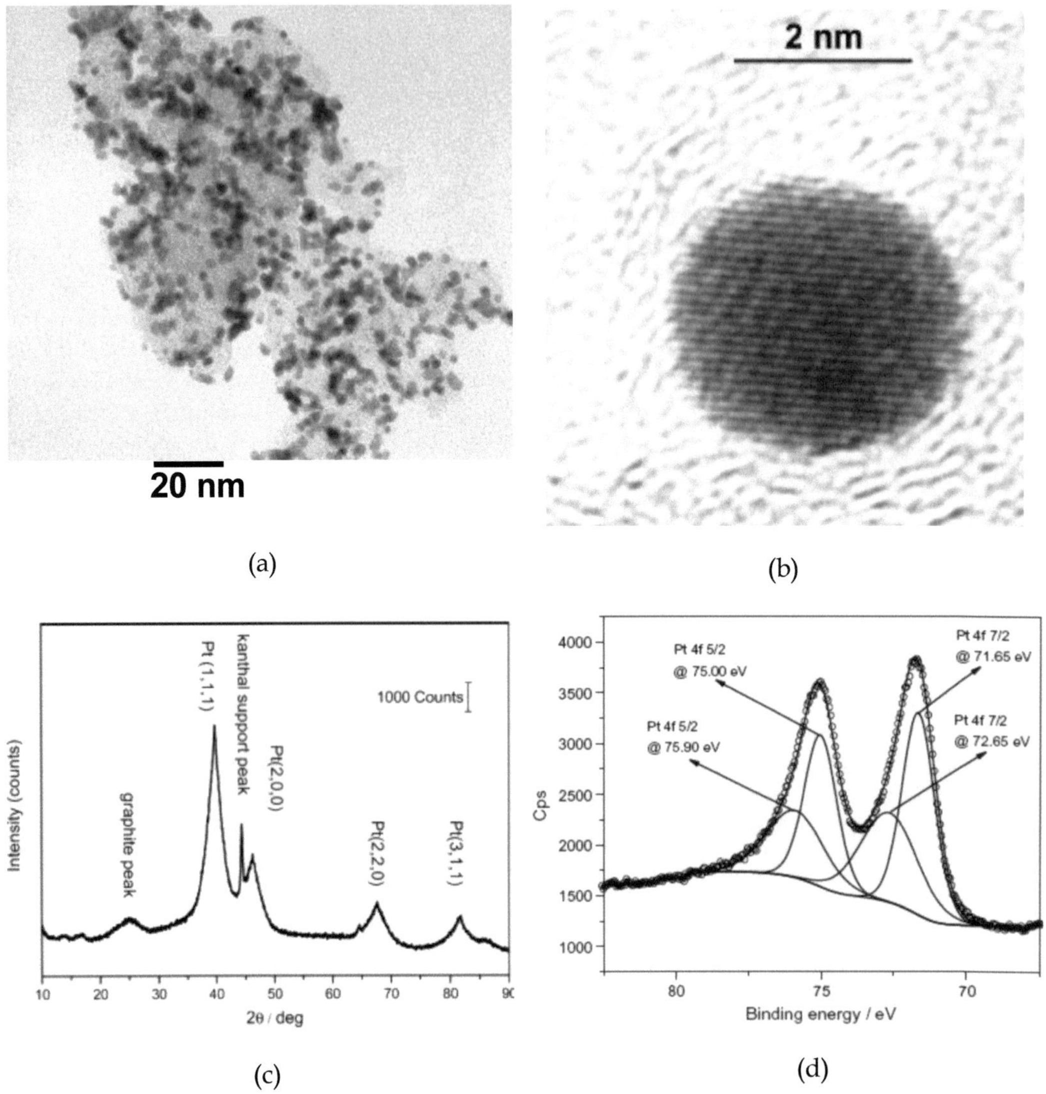

Fig. 2. **Charaterization of a Pt(40 wt.%)/C powder prepared by the Bönnemann method.** (a) TEM image, (b) HRTEM micrograph, (c) XRD pattern and (d) Pt 4f XPS core level spectrum recorded by XPS.

present case, the formation of water nano-droplets (where H_2PtCl_6 is dissolved) in the oil phase is the considered system, which will act as microreactor for the metal salt reduction to platinum nanoparticles. The reduction of metallic salt present in the microreactor is carried out by addition of a reducing agent ($NaBH_4$ for instance). Different chemical reactions were suggested in order to illustrate the reduction mechanism of H_2PtCl_6 by sodium borohydride; however, Ingelsten and coworker [Ingelsten et al., 2001] proposed the following one:

$$PtCl_6^{2-}(aq.) + BH_4^-(aq.) + 3H_2O \rightarrow Pt^0 + H_2BO_3^-(aq.) + 4H^+(aq.) + 6Cl^-(aq.) + 2H_2(g) \quad (5)$$

The synthesis protocol is represented on the scheme in Fig. 3. The desired amount of H_2PtCl_6, $6H_2O$ (from Alfa Aesar) is dissolved in the appropriate water volume in order to obtain a concentration of 0.2 mol L^{-1} in Pt^{4+}. The ternary system is realized by mixing the aqueous solution in presence of Brj®30 (water/Brij®30 molecular ratio = 4) with n-heptane; the reduction reaction is carried out by adding a large excess of solid $NaBH_4$ (n_{NaBH4}/n_{Pt} =15).

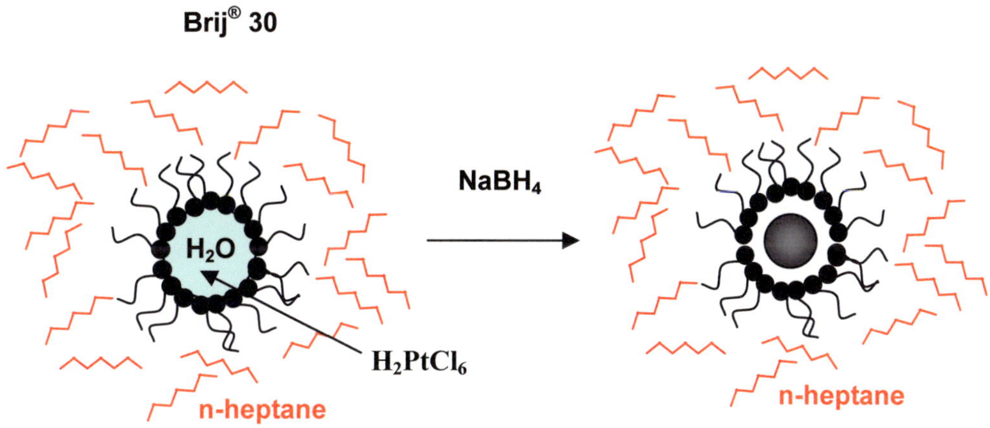

Fig. 3. **Synthesis by the "water in oil" microemulsion.** Scheme of the platinum nanoparticle synthesis via the "water in oil" microemulsion route.

The size of the microreactors was determined using a Laser DL135 granulometer from Cordouan Technologies; the hydrodynamic diameter was found to be ca. 16 nm; considering a size of 2 nm for a surfactant molecule, a microreactor with an internal diameter of 12 nm containing an aqueous solution of 0.2 mol L^{-1} H_2PtCl_6 should lead to platinum nanoparticles of ca. 1.5 nm.

After addition of the carbon powder (in a suitable amount to obtain a platinum loading of 40wt% assuming a total reduction of the platinum salt) and an abundant cleaning step with acetone, ethanol and water, the catalytic powder was characterized. TGA measurement indicated a platinum loading of ca. 37 wt%. TEM and XRD characterizations as shown in Fig. 4a and 4c, respectively, indicated that the mean platinum particle size was in the range of 3.0 nm, whereas the mean crystallite size was ca. 2.9 nm. It can be noted that TEM conical dark field clearly shows that the platinum particles obtained from "water in oil" microemulsion were composed by spherical crystallites (Fig. 4a). This observation is also confirmed by HRTEM of unsupported nanoparticles prepared by "water in oil" microemulsion on Fig. 4b.

In such synthesis route, the end of the growth process could be due to the depletion of metallic ions in microreactors. However, the mean particle size and crystallite size obtained from TEM and XRD measurements are significantly higher than that calculated from hydrodynamic diameter as determined by laser granulometry. Then, It seems that a second grain growth step is involved, which can be explained in terms of Brownian interparticles collision [Bradley, 1994] or Ostwald ripening [Peng et al., 1998].

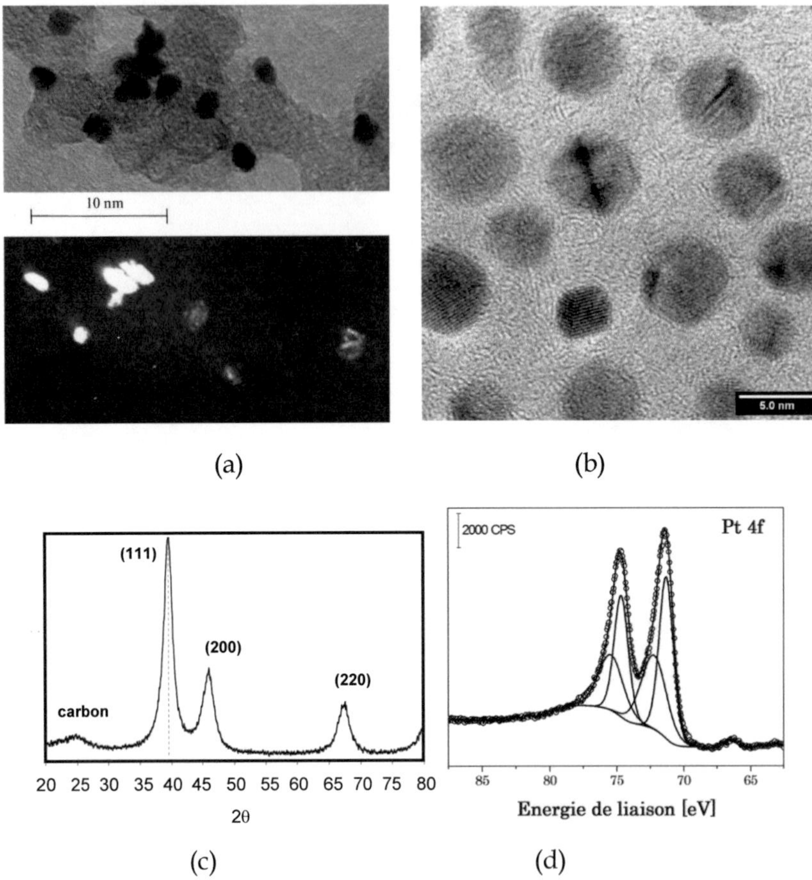

Fig. 4. **Characterization of a Pt(40 wt.%)/C powder prepared by the microemulsion route**. (a) TEM (top) and TEM conical dark field (down) images, (b) HRTEM micrograph, (c) XRD pattern and (d) Pt 4f XPS core level spectrum recorded.

Therefore, by correlating the TEM, HRTEM and XRD data, it can be proposed that isolated Pt nanoparticles prepared via the microemulsion method are *a priori* polycrystalline. XPS measurements carried out on a Pt/C powder (Fig. 4d) confirmed that platinum was also in that case present under 0 and +II oxidation states. From the particle size determined by TEM measurements, a real platinum surface area of ca. 55 m^2 g^{-1} could be calculated. It has also been estimated that 23% of the platinum entities were particle agglomerates with a size higher than 14 nm.

2.3 Synthesis of Pt/C by the polyol method

First described by Fievet et al. [Fievet et al., 1989], this synthesis method consists in the reduction of a metallic salt by using the reductive properties of polyol which also acts as solvent. Several polyol compounds were studied, particularly ethylene glycol, diethylene glycol and glycerol [Larcher et al., 2000]. In the present book chapter, we will focus on polyol synthesis methods using ethylene glycol as solvent and reducing agent.

Once the metallic salt is completely dissolved in an alkaline solution of ethylene glycol, intermediate phase of metal oxides or hydroxides are created. Then, the dehydration reaction of glycerol into acetaldehyde occurs, which allows the reduction reaction of metal oxides or hydroxides. The reaction mechanism is summarized in equations 6 to 8 [Larcher et al., 2000].

$$CH_2OH\text{-}CH_2OH \rightarrow CH_3\text{-}CHO + H_2 \quad \text{dehydration reaction} \quad (6)$$

$$CH_2OH\text{-}CH_2OH + 2OH^- \rightarrow CH_2O^-\text{-}CH_2O^- + H_2O \quad \text{acid-base reaction (pKa = 6.5)} \quad (7)$$

$$MO_2 + 4\ CH_3\text{-}CHO \rightarrow M + 2\ CH_3\text{-}CO\text{-}CO\text{-}CH_3 + 2\ H_2O \quad \text{reduction reaction} \quad (8)$$

This reduction reaction under relatively soft conditions favours low grain growth kinetics. The grain growth process is stopped due to the combination of two phenomena: adsorption of reaction by-products on the metal surface (particularly glycolates formed in alkaline medium according to equation 7) and depletion of metal salt to be reduced. Moreover, glycolate ligands, which also act as surfactant, do interact weakly and adsorbs preferentially on certain crystalline surface domains of platinum particles, so that nanoparticles with surface orientations (facetted) can be created.

Because ethylene glycol (EG) is a soft reducing agent, the reduction reaction of $H_2PtCl_6,6H_2O$ (1 g_{Pt} L^{-1}) in alkaline EG solution (pH = 11), in presence of Vulcan XC72 carbon powder (in appropriate amount for a nominal Pt loading of 40 wt%) needs to be activated either by temperature (at reflux, ca. 200°C for 2 hours) or by microwave irradiation [Lebegue et al., 2010].

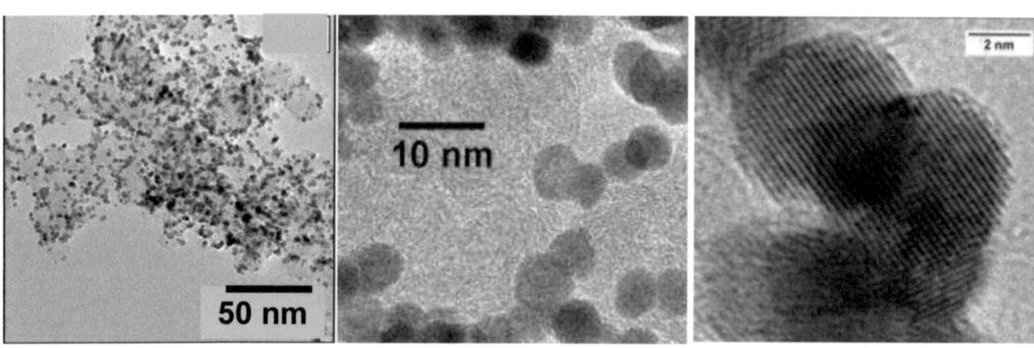

(a) (b) (c)

Fig. 5. Characterization of a Pt(40 wt.%)/C powder prepared by the polyol method at 200°C for 2 hours. (a) and (b) TEM images at different magnifications, and (c) HRTEM micrograph.

The TEM characterizations presented in Fig. 5 correspond to a Pt/C catalyst prepared by thermal treatment at reflux for 2 hours, leading to a platinum loading of ca. 36 wt%, as determined by TGA. The dispersion of platinum particle on the carbon support is relatively homogeneous, and the mean particle size was found to be ca. 4.0 nm. A majority of particles observed with higher magnification in Fig. 5b are clearly not round shaped, and facetted crystallites are formed. Concerning the crystalline structure of the Pt/C catalyst synthesized by the polyol method, XRD patterns have evidenced the classical fcc structure of platinum, from which the determined cell parameter was found to be close to that of bulk platinum (i.e. 0.392 nm) for a mean crystallite size of ca. 4 nm. The isolated nanoparticles would then correspond to single nanocrystals dispersed on the carbon substrate. This assumption is confirmed by the HRTEM image presented in Fig. 5c, where isolated particles appear really as single crystals.

2.4 Synthesis of Pt/C by the "instant method"

In order to obtain a high dispersion and prevent any undesirable nanoparticle agglomeration, a surfactant is commonly used in the chemical synthesis methods of platinum catalysts. This molecule controls the seed formation as well as its growth through the reduction of the metal salt. But its stabilizing effect depends on the pH and the temperature of the solution. As can be seen in the methods cited above, various molecules are proposed as surfactants for catalyst preparation. Reetz et al. [Reetz and Koch, 1998; Reetz and Koch, 1999] have shown that the use of a surfactant permits to produce metal oxides as intermediate species before their reduction to catalysts. Indeed, alkaline solutions such as NaOH, Na_2CO_3, or Li_2CO_3 are added to the platinum salts ($PtCl_4$ or H_2PtCl_6) in presence of water-soluble betaine stabilizer (3-(*N,N*-dimethyldodecylammonio)propane sulfonate $[R(CH_3)_2N^+(CH_3)_3SO_3^-]$). After 5 to 6 hours at 60°C, stable PtO_2 particles were obtained. The pH of the solution is one of the key parameter. Lithium carbonate permits to keep the pH close to 9, value at which chlorine are replaced by the hydroxide leading to PtO_2. The role of the surfactant in colloidal synthesis modes of Pt/C catalysts for fuel cells application is well demonstrated in literature. However, the synthesis methods of Pt/C catalysts described above require a heat treatment for removing this organic molecule. In order to avoid any use of surfactant and/or heat treatment, Reetz et al. [Reetz and Lopez, 2002, Reetz et al. 2004] developed a simple way called "instant method" for preparing PtO_x colloids free of surfactant. A catalysts support is used in place of the stabilizer. Therefore, the PtO_x colloids are immobilized on the support according to:

$$H_2PtCl_6 + C \xrightarrow{Li_2CO_3/H_2O} PtO_x/C \tag{9}$$

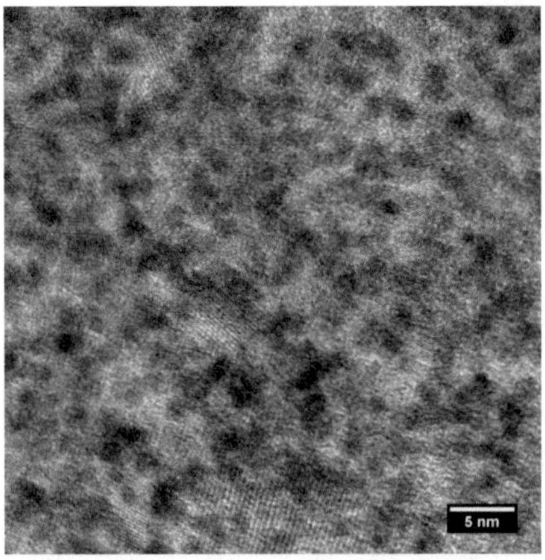

Fig. 6. **PtO_x particles from "Instant method".** TEM image of PtO_x colloids particles on carbon Vulcan XC72 (metal loading is 40%).

The pH of the solution is kept at 9 by the presence of lithium carbonate. The PtO_x colloid particles obtained in our group (Fig. 6) are small (1-2 nm). The reduction of the PtO_x colloids to Pt/C can be made by various reducing agents: hydrogen, as suggested by Reetz, sodium citrate or sodium borohydrate in a modified method. Fig. 7 shows high dispersion of Pt particles on carbon obtained by reducing PtOx colloids with an ice-cooled solution of sodium borohydrate.

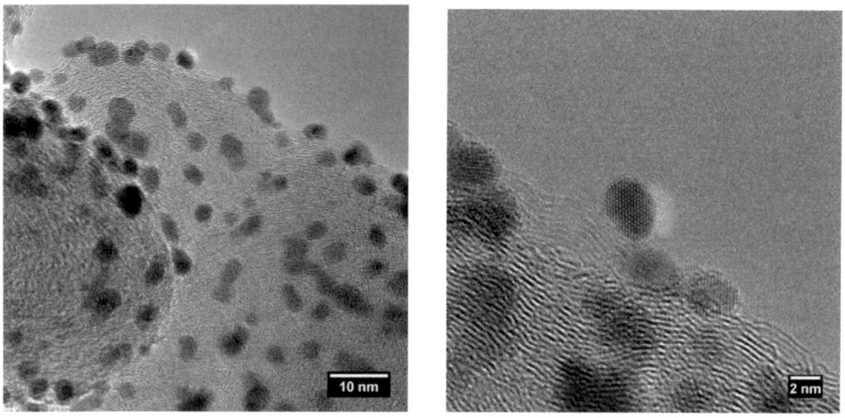

(a) (b)

Fig. 7. Characterization of a Pt(40wt%)/C catalysts obtained by "instant method" at 40°C.
(a) TEM image and (b) HRTEM image.

With the experimental conditions fixed above, the TEM characterizations show a mean Pt particle size about 3nm. The Thermogravimetry analysis confirmed the nominal Pt loading (38%). It is also observed that the particles size depends on the temperature of the colloidal solution and the metal loading. The higher dispersion and smaller crystallites were obtained with 40 wt% of Pt loading at 60°C.

3. Electrochemical characterization of the Pt/C catalysts

The working electrode was prepared by deposition of a catalytic ink on a glassy carbon disc (0.126 cm^2 geometric surface area) according to a method proposed by Gloaguen et al. [Gloaguen et al., 1994]. Catalytic powder was added to a mixture of 25 wt% (based on the powder content) Nafion® solution (5 wt.% from Aldrich) and ultra-pure water (MilliQ, Millipore, 18.2 MΩ cm). After ultrasonic homogenization of the Pt/C-Nafion® ink, a given volume was deposited from a syringe onto the fresh polished glassy carbon substrate yielding in the case of a Pt(40 wt.%)/C to a 60 μg cm^{-2} catalytic powder loading (i.e. 24 μg_{Pt} cm^{-2}). The solvent was then evaporated in a stream of ultra-pure nitrogen at room temperature. By this way, a catalytic layer was obtained with a thickness lower than 1μm.

The electrochemical active surface area (EASA) of Pt/C catalysts is determined from cyclic voltammograms (CV) by integrating the charge in the hydrogen desorption region corrected from the double layer capacity contribution [Coutanceau et al., 2000; Grolleau et al., 2008]. For each catalyst, the average active surface area is determined by recording the CV on four different ink depositions. Measurements are carried out in a N_2-saturated electrolyte at a scan rate of 50 mV s^{-1} after 10 voltammetric cycles performed between 0.05 V and 1.25 V vs. RHE in order to clean the platinum surface from remaining organic molecules coming from the synthesis and to obtain quasi-constant voltammograms [Maillard et al., 2007]. Fig. 8 displays a typical CV recorded on a commercially available e-Tek Pt(40wt%)/C catalyst in a 0.5 M H_2SO_4 electrolyte, as an example.

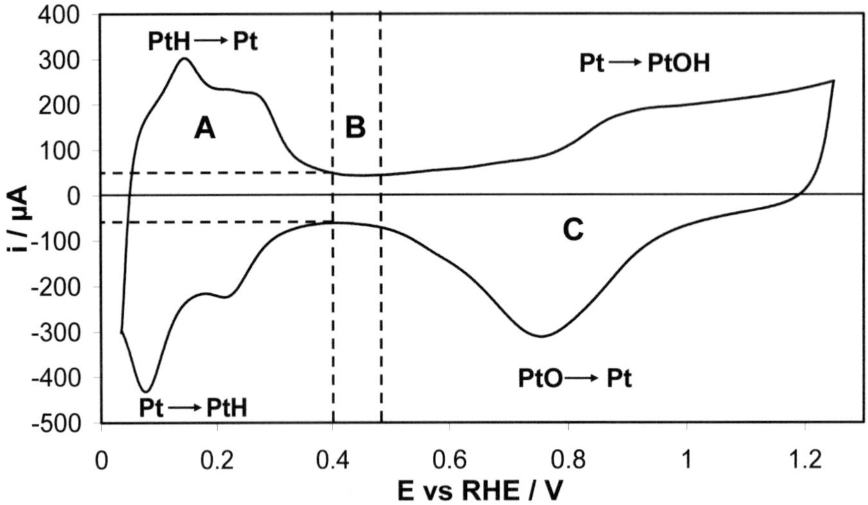

Fig. 8. **Electrchemical characterization of a Pt(40wt%)/XC72 e-tek® catalyst.** Voltammogram recorded in N_2-saturated 0.5 M H_2SO_4 electrolyte (v=20 mV.s^{-1}; T = 25 °C,).

The potential region (A) corresponds to the hydrogen adsorption (i < 0) and desorption (i > 0), the potential region (B) corresponds to the double layer capacity and the potential region (C) corresponds to the platinum surface oxide formation (i > 0) and reduction (i < 0). In the hydrogen region (A), by analogy with studies performed on low index single crystals [Solla-Gullon et al., 2006; Solla-Gullon et al., 2008; Attard et al., 2001], the oxidation peaks located at ca. 0.12 V and ca. 0.26 V can be assigned to sites of (110) and (100) symmetries, respectively. Changes in the height and the width of these peaks are representative of different surface states of platinum nanoparticles. The electrochemical active surface area is usually estimated from the determination of the charge associated to hydrogen desorption in the potential range from 0.05 V to 0.4 V vs RHE [Biegler et al., 1971], after correction of the double layer capacitive contribution. It is then assumed that a hydrogen monolayer is formed at the platinum surface and that the adsorption stoechiometry is one hydrogen atom per platinum surface atom. The charge involved for the hydrogen desorption reaction is calculated using the following equation:

$$Q_{H_{des}} = \int i(t)\,dt = \frac{1}{v}\int i(E)\,dE \qquad (10)$$

where $Q_{H_{des}}$ is the charge corresponding to the integrated current i(t) for a duration t, which corresponds to the integrated current in the potential range considered, i(E), divided by the potential scan rate, v.

From measurement on low index platinum single crystals, the charge involved during the hydrogen desorption could be determined [Markovic et al., 1997]:

- a Pt(111) surface exhibits 1.5 10^{15} atom cm^{-2}, leading to an associated theoretical charge of 240 $\mu C\ cm^{-2}$,
- a Pt(100) surface exhibits 1.3 10^{15} atom cm^{-2}, leading to an associated theoretical charge of 225 $\mu C\ cm^{-2}$,
- a Pt(110) surface exhibits 4.6 10^{14} atom cm^{-2}, leading to an associated theoretical charge of 147 $\mu C\ cm^{-2}$,

By weighting the different surface domains, a global charge of 210 $\mu C\ cm^{-2}$ was found for the adsorption of a hydrogen monolayer on a polycrystalline platinum surface ($Q_{monolayer}$). This value is now generally accepted. From this value, the accessible platinum surface per mass unit, named electrochemically active surface area (EASA), can be calculated by using equation (11):

$$EASA = \frac{Q_{H_{des}}}{m_{Pt} \times Q_{monolayer}} = \frac{\frac{1}{v}\int i(E)\,dE}{m_{Pt} \times Q_{monolayer}} \qquad (11)$$

where Q_{Hdes} is the charge determined in the hydrogen desorption region of the voltammogram recorded on the Pt/C catalyst in supporting electrolyte, $Q_{monolayer}$ is the charge related to the adsorption or desorption of a hydrogen monolayer on a polycrystalline Pt surface ($Q_{monolayer}$ = 210 $\mu C\ cm^{-2}$), i(E) is the current (in μA) recorded in the hydrogen desorption region, E is the electrode potential (in V vs RHE), v is the linear potential variation (in V s^{-1}) and m_{Pt} is the platinum loading on the electrode (in g). In the case of the catalyst used to record the voltammogram in Fig. 8, the measurement performed on five different electrodes prepared successively led to an EASA of 42 $m^2\ g_{Pt}^{-1}$, with a standard deviation of 3.7 $m^2\ g_{Pt}^{-1}$.

As it was shown in the previous part of this chapter, each synthesis method led to platinum particles having different structures and catalytic powders with different morphologies. Those changes will have an important effect on the structure of exhibited platinum surface and on the electrochemically active surface area, as it is shown in Fig. 9 where different Pt/C catalysts are compared with a commercial one (e-Tek catalyst).

The shape of the oxidation current peak in the hydrogen desorption region (i > 0) between 0.05 and 0.4 V vs RHE gives information on the catalyst surface. Considering for example the voltammogram of a Pt/C catalyst prepared by the polyol method, two well defined oxidation peaks are clearly visible, the first one centered at ca. 0.12 V and the second one at ca. 0.26 V were assigned to surface defects and to short range order (100) surface domains, respectively [Markovic et al., 1995; Attard et al., 2001]. Both those peaks are lower in intensity on the voltammogram recorded with a Pt/C catalyst prepared by the Bönnemann synthesis method. On the other hand, a third peak at ca. 0.2 V arises in this latter case, which was assigned to defective (110) planes [Gomez and Clavilier, 1993]. Commercial and "water in oil" microemulsion Pt/C catalysts exhibit the same general shape of the hydrogen region in the voltammograms, with a well-marked oxidation peak at ca. 0.12 V and a less-marked one at ca. 0.26 V. In the case of a Pt/C catalyst prepared via the "instant method" the three peaks are clearly visible as it can be seen in Fig. 10. This comparison indicates that the synthesis method of platinum nanoparticles, through the related Pt salt reduction kinetics and mode of growth stopping, has a great influence on the surface structure and the morphology of the catalyst, and can induce more or less surface defects, surface domains

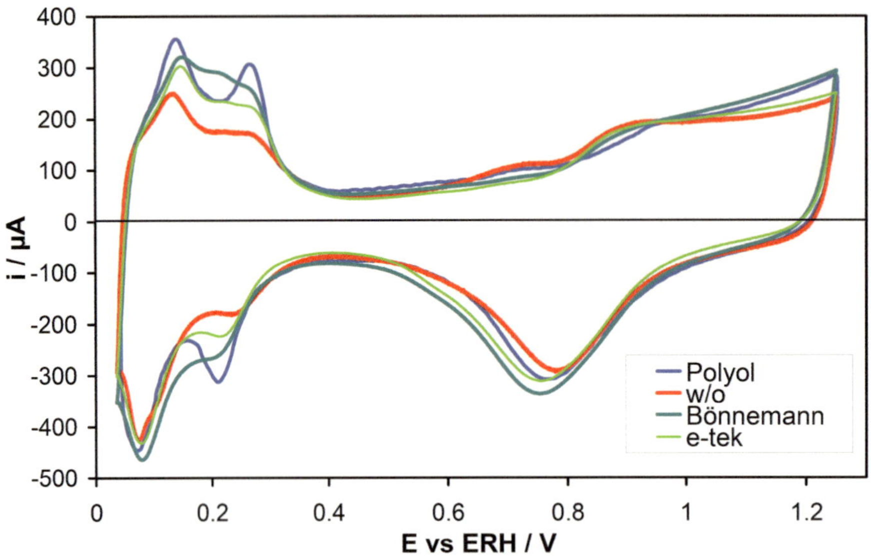

Fig. 9. **Electrochemical characterization of different Pt(40wt%)/XC72 catalysts.** Voltammograms recorded in N_2-saturated 0.5 M H_2SO_4 electrolyte (v=20 mV.s^{-1}; T = 25°C,).

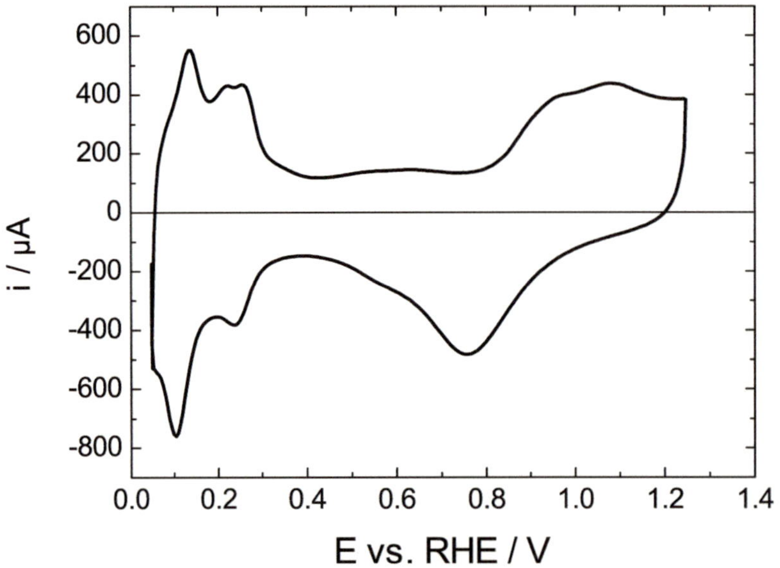

Fig. 10. Electrochemical characterization of a Pt(40wt%)/XC72 catalyst prepared via the "instant method". Voltammogram recorded in N_2-saturated 0.5 M H_2SO_4 electrolyte (v=20 mV.s^{-1}; T = 25°C,).

and surface heterogeneities. It is also likely that such changes in morphology and structure could lead to changes in the catalytic activity of the different Pt/C catalysts.

The electrochemical active surface area (EASA) of the different catalyst was also determined using equation (11) and was compared to the active surface area as determined from data obtained by TEM and XRD measurements (Table 1 summarizes the results). Assuming spherical particles of similar radius, an "electrochemical" mean particle size can be calculated from the following equations:

- the geometric surface area of a spherical particle is:

$$S_{geom} = 4\pi r^2 \tag{12}$$

- the volume of a spherical particle is:

$$V_p = \frac{4}{3}\pi r^3 \tag{13}$$

where r is the particle radius.

- the electrochemical active surface area of the particle is:

$$EASA = \frac{S_{geom}}{V_p \times \rho} = \frac{3}{\rho \times r} = \frac{6}{\rho \times d} \tag{14}$$

where ρ is the platinum density (21 450 kg m^{-3}).

- the electrochemical particle size d can be expressed as:

$$d = \frac{6}{\rho \times EASA} \tag{15}$$

	Bönnemann	Microemulsion w/o	Polyol	Instant Method	Commercial e-Tek
Pt loading (%)	37.5	37	36	38	40
Particle shape	Spherical	Spherical	spherical + facetted	Spherical	Spherical
d_{XRD} / nm	2.5	2.9	4.0	-	3.2
d_{TEM} / nm	2.5	3.0	4.0	2.8	3.5
$S_{real,TEM}$ / m^2 g^{-1}	110	91.7	68.8	98.3	78.6
EASA / m^2 g^{-1}	46	34	42	61	42
d / nm	6.0	7.6	6.5	4.5	6.5

Table 1. Structural and electrochemical characterization of different Pt(40 wt%)/C catalysts; EASA and d were determined from the voltammogram presented in Fig. 8, 9 and 10.

The values obtained from equation 15 are higher than that obtained by TEM or XRD, which seems to indicate that the whole surface of platinum particles is not used for electrochemical reactions. Considering the mean particle sizes determined by TEM for calculating the corresponding real surface areas ($S_{real,TEM}$), it appears that the electrochemical active surface area represents only between 37% (for the "water in oil" microemulsion" Pt/C catalyst) and 62% (for the Instant Method Pt/C catalyst) of the real surface areas as determined by TEM. This can be due to the non-utilization of the contact surface between platinum and carbon in the electrochemical reaction, to the aggregation of crystallites and to the presence of Nafion in the catalytic layer which can block some platinum adsorption sites. This discrepancy can also be due to the adsorption of species coming from the synthesis procedure, such as chloride (from used platinum salts), carbonate (for the instant method), traces of remaining surfactant, adsorbed ethylene glycol, etc. Considering the "water in oil" microemulsion catalyst, thermogravimetric analyses under air flow have shown that about 2 wt% Brij®30 remained on the catalyst surface, which could be desorbed from ca. 150°C (Fig. 11). The presence of traces of surfactant on the catalytic surface before calcination step, even with so low amount, can explain the very high difference between the electrochemical active surface area (34 $m^2\ g^{-1}$) and that calculated from d_{TEM} (ca. 92 $m^2\ g^{-1}$). The proportion of agglomerates with sizes higher than 14 nm (23 %) did not likely explain alone this very high difference. So, it can be proposed that a thermal treatment at ca. 200°C could clean the platinum particle surface from adsorbed surfactant. However, such treatment is expected to induce some changes in the platinum particle structure, and further on their electrocatalytic activity; nethertheless, it has been shown that the thermal treatment under air at 250°C of "water in oil" microemulsion Pt/C catalyst has no significant effect on the electrochemical active surface area (35 $m^2\ g^{-1}$).

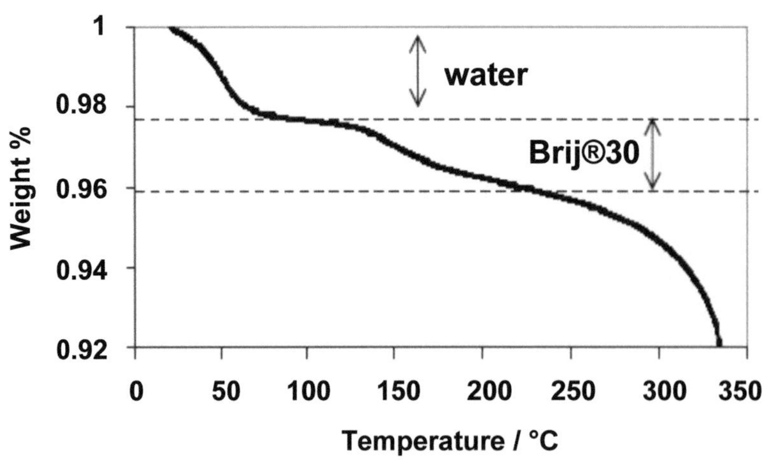

Fig. 11. TGA measurement of a Pt(40 wt%)/C catalyst prepared by the "water in oil" microemulsion method, recorded under air flow at 10°C min^{-1}.

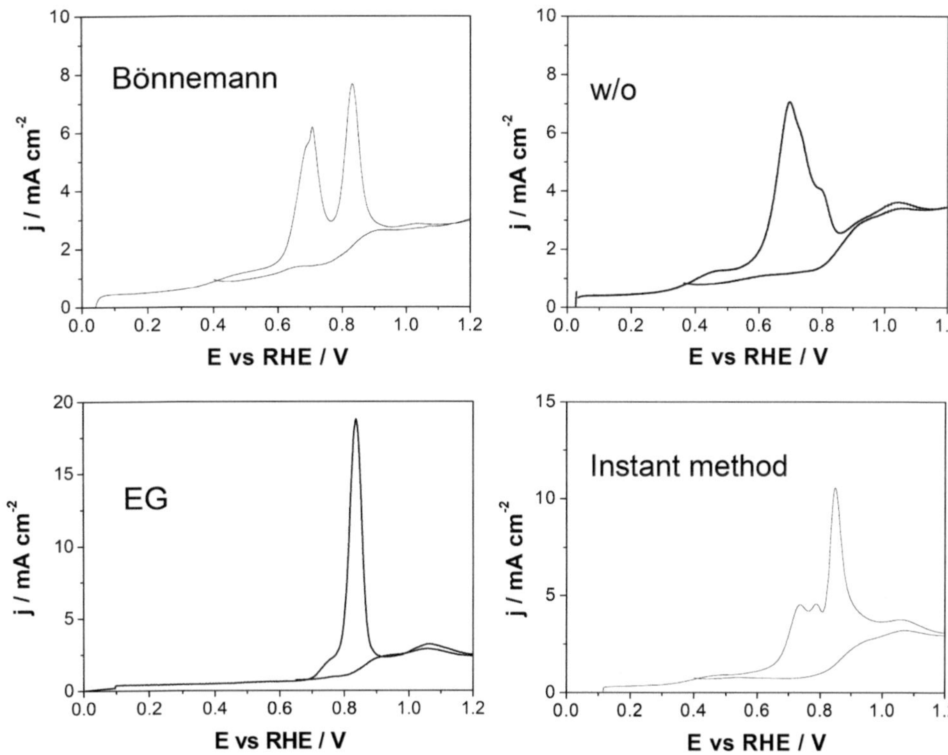

Fig. 12. **CO stripping measurements.** Voltammograms recorded on different Pt(40 wt %)/C in a deaerated H_2SO_4 0.5 M electrolyte at 20 mV s^{-1} (T = 293 K).

Interactions between platinum surfaces and carbon monoxide (CO) have been extensively studied in the few past decades [Iwasita, 2003]. CO is considered as a poisoning species for platinum-based anode catalysts. The strong affinity of CO for platinum and its sensitivity to surface structure originate its use in surface science as probe molecule for studying surface structure [Villegas and Weaver, 1994; Yoshimi et al., 1996] and eventually characterizing the electrocatalytic activity of platinum nanoparticles. On platinum nanoparticles, the most recognized CO electrooxidation mechanism is described by a Langmuir-Hinshelwood reaction mechanism, involving a water dissociative adsorption step:

$$H_2O \leftrightarrows OH_{ads} + H^+ + e^- \tag{16}$$

$$CO_{ads} + OH_{ads} \rightarrow CO_2 + H^+ + e^- \tag{17}$$

One of the main methods used for studying CO electrooxidation on platinum surface is the so-called CO-stripping voltammetry. Then, in order to get information on structure and morphology of nanoparticles, the complete surface coverage of platinum catalysts with CO is performed at 0.1 V vs. RHE for 5 min, then CO is removed from the bulk electrolyte by N_2 bubbling for 15 min maintaining the electrode under potential control at 0.1 V, at last CO stripping measurements are recorded by linearly varying the potential at 20 mV s^{-1} from 0 to 1.25 V vs RHE. Fig. 12 gives the recorded CO stripping voltammograms on the different synthesized catalysts. Clearly, the shape of the CO oxidation current feature is completely different according to the catalyst. The Bönnemann, w/o microemulsion and instant method catalysts, which are round shaped catalyst, lead to several oxidation peaks or shoulders, with a more or less important pre-peak from 0.4 to 0.6 V vs. RHE, whereas the polyol catalyst leads to a well defined single oxidation peak (and a small shoulder at lower potentials) and no pre-peak. The electrooxidation of carbon monoxide is often characterized by the existence of multiple oxidation peaks, which was explained either by a platinum particle size effect [Friedrich et al., 2000; Maillard et al., 2004], or by the presence of agglomerates [Maillard et al., 2005; Lopez-Cudero et al., 2010], or by the presence of grain boundaries [Plyasova et al., 2006] or again by surface crystallographic domain structures [Solla-Gullon et al., 2006; Solla-Gullon et al., 2008; Brimaud et al., 2008]. However, in the case of the catalysts presented here, the absence of an oxidation peak at ca. 0.7 V in the CO stripping voltammogram of the polyol catalyst could be interpreted by the absence of particle agglomeration according to other authors [Maillard et al., 2005; Lopez-Cudero et al., 2010], although TEM image in Fig. 5a clearly shows the presence of aggregates. Size effect could also hardly be proposed for the existence of the oxidation peak at ca. 0.85 V which was attributed by certain authors to CO oxidation on smaller particles [Friedrich et al., 2000; Maillard et al., 2004], as means particles size are in the same range for Bönnemann, w/o microemulsion and instant method catalysts. These experiments evidenced that the history of the Pt/C catalyst has a great importance on the electrochemical behaviour towards pre-adsorbed CO layer oxidation reaction, which is a well know surface structure sensitive reaction. However, based on current data and knowledge concerning this reaction, it is yet very difficult and too early to rely the electrochemical behavior to a given structure or morphology of a Pt/C catalyst.

4. Electrocatalytic activity towards the oxygen reduction reaction in acid medium

The oxygen reduction reaction is the limiting process in proton exchange fuel cells [Ralph and Hogarth, 2002]; this very irreversible reaction involves indeed the exchange of several electrons, and as a consequence, occurs with a high overvoltage. For this reason it has been extensively studied for the last few decades.

The reduction of dioxygen dissolved at saturation in 0.5 M H_2SO_4 supporting electrolyte was investigated at the Pt(40 wt%)/C deposited on a glassy carbon rotating disc electrode for several rotation rates Ω (from 0 to 3600 round per minute). The current vs potential curves were recorded during a slow voltammetric sweep (sweep rate of 2 mV s^{-1}) between 1.1 and 0.2 V vs RHE, so that the recorded curves are quasi-stationary. As a typical example, the current density (referred to the geometric surface area A_g) vs potential curves are shown in Fig. 13, for a platinum catalyst prepared by the polyol method.

A more general analysis of data was carried out by separating the contribution of the diffusion of molecular dioxygen from that of the surface processes involved in the oxygen reduction reaction using the $1/j$ vs $1/\Omega^{-1/2}$ Koutecky-Levich plots. For that purpose, a detailed reaction mechanism, involving the formation of H_2O_2, is written as follows [Tarasevich et al. 1983]:

$O_{2,sol} \rightarrow O_{2,surf}$ diffusion in the bulk electrolyte (diffusion coefficient D) (18)

$O_{2,surf} \rightarrow O_{2,cata}$ diffusion in the catalytic film (limiting current density j_l^{film}) (19)

$O_{2,cata} \rightarrow O_{2,ads}$ adsorption process (limiting current density j_l^{ads}) (20)

$O_{2,ads} + 2e^- \rightarrow [O_{2,ads}]^{2-}$ electron transfer rds (exchange current density j_0, Tafel slope b) (21)

$$[O_{2,ads}]^{2-} + H^+ \rightarrow [HO_{2,ads}]^- \quad (22)$$

$$[HO_{2,ads}]^- + H^+ \rightarrow [H_2O_2]_{ads} \rightarrow H_2O_2 \quad (23a)$$

$$\text{or } [HO_{2,ads}]^- + H^+ + e^- \rightarrow H_2O_2 \quad (23b)$$

followed by $H_2O_2 + 2\,H^+ + 2\,e^- \rightarrow 2\,H_2O$

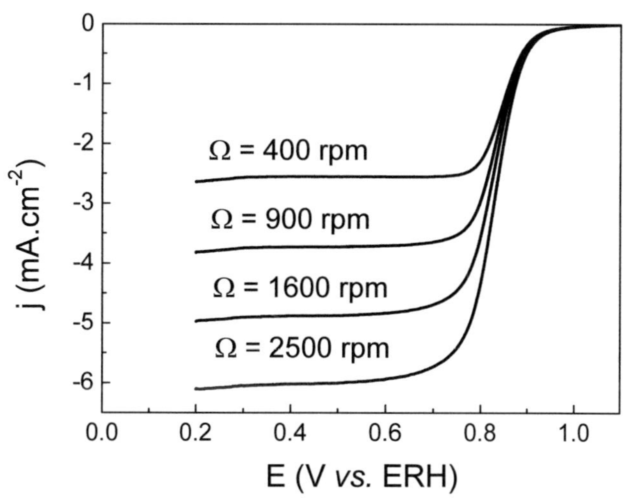

Fig. 13. **ORR study at a Pt/C catalyst prepared by a polyol method.** j(E) polarization curves at different electrode rotation rates Ω in revolution per minute(rpm) recorded in a O_2-saturated 0.5 M H_2SO_4 electrolyte. (T=20°C, scan rate = $2mVs^{-1}$).

Assuming that the reaction is first order with respect to oxygen and that the first electron transfer is the rate determining step, Koutecky-Levich plots can be drawn from these polarization curves, as it is shown in Fig. 14, using Equations (24) and (25) [Schmidt et al., 1999; Coutanceau et al., 2000]:

$$\frac{1}{j} = \frac{1}{j_0(\theta/\theta_e)e^{\eta/b}} + \frac{1}{j_l^{ads}} + \frac{1}{j_l^{film}} + \frac{1}{j_l^{diff}} \quad (24)$$

where j_l^{film} and j_l^{ads} correspond to the diffusion limiting current density in the catalytic film and to the adsorption limiting current density of dioxygen, respectively; $\eta = E - E_{eq}$ is the overpotential; b is the Tafel slope; j_0 is the exchange current density; θ and θ_e are the coverage of platinum surface by species coming from oxygen adsorption at potential E and at the equilibrium potential E_{eq} (1.185V vs RHE), respectively, and j_l^{diff} is the diffusion limiting current density which can be calculated from the Levich law:

$$j_l^{diff} = 0.2nF\left(D_{O_2}\right)^{2/3} \nu^{-1/6} C_{O_2} \Omega^{-1/2} \quad (25)$$

with n the number of electron exchanged per oxygen molecule, F the Faraday constant (96,500 C mol^{-1}), D_{O2} the coefficient diffusion of oxygen molecular in 0.5 M H_2SO_4 (2.1×10^{-5} cm^2 s^{-1}), ν the kinematic viscosity of the electrolyte (1.07×10^{-2} cm^2 s^{-1}), C_{O2} the bulk concentration of oxygen in a saturated electrolytic solution (1.03×10^{-3} mol L^{-1}) [Jakobs et al., 1985]. The coefficient 0.2 is used when the working electrode rotation rate Ω is expressed in revolutions per minute (rpm) [Bard et al., 2001; Zagal et al., 1980].

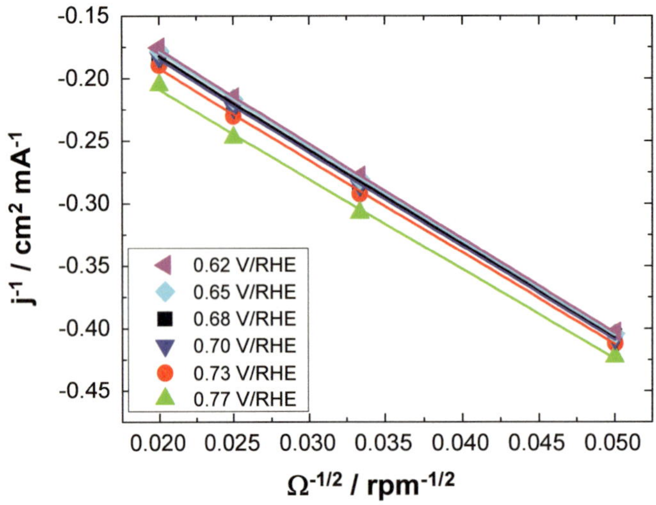

Fig. 14. **ORR study at a Pt/C catalyst prepared by a polyol method.** Koutecky-Levich plots at different potentials determined from j(E) curves presented in Fig. 12.

The film diffusion limiting current density and the adsorption limiting current density are both independent on disk electrode rotation rates and applied potential (E), thus it is impossible to dissociate them. Then, the current density j can be expressed as follows:

$$\frac{1}{j} = \frac{1}{j_0(\theta/\theta_e)e^{\eta/b}} + \frac{1}{j_l} + \frac{1}{j_l^{diff}} \text{ with } \frac{1}{j_l} = \frac{1}{j_l^{ads}} + \frac{1}{j_l^{film}} \quad (26)$$

Assuming that in the considered potential range the adsorption process of oxygen is more rapid than the charge transfer step, i. e. $\theta \approx \theta_e$ for the whole electrode potential range [Coutanceau et al., 2001]. The kinetic current density j_k can be expressed as:

$$\frac{1}{j_k} = \frac{1}{j_0 e^{\eta/b}} + \frac{1}{j_l} \quad (27)$$

Therefore, the slopes of the Koutecky-Levich straight lines will lead to the determination of the number of exchanged moles of electron per mole of oxygen, whereas their intercept with the Y-axis will lead to the determination of the kinetic current values as a function of potential. Equation 25 indicates that when the overpotential η tends to ∞ then $1/j_k$ tends to $1/j_l$, so that one can obtain the limiting current density j_l by extrapolating equation 25 at high η (Fig. 15a). This allowed to transform equation 25 as follows and to access the Tafel slope b (Fig. 15b) as well as the exchange current density j_0:

$$\eta = E - E_{eq} = -b\left(\ln\frac{j_l}{j_0} + \ln\left|\frac{j_k}{j_l - j_k}\right|\right) \quad (28)$$

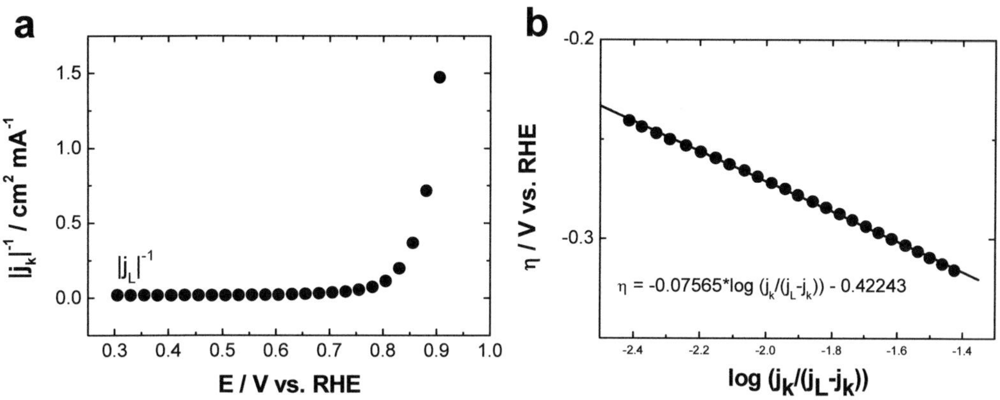

Fig. 15. **ORR study at a Pt/C catalyst prepared by a polyol method.** Plot of 1/jk as a function of the electrode potential and (d) related Tafel plot determined from Koutecky-Levich curves presented in Fig. 13.

	Bönnemann	Microemulsion w/o	Microemulsion w/o heat treated @ 250°C	Polyol	Commercial e-tek
Pt loading (%)	37.5	37	-	36	40
Particle shape	Spherical	Spherical	Spherical	Spherical +facetted	Spherical
EASA / $m^2\ g^{-1}$	46	34	35	42	42
j_k @ 0.95 V / mA cm^{-2}	0.82	0.17	0.81	0.74	0.014
j_{kw} @ 0.95 V / μA cm_{Pt}^{-2}	22	6	22	22	0.4

Table 2. **ORR kinetics data.** Values determined from RDE measurements recorded on different Pt(40wt%)/C catalysts in H_2SO_4 electrolyte according to the mathematical treatment of the Koutecky-Levich equation presented above.

Equation 26 not only allows determining the exchange current density j_0 but also the kinetic current densities j_k at a given potential for the different catalysts. An electrode potential of 0.95V vs. RHE for the determination of the diffusion corrected kinetic current density j_k is close to that of a cathode working in a PEMFC. So, the kinetic current at a potential convenient with that of a working electrode and the limiting current in the catalytic film can be determined; both parameters allow for well characterizing the catalytic film and selecting the best catalyst toward oxygen reduction reaction. However, in order to compare the catalytic activity, the kinetic current density at 0.95 V vs RHE may also be assessed by the active surface area expressed in $\mu A\ cm_{Pt}^{-2}$. Table 3 gives the kinetics parameters obtained for the different platinum catalysts prepared via the different methods presented previously.

Although platinum materials synthesized by the Bönnemann method, the "water in oil" microemulsion way and the polyol route, display a fcc structure with a cell parameter very close to that of bulk platinum (0.392 nm), these results indicate an important dependence of the catalytic activity on the catalyst synthesis method. However, due to the invariance of the cell parameter for all prepared catalyst, it is unlikely that the change in catalytic activity of the platinum material is related to geometric effect (change in Pt-Pt interatomic distance).

The catalyst prepared by the "water in oil" microemulsion leads to the lower activity. TEM images (Fig. 4) clearly show that platinum species are polycrystalline in the case of the w/o microemulsion Pt/C catalyst. A thermal treatment under air at 250°C of a w/o microemulsion Pt(40wt%)/C material allowed obtaining a catalyst leading to a kinetic current density of 0.81 mA cm^{-2} (corresponding to a j_{kw} of 22 $\mu A\ cm^{-2}$), whereas no significant change was observed by TEM and XRD measurements. Such improvement could be due to the removal of remaining surfactant as it was proposed in Fig. 11. However, no significant increase of the active surface area was recorded (35 $m^2\ g^{-1}$ after the thermal treatment against 34 $m^2\ g^{-1}$ for the fresh catalyst). Then, the low activity of the w/o microemulsion Pt/C catalyst can be attributed to a higher surface defect concentrations, i. e. higher density of low coordinated platinum atoms, which is not in favour of the activation of the oxygen reduction reaction [Kinoshita, 1990]. Indeed, conversely to the catalysts prepared by the Polyol route and the Bönnemann route, the growth stopping in the microemulsion synthesis is mainly due to the platinum cation depletion rather than to blocking of the platinum surface by surfactant adsorption. Moreover, no crystalline reconstruction occurs due to the absence of thermal treatment. Indeed it is known that a thermal treatment could improve the catalytic performance of PEM fuel cell catalysts for oxygen reduction reaction [Bezerra et al., 2007]. The residual constrains decrease with the thermal treatment, which leads to non negligible changes in the electronic effect.

The catalyst prepared by the polyol method display a high activity (j_k = 0.74 mA cm^{-2} and j_{kw} = 22 $\mu A\ cm_{Pt}^{-2}$), almost reaching that of the catalyst prepared by the Bönemann method (j_k = 0.82 mA cm^{-2} and j_{kw} = 22 $\mu A\ cm_{Pt}^{-2}$), although a higher mean particle size (and mean crystallite size) than that determined for the Bönnemann Pt/C, with ca. 4.0 nm against ca. 2.5 nm. This high activity of the Polyol Pt/C catalyst may be related to a size effect, a mean particle size of ca. 4.0 nm could be more favourable for the oxygen reduction reaction

[Kinoshita, 1990]. It could also be related to the presence of facets on the platinum surface, which could also favour the ORR. Indeed, facets display less low-coordinated surface atoms than edges or corners, which could be blocked by strongly adsorbed oxygen containing species difficult to reduce.

5. Conclusion

This book chapter intended to point out the importance of the Pt/C history on the electrocatalytic activity. For this purpose, different colloidal methods have been developed for the synthesis of platinum-based catalysts for Proton Exchange Membrane Fuel Cell (PEMFC) applications. Such methods are expected to produce metallic nano-sized particles with a narrow size distribution. However, the germination process, the grain growth and the metallic particles stabilization steps leading to Pt-surfactant stabilized colloids were different according to the synthesis route. For example, synthesis performed in organic or aqueous media could sometimes require thermal treatment (Bönnemann method) or not (Instant method). The end of the grain growth mechanism could occur either from depletion of metallic salts in surfactant stabilized nanoreactors ("water in oil" microemulsion method) or from the blocking of the platinum particle surface by adsorption of a surfactant (Bönnemann and polyol methods). All these mechanisms led to reach different catalyst morphologies and structures. The different methods of metal nanocatalyst synthesis via colloidal routes were first described, particularly Bönnemann method, "water in oil" microemulsion, polyol methods with thermal or microwave activation and instant method, in which different mechanisms are involved for the metal nanoparticle formation. Because the reactions occurring in Fuel Cells are known to be sensitive to the catalyst structure, the present book chapter has then focused on the relationship between synthesis methods, catalyst structure and morphology, crystallite microstructure, electrochemical properties and electrocatalytic behaviour of the materials.

The influence on the composite material structure and morphology has been discussed on the basis of microscopy observations (TEM, HRTEM) and X ray measurements (XRD, XPS). The characterization of the catalyst structure (mean sizes and size distribution of isolated particles), of the composite material morphology (mean size and size distribution of aggregates), and the platinum microstructure (crystallinity, crystallite sizes), etc. has also been performed and discussed.

Then, the electrochemical active surface areas, the behavior toward pre-adsorbed CO saturating layer oxidation and the activity toward the oxygen reduction reaction (ORR) have been compared using electrochemical methods (Cyclic voltammetry, rotating disc and ring disc electrodes). In the case of ORR studies, the number of exchanged electrons and the kinetic current densities at 0.95 V vs RHE, as determined from mathematical treatment of the Koutecky-Levich equation, have been used to compare the activity and selectivity of the different Pt/C catalysts. An important dependence of the catalytic activity on the catalyst synthesis method has been evidenced. However, due to the invariance of the cell parameter for all prepared catalyst, it is unlikely that the change in catalytic activity of the platinum material is related to geometric effect (change in Pt-Pt interatomic distance). But, the surface structure, presence of surface domains (facets) or of surface defects, seems also to have an important effect on the catalytic activity. It also seems that a thermal treatment, leading to crystallite reconstruction and decrease of residual strains, could lead to an enhancement of the catalytic activity toward orr.

Although the final objective, which was to correlate the activity of Pt/C catalysts to their morphology and structure, is not fully achieved, this book chapter shows unambiguously that the microstructure, macrostructure and morphology of a Pt/C catalyst has a great effect on its electrochemical behaviour toward molecules having energetic interest in fuel cell technology.

6. References

Attard, G. A. Gillies, J. E. Harris, C. A. Jenkins, D. J. Johnston, P. Price, M. A. Watson, D. J. Wells, P. B. (2001). Electrochemical evaluation of the morphology and enantioselectivity of Pt/graphite. *Appl. Catal. A : Gen.*, Vol. 222, N° 1-2, (December 2001) pp. 393-405.

Ayyadurai, S. M. Choi, Y-S. Ganesan, P. Kumaraguru, S. P. Popov. B. N. (2007). Novel PEMFC Cathodes Prepared by Pulse Deposition. *J. Electrochem. Soc.*, Vol. 154, N° 10 (August 2007) pp. B1063-B1073.

Bard, A. J. Faulkner, L. R.. in: Electrochemical Methods: Fundamentals and Applications, 2nd ed., John Wiley & Sons Inc., New York, 2001,.

Bezerra, C. W. B. Zhang, L. Liu, H. Lee, K. Marques, A. L. B. Marques, E. P. Wang, H. Zhang, J. (2007). A review of heat-treatment effects on activity and stability of PEM fuel cell catalysts for oxygen reduction reaction. *J. Power Sources*, Vol. 173, N° 2 (November 2007) pp. 891-908.

Biegler, T. Rand, D.A.J. Woods, R. (1971). Limiting Oxygen Coverage on Platinized Platinum; Relevance to Determination of Real Platinum Area by Hydrogen Adsorption. J. Electroanal. Chem., Vol. 29, N° 2, (February 1971) pp. 269-277.

Billy, E. Maillard, F. Morin A. Guetaz, L. Emieux, F. Thurier, C. Doppelt, P. Donet, S. Mailley, S. (2010). Impact of ultra-low Pt loadings on the performance of anode/cathode in a proton-exchange membrane fuel cell. *J. Power Sources*, Vol. 195, N° 9, (May 2010), pp. 2737-2746.

Bönnemann, H. Brijoux, W. (1995). The preparation, characterization and application of organosols of early transition metals. NanoStructured Materials, Vol. 5, N° 2, (February 1995) pp. 135-140.

Bönnemann, H. Brijoux, W. Brinkmann, R. Dinjus, E. Joussen, T. Korall, B. (1991). Formation of Colloidal Transition Metals in Organic Phases and Their Application in Catalysis. *Angew. Chem. Int. Engl.*, Vol. 30, N° 10, (October 1991) 1312-1314.

Bönnemann, H. Brijoux, W. Brinkmann, R. Fretzen, R. Joussen, T. Köppler, R. Korall, B. Neiteler, P. Richter, J. (1994). Preparation, characterization, and application of fine metal particles and metal colloids using hydrotriorganoborates. *J. Mol. Catal.*, Vol. 86, N° 1-3, (January 1994) 129-177.

Boulmer Leborgne, C. Benzerga, R. Scuderi, D. Perriere, J. Albert, O. Etchepare, J. Millon, E. (2006). Femtosecond laser beam in interaction with materials for thin film deposition. *SPIE Proceeding on High Power Laser Ablation VI*, Vol. 6261, Taos-New Mexico (USA), May 2006.

Boutonnet, M. Kizling, J. Stenius, P. Maire, G. (1982). The preparation of monodisperse colloidal metal particles from microemulsions. *Colloid and Surfaces*, Vol. 5, N° 3, (November 1982) pp. 209–225.

Bradley, J. S. (1994). The Chemistry of Transition Metal Colloids, In: *Clusters and Colloids, From Theory to Applications*, G. Schmid, pp. 459–544, Springer, ISBN 9783527290437, Weinheim, Germany.

Brault, P. Caillard, A. Thomam, A. L. Mathias, J. Charles, C. Boswell, R. W. Escribano, S. Durand, J. Sauvage, T. (2004). Plasma sputtering deposition of platinum into porous fuel cell electrode. *J. Phys. D : Appl. Phys.*, Vol. 37, N° 24, (December 2004), pp. 3419-3423.

Brimaud, S. Coutanceau, C. Garnier, E. Léger, J.-M. Gérard, F. Pronier, S. Leoni, M. (2007). Influence of surfactant removal by chemical or thermal methods on structure and electroactivity of Pt/C catalysts prepared by water-in-oil microemulsion. *J. Electroanal. Chem.*, Vol. 602, N° 2, (April 2007) pp. 226-236.

Brimaud, S. Pronier, S. Coutanceau, C. Léger, J.-M. (2008). New findings on CO electrooxidation at platinum nanoparticle surfaces. *Electrochem. Comm.*, Vol. 10, N° 11, (November 2008), pp. 1703-1707.

Caillard, A. Charles, C. Boswell, R. Brault, P. Coutanceau, C. (2007). Plasma based platinum nanoaggregates deposited on carbon nanofibers improve fuel cell efficiency. *Appl. Phys. Lett.*, Vol. 90, N° 22, (May 2007), pp. 223119-1-223119-3.

Chen, S. Kucernak. A. (2003). Electrodeposition of Platinum on Nanometer-Sized Carbon Electrodes. *J. Phys. Chem. B*, Vol. 107, N° 33, (August 2003) pp. 8392-8402.

Cho, Y-H. Yoo, S. J. Cho, Y-H. Park, H-S. Park, I-S. Lee, J.K. Sung, Y-E. Enhanced performance and improved interfacial properties of polymer electrolyte membrane fuel cells fabricated using sputter-deposited Pt thin layers. *Electrochim. Acta*, Vol. 53, N° 21,(September 2008) pp. 6111-6116.

Coutanceau, C. Croissant, M. J. Napporn, T. Lamy, C. (200). Electrocatalytic reduction of dioxygen at platinum particles dispersed in a polyaniline film. *Electrochim. Acta*, Vol. 46, N° 4, (December 2000) pp. 579-588.

Coutanceau, C. Rakotondrainibe, A. Lima, A. Garnier, E. Pronier, S. Léger, J. M. Lamy, C. (2004). Preparation of Pt-Ru bimetallic anodes by galvanostatic pulse electrodeposition : characterization and application to the direct methanol fuel cell. *J. Appl. Electrochem.*, Vol. 34, N° 1, (January 2004), pp. 61-66.

Coutanceau, C. Brimaud, S. Dubau, L. Lamy, C. Léger, J.-M. Rousseau, S. Vigier, F. (2008). Review of different methods for developing nanoelelectrocatalysts for the oxidation of organic compounds. *Electrochim. Acta*, Vol. 53, N° 23, (October 2008) pp. 6865-6880.

Devadas, A. Baranton, S. Napporn, T. W. Coutanceau, C. (2011). Tailoring of RuO_2 nanoparticles by microwave assisted "Instant method" for energy storage applications. *J. Power Sources*, Vol. 196, N° 8, (April 2011) pp. 4044-4053.

Eriksson, S. Nylén, U. Rojas, S. Boutonnet, M. (2004). Preparation of catalysts from microemulsions and their applications in heterogeneous catalysis. *Appl. Catal. A: Gen.*, Vol. 265. N° 2, (July 2004) pp. 207-219.

Fievet, F. Lagier, J. P. Blin, B. Beaudoin, B. Figlarz, M. (1989). Homogeneous and heterogeneous nucleations in the polyol process for the preparation of micron and submicron size metal particles. *Solid State Ionics*, Vol. 32-33, N° 1, (February-March 1989) pp. 198-205.

Friedrich, K. A. Henglein, F. Stimming, U. Unkauf, W. (2000). Size dependence of the CO monolayer oxidation on nanosized Pt particles supported on gold. *Electrochim. Acta*, Vol. 45, N° 20, (Juin 2000) pp. 3283-3293.

Gloagen, F. Andolfatto, N. Durand, R. Ozil, P. (1994). Kinetic study of electrochemical reactions at catalyst-recast ionomer interfaces from thin active layer modeling. *J. Appl. Electrochem.*, Vol. 24, N° 9, (September 1994) pp. 863-869. ISSN 0021-891X.

Gómez, R. Clavilier, J. (1993). Electrochemical behaviour of platinum surfaces containing (110) sites and the problem of the third oxidation peak . *J. Electroanal. Chem.*, Vol. 354, N° 1-2, (August 1993) pp. 189-208.

Grolleau, C. Coutanceau, C. Pierre, F. Leger, J. M. (2010). Optimization of a surfactant free polyol method for the synthesis of Platinum-Cobalt electrocatalysts using Taguchi design of experiments. *J. Power Sources*, Vol. 195, N° 6, (March 2010) pp. 1569-1576.

Grolleau, C. Coutanceau, C. Pierre, F. Léger. J. M. (2008). Effect of potential cycling on structure and activity of Pt nanoparticles dispersed on different carbon supports. *Electrochim. Acta*, Vol 53, N° 24, (October 2008) pp. 7157–7165.

Habrioux, A. Vogel, W. Guinel, M. Guetaz, L. Servat, K. Kokoh, B. Alonso-Vante, N. (2009). Structural and electrochemical studies of Au–Pt nanoalloys. *Phys. Chem. Chem. Phys.*, Vol. 11, N° 18, (May 2009), pp. 3573-3579. ISSN 1463-9076.

Henry, C. R. (2005). Morphology of supported nanoparticles. Progress in Surf. Sci. Vol. 80, N° 3-4, pp. 92–116.

Hwang B. J. Kumar, S. M. S. Chen, C-H. Monalisa. Cheng, M-Y. Liu, D-G. Lee J-F. (2007). An Investigation of Structure-Catalytic Activity Relationship for Pt-Co/C Bimetallic Nanoparticles toward the Oxygen Reduction Reaction. *J. Phys. Chem. C* , Vol. 111, N° 42,(October 2007) pp. 15267-15276.

Ingelsten, H. H. Ggwe, R. Palmqvist, A. Skoglundh, M. Svanberg, C. Hlmberg, K. Shah, D. O. J. (2001). Kinetics of the Formation of Nano-Sized Platinum Particles in Water-in-Oil Microemulsions. *J. Colloid Interface Sci.*, Vol. 241, N° 1, (Septembre 2001) pp. 104–111.

Iwasita, T. (2003). Methanol and CO électrooxydation. in *Handbook of Fuel Cells – Fundamentals, Technology and Applications*, Vielstich, W., Gasteiger, H. A., Lamm, A. John Wiley & Sons, Ltd.; New York, 2003, Vol. 2: Electrocatalysis, pp. 603-624.

Jakobs, R. C. M. Janssen, L. J. J. Barendrecht, E. (1985). Oxygen reduction at polypyrrole electrodes—I. Theory and evaluation of the rrde experiments. *Electrochim. Acta*, Vol. 30, N° 8, (August 1985) pp. 1085-1091.

Kinoshita, K. (1990). Particle size effects for oxygen reduction on highly dispersed platinum in acid electrolytes. J. Electrochem Soc., Vol. 137, N° 3, (March 1990) pp. 845-848.

Larcher, D. Patrice. R. (2000). Preparation of Metallic Powders and Alloys in Polyol Media: A Thermodynamic Approach. *J. Solid State Chem.*, Vol. 154, N° 2, (Novembre 2000) pp. 405-411.

Lebègue, E. Baranton, S. Coutanceau, C. (2011). Polyol synthesis of nanosized Pt/C electrocatalysts assisted by pulse microwaves activation. *J. Power Sources*, Vol. 196, N° 3, (February 2011) pp. 920-927.

Liu, Z. Gan, L. M. Hong, L. Chen, W. Lee, J. Y. (2005). Carbon supported Pt nanoparticles for PEMFC. *J. Power Sources*, Vol. 139, N° 1-2, (January 2005) 73-78 : ""

Lopez-Cudero, A. Solla-Gullon, J. Herrero, E. Aldaz, A. Feliu, J. M. (2010). CO electrooxidation on carbon supported platinum nanoparticles: Effect of aggregation. *J. Electroanal. Chem.*, Vol. 644, N° 2 (June 2010) pp. 117-126.

Maillard, F. Savinova, E. R. Simonov, P. A. Zaikovskii, V. I.. Stimming, U. (2004). Infrared Spectroscopic Study of CO Adsorption and Electro-oxidation on Carbon-Supported Pt Nanoparticles: Interparticle versus Intraparticle Heterogeneity. *J. Phys. Chem. B*, Vol. 108, N° 46, (November 2004) pp. 17893-17904.

Maillard, F. Schreier, S. Hanzlik, M. Savinova, E. R. Weinkauf, S. Stimming, U. (2005). Influence of particle agglomeration on the catalytic activity of carbon-supported Pt nanoparticles in CO monolayer oxidation. *Phys. Chem. Chem. Phys.*, Vol. 7, N° 2, (January 2005) pp. 385-393.

Maillard, F. Savinova, E. R. Stimming, U. (2007). CO monolayer oxidation on Pt nanoparticles: Further insights into the particle size effects. *J. Electroanal. Chem.*, Vol. 599, N° 2, (January 2007) pp. 221-232.

Markovic, N. M. Gasteiger, H. A. Ross, P. N. Oxygen Reduction on Platinum Low-Index Single- Crystal Surfaces in Sulfuric Acid Solution: Rotating Ring - Pt(hkl) Disk Studies. J. Phys. Chem., Vol. 99, N° 11, (March 1995) pp. 3411-3415.

Markovic, N. M. Grgur, B. N. Ross, P. N. (1997). Temperature-Dependent Hydrogen Electrochemistry on Platinum Low-Index Single-Crystal Surfaces in Acid Solutions. *J. Phys. Chem. B*, Vol. 101, N° 27, (July 1997) pp. 5405-5413.

Mottet, C. Gonialowski, J. Baletto, F. Ferrando, R. Tréglia, G. Modeling free and supported metallic nanoclusters: structure and dynamics. *Phase Transitions*, Vol. 77, N° 1-2, (January-February 2004) pp. 101–113. ISSN 0141-1594.

Oh, H-SOh, . J-G. Hong, Y-G. Kim, H. (2007). Investigation of carbon-supported Pt nanocatalyst preparation by the polyol process for fuel cell applications. *Electrochim. Acta*, Vol. 52, N° 25, (September 2007) pp. 7278–7285.

Park, J. Joo, J. Kwon, S. G. Jang, Y. Hyeon, T. (2007). Synthesis of monodisperse spherical nanoparticules. *Angewandte Chem. - Int. Ed.*, Vol. 46, N°25, (June 2007) pp. 4630-4660. ISSN 1521 3773.

Peng, X. Wickham, J. Alivisatos, A. P. (1998). Kinetics of II-VI and III-V Colloidal Semiconductor Nanocrystal Growth: "Focusing" of Size Distributions. *J. Am. Chem. Soc.*, Vol. 120, N° 21, (June 1998) pp. 5343-5344.

Perrière, J. Millon , E. Chamarro , M. Morcrette, M. Andreazza, C. (2001). Formation of GaAs nanocrystals by laser ablation. *Appl. Phys. Lett.*, Vol. 78, N° 19, (May 2001) 2949-1-2949-3.

Pieck, C. L. Marecot, P. Barbier, J. (1996). Effect of Pt-Re interaction on sulfur adsorption and coke deposition. *Appl. Catal. A: Gen.*, Vol. 145. N° 1-2, (October 1996) pp. 323-334.

Plyasova, L. M. Molina, I. Y. Gavrilov, A. N. Cherepanova, S. V. Cherstiouk, O. V. Rudina, N. A. Savinova, E. R. Tsirlina, G. A. (2006). Electrodeposited platinum revisited: Tuning nanostructure via the deposition potential. *Electrochim. Acta*, Vol. 51, N° 21 (June 2006) pp. 4477-4488.

Ralph, T. R. Hogarth, M. P. (2002). Catalysts for low temperature fuel cell. Part I: the cathode challenges. *Platinum Metal Rev.*, Vol. 46, N° 1 (January 2002) pp. 3-14

Richard, D. Gallezot, P. (1987). Preparation of highly dispersed carbon supported platinum catalysts. In: *Preparation of Catalysts IV*. B. Delmon, P. Grange, P.A. Jacobs, G. Poncelet, pp. , Elsevier Science Publishers B.V., Amsterdam.

Roman-Martinez, C. Cazorla-Amoros, D. Yamashita, H. de Miguel, S. Scelza, O. A. (2000). XAFS Study of Dried and Reduced PtSn/C Catalysts: Nature and Structure of the Catalytically Active Phase. *Langmuir*, Vol. 16, N° 3, (February 2000) pp. 1123-1131.

Reetz, M. T. Koch, M. G. Patent application DE-A 19852547.8, 13.11.1998.

Reetz, M. T. Koch. M. G. (1999). Water-Soluble Colloidal Adams Catalyst: Preparation and Use in Catalysis. *J. Am. Chem. Soc.*, Vol. 121, N° 34 (September 1999) pp. 7933–7934.

Reetz, M. T. Lopez, M. Patent Application DE-A 102 11701.2, 16.03.2002.

Reetz, M. T. Schulenburg, H. Lopez, M. Splienthoff, B. Tesche, B. (2004). Platinum-Nanoparticles on Different Types of Carbon Supports: Correlation of Electrocatalytic Activity with Carrier Morphology. *Chimia*, 58, N°12, (December 2004), pp. 896-899.

Sellin, R. Grolleau, C. Coutanceau, C. Léger, J.-M. Arrii-Clacens, S. Pronier, S. Clacens, J.-M. (2009). Effects of Temperature and Atmosphere on Carbon-Supported Platinum Fuel Cell Catalysts. *J. Phys. Chem. C,* Vol. 113, N° 52, (December 2009) pp. 21735-21744.

Sellin, R. Clacens, J-M. Coutanceau, C. (2010). A thermogravimetric analysis/mass spectroscopy study of the thermal and chemical stability of carbon in the Pt/C catalyst system. *Carbon,* Vol. 48, N° 8, (July 2010) pp. 2244-2254.

Shevchenko, E. V. Talapin, D. V. Schnablegger, H. Kornowski, A. Festin, O. Svedlindh, P. Haase, M. Weller H. (2003). Study of nucleation and growth in the organometallic synthesis of magnetic alloy nanocrystals : the role of nucleation rate in size control of $CoPt_3$ nanocrystals. *J. Am. Chem. Soc.,* Vol. 125, N° 30, (July 2003) pp. 9090-9101.

Schmidt, T. J. Gasteiger, H. A. Behm, R. J. (1999). Rotating Disk Electrode Measurements on the CO Tolerance of a High-Surface Area Pt/Vulcan Carbon Fuel Cell Catalyst. *J. Electrochem. Soc.,* Vol. 146, N° 4, (April 1999) pp. 1296-1304.

Solla-Gullon, J. Montiel, V. Aldaz, A. Clavilier. J. (2000). Electrochemical characterisation of platinum nanoparticles prepared by microemulsion: how to clean them without loss of crystalline surface structure. *J. Electroanal. Chem.,* Vol. 491, N° 1-2,(September 2000) pp. 69-77.

Solla-Gullón, J. Vidal-Iglesias, F. J. Herrero, E. Feliu, J. M. Aldaz, A. (2006). CO monolayer oxidation on semi-spherical and preferentially oriented (1 0 0) and (1 1 1) platinum nanoparticles. *Electrochem. Comm.,* Vol. 8, N° 1, (January 2006) pp. 189–194.

Solla-Gullón, J. Rodríguez, P. Herrero, E. Aldaz, A. Feliu, J. M. (2008). Surface characterization of platinum electrodes. *Phys. Chem. Chem. Phys.,* Vol. 10, N° 10, (March 2008) pp. 1359-1373. ISSN 1463-9076.

Tarasevich, M. R. Sadkowski, A. Yeager, E. (1983). Oxygen Electrochemistry. In *Kinetics and Mechanisms of Electrode Processes;* Conway, in: B.E. Conway, J.O'M. Bockris, E. Yeager, S.U.M. Khan (Eds.), Comprehensive Treatise of Electrochemistry, vol. 7, Plenum Press, New York, 1983, p. 301-398.

Vigier, F. Coutanceau, C. Perrard, A. Belgsir, E. M. Lamy, C. (2008). Development of anode catalyst for Direct Ethanol Fuel Cell. *J. Appl. Electrochem.,* Vol. 34, N° 4, (April 2004) pp. 439-446.

Villegas, I.. Weaver, M. J. Carbon monoxide adlayer structures on platinum (111) electrodes: A synergy between in situ scanning tunneling microscopy and infrared spectroscopy (1994). *J. Chem. Phys.,* Vol. 101, N° 2, (July 1994) pp. 1648-1660.

Vogel, W. Bradley, J. Vollmer, O. Abraham, I. (1998). Transition from Five-Fold Symmetric to Twinned FCC Gold Particles by Thermally Induced Growth. *J. Phys. Chem. B,* Vol. 102, N° 52, (December 1998) pp. 10853-10859.

Wang, J. Xue, J. M. Wan, D. M. Gan, B. K. (2000). Mechanically activating nucleation and growth of complex perovskites. *J. Solid State Chem.,* Vol. 154, N° 2, (November 2000) pp. 321-328.

Wilson, G. J. Matijasevitch, A. S. Mitchell, D. R. G. Schulz, J. C. Will, G. D. (2006). Modification of TiO2 for enhanced surface properties : finite Ostwald ripening by a microwave hydrothermal process. *Langmuir,* Vol. 22, N° 5, (February 2006) pp. 2016-2027.

Yoshimi, K. Song M. -B. Ito, M. (1996). Carbon monoxide oxidation on a Pt(111) electrode studied by in-situ IRAS and STM: coadsorption of CO with water on Pt(111). *Surf. Sci.,* Vol. 368, N° 1-3 (December 1996) pp. 389-395.

Zagal, J. Bindra, P. Yeager, E. (1980). A Mechanistic Study of O_2 Reduction on Water Soluble Phthalocyanines Adsorbed on Graphite Electrodes . *J. Electrochem. Soc.,* VoL 127, N° 7, (July 1980) pp. 1506-1517.

20

Characteristics of the Laser-Induced Breakdown Detection of Colloidal Nanoparticles for Determining Particle Size

E.C. Jung and H.R. Cho
Nuclear Chemistry Research Division, Korea Atomic Energy Research Institute
Republic of Korea

1. Introduction

Laser-induced breakdown detection (LIBD) of colloidal nanoparticles has been investigated in various fields, such as the measurement of natural colloids in drinking water (Bundschuh et al., 2005; Kaegi et al., 2008; Wagner et al., 2005; Walther et al., 2006), the in-situ observation of colloid mediated pollutant transport (Hauser et al., 2002; Möri et al., 2003) and the real-time measurement of the solubility of radioactive elements (Bundschuh et al., 2000; Cho et al., 2008; Knopp et al., 1999; J.I. Kim, 2006; Neck et al., 2001, 2003; Opel et al., 2007; Walther et al., 2007). LIBD is an established technique that measures the size and concentration of colloidal particles in aqueous media using plasma formation, which is induced by focusing a short pulse laser beam into the solution (Kim & Walther, 2007). LIBD is especially efficient for detecting small particles less than 100 nm in diameter, which are not easily detectable using commercially available devices that adopt the measurement of a scattered light intensity, such as photon correlation spectroscopy (PCS) (Bundschuh et al., 2001b). Although several different LIBD systems have been developed over the last decades, a commercial instrument that adopts the LIBD method is currently unavailable.

This chapter is composed of a brief review of LIBD systems that adopt different detection schemes (Section 2), followed by the particle size determination methods (Section 3). Although these methods in principle allow the determination of particle size, until now, particle sizing capabilities in most LIBD experiments have been tested only for polystyrene reference particles with a well-defined size. Thus it should be verified that these LIBD methods are suitable for determining the particle size of different materials (Jung et al., 2011). Newly investigated characteristics of LIBD of colloidal uranium and alumina particles were reported in the last section (Section 4) followed by the summary (Section 5).

2. LIBD systems

In a recent publication, two different LIBD systems and their applications were reviewed in detail (Kim & Walther, 2007). One was based on the optical detection of plasma flash using either a photomultiplier tube (Ajiro et al., 1992; Fujimori et al., 1992) or a charge-coupled device (CCD) camera (Bundschuh et al., 2001a, 2005; Hauser et al., 2002; Jung et al., 2006, 2007; J.W. Kim et al., 2008; Walther et al., 2002). A second system was based on the acoustic detection of a shock wave using a piezoelectric transducer (PZT) (Bundschuh et al., 2001b; Izumida et al., 1998; Kitamori et al., 1988, 1989; Satio et al., 1999; Scherbaum et al., 1996; Walther et al., 2004). Recently, we developed a modified LIBD system based on the optical detection of a shock wave; this system measures the deflection of a probe laser beam due to a shock wave-induced change in the refractive index of the liquid medium (Cho et al., 2008; Jung et al., 2009). In the next subsections, these different LIBD systems are described briefly based on the experimental apparatuses installed in our laboratory (Jung et al., 2006, 2007, 2009, 2011; Cho et al., 2008).

2.1 Optical detection of plasma flash using a CCD camera

A schematic diagram of the experimental arrangement of the LIBD system using a CCD camera is shown in Fig. 1 (Jung et al., 2006, 2007, 2011). The second harmonic (wavelength=532 nm, pulse width=6 ns, and pulse repetition rate=20 Hz) of the pulsed Nd:YAG laser (Continuum, Surelite II) was used for laser-induced breakdown. In order to adjust the laser beam diameter to approximately 4 mm, an iris diaphragm was installed on the travel path of the pulsed laser beam. Two Glan-Thompson polarizers, required to adjust the laser pulse energy, were arranged downstream of the iris diaphragm. The first polarizer was installed in such a fashion as to allow it to rotate; the second polarizer was installed to fix the direction so that only a polarized laser beam perpendicular to a bottom plane could pass through it. Accordingly, the laser pulse energy incident on the sample could be adjusted by rotating the first polarizer. The laser beam profile measured by a laser beam profiler (Newport, LBP-1) showed an approximately 80% Gaussian profile and the pulse-to-pulse fluctuation of the laser pulse energy was within 2.5% for 3,000 laser pulses. The laser beam was focused on the interior of a sample cell (Hellma, 111-QS) using a plano-convex

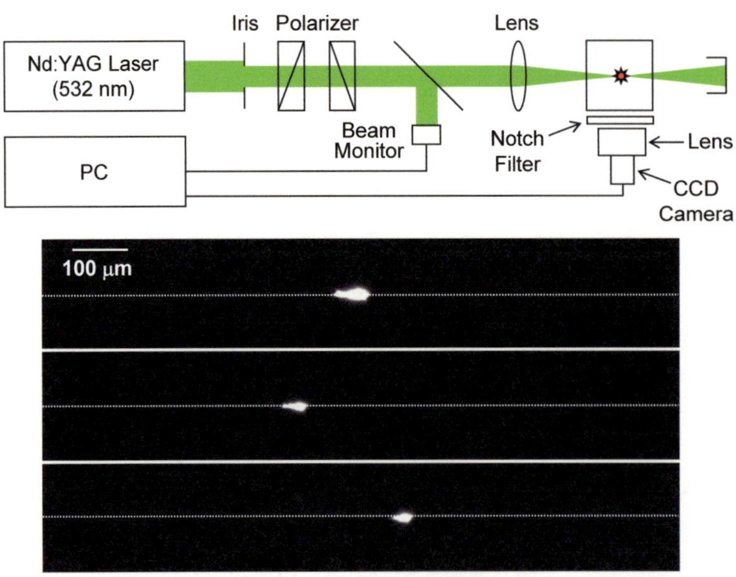

Fig. 1. Experimental setup and examples of plasma flashes acquired for each laser shot.

lens with a focal length of 40 mm. Laser-induced breakdowns occurred at the location on which the pulsed laser beam travelling in the Z-axis direction was focused.

The plasma flashes of the breakdown events were detected by a CCD camera (Hitachi Kokusai Electric, KP-F100BCL) with a variable macro-microscope. To prevent scattered laser light from reaching the CCD camera, a notch filter of ~1% transmission at 532 nm was inserted between the macro-microscope and the sample cell. Data from the CCD camera were recorded by a frame grabber card and processed by a home-made Labview software program. A typical example of the plasma flash is demonstrated in the lower part of Fig. 1, in which the optical image of a breakdown event is shown. The breakdown events occurred here and there along the laser beam propagation axis because of the Brownian motion of colloidal particles. In this study, the spatial distributions of 3,000 breakdown events were imaged on a pulse-to-pulse basis. A computer recorded the coordinates of plasma emission in the *X-Z* plane, the location and time of each breakdown event and the breakdown probability.

2.2 Acoustic detection of a shock wave using a PZT

A schematic diagram of the experimental arrangement of the LIBD system using a PZT is shown in Fig. 2. The role of each optical component was described in detail in previous subsection 2.1. When a laser-induced plasma is generated, a strong shock wave is propagated inside the liquid medium due to explosive expansion. To detect this shock wave, a PZT encapsulated in a cylindrical metal tube was attached to the wall of the sample cell. The coupling of metal tube housing to quartz sample cell was achieved by spiral springs. The waveform of the PZT signal was monitored using a digital oscilloscope (Tektronix TDS460A). The magnitude of the PZT signal was measured using a gated integrator (SRS SR250) and a boxcar averaging system (SRS SR280). The output was collected with a computer.

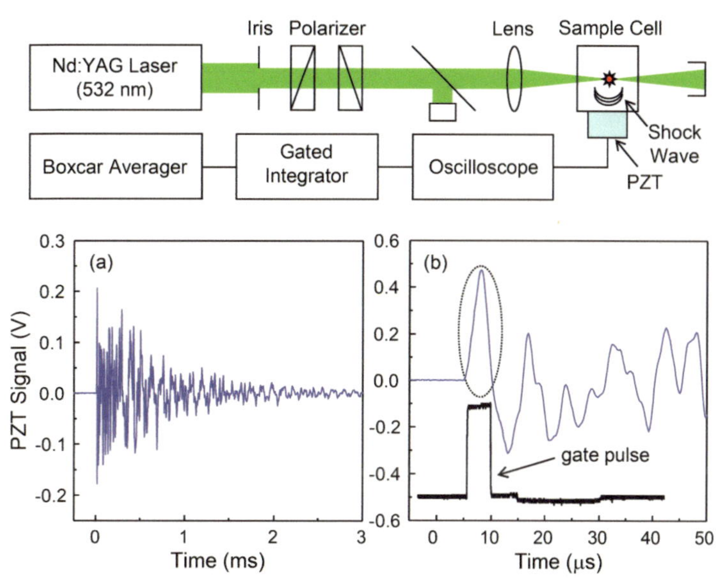

Fig. 2. Experimental setup and (a) PZT signal waveform acquired for a single laser shot, (b) early part of waveform (upper trace) and boxcar gate for signal detection (lower trace).

Fig. 2(a) and (b) shows the time trace as recorded by the PZT for a single laser shot and the early part of the waveform, respectively. The signal begins at a well-defined delay time *t* after the occurrence of the breakdown at $t = 0$, as shown in Fig. 2(b). This delay time was found to be linearly dependent on the distance between the focal point of the laser beam and the PZT surface. From these results, we are able to obtain a shock wave velocity of $\sim 1.48 \times 10^5$ cm/s. This is in good agreement with the established ultrasonic velocity at room temperature in water, which is $\sim 1.486 \times 10^5$ cm/s (Zapka et al., 1982).

In this experiment, the pulse height of the first signal in the waveform of the PZT signal, designated as a dotted circle in Fig. 2(b), showed a tendency to be proportional to the particle size (Kitamori et al., 1988, 1989). Thus the magnitude of the PZT signal was measured using a gated integrator with appropriate time delay and gate width as shown in the lower trace in Fig. 2(b). Because this signal magnitude was much higher than the magnitude of the electronic noise, each PZT signal magnitude greater than a certain threshold level was used to measure the breakdown event.

2.3 Optical detection of a shock wave using a probe beam

A schematic diagram of the experimental arrangement of the LIBD system using a probe beam is shown in Fig. 3 (Cho et al., 2008, Jung et al., 2009). The role of each optical component was described in detail in previous subsection 2.1. To detect the shock wave, a CW He-Ne laser (Uniphase, 20 mW) was used as a probe beam. When a laser-induced shock wave propagates through the medium, the refractive index of that medium changes. This change results in the deflection of the probe beam passing through the medium, and therefore the probe beam intensity reached at a high speed silicon photodiode (Thorlabs Det110) changes. A notch filter at 532 nm was used in order to prevent the photodiode from receiving the scattered light of the Nd:YAG laser. A pinhole (diameter of 0.5 mm) was used to detect the change in probe beam intensity for better sensitivity. The waveform of the probe beam signal was monitored using a digital oscilloscope (Tektronix TDS460A). The signal magnitude of the photodiode was measured using a gated integrator (SRS SR250) and a boxcar averaging system (SRS SR280). The output was collected with a computer.

Fig. 3(a) and (b) shows a typical waveform of a probe beam signal generated by a laser-induced shock wave and the early part of the waveform, respectively (Cho et al., 2008, Jung et al., 2009). The probe beam passed the 9.1 mm position below the focal point of the Nd:YAG laser beam, at which laser-induced breakdown occurred. The first peak at ~6.2 μs, which appears after the occurrence of the breakdown at $t = 0$, represents the deflected signal of the probe beam due to the shock wave. The second peak at ~9 μs results from the shock wave being reflected at the wall of the sample cell. The magnitude of the probe beam signal was measured using a gated integrator of appropriate time delay and gate width, as can be seen in the lower trace in Fig. 3(b). Comparing this technology with previously developed acoustic detection technology that used a PZT enables us to obtain a remote measurement in a non-contact manner. The probe beam signal begins at a well-defined delay time *t* after the occurrence of the breakdown at $t = 0$, as can be seen in Fig. 3(a) and (b). The delay time increases as the distance between the focal point of the laser beam and the probe beam position increases.

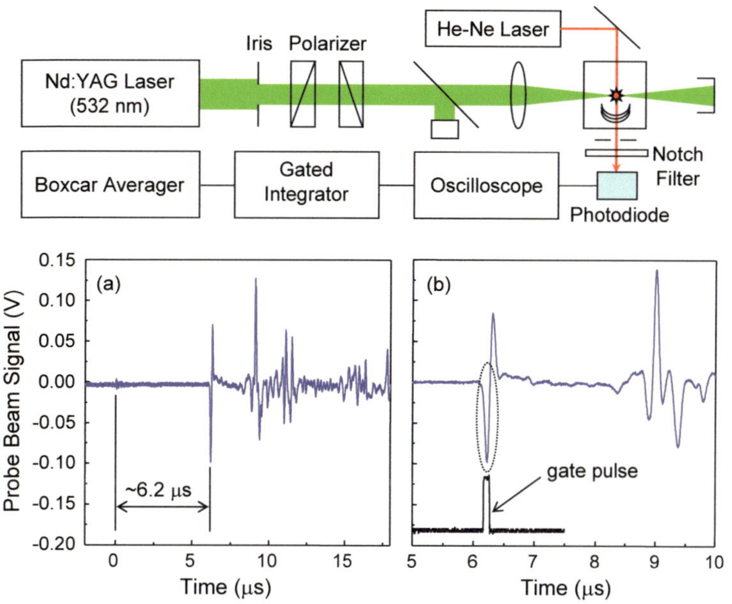

Fig. 3. Experimental setup and (a) probe beam signal waveform acquired for a single laser shot, (b) early part of waveform (upper trace) and boxcar gate for signal detection (lower trace).

Both a plasma emission and a shock wave can be measured simultaneously by constructing the detection components as shown in Fig. 4 (Jung et al., 2009). The laser beam propagated along the *Z*-axis direction was focused on an internal location of a sample cell. The probe beam propagated along the *Y*-axis direction passed by that location, which was lower than the focus of the pulsed laser beam by a distance "*d*" in the *X*-axis direction. By adjusting the distance downward from the focal point of the pulsed laser beam by approximately 12 mm, we were able to install the CCD camera and the PZT as can be seen in Fig. 4, to allow for simultaneous measurements.

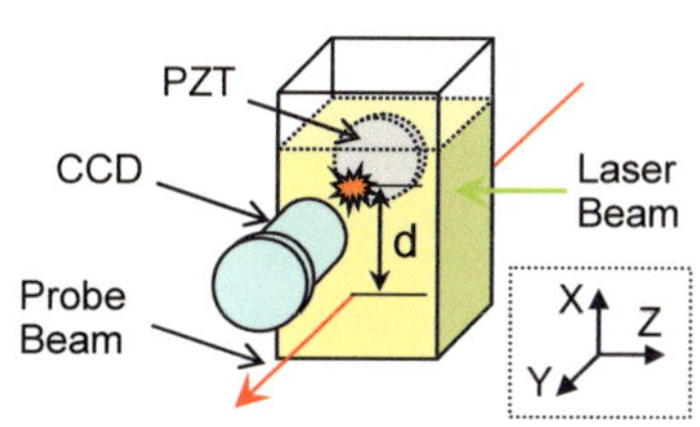

Fig. 4. Simultaneous detection using CCD camera, PZT and probe beam.

3. Particle size determination methods

The information on particle size is obtained by measuring the breakdown threshold energy which is defined as the minimum laser pulse energy required to generate a laser-induced plasma (Walther et al., 2002; Bundschuh et al., 2001a; Yun, 2007; Jung et al., 2011). The threshold energy can be determined from the breakdown probability measured as a function of the laser pulse energy (denoted as the "s-curve" in this chapter) (Walther et al.,

2002). Here, the breakdown probability is defined as the number of measured breakdown events divided by the total number of incident laser pulses.

Fig. 5 shows the breakdown probabilities as a function of the laser pulse energy for the 21 nm polystyrene particles at a laser wavelength of 532 nm. The empty and filled symbols represent the data derived from two different particle number densities, $5x10^8/cm^3$ and $5x10^7/cm^3$, for the same size particles. At a fixed particle number density, higher laser pulse energy resulted in higher breakdown probability. The threshold energy, corresponding to the laser pulse energy at which the breakdown probability is 0.01 in the s-curve, was determined in order to obtain the calibration curve for determining the particle size (Walther et al., 2002; Jung et al., 2011). At a fixed laser pulse energy, higher particle number density led to higher breakdown probability. As is apparent from the data shown in Fig. 5, the threshold energy is dependent on the particle concentration. In this study, the threshold energy was determined under experimental conditions in order to show that the increase in particle number density did not result in any further change of the threshold energy.

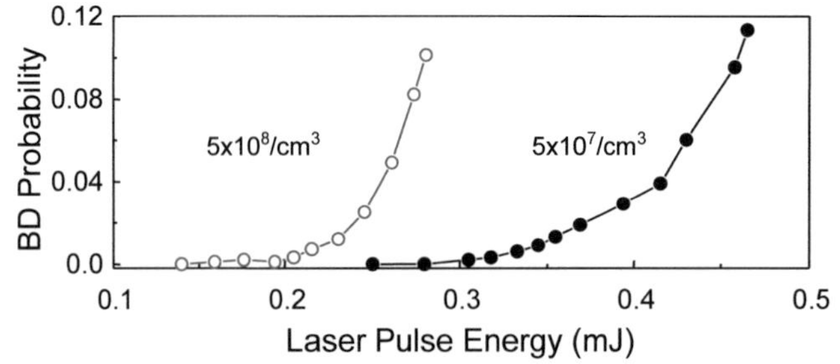

Fig. 5. Breakdown (BD) probability as a function of laser pulse energy for two different particle number densities. Data were obtained from polystyrene particles (size of 21 nm) at a laser wavelength of 532 nm.

Fig. 6 shows the dependence of the threshold energy on polystyrene particle size measured at a laser wavelength of 532 nm (Jung et al., 2011). At the fixed particle size, the threshold energy decreases with increasing particle number density, as depicted by the empty symbols in Fig. 6. The filled symbols indicate the minimum value of the threshold energy for each particle size. It is apparent that the threshold energy decreases with increasing particle size in this calibration curve.

The aforementioned particle sizing method, implemented by measuring the breakdown threshold energy, should be performed with a variation of the laser pulse energy. Under experimental conditions in which the laser pulse energy is fixed, the information on particle size can be obtained by measuring the spatial distribution of breakdown events (Bundschuh et al., 2001a; Hauser et al., 2002; Jung et al., 2007) and the frequency distribution curve of the magnitude of a laser-induced shock wave (Kitamori et al., 1989, Jung et al., 2009). In the next subsections, these methods are reported based on the experimental results obtained in our laboratory. In order to count only breakdown events induced by colloidal particles, the laser pulse energy was set to such a value that no breakdown event is observed in pure water.

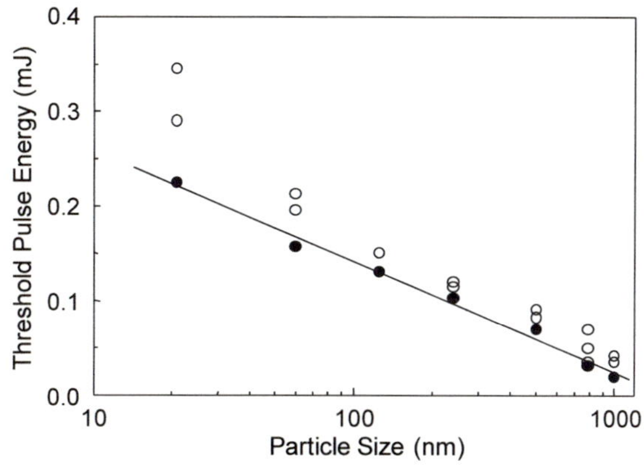

Fig. 6. Dependence of the breakdown threshold energies, which are defined as the energies at which the breakdown probability amounts to 0.01, on the particle size. Empty symbols represent the threshold energies determined from samples with different concentrations.

3.1 Particle sizing by measuring spatial distribution of breakdown events

The hyperbolic lines in Fig. 7 denote a focused laser beam propagated along the *Z*-axis; the spatial mode of the laser beam has a Gaussian profile with a beam waist of 5 μm at *Z*=0 (Jung et al., 2007). The exact focal point of the laser beam is designated as the origin of the coordinate axes. The contour lines in Fig. 7 represent the irradiance (equivalent to the power density) distribution of the laser beam. The regions indicated as "A" and "C" correspond to the upstream and downstream of the irradiance. The small ellipse indicated as "B" corresponds to the central region of the highest irradiance. As the laser beam waist increases along the *Z*-axis, the irradiance decreases, as depicted by the outer peanut-like shape.

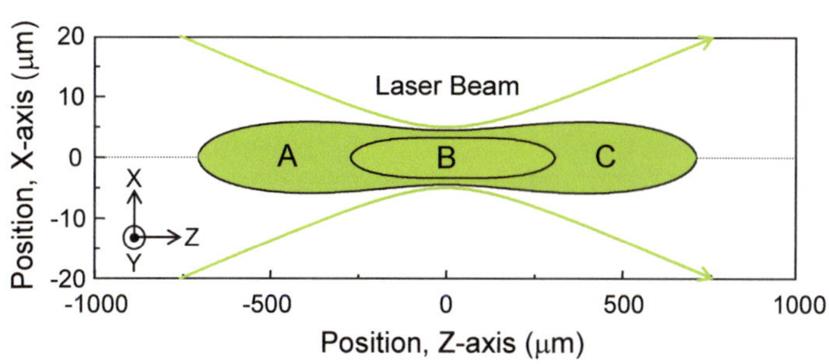

Fig. 7. Hyperbolic lines indicate the focused laser beam and contour lines indicate the laser irradiance distribution (A: upstream, B: central region of the highest irradiance, C: downstream).

When the CCD camera detects the plasma flash in the *X*-*Z* plane, as can be seen in Fig. 1, the spatial distribution of the breakdown events for small particles resembles the small ellipse shown in Fig. 7 because a breakdown is only induced at a region of a high irradiance, whereas the spatial distribution for large particles resembles the outer peanut-like shape due to their relatively low threshold energy for a breakdown (Bundschuh et al., 2001a).

Fig. 8 shows spatial distributions of 3,000 breakdown events for two different polystyrene particles (diameters of 21 nm and 60 nm) measured with a laser wavelength of 532 nm. Each data point represents the exact position on the *X-Z* plane at which a breakdown occurred. The length of the spatial distribution determined for 95% of the breakdown events (denoted the "effective focal length" in this manuscript) was measured in order to obtain the calibration curve for determining the particle size. At an incident laser pulse energy of 0.5 mJ, the breakdown probabilities were 0.2 for the 21 nm particles and 0.35 for the 60 nm particles, of which the concentrations were 1 parts per billion (ppb) and 7 ppb, respectively. The effective focal lengths were approximately 200 μm for the 21 nm particles and approximately 290 μm for the 60 nm particles. The spatial distributions of all data reported in this study exhibited good rotational symmetries centered on the origin of the *Z*-axis. Because the spatial distribution of the breakdown events along the *Z*-axis expands in proportion to a particle's diameter, a calibration for the particle sizing can be established by using these effective focal lengths (Bundschuh et al., 2001a).

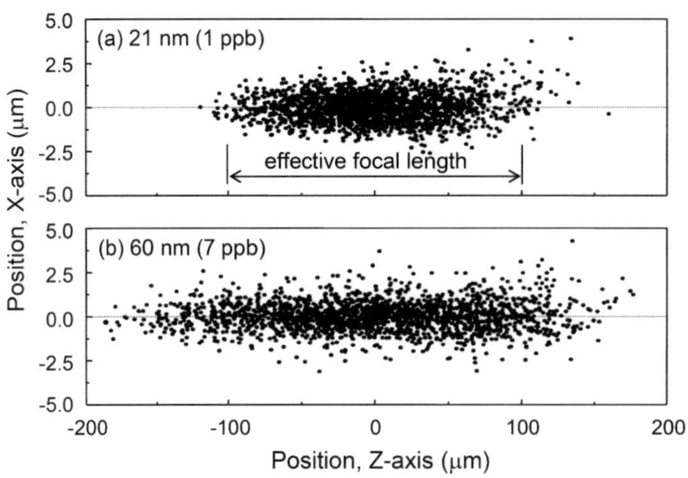

Fig. 8. Spatial distributions of 3,000 breakdown events for two different polystyrene particles (a) diameters of 21 nm (concentration of 1 ppb) and (b) 60 nm (concentration of 7 ppb). The length determined for the 95% of breakdown events was defined as the effective focal length.

3.2 Particle sizing by measuring frequency distribution of shock wave magnitude

Since each data point in Fig. 8 has its own acoustic signal magnitude, the main issue of this subsection is to investigate the size determination method by analyzing the acoustic signal magnitudes. Fig. 9 shows the distribution of PZT signal magnitude measured by the gated integrator and the boxcar averaging system, as explained in subsection 2.2. When a total of 3,000 laser pulses were incident to the sample cell containing the 21 nm polystyrene particles, 684 breakdown events, corresponding to a breakdown probability of 0.23, were measured. As is apparent from the data in Fig. 9, the signal magnitudes for the breakdown events were much higher (> 0.5 V) than the magnitude of the electrical noise of the PZT signal (< 0.2 V).

The frequency distribution curve of the PZT signal magnitudes, which were distributed in a wide voltage range, showed a Gaussian distribution, as plotted by solid line in Fig. 9. As an example of the frequency distribution curve, the data obtained from two different particles with diameters of 21 nm and 60 nm are shown in Fig. 10. Solid lines illustrate the Gaussian curves fitted to the experimental data points. The dependence of the peak and width of the distribution curve on the particle size is distinctive. The peak and width of the distribution curve increase as the particle size increases (Kitamori et al., 1989).

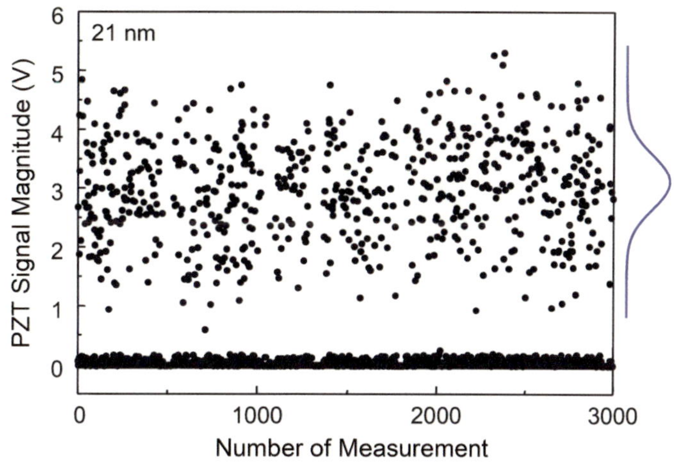

Fig. 9. Distribution of PZT signal magnitude for a total of 3,000 incident laser pulses.

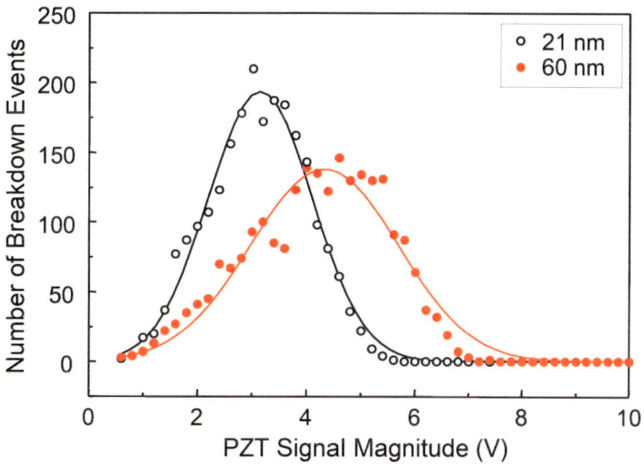

Fig. 10. The number of breakdown events as a function of the PZT signal magnitude for two different polystyrene particles (21 and 60 nm). Solid lines illustrate the Gaussian curves fitted to the experimental data points.

Although the detection method using PZT is very simple, this method has a drawback. It is difficult to obtain reproducible data due to the physical characteristics that are present when a PZT is directly attached to a sample cell to measure shock waves. Because several samples having particles of different sizes are contained in different sample cells, it is not easy to firmly attach the PZT to each sample cell in exactly the same manner. Therefore, difficulty arises when the magnitudes of the PZT signals are compared with each other for a plurality of different sample cells (Jung et al., 2009).

Comparing the optical detection method using a probe beam with the acoustic detection method using a PZT, the former method is advantageous for detecting harmful colloidal particles (such as radioactive substances), which must be located in a special environment isolated from their surroundings (in a glove box or in a cleaning booth), because of its capability for remote measurement of a shock wave in a non-contact manner (Cho et al., 2008). At first, the effect of the particle size on the waveform of the probe beam signal was investigated. The pulse height and width of the first signal in the waveform, designated as a dotted circle in Fig. 3(b), were different for particles with different sizes (data not shown). It was observed that the pulse height and width increased as the particle size increased (Jung et al., 2009).

Fig. 11 shows the results obtained by measuring the frequency distribution of the probe beam signal magnitude with respect to standard polystyrene nanoparticles having different sizes under conditions in which the delay and width of the gate pulse were fixed, as can be seen in Fig. 3(b). The concentrations of the particles having sizes of 21, 33, and 60 nm were 1, 2 and 7 ppb, respectively. In Fig. 11, the *X*-axis denotes the magnitude of the probe beam signal; the *Y*-axis denotes the number of breakdown events of the respective polystyrene nanoparticles having different sizes. The interval of the *X*-axis during which the data was processed to form a frequency distribution curve is 0.1 V. As the size of the polystyrene nanoparticles increases, the peak of the frequency distribution curve increases. Therefore, the calibration curve required for the determination of the particle size can be obtained by using the peaks of different frequency distribution curves (Jung et al., 2009).

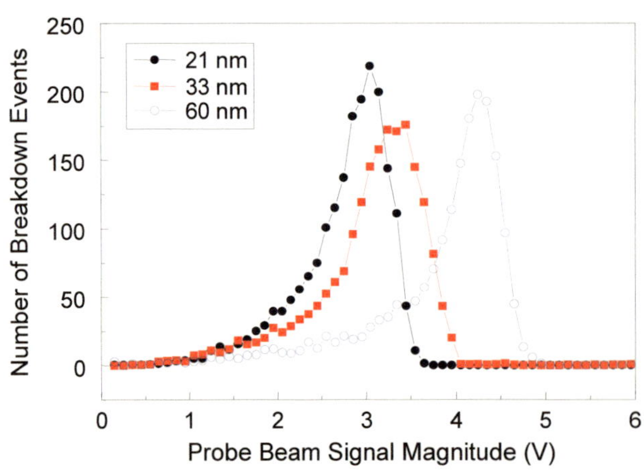

Fig. 11. The number of breakdown events as a function of the probe beam signal magnitude for three different polystyrene particles (21, 33 and 60 nm).

4. Material dependent characteristics of LIBD

Although the aforementioned LIBD methods in principle allow the determination of particle size, until now, particle sizing capabilities in most LIBD experiments have been tested only for polystyrene reference particles with a well-defined size. Because physical processes such as multiphoton ionization (MPI) and electron cascade growth affect the generation of laser-induced plasma, characteristics of LIBD (the breakdown threshold energy in the s-curve, the spatial distribution of breakdown events and the frequency distribution curve of the shock wave magnitude) may appear in different ways for different materials having the same particle size. Thus, it should be verified that these LIBD methods are suitable for determining the particle size of different materials (Jung et al., 2011).

In this section, LIBD experiments were performed for several different materials: polystyrene, uranium, silica and alumina nanoparticles. Colloidal uranium particles are generated from a spent nuclear fuel, thus, the measurement of their sizes is of interest to understand their behavior in groundwater, to allow for the safety assessment of nuclear waste disposal. We were also motivated to examine the size of silica and alumina nanoparticles, which can form natural colloids in groundwater, because these colloidal particles serve as crucial carriers for the migration of radionuclides in groundwater.

4.1 Sample preparation

Monodisperse polystyrene (21, 33, 60, 125, 240, 500, 800 and 1000 nm, Duke Scientific), silica (60, 238, 490, 730 and 990 nm, Corpuscular and Duke Scientific) and alumina (50, 240, 500, 800 and 1080 nm, Corpuscular) particles of a well-defined size were used in this study. Before sample preparation, the company-certified particle sizes of these reference particles were confirmed once again in our laboratory by TEM (Transmission Electron Microscopy, Jeol Ltd., JEM-2000FXII) and PCS (Malvern, Zetasizer, Nano ZS90). These particles were suspended in ultra-pure water from a water purification apparatus (Millipore, Milli-Q-element).

For the uranium sample, a pure U(VI) solution in $HClO_4$ (Merck, analysis grade) was prepared from natural UO_2 powder dissolved in $HClO_4$. A U(VI) stock solution of 1.0 mM at pH 3.8 in 0.1 M $NaClO_4$ was prepared by addition of a 1.0 M NaOH solution (Sigma-Aldrich, 99.99% decarbonated NaOH) at room temperature (Cho et al., 2008; Jung et al., 2009). The stock solution was equilibrated with air for one year. The uranium concentration of the stock solution was determined by using a kinetic phosphorescence analyzer (Chemcheck, KPA-11). U(VI) samples were prepared in a glove box purged with Ar gas. An aliquot of the stock solution was slowly titrated with four different 0.1 M $NaClO_4$ solutions at pH levels of 3.8, 9, 10, and 11 to maintain the ionic strength of the samples at 0.1 M $NaClO_4$. The uranium concentration of each sample was calculated from the measured concentration of the stock solution and the dilution factor during titration. The basic solutions (pH 9, 10, 11) of 0.1 M $NaClO_4$ were prepared with 99.99% decarbonated NaOH (Sigma-Aldrich) in a glove box. For purification, $NaClO_4{:}H_2O$ (Merck, analysis grade) was carefully recrystallized, diluted in pure water and filtered through a membrane filter (Advantec, mixed cellulose esters) with the pore size of 100 nm. By using LIBD system before the titration of the samples, it was proved that all the chemicals, including the uranium stock solution, contained no detectable colloidal particles. To minimize the adsorption of the formed uranium colloids on the surface of the test tubes, U(VI) samples were prepared in Teflon FEP tubes (Oak Ridge). U(VI) samples were equilibrated at 298±2 K during over one year in a glove box. The pH measurement was carried out with a glass combination pH electrode (Orion, Ross type, 6 mm diameter) calibrated with four pH buffer solutions (Mettler Toledo, pH 2.00-9.21). For the size measurement of uranium colloidal particles, a portion of the uranium sample was delivered in a sealable quartz cell out of a glove box.

4.2 Particle sizing of uranium colloids with a laser wavelength of 532 nm

Fig. 12 shows the breakdown probability as a function of the pH of the uranium sample at the concentration of approximately 1 parts per million (ppm). The breakdown probabilities were almost zero below a pH value of 5.5, at which the given uranium concentration exceeded the solubility limit of U(VI) hydrolysis compounds. At a pH of 5.6 there was an observable increase in breakdown probability and uranium colloids were formed. After the onset of breakdown, the breakdown probability increased with the increase of the pH value. Using LIBD, we determined the mean particle size to be approximately 600 nm in the sample at a pH of 7.0 just after preparing the sample. In this experiment, the particle size of colloidal uranium was determined by using the calibration curve made from the effective focal length of the spatial distribution of breakdown events of polystyrene nanoparticles.

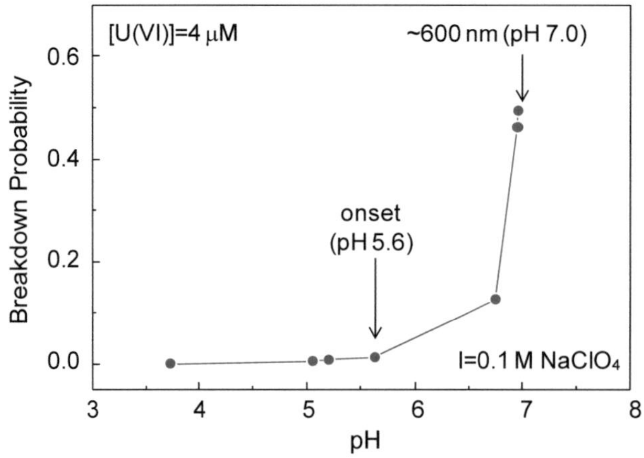

Fig. 12. Breakdown probability as a function of pH shows the formation of uranium colloidal particles.

One year after the sample preparation, the LIBD experiment was performed again to measure the size of the colloidal particles for the same sample. The solid symbol in Fig. 13 indicates the effective focal length obtained from the colloidal uranium particles, which were acidified to a pH of 5.8 after one year. When the effective focal length determined from the uranium sample was compared directly with the effective focal length values of the reference polystyrene particles, the mean particle size was found to be approximately 43 nm. From the change of the pH value from 7.0 to 5.8 and the change of the particle size from approximately 600 to 50 nm, we speculate that only small particles remained as suspended colloidal particles in the solution (Jung et al., 2009).

Because the characteristics of LIBD are dependent on not only the particle size but also the material properties, it is essential to compare the particle size determined using LIBD with the size determined using other methods, such as PCS and TEM. We tried to measure the particle size of uranium colloids, designated by the solid symbol in Fig. 13, using PCS, but we could not measure the particle size of uranium colloids because of their extremely low concentration. For the purpose of comparison, we prepared another sample that showed a strong enough light scattering intensity of uranium particles for PCS (Jung et al., 2009). The supernatant of the uranium sample (1.0 mM of U(VI) at pH 6.12) was carefully concentrated by ultrafiltration using a cellulose filter (10 kD, Centrion YM-10, Millipore). The concentrated supernatant was collected in a sealable quartz cell.

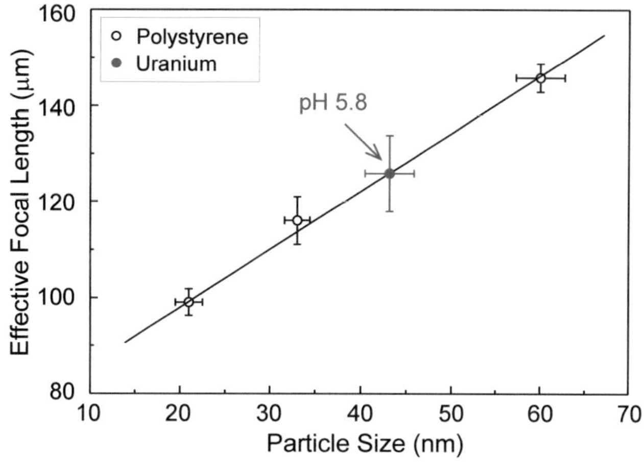

Fig. 13. Size measurement of colloidal uranium particles (solid symbol). The calibration curve was obtained from the effective focal lengths of the polystyrene reference particles with different sizes of 21, 33 and 60 nm in diameter (open symbols).

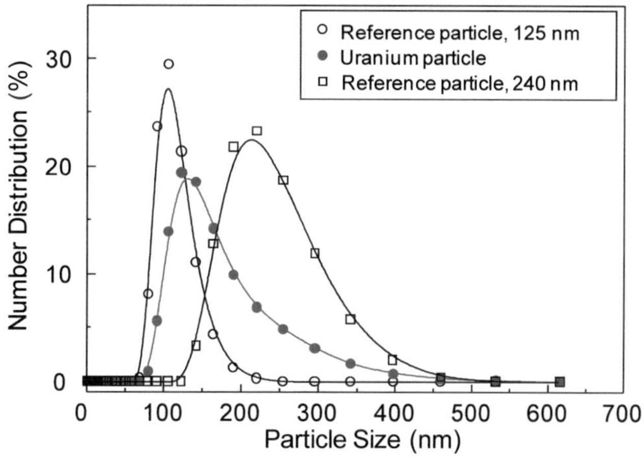

Fig. 14. Size measurement of colloidal uranium particles (solid symbol) and polystyrene reference particles (open symbols) using PCS.

The experimental number distributions for three different particles (two reference polystyrene samples and concentrated uranium sample) measured by PCS are presented in Fig. 14. The average sizes of the polystyrene particles were 143 and 257 nm with polydispersity indexes ranging from 0.05 to 0.08, respectively. Due to the hydrodynamic effect, these sizes were slightly larger than the sizes of 125 and 240 nm in diameter determined by the manufacturer using TEM. The average size of uranium colloids in Fig. 14 was 210 nm with a polydispersity index of 0.15. The polydispersity index of uranium colloids, higher than those of reference particles, indicates the heterogeneity of uranium colloids in size and morphology. Nevertheless, the cumulated mean sizes obtained by repeating the experiments for one week were in the range of 205-215 nm (Jung et al., 2009).

A TEM image of colloidal uranium particles and the number-weighted particle size distribution from the TEM image analysis are shown in Fig. 15. The average size of the uranium colloids shown in Fig. 15 is approximately 200 nm, with a high polydispersity. Even though the shapes of the uranium particles are not ideal spheres, the average size of the uranium particles obtained from TEM is found to reasonably agree with the size determined by PCS.

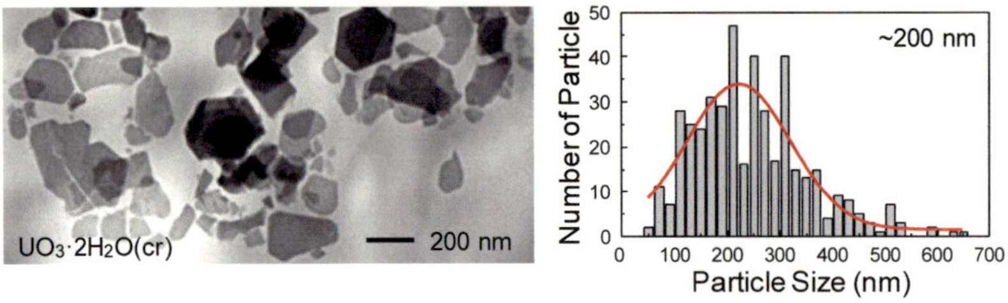

Fig. 15. TEM image of colloidal uranium particles and number-weighted particle size distribution obtained from TEM analysis.

Fig. 16 shows the particle sizes of concentrated uranium colloids determined by using the calibration curves made from the spatial distribution of 3,000 breakdown events for the reference polystyrene particles (21, 33, 60, 125 and 240 nm in diameter). The open and filled symbols in Fig. 16 represent the mean values obtained by repeating the experiments on the polystyrene and uranium particles, respectively. When the effective focal lengths determined in the uranium sample were compared directly with those of the reference particles, the estimated mean particle size was approximately 158 nm. Taking into account the polydispersity of the uranium colloids, the mean particle size determined using LIBD was in reasonable agreement with the sizes determined using PCS and TEM. Thus, we can obtain information on the relative differences or changes in the particle size of uranium colloids by using the present LIBD method (Jung et al., 2009).

To minimize the polydispersity of uranium particles, the large particles in the supernatant of the uranium sample were eliminated by filtration using a membrane filter (mixed cellulose ester, Advantec) having a pore size of 100 nm. The solid symbol designated by arrow in Fig. 16 represents the datum obtained from the filtrated sample. When the uranium datum was compared directly with the polystyrene data, the mean particle size of the filtrated uranium colloids was found to be approximately 31 nm. We tried to measure the particle size of the filtrated uranium colloids using PCS and TEM, but we could not measure the particle size of uranium colloids because of their extremely low concentration. Fig. 17 illustrates the spatial distributions of breakdown events for the same uranium particles as can be seen in Fig. 16. As is apparent from Fig. 17, the uranium colloids in the filtered sample show shorter effective focal length in the spatial distribution of breakdown events than that measured from the unfiltered, supernatant sample.

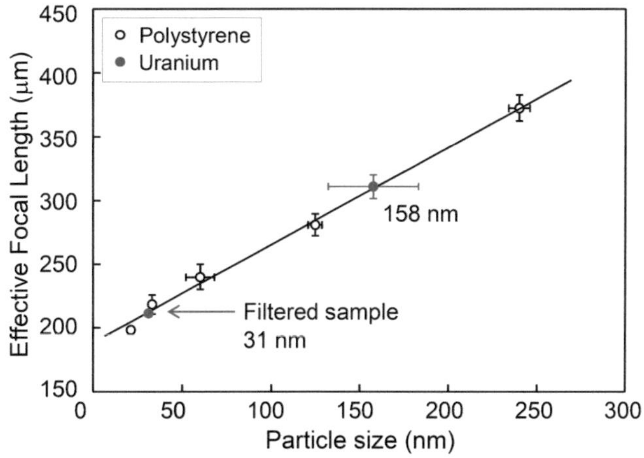

Fig. 16. Size measurements of colloidal uranium particles using effective focal lengths. The calibration curve was obtained from the size of the polystyrene reference particles having diameters of 21, 33, 60, 125 and 240 nm.

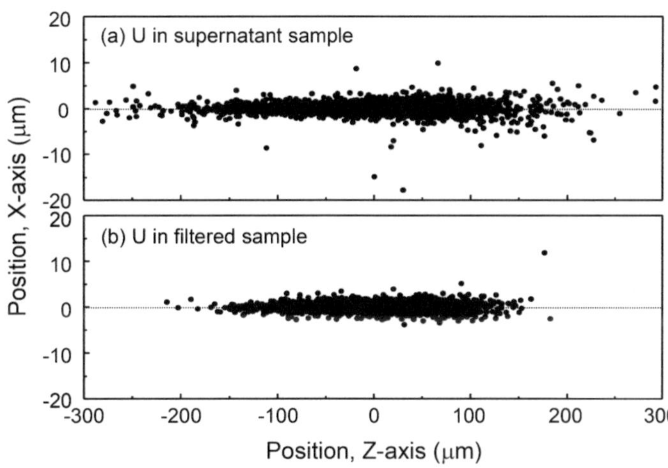

Fig. 17. Spatial distributions of 3,000 breakdown events for two different uranium particles in supernatant sample (a) and filtered sample (b).

4.3 Particle sizing of alumina colloids with a laser wavelength of 441 nm

Fig. 18 shows the effective focal lengths measured at a laser wavelength of 532 nm as a function of particle size for three different materials: polystyrene, silica and alumina colloidal particles. As already reported in the previous section, the effective focal length for polystyrene particles, designated as empty circles, is observed to be directly correlated to the particle size. In contrast, the effective focal lengths, designated as solid circles for alumina and square symbols for silica particles, are not correlated to the particle size. Although the solid lines fitted to the data points show slightly increasing behavior with poor correlation coefficients, it seems that the effective focal lengths for silica (Jung et al., 2011) and alumina are almost unchanged as the particle size increases.

The reason for these phenomena can be understood in terms of the difference in the ionization potentials (IPs) of these materials: ~7.8 eV for polystyrene, ~11.7 eV for silica and ~9.1 eV for alumina (Yun, 2007; Porter et al., 1955; B.H. Stephan, 1991). It is generally accepted that higher irradiance is required for the breakdown of a material with a higher IP. When a laser pulse at a wavelength of 532 nm (photon energy of ~2.33 eV) is used for the breakdown, simultaneous 4-photon absorption is required to induce MPI of polystyrene, while at least simultaneous 5-photon absorption is required for the MPI of silica and alumina. Thus, higher laser pulse energy is a prerequisite for the breakdown of silica and alumina, compared with that for polystyrene particles. It was reported that the breakdown threshold energy of silica was approximately 1.8-2.2 times higher than that of polystyrene particles (Bundschuh et al., 2001; Yun, 2007). The results shown in Fig. 18 imply that a size determination of silica and alumina particles using the spatial distribution of breakdown events measured at a laser wavelength of 532 nm is invalid under the present experimental conditions (Jung et al., 2011).

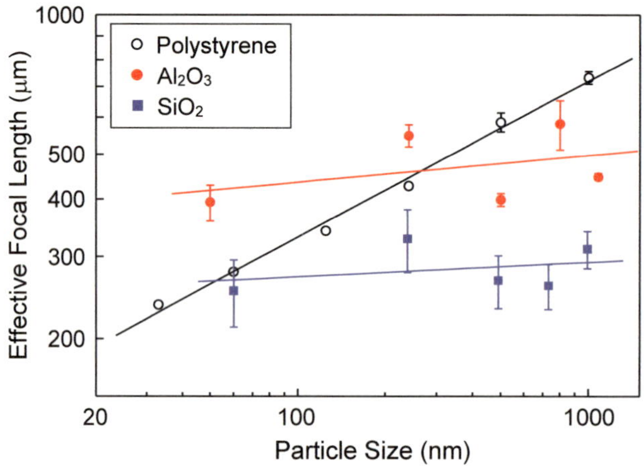

Fig. 18. Effective focal length as a function of particle size for polystyrene, silica and alumina particles at the laser wavelength of 532 nm.

As illustrated in Fig. 7, the irradiance distribution along the laser beam propagation axis can be divided into three distinct regions: the central region of highest irradiance near the focal point (region A), and the outside upstream and downstream regions of weak irradiance (regions B and C). The correlation between the effective focal length and the particle size for polystyrene reflects the fact that the breakdown of large particles is induced at the outside upstream and downstream regions, as well as at the central region, because the threshold energies of large particles are lower than those of small particles. The unchanged effective focal lengths for several silica and alumina particles indicate that breakdowns are only induced at the central region of high irradiance. It is speculated that these phenomena occur because the irradiance of the outside upstream and downstream regions is too low for breakdown, which requires simultaneous 5-photon absorption for the MPI (Jung et al., 2011).

The results shown in Fig. 19 illustrate the dependence of the effective focal length on the laser pulse energy, corroborating this speculation. Fig. 19 shows the effective focal length as a function of laser pulse energy for polystyrene, silica and filtrated uranium particles. For polystyrene particles with different diameters (21, 33 and 60 nm), the effective focal length increases almost linearly with the increase of the laser pulse energy, ranging from 0.4 to 0.6 mJ. Similar to that of polystyrene particles, the effective focal length of filtrated uranium shows increasing behavior with increasing laser pulse energy. Our recent measurement of solubility product of U(VI) hydrolysis compounds suggests that the colloidal particles are $UO_3 \cdot 2H_2O$(cr) (Cho et al., 2008). The information on the IP of $UO_3 \cdot 2H_2O$(cr) particles is unknown at this time. However, under the assumption that the IP of $UO_3 \cdot 2H_2O$(cr) is not much different from the IP of UO_2 (IP=~6.17 eV) (Han et al., 2003), the MPI of the uranium compound can easily occur with 3- or 4-photon absorption at the photon energy level of ~2.33 eV. For silica (60 nm in diameter), however, the effective focal length does not change as the laser pulse energy increases, ranging from 0.5 to 0.6 mJ (Jung et al., 2011). With laser pulse energy above 0.6 mJ, breakdown occurs even for ultra-pure water.

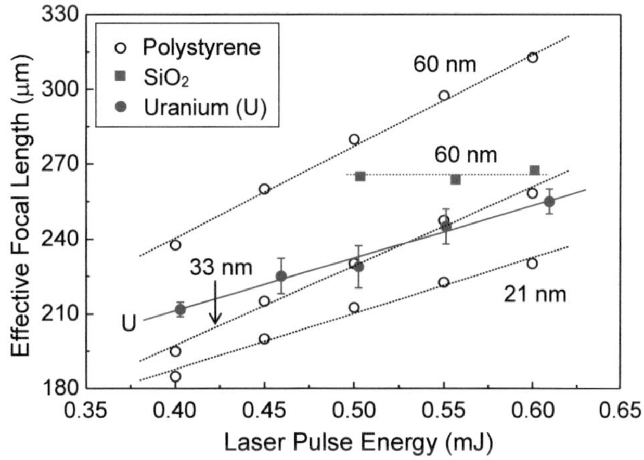

Fig. 19. Effective focal length as a function of laser pulse energy for polystyrene, uranium and silica particles at the laser wavelength of 532 nm.

When the laser photon energy is high enough, the MPI processes of silica and alumina may occur via simultaneous 4-photon absorption. As a result, breakdown events may occur in the upstream and downstream regions, as well as in the central region, because 4-photon absorption occurs at relatively low irradiance compared with 5-photon absorption. In this case, the correlation between the effective focal length in the spatial distribution of breakdown events and the particle size is expected even for silica and alumina particles.

To observe this effect, the dependence of the effective focal length on the particle size was investigated with a laser wavelength of 441 nm (photon energy of ~2.81 eV), and the results are shown in Fig. 20. It should be noted that the measurement of the incident wavelength-dependent characteristics of LIBD was performed by using a laser pulse generated from a wavelength tunable OPO (Optical Parametric Oscillator, OPOTEK, Vibrant) system. In this experiment, the OPO laser beam was focused on the interior of a sample cell using a bi-convex lens with a focal length of 15 mm. To prevent scattered laser light from reaching the CCD camera, a notch filter of ~1% transmission at 441 nm was inserted between the macro-microscope and the sample cell. When breakdown was induced by 441 nm laser radiation, the 4-photon absorption energy of approximately 11.2 eV exceeded the IP of ~9.1 eV for alumina. Therefore, the dependence of the effective focal length on the particle size was obviously observed for alumina, which is designated as filled symbols in Fig. 20, as expected. More recently, the results on the calibration curve obtained with 355 nm laser radiation (photon energy of ~3.49 eV) was reported for silica particles and the similar dependence of the effective focal length on the particle size was observed (Jung et al., 2011).

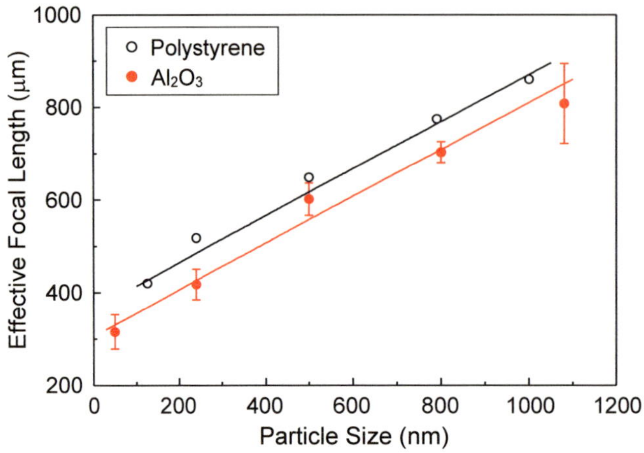

Fig. 20. Effective focal length as a function of particle size for polystyrene and alumina particles at the laser wavelength of 441 nm.

5. Summary

Several different LIBD systems using different detection devices, such as a PZT, a CCD camera and an optical probe beam, were described with their own particle size determination methods. In this chapter, material-dependent characteristics of LIBD were discussed for four different colloidal particles: polystyrene, $UO_3 \cdot 2H_2O$(cr), silica, and alumina. When a focused laser pulse of 532 nm wavelength is used to generate laser-induced plasma, simultaneous 4-photon absorption is required to induce MPI of polystyrene and $UO_3 \cdot 2H_2O$(cr); however, at least 5-photon absorption is required to induce MPI of silica and alumina. Thus, it was observed that the spatial distribution of the breakdown events measured with 532 nm laser radiation increased with increasing particle size for polystyrene and uranium compound, while spatial distribution of breakdown events was nearly unchanged for silica and alumina. The unchanged effective focal length with increase of particle size for silica and alumina resulted from the laser irradiance in the outside upstream and downstream regions, which was insufficient to induce MPI of these particles via simultaneous 5-photon absorption. When breakdown was induced by a laser wavelength of 441 nm, a correlation between the effective focal length and the particle size was observed for alumina because the simultaneous 4-photon absorption energy exceeds the IP of this material. Thus, the calibration curve obtained with an appropriate laser wavelength can be used to determine unknown particle sizes, even for materials with a high IPs.

6. Acknowledgment

This work was supported by the nuclear research and development program (2007-2011) through the National Research Foundation (NRF) of Korea funded by the Ministry of Education, Science and Technology (MEST). Part of this work was supported by the research and education program (2007) through the NRF of Korea funded by the MEST. The authors thank Mr. Hyungsoo Jung, Mr. Kilho Lee, and Mr. Sanghyun Park for obtaining data in section 3.1 and 3.2. The authors thank Ms. M.R. Park for obtaining data in section 4.3. The authors thank Dr. H.M. Kim for analyzing TEM results in section 4.2.

7. References

Ajiro, T.; Fujimori, H.; Matsui, T. & Izumi, S. (1992). Particle Size Dependence of Correlation Between Plasma Emission Delay Time and Plasma Emission Intensity of Laser Breakdown Induced by a Particle. *Japanese Journal of Applied Physics*, Vol. 31, No. 9, pp. 2760-2761, ISSN 0021-4922

Bundschuh, T.; Knopp, R.; Müller, R.; Kim, J.I.; Neck, V. & Fanghänel, Th. (2000). Application of LIBD to the Determination of the Solubility Product of Thorium(IV)-colloids. *Radiochimica Acta*, Vol. 88, No. 9-11, pp. 625-629, ISSN 0033-8230

Bundschuh, T.; Hauser, W.; Kim, J.I.; Knopp, R. & Scherbaum, F.J. (2001a). Determination of Colloid Size by 2-D Optical Detection of Laser Induced Plasma. *Colloids and Surfaces A: Physicochemical and Engineering Aspects*, Vol. 180, pp. 285-293, ISSN 0927-7757

Bundschuh, T.; Knopp, R. & Kim, J.I. (2001b). Laser-induced Breakdown Detection (LIBD) of Aquatic Colloids with Different Laser Systems. *Colloids and Surfaces A: Physicochemical and Engineering Aspects*, Vol. 177, pp. 47-55, ISSN 0927-7757

Bundschuh, T.; Wagner, T. U. & Koster, R. (2005). Laser-induced Breakdown Detection (LIBD) for the Highly Sensitive Quantification of Aquatic Colloids. Part I: Principle of LIBD and Mathematical Model. *Particle and Particle Systems Characterization*, Vol. 22, No. 3, pp. 172-180, ISSN 1521-4117

Cho, H.-R.; Jung, E.C. & Jee, K.Y. (2008). Probe Beam Detection of Laser-induced Breakdown for Measuring Solubility of Actinide Compounds. *Japanese Journal of Applied Physics*, Vol. 47, No. 5, pp. 3530-3532, ISSN 0021-4922

Fujimori, H.; Matsui, T.; Ajiro, T.; Yokose, K.; Hsueh, Y-M. & Izumi, S. (1992). Detection of Fine Particles in Liquids by Laser Breakdown Method. *Japanese Journal of Applied Physics*, Vol. 31, No. 5, pp. 1514-1518, ISSN 0021-4922

Han, J.; Kaledin, L.A.; Goncharov, V.; Komissarov, A.V.; Heaven, M.C. (2003). Accurate Ionization Potentials for UO and UO_2: A Rigorous Test of Relativistic Quantum Chemistry Calculations. *Journal of American Chemical Society*, Vol. 125, No. 24, pp. 7176-7177, ISSN 0002-7863

Hauser, W.; Geckeis, H.; Kim, J.I. & Fierz, Th. (2002). A Mobile Laser-induced Breakdown Detection System and Its Application for the in situ-monitoring of Colloid Migration. *Colloids and Surfaces A: Physicochemical and Engineering Aspects*, Vol. 203, 37-45, ISSN 0927-7757

Izumida, S.; Onishi, K. & Saito, M. (1998). Estimation of Laser-induced Breakdown Threshold of Microparticles in Water. *Japanese Journal of Applied Physics*, Vol. 37, No. 4, pp. 2039-2042, ISSN 0021-4922

Jung, E.C.; Yun, J.-I.; Kim, J.I.; Park, Y.J.; Park, K.K.; Fanghänel, Th. & Kim, W.H. (2006). Size Measurement of Nanoparticles Using the Emission Intensity Distribution of Laser-induced Plasma. *Applied Physics B*, Vol. 85, pp. 625-629, ISSN 0946-2171

Jung, E.C.; Yun, J.-I.; Kim, J.I.; Bouby, M.; Geckeis, H.; Park, Y.J.; Park, K.K.; Fanghänel, Th. & Kim, W.H. (2007). Measurement of Bimodal Size Distribution of Nanoparticles by Using the Spatial Distribution of Laser-induced Plasma. *Applied Physics B*, Vol. 87, pp. 497-502, ISSN 0946-2171

Jung, E.C.; Cho, H.-R.; Park, K.K.; Yeon, J.-W. & Song, K. (2009). Nanoparticle Sizing by a Laser-induced Breakdown Detection Using an Optical Probe Beam Deflection. *Applied Physics B*, Vol. 97, pp. 867-875, ISSN 0946-2171

Jung, E.C.; Cho, H.-R.; Park, M.R. (2011). Laser-Induced Breakdown Detection of Colloidal Silica Nanoparticles. Sumitted to *Applied Physics B*, ISSN 0946-2171

Kaegi, R.; Wagner, T.; Hetzer, B.; Sinnet, B.; Tzvekov, G. & Boller, M. (2008). Size, Number and Chemical Composition of Nanosized Particles in Drinking Water Determined by Analytical Microscopy and LIBD. *Water Research*, Vol. 42, pp. 2778-2786, ISSN 0043-1354

Kim, J.I. (2006). Significance of Actinide Chemistry for the Long-term Safety of Waste Disposal. *Nuclear Engineering and Technology*, Vol. 38, No. 6, pp. 459-482, ISSN 1738-5733

Kim, J.I. & Walther C. (2007). Laser-induced Breakdown Detection, In *Environmental Colloids and Particles: Behaviour, Separation and Characterisation*, Wilkinson, K.J. & Lead, J.R. (Ed.), pp. 555-612, Wiley, ISBN 978-0-470-02432-4, Chichester, U.K.

Kim, J.W.; Son, J.A.; Yun, J.-I.; Jung, E.C.; Park, S.H. & Choi, J.G. (2008). Analysis of Laser-induced Breakdown Images Measuring the Sizes of Mixed Aquatic Nanoparticles. *Chemical Physics Letters*, Vol. 462, No. 1-3, pp. 75-77, ISSN 0009-2614

Kitamori, T.; Yokose, K.; Suzuki, K.; Sawada, T. & Gohshi, Y. (1988). Laser Breakdown Acoustic Effect of Ultrafine Particle in Liquids and Its Application to Particle Counting. *Japanese Journal of Applied Physics*, Vol. 27, No. 6, pp. L983-L985, ISSN 0021-4922

Kitamori, T.; Yokose, K.; Sakagami, M. & Sawada, T. (1989). Detection and Counting of Ultrafine Particles in Ultrapure Water Using Laser Breakdown Acoustic Method. *Japanese Journal of Applied Physics*, Vol. 28, No. 7, pp. 1195-1198, ISSN 0021-4922

Knopp, R.; Neck, V. & Kim, J.I. (1999). Solubility, Hydrolysis and Colloid Formation of Plutonium(IV). *Radiochimica Acta*, Vol. 86, pp. 101-108, ISSN 0033-8230

Möri, A.; Alexander, W.R.; Geckeis, H.; Hauser, W.; Schäfer, T.; Eikenberg, J.; Fierz, T.; Degueldre, C. & Missana, T. (2003). The Colloid and Radionuclide Retardation Experiment at the Grimsel Test Site: Influence of Bentonite Colloids on Radionuclide Migration in a Fractured Rock. *Colloids and Surfaces A: Physicochemical and Engineering Aspects*, Vol. 217, No. 1-3, pp. 33-47, ISSN 0927-7757

Neck, V.; Kim, J.I.; Seidel, B.S.; Marquardt, C.M.; Dardenne, K.; Jensen, M.P. & Hauser, W. (2001). A Spectroscopic Study of the Hydrolysis, Colloid Formation and Solubility of Np(IV). *Radiochimica Acta*, Vol. 89, No. 7, pp. 439-446, ISSN 0033-8230

Neck, V.; Altmaier, M.; Müller, R.; Bauer, A. & Fanghänel, Th. (2003). Solubility of Crystalline Thorium Dioxide. *Radiochimica Acta*, Vol. 91, No. 5, pp. 253-262, ISSN 0033-8230

Opel, K.; Weiss, S.; Hübener, S.; Zänker, H. & Bernhard, G. (2007). Study of the Solubility of Amorphous and Crystalline Uranium Dioxide by Combined Spectroscopic Methods. *Radiochimica Acta*, Vol. 95, No. 3, 143-149, ISSN 0033-8230

Porter, R.F.; Chupka, W.A. & Inghram, M.G. (1955). Mass Spectrometric Study of Gaseous Species in the Si-SiO_2 System. *The Journal of Chemical Physics*, Vol. 23, pp. 216-217, ISSN 0032-9606

Saito, M.; Izumida, S.; Onishi, K. & Akazawa, J. (1999). Detection Efficiency of Microparticles in Laser Breakdown Water Analysis, *Journal of Applied Physics*, Vol. 85, No. 9, pp. 6353-6357, ISSN 0021-8979

Scherbaum, F.J.; Knopp, R. & Kim, J.I. (1996). Counting of Particle in Aqueous Solutions by Laser-induced Photoacoustic Breakdown Detection. *Applied Physics B*, Vol. 63, pp. 299-306, ISSN 0946-2171

Stephan, B.H.; Stephen, W.M. (1991). Determination of the Ionization Potentials of Aluminum Oxides via Charge Transfer. *The Journal of Physical Chemistry*, Vol. 95, No. 23, pp. 9091-9094, ISSN 0022-3654

Wagner, T. U.; Bundschuh, T. & Koster, R. (2005). Laser-induced Breakdown Detection (LIBD) for the Highly Sensitive Quantification of Aquatic Colloids. Part II: Experimental Setup of LIBD and Applications. *Particle and Particle Systems Characterization*, Vol. 22, No. 3, pp. 181-191, ISSN 1521-4117

Walther, C.; Bitea, C.; Hauser, W.; Kim, J.I. & Scherbaum, F.J. (2002). Laser Induced Breakdown Detection for the Assessment of Colloid Mediated Radionuclide Migration. *Nuclear Instruments and Methods in Physics Research B*, Vol. 195, pp. 374-388, ISSN 0168-583X

Walther, C.; Cho, H.-R. & Fanghänel, Th. (2004). Measuring Multimodal Size Distributions of Aquatic Colloids at Trace Concentrations. *Applied Physics Letters*, Vol. 85, No. 26, pp. 6329-6331, ISSN 0003-6951

Walther, C.; Büchner, S.; Filella, M. & Chanudet, V. (2006). Probing particle Size Distributions in Natural Surface waters from 15 nm to 2 mm by a Combination of LIBD and Single-particle Counting. *Journal of Colloid and Interface Science*, Vol. 301, No. 2, pp. 532-537, ISSN 0021-9797

Walther, C.; Cho, H.-R.; Marquardt, C.M.; Neck, V.; Seibert, A.; Yun, J.-I. & Fanghänel, Th. (2007). Hydrolysis of Plutonium(IV) in Acidic Solutions: No Effect of Hydrolysis on Absorption-spectra of Mononuclear Hydroxide Complexes. *Radiochimica Acta*, Vol. 95, No. 1, pp. 7-16, ISSN 0033-8230

Yun, J.-I. (2007). Material Dependence of Laser-induced Breakdown of Colloidal Particles in Water. *Journal of the Optical Society of Korea*, Vol. 11, No. 1, pp. 34-39, ISSN 1226-4776

Zapka, W.; Tam, A.C. (1982). Photoacoustic pulse generation and probe-beam deflection for ultrasonic velocity measurements in liquids. *Applied Physics Letters*, Vol. 40, No. 4, pp. 310-312, ISSN 0003-6951

21

Bulk Nanocrystalline Thermoelectrics Based on Bi-Sb-Te Solid Solution

L.P. Bulat[1], D.A. Pshenai-Severin[2], V.V. Karatayev[3], V.B. Osvenskii[3], Yu.N. Parkhomenko[3,5], M. Lavrentev[3], A. Sorokin[3], V.D. Blank[4], G.I. Pivovarov[4], V.T. Bublik[5] and N.Yu. Tabachkova[5]

1. Introduction

Thermoelectric energy conversion represents one of ways of direct conversion of the thermal energy to the electric energy. The thermoelectric converters - a thermoelectric power generator or a thermoelectric cooler are solid-state devices, therefore it possesses following advantages: environmentally cleanliness, simplicity of management and convenience of designing, high reliability and possibility of longtime operation without service, absence of moving parts, absence of noise, vibration and electromagnetic noise, compactness of modules, independence of space orientation, ability to carry out of considerable mechanical overloads. Despite obvious advantages of thermoelectric conversion it has the important lack - rather small value of the efficiency η; in the best cases η = (5 - 8) %. Therefore thermoelectric generators are used today, as a rule, only in «small power», where it is impossible or is economically inexpedient to bring usual electric mains: for power supply of space missions, at gas and oil pipelines, for power supply of sea navigating systems, etc. Nevertheless, the thermoelectric method of utilization of waste heat from units of cars and vessels is unique technically possible. It appears the thermoelectric generators can save up to 7 % of automobile fuel. The thermoelectric coolers become economically justified at enough small cooling power Qc as a rule no more than 10 - 100 W. However the thermoelectric coolers are widely applied in the most different areas: domestic refrigerators, water-chillers, picnic-boxes; coolers for medicine and biology, for scientific and laboratory equipment, refrigeration systems for transport facilities. Very important area connects with strong up-to-date requirements for the thermal management of micro- and optoelectronics elements, including microprocessors and integrated circuit; the requirements have essentially increased owing to the increase in their speed and miniaturization. And the desired value of local heat removal from concrete spots of chips can be realized only by means of the thermoelectric cooling.

[1]*National Research University ITMO, St. Petersburg, Russia*
[2]*Ioffe Physical Technical Institute, St Petersburg, Russia*
[3]*GIREDMET Ltd., Moscow, Russia*
[4]*Technological Institute of Superhard and New Carbon Materials, Troitsk, Russia*
[5]*National University of Science and Technology "MISIS", Moscow, Russia*

It is known, that the efficiency of thermoelectric generators and the coefficient of performance (COP) of thermoelectric coolers are defined by the dimensionless parameter of a thermoelectric material $ZT = \frac{\sigma\alpha^2}{\kappa}T$, where T - is the absolute temperature, Z - thermoelectric figure of merit, σ, κ, α - accordingly electric conductivity, thermal conductivity and Seebeck coefficient (thermoelectric power) of a used material.

Today the best commercial thermoelectric materials (thermoelectrics) has the efficiency ZT =1.0. Let us underline that ZT increased from 0.75 only to 1.0 during lust five decades. Obviously, competitiveness of thermoelectric generators and coolers will rise if it will be possible to increase the figure of merit. Thus the thermoelectric generation and cooling can provide a weighty contribution to a decision of the problem of utilization of renewed energy sources, a recycling of a low potential heat, and maintenance of storage of a foodstuff. Recently the increasing attention is involved to the thermoelectric refrigeration and the electric power generation as to environmentally clean methods. It is caused by several reasons. The main of them is caused by new scientific results on improve of the thermoelectric figure of merit. Important results in the development of highly effective nanostructured thermoelectric materials have been published last decade; see for example the reviews (Dresselhaus et al., 2007; Minnich et al., 2009; Dmitriev & Zvyagin, 2010; Lan et al., 2010). The thermoelectric efficiency ZT = 2.4 has been reached at T = 300K in the p-type semiconductor Bi_2Te_3/Sb_2Te_3 in superlattices with quantum wells (Venkatasubramanian et al, 2001); the estimation indicates that the value ZT ~ 3.5 has been received at T = 575 K in nanostructured n-type PbSeTe/PbTe with quantum dots (Harman et al., 2000, 2005). It is possible also to include thermotunnel elements (thermal diodes) to nanostructured thermoelectrics in which there exists the electron tunneling through a narrow vacuum or air gap (Tavkhelidze et al., 2002). The efficiency ZT=1.7 at the room temperature was received experimentally in thermoelements with cold junctions, consisting from the semiconductor branches of p-type $Bi_{0.5}Sb_{1.5}Te_3$ and n-type $Bi_2Te_{2.9}Se_{0.1}$ (Ghoshal et al., 2002a, 2002b). Let us note also the important results received in a set of papers, for example (Shakouri & Bowers, 1997), which specify in perspectives of the use of the emission nanostructures for creation of effective thermoelectric energy converters and coolers. The nanostructuring gives a new way of improvement of the thermoelectric efficiency because the governance of the sizes of nanostructured elements is a new important parameter for influence on the thermoelectric properties of a material.

Unfortunately the best values of ZT that were specified in nanostructures based on superlattices with quantum wells and quantum dots have not been reproduced in one laboratory of the world. On the other hand fabrication of such superlattices uses very expensive technologies; therefore industrial manufacture of such nanostructures is very problematic from the economical point of view. Good values of the thermoelectric figure of merit in thermotunnel devices and in thermoelements with point contacts also have not been reproduced. Therefore the special interest represents a creation of thermoelectric nanostructures by means of an adaptable to streamlined production and a cheap technique. An example of such technology is fabrication of bulk nanostructured thermoelectric samples by ball milling of initial materials with subsequent hot pressing (Poudel et al., 2008; Bublik et al., 2009; Bulat et al., 2008a, 2008b, 2009, 2011b; Minnich et al., 2009; Lan et al., 2010), spark plasma sintering (Bublik et al., 2010a, 2010b) or extrusion (Vasilevskiy et al., 2010). In Ref.

(Poudel et al., 2008) the value ZT = 1.4 have been received at T=100^{0}C and ZT = 1.2 at the room temperature in such bulk nanostructured thermoelectrics fabricated from the solid solutions based on p-type Bi-Sb-Te.

Thus from the set forth above it can be concluded that the following reliable preconditions of the obtaining of the high thermoelectric figure of merit in the nanostructured thermoelectrics take place: (a) the experimental results specify a possibility for the achievement of a high thermoelectric figure of merit in nanostructured thermoelectrics of various types; (b) in particular some experimental results confirm the possibility of the obtaining of high figure of merit in bulk nanostructured semiconductors. Experimental and theoretical results that obtained by the authors during lust few years on investigation of bulk nanocrystalline thermoelectrics based on Bi-Sb-Te solid solution including nanocomposites are summarized, systematized and analyzed in the present chapter.

2. Experiment

2.1 Fabrication of bulk nanocrystalline thermoelectrics

Two stages should be executed for preparation of bulk nanocrystalline materials. At first a powder from nanoparticles should be fabricated, and then it should be consolidated into a bulk sample. A crystalline thermoelectric material with high thermoelectric efficiency should be chosen as an initial material for the nanopowder preparation. In our case the solid solution based on *p*-type $Bi_xSb_{2-x}Te_3$ was selected as the initial material (Bublik et al., 2009, 2010a; Bulat et al., 2008a, 2008b, 2009a, 2009b, 2011b). It was grown up by zone melting method; and the dimensionless figure of merit ZT = 1.0 was measured along the C axis at the room temperature in primary samples. The initial crystalline material was grinded and purifying. The mechanoactivation process (the ball milling) is the most convenient and cheap way for fabrication of a nanopowder. We used the high-speed planetary mill AGO-2U to achieve the further superthin crushing and to prepare the nanopowder. Other types of mills: the Activator 2S, Retsch PM 400 also were applied at different stages of the nanopowder preparation. The processing of the powders fabrication in the mill is made by steel spheres which were collided with acceleration up to 90g. Tightly closed containers of the mill rotate in flowing water that protects a material from a warming up. It is necessary to provide absolute absence of the oxidation of the nanopowder. Therefore all operations were spent in the boxing filled with argon.

The duration of the mechanoactivation processing was varied from 30 min till 2 hours. The diffraction analysis has shown that the main sizes of nanoparticles of the powder are 8-10 nm. The following methods of pressing for fabrication of compact samples from highly active ultradisperse powder have been used (Bublik et al., 2009, 2010a; Bulat et al., 2009a): cold pressing of powders with the subsequent sintering in inert gas; sintering in graphite compression moulds; sintering in steel compression moulds (at more high pressure in comparison with the previous variant). Hot pressing of the nanopowder was made under the pressure in the range from 35 MPa to 3.3 GPa in the range of temperatures from 250 to 490 °C. To prevent the oxidation of nanoparticles all basic operations are made in the atmosphere of argon. As a result, series of compact *p*-type $Bi_xSb_{2-x}Te_3$ samples were produced. The method of spark plasma sintering (SPS) with the equipment SPS–511S for preparation of bulk nanostructure was also used (Bublik et al., 2010b, 2010c).

2.2 Methods of experimental investigation

For investigation of thermoelectric properties it is necessary to know values of four material parameters: Seebeck coefficient (thermoelectric power), electric conductivity, heat conductivity and thermoelectric figure of merit. The heat conductivity measurement of small samples is the most difficult because all traditional techniques of direct measurement are based on passing of a calibrated thermal flow through a sample; but a thermal flow measurement with sufficient accuracy and consideration of all losses on small samples is very complicated. Therefore we had been used the Harman method (Bublik et al., 2009, 2010a) which allows to fulfill the measurement of thermoelectric figure of merit Z directly by measurement of only electric parameters, not mentioning about thermal flows. Besides, the technique allows receiving in the same cycle of the measurement the values of Seebeck coefficient and the electric conductivity also. Then the value of heat conductivity can be calculated using the known value of Z. A mathematical model for calculations of thermoelectric parameters on the Harman method measurement has been developed, and processing of results of the measurement was carried out under specially developed soft.

For determining of the speed of longitudinal acoustic waves and for subsequent calculation of modules of elasticity and the modulus of dilatation the modified echo-pulse method with application of focusing system of an acoustic microscope has been used. The mode of ultrashort probing impulses has been utilized; it has given the opportunity to register separate signals caused by the reflexion of the impulse from walls of a sample. The microstructure of samples was investigated at metallographic sections made on grinding-and-polishing machine "Struers". Microhardness was measured on microhardness gauge PMT-3M by the method of cave-in of diamond tips. The microscope Olympus BX51 was used for the metallographic analysis.

The working capacity and reliability of thermoelectric devises are substantially caused by their strength characteristics. The strength at the extension occupies a special place among them. However for investigated materials the method of direct test for the extension is the extremely inexact for some reasons. Therefore the method of diametrical compression of disk or cylindrical samples was used; the advantage of the method consists that the extension pressure destruction begins inside a sample instead of its surface. Determination of the density of samples was made by the method of hydrostatic weighing. Laboratory analytical electronic scales "KERN", model 770-60 were used. The option "Sartorius" was applied for determination of the weight of a solid in a liquid.

The X-ray diffractometer methods were used for investigation of structure of nanopowder and bulk samples. The phase analysis was carried out by the method of X-rays diffraction with the diffractometer Bruker D8, equipped by the scintillation detector Bruker. The lattice constant of a solid solution of a thermoelectric material was determined by shooting of a diffractograms in the standard symmetric scheme of reflexion. A composition of the solid solution on the basis of measurements of the lattice constant was estimated. Values of nanograins were estimated by sizes of coherent dispersion areas (CDA) determined by the method of X-ray diffractometry on broadening of diffraction maxima. Calculation of CDA and estimation of microdeformation were spent by means of Outset program. The received values of CDA size were compared with the data received by a method of high resolution transmission electron microscopy (HRTEM). The structure of a sample was analyzed by means of creation of return polar figures. They were carried out by shooting of diffractograms in the standard symmetric scheme of reflexion. The following microscope equipment was used: the scanning electron microscope JSM-6480LV with the option for the energy-dispersive analysis INCA DRY; transmission electron microscope JEM 2100 with ultrahigh resolution and X-ray photoelectron spectroscope.

2.3 Structure and mechanical properties of nanopowder

Different types of nanopowder from $Bi_xSb_{2-x}Te_3$ solid solutions with different value of x were prepared with the following duration of the mechanoactivation process (the ball milling): 15 min, 30 min, 60 min and 120 min. For each type of powder the X-ray diffractograms, a distribution of CDA size and HRTEM images were received. A typical diffractogram of $Bi_{0,5}Sb_{1,5}Te_3$ powder is presented in (Bulat et al., 2009b). Examples of distribution of CDA size of the nanopowder prepared from $Bi_{0,5}Sb_{1,5}Te_3$ during 60 min ball milling and the correspondent HRTEM image are shown in Fig.1.

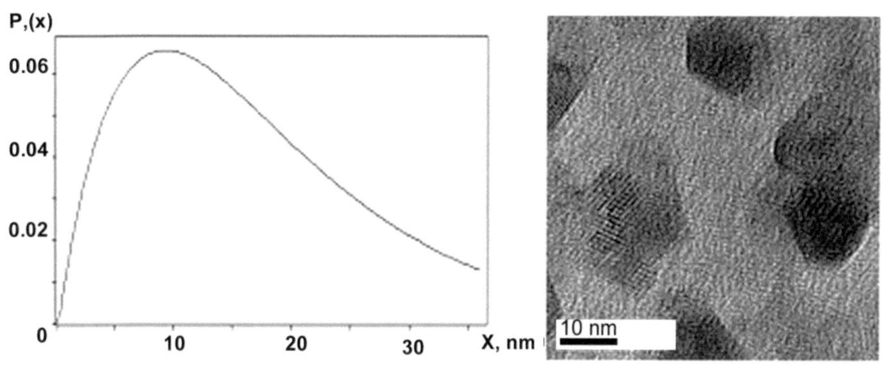

Fig. 1. Distribution of CDA size and TEM image of nanoparticles for $Bi_{0,5}Sb_{1,5}Te_3$ (60 min ball milling)

The nanopowder is the single-phase solid solution of $Bi_{0,5}Sb_{1,5}Te_3$ at each duration of the mechanoactivation process. Microdeformations of nanoparticles did not reveal. The electron microscopic data on the average size of nanoparticles confirm the calculation of CDA size determined by broadening of the diffraction maxima. In particular for 2 hours of ball milling the average value of nanoparticles was 8.5 nm and the greatest size was 35 nm. The distribution of sizes is homogeneous enough. An insignificant increase of the average size of CDA in comparison with the powder passed processing during 60 min is observed. A monotonous reduction of the average size of CDA was observed according to increase of the duration of the ball milling processing from 15 min to 60 min, and the size distribution of nanoparticles became more homogeneous. However the further increase of duration of milling leads any more to a reduction but to an increase of the average CDA size. Microstrains have been found out in the powder after 120 min ball milling. The mean-square microstrains are equal to 0.144%. So small particles cannot contain a dislocation therefore the presence of microstrains in the powder can be connected with heterogeneity of a structure of the solid solution, arising at long processing.

The lattice constants of the solid solution for all duration of processing are: a=0.4284 nm and c=3.0440 nm.

2.4 Structure and mechanical properties of bulk nanocrystalline samples

Taking into account set forth above the 60 min duration of the milling have been chosen for prepare of bulk samples. Therefore the average size of nanoparticles in the starting powder from $Bi_{0,5}Sb_{1,5}Te_3$ was equal to 8 - 10 nm. The cold and hot pressing and SPS method were applied for preparation of the bulk nanostructures.

The cold pressure under 1.5 GPa was carried out within 60 min (without sintering). The correspondent diffractogram shows that the sample is single-phase one, and it does not contain an exudation of moisture of another phase. The diffraction peaks belong to threefold solid solution of $Bi_{0,5}Sb_{1,5}Te_3$ with the lattice constants: a=0.4284 nm and c=3.0440 nm. The diffraction streaks remained strongly blurring as well as in a powder after the milling. The distribution of CDA size after the cold pressure illustrates Fig.2. The average size of CDA after the cold pressure is equal to 12 nm. As follows from Fig.2 at the cold pressing CDA size does not increase practically, and also uniformity of the nanoparticles size distribution increases. Microdeformations did not reveal.

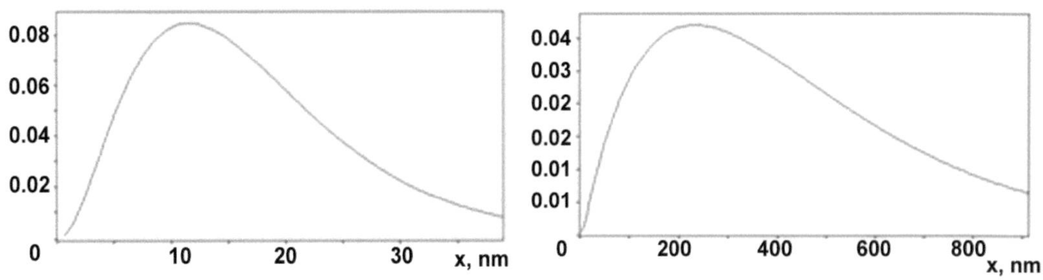

Fig. 2. Distribution of CDA size for bulk $Bi_{0,5}Sb_{1,5}Te_3$ after cold pressure; (a) – without sintering, (b) – after sintering

The investigation has shown that the sintering after cold pressing at temperatures ~ 300^0C and above leads to two effects: (a) to occurrence of the second phase (tellurium) and (b) to origination of microdeformations. For example after the cold pressing a sample was sintered at the temperature 350^0C within 25 min in argon atmosphere. It contains two phases $Bi_{0,5}Sb_{1,5}Te_3$ and tellurium. The lattice constants of the solid solution a=0.4296 nm and c=3.0447 nm are increased in comparison with the lattice constants in the initial sample before sintering. Root-mean-square microdeformation was equal to 0,086 %. Increase of the lattice constants of the solid solution as well as occurrence of microdeformations are apparently results of the excretion of tellurium from the solid solution. Thus as it was marked above microdeformations can be caused by heterogeneity of the solid solution arising at the raised temperature.

Fig.2 shows also the distribution of CDA size for the cold pressed bulk sample of $Bi_{0,5}Sb_{1,5}Te_3$ with the subsequent sintering. The average size of CDA is equal to 230 nm and the maximum size - 900 nm. Character of the size distribution in comparison with the cold pressing without the sintering has not changed, but the curve was displaced towards the big size. The increase of the CDA size testifies that there have passed processes of recrystallization during the sintering at 350 ^{0}C.

The SEM images of the sample received by cold pressing are characterized by high porosity (Fig.3). Pores are dark formations as they do not reflect electrons. The size of pores reaches 5 μm. It is visible at a big resolution that pores are not spherical; and coagulation of small pores takes place, therefore a facet is formed. The pores are faceted as a result of the diffusion processes during the sintering.

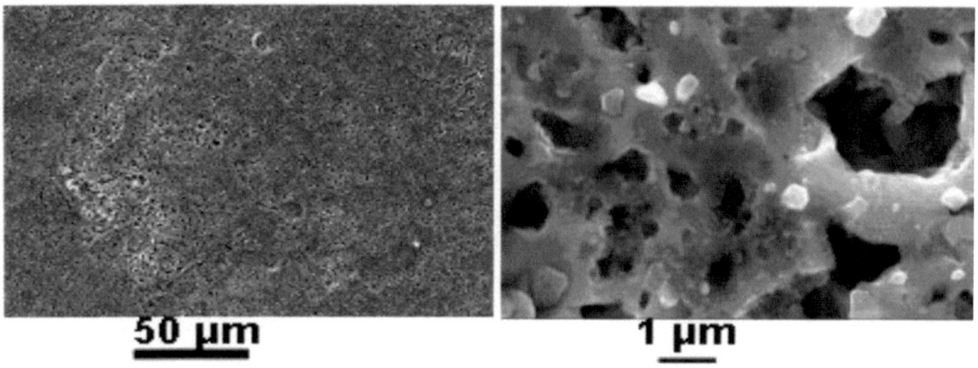

Fig. 3. The SEM images of the bulk $Bi_{0,5}Sb_{1,5}Te_3$ (cold pressing with sintering). The resolution is 500 and 10000 accordantly

The CDA size and structure of samples fabricated by the hot pressing are defined by three factors: a temperature of pressing, duration of stand-up under the loading and a value of pressure. But in any case the hot pressing as like as the cold pressing with the subsequent sintering leads to an occurrence of the second phase and to microdeformations. The increase of the CDA size due to the processes of recrystallization also takes part at the hot pressing. Fig.4 shows a typical distribution of CDA size for the hot pressed bulk sample; it was pressed during 20 min at 0.2 GPa and 289 ^{0}C. The sample contains two phases: $Bi_{0,5}Sb_{1,5}Te_3$ and tellurium. The lattice constants are increased in comparison with the lattice constants in the initial nanopowder. Root-mean-square microdeformation was equal to 0,055 %, the average CDA size ~ 85 nm, the biggest CDA has the size ~ 300 nm.

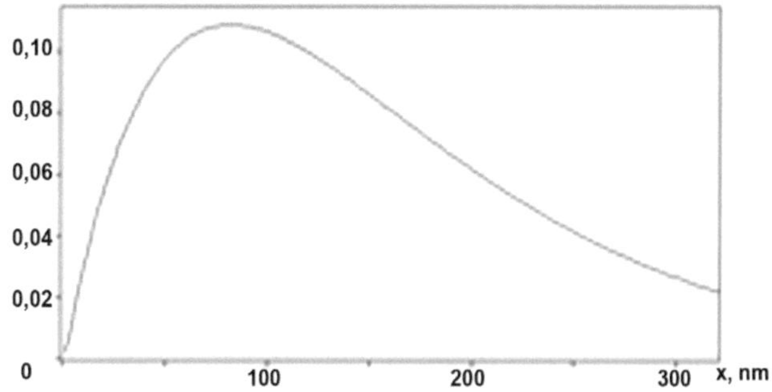

Fig. 4. Distribution of CDA sizes of the bulk $Bi_{0,5}Sb_{1,5}Te_3$ (hot pressing)

The SEM images of the surfaces of this sample are shown at Fig.5. A relief specified to different speed of dissolution arises after chemical polishing. The observable elements of structure: consertal formation, micropores, cracks, allocation of a second phase, microdeformations of CDA, connected with a method of hot pressing, are the factors reducing thermoelectric properties of the sample, first of all, the electric conductivity.

Fig. 5. The SEM images of the sample of bulk $Bi_{0,5}Sb_{1,5}Te_3$ (hot pressing). The resolution is 500 and 10000 accordantly

The samples for SPS were prepared by the cold pressure from the nanopowder at the room temperature. Then the SPS processing in a graphite press mould was made by passing of a pulse electric current under the pressure 50 MPa and the temperature 250-400 ^{0}C up to achievement of the hundred-percent density (from theoretical value for the given material). The correspondent SEM immerges can be seen in Fig.6.

At the temperatures of sintering 350 ^{0}C and 400 ^{0}C the grains grow and facet; that testifies about the active process of recrystallization. For samples that were sintered at a lower temperature the finely divided structure is typical, fragments of the fracture surface are not faceted, i.e. the grains have not recrystallized yet.

The average density of the samples fabricated from $Bi_{0,5}Sb_{1,5}Te$ solid solution by cold and hot pressing (plus 4 mass % of Te) are presented in table 1; the accuracy is ±0.02 g/cm^3. We see that increase of the temperature and increase of the pressure lead to gain of the density almost to the density of initial samples. However according to the ultrasonic microscopy microdefects in the form of separate cracks are found out even in fabricated at the high temperature and pressure samples. Such defects can lead to the decrease of the strength of nanostructured samples and can lead to reduce of the density. The values of elastic modules are presented in table 2.

Initial samples	Cold pressing at 1.5 GPa without sintering	Cold pressing at 1.5 GPa with subsequent sintering at 350°C	Hot pressing at 35 MPa and 470°C	Hot pressing at 250 MPa and 490°C
6.71	5.02	5.60	6.62	6.69
6.45	5.00	5.67	6.41	6.64
6.69	5.12	5.86	6.48	6.70

Table 1. Density of samples fabricated under different modes, g/cm^3

The change of the elasticity module E (Young's module) can be caused by changes of the concentration and sizes of defects in the type of micro- and sub-microcracks, formed at consolidation of nanostructured materials. It proves to be true according to the direct supervision of the samples structure (Fig.6).

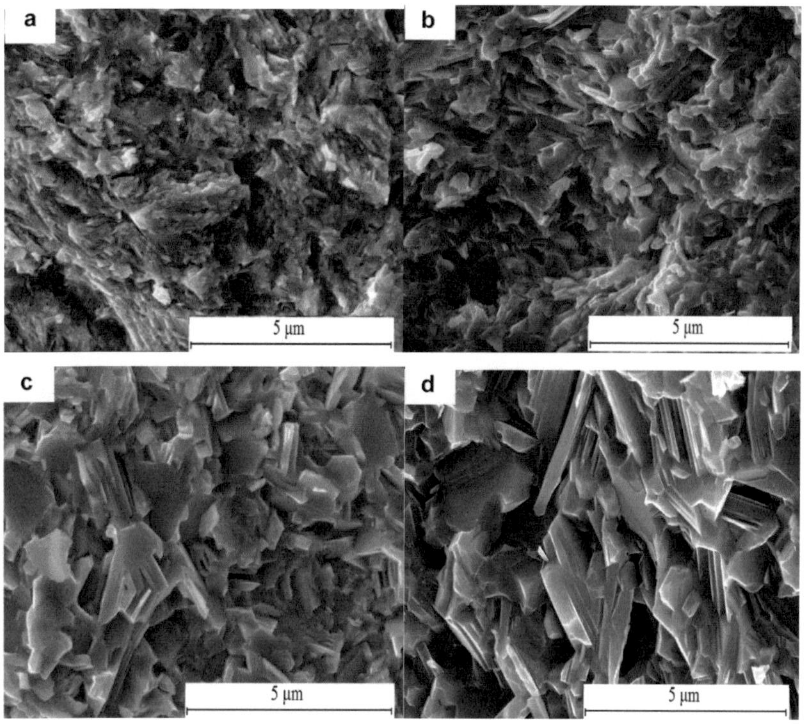

Fig. 6. SEM fractographs of sintered $Bi_{0,4}Sb_{1,6}Te_3$ samples at the pressure 50 MPa. SPS temperatures: a) 240 °C, b) 300 °C, c) 350°C, d) 400 °C.

Mode of sample fabrication	Elastic modules						
	V_L	V_T	ρV_L^2	ρV_T^2	B	E	Σ
Initial sample	3,26	1,77	71,84	21,1	43,7	54,5	0,292
Cold pressing at 1.5 GPa with subsequent sintering at 350°C	2,94	1,75	58,17	20,61	30,7	50,5	0,226
Hot pressing at 350 MPa and 470°C	3,46	2,16	80,09	31,21	38,48	73,65 73,71	0,181
Hot pressing at 250 MPa and 490°C	2,64	1,55	41,96	14,46	22,68	66,19	0,237
Hot pressing at 250 MPa and 490°C	2,72	1,59	44,54	15,22	24,25	69,36	0,240

Table 2. Elastic modules of samples fabricated under different modes, GPa

The received by SPS method samples were strong mechanically at all temperatures of sintering. Pores were absent. Results of the samples testing on diametrical compression are presented in Table 3.

Samples cut out from an ingot	Samples after cold pressing and sintering	Samples after hot pressing at 350 MPa and 470°C
6.8	20.3	27.3
11.4	15.0	31.0
4.5	22.6	29.5
	19.7	

Table 3. Compressive at diametrical strength (MPa)

2.5 Structure of bulk nanocomposites

Thermoelectric properties of materials with nanocrystalline structure should depend essentially on the size of nanograins (CDA) in a bulk sample. In turn the size of grain is defined by a number of factors: a temperature of the sintering or the hot pressing, duration of the hot pressing, value of the pressure, composition of materials, including presence of nano-adding of a second phase in a composite material. It is possible to ascertain that fabrication of initial nanopowder is less complex technological problem than maintain of the bulk nanostructure during the hot pressing that is caused by growth of initial nanoparticles due to recrystallization.

We see from Sec.2.4 that that the samples sintered after the cold as like as the hot pressure leads to increase of CDA size (or nanograins' size). In general nanostructured material is nonequilibrium by its nature, therefore thermal influences (at a manufacturing or an operation) are usually accompanied with the recrystallization of a compact material and a degradation of its properties. A possible way to reduce the average size of nanograins can be an inclusion of nanoparticles from another chemical composition, it means fabrication of nanocomposites. To investigate the relative change of nanograin size we added another nanoparticle-phase to the same solid solution matrix. They were added before the mechanical activation process. Three types of the extra nanoparticles were used for fabrication of nanocomposites: (a) MoS_2 with a laminated structure; (b) fullerene C_{60}, and (b) thermally expanded graphite (TEG). Values of nanograins was estimated by sizes of coherent dispersion areas (CDA) determined by the method of X-ray diffractometry on broadening of diffraction maxima. The received values of CDA sizes were compared to the data, obtained by the method of high resolution transmission electron microscopy (HRTEM). Both methods have shown a good consent of results at least at the size of grains up to several tens in nanometer. Larger grains also will consist from CDA with various crystallographic orientations which still influence on physical properties.

The content of MoS_2 was varied from 0.1 to 0.4 mass %. Only the peaks belonging to the triple solid solution $Bi_{0,4}Sb_{1,6}Te_3$ can be seen in the X-ray diffractogram of such a nanocomposite. The lattice constant of the nanocomposite does not change. Such a situation is repeated regardless of the pressing temperature. TEM study shows that MoS_2 nanoparticles are situated at the grain boundary, and do not dissolve in the matrix (Fig.7).

The MoS_2 particles have sizes ~ 20nm, and they have a crystalline structure. The introduction of MoS_2 greatly reduces the average size of nanograins and makes their size distribution more uniform. The maximum size of the nanograins decreases from 180 nm (in the solid solution without additives) to 80 nm (at a content of 0.1 mass % of MoS_2). The increase of the contents of MoS_2 up to 0.4 mass % leads to a future reduction of the average size of the nanograins and also leads to a more uniform size distribution. The toughness of the sample of the same composition was not less than 150 MPa. The addition of 0.1 mass % of MoS_2 brought about an increase of the toughness of 20-30 %.

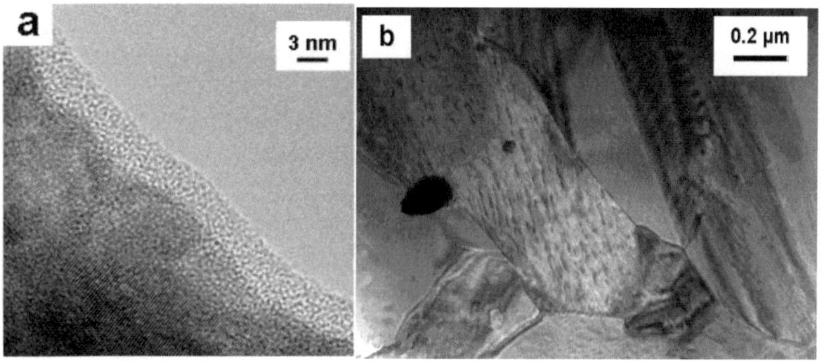

Fig. 7. (a) - HRTEM image of $Bi_{0.4}Sb_{1.6}Te_3$ nanoparticle covered by levels of MoS_2; (b) - image of the $Bi_{0.4}Sb_{1.6}Te_3$ sample fabricated at 350 MPa and 350ºC with 0.1 mass % of MoS_2

The fullerene C_{60} (1,5 mass %) or the thermally expanded graphite (TEG) (0.1 mass %) were added to the micropowder from initial crystalline material $Bi_{0,5}Sb_{1,5}Te_3$ of p-type. Then the mechanoactivation processing was made at different temperatures under pressure 350 MPa during 20 min in the argon atmosphere. The nanopowder received without carbon additives represented 100÷300 nm units consisting in turn from nanoparticles. The average CDA size was 8÷10 nm.

The mechanoactivation of $Bi_{0,5}Sb_{1,5}Te_3$ samples in the presence of TEG was accompanied by a stratification of the graphite and a formation of flakes with the size of few nanometers; layers from the graphite flakes cover the semiconductor nanoparticles. Fig.8 shows the size and a the configuration of carbon layers. The received layered covers on the semiconductor nanoparticles had as the ordered (similarly as layer of graphite on a surface) and the disorder structure. Let us notice that formation in the same process of the ordered and the disordered carbon covers is undesirable as they have different type of conductivity. This factor can cause a bad reproducibility of the properties of thermoelectric nanocomposites. Unlike TEG the fullerenes possess strongly pronounced electrophilic properties; therefore it would be interesting to track a combination of this form of carbon with the semiconductors' nanoparticles. The state of the interface «semiconductor - C_{60} - semiconductor» can make an essential impact on the transport properties at the expense of change of an electronic condition in thin layers of nanoparticles without chemical doping (Bulat et al., (2006). Nanoparticles from the semiconductor covered by layers with disorder structure from C_{60} molecules have been received by the mechanoactivated processing of $Bi_{0,5}Sb_{1,5}Te_3$ together with the fullerenes. The typical structure of such particle is shown at Fig. 9.

It has been determined, that at mechanoactivation processing of $Bi_{0,5}Sb_{1,5}Te_3$ solid solution the additive of nanocarbon do not influence to the average size of CDA; in all cases it was 8÷10 nm. However the application of nanocarbon has allowed to reduce essentially disorder of CDA size, and to reduce in 1.5÷2 times a share of concerning large (more then

30 nm) particles. Apparently the received carbon covers effectively break the recrystallization of nanoparticles: the CDA size have decreased in 1.7÷1.9 times at the temperature 400÷450°C.

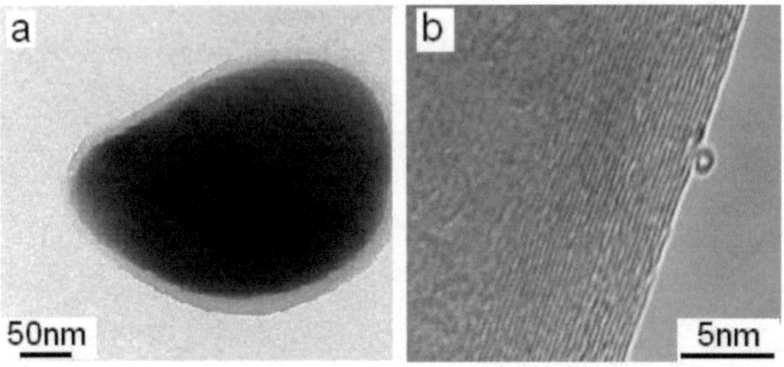

Fig. 8. Carbon covers from mechanoactivated TEG on the surface of $Bi_{0,5}Sb_{1,5}Te_3$ nanoparticles: (a) – ordered and (b) - disordered structure

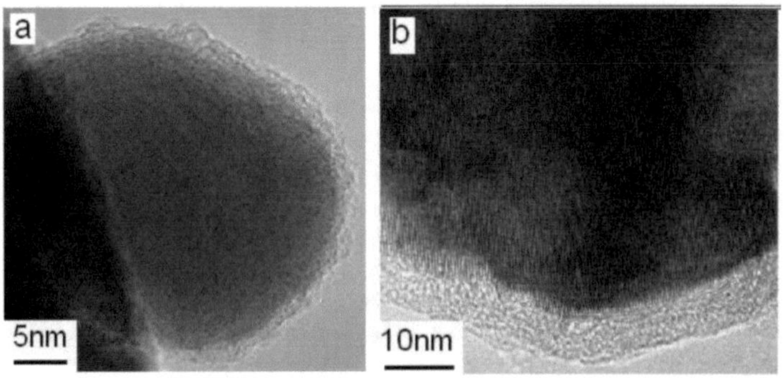

Fig. 9. HRTEM images of a semiconductor particle in a cover from molecules C_{60}: (a) - $Bi_{0,5}Sb_{1,5}Te_3$; (b) - $Bi_{0.4}Sb_{1.6}Te_3$

The samples cut out from an initial ingot have shown a considerable disorder of the strength σ_p=0.5÷2.5 MPa. The nanostructures samples from $Bi_{0,5}Sb_{1,5}Te_3$ had the strength σ_p=18.5÷20 MPa, and for $Bi_{0,5}Sb_{1,5}Te_3$ samples with TEG and with C_{60} the value of strength 26.3 and 31.0 MPa accordingly have been received.

Fig. 10 shows the consolidated data of the influence of different factors on the temperature dependencies of the average nanograins size for p-type nanocrystalline samples that were fabricated under different pressing mode (Bublik et al., 2009; Bulat et al., 2010b). We see that the main factors allowing slow growth of nanograins as a result of recrystallization are the reduction of the temperature and of the duration of the process, the increase of pressure, as well as the addition of small amount of additives (like MoS_2, TEG or fullerenes). In the case of additives the accidental particles in a nanocomposite settle down on borders of particles of the basic solid solution creating the structure like "core - cover". Let us underline that the CDA size coincides with the size of grains revealed on the SEM image of the break of surface in a compact sample at the sizes of grains to several tens nm. In the larger grains ~ 1-2 μm CDA are a part of the internal structure of a grain.

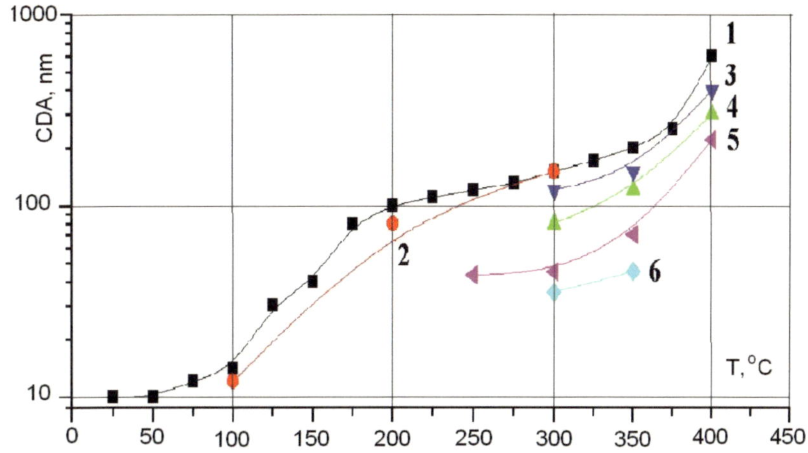

Fig. 10. Temperature dependence of the average size of nanograins of samples fabricated under different modes of pressing. 1 – in situ heating of $Bi_{0,4}Sb_{1,6}Te_3$ samples pressed at 25ºC and 1.5GPa in the thermocamera of the diffractometer; 2 – vacuum annealing at different temperatures of $Bi_{0,4}Sb_{1,6}Te_3$ samples pressed at 25ºC and 1.5GPa; 3 – $Bi_{0,4}Sb_{1,6}Te_3$ samples hot pressed at 35MPa; 4 – $Bi_{0,4}Sb_{1,6}Te_3$ samples hot pressed at 350MPa; 5 – $Bi_{0,4}Sb_{1,6}Te_3$ samples plus 0.1 mass% MoS_2 pressed at 350MPa; 6 – $Bi_{0,4}Sb_{1,6}Te_3$ samples plus 0.4 mass % MoS_2 pressed at 350MPa

As properties of a material to a great extent depend on its structure in micro- and nanoscale, the comparative analysis of the structure of the bulk samples received by SPS methods and traditional hot pressing has been carried out; these results are presented at Fig. 11, 12. The analysis of the received results shows that unlike the method of hot pressing the SPS method allows to receive at rather low pressure 50 MPa mechanically strong well sintered nanostructured materials. It does not contain pores even at temperatures more low then 300°C. The explanation of this result is that in SPS process the high density of allocated energy in contact zones between the powder particles causes the very strong local warming up (up to fusion of a grain blanket) whereas the basic volume of a the material remain at lower temperature. The CDA size for both methods of consolidating up to temperatures

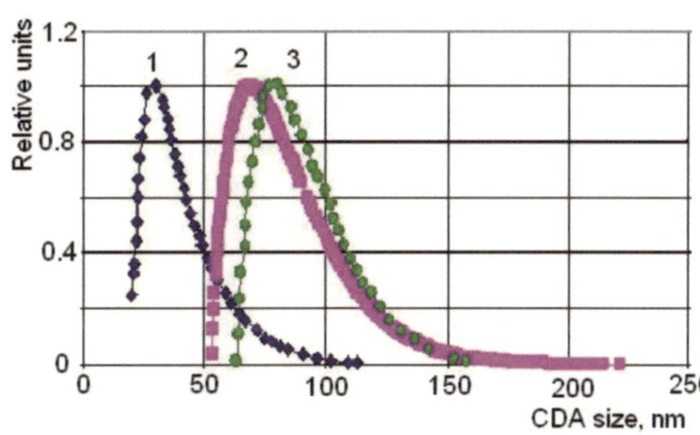

Fig. 11. Relative quantitative portion of different CDA size in nanostructured sintered $Bi_{0,4}Sb_{1,6}Te_3$. SPS temperatures: 1 - 240 °C; 2 - 300 °C; 3 - 350 °C; pressure 50 MPa

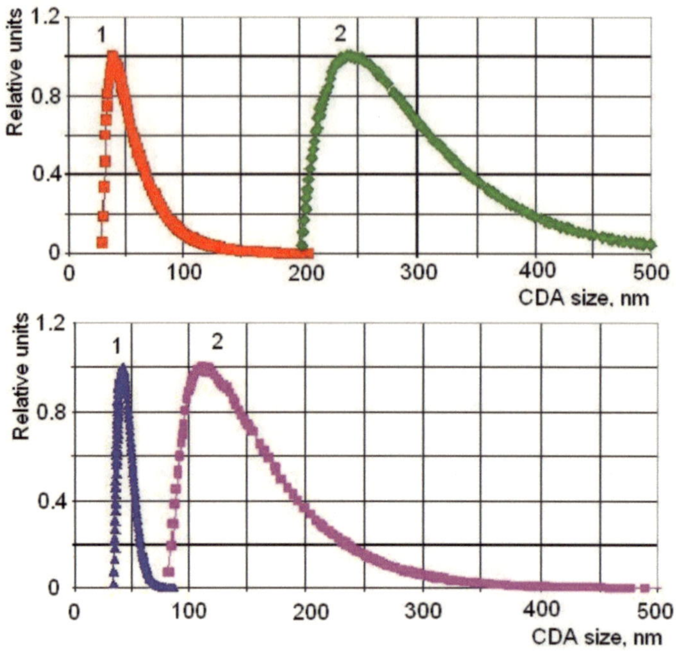

Fig. 12. Relative quantitative portion of different CDA size in hot pressed nanostructured $Bi_{0,4}Sb_{1,6}Te_3$ (350 MPa; 1-300°C &, 2- 400°C). a) $Bi_{0,4}Sb_{1,6}Te_3$, b) $Bi_{0,4}Sb_{1,6}Te_3$+0,1 mass % MoS_2

350˚C are comparable, whereas at sintering temperatures 400˚C and above (that in practice corresponds to temperatures of hot pressing) the CDA size in SPS method increases much less.

2.6 Thermoelectric properties of bulk nanostructures and nanocomposites

The main transport properties of nanocrystalline materials fabricated under different conditions were investigated. Dependences of the transport properties on average nanograins size were also analyzed.

The temperature dependences of the thermoelectric parameters of typical hot pressed nanostructured p-$Bi_{0,3}Sb_{1,7}Te_3$ sample was published and discussed in Ref. (Bulat et al., 2010b). The correspondent maximum value ZT=1.12 takes place at the temperatures ~ 350÷375 K. The same maximum efficiency ZT=1.1 in the same nanostructured material at the same temperature was measured in Ref. (Vasilevskiy et al., 2010) but the extrusion instead of hot pressing for consolidation of samples was used here.

Let as consider more in detail our investigation of thermoelectric properties of samples fabricated by the SPS method. Thermoelectric properties were studied depending on sintering temperature on samples of p-type $Bi_{0,5}Sb_{1,5}Te_3$ and $Bi_{0,4}Sb_{1,6}Te_3$. All samples have been received by SPS method at pressure 50 MPa, temperature from 250 to 500 ^{0}C, the duration of sintering was 5 min (for $Bi_{0,4}Sb_{1,6}Te_3$) and 20 min (for $Bi_{0,5}Sb_{1,5}Te_3$). The samples sintered at 300 ^{0}C have the maximum value of thermoelectric power. A small distinction exists between values of thermoelectric power for samples of various compositions. The electric conductivity increases with the raise of sintering temperature. The similar dependence is observed also for the heat conductivity; increase of the heat conductivity with rise of the temperature sintering is caused by increase in the electronic heat conductivity which is proportional to the electric conductivity.

It follows from experimental results that the samples received by SPS method at sintering temperature 450 ^{0}C have the greatest value of the efficiency ZT. The samples fabricated from the 0.5 μm powder have lower efficiency ZT in comparison with the nanostructured material. It was established that the pressure 50 MPa is the optimum one for obtain the high thermoelectric efficiency. Samples of $Bi_{0,4}Sb_{1,6}Te_3$ composition obtained at the sintering temperatures ~ 350 ^{0}C have higher ZT than $Bi_{0,5}Sb_{1,5}Te_3$ samples. The temperature dependence of thermoelectric parameters in nanostructured samples $Bi_{0,4}Sb_{1,6}Te_3$ received at sintering temperature 400 ^{0}C and pressure 50 MPa is presented at Fig. 13. Peak efficiency is reached at 90 ^{0}C and makes ZT=1.22.

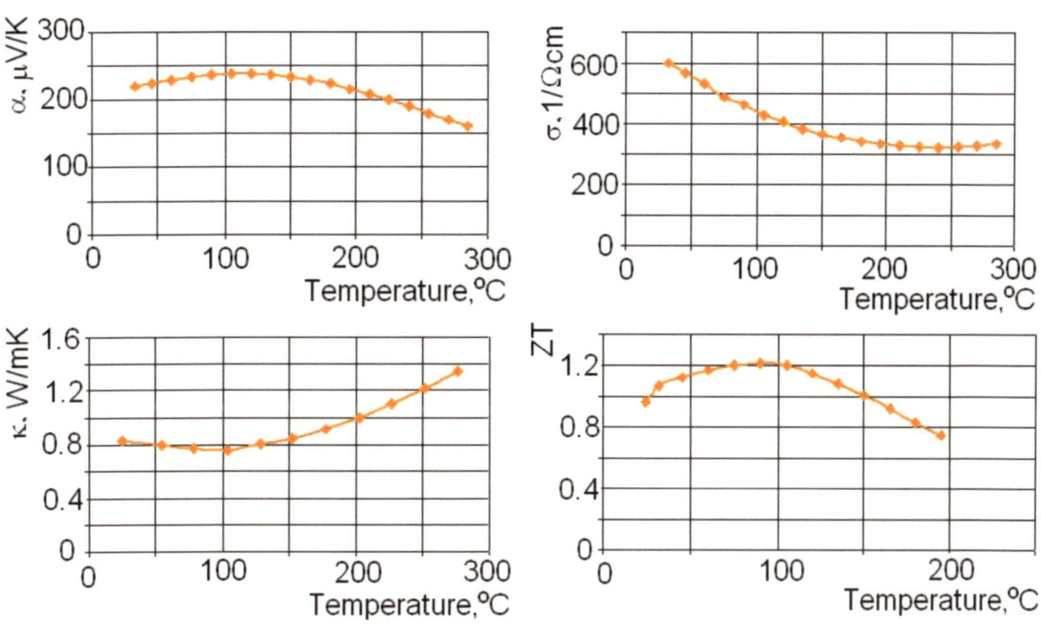

Fig. 13. Thermoelectric properties of sintered bulk nanostructured materials $Bi_{0,4}Sb_{1,6}Te_3$ as a function of measurement temperature: a) electrical conductivity, b) Seebeck coefficient, c) thermal conductivity, d) figure of merit, ZT.

Some dependences of measured thermoelectric coefficients in bulk nanostructured materials on the grain size for solid solution $Bi_xSb_{2-x}Te_3$ (Bulat et al., 2010c, 2011a) will be presented in Sec. 3.2 and 3.3.

3. Theory

Three mechanisms that can improve the thermoelectric efficiency are studied theoretically and compared with the experiment in Sec.3.

3.1 Electron tunneling

One of the possible mechanisms of electric transport in nanostructured materials is the tunneling of charge carriers through the intergrain barriers. This effect is similar to the thermionic or field emission thought the vacuum gap. The studies of the thermionic emission applied to the field of energy conversion began in the 1960th (Anselm, 1951). Though the efficiency of thermionic generators can reach 20% their working temperatures are about 1000K because of the large values of work function in metals and semiconductors.

To use thermionic devices at lower temperatures one should decrease the work function, e.g. by applying high electric field (Fleming & Henderson, 1940; Murphy & Good, 1956), by using special cathode coatings that can decrease the work function down to 0.8eV (Sommer, 1980) or by utilizing the tunneling effect through the thin vacuum gap (Hishinuma et al., 2001; Tavkhelidze et al., 2002). As was shown by Mahan (Mahan, 1994) to use thermionic devices for refrigeration at room temperature one needs to decrease the work function down to the values of 0.3-0.4eV that are not available at the present time. But in the case of nanoscale tunneling junction the tunneling probability increases and the noticeable cooling power can be reached even at the work functions of about 0.8eV (Hishinuma et al., 2001). One of the possible cooling applications of such device that consisted of metallic tip over the semiconducting plate was described in (Ghoshal, 2002b). Alternatively Schottky barriers or semiconductor heterostructures can be used instead of vacuum barriers (Mahan & Woods, 1998; Mahan et al., 1998). In such structures the barrier energy height can be as low as 0.1eV but the phonon thermal conductivity of semiconducting barrier will increase the total thermal conductivity and produce negative influence on the figure of merit.

In this section the bulk nanostructured material that consists of grains separated by tunneling junctions is considered. The influence of the charge carrier tunneling on the thermoelectric figure of merit of such material is theoretically investigated. The shape of nanoparticles is modelled by two truncated cones with the same base (Fig. 14) that allows one to perform calculations in an analytical way (Bulat & Pshenai-Severin, 2010a). The calculations of the heat flow inside the nanoparticles take into account the difference between the electron and phonon temperatures in the limiting case of vacuum gap when phonons cannot tunnel through the barriers.

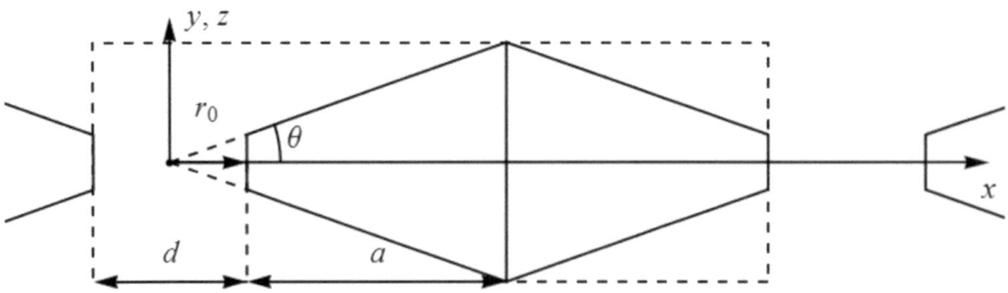

Fig. 14. The cross section of nanoparticle modeled by two truncated cones with the same base. $2a$ is the size of nanoparticle, 2θ is the cone aperture angle, the radius r_0 determines the size of truncated part, d is the tunneling junction width. Dashed rectangle represents the single period of the whole structure.

In order to estimate the thermoelectric figure of merit of such material the transport coefficients were calculated. The use of the linear approximation for tunneling coefficients in bulk nanostructured materials seems to be quite reasonable. Indeed if the typical size of samples are several millimeters and the grain size is about 10-40 nm then the voltage drop on single junction is 10^5 times smaller than on the whole sample. So even at several hundreds of volts bias on the sample the voltage drop on the single junction will be $\Delta V \sim 10^{-3}\,\mathrm{V}$. Similarly the temperature difference on single junction is about 10^{-3}K at the total temperature difference of 100K. So at the room temperature if the barrier energy height $\varepsilon_b \sim 0.1\mathrm{eV}$ than $|k_0 \Delta T|$ and $|q_0 \Delta V|$ are much less than both the thermal energy $k_0 T$ and barrier height ε_b. In this case one can use linear transport coefficient theory for the tunneling junction. The total current density through the tunneling junction is determined by the difference of emission currents from two electrodes (Burstein, 1969)

$$j_x = \sum_{\mathbf{k}, v_x > 0} 2(q_0 v_x)\, D(\varepsilon_x) \left[f_0\left(\frac{\varepsilon - \mu}{k_0 T}\right) - f_0\left(\frac{\varepsilon - (\mu + q_0 \Delta V)}{k_0 (T + \Delta T)}\right) \right]. \tag{1}$$

In this expression x-axis is directed at the right angle to the junction cross section (Fig. 14), ε is the total energy of electron with the wave vector $\mathbf{k}$, μ is the chemical potential of the left electrode, f_0 is the Fermi-Dirac distribution function, v_x and ε_x are velocity and kinetic energy of electron corresponding to its motion along x direction and $D(\varepsilon_x)$ is the tunneling probability. In order to obtain linear transport coefficients the expression for the total current density (1) was linearized with respect to small voltage ΔV and temperature ΔT differences. Finally the barrier electric conductivity can be obtained as $\sigma_b = -j_x / \Delta V$ (Bulat & Pshenai-Severin, 2010a)

$$\sigma_b = \frac{q_0^2\, m k_0 T}{2\pi^2 \hbar^3} \int_0^{\infty} D(\varepsilon_x^*)\, f_0(\varepsilon_x^* - \mu^*)\, d\varepsilon_x^* , \tag{2}$$

where energy ε_x^* and chemical potential μ^* with asterisks are measured in $k_0 T$ units and the effective mass of electron m is assumed to be the same inside nanoparticle and barrier. The Seebeck coefficient can be obtained from the zero current condition $\alpha_b = -(\Delta V / \Delta T)_{j_x = 0}$ and it was expressed as $\alpha_b = \beta_b / \sigma_b$, where (Bulat & Pshenai-Severin, 2010a)

$$\beta_b = \frac{k_0}{q_0} \frac{q_0^2\, m k_0 T}{2\pi^2 \hbar^3} \int_0^{\infty} D(\varepsilon_x^*) \left[u f_0(u) + ln\left(1 + e^{-u}\right) \right] d\varepsilon_x^* , \tag{3}$$

and the following notation was introduced $u = \varepsilon_x^* - \mu^*$.

The expression for electronic heat flow through the junction can be obtained from (1) after replacing $q_0 v_x$ with $(\varepsilon - \mu) v_x$. The value of barrier thermal conductivity measured at zero current κ_b can be expressed through the thermal conductivity at zero voltage drop $\kappa_{b,\Delta V=0}$ as (Bartkowiak & Mahan, 1999)

$$\kappa_b = \kappa_{b,\Delta V=0} - \alpha_b^{\ 2}\sigma_b T\ , \tag{4}$$

where (Bulat & Pshenai-Severin, 2010a)

$$\kappa_{b,\Delta V=0} = \frac{m k_0^{\ 3} T^2}{2\pi^2 \hbar^3} \int_0^{\infty} D(\varepsilon_x^*) \left[u^2 f_0(u) + 2u\,ln\left(1+e^{-u}\right) - 2Li_2\left(-e^{-u}\right) \right] d\varepsilon_x^*, \tag{5}$$

and the dilogarithm function is denoted as $Li_2(x)$. It is worth to note that the barrier electrical and thermal conductivities are determined with respect to voltage and temperature difference instead of their gradients as in the bulk case. Hence for the case of comparison with the bulk values it is more convenient to use the values of $\sigma_b d$ and $\kappa_b d$.

In the present calculations the intergrain barrier shape was assumed to be rectangular. In linear operating region the change of the tunneling barrier shape under applied field can be neglected. So the well-known expression for tunneling probability of rectangular barrier was used

$$D(\varepsilon_x) = \left(1 + \frac{\left(k_x^{\ 2} + k_b^{\ 2}\right)^2}{4 k_x^{\ 2} k_b^{\ 2}} sinh^2\left(k_b d\right) \right)^{-1}, \tag{6}$$

where $k_x = \sqrt{2m\varepsilon_x} / \hbar$ and $k_b = \sqrt{2m(\varepsilon_b - \varepsilon_x)} / \hbar$. Note that if $\varepsilon_x > \varepsilon_b$ the wave vector became pure imaginary $k_b = i\,|\,k_b\,|$ and hyperbolic sine should be changed to $sin(|\,k_b\,|d)/i$. Very often instead of exact expression for tunneling probability the WKB approximation is used $D_{WKB}(\varepsilon_x) = exp(-2k_b d)$. In WKB approximation tunneling probability for $\varepsilon_x > \varepsilon_b$ is equal to unity. In the following the values of tunneling transport coefficients calculated using these two approximations will be compared.

Fig. 15-17 show the dependencies of barrier electrical conductivity, Seebeck coefficient and Lorenz number on the size of tunneling junction for different barrier heights 0.4 and 0.8eV. In these estimations the effective mass was equal to $m = 0.7 m_0$ that corresponds to the typical hole effective mass in Bi_2Te_3 (Goltsman et al., 1972). Doping impurity concentration was equal to 10^{19}cm^{-3} for chemical potential close to the band edge. The curves plotted using exact expression for tunneling probability (6) and obtained in WKB approximation illustrate noticeable difference of two approaches. For metallic electrodes electron energies close to the Fermi level are important and WKB approximation can be used for small tunneling probabilities $D(\varepsilon_x) < 1/e$ as was stated in (Stratton, 1962). For semiconducting electrodes when the charge carrier energies are close to the band edge the preexponential factor can also be important because it approaches zero for small carrier energies. When the tunneling junction width becomes larger these difference decreases because the contribution of small energy carriers is less important.

As can be seen from Fig. 15 at the junction thicknesses smaller than 2nm the Seebeck coefficient can reach the values of about 300-350μV/K and slowly varies with the barrier thickness. The electrical conductivity is small (Fig. 16) and decreases exponentially with the barrier thickness. At larger junction thicknesses *d*>2nm the thermionic emission becomes more important than tunneling. Charge carriers with small energies are filtered out of the current that leads to the sharp increase in Seebeck coefficient and Lorenz factor (Fig. 15, 17). The dimensionless figure of merit for single junction is rather high $ZT \approx 3-4$.

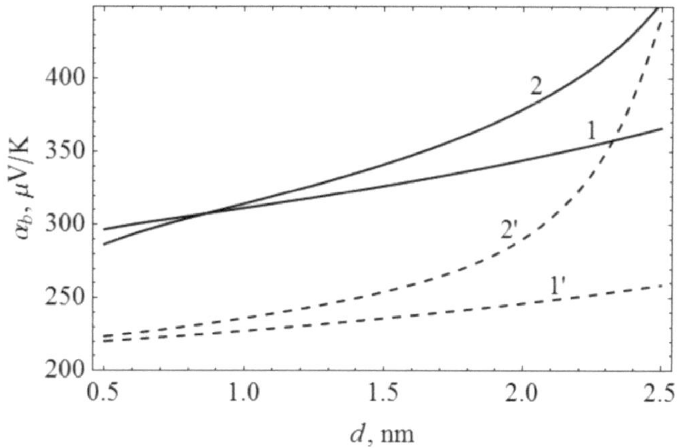

Fig. 15. The dependence of the barrier Seebeck coefficient on the tunneling junction size at room temperature for $\varepsilon_b = 0.8$eV (1, 1′) and 0.4eV (2, 2′) calculated using exact expression for tunneling probability (1, 2) and WKB approximation (1′, 2′).

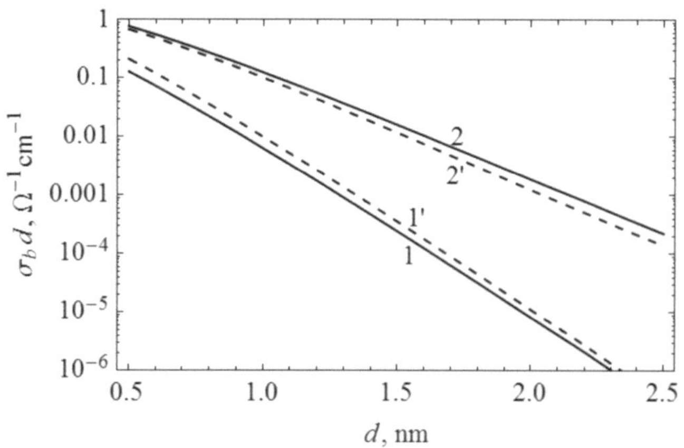

Fig. 16. The dependence of the barrier electrical conductivity on the tunneling junction size (see Fig. 15 for notation).

In order to calculate effective transport coefficients in the whole structure charge and heat flow inside nanoparticles should be taken into account. As the approximation of zero phonon thermal conductivity of the barrier is considered the heat flow through the junction is only due to charge carriers. Hence the equations for the heat flow should take into account the differences in electron T_e and phonon T_p temperatures (Ghoshal, 2002b;

Bartkowiak & Mahan, 1999). The electron-phonon scattering inside nanoparticle leads to the equilibrating of their temperatures on the length scale l_c which is called cooling length. The general solution for conical geometry was obtained in (Ghoshal, 2002b) where the limiting case $l_c << a$ was analyzed. In the considered materials based on bismuth antimony telluride solid solutions the values of cooling length are 66nm for Bi_2Te_2 and 156nm for Sb_2Te_3 (da Silva & Kariany, 2004). So for nanoparticle size of 10-20nm the limit of $l_c >> a$ can be considered. For this case the heat transfer equations were solved for each of two truncated cones representing the nanoparticle in (Bulat & Pshenai-Severin, 2010a). As a result the equations for total resistance and thermal conductance of nanoparticle were obtained (Bulat & Pshenai-Severin, 2010a)

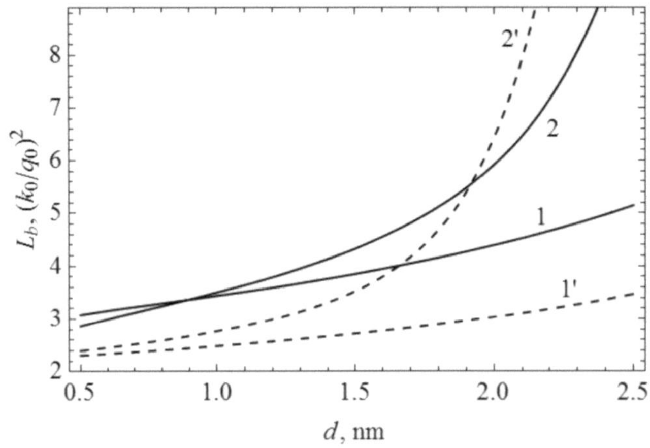

Fig. 17. The dependence of the barrier Lorenz number on the tunneling junction size (see Fig. 2 for notation).

$$R_n = \left(\sigma \gamma \Omega r_0^2\right)^{-1}, \quad K_n = \kappa_e \xi \gamma \Omega r_0^2, \tag{7}$$

where $\gamma = (r_0 + a) / r_0 L_n$, $L_n = 2a$ is the total length of nanoparticle in the x-axis direction, r_0 determines the size of truncated part (Fig. 14), $\Omega = 4\pi sin^2 \theta / 2$ is the cone solid angle and

$$\xi = 1 + \frac{\kappa_p}{\kappa_e + \kappa_p} \frac{(a / l_c)^2}{3\gamma L_n}. \tag{8}$$

From these equations it can be seen that in the limit of the small nanoparticle size compared to the cooling length the electrical resistance does not change due to the difference in T_e and T_p. The correction to the thermal conductance due to this effect is only second order of magnitude with respect to small parameter a / l_c and K_n is determined mainly by electronic contribution.

It is interesting to note that the transition to the layered geometry can be obtained if $r_0 \to \infty$ and $\theta \to 0$ in such a way that the area Ωr_0^2 is constant. In this limit $\gamma L_n = 1$ and from (7) it is easy to get corresponding equation for the layered system (Anatychuk & Bulat, 2001).

Though in real nanostructured material the size on nanoparticles and their positions are randomly distributed here for estimations of effective transport coefficient the material is modeled as an ordered set of primary cells outlined by dashed lines on the Fig. 14. In this case the total current flow is directed along x-axis and effective transport coefficients can be calculated based on equations for layered medium (see, e.g., Snarsky et al., 1997). The effective transport coefficients for the present case were calculated in (Bulat & Pshenai-Severin, 2010a). The thermal conductivity can be obtained as a series connection of barrier and nanoparticle thermal conductivities

$$\kappa_{eff} = \frac{\kappa_b \kappa_e \xi \gamma}{\kappa_b + \kappa_e \xi \gamma} \gamma_t , \tag{9}$$

where geometric factor $\gamma_t = r_0^2 (d + 2a) / (r_0 + a)^2$ was introduced. The effective Seebeck coefficient can be obtain as a sum of Seebeck coefficients of barrier and nanoparticle taking into account corresponding temperature differences on each part

$$\alpha_{eff} = \frac{\alpha_n \kappa_b + \alpha_b \kappa_e \xi \gamma}{\kappa_b + \kappa_e \xi \gamma} . \tag{10}$$

In calculations of electrical conductivity the average sample temperature T_{av} is assumed to be constant. But due to Peltier effect the temperatures of neighboring contacts are different. So in the equation for effective electrical conductivity in addition to common expression for series resistance the factor due to Peltier effect induced thermopower should be taken into account

$$\sigma_{eff} = \frac{\sigma_b \sigma \gamma \gamma_t}{\sigma_b + \sigma \gamma} \left(1 + \frac{(\alpha_b - \alpha_n)^2}{\kappa_b + \kappa_e \xi \gamma} \frac{\sigma_b \sigma \gamma T_{av}}{\sigma_b + \sigma \gamma} \right)^{-1} , \tag{11}$$

where T_{av} is the average temperature of sample.

The effective figure of merit of bulk nanostructured material can be calculated using equation (9)-(11) as $Z_{eff} = \alpha_{eff}^2 \sigma_{eff} / \kappa_{eff}$. For estimations the typical room temperature parameter for Bi_2Te_3 from (Goltsman et al., 1972) were used: α_n=200 μV/K, σ=830 $\Omega^{-1}cm^{-1}$ and κ_p=1 W/m K. Though Bi_2Te_3 is anisotropic material nanocrystals inside the sample are randomly oriented. So for the estimations the values of thermal and electrical conductivities were average over all directions.

On Fig. 18 the dependencies of effective electrical conductivities on the tunneling junction thickness are plotted. It is interesting to note that effective transport coefficients are independent of cone aperture angle θ because only cross-section areas depend on it and these dependences are canceled out. For larger r_0 the electrical conductivity increases due to the increase of the smaller cross-section of the cone. As was noted above in the limit of large r_0 the transport coefficients approach the values for layered geometry (compare i, i' and i'' with L_i on Fig. 18 for i=1, 2). For considered parameter range the electrical conductivity of the tunneling junction is much less than the usual values in semiconductors. Hence the effective electrical conductivity is determined mainly by barrier part but it is related to the total period of the structure. For example for layered geometry $\sigma_{eff} \approx (\sigma_b d)(L_n + d) / d > \sigma_b d$ (compare L_i with B_i on the Fig. 18 for i=1, 2). For the case of conical geometry the factor γ_t should be taken into account that diminishes σ_{eff} for small r_0.

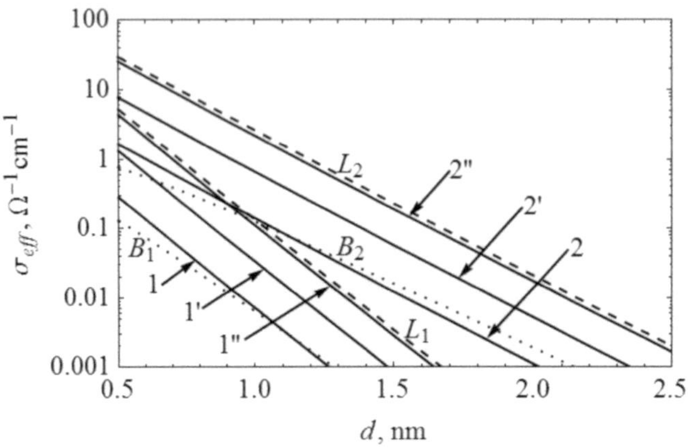

Fig. 18. The dependence of effective electrical conductivity on the tunneling junction thickness for barrier height $\varepsilon_b = 0.8$ (1, 1′, 1′′) and 0.4eV (2, 2′, 2′′) for cone-shaped nanoparticles. The same dependences for single barrier and layer structure are plotted as B_1, B_2 and L_1, L_2 correspondingly. The ratio r_0/a=0.3 (1,2), 1 (1′, 2′) and 10 (1′′,2′′); $2a$=20nm.

On Fig. 19 the dependence of effective thermoelectric figure of merit on the tunneling junction thickness is plotted. The estimations showed that in the absence of the phonon thermal conductivity in the barrier for all considered ranges of tunneling junction parameters (see Fig.16, 17) the barrier thermal conductivity is much smaller than the thermal conductivity of nanoparticle. Hence relatively high values of the effective Seebeck coefficient are determined mainly by large α_b (Fig. 15) and the ratio $\sigma_{eff} / \kappa_{eff}$ in Z_{eff} is determined by the effective Lorenz number that has usual values for small d and begins to increase with the increase of d (see Fig. 17). So in the present case the large values of $Z_{eff}T \approx 2.5-4$ are determined by large barrier Seebeck coefficient and the decrease of $Z_{eff}T$ for larger d is due to the increase of barrier Lorenz number. Simple estimations of the effect of phonon thermal conductivity of the barrier performed in one-temperature approximation showed that to increase the thermoelectric figure of merit compared to initial semiconducting material the phonon barrier thermal conductivity $\kappa_{b,ph}$ should be about 4 time smaller than the electronic contribution.

To conclude this section it can be said that the thermoelectric figure of merit of the structures with tunneling junctions can be quite large $Z_{eff}T \approx 2.5-4$ if the barrier phonon thermal conductivity is negligible. These large values are determined by the large values of the barrier Seebeck coefficient and greatly reduced in the presence of $\kappa_{b,ph}$. In addition irregularities in the tunneling junction width or the size of nanoparticles can also lead to the decrease of the figure of merit in real structure. The comparison with the experimental data from Sec.2.6 and Ref. (Poudel et al., 2008; Bulat et al., 2010b; Vasilevskiy et al., 2010) showed that the increase of the figure of merit in these materials is hardly connected with the tunneling effect because of the large difference between measured electrical conductivity (of the order of $1000\ \Omega^{-1}\,\text{cm}^{-1}$) and estimated values that are much less than $100\ \Omega^{-1}\,\text{cm}^{-1}$.

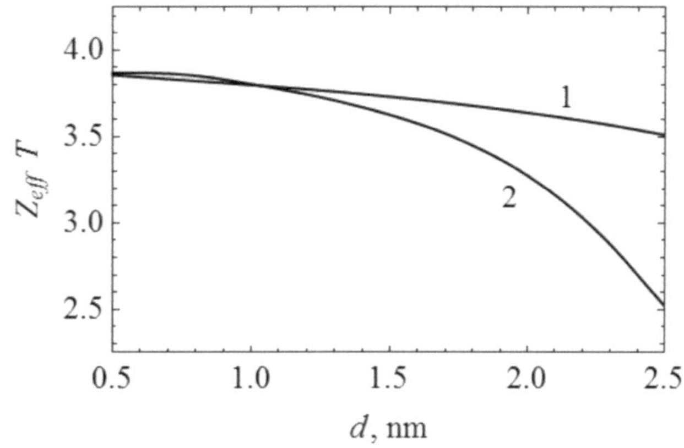

Fig. 19. The dependence of effective thermoelectric figure of merit on the tunneling junction thickness for barrier height $\varepsilon_b = 0.8$ (1) and 0.4eV (2).

3.2 Boundary scattering

In this section the influence of boundary scattering on the thermal conductivity of bulk nanostructured materials obtained by ball-milling with subsequent hot pressing is considered following (Bulat et al., 2010c). These materials are polycrystalline with small grain sizes in the range from 10 nm to several hundreds on nanometers depending on the temperature of hot pressing. The most common way to estimate the influence of grain boundary scattering on the thermal and electrical conductivities is to include additional scattering mechanism with the mean free path equal to the grain size L_n. The theory of this effect applied to thermal conductivity in different polycrystalline solid solutions was described in (Goldsmid et al., 1995) but due to relatively large grain size considered there it was predicted that the effect of boundary scattering on thermal conductivity in bismuth telluride alloys is negligible.

Usually the grain boundary effect is considered to be important for thermal conductivity only at low temperatures when the probability of phonon-phonon scattering decreases. But it is related mainly to pure single crystals (Goldsmid et al., 1995). In solid solutions at high temperatures the contribution of short wavelength phonons to the thermal conductivity is reduced due to the point defect. So in solid solutions the contribution of long wavelength phonons to thermal conductivity is relatively more important than in pure crystals. This contribution can be effectively reduced by introducing boundary scattering.

The estimations performed here are based on Debye model for acoustic phonons with linear spectrum up to Debye frequency ω_D. The following scattering mechanisms are taken into account. In pure single crystals the most important scattering mechanism at room temperature is phonon-phonon umklapp scattering with the relaxation time $\tau_U = A_U / \omega^2$. The thermal conductivity in this case can be written as (Goldsmid et al., 1995)

$$\kappa_0 = \frac{1}{3} c_V \, v_D \, \overline{l_0}, \tag{12}$$

where $c_V = 3k_0 N_V$ is the heat capacity associated with acoustic modes of the crystal containing N_V primary cells, v_D is the mode averaged Debye speed of sound and $\overline{l_0}$ is the mean free path associated with the umklapp scattering. Knowing the value of thermal conductivity in single crystal κ_0 the mean free path $\overline{l_0}$ and the constant A_U can be obtained.

When the second component is added forming solid solution the thermal conductivity κ_s becomes less than κ_0 due to the point defect scattering $\tau_P = A_P / \omega^4$. The constant A_P can be deduced from experimental value of κ_s assuming that A_U is the same as in initial single crystal. Finally the boundary scattering is described by frequency independent relaxation time $\tau_b = L_n / v_D$. In the simplified treatment (Goldsmid et al., 1995) it was proposed to divide the total range of phonon frequencies into three parts. For each part of the spectrum only the most important relaxation time is considered: τ_b, τ_U and τ_P for lower, medium and high frequency parts correspondingly. Then the simple equation for phonon thermal conductivity in polycrystalline material was obtained (Goldsmid et al., 1995)

$$\kappa_{ph} = \kappa_s - \frac{2}{3} \kappa_0 \sqrt{\frac{\overline{l_0}}{3L_n}} \, . \tag{13}$$

In order to compare the values of κ_{ph} with experiment for nanostructured material based on p-$Bi_xSb_{1-x}Te_3$ (Bulat et al., 2010c) the hole contribution should be subtracted from experimental values of thermal conductivity. So the proper estimations of electrical conductivity and hole thermal conductivity are necessary. The electrical conductivity in initial solid solution is anisotropic but after ball-milling and hot pressing the samples became isotropic on average. To take the anisotropy into account it was assumed that it is connected mainly with the anisotropy of effective masses and the relaxation time is a scalar. Then using the effective medium theory for average electrical conductivity the effective mass of conductivity in polycrystalline material can be expressed as (Bulat et al., 2010c)

$$m_c = \frac{4m_{c11}}{1 + \sqrt{1 + 8m_{c11} / m_{c33}}}, \tag{14}$$

where m_{cii} are effective conductivity masses along main crystalline directions ($i = 1,2,3$).

The boundary scattering of holes was taken into account using relaxation time in the form $\tau_{b,h} = L_n / v$. The relaxation time energy dependence for acoustic scattering is $\tau_a \sim \varepsilon^{-1/2}$. It is the same as that for point defect scattering or alloy scattering in solid solution. It appears that this energy dependence is the same also for boundary scattering of holes. So the change of mobility in nanostructured material can be describe as (Bulat et al., 2010c)

$$u = \frac{L_n / l_s}{1 + L_n / l_s} u_s , \qquad (15)$$

where l_s and u_s are the mean free path and mobility in initial solid solution. The Lorenz number and the Seebeck coefficient in this case are the same as for acoustical scattering due to the same energy dependencies of the relaxation times.

The experimental values of electrical conductivity together with estimations based on equation (15) are shown on Fig. 20. In the initial solid solution $Bi_{0.4}Sb_{1.6}Te_3$ the values of electrical conductivity in the cleavage plane and the Seebeck coefficient were equal to $1000\ \Omega^{-1}cm^{-1}$ and $195\ \mu V/K$ correspondingly. In $Bi_{0.3}Sb_{1.7}Te_3$ these values were equal to $1387\ \Omega^{-1}cm^{-1}$ and $187\ \mu V/K$. The experimental values of mobility in $Bi_{0.3}Sb_{1.7}Te_3$ were 15% higher than in $Bi_{0.4}Sb_{1.6}Te_3$. The values of effective masses were taken from two different sources (Luk'yanova et al., 2010) and (Stordeur et al., 1988). The effective masses of the density of state per one ellipsoid m_{d1} and of conductivity m_c obtained using (14) were equal to $0.069m_0$ and $0.054m_0$ (Luk'yanova et al., 2010) and $0.305m_0$ and $0.186m_0$ (Stordeur et al., 1988). Due to the wide spread of the effective mass values the estimations of the mean free path in the initial solid solution $Bi_{0.4}Sb_{1.6}Te_3$ were quite different l_a=23 nm and 4nm correspondingly. This is reflected on the Fig. 20 where the effect of boundary scattering is more prominent for the estimations with larger l_a (compare curves 1 and 1′).

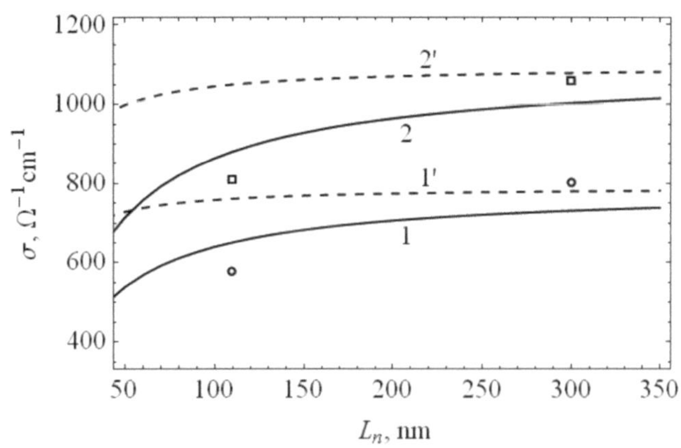

Fig. 20. The dependence of electrical conductivity of bulk nanostructured materials on the grain size for $Bi_{0.4}Sb_{1.6}Te_3$ (circles– experimental data; 1, 1′ – estimations) and for $Bi_{0.3}Sb_{1.7}Te_3$ (squares - experimental data; 2, 2′ – estimations). Estimations use effective mass values from (Luk'yanova et al., 2010) - 1, 2 and from (Stordeur et al., 1988) – 1′, 2′.

For estimations of the influence of boundary scattering on the phonon thermal conductivity the following material parameters were used. The lattice thermal conductivity in Sb_2Te_3 in the cleavage plane at room temperature is equal to $\kappa_{0,11} = 1.9$ W/m K (Goltsman et al., 1972) and the anisotropy of the thermal conductivity is equal to 2.38 (Madelung et al., 1998). Averaging similar to (14) gives the thermal conductivity $\kappa_0 = 1.47$ W/m K. In $Bi_{0.4}Sb_{1.6}Te_3$ solid solution the thermal conductivity in the cleavage plane is 1.2 W/m K (Goltsman et al., 1972) that after averaging using anisotropy value of 2.22 (Madelung et al., 1998) gives

$\kappa_s = 0.94$ W/m K. The Debye temperature in Sb_2Te_3 is about 160K, Debye velocity was estimated as $v_D = 3.6 \cdot 10^5$ cm/s and the heat capacity at room temperature is close to usual value 24.9 J/mol K (Goltsman et al., 1972). This data allowed estimating the average mean free path in the pure crystal as $\bar{l}_0 = 4.7$ nm.

The comparison of the estimated thermal conductivity of nanostructured material with the experimental data is presented on Fig. 21. The electronic contribution to the thermal conductivity was subtracted using Lorenz factor calculated as described above. The results of estimations are quite well correlate with the experimental data. This allows one to conclude that the boundary scattering is important mechanism of reduction of phonon thermal conductivity in bulk nanostructured materials. The estimation of the decrease of the lattice thermal conductivity due to boundary scattering is shown on Fig. 22. It can be seen that the decrease can reach the values of 30-40% at the grain size of about 10-20 nm.

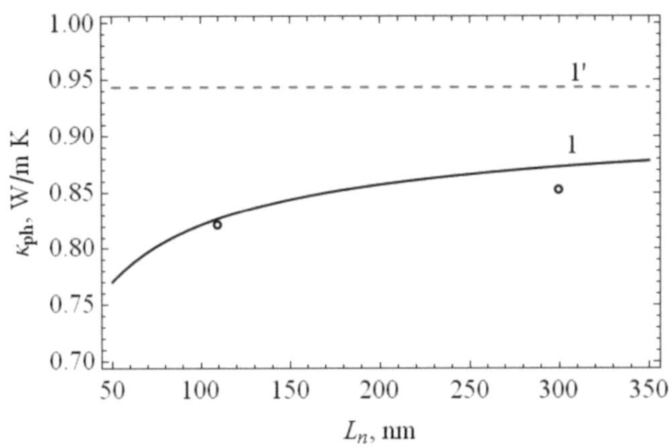

Fig. 21. The dependence of phonon thermal conductivity of bulk nanostructured materials on the grain size for $Bi_{0.4}Sb_{1.6}Te_3$ (circles – experimental data; 1 – estimations; 1′ – the value in initial solid solution).

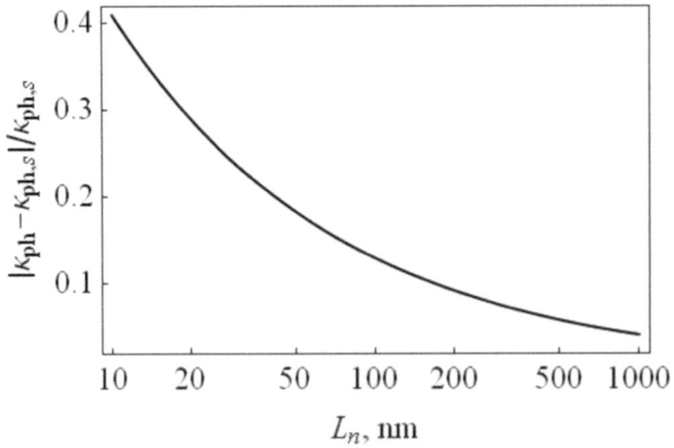

Fig. 22. The dependence of the relative decrease of lattice thermal conductivity on the grain size in nanostructured material based on $Bi_{0.4}Sb_{1.6}Te_3$ solid solution.

3.3 Energy filtering

In the previous section the influence of the boundary scattering on the electrical conductivity of nanostructured material was considered. The boundary scattering was described using constant mean free path equal to the size of grains L_n . As was noticed above the energy dependence of the relaxation time for boundary scattering in this approximation is the same as for acoustical scattering. So the Lorenz factor and the Seebeck coefficient should not differ from that in initial solid solution if the concentration remains the same. On the other hand the experimental data (Bulat et al., 2011a) showed the increase of the Seebeck coefficient in the samples with smaller grain size. To describe this effect the more detailed study of the scattering process was performed. The energy dependence of the probability of carrier scattering on the potential barrier at the grain boundary was taken into account. As the charge carriers with smaller energy scatter more intensively their contribution to the electrical and heat current decrease. This energy filtering can lead to the increase of the Seebeck coefficient if the energy relaxation length l_ε is much greater than the momentum mean free path l_p (Moizhes & Nemchinsky, 1998). At the temperatures much higher than Debye temperature the estimations for typical parameters of semiconductors (Moizhes & Nemchinsky, 1998) showed that as $l_p \sim 50$nm the energy relaxation length is about 500 nm that is much greater than the grain size considered in the present section.

There are several approaches that take into account the influence of the energy filtering on the transport coefficient. In (Ravich, 1995) the scattering on the single barrier was considered. In (Popescu et al., 2009) the exact expression for scattering probability was used but it was not taken into account that it should depend on the part of the kinetic energy corresponding to the motion normal to the boundary rather than the total energy. In (Mayadas & Shatzkes, 1970; Gridchin et al., 2005) the boundary scattering in polycrystalline thin films was considered but the relaxation time was anisotropic.

In the bulk nanostructured samples considered in this section the electrical conductivity appears to be isotropic due to random grain orientation and the following approach for calculation of relaxation time was used (Bulat et al., 2011a). In this approach the boundary scattering is modeled through the specular scattering on the randomly oriented planes representing grain boundaries and the inter plane distance is equal to the grain size L_n . The estimations of the mean free path in the previous section gave $l_p \sim 20$ nm. So the grain size is greater than the mean free path. In this case the multiple scattering can be taken into account through the summing up the probabilities of scattering rather than the matrix elements. In isotropic polycrystalline material with random grain orientation the summation of the probability of multiply scattering leads to the averaging over the boundary plane orientations. The total number of planes was estimated as $3L / L_n$, where L is the characteristic sample size. Due to the conservation lows only two final states in the individual scattering act are possible, namely forward scattering and reflection. In the relaxation time calculation only the second type gives contribution. For the probability of reflection the exact expression is used $W_r(k_n) = 1 - D(k_n)$ where tunneling probability $D(k_n)$ is defined by equation (6) and k_n is the wave vector normal to the grain boundary. As the number of incident electrons on the unit area of the boundary in one second is equal to the density of electron flow $j_i = \hbar k_n / mL$, the number of reflections in the unit time is equal to $j_i W_r(k_n)$. Finally the relaxation time can be calculated as

$$\tau_b^{-1} = \sum_{\mathbf{n}} \frac{\hbar k_n}{mL} W_r(k_n) \frac{-\Delta \mathbf{k_n} \mathbf{k}}{k^2}, \tag{16}$$

where the summation over $\mathbf{n}$ takes into account all possible boundary orientations. The summation can be replaced with the integration over polar and azimuthal angles θ and φ determining the direction of normal vector $\mathbf{n}$. Then the expression for relaxation time due to boundary scattering can be obtained in the following form (Bulat et al., 2011a)

$$\tau_b^{-1} = \frac{6\hbar k}{mL_n} \int_0^1 W_r(k\chi)\chi^3 \, d\chi, \tag{17}$$

where $\chi = cos\,\theta$.

The experimental data and theoretical estimations for electrical conductivity and Seebeck coefficient in the bulk nanostructured materials based on Bi_2Te_3-Sb_2Te_3 solid solutions are presented on Fig. 23, 24. The scattering on the grain boundaries including energy filtering (17) and the scattering on acoustic phonons were taken into account. Because the exact account of anisotropy in Bi_2Te_3 based materials is complicated in equation (17) the density of state effective mass m_{d1} was used. The unknown parameters in calculations were the width d and the energy height ε_b of the intergrain barrier. The estimations showed that quite good agreement with the experimental data can be obtained at the reasonable values of these parameters equal to $d = 5$ nm and $\varepsilon_b = 1.5 k_0 T$. The other parameters were the same as for the estimations of boundary scattering in the constant mean free path approximation discussed in the previous section. In order to check the applicability of relaxation time

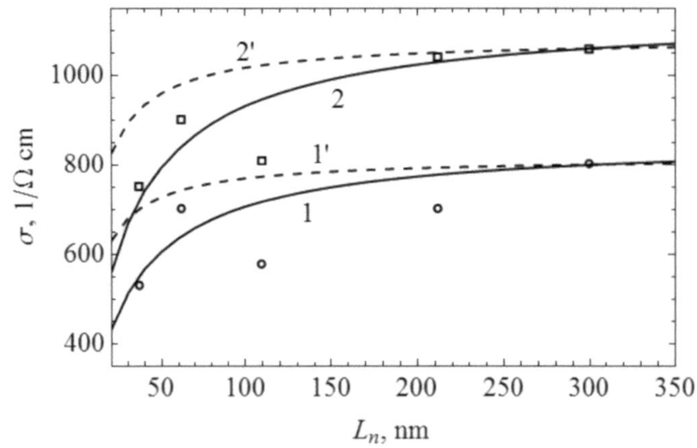

Fig. 23. The dependence of electrical conductivity of bulk nanostructured materials on the grain size for $Bi_{0.4}Sb_{1.6}Te_3$ (circles– experimental data; 1, 1′ – estimations) and for $Bi_{0.3}Sb_{1.7}Te_3$ (squares – experimental data; 2, 2′ – estimations). Estimations use effective mass values from (Luk'yanova et al., 2010) - 1, 2 and from (Stordeur et al., 1988) – 1′, 2′.

approximation the estimation of typical values of the relaxation time were made. It is known that relaxation time approximation is applicable if $\tau > \hbar / k_0 T$ and if the temperature difference on the length of mean free path is small compared to average temperature. If the temperature difference on the sample with $L \sim 0.1\,\text{cm}$ is 100K, then the temperature difference of the length of the order of mean free path about 10 nm is 10^{-3}K. The estimations gave $\tau \sim 10^{-13}$ s and $\hbar / k_0 T = 2.5 \cdot 10^{-14}$ s at room temperature so the both criteria are well satisfied.

Finally the conclusion can be made that the energy filtering effect quite well describes the change of both electrical conductivity and the Seebeck coefficient in nanostructured materials. The estimations showed that in the bulk nanostructured materials based on $Bi_xSb_{1-x}Te_3$ the increase of the Seebeck coefficient due to this effect can reach 10-20% at the grain size of 20-30nm. If the lattice thermal conductivity decrease is the same as that for electrical conductivity this can give the 20-40% increase in the figure of merit.

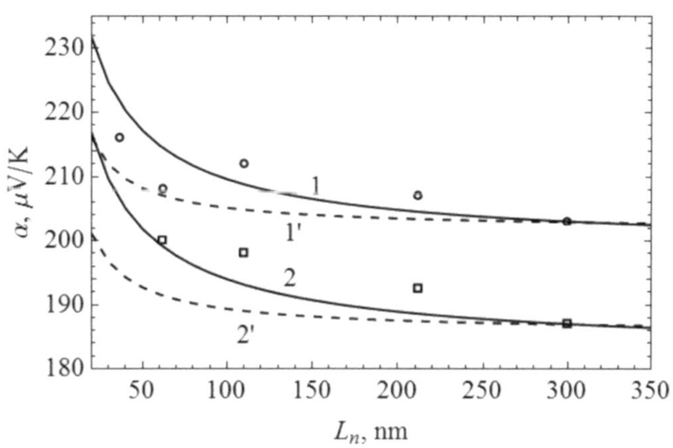

Fig. 24. The dependence of the Seebeck coefficient of bulk nanostructured materials on the grain size (see Fig. 23 for notation).

4. Conclusion

The nanopowder p-Bi-Sb-Te with particles ~ 10 nm were fabricated by the mechanical activation method (ball milling) using different technological modes. Cold and hot pressing at different conditions and also SPS process were used for consolidation of the powder into a bulk nanostructure and nanocomposites.

Nanoparticles keep composition of initial solid solution $Bi_xSb_{2-x}Te_3$. The change of the hot pressing temperature did not result in the change of phase composition and of the lattice parameter of samples. The main factors allowing slowing-down of the growth of nanograins as a result of recrystallization are the reduction of the temperature and of the duration of the pressing, the increase of the pressure, as well as addition of small value additives (like MoS_2, thermally expanded graphite or fullerenes). The SPS processing is also an effective way for reduction of the CDA (or nanograins) size, and as consequence for improvement of the figure of merit. The best value of the efficiency ZT=1.12 (at the temperature ~ 350÷375 K) was measured in the hot pressed bulk nanostructures $Bi_{0,3}Sb_{1,7}Te_3$, while it was reached ZT=1.22 (at 360 K) in the bulk nanostructure $Bi_{0,4}Sb_{1,6}Te_3$ fabricated by SPS method.

The theoretical dependence of the electric and heat conductivities and the thermoelectric power as the function of nanograins size L_n in $Bi_xSb_{2-x}Te_3$ bulk nanostructure are quite accurately correlates with the experimental data (see Sec. 3.2 & 3.3). It means that the phonons and holes scattering on the nanograin boundaries takes place. And the intensity of the scattering increases with reduction of L_n, that results in simultaneous decrease both phonons heat conductivity and electric conductivity. Our study shows that reduction of CDA size really can lead to improvement of the thermoelectric figure of merit.

Some theoretical results (Sec.3) on investigation of mechanisms of the thermoelectric efficiency improvement in bulk nanocrystalline semiconductors based on $Bi_xSb_{2-x}Te_3$ are summarized in Table 4.

Mechanism of improvement Z	Ways of realization	Probable value of increasing Z
Additional phonon scattering	L_n < (10 – 20) nm	(15 – 25)%
Tunneling of carriers	L_n < (10 – 20) nm Vacuum gaps between nanograins ~ 1 – 2 nm	ZT – up to 3,0 – 3,5
Energy filtering of carriers	L_n < (20 – 30) nm Decrease of electrical conductivity and lattice thermal conductivity compensate each other	(20 – 40)%

Table 4. Comparison of mechanisms of the figure of merit improvement

The increase of the thermoelectric power by 10-20 % at L_n =20 - 30 nm can lead to significant (20 - 40 %) increase of the thermoelectric efficiency provided that the reduction of the electric conductivity and the lattice heat conductivity compensate each other. In the investigated samples the full indemnification does not occur, however the thermoelectric efficiency nevertheless managed to be increased up to the values ZT =1.1 - 1.2 (see Sec.2.6).

Table 4 shows that it is necessary to provide the small nanograin size (be more exact – CDA size) ~ 10 – 20 nm for realization of all three mechanisms of the figure of merit improvement. It is difficult to create such nanostructure technologically; the reason is the growth of the initial nanoparticles due to the recrystallization processes. However technological conditions have been determined (see Sec.2.5) for fabrication of the bulk nanostructures and nanocomposites based on $Bi_xSb_{2-x}Te_3$ solid solution from nanopowder by hot pressing and SPS methods which have given the reliable opportunity to obtain the CDA sizes L_n ~ 40 nm.

Fabrication of the vacuum gap ~ 1 – 2 nm between the nanograins for realization of the tunneling mechanism of the improvement of the figure of merit and the cutting off the phonons transport hardly will be possible by the technology of ball milling with the hot pressing or SPS process. Moreover, the electronic microscopy research has not found out any gaps between grains (or CDA) in studied nanostructures - no vacuum, no oxide (Sec.

2.4, 2.5). The accomplishment of all listed in Table 4 requirements to the structure of the nano-thermoelectrics based on $Bi_xSb_{2-x}Te_3$ solid solution should provide the increase of ZT up to 3,5 at the room temperatures. If the vacuum gaps ~ (1 - 2) nm between the grains can not be created technologically, but if the bulk nanostructure with the grain sizes ~ (10 - 20) nm can be realized, the increase of ZT up to 1.5 can be expected.

5. Acknowledgment

The work was supported by the Ministry of Education and Science of the Russian Federation, contract № 16.523.11.3002 and partially by the grant of the President of the Russian Federation № MK-7419.2010.2.

6. References

Anatychuk L.I. & Bulat L.P. (2001). *Semiconductors in Extreme Temperature Conditions.* Nauka, ISBN 5-02-024960-2, St. Petersburg, Russia (In Russian)

Anselm A.I. (1951). *Thermionic Vacuum Thermoelement.* Academy of Science of USSR, Leningrad, USSA (In Russian)

Bartkowiak M. & Mahan G.D. (1999). Boundary effects in thin-film thermoelectrics. *MRS Proceedings,* Vol.545, pp.265-272, ISSN 1946-4274

Bublik V.T., Bulat L.P., Karataev V.V., Maronchuk I.I., Osvenskii V.B., Pivovarov G. I., Pshenai-Severin D.A., Sagalova T.B. & Tabachkova N.Yu. (2009). Connection between Properties and Composition Parameters of Thermoelectric Material Based on Chalcogenides of Bismuth and Antimony. *Materials for electronic technics,* No.4, pp.61-64, ISSN: 1609-3597 (In Russian)

Bublik V.T., Bulat L.P., Karataev V.V., Maronchuk I.I., Osvenskii V.B., Pivovarov G. I., Pshenai-Severin D.A. & Tabachkova N.Yu. (2010a). Possibilities of nanostructured state maintenance at Thermoelectric Material Based on Bismuth and Antimony Chalcogenides. News of Universities, Physics, No.3-2, pp. 37-41, ISSN 0021-3411 (In Russian)

Bublik V.T., Dashevsky Z.M., Drabkin I.A., Karataev V.V., Kasian V.A., Osvenskii V.B., Pivovarov G. I., Pshenai-Severin D.A., Tabachkova N.Yu. & Bohmsteon N. (2010b). Transport Properties in 10-300 K Temperature Range for Nanostructured p-$Bi_{0,5}Sb_{1,5}Te_3$ Obtained by Spark Plasma Sintering. *Thermoelectrics and their Application,* pp.47-52, ISBN 978-5-86763-272-4, Ioffe PTI, St. Petersburg, Russia (In Russian)

Bublik V.T., Dashevsky Z.M., Drabkin I.A., Karataev V.V., Kasian V.A., Lavrentiev M.G., Osvenskii V.B., Pivovarov G. I., Sorokin A.I., Tabachkova N.Yu. & Bohmsteon N. (2010c). Thermoelectric Properties of Nanostructured p-$Bi_{0,5}Sb_{1,5}Te_3$ Obtained by SPS Method. *Thermoelectrics and their Application,* pp.53-57, ISBN 978-5-86763-272-4, Ioffe PTI, St. Petersburg, Russia (In Russian)

Bulat L.P., Pivovarov G.I., Snarskii A.A. (2006), Thermoelectrics based on fullerenes. In: *Thermoelectrics and their Application,* pp.36-40, ISBN 5-86763-185-0, Ioffe PTI, St. Petersburg, Russia (In Russian)

Bulat L.P., Osvensky V.B., Pivovarov G.I., Snarskii A.A., Tatyanin E.V. & Tay A.A.O. (2008a), On the effective kinetic coefficients of thermoelectric nanocomposites.

Proceedings of 6th European Conference on Thermoelectrics, pp. I2-1 - I2-6. Paris, France, July 2-4, 2008

Bulat L.P., Drabkin I.A., Pivovarov G.I., Osvensky V.B., (2008b). On Thermoelectric Properties of Materials with Nanocrystalline Structure. *Journal of Thermoelectricity,* No.4, pp. 27-31, ISSN 1607-8829

Bulat L.P., Bublik V.T., Drabkin I.A., Karatayev V.V., Osvensky V.B., Pivovarov G.I., Pshenai-Severin D.A., Tatyanin E.V. & Tabachkova N.Yu. (2009a), Bulk Nanostructured Thermoelectrics Based on Bismuth Telluride. *Journal of Thermoelectricity,* No.3, pp. 67-72, ISSN 1607-8829

Bulat L.P., Bublik V.T., Karatayev V.V., Osvensky V.B., Pivovarov G.I. (2009b). Methods of Investigation of Mechanical Properties and Structure of Nanomaterials for Thermoelectric Coolers. Bulletin of the International Academy of Refrigeration, No.3, pp.4-7, ISSN 1606-4313 (In Russian)

Bulat L.P. & Pshenai-Severin D. A. (2010a), Effect of Tunneling on the Thermoelectric Efficiency of Bulk Nanostructured Materials, *Physics of the Solid State,* Vol.52, No. 3, pp.485–492, ISSN 1063-7834

Bulat L. P., Bublik V.T., Drabkin I.A., Karataev V.V., Osvenskii V.B., Parkhomenko Yu.N., Pivovarov G. I., Pshenai-Severin D.A. & Tabachkova N.Yu. (2010b), Bulk Nanostructured Polycrystalline p-Bi-Sb-Te Thermoelectrics Obtained by Mechanical Activation Method with Hot Pressing, *Journal of Electronic Materials,* Vol.39, No.9, (September, 2010), pp.1650-1653, ISSN 0361-5235

Bulat L. P., Drabkin I.A., Karataev V.V., Osvenskii V.B. & Pshenai-Severin D.A. (2010c), Effect of Boundary Scattering on the Thermal Conductivity of a Nanostructured Semiconductor Material Based on the $Bi_xSb_{2-x}Te_3$ Solid Solution, *Physics of the Solid State,* Vol. 52, No. 9, pp. 1836–1841, ISSN 1063-7834

Bulat L. P., Drabkin I.A., Karataev V.V., Osvenskii V.B., Parkhomenko Yu. N., Pshenai-Severin D. A., Pivovarov G. I., & Tabachkova N. Yu. (2011a), Energy Filtration of Charge Carriers in a Nanostructured Material Based on Bismuth Telluride, *Physics of the Solid State,* Vol. 53, No. 1, pp. 29–34, ISSN 1063-7834

Bulat L.P., Pshenai-Severin D.A., Drabkin I.A., Karatayev V.V., Osvensky V.B., Parkhomenko Yu.N., Blank V.D., Pivovarov G.I., Bublik V.T. & Tabachkova N.Yu. (2011b), Mechanisms of Increasing of Thermoelectric Efficiency in Bulk Nanostructured Polycrystals. *Journal of Thermoelectricity,* No.1, pp. 14-19, ISSN 1607-8829

Burstein E. (ed.) (1969). *Tunneling Phenomena in Solids.* New York, Plenum Press, 579 p., ISBN: 0306303620

da Silva L.W. & Kariany M. (2004). Micro-thermoelectric cooler: interfacial effects on thermal and electrical transport. *International Journal of Heat and Mass Transfer,* Vol.47, No.10-11, pp.2417-2435, ISSN 0017-9310

Dmitriev A.V., Zvyagin I.P. (2010). Current trends in the physics of thermoelectric materials. *Uspekhi Fizicheskikh Nauk,* Vol.180, No.8, pp. 821- 838, ISSN 0042-1294

Drabble J.R. & Wolfe R. (1956). Anisotropic galvanomagnetic effects in semiconductors. *Proc. Phys. Soc. (London),* Vol.B69, No.2, pp.1101-1108, ISSN 0370-1328

Dresselhaus M.S., Chen G., Tang M.Y., Yang R., Lee H., Wang D., Ren Z., Fleurial J.-P. & Gogna P. (2007). New Directions for Low-Dimensional Thermoelectric Materials. *Adv. Mater.* Vol.19, No.8, (April, 2007), pp.1043–1053, ISSN 1521-4095

Fleming G.M. & Henderson J.E. (1940). The energy losses attending field current and thermionic emission of electrons from metals. *Phys. Rev.* Vol.58, No.10, pp.887-894, ISSN 1943-2879

Ghoshal, U., Ghoshal, S., McDowell, C., Shi, L., Cordes, S. & Farinelli, M. (2002a). Enhanced thermoelectric cooling at cold junction interfaces. *Appl. Phys. Letters,* Vol.80, No.16, pp. 3006-3008, ISSN 0003-6951

Ghoshal U. (2002b). Design and Characterization of Cold Point Thermoelectric Coolers. *Proceedings of XXI International Conf. on Thermoelectrics,* pp. 540-543, ISBN: 0-7803-7683-8, Long Beach, California, USA, August 26-29, 2002.

Goldsmid H.J., Lyon H.B. & Volckmann E.H. (1995). A simplified theory of phonon boundary scattering in solid solutions. *Proc. of the XIV Int. Conf. on Thermoelectrics,* pp.16-19, ISBN: 5-86763-081-1, St. Petersburg, Russia, June 27-30, 1995.

Goltsman B.M., Kudinov B.A. & Smirnov I.A. (1972). *Semiconductor Thermoelectric Materials Based on* Bi_2Te_3. Moskow, Nauka, 320 p., (In Russian)

Gridchin V.A., Lyubimskii V.M. & Moiseev A.G. (2005). Scattering of charge carriers at the boundaries of crystallites in films of polycrystalline silicon. *Semiconductors,* Vol.39, No.2, pp.192-197, ISSN 1063-7826

Harman, T. C., P. J. Taylor, Spears D. L. & Walsh M. P. (2000). Thermoelectric quantum-dot superlattices with high ZT. *Journal of Electronic Materials,* Vol.29, No.1, pp. L1-L2, ISSN 0361-5235

Harman, T. C., M. P. Walsh, Lafarge B.E. & Turner G.W. (2005). Nanostructured thermoelectric materials. *Journal of Electronic Materials,* Vol.34, No.5, pp. L19-L22, ISSN 0361-5235

Hishinuma Y., Geballe T.H., Moyzhes B.Y. & Kenny T.W. (2001). Refrigeration by combined tunneling and thermionic emission in vacuum: Use of nanometer scale design. *Appl. Phys. Lett.,* Vol.78, No.17, pp.2572-2574, ISSN 0003-6951

Lan Y., Minnich A.J., Chen G. & Ren Z. (2010). Enhancement of Thermoelectric Figure-of-Merit by a Bulk Nanostructuring Approach. *Advanced Functional Materials,* Vol.20, No.3, pp. 357-376, ISSN 1616-301X

Madelung O., Rössler U. & Schulz M. (ed.). (1998). $(Bi_{(1-x)}Sb_{(x)})_2Te_3$ physical properties, In: *Landolt-Börnstein - Group III Condensed Matter. Numerical Data and Functional Relationships in Science and Technology. Vol. 41C: Non-Tetrahedrally Bonded Elements and Binary Compounds I.* Berlin, Springer-Verlag, ISBN: 978-3-540-64583-2

Luk'yanova L.N., Kutasov V.A., Konstantinov P.P. & Popov V.V. (2010). Thermoelectric figure-of-merit in p-type bismuth- and antimony-chalcogenide-based solid solutions above room temperature. *Physics of the Solid State,* Vol.52, No.8, pp.1599-1605, ISSN 1063-7834

Mahan G.D. (1994). Thermionic refrigeration. *J. Appl. Phys.,* Vol.76, No.7, pp. 4362-4366, ISSN 0021-8979

Mahan G. D. & Woods L. M. (1998). Multilayer Thermionic Refrigeration. *Phys. Rev. Lett.* Vol.80, No.18, (4 May 1998) pp.4016-4019, ISSN 0031-9007

Mahan, G. D., Sofo, J. O. Bartkowiak, M. (1998). Multilayer thermionic refrigerator and generator. *J. Appl. Phys.* Vol.83, No.9, pp. 4683-4689, ISSN 0021-8979

Mayadas A. F. & Shatzkes M. (1970). Electrical-resistivity model for polycrystalline films: the case of arbitrary reflection at external surfaces. *Phys. Rev. B,* Vol.1, No.4, pp.1382–1389, ISSN 1098-0121

Minnich A.J., Dresselhaus M.S., Ren Z.F., & Chen G. (2009). Bulk nanostructured thermoelectric materials: current research and future prospects, *Energy and Environmental Science*, Vol. 2, No.5, pp. 466-479, *ISSN* 1754-5692

Moizhes B.Ya. & Nemchinsky V. (1998). Thermoelectric figure of merit of metal–semiconductor barrier structure based on energy relaxation length. *Appl. Phys. Lett.*, Vol.73, No.13, pp.1895-1897, ISSN 0003-6951

Murphy E.L. & Good R.H. (1956). Thermionic emission, field emission, and the transition region. *Phys. Rev.*, Vol.102, pp.1464-1473, ISSN 1943-2879

Popescu A., Woods L.M., Martin J. & Nolas G.S. (2009). Model of transport properties of thermoelectric nanocomposite materials. *Phys. Rev. B*, Vol.79, No.20, pp.205302-205308, ISSN 1098-0121

Poudel Bed, Hao Qing, Ma Yi, Lan Yucheng, Minnich Austin, Yu Bo, Yan Xiao, Wang Dezhi, Muto Andrew, Vashaee Daryoosh, Chen Xiaoyuan, Liu Junming, Dresselhaus Mildred S., Chen Gang, Ren Zhifeng (2008). High-Thermoelectric Performance of Nanostructured Bismuth Antimony Telluride Bulk Alloys. *Science*, Vol. *320, No. 5876, (2 May 2008), pp. 634-638,* ISSN 0272-4634.

Ravich Yu.I. (1995). Selective scattering in thermoelectric materials. In: *CRC Handbook of Thermoelectrics. Ed. by Rowe D.M.* CRC Press, N.Y., pp.67-73, ISBN 0-8493-0146-7

Shakouri Ali & Bowers J. E. (1997). Heterostructure integrated thermionic coolers. *Appl. Phys. Lett.* Vol.71, No.9, pp.1234-1236, ISSN 1077-3118

Snarsky A. A., Palti A.M. & Ascheulov A.A. (1997). Anisotropic thermoelements. Review. *Fizika i Tekhnika Poluprovodnikov*, Vol.31, No.11, pp.1281-1298, (In Russian)

Sommer A. H. (1980). *Photoemissive Materials*. Krieger, New York, 270p., ISBN: 0898740096

Stordeur M., Stölzer M., Sobotta H. & Riede V. (1988). Investigation of the valence band structure of thermoelectric $(Bi_{1-x}Sb_x)_2Te_3$ single crystals. *phys. stat. sol. b*, Vol.150, No.1, pp.165–176, ISSN 0370-1972

Stratton R. (1962). Theory of field emission from semiconductors. *Phys. Ref.* Vol.125, No.1, pp.67-82, ISSN 1943-2879

Tavkhelidze, A., Skhiladze, G., Bibilashvili, A., Tsakadze, L., Jangadze, L., Taliashvili, Z., Cox, I. & Berishvili, Z. (2002) Electron Tunneling Through Large Area Vacuum Gap - Preliminary Results. *Proceedings of XXI International Conf. on Thermoelectrics*, pp. 435- 438, ISBN: 0-7803-7683-8, Long Beach, California, USA, August 26-29, 2002

Vasilevskiy, D., Dawood, M. S., Masse J.-P., Turenne, S. & Masut R. A. Generation of Nanosized Particles during Mechanical Alloying and Their Evolution through the Hot Extrusion Process in Bismuth-Telluride-Based Alloys. *Journal of Electronic Materials,* Vol.39, No.9, (September, 2010), pp. 1890-1896, ISSN 0361-5235

Venkatasubramanian R., Siivola E., Colpitts T. & O'Quinn B. (2001). Thin-film thermoelectric devices with high room-temperature figures of merit. *Nature,* Vol. 413 (11 October 2001), pp. 597-602, *ISSN* 0028-0836

22

Self-Assembling Siloxane Nanoparticles with Three Phases

Masatoshi Iji
NEC Corporation
Japan

1. Introduction

Inorganic nanometer-sized particles (nanoparticles) are attracting attention as reinforcing fillers for use in polymer-nanoparticle composites (nanocomposites) because they improve key characteristics of these composites at a relatively low content (Hussain et al., 2006, Jordan et al., 2005). For example, clay nanoparticles (Usuki et al, 2005), metal oxide nanoparticles such as silica (Rosso et al 2006) and titania nanoparticles (Zelikman et al., 2006), and carbon nanotubes (Moniruzzaman & Winey, 2006) increase the mechanical properties, especially the elasticity modulus, of many kinds of polymer composites. However, currently available nanoparticles insufficiently improve the tenacity (elongation at breaking point) of nanocomposites, which is necessary if nanocomposites are to be used in durable products such as electronic equipments and automobiles. This is mainly because these nanoparticles lack rubber-like elasticity although they have an affinity for a polymer matrix. Adding a typical elastomer such as rubber or plasticizer (e.g., a long chain alkyl ester) with a high affinity for a polymer matrix, however, reduces the breaking strength and elasticity modulus of the composite due to their lack of rigidity (Li & Turng, 2006, Shibata et al., 2006).

Nanoparticles with multiple-phases, a high-density phase (core) with rigidity and outside phases with rubber-like elasticity and affinity for matrix polymers, should improve the tenacity of nanocomposites without degrading their breaking strength. Metal oxide nanoparticles are typically formed by hydrolysis and condensation reactions of organic metal compounds, mainly silicon alkoxides (Ha & Cho, 2000, Chujo & Saegusa, 1992, Tamaki & Chujo,1999, Li et al., 2001, Kim et al., 2003) and perhydropolysilazane (Yamano, & Kozuka, 2009), in solvents or polymers, i.e., a sol-gel method. However, the formation of multiple-phased nanoparticles using these organic metal compounds is difficult due to their limited chemical structures. Moreover, the use of conventional surface treatment agents such as organic metal alkoxides to uniformly form multiple phases on nanoparticles while preventing their coagulation is practically difficult because such treatment is an extremely complex process.

Poly L-lactic acid (PLLA), a representative mass-produced biopolymer made of biomass (starch), is attractive for use in environmentally sensitive applications because its use prevents petroleum exhaust and reduces plastic waste due to its biodegradability after disposal. Although it has a relatively high breaking strength, its tenacity is extremely inadequate for it to be used in a variety of applications including durable products like those mentioned above because of its stiff structure, which is due to the hard crystalline region. Studies on PLLA nanocomposites have focused on the usual nanoparticles such as clay and calcium carbonate ones (Li & Turng, 2006, Petersson & Oksman, 2006, Jiang et al.,2007). However, to the best of the authors' knowledge, the use of multi-phased nanoparticles to increase the tenacity of PLLA has not been reported.

In this chapter, self-assembling siloxane nanoparticles with three phases that improve the tenacity of PLLA are reported. The particles consist of a high-density siloxane phase (plural cores), an elastomeric silicone phase, and a caprolactone oligomer phase. Self-assembly by aggregation and condensation of an organosiloxane with three units forms each phase. Testing showed that the use of these nanoparticles increases the tenacity (breaking strain) of PLLA while maintaining its relatively high breaking strength (Iji, 2011).

2. Self-assembling siloxane nano-particle with three phases

2.1 Preparation of siloxane nano-particles with three phases

Figure 1 illustrates the process used to prepare the organosiloxane with three units and the assumed self-assembly of the nanoparticles with three phases through aggregation and condensation of the organosiloxane, which consists of three units: isocyanatepropyltrimethoxysilane (IPTS), polymethylpropyloxysiloxane (PMPS), and a caprolactone oligomer (CLO).

IPTS was selected to form the high-density siloxane phase (plural cores) because it contains methoxy groups, which are highly polar and reactive, at a high molecular ratio, and thus preferentially aggregates and condensates, producing a rigidly cross-linked (high-density) siloxane network that forms more than one core in the particle. PMPS was selected to form the elastomeric silicone phase with an appropriate (nanometer) size around the cores because it contains propyloxy groups, which have moderate polarity and reactivity, at a low molecular ratio in an adequate-length siloxane chain (siloxane number: 8.0). Its use produces a loosely cross-linked (relatively low-density) siloxane network after the core formation. The CLO was selected to form the outside phase due to its low polarity compared with those of IPTS and PMPS and its high affinity for the PLLA matrix (it is highly soluble in melted PLLA).

The organosiloxane was synthesized by mixing PMPS with CLO at a molecular ratio of 1:1 and then mixing the resulting compound with IPTS at a molecular ratio of 1:1 (Figure 1 (A)). After the binding reactions had ceased, the unreacted IPTS, PMPS and CLO were removed using a column packed with polystyrene particles. The reactions were confirmed by hydrogen-nuclear magnetic resonance (H-NMR) analysis of the functional groups in the resulting compounds. Moreover, we performed gel permeation chromatography (GPC) analysis to determine the molecular weight, element analysis of carbon, hydrogen, and nitrogen, and Fourier transform infrared spectroscopy (FT-IR) to determine major bonds. These results showed that the IPTS, PMPS, and CLO were combined in almost equal molecular proportions as we intend. The CLO and IPTS were probably randomly located on the PMPS in the final compound but apart to some-extent because of the difference in their polarities. The reference organosiloxanes with two units (IPTS+PMPS or PMPS+CLO) were prepared and determined based on these methods.

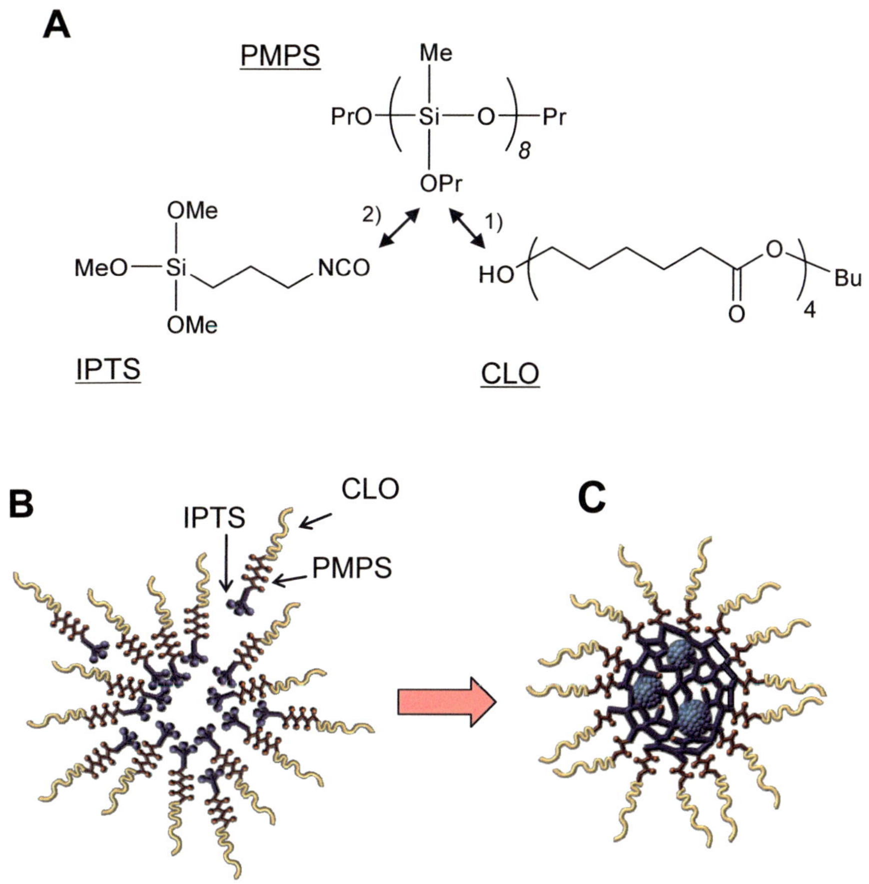

Fig. 1. Preparation of organosiloxane with three units (A) and self-assembly of three-phased nanoparticles through aggregation (B) and condensation (C) of organosiloxane.

We confirmed that the organosiloxane with three units dissolved in tetrahydrofuran (THF) aggregates and forms nanoparticles due to condensation by using water and ammonia as a base catalyst. To avoid confusing the formation of the high-density siloxane cores with the formation of the elastomeric silicone phase around them, we initiated two-step condensation of the organosiloxane by taking advantage of the higher reactivity of the methoxy groups in the IPTS unit than that of the propyloxy groups in the PMPS unit. The first step was core formation through hydrolysis and condensation of the methoxy groups in the IPTS unit at room temperature for 24 hours in THF. The second step was elastomeric silicone phase formation through hydrolysis and condensation of the propyloxy groups in the PMPS unit by heating at a high temperature (180°C) for 20 minutes after replacing the THF with dimethyl sulfoxide, which has a higher boiling point (189°C). These stepwise reactions were ascertained by H-NMR analysis for the methoxy, propyloxy and silanol. After the first step, the methoxy groups in the IPTS were almost completely hydrolyzed and condensed, forming siloxane, while the propyloxy groups in the PMPS unit did not hydrolyzed. After the second step, the propyloxy groups were almost completely hydrolyzed and condensed. These detail methods and results were shown in our paper (Iji, 2011).

2.2 Structures of siloxane nano-particles

As shown in Figure 2, the formed nanoparticles were observed with a scanning electron microscope (SEM). The size distribution, measured by light scattering analysis, indicated a relatively narrow size range, with an average diameter of 13 nm.

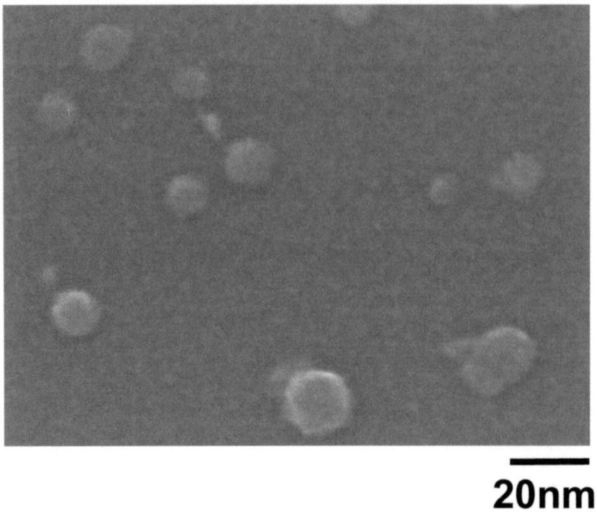

Fig. 2. Observation of three-phased nanoparticles using scanning electron microscope.

The inside structures of the nanoparticles were investigated by scanning electron microscopy and energy dispersive X-ray (SEM-EDX) analysis. Figure 3 shows representative results for the amounts (intensities) of silicon and carbon that were detected along cross-sections of the nanoparticles. The amount of silicon originating from the IPTS and PMPS units was remarkably higher at several points around the center. This indicates that the nanoparticle had plural high-density siloxane cores formed mainly from the IPTS unit and that, around the cores, there was a relatively low-density siloxane phase formed mainly from the PMPS unit. The amount of carbon, which originated from all the units, especially the CLO unit, did not significantly vary throughout the particle. This indicates that there was a relatively high concentration of carbon in the outer layer of the particle, meaning that the CLO unit formed the outside phase fairly well.

Figure 4 shows the thermo-gravity analysis of the nanoparticle formed by the organosiloxane with IPTS, PMPS, and CLO, the reference particle formed by the organosiloxane with PMPS and CLO (through the second step described above), and these organosiloxanes. The results indicated that the nanoparticle showed higher thermo-degradation resistance than the reference particle, which suggests the formation of the high density cross-linking phase to retard the thermo-degradation; we considered the phase to be the cores in the nanoparticle. Furthermore, these particles showed considerably higher thermo-degradation resistances than those of the organosiloxanes. These suggest that not only the high- density cores derived from IPTS but also the relatively low density cross-linking silicone phase derived from PMTS were formed.

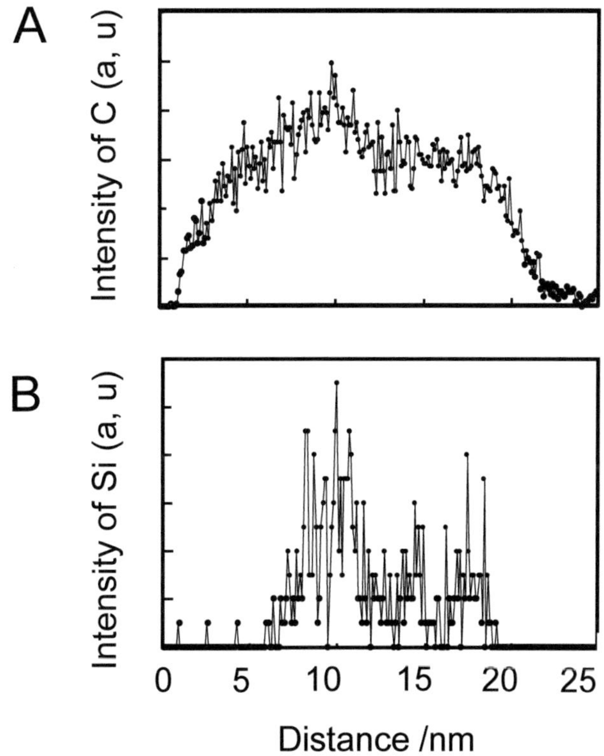

Fig. 3. Analysis of silicon and carbon in a three-phased nanoparticle by SEM-EDX: intensities of silicon (A) and carbon (B) detected along cross-section of the nanoparticle.

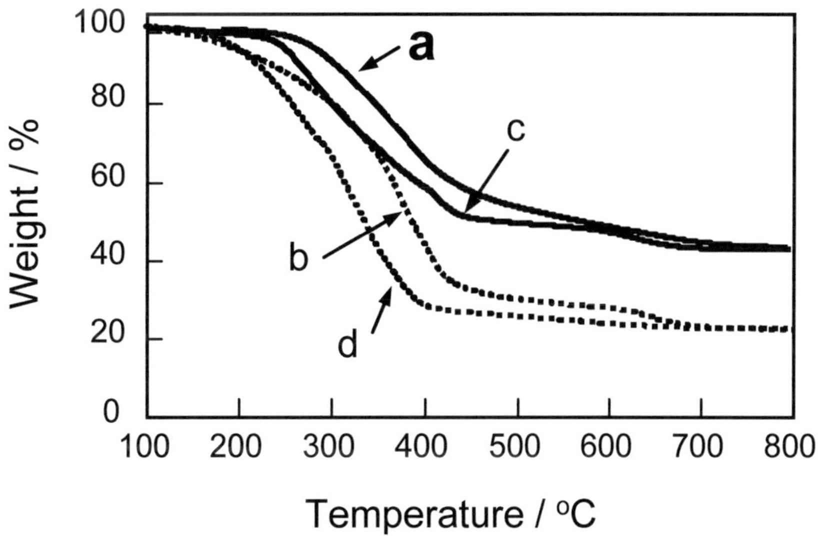

Fig. 4. Thermo-gravimetric analysis of nanoparticles and organosiloxanes (in nitrogen by heating at 10 centigrade per minute)
(a) Three-phased nanoparticle, (b) Organosiloxane consisting of IPTS, PMPS and CLO to

form (a), (c) Two-phased particle consisting of elastomeric silicone and CLO phases, (d) Organosiloxane consisting of PMPS and CLO to form (c)

From these results, it seems reasonable that three-phased nanoparticles can be formed by self-assembly of the organosiloxane with three units, as shown in Figure 1 (B, C). The IPTS unit mainly performs the aggregation of the organosiloxane. The aggregated IPTS unit mainly forms the high-density cross-linked siloxane phase (plural cores) through preferential hydrolysis and condensation reactions of its methoxy groups. After the core formation, the PMPS unit mainly forms the middle phase, the relatively low-density cross-linked siloxane network (elastomeric silicone) around the cores through hydrolysis and condensation reactions of its propyloxy groups. The CLO unit mainly forms the outside phase of the nanoparticles after the organosiloxane has aggregated because of its position and relatively low polarity.

2.3 Characteristics of nanocomposites consisting of siloxane nano-particles and polylactic acid

Using these three-phased siloxane nanoparticles at 5wt%, we prepared a molded PLLA nanocomposite to measure its tenacity by flexural and tensile testing. After the first step (core formation), the THF, water, and ammonia were removed by evaporation. The nanoparticles were mixed with PLLA in chloroform, followed by removing the solvent by evaporation. The resulting composite was extruded at 180°C for 10 minutes using a screw-type mixer and then molded by pressing while heating at 180°C for 10 minutes, followed by crystallization of the PLLA in the composite by heating at 100°C for 4 hours. It is likely that the condensation reaction to form the elastomeric silicone phase in the particle was mostly finished during the extruding and molding of the PLLA composite because the same heating condition as that for the solvent in the second step above resulted in the same condensation reaction.

As shown in Figure 5 (A, B), the three-phased nanoparticles greatly increased the PLLA's tenacity without degrading its high breaking strength. The elongation of the PLLA nanocomposite was more than twice that of PLLA while the elasticity modulus and breaking (maximum) strength were comparable to those of PLLA.

As references, PLLA composites containing commercial silica nanoparticles, two-phased nanoparticles (cores and elastomeric silicone phase), or the organosiloxane consisting of PMPS and CLO units were prepared using the method described above. The composite containing the commercial silica nanoparticles, average diameter 12 nm, showed a slight increase in the modulus, but its strength and elongation were less than those of PLLA (Figure 5A (c)). The two-phased nanoparticles, average diameter of 10 nm, were formed using the organosiloxane with IPTS and PMPS units through the same core formation step described above. The composite containing the nanoparticles showed only a slight increase in elongation compared with that of PLLA (Figure 5A (d)). The composite containing the organosiloxane with PMPS and CLO (PMPS might be cross-linked when mixing with PLLA during heating) showed substantial increase in elongation compared with that of PLLA, but its strength and modulus were less (Figure 5A (e)). The composite containing CLO alone decreased the elasticity modulus of the PLLA composite and insufficiently increased its elongation (Figure 5A (f)). The amount of CLO alone in the composite was 2 wt%, which is near to the amounts of the CLO parts of the three-phased nanoparticles and the organosiloxane with PMPS and CLO. Increasing the amount of CLO further decreased both the modulus and also the strength while it increasing the elongation. These mean that each phase in the three-phased nanoparticles is necessary to increase the PLLA's tenacity while maintaining its breaking strength and modulus.

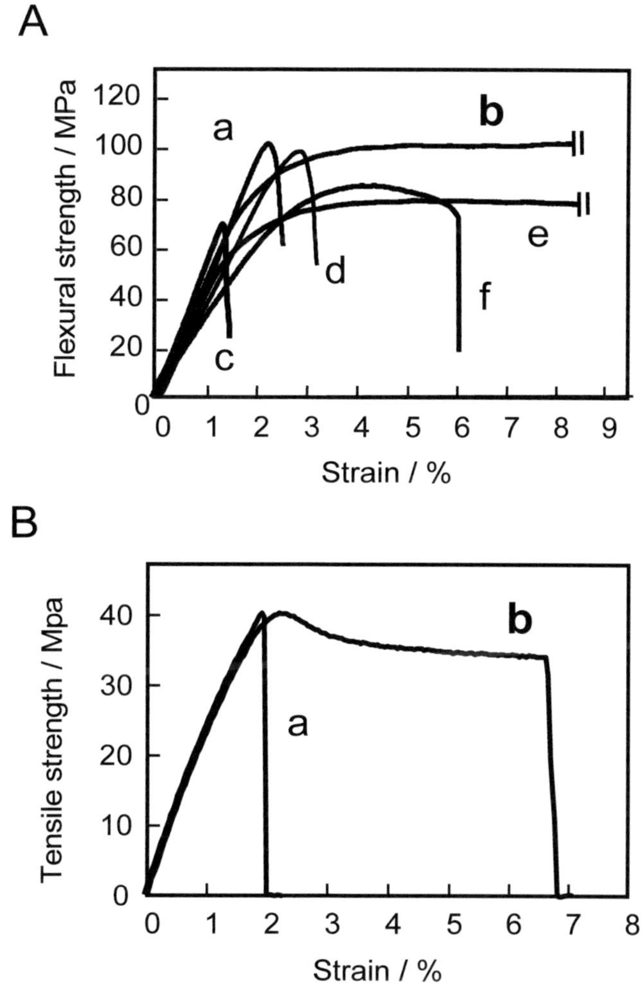

Fig. 5. Flexural (A) and tensile (B) testing of PLLA composites
(a) PLLA, **(b) PLLA nanocomposite with three-phased nanoparticles (5 wt%)**, (c) PLLA nanocomposite with silica nanoparticles (5 wt%), (d) PLLA nanocomposite with two-phased nanoparticles consisting of cores and elastomeric silicone phase (5 wt%), (e) PLLA composite with organosiloxane consisting of PMPS and CLO (5 wt%), and (f) PLLA composite with CLO (2 wt%).

These results suggest that the mechanism of the improved tenacity due to the use of the three-phased nanoparticles is as follows. During the initial period of the PLLA nanocomposite deformation, the nanoparticles create a high elasticity modulus and maximize the strength because of their core rigidity and the high affinity of the outside CLO phase for the PLLA matrix. This idea is supported by the results that the reference organosiloxane with PMPS and CLO, not forming cores and also, the nano-silica, which aggregated in the PLLA composites due to its low affinity with PLLA, did not maintain such a high modulus and strength, simultaneously. In the middle and final periods of the deformation, the three-phased nanoparticles elongated the composite due to the rubber-like elasticity of the elastomeric silicone phase derived from PMPS and the plasticity of the outside CLO phase. While the nano-silica, the reference two-phased particles without the CLO phase, and CLO alone did not perform such elongation of the PLLA composites, the organosiloxane with PMPS and CLO elongated the composite, which can supports the proposed mechanism.

Furthermore, we have cleared that the influence of adding the three-phased nanoparticles on the heat resistance of PLLA (Table 1). The glass transition temperature and heat distortion temperature of the PLLA composite with the nanoparticles (5wt%) slightly decreased comparing with PLLA, but these levels were fairly kept. The decomposition temperature measured by TGA increased. Adding typical elastomers such as rubber or plasticizer reduces heat resiatance of the composites. However, the nanoparticles maintained the heat resistance because of its core rigidity and its higher thermo-degradation resistance as above mentioned.

	Glass transition temp. (°C)	Heat distortion temp.(°C) Load: 0.45MPa / 1.80MPa	Decomposition temp.(°C) / 10% weight loss
PLLA	63	124 / 66	337
PLLA with three-phased nanoparticle (5wt%)	59	119 / 64	344

Table 1. Heat resistance of PLLA and PLLA composites with three-phased nanoparticle

3. Conclusion

In conclusion, we developed self-assembling siloxane nanoparticles with three phases: a high-density cross-linked siloxane phase (plural cores), an elastomeric silicone phase around the cores, and an outside CLO phase with a high affinity for the PLLA matrix. These nanoparticles self-assemble by aggregation and condensation of the organosiloxane with three units, IPTS, PMPS, and CLO that respectively form each phase. Adding these nanoparticles to PLLA increases the tenacity of the PLLA while maintaining its high breaking strength. Their use will expand the use of PLLA in durable product applications and other new applications. These nanoparticles can also be applied to various other brittle polymers by modifying the structure of the outside phase to achieve a high affinity with these polymers.

4. References

Hussain, F., Hojjati, M., Okamoto, M., & Gorga, R. E. (2006). Polymer-Matrix Nanocomposites, Processing, Manufacturing, and Application. *J. Compos. Mat.* Vol.40, pp.1511-1575.

Jordan, J., Jacob, K. I., Tannenhaum, R., Sharif, M. A., & Jasiuk, I. (2005). Experimental Trends in Polymer Nanocomposites- A Review. *Mat. Sci. Eng.* Vol.A 393, pp.1-11.

Usuki, A., Hasegawa, N., & Kato, M. (2005).Polymer-Clay Nonocomposites. *Adv. Polym. Sci.* Vol.179, pp.135-195.

Rosso, P., Ye, L., Friedrich, K., & Sprenger, S. (2006). A Toughened Epoxy Resin by Silica Nanoparticle Reinforcement. *J. Appl. Polym. Sci.* Vol. 100, pp.1849-1855.

Zelikman, E., Tchoudakov, R., & Moshe, N. (2006).Particulate Multi-Phase Polymeric Nanocomposites. *Polym. Comp.* Vol. 27, No. 4, pp. 425-430.

Moniruzzaman, M., & Winey, K. I. (2006). Polymer Nanocomposites Containing Carbon Nanotubes. *Macromolecules.* Vol. 39, pp.5194-5205.

Li, T., & Turng, L. (2006). Polylactide, Nanoclay, and Core-Shell Rubber. *Polym. Eng. Sci.* pp.1419-1427 .

Shibata, M, Someya, Y., Orihara, M., & Miyoshi, M. (2006). Thermal and Mechanical Properties of Plasticized Poly (L-lactide) Nanocomposites with Organo-Modified Montmorillonites. *J. Appl. Polym. Sci.* Vol.99, pp.2594-2602.

Ha, C. S. & Cho, W. (2000). Microstructure and Interface in Organic/Inorganic Hybrid Composites. *J. Polym. Adv. Technol.* Vol. 11, pp.145-150.

Chujo, Y. & Saegusa T. (1992). Organic Polymer Hybrids with Silica Gel Formed by Means of the Sol-Gel Method. *Adv. Polym.* Sci. Vol. 100, pp.11-29.

Tamaki, R. & Chujo, Y. (1999). Synthesis of Polystyrene and Silica Gel Polymer Hybrids Utilizing Ionic Interactions. *Chem. Mater.* Vol.11, pp.1719-1726.

Li, G. Z., Wang, L., Toghiani, H., Daulton, T. L., Koyama, K., & Pittman, C. U. (2001).Viscoelastic and Mechanical Properties of Epoxy/Multifunctional Polyhedral Oligomeric Silsesquioxane Nanocomposites and Epoxy/Ladderlike Polyphenylsilsesquioxane Blends. *Macromolecules.* Vol.34, pp.8686-8693.

Kim, G. M., Qin, H., Fang, X., Sun, F.C., & Mather, P.T. (2003).Hybrid Epoxy-Based Thermosets Based on Polyhedral Oligosilsesquioxane: Cure Behavior and Toughening Mechanisms. *J. Polym. Sci. B, Polym. Phys.* Vol.41, No.24, pp. 3299-3313.

Yamano, A., & Kozuka, H. (2009).Preparation of Silica Coatings Hevily Doped with Spiropyran Using Perhydropolysilazane as the Silica Source and Their Photochromic Properties. *J. Phys. Chem. B*, Vol. 113, pp. 5769-5776.

Petersson, L., & Oksman, K. (2006). Biopolymer Based Nanocomposites: Comparing Layered Silicates and Microcrystalline Cellulose as Nanoreinforcement. *Com. Sci, Tech.* Vol.66, pp.2187-2196.

Jiang, L., Zhang, J., & Wolcott, M. P. (2007). Comparison of Polylactide/Nano-Sized Calcium Carbonate and Polylactide /Montmorillonite Composites: Reinforcing Effects and Toughening Mechanism. *Polymer*, Vol.48, pp.7632-7644.

Iji, M., Morishita, N. & Kai,H.(2011). Self-assembling siloxane nanoparticles with three phases that increase tenacity of poly L-lactic acid", *Polymer Journal,* Vol. 43, pp.101-104.

One-Step Synthesis of Oval Shaped Silica/Epoxy Nanocomposite: Process, Formation Mechanism and Properties

Nopphawan Phonthammachai[1], Hongling Chia[1] and Chaobin He[1,2]
[1]*Institute of Materials Research and Engineering, Singapore, A*STAR (Agency for Science, Technology and Research),*
[2]*Department of Materials Science & Engineering, National University of Singapore, Singapore*

1. Introduction

Silica/epoxy nanocomposites have been widely employed as high strength material for aerospace, automobile, electronic and sporting equipment industries (Deng et al., 2007; Preghenella et al., 2005; Hsiue et al., 2001; Fu et al., 2008). The thermal mechanical properties of nanocomposites were reported to be influenced by the shape and size of nano-filler, volume fraction, quality of dispersion, and the interaction between filler and matrix (Ragosta et al., 2005; Kwon et al., 2008; Adachi et al., 2008; Zhang et al., 2006). These parameters determine the molecular mobility of matrix and the amount of energy dissipation at crack initiation and propagation as the stress transfer from matrix can be promoted by these high specific surface area particles.

To obtain silica/epoxy nanocomposites with good properties, many approaches have been devised to improving the dispersion of silica and the interaction between silica and epoxy (Fu et al., 2008; Kwon et al., 2008; Zhang et al., 2006; Chen et al., 2008; Wang et al., 2005; Zhang et al., 2008). The solution blending process has been reported as a method providing good silica dispersion and silica-epoxy bonding (Hsiue et al., 2001; Zhang et al., 2008; Deng et al., 2007; Liu et al., 2003; Mascia et al., 2006; Huang et al., 2005; Araki et al., 2008). However, the multiple steps are involved in this process. The spherical shaped silica synthesized by sol-gel process is normally used as a silica source due to its availability in solvents. The further step to functionalize silica surfaces by amine-terminated coupling agents is required to improve the dispersion of silica nanoparticles and their adhesion with matrix (Mascia et al., 2006). Then, the high pressure and temperature mixing with epoxy compositions is required after the above process. As the solvents are involved in this process, the steps to remove, recycle and dispose solvents are needed. This poses the environmental, health and safety issues, in addition to the additional costs involved in solvent removal/disposal. Moreover, large amounts of silica at 5-30 % by weight of total composite composition are required to obtain nanocomposite with good mechanical and thermal properties (Deng et al., 2007; Preghenella et al., 2005; Zhang et al., 2008). The high percentages of silica significantly increase the viscosity of compositions and influence many intrinsic properties of epoxy matrix such as weight, ductility, processability and transparency. Therefore, there is a need to develop an effective, convenient and low-cost process to prepare high performance silica/epoxy nanocomposite that exhibits uniform dispersion of silica with great silica- epoxy adhesion, to target a wide range of applications.

In the present work, a "Solvent-Free One-Pot Synthesis" method was developed for the preparation of high performance nanocomposites with uniform dispersion of oval shaped silica in epoxy and strong silica-epoxy bonding. The silica formation, surface functionalization and dispersion in epoxy compositions are combined into one step at 50 °C under mechanical stirring. In this process, the solvent was not involved in the nanocomposite preparation. Thus, it is friendly to the environment and benefits to the formation of oval shaped silica because high shear rate was applied for the mixing of viscous mixture. The details on the synthesis process, the chemical composition and morphology, the thermal mechanical properties of silanized silica/epoxy nanocomposites were studied. The properties of prepared nanocomposite were compared with those of neat epoxy, non-functionalized silica/epoxy and commercial available silica/epoxy systems.

2. Experimental

2.1 Materials

Diglycidyl ether of bisphenol A (DGEBA, D.E.R.™ 332) was supplied by Dow Chemicals. Diethyltoluenediamine (Ethacure 100-LC) was obtained from Albemarle. Tetraethylorthosilicate (TEOS, ≥ 99%) and (3-aminopropyl)trimethoxysilane (APTMS, 97%) were purchased from Sigma-Aldrich. Ammonia solution (25 wt-%) was supplied by Merck. The commercial available silica/epoxy (Nanopox F400) was supplied by Nanoresins AG.

2.2 Preparation of silanized silica/epoxy nanocomposite

A mixture of epoxy, Ethacure 100-LC, TEOS and APTMS was stirred vigorously at 50 °C. The weight ratio of epoxy: Ethacure 100-LC was fixed at 3.8:1. The amount of TEOS was varied to obtain 1-4 wt-% silica in epoxy composition. The amount of APTMS was fixed at 10 wt-% APTMS to silica amount. An ammonia solution with NH_3:TEOS molar ratio of 2.3:1 was injected into the above solution and aged for 60 min. The mixture was degassed under vacuum at 75 °C and poured into a mold coated with releasing agent. The sample was then cured in an air purged oven at 130 °C for 1 h, 160 °C for 2 h and 270 °C for 4 h.

2.3 Preparation of comparative samples

The neat epoxy sample was prepared by mixing an epoxy resin with Ethacure 100-LC at weight ratio of 3.8:1. A commercial available silica/epoxy sample (Nanopox F-400, 2 wt-% silica in epoxy composition) was formed by mixing epoxy, Ethacure 100-LC and Nanopox F-400 at the weight ratio of 3.6:1:0.24. The non-functionalized silica/epoxy (2 wt-% silica in epoxy composition) was prepared following the process shown in 2.2 without adding APTMS. These mixtures were degassed under vacuum at 75 °C before transferred into a mold and cured following the same curing process as that for silanized silica/epoxy nanocomposite.

2.4 Material characterization

2.4.1 Morphology and chemical composition

The morphology, size and dispersion of silica were investigated using transmission electron microscope (TEM) that was conducted in high resolution mode using a JEOL 2100F instrument and operated at 200 kV. The samples were cut using a Leica Ultracut UCT ultramicrotome and placed on 200 mesh copper grids. The chemical compositions of nanocomposites were analyzed using an energy dispersive X-ray spectroscopy in a transmission electron microscope (EDX-equipped TEM). The chemical state of elements in nanocomposite was determined by X-ray photoelectron spectroscopy (XPS) that was conducted using a VG Escalab 220i instrument with monochromatic Al radiation and spot size of 700 μm.

2.4.2 Thermal properties

Single-cantilever mode of the dynamic mechanical analyzer (DMA Q800, TA Instruments) was used to measure the dynamic modulus (E′) and glass transition temperature (T_g) of materials by heating the samples from 25 to 250 °C with a ramping rate of 3 °C/min, frequency of 1 Hz and oscillation amplitude of 20 μm. Thermogravimetric analysis (TGA) was performed with a TA instrument Q500 thermogravimetric analyzer. The degradation temperature (T_d) of materials was measured under nitrogen atmosphere by heating the samples to 800 °C at a ramping rate of 5 °C/min. The temperature at the middle of thermal transition of composites was defined as the degradation temperature (T_d).

2.4.3 Mechanical properties

The flexural strength and modulus of nanocomposites were determined by 3-point bending test according to the ASTM Standard D 790-96, with specimens of 55 × 13 × 2.2 mm^3. The tests were conducted with crosshead speed of 1 mm/min, at a span length of 40 mm. The tensile tests were carried out according to the ASTM Standard D 638-03 using an Instron 5569 testing machine at tensile speed of 1 mm/min. The specimens were cut into dog-bone shape with dimension of 55 × 3 × 2.2 mm^3. The fracture toughness was measured using the Single-Edge-Notch 3-Point-Bend (SEN-3PB) Tests. The Mode-I critical stress intensity factor (K_{Ic}) was measured using SEN-3PB geometry (span = 50.8 mm) and single-edge-notched (SEN) specimens of 60 × 12.7 × 3.0 mm^3, which meets the plane strain condition requirements. A sharp notch was introduced by pressing afresh razor blade at the bottom of a saw-slot in the middle of the rectangular bar with the Instron 5569 at a crosshead speed of 0.5 mm/min. The tests were conducted on the same Instron 5569 at a crosshead speed of 1 mm/min.

3. Results and discussion

3.1 Synthesis of silica/epoxy nanocomposite

In "Solvent-Free One-Pot Synthesis" method, silanized silica nanoparticles were synthesized through the sol-gel process of TEOS, APTMS and ammonia solution (25 wt-% ammonia in water) in epoxy resin (Scheme 1).

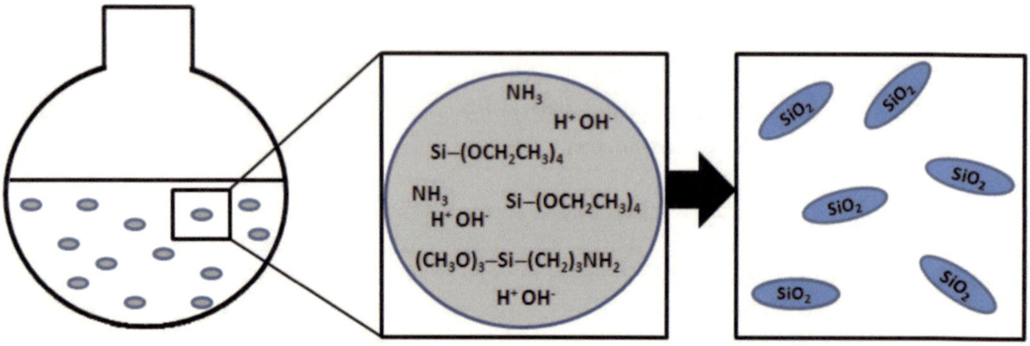

Scheme 1. The formation of silanized silica/epoxy nanocomposite prepared using the "Solvent-Free One-Pot Synthesis" method.

In this process, the water molecules conduct the hydrolysis of TEOS and APTMS, while NH_3 is a basic catalyst that accelerates the reaction (reaction a and b, Scheme 2). The condensation of hydrolyzed TEOS was subsequently occurred to form silica nanoparticles (reaction c, Scheme 2). The hydrolyzed APTMS was functionalized onto the silica surfaces through the reaction between silanol groups of silica and hydroxyl groups of hydrolyzed APTMS to form silanized silica with $-NH_2$ functional groups those can be reacted with epoxy resin to form strong silica-epoxy bonding (reaction d and e, Scheme 2).

The formation of silanized silica as proposed mechanism was confirmed by the transmission electron microscope (TEM) with an energy dispersive X-ray spectroscopy (EDX) and the X-ray photoelectron spectroscopy (XPS). From the TEM image of 4 wt-% silanized silica/epoxy nanocomposite (Fig. 1a), the oval-shaped nanoparticles with diameter of 65-140 nm were uniformly dispersed inside epoxy resin. The preferable oval-shape of these particles (aspect ratio > 1) is possibly occurred through a high shear rate of viscous epoxy\TEOS\APTMS\$NH_3$\$H_2O$ mixture under vigorous stirring. These oval shaped silica nanoparticles could provide added benefit to the mechanical properties of the resulting nanocomposite system as it has higher aspect ratio than those spherical shaped silica. The chemical composition of dispersed particles was confirmed by EDX analysis to be Si, C and O atoms (Fig. 1b).

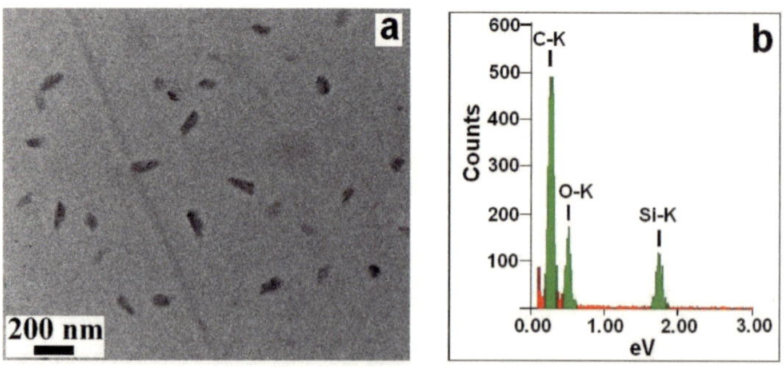

Fig. 1. Homogeneous dispersion of oval shaped silica nanoparticles in epoxy matrix prepared using the "Solvent-Free One-Pot Synthesis" method (a), the chemical composition of silanized silica/epoxy nanocomposite detected by EDX-equipped TEM (b).

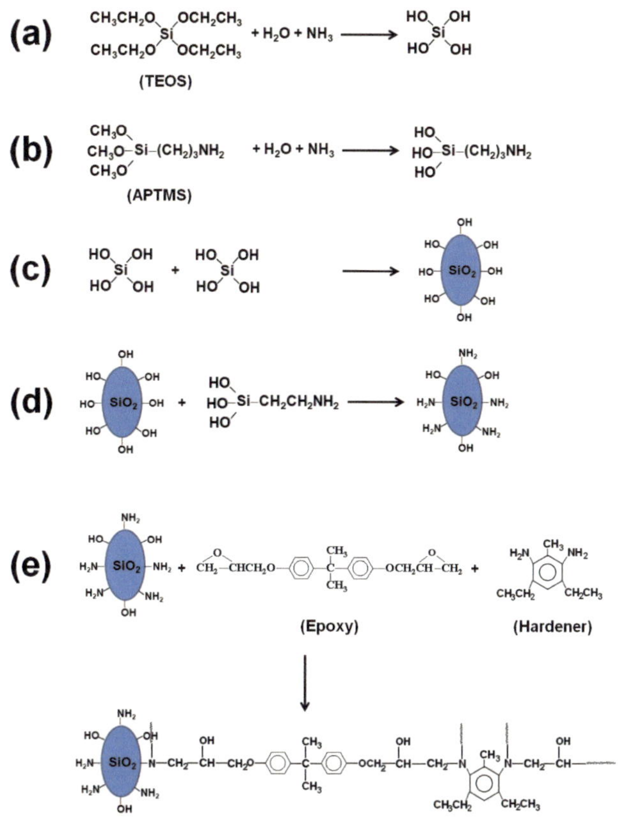

Scheme 2. Proposed mechanisms for the formation of silanized silica/epoxy nanocomposite prepared using the "Solvent-Free One-Pot Synthesis" method.

The XPS of silanized silica/epoxy nanocomposite showed broad peak of Si2p corresponding to three Si species (Fig. 2a). The part at highest binding energy (B.E.) of 103.8 eV represents the silanol groups (Si-OH) on silica surfaces. The Si-O-Si structure of silica particles and Si-alkylamine of functionalized APTMS was found at 102.5 and 101.7 eV respectively. The XPS result of non-functionalized silica/epoxy nanocomposite was compared with the above sample to confirm the present of APTMS functionalized on the silica surfaces (Fig. 2b). As expected, only two Si species were obtained from the non-functionalized silica/epoxy sample at 103.1 eV and 102.1 eV, respectively. The species at high B.E. refers to the silanol groups (Si-OH) on silica surfaces, while the one at lower B.E. is Si-O-Si structure of silica particles.

In accordance with the Si2p spectra, the N1s of silanzied silica/epoxy nanocomposite showed two Ni species represented the N1s of Ethacure 100-LC at B.E. 398.9 eV and APTMS at B.E. of 400.3 eV (Fig. 2c). In Etacure 100-LC, N atoms are connected with a stronger electron donor group (benzene ring) than N atoms of APTMS, therefore, they showed a signal at lower binding energy (Liu et al., 2006). However, only one N from C-N linkage between epoxy and Etacure 100-LC at 399.9 eV was observed from the non-functionalized sample (Fig. 2d).

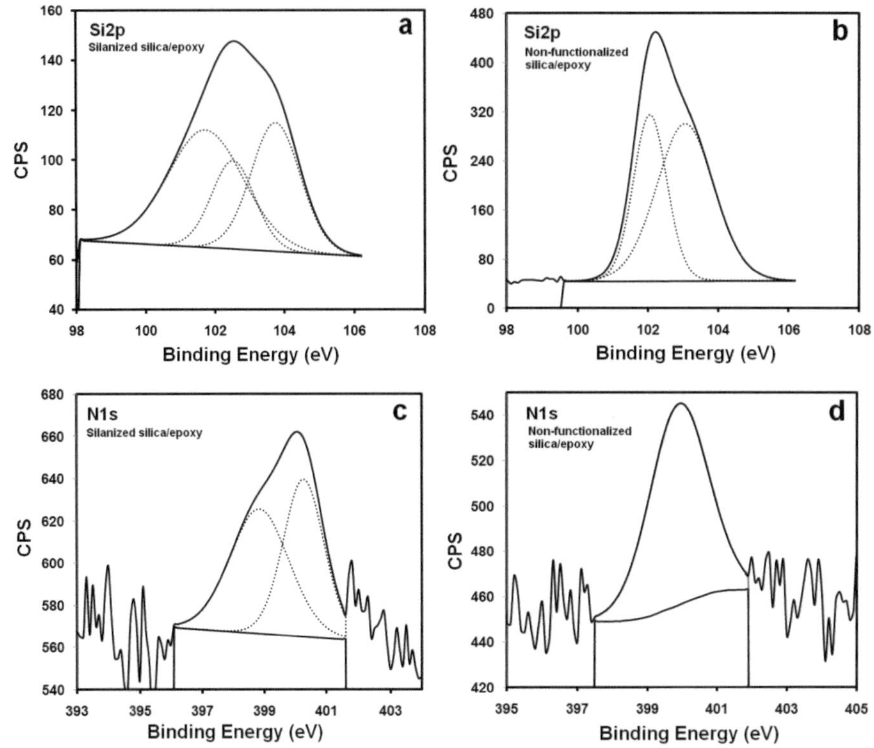

Fig. 2. XPS shows Si2p and N1s curve fit of 2 wt-% silanized silica/epoxy (a and c) and 2 wt% non-functionalized silica/epoxy nanocomposites (b and d).

Therefore, it can be concluded that the oval shaped silica nanoparticles were formed by the sol-gel process of TEOS in the presence of ammonia solution. The APTMS was functionalized on silica surfaces and acts as a linker to form a strong filler-matrix bonding during curing process at elevate temperature of 130-270 °C.

3.2 Thermal mechanical properties of silanized silica/epoxy nanocomposite

The thermal mechanical properties of silanized silica/epoxy nanocomposites were studied at various silica contents of 1-4 wt-% in epoxy composition. The thermal properties of nanocomposites were tested by DMA and TGA. The mechanical properties such as tensile modulus, flexural modulus and fracture toughness were done by tensile, 3-point bending and single-edge-notch 3-point-bend (SEN-3PB) tests.

From the DMA measurement, the glass transition temperature (T_g) of nanocomposites with 0-2 wt-% silanized silica were comparable at 205-210 °C (Fig. 3). The depletion of T_g to 185 °C and 161 °C was occurred when incorporated 3-4 wt-% silica into epoxy resin. Similar trend of result was achieved from the degradation temperature (T_d), in which the comparable T_d at 375-376 °C was shown in 0-2 wt-% silanized silica. However, the T_d of composite became lower to 374 and 360 °C at 3 and 4 wt-% silica contents. The depletion of T_g and T_d at high silica content may occurred from the retardant of cross-linked reaction of epoxy network by large amount of silica particles (Chen et al., 2008).

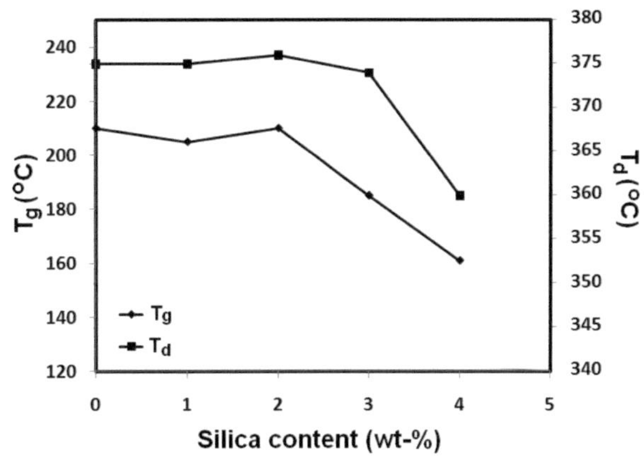

Fig. 3. T_g and T_d of neat epoxy and silanized silica/epoxy nanocomposites prepared at different silica contents of 1-4 wt-% in epoxy composition.

The storage modulus of nanocomposite was increased in accordance with the amount of silica incorporated into epoxy resin (Fig. 4). The similar trend of result was obtained from the fracture toughness, represented as the mode-I critical stress intensity factor (K_{Ic}) (Fig. 5). The K_{Ic} was dramatically improved for 54 % by incorporated 4 wt-% silanized silica into epoxy. The value was increased from 0.35±0.20 MPa.m$^{1/2}$ at 0 wt-% (neat epoxy) to 0.76±0.13 MPa.m$^{1/2}$ at 4 wt-%. It means that the solid silica nanoparticles those uniformly dispersed and formed strong bond with epoxy provide extra reinforcement to the composite structure.

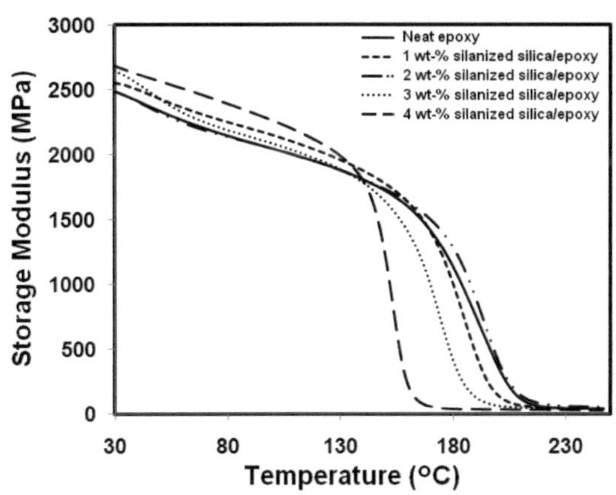

Fig. 4. Plot of storage modulus against temperature of neat epoxy and silanized silica/epoxy nanocomposites prepared at different silica contents of 1-4 wt-% in epoxy composition.

The enhancement of flexural and tensile modulus emphasizes the advantage of oval-shaped silica of the present method to the reinforcement of epoxy resin (Fig. 6). 20% improvement on the flexural and tensile modulus was achieved when introduced few percentages of oval shaped silica (2-3 wt-%) into epoxy, whereas the modulus became maintained at silica content greater than 3 wt-%. As the oval-shaped silica nanoparticles of the present method are uniformly dispersed in epoxy matrix with very low degree of aggregation between particles and form strong bond with epoxy, only few percentages of silica were required to improve the properties of neat resin. No different on the flexural property was observed when added too small quantity of silica (1 wt-%) into epoxy because the number of reinforced silica particles is not enough to prevent the fracture of brittle matrix. At the optimum ranges where the number of silica nanoparticles is enough to reinforce epoxy structure, the improvements of properties were achieved.

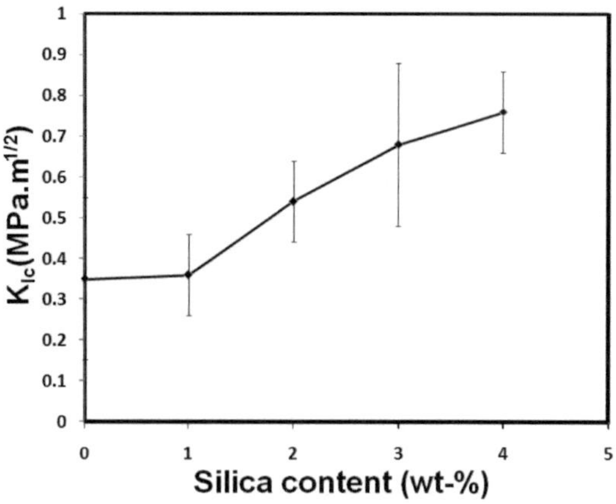

Fig. 5. Mode-I critical stress intensity factor (K_{Ic}) represents the fracture toughness of neat epoxy and silanized silica/epoxy nanocomposites prepared at different silica contents of 1-4 wt-% in epoxy composition. The K_{Ic} value is increased at higher amount of silica content in composite.

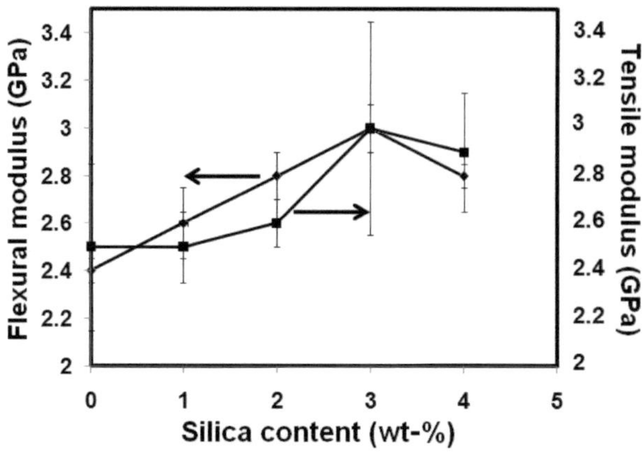

Fig. 6. Flexural and tensile modulus of neat epoxy and silanized silica/epoxy nanocomposites prepared at different silica contents of 1-4 wt-% in epoxy composition.

3.3 Comparison between the thermal mechanical properties of silanized silica/epoxy nanocomposite and comparative references

The silanized silica/epoxy nanocomposite prepared using the present method was compared with three comparative references, which are the neat epoxy resin, non-functionalized silica/epoxy and commercial silica/epoxy. The silica content of nanocomposite systems was fixed at 2 wt-% in epoxy composition.

From DMA and TGA results, it was shown that the T_g and T_d of neat epoxy resin and silanized silica/epoxy were comparable at T_g of 210 °C and T_d of 376 °C (Fig. 7). The commercial silica/epoxy showed slightly lower temperature at T_g of 209 °C and T_d of 374 °C. The lowest T_g was obtained from non-functionalized silica/epoxy nanocomposite at T_g of 207 °C and T_d of 372 °C. It was found from these results that the surface functionalization of silica particles is important to the strength of composite network, in which the thermal stability of nanocomposite structure can be promoted by strong silica-epoxy bonding.

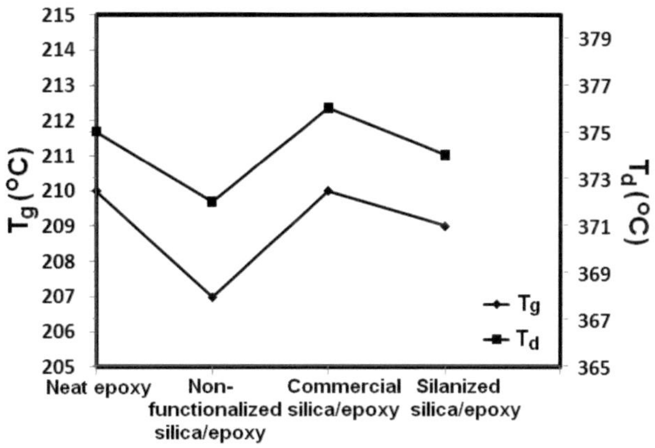

Fig. 7. Comparison between the T_g and T_d of 2 wt-% silanized silica/epoxy nanocomposites and comparative references (neat epoxy, 2 wt-% non-functionalized silica/epoxy and 2 wt-% commercial silica/epoxy).

The comparative results of storage modulus were correspondence with the T_g (Fig. 8.). The highest value was achieved from silanized silica/epoxy at 2510 MPa, while the lower modulus was obtained from non-functionalized silica/epoxy and commercial silica/epoxy nanocomposites at 2330 and 2370 MPa, respectively.

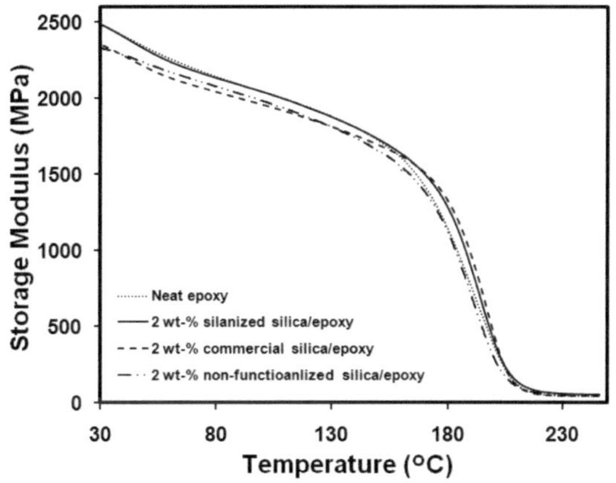

Fig. 8. Comparison between the storage modulus of 2 wt-% silanized silica/epoxy nanocomposites and comparative references (neat epoxy, 2 wt-% non-functionalized silica/epoxy and 2 wt-% commercial silica/epoxy).

The flexural and tensile modulus of neat epoxy resin were found to significantly be enhanced by incorporating silanized silica of the present method (Fig. 9.). In contrast, the poorest properties were achieved when mixed commercial silica with epoxy resin. This may occurred through the aggregation of high concentrated masterbatch of commercial silica (40 wt-%). The size and shape of silica also influence the mechanical properties of nanocomposite. As the silanized silica is oval in shape and its size is suitable for good composite reinforcement, only few percentages of silica are required for the property enhancement. Therefore, the special mixing process may require for the dispersion process of commercial silica in epoxy. Larger amount of commercial silica is also required to achieve a similar range of properties as our silanized silica/epoxy system. Therefore, it can clearly be seen from the comparison that the silanized silica/epoxy nanocomposite prepared by the present method is a competitive method where good thermal mechanical properties of epoxy can be enhanced through one-pot, solvent-free process and required only few percentages of silica.

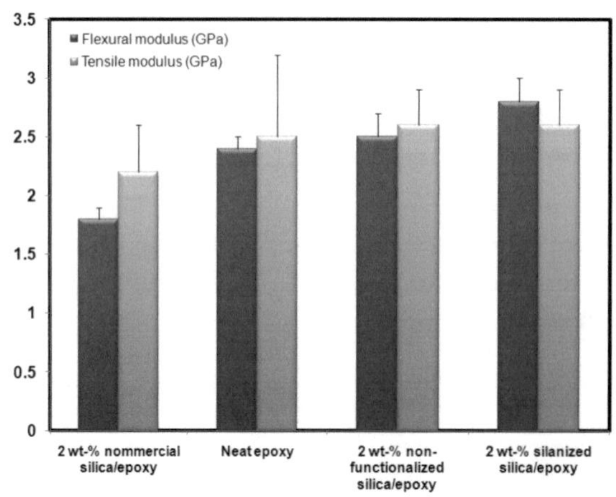

Fig. 9. Comparison between the flexural and tensile modulus of 2 wt-% silanized silica/epoxy nanocomposite and comparative references (neat epoxy, 2 wt-% non-functionalized silica/epoxy and 2 wt-% commercial silica/epoxy).

4. Conclusion

Oval shaped silica/epoxy nanocomposites with uniform silica dispersion and strong silica-epoxy adhesion were effectively and conveniently synthesized via the "Solvent-Free One-Pot Synthesis" method from TEOS, APTMS, ammonia solution, and epoxy compositions at 50°C. Small amount of silica incorporation in the nanocomposite (1-4 wt-% silica in epoxy composition) could enhance the property of silica/epoxy composite significantly. The lower T_g at higher silica loading is due to the decrease of the cross-linked density of epoxy network. The high performance silica/epoxy nanocomposites prepared using the present method exhibit better mechanical properties over neat epoxy (20% and 17% improvements on the flexural and tensile modulus) and commercial available silica/epoxy nanocomposite systems (36% improvement on the flexural modulus).

5. Acknowledgment

This work is funded by the Science and Engineering Research Council (SERC), A*STAR (Agency for Science, Technology and Research) under Grant No. 092 137 0013.

6. References

Deng, S., Ye, L., & Friedrich, K.J. (2007). Fracture Behaviours of Epoxy Nanocomposites with Nano-Silica at Low and Elevate Temperatures. *Journal of Materials Science*, Vol.42, No.8, (April 2007), pp. 2766-2774, ISSN 0022-2461

Preghenella, M., Pegoretti, A., & Migliaresi, C. (2005). Thermo-Mechanical Characterization of Fumed Silica-Epoxy Nanocomposites. *Polymer*, Vol.46, No.26, (December 2005), pp. 12065-12072, ISSN 0032-3896

Hsiue, G.H., Liu, Y.L., & Liao, H.H. (2001). Flame-Retardant Epoxy Resins: An Approach from Organic-Inorganic Hybrid Nanocomposites. *Journal of Polymer Science*, Vol.39, No.7, (April 2001), pp. 986-996, ISSN 1099-0518

Fu, S.-Y., Feng, X.-Q., Lauke, B., & Mai, Y.-W. (2008). Effects of Particle size, Particle/Matrix Interface Adhesion and Particle Loading on Mechanical Properties of Particulate-Polymer Composites. *Composites Part B*, Vol.39, No.6, (September 2008), pp. 933-961, ISSN 1359-8368

Ragosta, G., Abbate, M., Musto, P., Scarinzi, G., & Mascia, L. (2005). Epoxy-Silica Particulate Nanocomposites:Chemical Interactions, Reinforcement and Fracture Toughness. *Polymer*, Vol.46, No.23, (November 2005), pp. 10506-10516, ISSN 0032-3896

Kwon, S-C., Adachi, T., & Araki, W. (2008). Temperature Dependence of Fracture Toughness of Silica/Epoxy Composites: Related to Microstructure of Nano- and Micro-Particles Packing. *Composites Part B*, Vol.39, No.5, (July 2008), pp. 773-781, ISSN 1359-8368

Adachi, T., Osaki, M., Araki, W., & Kwon, S.-C. (2008). Fracture Toughness of Nano- and Micro-Spherical Silica-Particle-Filled Epoxy Composites. *Acta Materialia*, Vol.56, No.9, (May 2008), pp. 2101-2109, ISSN 1359-6454

Zhang, H., Zhang, Z., Friedrich, K., & Eger, C. (2006). Property Improvements of In Situ Epoxy Nanocomposites with Reduced Interparticle Distance at High Nanosilica Content. *Acta Materialia*, Vol.54, No.7, (April 2006), pp. 1833-1842, ISSN 1359-6454

Chen, Q., Chasiotis, I., Chen, C., & Roy, A. (2008). Nanoscale and Effective Mechanical Behavior and Fracture of Silica Nanocomposites. *Composites Science and Technology*, Vol.68, No.15-16, (December 2008), pp. 3137-3144, ISSN 0266-3538

Wang, K., Chen, L., Wu, J., Toh, M.L., He, C., & Yee, A.F. (2005). Epoxy Nanocomposites with Highly Exfoliated Clay: Mechanical Properties and Fracture Mechanisms. *Macromolecules*, Vol.38, No.3, (January 2005), pp. 788-800, ISSN 0024-9297

Zhang, H., Tang, L.C., Zhang, Z., Friedrich, K., & Sprenger, S. (2008). Fracture Behaviours of In Situ Silica Nanoparticle-Filled Epoxy at Different Temperatures. *Polymer*, Vol.49, No.17, (August 2008), pp. 3816-3825, ISSN 0032-3896

Deng, S., Hou, M., & Ye, L. (2007). Temperature-Dependent Elastic Moduli of Epoxies Measured by DMA and Their Correlations to Mechanical Testing Data. *Polymer Testing*, Vol.26, No.6, (September 2007), pp. 803-813, ISSN 0142-9418

Liu, Y.-L., Hsu, C.Y., Wei, W.-L., & Jeng, R.-J. (2003). Preparation and Thermal Properties of Epoxy-Silica Nanocomposites from Nanoscale Colloidal Silica. *Polymer*, Vol.44, No.18, (August 2003), pp. 5159-5167, ISSN 0032-3896

Mascia, L., Prezzi, L., & Haworth, B. (2006). Substantiating the Role of Phase Bicontinuity and Interfacial Bonding in Epoxy-Silica Nanocomposites. *Journal of Materials Science*, Vol.41, No.4, (February 2006), pp. 1145-1155, ISSN 0022-2461

Huang, C.J., Fu, S.Y., Zhang, Y.H., Lauke, B., Li, L.F., & Ye, L. (2005). Cryohenic Properties of SiO_2/Epoxy Nanocomposites. *Cryogenics*, Vol.45, No.6, (June 2005), pp. 450-454, ISSN 0011-2275

Araki, W., Wada, S., & Adachi, T. (2008). Viscoelasticity of Epoxy Resin/Silica Hybrid Materials with an Acid Anhydride Curing Agent. *Journal of Applied Polymer Science*, Vol.108L, No.4, (May 2008), pp. 2421-2427, ISSN 0021-8995

Li, X., Cao, Z., Zhang, Z., & Dang, H. (2006). Surface-Modification In Situ of Nano-SiO_2 and Its Structure and Tribological Properties. *Applied Surface Science*, Vol.252, No.22, (September 2006), pp. 7856-7861, ISSN 0169-4332

24

Low Energy Emulsification Methods for Nanoparticles Synthesis

Veronique Sadtler[1], Johanna M. Galindo-Alvarez[1]
and Emmanuelle Marie -Bégué[2]
[1]*Laboratoire Réactions et Génie des Procédés - GEMICO, CNRS-Nancy Université, Nancy*
[2]*UMR 8640 CNRS-ENS-UPMC, Ecole Normale Supérieure, Paris, France*

1. Introduction

Nanoparticles synthesis by miniemulsion polymerization produces materials that are not obtainable by means of other techniques such as conventional emulsion polymerization. The reason is that, in miniemulsion polymerization, particles are mainly formed by droplet nucleation (Asua, 2002). However, the high energy requirement for preparation of nano-emulsions by traditional methods (Mason et al., 2006; Solans et al., 2005) has precluded widespread use and commercialization.

Nanoemulsions, also referred as miniemulsions or ultrafine emulsions, compose a particular class of emulsions consisting of colloidal dispersions, transparent or bluish for the smallest droplet sizes between 20–100 nm, or milky for sizes up to 500 nm (Solans et al., 2002). In opposition to microemulsions, these systems are thermodynamically unstable, and the droplet size tends to increase with time before phase separation. Nevertheless, the very small initial droplet size makes them kinetically stable (Tadros et al., 2004).

As nanoemulsions are non-equilibrated systems, external energy is required for their preparation. Two generating processes are reported in the literature. In the first case, high mechanical energy is applied during emulsification, generally by using high shear stirring, high pressure homogenizers and/or ultrasound generators. On the contrary, the lower energy method, or condensation method, is based on the phase transitions taking place during the emulsification process (Lamaallam et al., 2005; Solans, et al., 2002; Tadros, et al., 2004). These phase transitions result from changes in the spontaneous curvature of the surfactant and can be achieved (i) at constant composition by changing the spontaneous curvature of non-ionic surfactants with temperature, the well-known Phase Inversion Temperature, PIT, widely used in industry (Izquierdo et al., 2005; K. Shinoda & Saito, 1968)or (ii) at constant temperature by varying the composition of the system by the Emulsion Inversion Point (EIP) method (Forgiarini et al., 2001; Pey et al., 2006; Porras et al., 2008).

Thus, nanoemulsions are specially formulated heterophase systems where stable nanodroplets (with a diameter lower than 500 nm) of one monomer phase are dispersed in a second continuous phase before polymerization takes place, often following a radical mechanism. Since its introduction, this approach has extended the classical emulsion radical polymerization, as ideally each nanodroplet could be regarded as an individual batch reactor, a nanoreactor. Indeed, when (oligo) radicals are generated in the continuous phase, nanodroplets compete with micelles for their capture. In addition, the amount of surfactant added in the feed is usually adjusted so as to minimize (or avoid) the presence of micelles in the continuous phase (Anton et al., 2008; Antonietti & Landfester, 2002).

In those conditions, it could be a good approximation to consider that each droplet behaves as an independent reaction vessel, a hypothetical bulk state where the continuous phase may still transport initiators, side products and heat. Thus, miniemulsion polymerization allows preparing water-based formulated polymers with high solid contents. Additionally this particular mode of design of nanoparticles becomes an advantage, since the chemical composition and colloidal characteristics of the initial nanoemulsion can be used to prepare polymer nanoparticles by "miniemulsion polymerization" of the monomer contained in the oil droplets. The nanoemulsions used for that purpose are mainly prepared by high-energy emulsification methods (Asua, 2002). The aim of this chapter is to show the EIP Method and the Near - PIT concept as a tool to produce miniemulsion templates for miniemulsion polymerization.

2. Nanoparticles by Emulsion Inversion Point (EIP) method

Studies showed that nanoemulsions with very small droplet sizes can be obtained through low-energy methods if, during the emulsification process, the oil is completely dissolved in a single phase, like a bicontinuous microemulsion or a lamellar crystalline phase (Mohlin et al., 2003; Rang & Miller, 1999).The further evolution of the system led to the dislocation of this continuous phase into small nanodroplets. For instance, in the EIP method, the addition of water to a system of water/oil/surfactant forming a lamellar phase increases the hydratation degree of the surfactant polar head thereby increasing its spontaneous curvature. The lamellar phase is disrupted, and the oil, which was initially dissolved, forms small droplets in the size order of the thickness of the hydrophobic layer (See Figure 1). Such methods of nanoemulsion preparation have received increasing attention (Maestro et al., 2008), since even active molecule (*i.e.* lidocaïne) encapsulation in emulsions is achievable by these protocols (Sadurní et al., 2005).

As already mentioned, nanoemulsions can be used to prepare polymer nanoparticles by miniemulsion polymerization of the monomer contained in the oil droplets. The nanoemulsions used for that purpose are mainly prepared by sonifiers and high-pressure homogenizers (Asua, 2002). Only a few studies described the preparation of nanoparticles from nanoemulsions obtained by condensation methods (Calderó et al., 2011; Isabel Solè et al., 2010; Liat Spernath & Magdassi, 2007; L. Spernath et al., 2009).

In this section the formation by EIP, of monomer-in-water nanoemulsions, followed by their conversion in polymer nanoparticles will be considered (Sadtler et al., 2010). For this purpose, the water/Brij 98/styrene system was chosen. Brij compounds are POE-based non ionic surfactants which are commonly used for biomedical applications. Water always

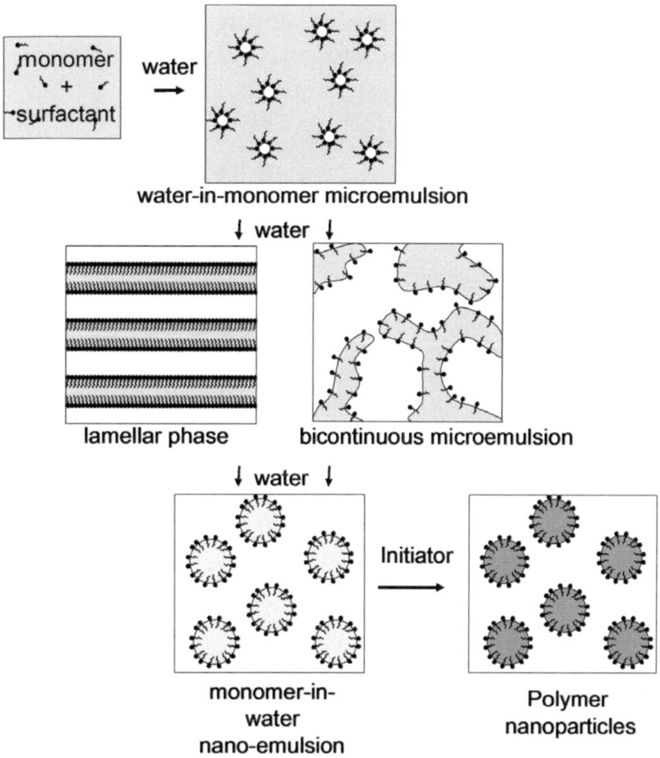

Fig. 1. Schematic representation of the polymer nanoparticles synthesis from nanoemulsions prepared by the EIP method (Adapted from Sadtler, et al., 2010)

contained 0.1 M of NaCl and a small quantity of hexadecane was always solubilized in styrene (in the 5:95 ratio) to avoid Ostwald ripening after nanoemulsion formation (Kabal'nov et al., 1987; Marie et al., 2007). Because lamellar liquid crystalline phase and/or bicontinuous microemulsions are necessary to generate nanoemulsions, the partial phase diagram of the system was determined prior to the nanoemulsion preparation and its miniemulsion polymerization. The phase diagram has been carried out at 50°C due to the suitable temperature to styrene polymerization (Figure 2).

The phase behaviour was found to be in good agreement with the one of other systems containing polyoxyethylene alkyl ether non-ionic surfactant of technical grade. A domain of liquid isotropic phase extends along the surfactant/styrene axis solubilizing up to 10 % water. According to the literature, the most probable structures are inverse micelles or W/O microemulsion (Om). Higher amounts of water (up to 20% approximately) led to the appearance of the lamellar crystalline phase (Lα) that coexists with the Om phase. A wide multiphasic region, with two or three phases comprising liquid crystalline phases (equilibrium not determined), occupies the centre of the diagram for water composition from 20% to 60%. The lamellar crystalline phase in equilibrium with water, (W + Lα) is observed at high-medium surfactant concentration. For the higher amounts of water (up to 90%), the two phase region is present: oil-in-water microemulsion (Wm) and free oil (O). Above this area, by increasing the surfactant concentration, the oil is completely incorporated in the oil-in-water microemulsion. Finally, at low Brij 98 concentrations (O:S ratio above 80:20), appears the three-phase region (Forgiarini, et al., 2001; I. Solè et al., 2006).

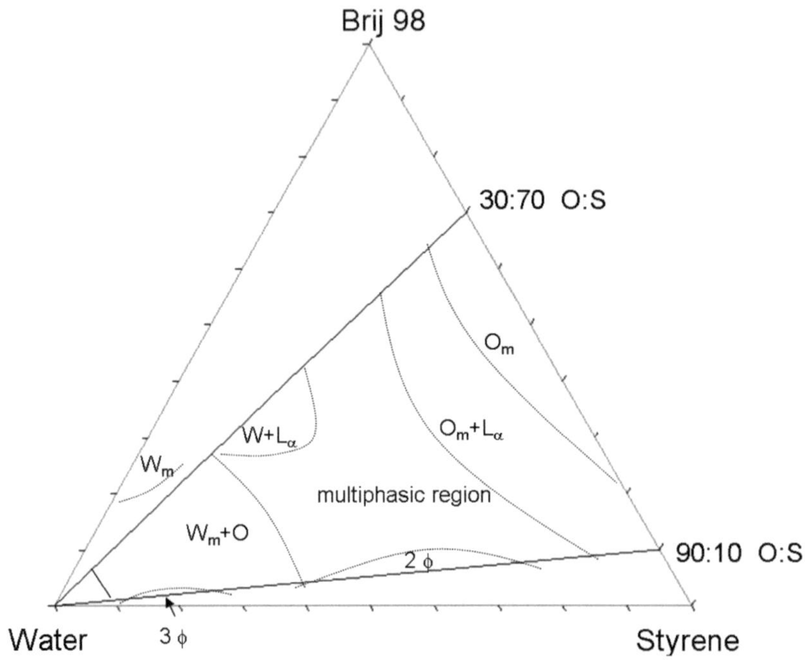

Fig. 2. Partial phase diagram of the water/Brij 98/styrene system, at 50° C. Wm: direct micellar solution or O/W microemulsion; Om: reverse micellar solution or W/O microemulsion); Lα: anisotropic phase (lamellar liquid crystalline phase); W: water phase; O: oil phase. 2ϕ: two isotropic liquid phases; 3ϕ: three isotropic liquid phases. (Adapted from Sadtler, et al., 2010)

The equilibrium phase diagram (figure 2), allows identifying a suitable region for nanoemulsion formation. This domain corresponds to the two-phase region, Wm + O, for O/S ratio between 30:70 and 80:20. The emulsification process path is schematically represented by an arrow on the phase diagram (figure 3). Thus, the addition of water at constant rate to different mixtures of Brij 98 and styrene (inside the suitable region), allowed the system to cross the multiphasic central region, with two or three phases comprising liquid crystalline phases, favouring nanoemulsion formation. The final water concentration was fixed at 80 wt. %, to keep a relatively high percentage of dispersed phase. These aqueous dispersions can be regarded as O/W nanoemulsions (and not microemulsions) because they are formed in the multiphase region (W_m + O). Bluish dispersions were obtained. After addition of KPS solution (the water soluble thermal initiator), miniemulsion polymerization was carried out at 50° C for 24 h.

Figure 4 shows the evolution of nanoparticle sizes as a function of O:S ratio for nanoemulsion containing 80 wt.% water. Results from Sadtler *et al.*, (2010), showed that nanoparticle sizes were clearly dependent on O:S ratio and increased with O:S ratio. After polymerization, nanoparticle sizes varied between 36 nm (O:S ratio of around 0.5) and

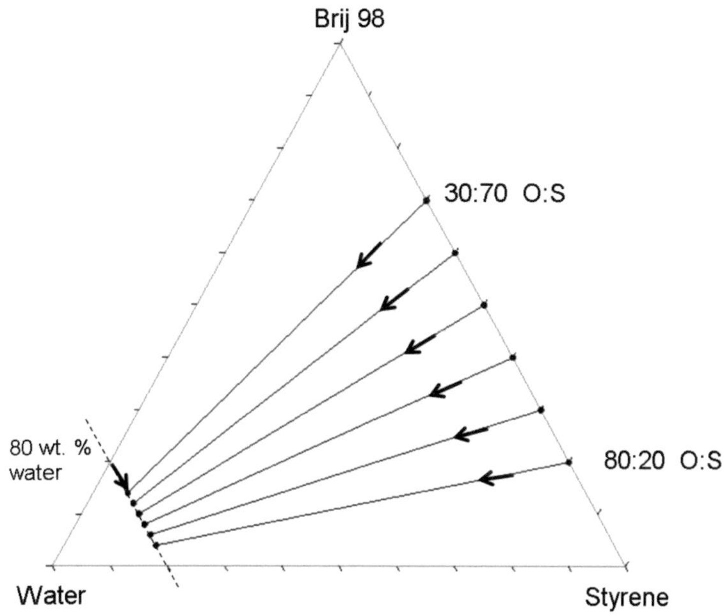

Fig. 3. Schematic representation of the emulsification paths: stepwise addition of 80 wt. % water to different oil:surfactant ratios mixtures

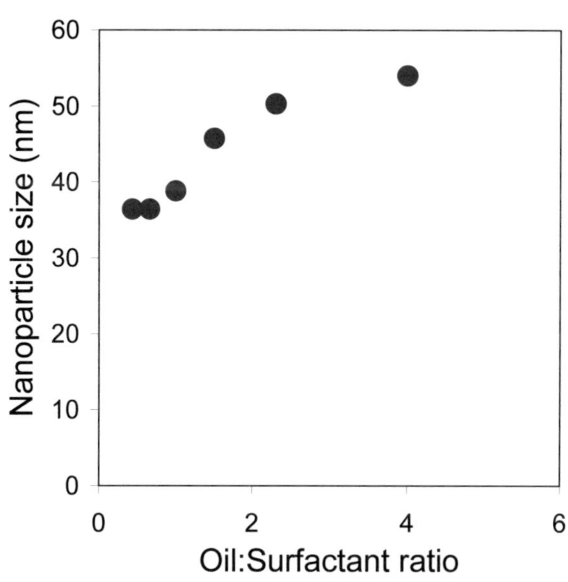

Fig. 4. Polystyrene nanoparticle diameter as function of O:S initial ratio (water addition rate = 4.6 ml/h). (Adapted from Sadtler, et al., 2010)

50 nm (O:S ratio = 4). The fact that the nanoparticle diameter progressively increased with the oil:surfactant ratio suggests that the styrene constituted the inner core of the nanodroplets, which was consistent with a direct O/W -type structure. It should be noticed that the polystyrene nanoparticles sizes obtained by this emulsification path, were

exceptionally small for the water/Brij 98/styrene system, compared to those reported in the literature from high-energy emulsification method (Antonietti & Landfester, 2002; Asua, 2002; Bouanani et al., 2008; Marie, et al., 2007).

Figure 5 shows the size of the nanoparticles synthesized from nanoemulsions prepared at different water flow rates, ranging from 4 ml/h to 150 ml/h, for a O:S ratio of 30:70 and final water composition of 80 wt. %. As predicted by Pey (2006), the polystyrene particle sizes increased with water addition rate. This could be related by the crossing rate of the phases along the emulsification paths, *i.e.* the kinetic of the whole emulsification process.

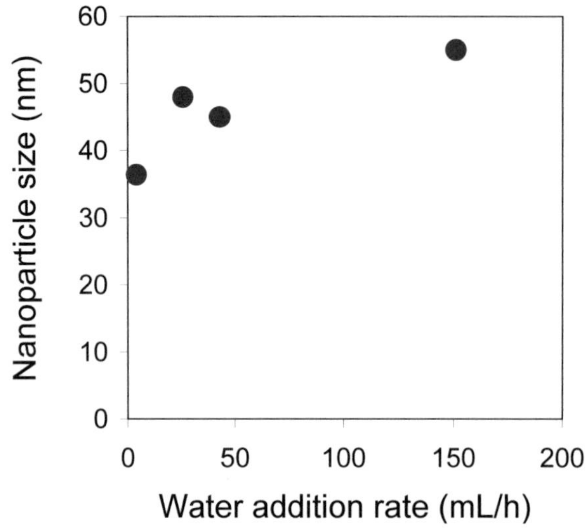

Fig. 5. Polystyrene nanoparticle size as function of the water addition rate (O:S ratio = 0.42). (Adapted from Sadtler, et al., 2010)

To confirm this point, the same path was followed as in the first experiment by adding water at once. In the second experiment, the order of the addition was modified: the oil was added at once to the mixture of water and surfactant (the sample compositions were identical in the two cases).

Figure 6 presents the particle sizes after polymerization obtained following these different pathways. Particles resulting from stepwise addition of water over the mixture of oil and surfactant were smaller than the one obtained by water addition at once (36 and 65 nm respectively).

Phase transitions that take place during the emulsification process (as result of the change in the spontaneous curvature of surfactant), allow to low energy emulsification methods make use of stored chemical energy to get a small drop size distribution. However, when styrene was added to the water and Brij 98 mixture (at once), milky emulsions were formed and the polymerization of the oil droplets did not produced small nanoparticles. Hence the resulting polystyrene dispersion presented an average size of 420 nm (figure 6). The polymerization process might even be totally different in this case, switching from real miniemulsion polymerization to a "simple" emulsion polymerization process. Indeed, miniemulsion polymerization is only possible if the droplets are nucleated thereby leading to polymerization inside the droplets. The smaller the droplet size, the higher the probability of radical entry into the droplets because of the higher interface area. When the emulsion droplets are bigger, the radical entry probability decreases while the micellar and/or homogeneous nucleation increased. Theses nucleation processes are found in the emulsion polymerization process.

Concerning the emulsification protocol, internal phase addition should favour a proper mixing, to assure to reach the equilibrium with all the oil dissolved into the critical phases (*i.e.*, cubic liquid crystal or lamellar phase) (Isabel Solè, et al., 2010).

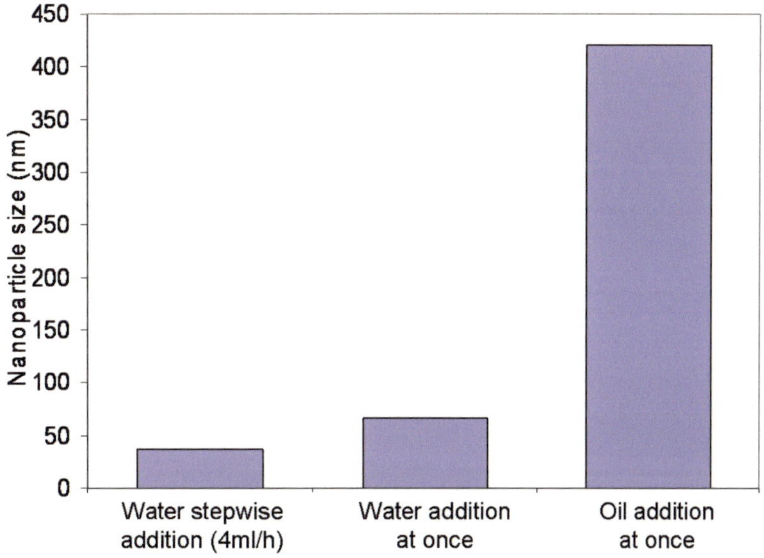

Fig. 6. Polystyrene nanoparticle diameter as function of the emulsification process. The nanoemulsions were prepared at 30:70 O:S ratio, with a final water concentration of 20 wt.%.

3. Nanoparticles by near-PIT method

The formulation-composition map is the graphical representation of the so-called generalized formulation (see Figure 7). The middle shaded zone corresponds to the three phase behavior at or near the optimum formulation. The formulation variable scale is such that the hydrophilicity increases from top to bottom and the stair like bold line is the standard inversion frontier. This line separates the regions in which O/W and W/O emulsions are formed as the result of the stirring of an equilibrate surfactant-oil-water system (Salager, 2000b; Salager et al., 1983). The crossing over through the inversion frontier represents a dynamic phase inversion, since the curvature of the liquid-liquid interface swaps its bending from one way to the other. This change is the consequence of the variation in one of formulation variables (*i.e.*, surfactant affinity) or composition variables (*i.e.*, oil/water ratio) during the stirring process.

If the change is rendered in the map as a vertical shift (crossing through the horizontal branch of inversion line), as for instance in the continuous change in temperature (in the case of non-ionic surfactant), the inversion will always take place under the same conditions (at so-called optimum formulation). Such a dynamic inversion, which is found to be reversible, has been called transitional because it is linked to a phase behavior transition (Salager, 2000a). On the other hand, when the inversion takes place by crossing through a vertical branch of the inversion line, it is called catastrophic because it may be modeled by using catastrophe theory (Salager, 1988; Salager, et al., 1983).

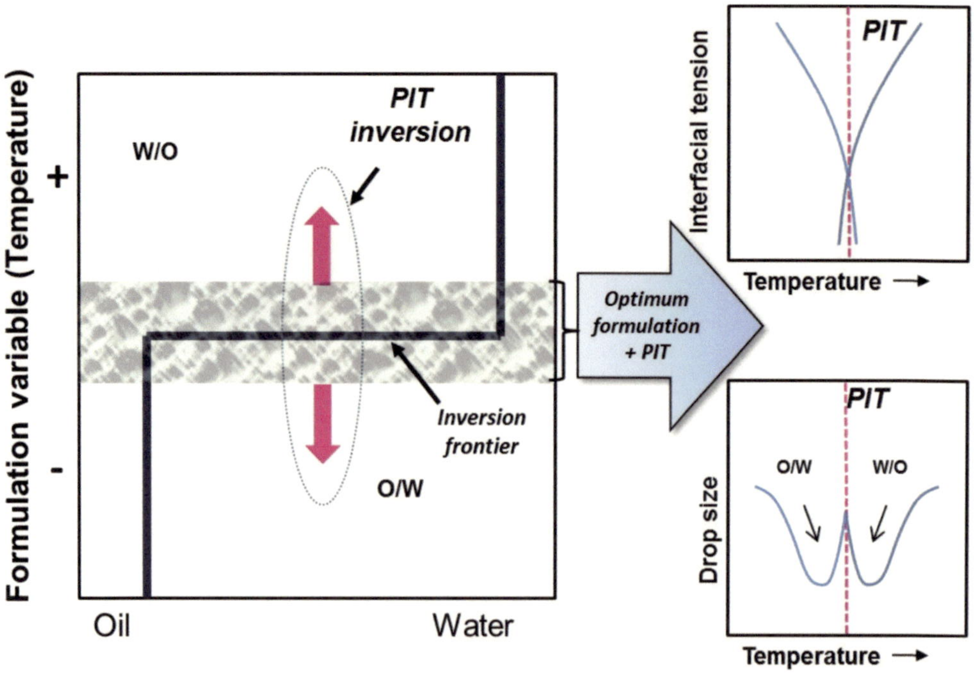

Fig. 7. Formulation - composition map. The bold line is the standard inversion frontier. At right side upper schema illustrates the minimum of interfacial tension obtained near to PIT value. Bottom schema show the regions near to PIT value where droplet size presents a minimum.

The transitional phase inversion is based on the particular ability of emulsions stabilized by poly(ethylene oxide) (PEO)-based non-ionic surfactants to undergo a phase inversion upon temperature variation (Kōzō Shinoda & Arai, 1964). A change of formulation (i.e. induced by temperature increase) along a vertical line, results in a minimum of both the interfacial tension and the emulsion stability at the optimum formulation (see Figure 5). The minimum of stability at optimum formulation has been attributed either to the percolation through liquid crystals located across the thin film, or to the trapping of all surfactant in the microemulsion (Antón et al., 1986). Thus as optimum formulation is approached (either from above or from below the standard inversion line) both the interfacial tension and the emulsion stability decrease. As far as the emulsion droplet size is concerned, the two resulting effects are opposite. The weakening of interfacial tension tends to enhance the efficiency of stirring-mixing process and thus produces smaller droplets, while the decrease in emulsion stability favors the occurrence of coalescence events, and thus results in larger droplets (Salager et al., 1996) (see Figure 7).

The use of the low-energy PIT method has been reported for miniemulsion polymerization by heating above PIT temperature to inverse the emulsion and then cooling to induce the re-inversion followed sometimes by a rapid cooling in an ice bath to set droplet size within the submicronic range (Jahanzad et al., 2007; Liat Spernath & Magdassi, 2007; L. Spernath & Magdassi, 2010). In this section the Near - PIT method (Galindo-Alvarez et al., 2011), for which the strong decrease in interfacial tension near to optimum formulation is used to form submicronic droplets, will be discussed. In contrast to other protocols, Near-PIT method does not reach and cross temperatures aboves PIT, thus temperature sensitive molecules can be use through a carefully match of surfactant system.

It has been reported that stable O/W nanoemulsions can be produced by the PIT method if the dispersed system is rapidly cooled by about 30°C away from its temperature of transitional phase inversion (Solans, et al., 2005). In those conditions, droplet coalescence becomes negligible because the non-ionic surfactant molecules provide an efficient steric barrier. Therefore the miniemulsion templates should exhibit a PIT value about 30 °C higher than the targeted polymerization temperature. PIT value results from interaction between overall surfactant concentration, surfactant mixing ratio and weight fraction oil (K. Shinoda & Arai, 1967). Thus, the PIT value of the studied system was tuned by the appropriate selection of the constituents.

In the case of non-ionic surfactant mixtures, it is well-known that increasing the length of the poly(ethylene oxide) chain results in higher HLB numbers and thus the increase in PIT. Two non-ionic surfactants, PEO stearyl ethers (Brij 78 and Brij 700), differing by the length of the PEO chain (20 and 100 repeat units, respectively) were used for formulating the nanoemulsions allowing a certain adjustment of the PIT value within the convenient range. Figure 8 shows the influence of surfactant mixing ratio and weight fraction of dispersed phase over PIT value. Thus a water/Brij 78 + Brij 700/styrene system containing 1%w/v of NaCl, a surfactant mixing ratio of 0.35/0.65 Brij 700/Brij 78 and 35 wt% of dispersed phase with PIT value around 80°C has been chosen to carry out the miniemulsion polymerization at 50°C using potassium persulfate (KPS) as water-soluble initiator.

On the basis of the previously selected formulation, the Near-PIT emulsification procedure is designed and compared to classical sonification and emulsion polymerization with regard to the final nanoparticle size obtained after reaction completion. For used conditions, a polymerization temperature of 50 °C ensures fast enough initiator decomposition so that no limitation by the polymerization reaction is considered. About particle nucleation mechanism, droplet size distribution of miniemulsion polymerization templates was similar to droplet size distribution of latex particles, suggesting predominance of droplet nucleation mechanism.

Near-PIT protocol, as discussed at the beginning of this section, is based on the effect that droplet size decreases when PIT temperature is approached as the result of an enhanced stirring efficiency due to the very low interfacial tension. Nevertheless, in that temperature range, close to the PIT, resulting emulsions turn out to be very unstable; and no theoretical relationships are available to discriminate zones of minimum droplet size from unstable emulsion. In a general way, in Near-PIT protocol the system is heated until a temperature close to PIT value, equivalent to: -5°C or – 10°C below PIT temperature. As this value is particular for each system formulation, Figure 9 illustrates the final average particle diameters obtained for suspensions resulting from miniemulsion polymerizations after Near-PIT protocols carried out at PIT-10°C, PIT-5°C and PIT-0.5 ◦C values. The used formulation was ϕ_{oil} = 0.2, $X_{Brij700}$ = 0.35 and the overall surfactant concentration equal to 5.2 wt%. For the three examined conditions, particles with diameters lower than 100 nm were obtained after polymerization. In addition, polydispersity indices were relatively low (<0.22) indicating reasonably narrow size distributions. Even if the average diameters were similar for the different temperatures, the better compromise was obtained for the experience carried out at PIT – 5°C, since its polydispersity index was the lowest.

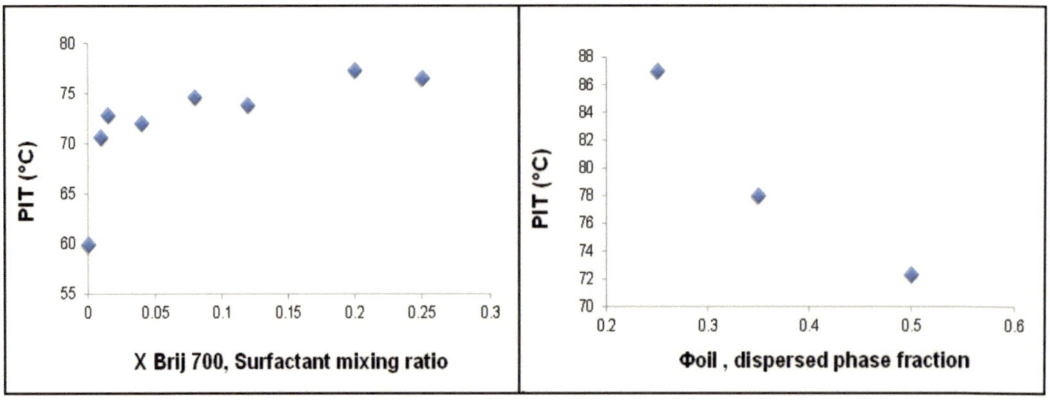

a) b)

Fig. 8. a) Increase of PIT value as function of Brij® 700 content, XBrij700, ϕ_{oil} = 0.5 and 8wt% of surfactant concentration. b) Variation of PIT value with the weight fraction of dispersed phase, ϕ_{oil}, $X_{Brij\ 700}$ = 0.25 and 8wt% of surfactant mixture.
(Adapted from Galindo-Alvarez, et al., 2011)

In miniemulsion polymerization, the use of an effective surfactant system may give very small (20–300 nm) monomer droplets with very large surface area and almost all the surfactant adsorbed at the droplet surface (the concept of critically "stabilized miniemulsion"). Particle nucleation occurs primarily via radical (primary or oligomeric) entry into monomer droplets, since little or no surfactant is present in the form of micelles. The reaction proceeds by polymerization of the monomer in these small droplets, since the loci of polymerization become the monomer droplets and ends when all monomer in droplet is consumed (Schork et al., 2005).

In contrast for macroemulsion polymerization, polymerization starts with large monomer droplets (diameters higher than 10 µm) stabilized by surfactant and coexisting with empty or monomer-swollen surfactant micelles. The water-soluble initiator forms oligoradicals with the slightly water-soluble monomer molecules and these oligoradicals go inside the micelles (heterogeneous nucleation) or start nucleate particles in the continuous phase after reaching a critical degree of polymerization (homogeneous nucleation). During polymerization, the monomer diffuses from the large monomer droplets through the continuous phase to the polymer particles and sustain polymer particle growth until the monomer droplets have vanished (Antonietti & Landfester, 2002). Thus in miniemulsion polymerization latex particles size distribution are expected to correspond to the primary emulsion droplets. On the contrary, in macroemulsion polymerization the particle size distribution is established by the contribution of several nucleation processes leading to average diameters usually larger than 100nm and sometimes to the formation of several populations within the final sample.

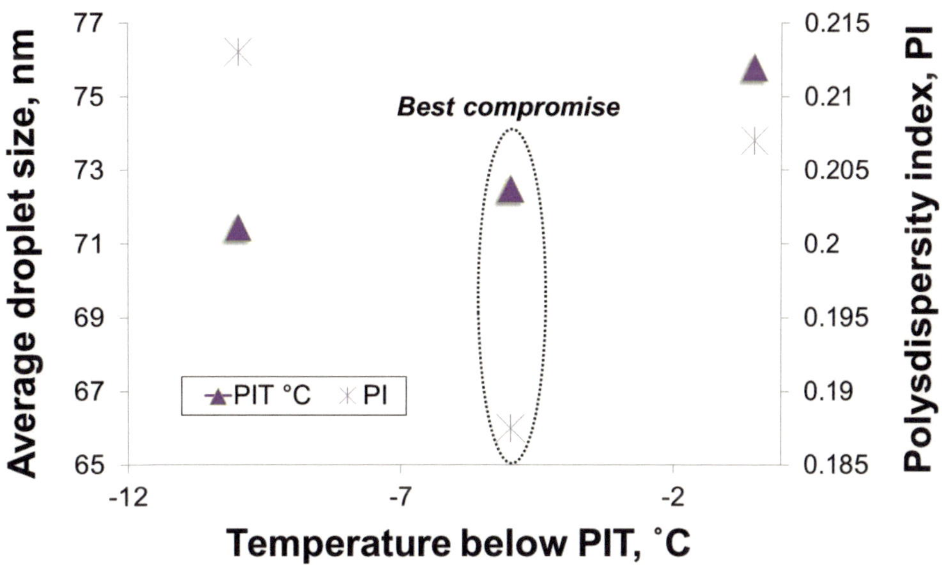

Fig. 9. Influence of polymerization temperature below PIT on final average particle size. Nanoemulsion formulation: $\phi_{oil} = 0.2$, $X_{Brij700} = 0.35$ and 5.2 wt% surfactant concentration. (Adapted from Galindo-Alvarez, et al., 2011)

Figure 10 illustrates the particle size obtained from two miniemulsion polymerization methods (low energy Near-PIT and ultrasound emulsification) and one coarse-emulsion polymerization (standard mechanical emulsification) with various surfactant to oil weight ratios and composition of surfactant mixture. In macroemulsion polymerization protocol, coarse-emulsion is agitated at 800rpm as in Near-PIT method, but the system is heated only until polymerization temperature and not 25 ◦C beyond as in Near-PIT. As expected, macroemulsion polymerization from coarse-emulsion gave the highest particle diameters and polydispersity indices which indicate a large and probably multimodal particle size distribution (see figures. 10 and 11). Thus, the viability of Near-PIT method to produce submicronic droplets as templates for miniemulsion polymerization, with slightly better efficiency than that found for ultrasonic emulsification method, has been confirmed.

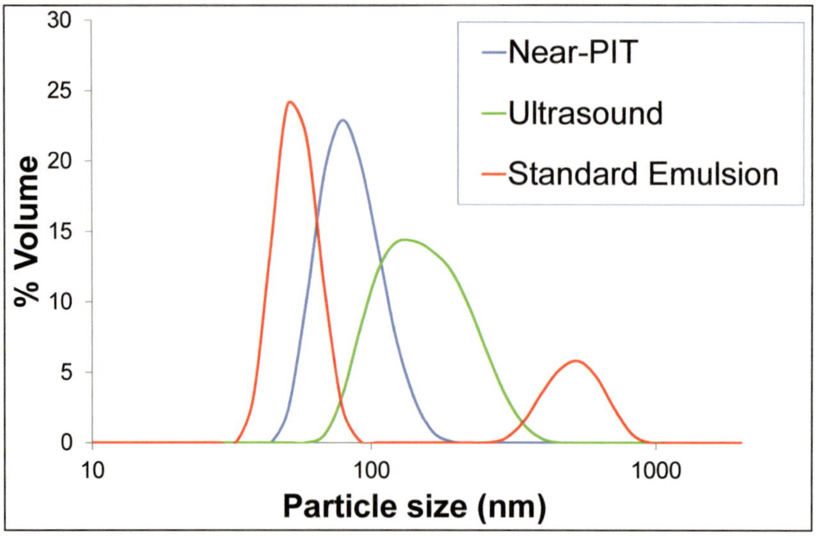

Fig. 10. Influence of emulsification process in particle size distribution, nanoemulsion formulation: °oil = 0.35, $X_{Brij\ 700}$ = 0.4 and 5.2 wt% surfactant concentration. (Adapted from Galindo-Alvarez, et al., 2011)

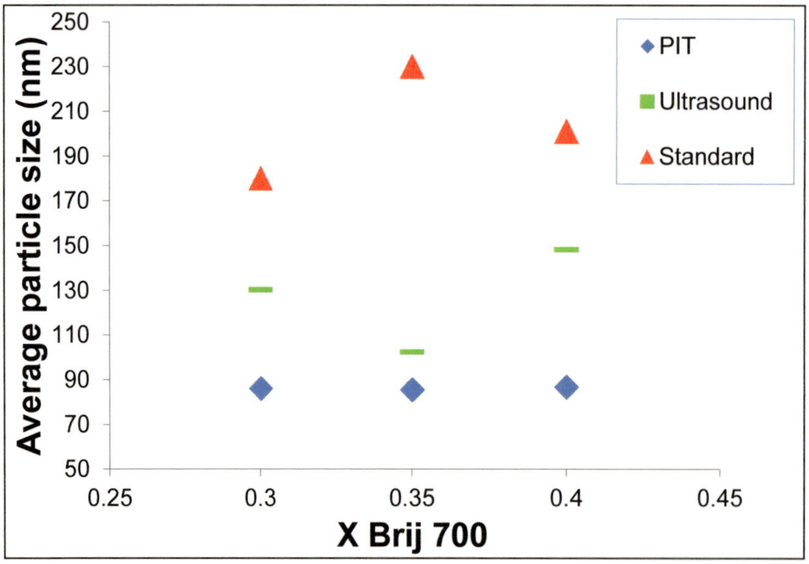

Fig. 11. Influence of emulsification process in average particle size, miniémulsion formulation: °oil = 0.35 and 5.2 wt% surfactant concentration. (Adapted from Galindo-Alvarez, et al., 2011)

4. Conclusion

This chapter has showed the viability to produce polystyrene nanoparticles by two different types of low energy emulsification methods: EIP, emulsion inversion point and Near-PIT, near phase inversion temperature.

The Emulsion Inversion Point technique was used on the water/Brij 98/styrene system to the formation of direct styrene-in-water nanoemulsions. After miniemulsion polymerization, particle sizes as low as 36 nm were obtained. These values are much lower than the one classically reached by high-energy emulsification methods. Thus Emulsion Inversion Point method is very attractive in industrial applications for nanoparticle synthesis, since the nano-emulsion formation does not require high concentration of surfactant, as in the case of microemulsion, or special high-shear equipments as in the case of most reported miniemulsion polymerization

The Phase Inversion Temperature concept as a tool to produce miniemulsion templates for miniemulsion polymerization is a promising methodology in polymerization field to obtain monodisperse and aggregate-free nanoparticle suspensions. In this review a low-energy emulsification method has been designed and showed to allow the preparation of polymeric particles smaller than usual ultrasound miniemulsification methods (about 75 nm) in a water/Brij 78 + Brij 700/styrene system. The operating conditions were adjusted so as to conciliate particle size distribution, colloidal stability and polymerization kinetics. Contrary to usual PIT methods, the Near-PIT procedure did not imply heating the samples at temperatures higher than PIT. Final particles had comparable characteristics to those obtained by traditional PIT methods. In addition, we showed the relevance of temperature control (PIT or PIT-ΔT) over nanoparticle size to obtain even slightly smaller particles than those obtained after ultrasound emulsification. Finally it is possible to vary the composition of the surfactant mixture, within a certain range, without strongly modifying nanoparticles final characteristics, but in way to control the thickness of the hydrophilic superficial layer.

5. Acknowledgments

This work has been funded by the BLAN-06-0174 ANR (National Research Agency) program. The authors thank Prof. Alain Durand, Dr. Marianna Rondon Gonzales, Mrs. Audrey Acrement, Mrs. Ioulia Habipi and Mr. David Boyd for their contributions to this study.

6. References

Anton, N., Benoit, J. P., & Saulnier, P. (2008). Design and production of nanoparticles formulated from nano-emulsion templates-A review. *Journal of Controlled Release, 128*(3), 185-199.

Antón, R. E., Castillo, P., & Salager, J.-L. (1986). Surfactant-oil-water systems near the affinity inversion part IV: emulsion inversion temperature. *Journal of Dispersion Science and Technology 7*(3), 319-329.

Antonietti, M., & Landfester, K. (2002). Polyreactions in miniemulsions. *Progress in Polymer Science (Oxford), 27*(4), 689-757.

Asua, J. M. (2002). Miniemulsion polymerization. *Progress in Polymer Science (Oxford), 27*(7), 1283-1346.

Bouanani, F., Bendedouch, D., Hemery, P., & Bounaceur, B. (2008). Encapsulation of montmorillonite in nanoparticles by miniemulsion polymerization. *Colloids and Surfaces A: Physicochemical and Engineering Aspects, 317*(1-3), 751-755.

Calderó, G., García-Celma, M. J., & Solans, C. (2011). Formation of polymeric nano-emulsions by a low-energy method and their use for nanoparticle preparation. *Journal of Colloid And Interface Science, 353*(2), 406-411.

Forgiarini, A., Esquena, J., Gonzalez, C., & Solans, C. (2001). Formation of nano-emulsions by low-energy emulsification methods at constant temperature. *Langmuir, 17* 2076-2083.

Galindo-Alvarez, J., Boyd, D., Marchal, P., Tribet, C., Perrin, P., Marie-Bégué, E., et al. (2011). Miniemulsion polymerization templates: A systematic comparison between low energy emulsification (Near-PIT) and ultrasound emulsification methods. *Colloids and Surfaces A: Physicochemical and Engineering Aspects, 374*(1-3), 134-141.

Izquierdo, P., Feng, J., Esquena, J., Tadros, T. F., Dederen, J. C., Garcia, M. J., et al. (2005). The influence of surfactant mixing ratio on nano-emulsion formation by the pit method. *Journal of Colloid And Interface Science, 285*(1), 388-394.

Jahanzad, F., Chauhan, G., Mustafa, S., Saha, B., Sajjadi, S., & Brooks, B. W. (2007). Composite Polymer Nanoparticles via Transitional Phase Inversion Emulsification and Polymerisation. *Macromolecular Symposia, 259*(1), 145-150.

Kabal'nov, A. S., Pertzov, A. V., & Shchukin, E. D. (1987). Ostwald ripening in two-component disperse phase systems: Application to emulsion stability. *Colloids and Surfaces, 24*(1), 19-32.

Lamaallam, S., Bataller, H., Dicharry, C., & Lachaise, J. (2005). Formation and stability of miniemulsions produced by dispersion of water/oil/surfactants concentrates in a large amount of water. *Colloids and Surfaces A: Physicochemical and Engineering Aspects, 270-271*(1-3), 44-51.

Maestro, A., Solè, I., González, C., Solans, C., & Gutiérrez, J. M. (2008). Influence of the phase behavior on the properties of ionic nanoemulsions prepared by the phase inversion composition method. *Journal of Colloid And Interface Science, 327*(2), 433-439.

Marie, E., Rotureau, E., Dellacherie, E., & Durand, A. (2007). From polymeric surfactants to colloidal systems. 4. Neutral and anionic amphiphilic polysaccharides for miniemulsion stabilization and polymerization. *Colloids and Surfaces A: Physicochemical and Engineering Aspects, 308*(1-3), 25-32.

Mason, T. G., Graves, S. M., Wilking, J. N., & Lin, M. Y. (2006). Extreme emulsification: Formation and structure of nanoemulsions. *Condensed Matter Physics, 9*(1), 193-199.

Mohlin, K., Holmberg, K., Esquena, J., & Solans, C. (2003). Study of low energy emulsification of alkyl ketene dimer related to the phase behavior of the system. *Colloids and Surfaces A: Physicochemical and Engineering Aspects, 218*(1-3), 189-200.

Pey, C. M., Maestro, A., Solé, I., González, C., Solans, C., & Gutiérrez, J. M. (2006). Optimization of nano-emulsions prepared by low-energy emulsification methods at constant temperature using a factorial design study. *Colloids and Surfaces A: Physicochemical and Engineering Aspects, 288*(1-3), 144-150.

Porras, M., Solans, C., González, C., & Gutiérrez, J. M. (2008). Properties of water-in-oil (W/O) nano-emulsions prepared by a low-energy emulsification method. *Colloids and Surfaces A: Physicochemical and Engineering Aspects, 324*(1-3), 181-188.

Rang, M. J., & Miller, C. A. (1999). Spontaneous emulsification of oils containing hydrocarbon, nonionic surfactant, and oleyl alcohol. *Journal of Colloid And Interface Science, 209*(1), 179-192.

Sadtler, V., Rondon-Gonzalez, M., Acrement, A., Choplin, L., & Marie, E. (2010). PEO-Covered Nanoparticles by Emulsion Inversion Point (EIP) Method. *Macromolecular Rapid Communications, 31*(11), 998-1002.

Sadurní, N., Solans, C., Azemar, N., & García-Celma, M. J. (2005). Studies on the formation of O/W nano-emulsions, by low-energy emulsification methods, suitable for pharmaceutical applications. *European Journal of Pharmaceutical Sciences, 26*(5), 438-445.

Salager, J.-L. (1988). 2 Phase transformation and emulsion inversion on the basis of catastrophe theory. In P. Becher (Ed.), *Encyclopedia of Emulsion Technology: Basic Theory, Measurement, Applications* (Vol. 3, pp. 79 - 134). New York - Basel: Marcel Dekker.

Salager, J.-L. (2000a). Emulsion properties and related know-how to attain them. In F. Nielloud & G. Marti-Mestres (Eds.), *Pharmaceutical Emulsions and Suspensions* (pp. 73 - 125). New York: Marcel Dekker.

Salager, J.-L. (2000b). Formulation concepts for the emulsion maker. In F. Nielloud & G. Marti-Mestres (Eds.), *Pharmaceutical Emulsions and Suspensions* (pp. 19 - 72). New York: Marcel Dekker.

Salager, J.-L., Miñana-Perez, M., M. Pérez-Sánchez, Ramirez-Gouveia, M., & Rojas, C. I. (1983). Surfactant-oil-water systems near the affinity inversion part III: the two kinds of emulsion inversion. *Journal of Dispersion Science and Technology 4*(3), 313 - 329.

Salager, J.-L., Perez-Sanchez, M., & Garcia, Y. (1996). Physicochemical parameters influencing the emulsion drop size. *Colloid and Polymer Science, 274*(1), 81-84.

Schork, F. J., Luo, Y., Smulders, W., Russum, J. P., Butté, A., & Fontenot, K. (2005). Miniemulsion Polymerization *Polymer Particles* (pp. 129-255).

Shinoda, K., & Arai, H. (1964). The correlation between phase inversion temperature in emulsion and cloud point in solution of nonionic emulsifier. *Journal of Physical Chemistry, 68*(12), 3485-3490.

Shinoda, K., & Arai, H. (1967). The effect of phase volume on the phase inversion temperature of emulsions stabilized with nonionic surfactants. *Journal of Colloid And Interface Science, 25*(3), 429-431.

Shinoda, K., & Saito, H. (1968). The effect of temperature on the phase equilibria and the types of dispersions of the ternary system composed of water, cyclohexane, and nonionic surfactant. *Journal of Colloid and Interface Science, 26*(1), 70-74.

Solans, C., Esquena, J., Forgiarini, A. M., Uson, N., Morales, D., Izquierdo, P., et al. (2002). Nanoemulsions: Formation and Properties. In K. L. Mittal & D. O. Shah (Eds.), *Surfactants in Solution: Fundamentals and Applications* (pp. 525). New York: Marcel Dekker.

Solans, C., Izquierdo, P., Nolla, J., Azemar, N., & Garcia-Celma, M. J. (2005). Nano-emulsions. *Current Opinion in Colloid & Interface Science, 10*(3-4), 102-110.

Solè, I., Maestro, A., González, C., Solans, C., & Gutiérrez, J. M. (2006). Optimization of nano-emulsion preparation by low-energy methods in an ionic surfactant system. *Langmuir, 22*(20), 8326-8332.

Solè, I., Pey, C. M., Maestro, A., González, C., Porras, M., Solans, C., et al. (2010). Nano-emulsions prepared by the phase inversion composition method: Preparation variables and scale up. *Journal of Colloid And Interface Science, 344*(2), 417-423.

Spernath, L., & Magdassi, S. (2007). A new method for preparation of poly-lauryl acrylate nanoparticles from nanoemulsions obtained by the phase inversion temperature process. *Polymers for Advanced Technologies, 18*(9), 705-711.

Spernath, L., & Magdassi, S. (2010). Formation of silica nanocapsules from nanoemulsions obtained by the phase inversion temperature method. *Micro and Nano Letters, 5*(1), 28-36.

Spernath, L., Regev, O., Levi-Kalisman, Y., & Magdassi, S. (2009). Phase transitions in O/W lauryl acrylate emulsions during phase inversion, studied by light microscopy and cryo-TEM. *Colloids and Surfaces A: Physicochemical and Engineering Aspects, 332*(1), 19-25.

Tadros, T., Izquierdo, P., Esquena, J., & Solans, C. (2004). Formation and stability of nano-emulsions. *Advances in Colloid and Interface Science, 108-109*, 303-318.

25

Nanoparticles in Ancient Materials: The Metallic Lustre Decorations of Medieval Ceramics

Philippe Sciau
CEMES-CNRS, Université de Toulouse
France

1. Introduction

The scientific community has begun to focus on optical properties of metallic colloids since the early twentieth century with Gustav Mie's works (Mie 1908). However the use of their outstanding properties is much older and dates back to several millennia ago (Colomban 2009; Garcia 2011).

Investigations using various techniques showed that red glasses of the late Bronze Age (1200-1000 BCE) from Frattesina di Rovigo (Italy) were coloured thanks to the excitation of phasmon surface modes of copper nanoparticles (Angelini *et al.* 2004; Artioli *et al.* 2008). The protohistoric community of this region developed advanced glass-manufacturing technology and was able to induce the exsolution of metallic copper crystals in the top layer of glass by exposing the material to reducing conditions. The presence of copper nanoparticles and cuprous oxide (cuprite Cu_2O) had already been reported in Celtic red enamels dated from 400 to 100 BCE (Brun *et al.* 1991). The use of metallic particles for colouring glass spread during the Roman period. Most of the red tesserae used in Roman mosaics were made of glass containing a dispersion of copper nanocrystals (Brun, Mazerolles *et al.* 1991; Colomban *et al.* 2003; Ricciardi *et al.* 2009). In addition to the copper crystals, gold nanoparticles were identified in some red tesserae showing that other metallic nanocrystals were used during Roman times (Colomban, March *et al.* 2003). It is precisely the case of the well-known Roman Lycurgus Cup in glass dated from the 4th century CE and currently exhibited in the British Museum (Freestone *et al.* 2007). The glass of this cup is dichroic and resembles jade with an opaque greenish-yellow tone, but when light shines through the glass (transmitted light) it turns into a translucent ruby colour. It has been demonstrated that the spectacular colour change is caused by colloidal metal and more precisely by nanocrystals of a silver-gold alloy dispersed throughout the glassy matrix (Barber & Freestone 1990). A handful of other Roman glasses showing a dichroic effect were also reported and although the colour change is not so spectacular, the Lycurgus Cup is obviously the result of a good technical mastery of Roman glass-workers (Freestone, Meeks *et al.* 2007). The Roman craftsmen knew that glass could be red coloured and that unusual colour change effect generated by the addition of noble metal bearing material when the glass was molten could be engineered. Nevertheless, the difficulties in controlling the coloration process meant that relatively few glasses of this type were produced, and even fewer have survived.

During the Middle Ages, glass manufacturing expanded considerably, especially to address the demand for stained glass (Kurmann-Schwarz & Lautier 2009). This development was accompanied by an increase in the type of colloidal metal used for colouring glass (Perez-Villar *et al.* 2008; Rubio *et al.* 2009; Gimeno *et al.* 2010). This age also saw the emergence of lusterware, a special type of glazed ceramics, with striking optical effects again obtained from metallic nanoparticles (Caiger-Smith 1991; Pérez-Arantegui *et al.* 2001). Then the progress in glass chemistry during the Renaissance period (Simmons & Mysak 2010) and especially in modern times allowed for better tuning of coloration effects based on the surface plasmons of metallic nanoparticles (Gil *et al.* 2006; Hartland 2011).

The manufacturing process of red glass was used worldwide. The famous Satsuma glasses produced in Japan in the mid-19th century were obtained using a similar technique and their ruby colour comes also from the absorption properties of copper nanocrystals (Nakai *et al.* 1999). It is also the case of the famed red *flambé* and mixed blue-red Jun glazed porcelains from Song and Ming to Qing Chinese Dynasties (Wood 1999).

2. Lustre decorations of medieval ceramics

The lustre is a variety of glaze decoration on ceramics, which appears in medieval times as mentioned in the introduction. Like the ruby glass, the colour of the lustre decorations has a physical basis coming from metallic nanoparticles (Bobin *et al.* 2003; Colomban 2009; Lafait *et al.* 2009). However, lustres possess the particularity of having a colour which can change depending on the angle from which it is observed. An example of these types of ceramic decorations is given in Figure 1 (Mirguet, Roucau *et al.* 2009). The colour change under specular reflection is often spectacular and produces a very intense coloured metallic shine, which can be golden-yellow, blue, green, pink, etc … The density of nanoparticles in the top layers of glaze is higher than that for ruby glass and shows a structuration in depth, which can be more or less complex, as in the lustre of figure 1. This multi-layer structuration on the scale of wavelengths of visible light gives rise to interference phenomena and scattering through rough interfaces, which adds to the surface plasmon effect and strongly contributes to the observed colour. As pointed out by Jacques Lafait *et al.* in their paper concerning the physical colours in cultural heritage, the colours with structural origin are particularly striking and very brilliant. The understanding of these structural effects on optical properties (photonic crystals) is very recent and it is fascinating to see that Islamic potters were able to create such complex structure through empirical chemical means in order to exploit their outstanding optical properties. However before going into detail about these various aspects, a word about the historical context in order to have a few chronological and geographic references.

2.1 Historical context

The earliest lustred potteries were found in Mesopotamia and most of them originate from the site of the Abbasid Caliphs' palace of Samarra in present-day Iraq (Caiger-Smith 1991). This monumental palace-complex whose building was begun by Caliph Mu'tasim in 836 CE, was abandoned in 883 CE.

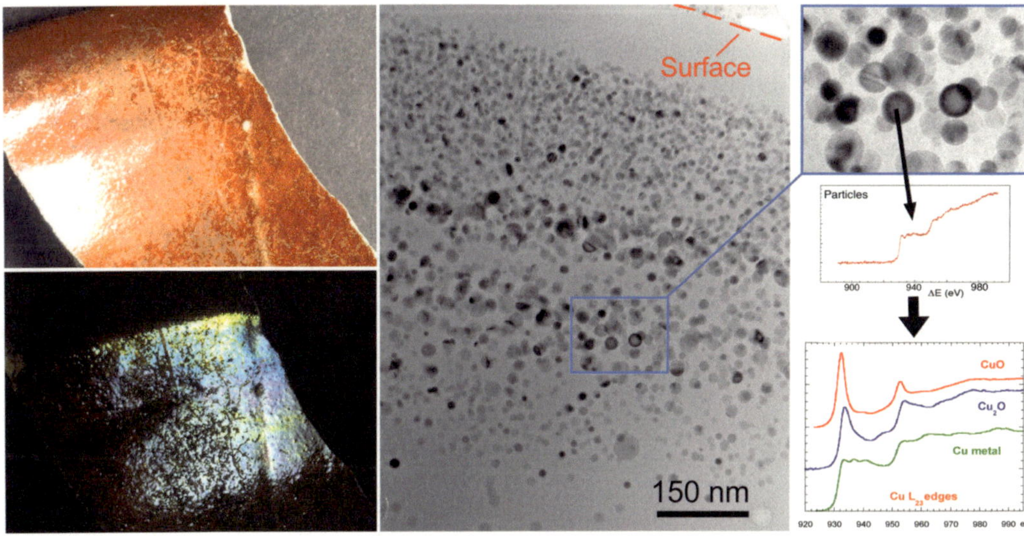

Fig. 1. 9th century CE lusterware from Mesopotamia (Susa). The colour changes from red (top left) to blue/green (bottom left) (Micrographs, courtesy D. Chabanne), nanoparticle distribution observed by TEM (in the centre) and on the right, the identification of particle nature (copper metal) by EELS. Details can be found in references (Chabanne 2005) and (Mirguet *et al.* 2009).

The early lustre manufacturing is therefore rather well dated, though it is currently assumed that first experiments may have occurred earlier, possibly in the time of Harun-al-Rashid (766-809 CE). Abbasid lustres were also found in some quantity in other Mesopotamian cities such as Baghdad, Basra, Kufa or Susa in present-day Iran. Theses cities are often presented as potential production centres, but it is yet an open question. Mesopotamian lustres were discovered outside this geographic area. Tiles with lustre decorations from Iraq were used in the partial reconstruction of the Kairouan Great Mosque (Tunisia), in the 9th century. Fragments have been found at Fustat, which was the main citadel of Lower Egypt in the 9th century. Shards have also been excavated from the site of the palace of Qal'a in Algeria, which was until 1052 CE the capital of the Hammamid princes. In fact, lustre decorations were certainly created in the early 9th century for courts and courtiers and seldom appeared in any other setting. For several centuries the lusterware kept its status of luxury tableware for princely courts.

The annexation of Egypt by the Fatimids (969 CE) led to profound modifications, not only on a governmental level, but also in the population. The Fatimid capital was transferred from Tunisia to al-Qahira, modern Cairo, and the old city of Fustat provided quarters for craftsman who worked for the new capital a few kilometres to the north. The demand for lustre by the new court led to the development of the local production. It is now attested that lustres were made in Fustat before the Fatimid period. However this production, often called pre-Fatimid, seems to have been very limited and of poor quality. The Egyptian production actually began with the arrival of the Fatimids, and during two centuries, a great deal of good quality lustre was being made reflecting the interests and cultural traditions of the new dynasty and its courtiers.

During the 12th century, lustre technique began to extend from Egypt to Syria and to Persia (present-day Iran). Craftsmen from Fustat allegedly brought the technique there during the decline of the Fatimid dynasty, which occurred in the middle 12th century. Concerning the diffusion in Persia, the subject is treated in detail in Oliver Watson's book (Watson 1985).

It seems that the technique appeared in the Occident (southern Spain) during the same period, as soon as the taifa emerged after the dissolution of the Spanish Umayyad caliphate. However, it is only under the Nasrid dynasty (1237-1492 CE) that the lustre technique really flourished in Spain. Its apogee, between the 14th–15th centuries, gave rise to the Hispano-Moorish ceramic, which was elaborated in the Valencia region up to the 18th century.

The technique found a new application during the Italian Renaissance (15th and 16th centuries) where Deruta and Gubbio became the most famous production centres of lustred glazed majolica (Padeletti *et al.* 2006). The main production centres with chronological data are summarized in the figure 2.

Fig. 2. Localization of the main centres of lustre productions.

2.2 Nanoparticle layer

Many studies were devoted to determining the elementary composition of glazes. The results obtained by the C2RMF lustre team, which analysed a significant corpus of specimens, are available online[1]. A review of the main results of the other studies can be found in Philippe Colomban's paper (Colomban 2009). The composition of glaze used in the lustre decoration is highly varied with alkaline and high lead glaze. There is not a specific composition. During the Abbasid times, the first productions used alkaline glazes but lead was then introduced with, in some cases, tin. The glazes used during the Fatimid epoch were mostly leaded alkaline. The Hispano-Moorish productions were characterized by high lead glaze containing a small amount of sodium and potassium. Renaissance decorations were also applied on leaded glazes. Significant composition variations were observed inside the same geographic area, which could be linked to a relative chronology, according to C2RMF's work. Several research groups are seeking to confirm this. The ceramic bodies are

[1] *Ceramics with metallic lustre decoration. A detailed knowledge of Islamic productions from 9th century until Renaissance.* D. Chabanne, M. Aucouturier, A. Bouquillon1, E. Darque-Ceretti, S. Makariou, X. Dectot, A. Faÿ-Hallé, D. Miroudot (2011), 2011arXiv1101.2321C

also various. Some lustre decorations were affixed on ceramic clay bodies, whereas others were deposited on siliceous pasta. The various associations of ceramic/glaze in relation to the different geographic areas and periods are listed in the online paper of the C2RMF lustre team (cf. note 1).

The composition of nanoparticles is much less diverse. Only copper and silver were used in all lustre decoration from the 9th century to now. On the other hand, the size, the shape and the spatial organisation can be very different (Fig. 3). One of the best tools for studying an organisation on a nanometre scale is the transmission electron microscopy (TEM). It is this technique which allowed for the first observation of nanocrystals in an archaeological lustre, a 13th century lustre of Hispano-Moorish period (Pérez-Arantegui, Molera *et al.* 2001). Although it is an abrasive technique, which is an obstacle for the investigation of museum pieces, several specimens from different periods and different geographic areas have been analysed by TEM (Borgia *et al.* 2002; Padeletti & Fermo 2003b; Pérez-Arantegui & Larrea 2003; Fredrickx *et al.* 2004; Padeletti & Fermo 2004; Roqué *et al.* 2007; Mirguet *et al.* 2008; Mirguet, Roucau *et al.* 2009; Sciau *et al.* 2009a). For some of them, sampling was limited to a few thousand cubic micrometres, using focused ion beam (FIB) techniques (Sciau *et al.* 2009b). However, since these ceramics are not conductors, a carbon deposit must be performed, which is not easy to carry out on museum pieces.

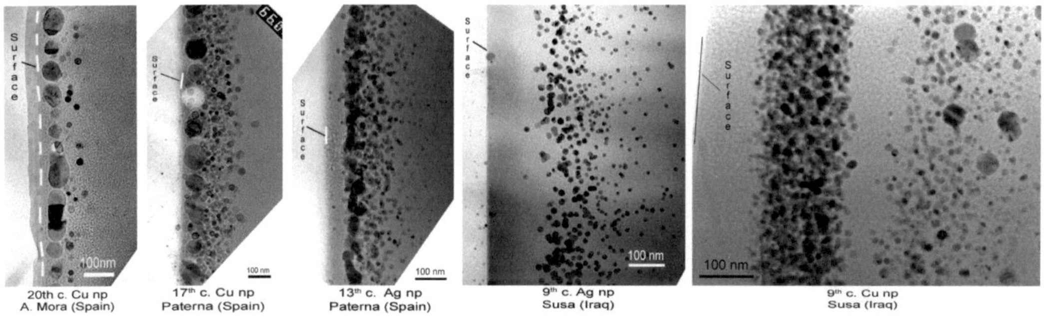

Fig. 3. A selection of bright field TEM images of various lustres from Mesopotamia (right) to Spain (left). More details can be found in the references (Chabanne 2005) and (Mirguet, Roucau *et al.* 2009).

Synchrotron radiation was used to obtain information on nanoparticles. In addition to the nature of nanoparticles, X-ray absorption fine structure (XAFS) measurements give interesting information on the presence of metallic ions (Ag^+, Cu^+, Cu^{2+}) in the glassy matrix (Padovani *et al.* 2006). In some cases, the size of the metallic oxide clusters can be estimated by fitting the extended X-ray absorption fine structure (EXAFS) spectra, whereas the average size of metallic nanoparticles can be deduced from the broadness of X-ray diffraction reflections. In addition, using glancing incidence X-ray diffraction (GIXRD) techniques, information can be obtained on structure and depth distribution of nanoparticles (Bontempi *et al.* 2006). Nevertheless, all these techniques give only partial information about the nanoparticle layers and require a synchrotron facility.

An alternative solution has been proposed using ion beam analysis (IBA) (Salomon *et al.* 2008; Pichon *et al.* 2010). The association of particle induced X-ray emission (PIXE) and elastic Rutherford backscattering spectrometry (RBS) allowed them to obtain significant data concerning the glassy matrix composition and the nature and the depth distribution of nanoparticles for a number of lustre ceramics, including valuable museum objects (Darque-Ceretti *et al.* 2005; Padeletti, Ingo *et al.* 2006; Chabanne *et al.* 2008; del Rio & Castaing 2010; del Rio *et al.* 2010). Whereas PIXE gives the chemical composition of the glassy matrix, RBS can provide detailed depth information. Thus, from a simulation of the experimental RBS spectrum, the nanoparticle distribution can be modelled. However, a calibration is necessary and for this, TEM observations are very useful (Chabanne, Bouquillon *et al.* 2008).

Several deductions can be made from all these investigations. The Abbasid and Fatimid samples observed by TEM showed a more complex structuration in depth than the Hispano-Moorish productions (Fig. 3). The particles of the first lustres are small with an average size of around 10-15 nm, and the thickness of the layer without particles below the glaze surface is superior to 100 nm (Chabanne 2005; Mirguet, Fredrickx *et al.* 2008). On the contrary, the Hispanic productions are characterized by a layer of big particles (50-100 nm) close to the glaze surface (Chabanne 2005; Mirguet, Roucau *et al.* 2009). This is even more pronounced for the 17th century lustres and modern replica made by Spanish artisans using a traditional process. Several Abbasid and Fatimid lustres present a partial structuring in depth such as shown in figure 3 and 4.

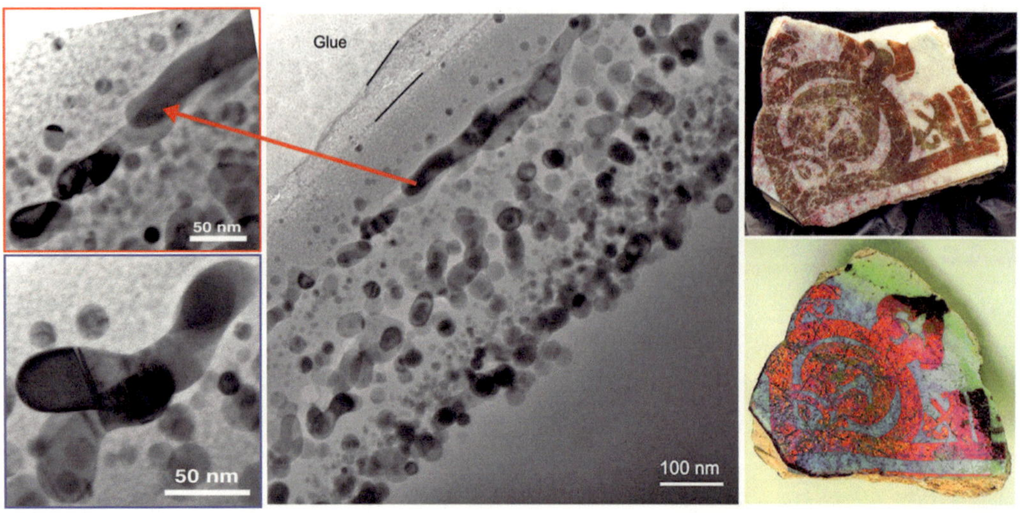

Fig. 4. Lustre from the Fatimid period (12th century CE) showing a partial multi layer silver nanoparticle organization (Mirguet, Roucau *et al.* 2009; Sciau, Mirguet *et al.* 2009a). On the right, the colour change from brown (scattering light) to pink (specular position) and on the left, magnifications of elongated silver particles ("metal worms").

TEM investigations even brought to light a lustre decoration from the Fatimid period with a very regular nanoparticle distribution in two well-separated layers (Fig. 5). Electron diffraction revealed that all particles were silver with a CFC structure (Sciau, Mirguet *et al.* 2009a). The electron energy loss spectroscopy (EELS) confirmed that silver is only present in the nanoparticle layers whereas copper is found everywhere in the glaze, however in the ionic form (Cu^{2+}). The green colour of the glaze indeed comes from the Cu^{2+} ions. The distance between the two layers is amazingly constant. With a value of around 430 nm, this distance is of the same order of magnitude as visible wavelengths. It results in that this double layer structure behaves as an optical network. A more in depth investigation of nanopaticles showed that the particles in the second layer are slightly larger and that their shapes are less spherical. Some particles have even coalesced forming larger particles. High resolution electron microscopy (HREM) showed that many of them have structural defects of a stacking fault type (Sciau, Mirguet *et al.* 2009a).

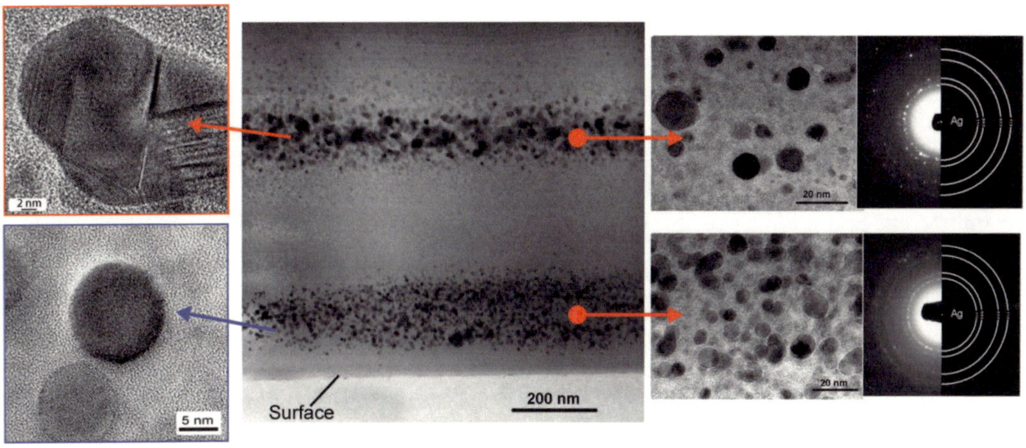

Fig. 5. Lustre from the Fatimid period (12th century CE) showing a well defined multi layer silver nanoparticle organization, with on the left, HREM images of nanoparticles of each layer and on the right, electron diffraction identification (Mirguet, Roucau *et al.* 2009; Sciau, Mirguet *et al.* 2009a).

2.3 Optical properties

Since the discovery of their astonishing colorimetric properties, several studies have sought to model the optical behaviour of lustre decorations.

Olivier Bobin was the first to carry out a theoretical investigation of optical properties of lusterware and thus to prove the role of the surface phasmon resonance of silver and/or copper nanocrystals in the colouring process (Bobin, Schvoerer *et al.* 2003). His modelling, using Mie's theory, and based on the copper-silver ratio, the particle size, the particle density and the nature of embedded glaze give rather good results for the colours observed in scattering light. In the modelling of the size effect on the surface plasmon resonance, two regimes are usually distinguished depending on the nanoparticle size range (Garcia 2011). For the small particles (smaller than light wavelength i.e. with a radius up to 50 nm), the particle can be properly described by a dielectric dipole. The size variation affects mainly the width and the intensity of the resonance band. On the other hand, the resonance wavelength is only slightly shifted. For the larger particles with a size comparable to the wavelength i.e. with a radius superior or equal to 50 nm, the dipole approximation is not sufficient and multipolar terms must be added leading to the splitting of the resonance band into several peaks: two peaks for quadrupole, three peaks for an octopole, etc … (Kreibig *et al.* 1987). The metallic particles present in the lusterware are seldom superior to 50 nm in radius (cf. § 2.2); also the dipole approximation is sufficient to describe the surface plasmon resonance. Nevertheless, the size dispersion is large and that has as a consequence a significant broadening of the absorption band. Since the restoring force for surface plasmons is related to the charge accumulated at the surface, it is influenced as well by the particle shape. With elongated particles, the absorption band is split into two bands: the transversal and longitudinal bands. The frequency shift is proportional to the ratio between longitudinal and transversal lengths. While the resonant frequency of transversal plasmons falls at about the same position as for spherical particles (actually, at wavelengths slightly smaller), the resonance of longitudinal plasmons shifts towards larger wavelengths when the ratio increases. The intensity of the longitudinal plasmon band increases with the ratio while the one of the transversal band decreases. For the lustre of figure 4, the shape effect must be taken into account, but in this case other effects such as the interference phenomena must also be considered. It is obvious that the interferences have a significant influence on the colour of metallic shine (the specular position) for the lustres having a partial multi-layer structuration. The study of the lusterware with the double layer (Fig. 5) demonstrated that the interferences are at the origin of the colour variation of the metallic shine from blue to green (Sciau, Mirguet *et al.* 2009a).

The first model taking into account the interference phenomena was proposed by Vincent Reillon (Reillon & Berthier 2006). However, the modelling of such a complex system was not easy and it was only recently that a model integrating all phenomena (surface plasmon absorption, interference and scattering) was published (Reillon 2008; Reillon *et al.* 2010). From this model, it is now possible to correctly simulate the reflection spectra recorded as well in the specular direction as in the scattering directions. The evolution of the colour between the specular and the diffusion directions can be perfectly calculated. Thus, the key parameters determining colour behaviour are (Lafait, Berthier *et al.* 2009):

- in the specular direction, interference phenomena play a major role with the key parameters being the number of layers, the optical index and the thickness of each layer,
- in the scattering cone, plasmon absorption is predominant and the key parameters are the kind of metal, the metal volume fraction, the particle size and shape, and the glass matrix composition,
- in the intermediate cone, there is a transition between a coherent component (dominant close to the specular direction) and a scattered incoherent component (dominant close to the scattering cone).

The colour behaviour of lustre is schematized in figure 6. To simulate the experimental spectra, the modelling uses a schematic representation of the multilayer structure of lustre decoration. Hence inversely from a modelling of a set of experimental spectra collected from different directions, it is possible to obtain significant information on the nanoparticle distribution.

2.4 Manufacturing process

Several descriptions of the glazing technique were proposed (Pérez-Arantegui, Molera *et al.* 2001; Padeletti & Fermo 2003a; Colomban & Truong 2004; Pradell *et al.* 2005; Roqué *et al.* 2005; Pradell *et al.* 2006; Roqué *et al.* 2008; Colomban 2009) on the basis of experimental evidence and on information extracted from the transcription of ancient recipes (Abu al Qâsem 14th century, Picolpasso 16th century, Deck and Bertan 19th century and Artigas 20th century).

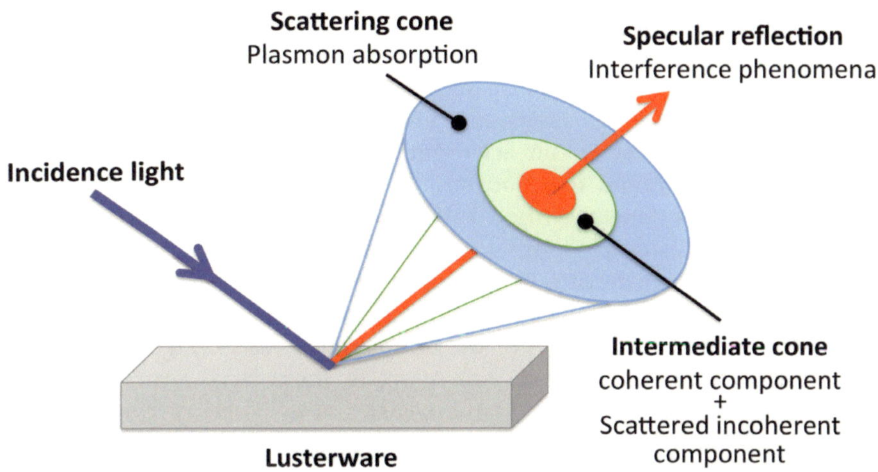

Fig. 6. Schematic representation of the light scattered and reflected by a lustre glaze decoration from Reillon's modelling (Reillon 2008; Reillon, Berthier *et al.* 2010).

Lustre decorations were obtained by applying a mixture of a paint, which contained copper and silver salts, water and more or less vinegar and lye, onto a glazed ceramic, which was subsequently annealed in a reducing atmosphere. Inside the kiln, the raw paint reacted with the glaze surface, and after firing, the remaining paint was washed off, revealing the lustre decoration beneath.

The role of ionic exchange was first identified in the lustre formation process (Smith *et al.* 2003; Pradell, Molera *et al.* 2005). From the analysis of medieval ceramics, some interesting trends and information were obtained concerning the driving force responsible for the diffusion of silver and copper ions into the glaze, consisting of an "ionic exchange" of Ag^+ and Cu^+ with Na^+ and K^+. This type of ionic exchange is a well-known mechanism in glasses and glazes (Pradell, Molera *et al.* 2006). When glasses are immersed in a solution containing copper and/or silver molten salt (typically sulphates or nitrates), atomic exchanges take place and the alkalis (K^+ and Na^+) of the glaze are replaced by Ag^+ and Cu^+ ions of the solution. Accurate chemical analysis of medieval ceramics showed a clear inverse correlation between the metal components of the lustre decorations (Cu and Ag) and the amount of Na and K in the glaze. This correlation has been found in early Islamic lusterwares from Iraq (9th century CE) as well as in late Hispano-Moorish lusterwares from Paterna (13th-17th centuries CE). In addition, reproductions and ancient lustre surfaces were observed by means of white light interferometry, atomic forces microscopy, X-ray diffraction and electron microprobe (Roqué, Pradell *et al.* 2005). These observations showed that lustre layers do not appear as superimposed layers on the top of the glaze, but rather as a surface roughness resulting in the nanocrystals growth inside the glassy matrix. The surface roughness increases during the formation process as a result of metal nanoparticle growth. The lustre formation process involves a two step process: ion exchange and crystallization (nucleation and crystal growth) of copper and silver metallic nanoparticles inside the glassy matrix (Roqué, Molera *et al.* 2008).

More recently, it has been assumed that the burning of organic residues could be used to control the surface temperature and embedded metal dispersion allowing one to set the final lustre colour in different places of the same item (Mirguet, Fredrickx *et al.* 2008). The strong temperature gradient arising from the combustion of surface acetate residues could control the self-organization of the metal particles leading to light diffraction. The multilayer particle distribution of lustre could be explained by special firing cycles where repeated heat flashes provoked by surface organic residue combustion make it possible to control the size, the shape and the distribution of nanoparticles. In several cases, very elongated silver particles ("metal worms") were observed by TEM. It is particularly well marked in the Fatimid lustre of Figure 4, which shows 3 layers of metal worm particles separated by areas of lower density in nanoparticles. The particles of these intermediate zones are smaller and spherical. The nanoparticle coalescences forming larger particles are also observed in the second layer of the double layer lustre (Fig. 5). High-resolution electron microscopy (HREM) showed that many of them have structural defects of a stacking fault type. It is not the case of nanoparticles of the first layer, which are smaller and quite spherical. The nanoparticles of the two layers are different enough to conclude that they were not formed under the same thermal conditions. Nanoparticle coalescences were also found in the two other lustres of figure 3 with silver particles (Abbasid from Susa and 13th century from Paterna).

The elongated shape of silver particles proves that the temperatures close to that of the silver melting point (~960°C) were reached. Diffusion-controlled phenomena cannot lead to such worm-shapes. The Melting of metallic silver demonstrates that glaze surface temperatures close to 1000°C were achieved at the peak temperature cycle. This is consistent with an increase in temperature obtained thanks to the combustion of organic residue. Except for some modern replica (Mora productions, Fig. 3), worm-shapes were not observed for copper particles, which indicated that the melting point of copper (~1080°C) was not reached.

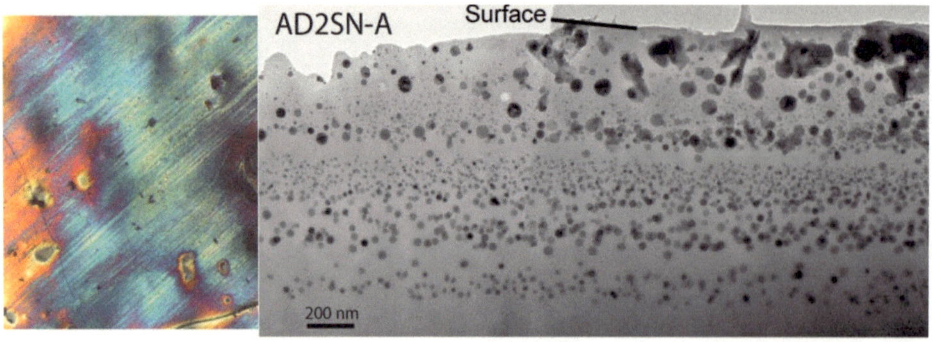

Fig. 7. Lustre creation made by Eva Haudum (Colomban 2009).

3. Conclusion

The review of present knowledge on ancient materials, in which the size and distribution as well as the reduction of metallic nanoprecipitates were organized, shows the high level of empirical control carried out by ancient potters and glass-makers. The ancient potters were certainly not aware of material nanostructuration, since they did not have any nanostructure checking facilities. However, their know-how allowed for the realization of two perfectly well separated layers (Sciau, Mirguet *et al.* 2009a). This special nanostructuration has been up to now observed on only one sample. It is important to keep in mind that this sample might have been accidently obtained or that the first layer (uppermost one) was created because the first visual aspect did not suit the craftsman. However, other lustres of the Fatimid (12th century CE) and Abbasid (9th - 10th centuries CE) periods show a partial organisation in multi-layered nanoparticles. So there is a good chance that this sample is not an isolated case and that this technology was developed by the Islamic craftsmen to exploit the complex nano-optical properties of multi layered particles.

The optical modelling performed by Vincent Reillon confirmed the role of interference in the colour of metallic shine reflection (specular direction). It is obvious that the multi-layer organisation of nanoparticles, with distances between layers comparable to the visible wavelengths, strengthens the interference effects allowing for the obtainment of very bright iridescent colours from a large palette of hues. It is likely that it was the aim of the Abbasid and Fatimid potters to develop empirical processes for creating such multi-layered nanoparticles through a know-how, which has nowadays partially been lost. Modern artists (S. Çizer, E. Haudum ...) search to recreate lustre decorations with strong iridescent effects and bright colours using modern kilns permitting alternative oxidising (oxygen flux) and reducing (CO flux) phases during the firing (Colomban 2009). Their best recreations exhibit the partial multi-layer structure observed in the Abbasid and Fatimid artefacts, but maybe with a weaker organisation and sometimes residues of raw mixture as observed in figure 7.

4. Acknowledgments

I would like to thank A. Bouquillon and D. de Camaret for their help in the writing of this chapter as well as colleagues who allowed me to develop this activity in CEMES-CNRS lab (G. BenAssayag, B. Brunetti, R. Carles, Ph. Colomban, H. Dexpert, C. Roucau). I thank all under-graduate and PhD students who experimentally contributed to this paper, especially D. Chabanne, C. Dejoie, Y. Leon, C. Mirguet and J. Roqué-Rosell. Research supported by the *Conseil Régional de Midi-Pyrénées* under several contracts.

5. References

Angelini, I.; Artioli, G.; Bellintani, P.; Diella, V.; Gemmi, M.; Polla, A. & Rossi, A. (2004). Chemical analyses of bronze age glasses from Frattesina di Rovigo, northern Italy. *Journal of Archaeological Science*, Vol.31, No.8, pp. 1175-1184, ISSN 0305-4403

Artioli, G.; Angelini, I. & Polla, A. (2008). Crystals and phase transitions in protohistoric glass materials. *Phase Transitions*, Vol.81, No.2-3, pp. 233-252, ISSN 0141-1594

Barber, D. J. & Freestone, I. C. (1990). An investigation of the origin of the color of Lycurgus cup by analytical transmission electron-microscopy. *Archaeometry*, Vol.32, (February 1990), pp. 33-45, ISSN 0003-813X

Bobin, O.; Schvoerer, M.; Miane, J. L. & Fabre, J. F. (2003). Coloured metallic shine associated to lustre decoration of glazed ceramics: a theoretical analysis of the optical properties. *Journal of Non-Crystalline Solids*, Vol.332, No.1-3, (December 2003), pp. 28-34, ISSN 0022-3093

Bontempi, E.; Colombi, P.; Depero, L. E.; Cartechini, L.; Brunetti, B. G. & Sgamellotti, A. (2006). Glancing-incidence X-ray diffraction of Ag nanoparticles in gold lustre decoration of Italian Renaissance pottery. *Applied Physics A: Materials Science & Processing*, Vol.83, No.4, (Jun 2006), pp. 543-546, ISSN 0947-8396

Borgia, I.; Brunetti, B.; Mariani, I.; Sgamellotti, A.; Cariati, F.; Fermo, P.; Mellini, M.; Viti, C. & Padeletti, G. (2002). Heterogeneous distribution of metal nanocrystals in glazes of historical pottery. *Applied Surface Science*, Vol.185, No.3-4, (January 2002), pp. 206-216, ISSN 0169-4332

Brun, N.; Mazerolles, L. & Pernot, M. (1991). Microstructure of opaque red glass containing copper. *Journal of Materials Science Letters*, Vol.10, No.23, (December 1991), pp. 1418-1420, ISSN 0261-8028

Caiger-Smith, A. (1991). *Lustre pottery. Technique, tradition and innovation in Islam and the Western World*, New Amsterdam Books, ISBN 1-56131-030-1, New York, USA.

Chabanne, D. (2005). *Le décor de lustre métallique des céramiques glaçurées (IXème-XVIIème siècles). Matériaux, couleurs et techniques*. PhD thesis, University Bordeaux 3.

Chabanne, D.; Bouquillon, A.; Aucouturier, M.; Dectot, X. & Padeletti, G. (2008). Physico-chemical analyses of Hipano-Moresque lustred ceramic: a precursor for Italian majolica? *Applied Physics A: Materials Science & Processing*, Vol.92, No.1, (July), pp. 11-18, ISSN 0947-8396

Colomban, P. (2009). The use of metal nanoparticles to produce yellow, red and iridescent colour, from bronze age to present times in lustre pottery and glass: solid state chemistry, spectroscopy and nanostructure. *Journal of Nano Research*, Vol.8, pp. 109-132, ISSN 1662-5250

Colomban, P.; March, G.; Mazerolles, L.; Karmous, T.; Ayed, N.; Ennabli, A. & Slim, H. (2003). Raman identification of materials used for jewellery and mosaics in Ifriqiya. *Journal of Raman Spectroscopy*, Vol.34, No.3, (March 2003), pp. 205-213, ISSN 0377-0486

Colomban, P. & Truong, C. (2004). Non-destructive raman study of the glazing technique in lustre potteries and faience (9-14th centuries): silver ions, nanoclusters microstructure and processing. *Journal of Raman Spectroscopy*, Vol.35, No.3, (March 2004), pp. 195-207, ISSN 0377-0486

Darque-Ceretti, E.; Hélary, D.; Bouquillon, A. & Aucouturier, M. (2005). Gold like lustre: nanometric surface treatment for decoration of glazed ceramics in ancient Islam, Moresque Spain and Renaissance Italy. *Surface Engineering*, Vol.31, No.5-6, (December 2003), pp. 352-358, ISSN 0267-0844

del Rio, A. P. & Castaing, J. (2010). Lustre decorated ceramics from a 15th-16th century production in Seville. *Archaeometry*, Vol.52, (February 2010), pp. 83-98, ISSN 0003-813X

del Rio, A. P.; Roehrs, S.; Aucouturier, M.; Castaing, J. & Bouquillon, A. (2010). Medinal Al-Zahra lustre ceramics: 10th century local nanotechnology or importation from middle east. *Arabian Journal for Science and Engineering*, Vol.35, No.1C, (Jun 2010), pp. 157-168, ISSN 1319-8025

Fredrickx, P.; Hélary, D.; Schryvers, D. & Darque-Ceretti, E. (2004). A TEM study of nanoparticles in lustre glazes. *Applied Physics A: Materials Science & Processing*, Vol.79, No.2, (July 2004), pp. 283-288, ISSN 0947-8396

Freestone, I.; Meeks, N.; Sax, M. & Higgitt, C. (2007). The Lycurgus Cup - A Roman nanotechnology. *Gold Bulletin*, Vol.40, No.4, pp. 270-277, ISSN 0017-1557

Garcia, M. A. (2011). Surface plasmons in metallic nanoparticles: fundamentals and applications. *Journal of Physics D: Applied Physics*, Vol.44, No.28, (July 2011), pp. 283001, ISSN 0022-3727

Gil, C.; Villegas, M. A. & Navarro, J. M. F. (2006). TEM monitoring of silver nanoparticles formation on the surface of lead crystal glass. *Applied Surface Science*, Vol.253, No.4, (December 2006), pp. 1882-1888, ISSN 0169-4332

Gimeno, D.; Aulinas, M.; Bazzocchi, F.; Fernandez-Turiel, J. L.; Garcia-Valles, M.; Novembre, D.; Basso, E.; Messiga, B.; Riccardi, M. P.; Tarozzi, C. & Mendera, M. (2010). Chemical characterization of the stained glass window from the rose window, Siena Duomo (Italy, 1288-1289). *Boletin De La Sociedad Espanola De Ceramica Y Vidrio*, Vol.49, No.3, (May-Jun 2010), pp. 205-213, ISSN 0366-3175

Hartland, G. V. (2011). Optical Studies of Dynamics in Noble Metal Nanostructures. *Chemical Reviews*, Vol.111, No.6, (Jun 2011), pp. 3858-3887, ISSN 0009-2665

Kreibig, U.; Schmitz, B. & Breuer, H. D. (1987). Separation of plasmon-polariton modes of small metal particles *Physical Review B*, Vol.36, No.9, (September 1987), pp. 5027-5030, ISSN 0163-1829

Kurmann-Schwarz, B. & Lautier, C. (2009). The Medieval stained-glass window in Europe: 10 years of abundant research. *Perspective-La Revue De L Inha*, No.1, (March 2009), pp. 99-130, ISSN 1777-7852

Lafait, J.; Berthier, S.; Andraud, C.; Reillon, V. & Boulenguez, J. (2009). Physical colors in cultural heritage: surface plasmons in glass. *Comptes Rendus Physique*, Vol.10, No.7, (September 2009), pp. 649-659, ISSN 1631-0705

Mie, G. (1908). Beiträge zur optik trüber medien, speziell kolloidaler metallösungen. *Annalen der Physik*, Vol.25, No.3, (March 1908), pp. 377-445, ISSN 0003-3804

Mirguet, C.; Fredrickx, P.; Sciau, P. & Colomban, P. (2008). Origin of the self-organisation of Cu°/Ag° nanoparticles in ancient lustre pottery. A TEM study. *Phase Transitions*, Vol.81, No.2-3, pp. 253-266, ISSN 0141-1594

Mirguet, C.; Roucau, C. & Sciau, P. (2009). Transmission electron microscopy a powderful means to investigate the glazed coating of ancient ceramics. *Journal of Nano Research*, Vol.8, pp. 141-146, ISSN 1662-5250

Nakai, I.; Numako, C.; Hosono, H. & Yamasaki, K. (1999). Origin of the red color of satsuma copper-ruby glass as determined by EXAFS and optical absorption spectroscopy. *Journal of the American Ceramic Society*, Vol.82, No.3, (March 1999), pp. 689-695, ISSN 0002-7820

Padeletti, G. & Fermo, P. (2003a). How the masters in Umbria, Italy, generated and used nanoparticles in art fabrication during the Renaissance period. *Applied Physics A: Solids and Surfaces*, Vol.76, pp. 515-525,

Padeletti, G. & Fermo, P. (2003b). How the masters in Umbria, Italy, generated and used nanoparticles in art fabrication during the Renaissance period. *Applied Physics a-*

Materials Science & Processing, Vol.76, No.4, (March 2003), pp. 515-525, ISSN 0947-8396

Padeletti, G. & Fermo, P. (2004). Production of gold and ruby-red lustres in Gubbio (Umbria, Italy) during the renaissance period. *Applied Physics A: Materials Science & Processing*, Vol.79, No.2, (July 2004), pp. 241-245, ISSN 0947-8396

Padeletti, G.; Ingo, G. M.; Bouquillon, A.; Aucouturier, M.; Roehrs, S. & Fermo, P. (2006). First-time observation of Mastro Giorgio materpieces by means of non-destructive techniques. *Applied Physics A: Materials Science & Processing*, Vol.83, No.4, (Jun 2006), pp. 475-483, ISSN 0947-8396

Padovani, S.; Puzzovio, D.; Mazzoldi, P.; Borgia, I.; Sgamellotti, A.; Brunetti, B. G.; Cartechini, L.; D'Acapito, F.; Maurizio, C.; Shokouhi, F.; Oliaiy, P.; Rahighi, J.; Lamehi-Rachti, M. & Pantos, E. (2006). XAFS study of copper and silver nanoparticles in glazes of medieval middle-east lustreware (10th-13th century). *Applied Physics A: Materials Science & Processing*, Vol.83, No.4, (Jun 2006), pp. 521-528, ISSN 0947-8396

Pérez-Arantegui, J. & Larrea, A. (2003). The secret of early nanomaterials is revealed, thanks to transmission electron microscopy. *Trends in Analytical Chemistry*, Vol.22, No.5, (May 2003), pp. 327-329, ISSN 0165-9936

Pérez-Arantegui, J.; Molera, J.; Larrea, A.; Pradell, T.; Vendrell-Saz, M.; Borgia, I.; Brunetti, B. G.; Cariati, F.; Fermo, P.; Mellini, M.; Sgamellotti, A. & Viti, C. (2001). Luster pottery from the thirteenth century to the sixteenth century: a nanostructured thin metallic film. *Journal of the American Ceramic Society*, Vol.84, No.2, (February 2001), pp. 442-446, ISSN 0002-7820

Perez-Villar, S.; Rubio, J. & Oteo, J. L. (2008). Study of color and structural changes in silver painted medieval glasses. *Journal of Non-Crystalline Solids*, Vol.354, No.17, (April 2008), pp. 1833-1844, ISSN 0022-3093

Pichon, L.; Beck, L.; Walter, P.; Moignard, B. & Guillou, T. (2010). A new mapping acquisition and processing system for simultaneous PIXE-RBS analysis external beam. *Nuclear Instruments & Methods in Physics Research, Section B: Beam Interactions with Materials and Atoms*, Vol.268, No.11-12, (Jun 2010), pp. 2028-2033, ISSN 0168-583X

Pradell, T.; Molera, J.; Bayes, C. & Roura, P. (2006). Luster decoration of ceramics: mechanisms of metallic luster formation. *Applied Physics A: Materials Science & Processing*, Vol.83, No.2, (May 2006), pp. 203-208, ISSN 0947-8396

Pradell, T.; Molera, J.; Roqué, J.; Vendrell-Saz, M.; Smith, A. D.; Pantos, E. & Crespo, D. (2005). Ionic-exchange mechanism in the formation of medieval luster decorations. *Journal of the American Ceramic Society*, Vol.88, No.5, (May 2005), pp. 1281-1289, ISSN 0002-7820

Reillon, V. (2008). *Caractérisation et modélisation des propriétés optiques des ceramiques lustrées.* PhD thesis, Université Pierre et Marie Curie – Paris 6.

Reillon, V. & Berthier, S. (2006). Modelization of the optical and colorimetric properties of lustred ceramics. *Applied Physics A: Materials Science & Processing*, Vol.83, No.2, (May 2006), pp. 257-265, ISSN 0947-8396

Reillon, V.; Berthier, S. & Andraud, C. (2010). Optical properties of lustred ceramics: complete modelling of the actual structure. *Applied Physics A: Materials Science & Processing*, Vol.100, No.3, (September 2010), pp. 901-910, ISSN 0947-8396

Ricciardi, P.; Colomban, P.; Tournié, A.; Macchiarola, M. & Ayed, N. (2009). A non-invasive study of Roman Age mosaic glass tesserae by means of Raman spectroscopy. *Journal of Archaeological Science*, Vol.36, No.11, (November 2009), pp. 2551-2559, ISSN 0305-4403

Roqué, J.; Molera, J.; Cepria, G.; Vendrell-Saz, M. & Perez-Arantegui, J. (2008). Analytical study of the behaviour of some ingredients used in lustre ceramic decorations following different recipes. *Phase Transitions*, Vol.81, No.2-3, pp. 267-282, ISSN 0141-1594

Roqué, J.; Molera, J.; Pérez-Arantegui, J.; Calabuig, C.; Portillo, J. & Vendrell-Saz, M. (2007). Lustre colour ans shine from the Olleries xiques workshop in Paterna (Spain), 13th century AD: Nanostructure, chemical composition and annealing conditions. *Archaeometry*, Vol.49, No.3, (August 2007), pp. 511-528, ISSN 0003-813X

Roqué, J.; Pradell, T.; Molera, J. & Vendrell-Saz, M. (2005). Evidence of the nucleation and growth of the metal Cu and Ag nanoparticles in lustre: AFM surface characterization. *Journal of Non-Crystalline Solids*, Vol.351, No.6-7, (March 2005), pp. 568-575, ISSN 0022-3093

Rubio, F.; Perez-Villar, S.; Garrido, M. A.; Rubio, J. & Oteo, J. L. (2009). Application of Gradient and Confocal Raman Spectroscopy to Analyze Silver Nanoparticle Diffusion in Medieval Glasses. *Journal of Nano Research*, Vol.8, pp. 89-97, ISSN 1662-5250

Salomon, J.; Dran, J.-C.; Guillou, T.; Moignard, B.; Pichon, L.; Walter, P. & Mathis, F. (2008). Ion-beam analysis for cultural heritage on the AGLAE facility: impact of PIXE/RBS combination. *Applied Physics A: Materials Science & Processing*, Vol.92, No.1, (July 2008), pp. 43-50, ISSN 0947-8396

Sciau, P.; Mirguet, C.; Roucau, C.; Chabanne, D. & Schvoerer, M. (2009a). Double nanoparticle layer in the 12th century lustreware decoration: accident or technological mastery. *Journal of Nano Research*, Vol.8, pp. 133-139, ISSN 1662-5250

Sciau, P.; Salles, P.; Roucau, C.; Mehta, A. & Benassayag, G. (2009b). Applications of focused ion beam for preparation of specimens of ancient ceramic for electron microscopy and synchrotron X-ray studies. *Micron*, Vol.40, No.5-6, (July-August 2009), pp. 597-604, ISSN 0968-4328

Simmons, C. T. & Mysak, L. A. (2010). Transmissive properties of Medieval and Renaissance stained glass in European churches. *Architectural Science Review*, Vol.53, No.2, (May 2010), pp. 251-274, ISSN 0003-8628

Smith, A. D.; Pradell, T.; Molera, J.; Vendrell-Saz, M.; Marcus, M. A. & Pantos, E. (2003). MicroEXAFS study into the oxidation states of copper coloured Hipano-Moresque Lustre decorations. *Journal de Physique IV*, Vol.104, (March 2003), pp. 519-522, ISSN 1155-4339

Watson, O. (1985). *Persian lustre ware*, Faber and Faber, ISBN 9780571132355, London, UK.

Wood, N. (1999). *Chinese glazes, their origins, chemistry and recreation,* University of Pennsylvania Press, ISBN 978-0-8122-3476-3, Philadelphia, USA.